INDEX OF TABLES

VOLUME ONE

UNIVERSITY PHYSICS

SEVENTH EDITION

The end-of-chapter Problem Sets were revised for the *Seventh Edition* by A. Lewis Ford, Texas A&M University.

Contributions to the Problem Sets were made by Lawrence B. Coleman, University of California, Davis; James L. Monroe, Pennsylvania State University, The Beaver Campus; Terry F. O'Dwyer, Nassau Community College.

Craig Watkins, Massachusetts Institute of Technology, provided assistance to both Professor Young and to Professor Ford in the development of the manuscript.

VOLUME ONE

UNIVERSITY PHYSICS

SEVENTH EDITION

Francis W. Sears

Late Professor Emeritus
Dartmouth College

Mark W. Zemansky

Late Professor Emeritus
City College of the City University of New York

Hugh D. Young

Professor of Physics
Carnegie-Mellon University

ADDISON-WESLEY PUBLISHING COMPANY
Reading, Massachusetts ▪ Menlo Park, California
Don Mills, Ontario ▪ Wokingham, England ▪ Amsterdam
Sydney ▪ Singapore ▪ Tokyo ▪ Madrid ▪ Bogotá
Santiago ▪ San Juan

This book is in the
Addison-Wesley Series in Physics

Sponsoring Editor: Bruce Spatz, Debra Hunter, Steve Mautner
Developmental Editor: David M. Chelton
Production Supervisor: Marion E. Howe
Copy Editor: Jacqueline M. Dormitzer
Text Designer: Catherine L. Dorin
Layout Artist: Lorraine Hodsdon
Illustrators: Oxford Illustrators, Ltd.
Art Consultant: Loretta Bailey
Manufacturing Supervisor: Ann DeLacey
Cover: Marshall Henrichs

Photo credits: page 1, NASA. 3, Lockheed Missiles and Space Company. 25, Education Development Center. 49 and 71, Dr. Harold Edgerton, M.I.T., Cambridge MA. 90, National Center for Atmospheric Research/National Science Foundation. 119, AP/Wide World Photos. 143, Dr. Harold Edgerton, M.I.T., Cambridge MA. 145, Matson Navigation Co. 182, Education Development Center. 210, R. V. Willstrop/Anglo-Australian Telescope Board. 249, Golden Gate Bridge, Highway, and Transportation District. 263, Dr. Harold Edgerton, M.I.T., Cambridge MA. 289, U.S. Geological Survey. 291, M. S. Paterson, Australian National University. 306, Department of Aeronautics, Imperial College of Science and Technology. 340, N.Y. State Department of Commerce. 357, Nancy Rodger/Exploratorium. 374, Lockheed Corp. 389, Barry L. Runk/Grant Heilman Photography. 403, Pacific Gas & Electric. 424, Kurt Rogers, San Francisco Examiner. 451, Lawrence Berkeley Laboratory. 473, M. P. Moller, Inc. 475, Fundamental Photographs. 496, The Exploratorium. 512, Steinway & Sons.

Library of Congress Cataloging-in-Publication Data

Sears, Francis Weston, 1898–
    University physics.

    Includes index.
    1. Physics. I. Zemansky, Mark Waldo, 1900–
II. Young, Hugh D. III. Title.
QC21.2.S36 1986   530   85–28801
ISBN 0-201-06682-3

ABCDEFGHIJ-MU-8987

PREFACE

In this new edition of *University Physics,* we have tried to preserve those qualities and features that users of previous editions have found useful. Yet this is the most comprehensive revision in the long, successful history of the book. Physics courses and physics students have changed substantially in recent years, and new editions must keep pace with these changes.

Our basic goals have not changed. Our objective is to provide a broad, rigorous introduction to physics at the beginning college level. This book is appropriate for students of science and engineering who are taking an introductory calculus course concurrently. We place primary emphasis on physical principles and the development of problem-solving ability, rather than on historical background or specialized applications. The complete text may be taught in an intensive two- or three-semester course, and the book is also adaptable to a wide variety of shorter courses. It is available as a single volume or as two volumes. Volume I includes mechanics, heat, and mechanical waves, and Volume II includes electricity and magnetism, optics, and atomic and nuclear physics.

Here are some of the most important new features in this edition:

Table of Contents. After careful consideration and consultation with many users of our book, we have reorganized the chapters on mechanics. We now conform to the usual order in introductory courses, beginning with kinematics and dynamics and treating statics later as a special case of dynamics. Getting students into the study of motion immediately helps to build motivation for the study of physics, and it also helps to tie the physics course in with the often-concurrent calculus course.

Problems. The end-of-chapter problem collections have been extensively revised and augmented. We now group each collection into three categories: *Exercises,* single-concept problems that are keyed to specific sections of the text; *Problems,* usually requiring two or more nontrivial steps for their solution; and *Challenge Problems,* intended to challenge the strongest students. The number of problems has grown by 12%; the total number is now approximately 1800, of which over 25% are new. We have also added to the lists of thought-provoking questions at the ends of the chapters, about 700 questions in all. The revision of the problem collections was carried out by Professor A.

Lewis Ford (Texas A. & M. University), with the assistance of Mr. Craig Watkins (Massachusetts Institute of Technology). Additional problems were contributed by Professors Lawrence B. Coleman (University of California, Davis), James L. Monroe (Pennsylvania State University, The Beaver Campus), and Terry F. O'Dwyer (Nassau Community College).

Problem-Solving Strategies. The remark heard most often in the freshman Physics classroom is: "I understood the material, but I couldn't do the problems!" To respond to this universal cry for help, we have included in each chapter one or more sections called *Problem-Solving Strategy,* where we list suggestions for developing a methodical and systematic approach to solving problems. Our most important objective in this book is to help students learn to apply physical principles to a wide variety of problems, and these new strategy sections should be a substantial help.

Chapter Summaries. Each chapter concludes with a list of *Key terms,* which have been highlighted in boldface type in the text, and a *summary,* in prose and equations, of the most important principles presented in the chapter. These will be a useful aid for the student, especially in identifying and emphasizing the concepts and relationships that are of central importance.

Chapter Introductions and Perspectives. Each chapter now begins with an introductory paragraph summarizing briefly the content of the chapter and relating it to what has come before. In addition, a *Prospectus* and seven *Perspectives* are included at intervals throughout the book. Their object is to enhance continuity by looking both backward and forward to show how various areas of physics are interrelated and to exhibit as clearly as possible the beauty and fundamental unity of all branches of physics.

Study Notes. Brief notes inserted in the margins act as references for the reader and identify key ideas and concepts in the text. They help the student locate quickly a particular discussion or example, and they provide capsule summaries of paragraphs and sections of text, useful for re-study and review of troublesome areas.

Mathematical Level. The level of mathematical sophistication has not changed substantially, but we have increased somewhat the use of unit vectors and calculus in worked-out examples. There are also more problems providing opportunities to use unit vectors and calculus.

Changes in subject matter. Many topics have been added or treated in greater depth than in previous editions. A partial list includes:

estimates and orders of magnitude
gravitational field
automotive power
damped and forced oscillations
Maxwell–Boltzmann distribution
sign conventions for Kirchhoff's rules
Maxwell's equations
circular apertures and resolving power
relativistic Doppler effect
Planck radiation law
superconductivity
band theory of solids

The elasticity chapter has been rewritten so that each type of stress is introduced along with its corresponding strain. The treatment of fluid mechanics has been condensed and reduced from two chapters to one. The material on acoustic phenomena has been reduced, as has the treatment of magnetic materials. The discussion of polarization of light has been shortened and incorporated into the chapter on Nature and Propagation of Light. The material on refracting surfaces has been reorganized so that the thin-lens equation can be presented earlier.

Units and Notation. We have moved closer to 100% SI units. English units are retained in a few examples and problems in the first half of the text, but SI units are used exclusively in the second half. We use the joule as the standard unit of energy of all forms, including heat. In examples, units are always carried through all stages of numerical calculations. As usual, boldface symbols are used for vector quantities, and in addition boldface +, −, and = signs are used to remind the student at every opportunity of the crucial distinctions between operations with vectors and those with numbers.

Supplements. A textbook should stand on its own feet. Yet some students benefit from supplementary materials designed to be used with the text. With this thought in mind, we offer a *Study Guide* and a *Solutions Manual*. The *Study Guide*, prepared by Professors James R. Gaines and William F. Palmer, includes for each chapter & statement of objectives, a review of central concepts, problem-solving hints, additional worked-out examples, and a short quiz. The *Solutions Guide*, prepared by Professor A. Lewis Ford, includes completely worked-out solutions for about one third of the problems in the book, drawn from odd-numbered problems only. Answers to all odd-numbered problems are listed at the end of the text, and a booklet containing answers to all even-numbered problems can be obtained by instructors from the publisher.

Reviewers. Many of the changes in this edition are a direct result of recommendations from colleagues who have used earlier editions with their students. The views and suggestions collected through reviews, written questionnaires, telephone surveys, and discussion groups have been invaluable. In addition to those named in connection with the problem revisions, I gratefully acknowledge the very helpful and valuable contributions of the following reviewers and discussion-group participants:

Alex Azima, Lansing Community College
Dilip Balamore, Nassau Community College
Arun Bansil, Northeastern University
Albert Bartlett, University of Colorado
Lev I. Berger, San Diego State University
James Brooks, Boston University
Nicholas E. Brown, California Polytechnic University–San Luis Obispo
Hans Courant, University of Minnesota
Gayl Cook, University of Colorado
Bruce A. Craver, University of Dayton
Steve Detweiler, University of Florida
Lewis Ford, Texas A & M University
Walter S. Gray, University of Michigan
Graham D. Gutsche, U. S. Naval Academy–Annapolis
Michael J. Harrison, Michigan State University
Howard Hayden, University of Connecticut
Lorella Jones, University of Illinois
Jean P. Krisch, University of Michigan
Alfred Leitner, Rensselaer Polytechnic University
David Markowitz, University of Connecticut
Joseph L. McCauley, University of Houston
T. K. McCubbin, Jr., Pennsylvania State University
Thomas Meyer, Texas A & M University
Herbert Muether, S.U.N.Y.–Stony Brook
Jack Munsee, California State University–Long Beach

Lorenzo Narducci, Drexel University
Van E. Neie, Purdue University
David A. Nordling, U. S. Naval Academy–Annapolis
W. F. Parks, University of Missouri
Arnold Perlmutter, University of Miami
John S. Risley, North Carolina State University
Richard Roth, Eastern Michigan University
Rajarshi Roy, Georgia Institute of Technology
Russell A. Roy, Santa Fe Community College
Stan Shepherd, Pennsylvania State University
Malcolm Smith, University of Lowell
James Smith, U. S. Military Academy–West Point
Conley Stutz, Bradley University
G. David Toot, Alfred University
George Williams, University of Utah
John Williams, Auburn University
D. H. Ziebell, Manatee Community College
George O. Zimmerman, Boston University

With the departure of Professors Sears and Zemansky from this life, I have assumed sole responsibility for the book. I feel a little like a violinmaker who is asked to take a Stradivarius apart and repair it. It is an honor to be asked to do it, but it is also an awesome responsibility. I have taken great care to remain true to the original spirit of this text, while making it as useful for today's students as the first edition was for its users a few decades ago.

Acknowledgments. A special debt of gratitude is owed to the author's colleagues at Carnegie-Mellon, especially Professors Robert Kraemer, Bruce Sherwood, and Helmut Vogel, for many stimulating discussions about physics pedagogy, and to Professor Kraemer for major contributions to the high-energy physics material. An equally important debt of a different kind is owed to Dr. Michael Schur for his support and encouragement when they were most needed. Finally and most important, I offer my gratitude to my wife Alice and our children Gretchen and Rebecca for their love, support, and emotional sustenance during a difficult period in my life. May all men be blessed with love such as theirs.

As always, I welcome communications from students and professors, especially when they concern errors or deficiencies that may be found in this edition. I have written the best book I know how to write; I hope it will help you to teach and learn physics. In turn, you can help me by letting me know what still needs to be improved!

Pittsburgh, Pennsylvania   H. D. Y.
November 1986

CONTENTS

PART TWO

MECHANICS—FURTHER DEVELOPMENTS 143

7

8

9

14

15

16

17

PART FOUR

WAVES 473

21

MECHANICAL WAVES   475

22

VIBRATING BODIES   496

23

ACOUSTIC PHENOMENA   512

APPENDIXES

ABRIDGED CONTENTS

PART SEVEN

OPTICS

PART EIGHT

MODERN PHYSICS

VOLUME ONE

UNIVERSITY PHYSICS

SEVENTH EDITION

MECHANICS—FUNDAMENTALS

PROSPECTUS

The study of physics is an adventure. It is challenging, sometimes frustrating, occasionally painful, and often richly rewarding and satisfying. It appeals to the emotions and the aesthetic sense as well as to the intellect. The achievements of such scientific giants as Galileo, Newton, Maxwell, and Einstein form the foundation for our present understanding of the physical world. You can share the excitement of discovery that they experienced when you learn the value of physics in solving practical problems and in gaining insight into everyday phenomena, and its significance as an achievement of the human intellect in its quest for understanding of the world we all live in.

We begin our study of physics with the subject of *mechanics:* the study of motion and its causes. This is a natural starting point; everyday experience offers abundant examples of mechanical principles, more than for any other area of physics. In the opening chapter we introduce several elements of the language of physics, including units, calculational techniques, and vector algebra. In the following chapters we develop detailed language for describing motion. We begin with motion of a single particle, which is a body with no size and no shape; we represent such a body as a geometric point. The simplest case is motion of a point along a straight line. We then progress to examples of motion in a plane, including motion in parabolic and circular paths. Next we consider the relationships between motion and the forces that are always associated with it; whenever a particle speeds up, slows down, or changes the direction of its motion, there is always an associated force. The relationships of force and motion are summarized neatly in Newton's three laws of motion.

In studying these laws, we also study the interplay of theory and experiment; every physical theory must be grounded in experimental observations of phenomena in the physical world. As we learn how to apply these laws to a variety of practical problems, we develop systematic problem-solving procedures that help up set up problems and carry out solutions efficiently and accurately. We also begin to appreciate the role of idealized models: approximate representations of physical situations that are simplifed to facilitate analysis and calculation. Equally important, we begin to develop a sense of the beauty of physics as we learn how the essential relationships in the entire area of mechanics are wrapped up in a wonderfully neat and compact package that we label "Newton's laws of motion."

1

UNITS, PHYSICAL QUANTITIES, AND VECTORS

WE BEGIN OUR STUDY OF PHYSICS WITH SEVERAL IMPORTANT MATTERS of language. The first of these is the concept of *units*. To describe physical phenomena in a quantitative way, we need to use numbers; we usually describe a quantity as a multiple of some standard unit of that quantity. In this chapter we study various unit systems, the standard notations used with them, and procedures for converting quantities from one set of units to another. Closely related to these concepts are the ideas of *precision* of a measurement and its number of *significant figures*. We also study examples of problems where no precise calculations are possible but where rough *estimates* can be interesting and useful. Finally, we study several aspects of *vector algebra*. Vectors are mathematical entities that we use to describe physical quantities having directions in space as well as numerical magnitudes. As such, vectors are an essential part of the language of all areas of physics; it is important to learn this language thoroughly, and the other items we mentioned, at the beginning.

1–1 INTRODUCTION

Physics is an experimental science. Everything we know about the physical world and about the principles that govern its behavior has been learned through experiment, that is, through observations of the phenomena of nature. The ultimate test of any physical theory is its agreement with experimental observations. These observations usually involve measurements; thus physics is inherently a science of *experiment* and *measurement*. Figure 1–1 shows two well-known experimental facilities.

To learn the laws of nature, we must *observe* nature.

Any number used to describe a physical phenomenon quantitatively is called a **physical quantity.** A physical quantity is defined in one of two ways: We may specify a procedure for *measuring* the quantity, or we may describe a way to *calculate* the quantity from other quantities that we can measure. For example, in the first case we might use a ruler to measure a distance or a stopwatch to measure a time interval. In the second case we might define the average speed of a moving object as the distance traveled (measured with a ruler) divided by the time of travel (measured with a stopwatch).

(a)

(b)

1–1 Two research laboratories. (a) The Cathedral of Pisa (Italy), with the Baptistry in the foreground, the Cathedral behind it, and the famous Leaning Tower at the far right. According to legend, Galileo studied the motion of freely falling bodies by dropping them from the tower. He is also alleged to have gained insights into pendulum motion by observing the swinging of the hanging chandeliers in the Cathedral. The nearly parabolic dome of the Baptistry produces interesting acoustical effects. (Art Resource) (b) The Hubble Space Telescope. When this 2.4-m reflecting telescope is placed in orbit 500 km above the surface of the earth, it will permit observation of celestial objects 100,000,000 times fainter in brightness than the faintest objects visible with the best earth-based telescope. (Courtesy Lockheed Missiles and Space Company)

Operational definitions: describing how to measure a quantity

A definition that gives a procedure for measuring the defined quantity is called an **operational definition.** Some quantities, such as mass, length, and time, are so fundamental that they can be defined only with operational definitions. Later we will encounter operational definitions of other fundamental quantities such as temperature (Chapter 14) and electric current (Chapter 31).

1–2 STANDARDS AND UNITS

Units: standards for describing magnitudes of physical quantities

When we measure a quantity, we always compare it with some reference standard. When we say a rope is 30 meters long, we mean that it is 30 times as long as an object, such as a meter stick, that is defined to be one meter long. Such a standard is called a **unit** of the quantity. Thus the meter is a unit of distance, and the second is a unit of time.

To make precise measurements we need definitions of the units of measurement that do not change and that can be duplicated by observers in various locations. When the metric system was established in 1791 by the Paris Academy of Sciences, the **meter** was originally defined as one ten-millionth of the distance from the equator to the North Pole, and the **second** as the time for a pendulum one meter long to swing from one side to the other.

The International System of units (SI)

These definitions were cumbersome and hard to duplicate precisely; in more recent years they have been replaced by more refined definitions. Since 1889 the definitions of the basic units have been established by an international organization, the General Conference on Weights and Measures. The system of units defined by this organization is based on the metric system, and since 1960 it has been known officially as the **International System,** or SI (the abbreviation for the French equivalent, Système International).

Until 1960 the unit of time was based on a certain fraction of the mean solar day, the average time interval between successive arrivals of the sun at its highest point in the sky. The present standard, adopted in 1967, is an atomic one, based on the two lowest energy states of the cesium atom. These two states have slightly different energies, depending on whether the spin of the outermost electron is parallel or antiparallel to the nuclear spin. Electromagnetic radiation (microwaves) of precisely the proper frequency causes transitions from one state to the other. We now define one second as the time required for 9,192,631,770 cycles of this radiation. Figure 1–2 shows the instrument currently used to establish this standard.

In 1960 the meter was defined by an atomic standard, in terms of the wavelength of the orange-red light emitted by atoms of krypton (^{86}Kr) in a glow discharge tube; one meter was defined as 1,650,763.73 of these wavelengths. In November 1983 the standard was changed again in a more radical way. The new definition of the meter is the distance light travels in 1/299,792,458 second. This has the effect of defining the speed of light to be precisely 299,792,458 m·s^{-1}; we then define the meter to be consistent with this number and with the definition of the second given above. The reason for this change is that at present we can measure the speed of light and intervals of time much more precisely than distances.

The standard of *mass* is the mass of a particular cylinder of platinum-iridium alloy. Its mass is defined to be one **kilogram,** and it is kept at the International Bureau of Weights and Measures at Sèvres, near Paris. An atomic standard of mass has not yet been adopted because at present we cannot measure masses on an atomic scale with as much precision as on a macroscopic scale.

Once the fundamental units are defined, it is easy to introduce larger and smaller units for the same physical quantities. In all versions of the metric system, including SI, these other units are always related to the fundamental units by multiples of 10 or 1/10. Thus one kilometer (1 km) is 1000 meters,

Definition of the second, the unit of time

Definition of the meter, the unit of length

Definition of the kilogram, the unit of mass

1–2 NBS-6 is the latest of six generations of primary atomic frequency standards developed by the National Bureau of Standards (NBS). Consisting of a 6-m cesium beam tube, NBS-6 achieves an accuracy of better than one part in 10^{13}, and when operated as a clock, can keep time to within 3 millionths of a second per year. (Courtesy National Bureau of Standards.)

TABLE 1–1 PREFIXES FOR POWERS OF TEN

Power of ten	10^{-18}	10^{-15}	10^{-12}	10^{-9}	10^{-6}	10^{-3}	10^{-2}	10^{3}	10^{6}	10^{9}	10^{12}	10^{15}	10^{18}
Prefix	atto-	femto-	pico-	nano-	micro-	milli-	centi-	kilo-	mega-	giga-	tera-	peta-	exa-
Abbreviation	a	f	p	n	μ	m	c	k	M	G	T	P	E

How to use unit prefixes

one centimeter (1 cm) is 1/100 meter, and so on. We usually express the multiplicative factors in exponential notation; thus $1000 = 10^3$, $1/1000 = 10^{-3}$, and so on. The names of the additional units are always derived by adding a **prefix** to the name of the fundamental unit. For example, the prefix "kilo-", abbreviated k, always means a unit larger by a factor of 1000; thus

$$1 \text{ kilometer} = 1 \text{ km} = 10^3 \text{ meters} = 10^3 \text{ m},$$
$$1 \text{ kilogram} = 1 \text{ kg} = 10^3 \text{ grams} = 10^3 \text{ g},$$
$$1 \text{ kilowatt} = 1 \text{ kW} = 10^3 \text{ watts} = 10^3 \text{ W}.$$

Table 1–1 lists the standard SI prefixes with their meanings and abbreviations. We note that most of these are multiples of 10^3.

When pronouncing unit names with prefixes, we always accent the first syllable; some examples are KIL-o-gram, KIL-o-meter, CEN-ti-meter, and MIC-ro-meter.

Here are several examples of the use of multiples of 10 and their prefixes. Some additional time units are also included.

$$1 \text{ nanometer} = 1 \text{ nm} = 10^{-9} \text{ m (a few times the size of an atom)}$$
$$1 \text{ micrometer} = 1 \text{ } \mu\text{m} = 10^{-6} \text{ m (size of some bacteria and cells)}$$
$$1 \text{ millimeter} = 1 \text{ mm} = 10^{-3} \text{ m (point of a ballpoint pen)}$$
$$1 \text{ centimeter} = 1 \text{ cm} = 10^{-2} \text{ m (diameter of your little finger)}$$
$$1 \text{ kilometer} = 1 \text{ km} = 10^{3} \text{ m (a 10-minute walk)}$$
$$1 \text{ microgram} = 1 \text{ } \mu\text{g} = 10^{-9} \text{ kg}$$
$$1 \text{ milligram} = 1 \text{ mg} = 10^{-6} \text{ kg}$$
$$1 \text{ gram} = 1 \text{ g} = 10^{-3} \text{ kg (mass of a paper clip)}$$
$$1 \text{ nanosecond} = 1 \text{ ns} = 10^{-9} \text{ s (time for light to travel 0.3 m)}$$
$$1 \text{ microsecond} = 1 \text{ } \mu\text{s} = 10^{-6} \text{ s}$$
$$1 \text{ millisecond} = 1 \text{ ms} = 10^{-3} \text{ s (time for sound to travel 0.35 m)}$$
$$1 \text{ minute} = 1 \text{ min} = 60 \text{ s}$$
$$1 \text{ hour} = 1 \text{ hr} = 3600 \text{ s}$$
$$1 \text{ day} = 1 \text{ da} = 86,400 \text{ s}$$

The British system of units

Finally, we should mention the British system of units. These units are used only in the United States and a few other countries, and they are rapidly being replaced by SI in the latter. British units are now officially defined in terms of SI units, as follows:

Length: 1 inch = 2.54 cm (exactly).
Force: 1 pound = 4.448221615260 newtons (exactly).

The fundamental British unit of time is the second, defined the same way as in SI. In physics, British units are used only in mechanics and thermodynamics; there is no British system of electrical units. In this book we use SI units for all examples and problems, but occasionally in the early chapters we give approximate equivalents in British units. A few of the exercises also use British units.

1–3 UNIT CONSISTENCY AND CONVERSIONS

We often use equations to express relations among physical quantities that are represented by algebraic symbols. An algebraic symbol always denotes both a number and a unit. For example, d might represent a distance of 10 m, t a time of 5 s, and v (for velocity) a speed of 2 m/s or 2 m·s^{-1}. (In this book we usually use negative exponents with units to avoid use of the fraction bar.)

An equation must always be **dimensionally consistent;** this means that two terms may be added or equated only if they have the same units. For example, if a body moving with constant speed v travels a distance d in a time t, these quantities are related by the equation

$$d = vt. \tag{1–1}$$

If d is measured in meters, then the product vt must also be expressed in meters. Using the numbers above as an example, we may write

$$10 \text{ m} = (2 \text{ m·s}^{-1})(5 \text{ s}).$$

Because the unit s^{-1} or 1/s cancels the unit s on the right side, the product vt is indeed expressed in meters, as it must be. In calculations, units are always treated just like algebraic symbols with respect to multiplication and division.

When a problem requires calculations using numbers with units, the numbers should always be written with the correct units, and the units should be carried through the calculation as in the example above. This provides a very useful check for calculations. If at some stage in a calculation you find that an equation or an expression has inconsistent units, you know you have made an error somewhere. In this book we will always carry units through all calculations, and we strongly urge you to follow this practice when you solve problems.

Unit consistency: You can't add apples and artichokes.

PROBLEM-SOLVING STRATEGY: *Unit conversions*

Units are multiplied and divided just like ordinary algebraic symbols. This fact provides a convenient procedure for converting a quantity from one set of units to another. The key to the procedure is the fact that we can use equality to represent the same physical quantity when we express it in two different units. For example, to say that 1 min = 60 s does not mean that the number 1 is equal to the number 60; it means that 1 min represents the same physical time interval as 60 s. Thus we may multiply a quantity by 1 min and then divide it by 60 s, or multiply by the quantity (1 min/60 s), without changing its physical meaning. To find the number of seconds in 3 min, we write

$$3 \text{ min} = (3 \text{ min})\left(\frac{60 \text{ s}}{1 \text{ min}}\right) = 180 \text{ s}.$$

This procedure is sometimes called the **factor-label method.**

EXAMPLE 1–1 American women in the age group 19 to 22 years have an average height of 5 ft, 4 in. What is this height in centimeters? In meters?

SOLUTION We first express the height in inches:

$$5 \text{ ft} = \left(\frac{12 \text{ in.}}{1 \text{ ft}}\right) 5 \text{ ft} = 60 \text{ in.}$$

$$5 \text{ ft, 4 in.} = 5 \text{ ft} + 4 \text{ in.} = 60 \text{ in.} + 4 \text{ in.} = 64 \text{ in.}$$

Then

$$64 \text{ in.} = \left(\frac{2.54 \text{ cm}}{1 \text{ in.}}\right) 64 \text{ in.} = 163 \text{ cm}$$

(The product has been rounded to the nearest centimeter.) Finally,

$$163 \text{ cm} = \left(\frac{1 \text{ m}}{100 \text{ cm}}\right) 163 \text{ cm} = 1.63 \text{ m}.$$

EXAMPLE 1–2   A woman drives a car in Germany at 50 km·hr^{-1} (50 kilometers per hour). Express this speed in meters per second.

SOLUTION

$$50 \text{ km·hr}^{-1} = (50 \text{ km·hr}^{-1})\left(\frac{1000 \text{ m}}{1 \text{ km}}\right)\left(\frac{1 \text{ hr}}{3600 \text{ s}}\right) = 13.89 \text{ m·s}^{-1}.$$

1–4 PRECISION AND SIGNIFICANT FIGURES

Precision: How exact is that number?

Measurements always have uncertainties. When we measure a distance with an ordinary ruler, it is usually reliable only to the nearest millimeter, while a precision micrometer caliper can measure distances dependably to 0.01 mm or even less. We often indicate the precision of a number by writing the number, the symbol ±, and a second number indicating the maximum likely uncertainty. If the diameter of a steel rod is given as 56.47 ± 0.02 mm, this means that the true value is unlikely to be less than 56.45 mm or greater than 56.49 mm.

Percent uncertainty

We can also express precision in terms of the maximum likely fractional or percent uncertainty. A resistor labeled "47 ohms, 10%" probably has a true resistance differing from 47 ohms by no more than 10% of 47 ohms, or about 5 ohms; that is, the resistance is between about 42 and 52 ohms. In the steel rod example above, the fractional uncertainty is (0.02 mm)/(56.47 mm), or about 0.00035; the percent uncertainty is (0.00035)(100%), or about 0.035%.

When we use numbers having uncertainties or errors to compute other numbers, the computed numbers are also uncertain. It is especially important to understand this when comparing a number obtained from measurements with a value obtained from a theoretical prediction. Suppose you want to verify the value of π, the ratio of the circumference to the diameter of a circle. The true value of this ratio, to ten digits, is 3.141592654. To make your own calculation, you draw a large circle and measure its diameter and circumference to the nearest millimeter, obtaining the values 135 mm and 424 mm. You punch these into your calculator and obtain the quotient 3.140740741. Does this agree with the true value or not?

Significant figures: an indication of precision

To answer this question we must first recognize that at least the last six digits in your answer are meaningless because they imply greater precision than is possible with your measurements. The number of meaningful digits in a number is called the number of **significant figures;** usually a numerical result has no more significant figures than the numbers from which it is computed. Thus your value of π has only three significant figures and should be stated simply as 3.14 or possibly as 3.141 (rounded to four figures). Within the limit of three significant figures, your value does agree with the true value.

In the examples and problems in this book, we usually assume that the numerical values we give are precise to three or at most four significant figures, and thus your answers should show at most four significant figures. You may do the arithmetic with a calculator having a display with five to ten digits. But you should not give a ten-digit answer for a calculation using numbers with three significant figures. To do so is not only unnecessary but it is also genuinely wrong because it misrepresents the precision of the results. Always round your answer to keep only the correct number of significant figures, or in doubtful cases one more at most. Thus in Example 1–2 the result should have been stated as 13.9, or 14 m·s⁻¹. Of course, significant figures provide only a crude representation of the reliability of a number. The fractional uncertainty of 104 is not appreciably less than that of 96, despite the difference in number of significant figures. When a better representation of uncertainty is needed, more sophisticated statistical methods are used.

In calculations with very large or very small numbers, we can show significant figures much more easily by using powers-of-ten notation, sometimes called **scientific notation.** The distance from the earth to the sun is about 149,000,000,000 m, but to write the number in this form gives no indication of the number of significant figures. Certainly not all 12 are significant! Instead, we move the decimal point 11 places to the left (corresponding to dividing by 10^{11}) and multiply by 10^{11}. That is,

$$149{,}000{,}000{,}000 \text{ m} = 1.49 \times 10^{11} \text{ m}.$$

Scientific notation: using powers of ten to represent very large and very small numbers

In this form it is clear that the number of significant figures is three. In scientific notation the usual practice is to express the quantity as a number between 1 and 10 multiplied by the appropriate power of ten.

We can use the same technique when we multiply or divide very large or very small numbers. For example, the energy E corresponding to the mass m of an electron is given by the equation

$$E = mc^2, \tag{1-2}$$

where c is the speed of light. The appropriate numbers are $m = 9.11 \times 10^{-31}$ kg and $c = 3.00 \times 10^8$ m·s⁻¹. We find

$$E = (9.11 \times 10^{-31} \text{ kg})(3.00 \times 10^8 \text{ m·s}^{-1})^2$$
$$= (9.11)(3.00)^2(10^{-31})(10^8)^2 \text{ kg·m}^2\text{·s}^{-2}$$
$$= (82.0)(10^{[-31+(2\times8)]}) \text{ kg·m}^2\text{·s}^{-2}$$
$$= 8.20 \times 10^{-14} \text{ kg·m}^2\text{·s}^{-2}.$$

Most calculators use scientific notation and do this addition of exponents automatically for you; but you should be able to do such calculations by hand when necessary. Incidentally, the value used for c has three significant figures even though two of them are zeros. To greater accuracy, $c = 2.997925 \times 10^8$ m·s⁻¹; thus it would *not* be correct to write $c = 3.000 \times 10^8$ m·s⁻¹.

1–5 ESTIMATES AND ORDERS OF MAGNITUDE

We have discussed the importance of knowing the precision of numbers that represent physical quantities. But there are also situations where even a very crude estimate of a quantity may provide useful information. We may know how to calculate a certain quantity if we are given the necessary input data, but

Estimates: A good guess is better than no information at all.

those data may be unavailable or may have to be guessed at. Or the calculation may be too complicated to carry out exactly. In any case, the result of such a calculation is also a guess, but in some situations a guess is useful, even if it is uncertain by a factor or two or ten or more. Such calculations are often called **order-of-magnitude** calculations or estimates.

EXAMPLE 1–3 You are writing an international espionage novel in which the hero escapes across the border with a billion dollars' worth of gold in a suitcase. Is this possible? Would it fit? Would it be too heavy to carry?

SOLUTION Gold sells for around $400 an ounce. On a particular day it may be $200 or $600, but never mind. An ounce is about 30 grams. Actually, an ordinary (avoirdupois) ounce is 28.35 g; an ounce of gold is a troy ounce, which is 9.45% more. Again, never mind. Ten dollars' worth of gold has a mass somewhere around one gram, so a billion (10^9) dollars' worth is a hundred million (10^8) grams, or a hundred thousand (10^5) kilograms. This corresponds to a weight in British units of around 200,000 lb, or a hundred tons. Whether the precise number is 50 tons or 200 does not matter; either way, our hero is not about to carry it across the border in a suitcase.

We can also estimate the *volume* of this gold. If its density were the same as that of water (1 g·cm^{-3}), the volume would be 10^8 cm^3, or 100 m^3. But gold is a heavy metal; we might guess its density as ten times that of water. It is actually 19.3 times as dense as water. But guessing ten, we find a volume of 10 m^3. Visualize ten cubical stacks of gold bricks, each one meter on a side, and ask whether it would fit in a suitcase!

Exercises 1–15 through 1–23 at the end of this chapter are of the estimating or "order-of-magnitude" variety. Some are silly, and most require much guesswork for the needed input data. Do not try to look up a lot of data; make the best guesses you can. Even when they are off by a factor of ten, the results can be useful and interesting.

1–6 VECTORS AND VECTOR ADDITION

Some physical quantities, such as time, temperature, mass, density, and electric charge, can be described completely by a single number with a unit. Many other quantities, however, have a *directional* quality that cannot be described by a single number. A familiar example is velocity. To describe the motion of a body we must say not only how fast it is moving but also in what direction. Force is another example. When we push or pull on a body, we exert a force on it. To describe a force we need to describe the direction in which it acts, as well as its magnitude, or "how hard" the force pushes or pulls.

A physical quantity that is described by a single number is called a **scalar quantity;** a quantity having both magnitude (the "how much" or "how big" part) and direction is called a **vector quantity.** Calculations with scalar quantities use the operations of ordinary arithmetic, but calculations with vector quantities are somewhat different. Vector quantities play an essential role in

Scalar and vector quantities: Does the quantity have direction or only a "how much" number?

all areas of physics, and so we turn now to a discussion of their nature and the operation of vector addition.

We begin with a vector quantity called **displacement.** When a particle, which we represent as a point, moves from one location in space to another, it undergoes a displacement. In Fig. 1–3a we represent the change of position from point P_1 to point P_2 by the directed line segment P_1P_2, with an arrowhead at P_2 to represent the direction of motion. Displacement is a vector quantity because we must state not only how far the particle moves but also in what direction. A displacement of 3 km north is not the same as a displacement of 3 km southeast.

We usually represent a displacement by a single letter, such as *A* in Fig. 1–3a. In this book we always print vector symbols in boldface type as a reminder that vector quantities have different properties from scalar quantities. In handwriting, vector symbols are usually underlined or written with an arrow above, as shown in Fig. 1–3c, to indicate that they represent vector quantities.

The vector from point P_3 to point P_4 in Fig. 1–3b has the same length and direction as the one from P_1 to P_2. These two displacements are equal, even though they start at different points. By definition, two vector quantities are equal if they have the same magnitude (length) and direction, no matter where they are located in space. The vector **B**, however, is not equal to **A** because its direction is opposite to that of **A.** We define the *negative* of a vector as a vector having the same magnitude as, but the opposite direction to, the original vector. The negative of vector quantity **A** is denoted as −**A,** and we use a boldface "minus" to emphasize the vector nature of the quantities. Thus the relation between **A** and **B** may be written as **A = −B** or **B = −A.** The vectors **A** and **B** are *antiparallel*. Note that a boldface "equals" sign is also used to emphasize that equality of two vector quantities is not the same relationship as equality of scalar quantities.

Displacement is always a straight-line segment, directed from the starting point to the endpoint, even though the path of the particle may be curved. Thus in Fig. 1–4, when the particle moves along the curved path shown from P_1 to P_2, the displacement is still the vector **A** shown. Also, when it continues on to P_3 and then returns to P_1, the displacement for the entire trip is zero.

We represent the **magnitude** of a vector quantity (its length, in the case of a displacement vector) by the same letter used for the vector, but in light italic type rather than boldface italic. An alternative notation is the vector symbol with vertical bars on both sides. Thus

$$\text{(Magnitude of } \mathbf{A}\text{)} = A = |\mathbf{A}|. \tag{1–3}$$

By definition, the magnitude of a vector quantity is a scalar quantity (a single number) and is always positive. We also note that a vector quantity can never be equal to a scalar one because they are different kinds of quantities. The expression *A* = 6 m is just as wrong as 2 oranges = 3 apples or 6 lb = 7 km!

Now suppose a particle undergoes a displacement **A,** followed by a second displacement **B,** as shown in Fig. 1–5a. The final result is the same as though it had started at the same initial point and undergone a single displacement **C,** as shown. We call displacement **C** the **vector sum** of displacements **A** and **B;** the relationship is expressed symbolically as

$$\mathbf{C} = \mathbf{A} + \mathbf{B}. \tag{1–4}$$

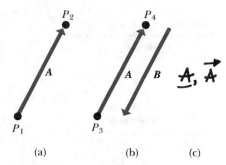

(a) (b) (c)

1–3 (a) Vector **A** is the displacement from point P_1 to point P_2. (b) The displacement from P_3 to P_4 is equal to that from P_1 to P_2, but displacement **B** is the negative of displacement **A.**

Equality of vector quantities

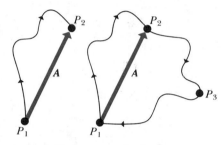

1–4 A displacement is always a straight line segment, directed from the starting point to the endpoint, even if the actual path is curved. When a point ends at the same place it started, the displacement is zero.

Magnitude of a vector quantity

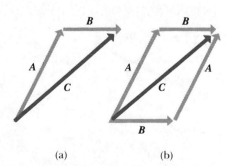

(a) (b)

1–5 (a) Vector **C** is the vector sum of vectors **A** and **B.** (b) The order in vector addition is immaterial.

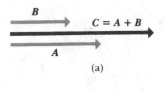

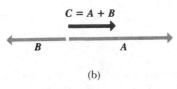

1–6 Vector sum of (a) two parallel vectors, and of (b) two antiparallel vectors.

Resultant: another name for vector sum

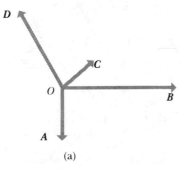

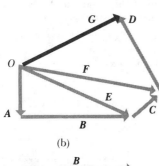

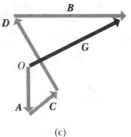

1–7 Polygon method of vector addition.

The boldface "plus" sign emphasizes that adding two vector quantities requires a geometrical process and is not the same operation as adding two scalar quantities, such as $2 + 3 = 5$.

If we make the displacements A and B in the reverse order, as in Fig. 1–5b, with B first and A second, the result is the same, as the figure shows. Thus

$$C = B + A \quad \text{and} \quad A + B = B + A. \quad (1–5)$$

This shows that vector addition obeys the *commutative law;* The order of the terms in the sum does not matter.

Figure 1–5b also suggests an alternative graphical representation of the vector sum: When vectors A and B are both drawn from a common point, vector C is the diagonal of a parallelogram constructed with A and B as two adjacent sides.

The vector sum is often called the **resultant;** thus the vector sum C of vectors A and B can be called the resultant displacement.

Figure 1–6 shows a special case in which two vectors A and B are parallel, as in (a), or antiparallel, as in (b). When they are parallel, the magnitude of the vector sum equals the *sum* of the magnitudes of A and B; when they are antiparallel, it equals the *difference* of their magnitudes. The vectors in Fig. 1–6 have been displaced slightly sidewise to show them more clearly, but they actually lie along the same geometric line.

When more than two vectors are to be added, we may first find the vector sum of any two, add this vectorially to the third, and so on. This process is shown in Fig. 1–7; part (a) shows four vectors A, B, C, and D. In Fig. 1–7b, vectors A and B are first added, giving a vector sum E; vectors E and C are then added by the same process, to obtain the vector sum F; finally F and D are added, to obtain the vector sum

$$G = A + B + C + D.$$

We do not need to draw vectors E and F; all we need do is draw the given vectors in succession, with the tail of each at the head of the one preceding it, and complete the polygon by a vector G from the *tail* of the first to the *head* of the last vector. The order makes no difference, as shown in Fig. 1–7c; we invite you to try other orders.

Diagrams for addition of displacement vectors do not need to be drawn actual size. It is often convenient to use a scale similar to those used for maps, where the distance on the diagram is proportional to the actual distance, such as 1 cm for 5 km. When we work with other vector quantities whose units are not distance units, we *must* use a scale. For example, in a diagram for force vectors we might use a scale in which a vector 1 cm long represents a force of magnitude 5 N. (The newton, abbreviated N, is the SI unit of force.) A 20-N force would then be represented by a vector 4 cm long with the appropriate direction.

A vector quantity such as a displacement can be multiplied by a scalar quantity (an ordinary number). The displacement $2A$ is a displacement (vector quantity) in the same direction as the vector A but twice as long. The scalar quantity used to multiply a vector may be a physical quantity having units. For example, you may be familiar with the relationship $F = ma$; force F (a vector quantity) is equal to the product of mass m (a scalar quantity) and acceleration *a* (a vector quantity). The magnitude of the force is equal to the mass multiplied by the magnitude of the acceleration, and the unit of the magnitude of force is the product of the unit of mass and that of the magnitude of acceleration.

We have already mentioned the special case of multiplication by -1: $(-1)A = -A$ is by definition a vector having the same magnitude as *A* but the opposite direction. This provides the basis for defining vector subtraction. We define the difference $A - B$ of the two vectors *A* and *B* to be the vector sum of *A* and $-B$:

$$A - B = A + (-B). \tag{1-6}$$

The boldface $+$, $-$, and $=$ signs remind us of the vector nature of these operations.

Subtraction of vector quantities

1–7 COMPONENTS OF VECTORS

Addition and subtraction of vectors are usually carried out by the use of **components.** To define components we use a rectangular (cartesian) coordinate-axis system as in Fig. 1–8. We can represent any vector lying in the *xy*-plane as the sum of a vector parallel to the *x*-axis and a vector parallel to the *y*-axis. These two vectors are labeled A_x and A_y in the figure; they are called the component vectors of vector *A.* The relation is expressed formally as

$$A = A_x + A_y. \tag{1-7}$$

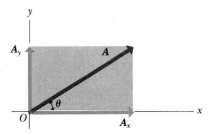

1–8 Vectors A_x and A_y are the rectangular components of *A* in the directions of the *x*- and *y*-axes.

By definition, each component vector lies along a coordinate-axis direction. Thus only a single number is needed to describe each one. For example, the number A_x is, apart from a possible negative sign, the magnitude of the component vector A_x. We further specify that A_x is positive when A_x points in the positive axis direction and negative when it points in the opposite direction. The two numbers A_x and A_y are called simply the components of *A.*

If we know the magnitude *A* of the vector *A* and its direction, given by angle θ in Fig. 1–8, we can calculate the components. From the definitions of the trigonometric functions,

$$\frac{A_x}{A} = \cos \theta \qquad \text{and} \qquad \frac{A_y}{A} = \sin \theta;$$

$$A_x = A \cos \theta \qquad \text{and} \qquad A_y = A \sin \theta. \tag{1-8}$$

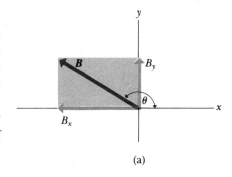

(a)

In Fig. 1–9, the component B_x is negative, because its direction is opposite to that of the positive *x*-axis. This is consistent with Eqs. (1–8); the cosine of an angle in the second quadrant is negative. The component B_y is positive, but both C_x and C_y are negative.

We can use either the magnitude and direction or the *x*- and *y*-components of a vector quantity to describe it completely. Equations (1–8) show how to obtain the components if the magnitude and direction are given. Or if we are given the components, we can find the magnitude and direction. Applying the Pythagorean theorem to Fig. 1–8, we find

$$A = \sqrt{A_x{}^2 + A_y{}^2}. \tag{1-9}$$

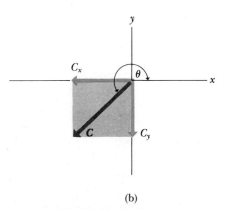

(b)

1–9 Components of a vector may be positive or negative numbers.

Also, from the definition of the tangent of an angle,

$$\tan \theta = \frac{A_y}{A_x} \quad \text{and} \quad \theta = \arctan \frac{A_y}{A_x}. \qquad (1\text{--}10)$$

There is one slight complication in using Eq. (1–10) to find θ. Suppose $A_x = 2$ m and $A_y = -2$ m; then $\tan \theta = -1$. But there are two angles having tangents of -1, namely 135° and 315° (or -45°). To decide which is correct we must look at the individual components; because A_x is positive and A_y is negative, the angle must be in the fourth quadrant; thus $\theta = 315°$ (or -45°) is the correct value. Most pocket calculators give arctan $(-1) = -45°$; in this case that is correct, but if instead we have $A_x = -2$ m and $A_y = 2$ m, then the correct angle is 135°. Thus you should always draw a sketch to check which of the two possibilities is the correct one.

Components provide an efficient means of calculating the vector sum of several vectors. In principle, such sums can always be carried out by using diagrams such as those in Figs. 1–5 and 1–7; but to do this we must either draw and measure a scale diagram (which is hard to do accurately) or carry out a trigonometric solution of oblique triangles (which can be very complicated). The component method, by contrast, requires only right triangles and simple computations, and it can be carried out with great accuracy.

Here is the basic idea of the component method. Figure 1–10 shows two vectors A and B and their vector sum (resultant) C, along with x- and y-components of all three vectors. You can see from the diagram that the x-component C_x of the vector sum is simply the sum $(A_x + B_x)$ of the x-components of the vectors being added. The same is true for the y-components.

$$C_x = A_x + B_x,$$
$$C_y = A_y + B_y. \qquad (1\text{--}11)$$

Thus once we know the components of A and B, perhaps by using Eqs. (1–8), we can compute the components of the vector sum. Then if the magnitude and direction of C are needed, we can obtain them from Eqs. (1–9) and (1–10).

This procedure for finding the sum of two vectors can easily be extended to any number. Let R be the vector sum of $A, B, C, D, E, \ldots$. Then

$$R_x = A_x + B_x + C_x + D_x + E_x + \cdots,$$
$$R_y = A_y + B_y + C_y + D_y + E_y + \cdots. \qquad (1\text{--}12)$$

When we have to add a large number of vectors, the component method is often the only reasonable method.

Using components to calculate vector sums

1–10 C_x is the x-component of the vector sum C of vectors A and B, and is equal to the sum of the x-components of A and B. The y-components are similarly related.

1–11 Three successive displacements $A, B,$ and $C,$ and the resultant or vector sum displacement $R = A + B + C.$

EXAMPLE 1–4 The pilot of a private plane flies 20.0 km in a direction 60° north of east, then 30.0 km straight east, then 10.0 km straight north. How far and in what direction is the plane from the starting point?

SOLUTION The situation is shown in Fig. 1–11. We have chosen the x-axis as east and the y-axis as north, the usual choice for maps. Let A be the first displacement, B the second, C the third, and R the vector sum or resultant displacement. We estimate from the diagram that R is about 50 km, at an angle of about 30°. We can later check this estimate against our calculated results.

The components of A are

$$A_x = (20.0 \text{ km})(\cos 60°) = 10.0 \text{ km},$$
$$A_y = (20.0 \text{ km})(\sin 60°) = 17.3 \text{ km}.$$

The components of all the displacements and the calculations can be arranged systematically, as in Table 1–2.

TABLE 1–2

Distance	Angle	x-component	y-component
$A = 20.0$ km	60°	10.0 km	17.3 km
$B = 30.0$ km	0°	30.0 km	0
$C = 10.0$ km	90°	0	10.0 km
		$R_x = 40.0$ km	$R_y = 27.3$ km

$$R = \sqrt{(40.0 \text{ km})^2 + (27.3 \text{ km})^2} = 48.4 \text{ km},$$

$$\theta = \arctan \frac{27.3 \text{ km}}{40.0 \text{ km}} = 34.3°.$$

Alternatively, we can find θ first, as above, then use Eqs. (1–8) to find R:

$$R = \frac{R_x}{\cos \theta} = \frac{40.0 \text{ km}}{\cos 34.3°} = 48.4 \text{ km},$$

or

$$R = \frac{R_y}{\sin \theta} = \frac{27.3 \text{ km}}{\sin 34.3°} = 48.4 \text{ km}.$$

This discussion has been confined to vectors lying in the xy-plane, but we can easily generalize it to vectors having any direction in space. We introduce a z-axis perpendicular to the xy-plane; then in general a vector A has components A_x, A_y, and A_z in the three coordinate directions. The magnitude A is given by

$$A = \sqrt{A_x^2 + A_y^2 + A_z^2}. \tag{1-13}$$

1–8 UNIT VECTORS

A **unit vector** is a vector having a magnitude of unity, with no units. Its only purpose is to describe a direction in space. Unit vectors provide a convenient notation in many expressions involving components of vectors.

Unit vectors: a handy way to describe directions in space

In an xy-coordinate system we can define a unit vector i that points in the direction of the positive x-axis, and a unit vector j in the direction of the positive y-axis. Then we can express the relationships between component vectors and components, described at the beginning of Section 1–7, as follows:

$$A_x = A_x i, \qquad A_y = A_y j. \tag{1-14}$$

Similarly, we write vector A in terms of its components as

$$A = A_x i + A_y j. \tag{1-15}$$

Equations (1–14) and (1–15) are *vector* equations; each term, such as $A_x i$, is a vector quantity, and the boldface = and + signs denote vector equality and addition.

When two vectors **A** and **B** are represented in terms of their components, we can express the vector sum using unit vectors, as follows:

$$A = A_x i + A_y j, \qquad B = B_x i + B_y j,$$
$$C = A + B$$
$$= (A_x i + A_y j) + (B_x i + B_y j)$$
$$= (A_x + B_x)i + (A_y + B_y)j$$
$$= C_x i + C_y j. \tag{1–16}$$

Equation (1–16) restates the content of Eqs. (1–11) in the form of a single vector equation rather than two ordinary equations.

If the vectors do not all lie in the *xy*-plane, then a third component is needed. We introduce a third unit vector **k** in the *z*-axis direction. The generalized forms of Eqs. (1–15) and (1–16) are

$$A = A_x i + A_y j + A_z k, \tag{1–17}$$
$$C = (A_x + B_x)i + (A_y + B_y)j + (A_z + B_z)k$$
$$= C_x i + C_y j + C_z k. \tag{1–18}$$

1–9 PRODUCTS OF VECTORS

Many physical relationships can be expressed concisely by the use of *products* of vectors. Because vectors are not ordinary numbers, ordinary multiplication is not directly applicable to vectors. Just as addition of vectors is different from addition of scalars, so it is with multiplication. In fact, there are two different kinds of vector products. The first, called the scalar product, yields a result that is a scalar quantity, while the second, the vector product, yields another vector.

Here is the definition of the **scalar product** of two vectors **A** and **B**. We draw the two vectors from a common point, as in Fig. 1–12a. The angle between their directions is θ, as shown. We define the scalar product, denoted by **A · B,** as

$$A \cdot B = AB \cos \theta = |A||B| \cos \theta \tag{1–19}$$

Because of this notation, the scalar product is also called the *dot product*. It is a scalar quantity, not a vector, and it may be positive or negative. When θ is between zero and 90°, the scalar product is positive; when θ is between 90° and 180°, it is negative; and when $\theta = 90°$, **A · B** = 0. *The scalar product of two perpendicular vectors is always zero.*

The scalar product obeys the *commutative* law of multiplication; the order of the two vectors doesn't matter. For any two vectors **A** and **B, A · B** = **B · A.** This property follows directly from the definition.

Scalar product: one of the ways to multiply vectors

The scalar product of two vectors is a scalar, not a vector.

1–12 (a) Two vectors drawn from a common starting point to define their scalar product. (b) $B \cos \theta$ is the component of **B** in the direction of **A**, and **A · B** is the product of this component with the magnitude of **A**.

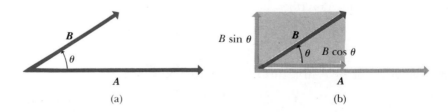
(a)   (b)

We can represent vector B in terms of a component parallel to A and a component perpendicular to A, as shown in Fig. 1–12b; the component parallel to A is $(B \cos \theta)$. Thus, from Eq. (1–19), $A \cdot B$ is equal to the component of B parallel to A, multiplied by the magnitude of A. Alternatively, it is also the component of A parallel to B, multiplied by the magnitude of B.

If we know the components of A and B, we can calculate their scalar product. The easiest procedure is to use the unit vector representation introduced in Section 1–8. First,

Calculating the scalar product from the components of the vectors

$$A \cdot B = (A_x i + A_y j + A_z k) \cdot (B_x i + B_y j + B_z k). \qquad (1\text{–}20)$$

We expand the product of the two sets of parentheses on the right, obtaining nine terms in all, as follows:

$$
\begin{aligned}
A \cdot B = \; & A_x i \cdot B_x i + A_x i \cdot B_y j + A_x i \cdot B_z k \\
& + A_y j \cdot B_x i + A_y j \cdot B_y j + A_y j \cdot B_z k \\
& + A_z k \cdot B_x i + A_z k \cdot B_y j + A_z k \cdot B_z k.
\end{aligned}
\qquad (1\text{–}21)
$$

Each of these terms contains the scalar product of two vectors that are either parallel or perpendicular. For example, in $A_x i \cdot B_x i$, the two vectors are parallel, the angle between them is zero, its cosine is unity, and the scalar product is simply the product $A_x B_x$ of the magnitudes. But in $A_x i \cdot B_y j$, the two vectors are perpendicular and the scalar product is zero. Thus six of the nine terms are zero, and the three that survive give simply

$$A \cdot B = A_x B_x + A_y B_y + A_z B_z. \qquad (1\text{–}22)$$

EXAMPLE 1–5   Find the angle between the two vectors

$$A = 2i + 3j + 4k, \qquad B = i - 2j + 3k.$$

SOLUTION   We have

$$
\begin{array}{ll}
A_x = 2 & B_x = 1 \\
A_y = 3 & B_y = -2 \\
A_z = 4 & B_z = 3
\end{array}
$$

The scalar product is given by either Eq. (1–19) or (1–22). Equating these two and rearranging, we obtain

$$\cos \theta = \frac{A_x B_x + A_y B_y + A_z B_z}{AB}. \qquad (1\text{–}23)$$

In our example,

$$A_x B_x + A_y B_y + A_z B_z = (2)(1) + (3)(-2) + (4)(3) = 8,$$

$$A = \sqrt{2^2 + 3^2 + 4^2} = \sqrt{29}, \qquad B = \sqrt{1^2 + (-2)^2 + 3^2} = \sqrt{14},$$

$$\cos \theta = \frac{8}{\sqrt{29}\sqrt{14}} = 0.397,$$

and

$$\theta = 66.6°.$$

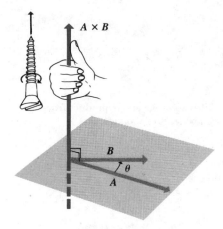

1–13 Vectors **A** and **B** lie in a plane; the vector product **A** × **B** is perpendicular to this plane, in a direction determined by the "right-hand rule."

Our first physical application of the scalar product will come in Chapter 7 with the concept of work. When a constant force **F** is applied to a body that undergoes a displacement **d,** the word W (a scalar quantity) done by the force is given by

$$W = \mathbf{F} \cdot \mathbf{d}. \tag{1–24}$$

The **vector product** of two vectors **A** and **B** is denoted by **A** × **B**. To define the vector product (also called the *cross product* because of this notation) we again draw **A** and **B** from a common point. The two vectors then lie in a plane. We define the vector product as a vector quantity with a direction perpendicular to this plane (i.e., perpendicular to both **A** and **B**) and a magnitude given by $AB \sin \theta$. That is, if **C** = **A** × **B,** then

$$C = AB \sin \theta. \tag{1–25}$$

We measure the angle θ from **A** toward **B** and take it always to be between 0 and 180°. Thus C in Eq. (1–25) is always positive, as a vector magnitude must be. We note also that when **A** and **B** are parallel or antiparallel, $\theta = 0$ or 180° and $C = 0$. That is, *the vector product of two parallel or antiparallel vectors is always zero.*

There are always *two* directions perpendicular to a given plane. To distinguish between these, we imagine rotating vector **A** about the perpendicular line until it is aligned with **B** (choosing the smaller of the two possible angles). We then curl the fingers of the right hand around this perpendicular line so that the fingertips point in the direction of rotation; the thumb then gives the direction of the vector product. This rule is shown in Fig. 1–13. Alternatively, the direction of the vector product is the direction a right-hand-thread screw advances if turned in the sense from **A** toward **B,** as shown in the figure.

Similarly, we determine the direction of the vector product **B** × **A** by rotating **B** into **A** in Fig. 1–13. This yields a result *opposite* to that for **A** × **B**. The vector product is not commutative! In fact, for any two vectors **A** and **B,**

$$\mathbf{A} \times \mathbf{B} = -\mathbf{B} \times \mathbf{A}. \tag{1–26}$$

If we know the components of **A** and **B,** we can calculate the components of the vector product by using a procedure similar to that for the scalar product. We expand the expression

$$\begin{aligned}
\mathbf{A} \times \mathbf{B} = {} & (A_x \mathbf{i} + A_y \mathbf{j} + A_z \mathbf{k}) \times (B_x \mathbf{i} + B_y \mathbf{j} + B_z \mathbf{k}), \\
= {} & A_x \mathbf{i} \times B_x \mathbf{i} + A_x \mathbf{i} \times B_y \mathbf{j} + A_x \mathbf{i} \times B_z \mathbf{k} \\
& + A_y \mathbf{j} \times B_x \mathbf{i} + A_y \mathbf{j} \times B_y \mathbf{j} + A_y \mathbf{j} \times B_z \mathbf{k} \\
& + A_z \mathbf{k} \times B_x \mathbf{i} + A_z \mathbf{k} \times B_y \mathbf{j} + A_z \mathbf{k} \times B_z \mathbf{k}.
\end{aligned} \tag{1–27}$$

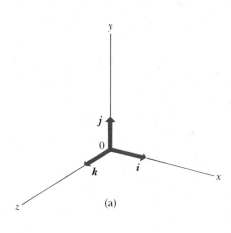

(a)

The individual terms may also be rewritten as $A_x \mathbf{i} \times B_y \mathbf{j} = A_x B_y \mathbf{i} \times \mathbf{j}$, and so on. Each term in which the same unit vector appears twice, such as $\mathbf{i} \times \mathbf{i}$, is zero because it is a product of two parallel vectors. To evaluate the others we refer to the axis system of Fig. 1–14a. We find, for example, $\mathbf{i} \times \mathbf{j} = \mathbf{k}$, and $\mathbf{j} \times \mathbf{i} = -\mathbf{k}$. Thus $A_x \mathbf{i} \times B_y \mathbf{j} = A_x B_y \mathbf{k}$, and so on. We obtain finally

$$\mathbf{A} \times \mathbf{B} = (A_y B_z - A_z B_y)\mathbf{i} + (A_z B_x - A_x B_z)\mathbf{j} + A_x B_y - A_y B_x)\mathbf{k}. \tag{1–28}$$

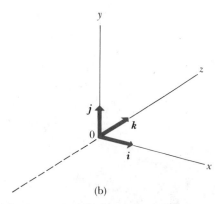

(b)

1–14 (a) A right-handed coordinate system, in which $\mathbf{i} \times \mathbf{j} = \mathbf{k}, \mathbf{j} \times \mathbf{k} = \mathbf{i}$, and $\mathbf{k} \times \mathbf{i} = \mathbf{j}$. (b) A left-handed coordinate system, in which $\mathbf{i} \times \mathbf{j} = -\mathbf{k}$, and so on. Usually only right-handed systems are used.

If **C** = **A** × **B,** the components of **C** are given by

$$\begin{aligned}
C_x &= A_y B_z - A_z B_y, \\
C_y &= A_z B_x - A_x B_z, \\
C_z &= A_x B_y - A_y B_x.
\end{aligned} \tag{1–29}$$

Our first physical application of the vector product will come in Chapter 9 with the definitions of torque and angular momentum. We will also use it extensively in the chapters on magnetic fields.

Here is an interesting quirk of the vector product. In Fig. 1–14a, suppose we reverse the direction of the z-axis, giving the axis system shown in Fig. 1–14b. Then, as you may verify, the definition of the vector product gives $i \times j = -k$ instead of $i \times j = k$. If two axis directions are reversed, we obtain again $i \times j = k$, and if all three are reversed, $i \times j = -k$. Thus there are two kinds of coordinate-axis systems, differing in the signs of the products of unit vectors. In using the vector product, we must specify which kind we are using, to avoid ambiguity.

An axis systems in which $i \times j = k$ is called a **right-handed system.** The usual practice is to use *only* right-handed systems; we will follow that practice throughout this book.

> Coordinate axis systems, like people, are either right-handed or left-handed.

EXAMPLE 1–6 Vector **A** has magnitude 6 units and is in the direction of the $+x$-axis; vector **B** has magnitude 4 units and lies in the xy-plane, making an angle of 30° with the $+x$-axis and an angle of 60° with the $+y$-axis. Find the vector product **A** × **B**.

SOLUTION   From Eq. (1–25), the magnitude of the vector product is

$$AB \sin \theta = (6)(4)(\sin 30°) = 12.$$

From the right-hand rule, the direction of **A** × **B** is that of the $+z$-axis. Alternatively, we may write the components of **A** and **B** and use Eqs. (1–29):

$$A_x = 6, \quad A_y = 0, \quad A_z = 0,$$
$$B_x = 4 \cos 30° = 2\sqrt{3}, \quad B_y = 4 \cos 60° = 2, \quad B_z = 0.$$

If **C** = **A** × **B,** then

$$C_x = (0)(0) - (0)(2) = 0,$$
$$C_y = (0)(2\sqrt{3}) - (6)(0) = 0.$$
$$C_z = (6)(2) - (0)(2\sqrt{3}) = 12.$$

The vector product **C** has only a z-component, and it lies along the z-axis. The magnitude agrees with the above result.

SUMMARY

The fundamental physical quantities of mechanics are mass, length, and time; the corresponding SI units are the kilogram, the meter, and the second, respectively. Other units are related to these by multiples of powers of ten and are identified by adding a prefix to the basic unit. Derived units for other physical quantities are products or quotients of the basic units.

Equations must be dimensionally consistent; two terms can be added or equated only when they have the same units. Treating units like algebraic symbols makes it possible to check this consistency and to convert units easily from one system to another.

The uncertainty of a number can be indicated by the number of significant figures or by an expressed uncertainty. The result of a calculation never has more significant figures than the input data. Even when only crude esti-

KEY TERMS

physical quantity

operational definition

unit

second

kilogram

meter

International System

prefix

dimensional consistency

factor-label method

significant figures

mates are available for the magnitudes of the input data, useful estimates (called order-of-magnitude estimates) can often be obtained.

Scalar quantities are numbers and are combined by the usual rules of arithmetic. Vector quantities have direction as well as magnitude and are combined according to the rules of vector addition. Vector addition can be carried out by using components of vectors. If A_x and A_y are the components of vector A, and B_x and B_y the components of vector B, the components of the vector sum $C = A + B$ are given by

$$C_x = A_x + B_x,$$
$$C_y = A_y + B_y. \tag{1-11}$$

Unit vectors describe directions in space. A unit vector has a magnitude of unity, with no units. Unit vectors aligned with the coordinate axes of a rectangular coordinate system are especially useful.

There are two kinds of products of vectors: the scalar product (or dot product) and the vector product (or cross product). The scalar product $C = A \cdot B$ of two vectors A and B is a scalar quantity, defined as

$$A \cdot B = AB \cos \theta = |A||B| \cos \theta. \tag{1-19}$$

The scalar product can also be expressed in terms of the components of the vectors:

$$A \cdot B = A_x B_x + A_y B_y + A_z B_z. \tag{1-22}$$

The scalar product is commutative; for any two vectors A and B, $A \cdot B = B \cdot A$. The scalar product of two perpendicular vectors is zero.

The vector product $C = A \times B$ of two vectors A and B is another vector C, with magnitude given by

$$C = AB \sin \theta. \tag{1-25}$$

Its direction is perpendicular to the plane of the two vectors, as given by the right-hand rule. The components of the vector product can be expressed in terms of the components of the two vectors being multiplied, as follows:

$$C_x = A_y B_z - A_z B_y,$$
$$C_y = A_z B_x - A_x B_z,$$
$$C_z = A_x B_y - A_y B_x. \tag{1-29}$$

The vector product is not commutative; the order of the factors must not be interchanged. For any two vectors A and B, $A \times B = -B \times A$. The vector product of two parallel or antiparallel vectors is zero.

QUESTIONS

1–1 What are the units of the number π?

1–2 The rate of climb of a mountain trail was described in the guidebook as 150 meters per kilometer. How can this be expressed as a number with no units?

1–3 Suppose you are asked to compute the cosine of 3 meters. Is this possible?

1–4 Hydrologists describe the rate of volume flow of rivers in "second-feet." Is this unit technically correct? If not, what would be a correct unit?

1–5 A highway contractor stated that in building a bridge deck he had poured 200 yards of concrete. What do you think he meant?

1–6 Does a vector having zero length have a direction?

1–7 What is your weight in newtons?

1–8 What is your height in centimeters?

1–9 What physical phenomena (other than a pendulum or cesium clock) could be used to define a time standard?

1–10 Could some atomic quantity be used for a definition of a unit of mass? What advantages or disadvantages would this have compared to the 1-kilogram platinum cylinder kept at Sèvres?

1–11 How could you measure the thickness of a sheet of paper with an ordinary ruler?

1–12 Can you find two vectors with different lengths that have a vector sum of zero? What length restrictions are required for three vectors to have a vector sum of zero?

1–13 What is the displacement when a bicyclist travels from the north side of a circular race track of radius 500 meters to the south side? When she makes one complete circle around the track?

1–14 What are the units of volume? Suppose a student tells you a cylinder of radius r and height h has volume given by $\pi r^3 h$. Explain why this cannot be right.

1–15 An angle (measured in radians) is a number with no units, since it is a ratio of two lengths. Think of other geometrical or physical quantities that are unitless.

1–16 Can you find a vector quantity that has components different from zero but a magnitude of zero?

1–17 One sometimes speaks of the "direction of time," evolving from past to future. Does this mean that time is a vector quantity?

1–18 Is the scalar product of two vectors commutative? Explain.

1–19 What is the scalar product of a vector with itself? The vector product?

EXERCISES

Section 1–2 Standards and Units

Section 1–3 Unit Consistency and Conversions

1–1 Starting with the definition 1.00 in. = 2.54 cm, compute the number of kilometers in 1 mi.

1–2 The density of water is $1 \text{ g} \cdot \text{cm}^{-3}$. What is this value in kilograms per cubic meter?

1–3 The Concorde is the fastest airliner used for commercial service and can cruise at $1450 \text{ mi} \cdot \text{hr}^{-1}$ (about two times the speed of sound, or in other words Mach 2).
a) What is the cruise speed of the Concorde in $\text{mi} \cdot \text{s}^{-1}$?
b) What is the cruise speed of the Concorde in $\text{m} \cdot \text{s}^{-1}$?

1–4 Compute the number of seconds in a day (24 hr), and in a year (365 da).

1–5 The piston displacement of a certain automobile engine is given as 2.0 liters (L). Using only the facts that $1.0 \text{ L} = 1000 \text{ cm}^3$ and 1.0 in. = 2.54 cm, express this volume in cubic inches.

1–6 If one Deutschmark (the West German unit of currency) is worth 40 cents and gasoline costs 1.30 Deutschmarks per liter, what is its cost in dollars per gallon? Use the conversion factors in Appendix E. How does your answer compare with the cost of gasoline in the United States?

1–7 The gasoline consumption of a small car is $17.0 \text{ km} \cdot \text{L}^{-1}$. How many miles per gallon is this? Use the conversion factors in Appendix E.

1–8 The speed limit on a highway in Lower Slobbovia was given as 150,000 furlongs per fortnight. How many miles per hour is this? (One furlong is 1/8 mile, and a fortnight is 14 days. A furlong originally referred to the length of a plowed furrow.)

1–9 One standard of frequency is the radio waves the hydrogen maser has as its output. These radio waves have a frequency of 1,420,405,751.786 hertz. (A hertz is just a special name for cycles per second.) A clock controlled by a hydrogen maser is off by only 1 s in 100,000 years. For the following questions use only three significant figures. (The large number of significant figures for the frequency was given to illustrate the remarkable precision to which it has been measured.)
a) What is the time for one cycle of the radio wave?
b) How many cycles occur in 1 hour?
c) How many cycles would have occurred during the age of the universe, which is estimated to be 10 billion years?
d) By how many seconds would a hydrogen maser clock be off during the lifetime of the earth, which is estimated to be 4600 million years?

Section 1–4 Precision and Significant Figures

1–10 What is the percent error in each of the following approximations to π?
a) 22/7   b) 355/113

1–11 What is the fractional error in the approximate statement $1 \text{ yr} = \pi \times 10^7$ s? (Assume that a year is 365 days.)

1–12 Estimate the percent error in measuring
a) a distance of about 50 cm with a meter stick;
b) a mass of about 1 g with a chemical balance;
c) a time interval of about 4 min with a stopwatch.

1–13 The mass of the earth is 5.98×10^{24} kg, and its radius is 6.38×10^6 m. Compute the density of the earth, using powers-of-ten notation and the correct number of significant figures. (The density of an object is defined as its mass divided by its volume. The formula for the volume of a sphere is given in Appendix B.)

1–14 An angle is given, to one significant figure, as 5°, meaning that its value is between 4.5° and 5.5°. Find the corresponding range of possible values of the cosine of the angle. Is this a case where there are more significant figures in the result than in the input data?

Section 1–5 Estimates and Orders of Magnitude

1–15 A box of typewriter paper is $11'' \times 17'' \times 9''$; it is marked "10 M." Does that mean it contains 10 thousand sheets or 10 million?

1–16 What total volume of air does a person breathe in a lifetime? How does that compare with the volume of the Houston Astrodome?

1–17 Could the water-supply needs of Los Angeles be met by hauling water in by truck? By railroad?

1–18 How many kernels of corn does it take to fill a 1-L soft-drink bottle?

1–19 How many hairs do you have on your head?

1–20 How many times does a human heart beat during a lifetime? How many gallons of blood does it pump?

1–21 How much would it cost to paper the continental United States with dollar bills?

1–22 How many dollar bills would have to be stacked up to reach the moon? Would that be cheaper than building and launching a spacecraft?

1–23 How many cars can pass through a two-lane tunnel through a mountain in 1 hour?

Section 1–6 Vectors and Vector Addition

Section 1–7 Components of Vectors

1–24 A bug starts at the center of a 12-in. phonograph record and crawls along a straight radial line to the edge. While this is happening, the record turns through an angle of 45°. Draw a sketch of the situation and describe the magnitude and direction of the bug's final displacement from its starting point.

1–25 Find the magnitude and direction of the vector represented by each of the following pairs of components:

a) $A_x = 3.0$ cm, $A_y = -4.0$ cm;

b) $A_x = -5.0$ m, $A_y = -12.0$ m;

c) $A_x = -2.0$ km, $A_y = 3.0$ km.

1–26 Vector A has components $A_x = 2.0$ cm, $A_y = 3.0$ cm, and vector B has components $B_x = 4.0$ cm, $B_y = -2.0$ cm. Find

a) the components of the vector sum $A + B$;

b) the magnitude and direction of $A + B$;

c) the components of the vector difference $A - B$;

d) the magnitude and direction of $A - B$.

1–27 A disoriented physics professor drives 5.0 km east, then 4.0 km south, then 2.0 km west. Find the magnitude and direction of the resultant displacement.

1–28 A postal employee drives a delivery truck 1 mi north, then 2 mi east, then 5 mi northwest. Determine the resultant displacement.

1–29 Find the magnitude and direction of

a) the vector sum $A + B$;

b) the vector difference $A - B$

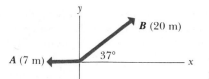

FIGURE 1–15

for the vectors A and B given in Fig. 1–15.

1–30 Vector A is 2.0 cm long and is 60° above the x-axis in the first quadrant. Vector B is 2.0 cm long and is 60° below the x-axis in the fourth quadrant. Find

a) the vector sum $A + B$;

b) the vector differences $A - B$ and $B - A$.

1–31 Vector M, of magnitude 5.0 cm, is at 36.9° counterclockwise from the $+x$-axis. It is added to vector N, and the resultant is a vector of magnitude 5.0 cm, at 53.1° counterclockwise from the $+x$-axis. Find

a) the components of N;

b) the magnitude and direction of N.

Section 1–8 Unit Vectors

1–32 Write each of the two vectors in Fig. 1–15 in terms of the unit vectors i and j.

1–33 Given two vectors $A = 2i + 3j$ and $B = i - 2j$, do the following:

a) Find the magnitude of each vector.

b) Write an expression for the vector sum, using unit vectors.

c) Find the magnitude and direction of the vector sum.

d) Write an expression for the vector difference $A - B$, using unit vectors.

e) Find the magnitude and direction of the vector difference $A - B$.

Section 1–9 Products of Vectors

1–34 Find the scalar product $A \cdot B$ of the two vectors in Fig. 1–15.

1–35 Write out a multiplication table for the scalar products of all possible pairs of unit vectors, such as $i \cdot i = ?$, $i \cdot j = ?$, and so on.

1–36 Find the scalar product of the two vectors given in Exercise 1–33.

1–37 Find the angle between the two vectors $A = -2i + 5j$ and $B = 3i - j$.

1–38

a) Find the magnitude and direction of the vector product $A \times B$ of the two vectors in Fig. 1–15.

b) Find the magnitude and direction of the vector product $B \times A$ of the two vectors in Fig. 1–15.

1–39 Write out a multiplication table for the vector products of all possible pairs of unit vectors, such as $i \times i = ?$, $i \times j = ?$, and so on, using a right-handed coordinate system.

1–40 Find the vector product of the two vectors given in Exercise 1–33. What is the magnitude of this vector product?

1–41

a) Find the scalar product and the vector product for the two vectors $A = 2i - 5j$ and $B = 5i + 2j$.

b) Find the scalar product and the vector product for the two vectors $A = 2i - 5j$ and $B = 4i - 10j$.

1–42 Which of the following are legitimate mathematical operations?

a) $A \cdot (B - C)$ b) $(A - B) \times C$

c) $A \cdot (B \times C)$ d) $A \times (B \times C)$

e) $A \times (B \cdot C)$

1–43

a) Show that Eq. (1–29) can be written in terms of a determinant as follows:

$$C = \begin{vmatrix} i & j & k \\ A_x & A_y & A_z \\ B_x & B_y & B_z \end{vmatrix}.$$

b) Show that

$$(A \times B) \cdot C = \begin{vmatrix} A_x & A_y & A_z \\ B_x & B_y & B_z \\ C_x & C_y & C_z \end{vmatrix}.$$

PROBLEMS

1–44 Physicists, mathematicians, and others often deal with large numbers. The number 10^{100} has been given the whimsical name of a *googol* by mathematicians. (See Edward Kasner and J. R. Newman in Vol. 3 of *The World of Mathematics*, ed. by J. R. Newman. New York: Simon and Schuster.) Let us compare some large numbers in physics with the googol. (*Note*. The following problem requires numerical values that can be found in the Appendixes of the book, with which you should become familiar.)

a) Approximately how many atoms make up the earth? For simplicity, take the average atomic mass of the atoms to be 14 g·mole^{-1}. Avogadro's number gives the number of atoms in a mole.

b) Approximately how many neutrons are in a neutron star? Neutron stars are made up of neutrons and have approximately twice the mass of the sun.

c) In one theory of the origin of the universe, the universe at a very early time had a density (mass divided by volume) of 10^{15} g·cm^{-3}, while its radius was approximately the present distance of the earth to the sun. Assuming $\frac{1}{3}$ of the particles were protons, $\frac{1}{3}$ of the particles were neutrons, and the remaining $\frac{1}{3}$ were electrons, how many particles then made up the universe?

1–45 A spelunker is surveying a cave. He follows a passage 100 m straight east, then 50 m in a direction 30° west of north, then 150 m at 45° west of south. After a fourth unmeasured displacement he finds himself back where he started. Determine the fourth displacement (magnitude and direction).

1–46 A sailboat sails 2.0 km east, then 4.0 km southeast, then an additional distance in an unknown direction. Its final position is 6.0 km directly east of the starting point. Find the magnitude and direction of the third leg of the journey.

1–47 Two points P_1 and P_2 are described by their x- and y-coordinates, (x_1, y_1) and (x_2, y_2), respectively. Show that the components of the displacement A from P_1 to P_2 are

$A_x = x_2 - x_1$ and $A_y = y_2 - y_1$. Also derive expressions for the magnitude and direction of this displacement.

1–48

a) Find the scalar product $A \cdot B$ of the two vectors A and B in Exercise 1–30.

b) Find the magnitude and direction of the vector product $A \times B$ of the two vectors A and B in Exercise 1–30.

1–49 When two vectors A and B are drawn from a common point, the angle between them is θ.

a) Show that the magnitude of their vector sum is given by

$$\sqrt{A^2 + B^2 + 2AB \cos \theta}.$$

b) If A and B have the same magnitude, under what circumstances will their vector sum have the same magnitude as A or B?

c) Derive a result analogous to that in (a) for the magnitude of the vector difference $A - B$.

d) If A and B have the same magnitude, under what circumstance will $A - B$ have this same magnitude?

1–50 Given two vectors $A = -i + 2j - 5k$ and $B = 2i + 3j - 2k$, obtain the following:

a) Find the magnitude of each vector.

b) Write an expression for the vector sum, using unit vectors.

c) Find the magnitude of the vector sum.

d) Write the expression for the vector difference $A - B$, using unit vectors.

e) Find the magnitude of the vector difference $A - B$. Is this the same as the magnitude of $B - A$? Explain.

1–51 Find the scalar product of the two vectors given in Problem 1–50.

1–52 Find the vector product of the two vectors given in Problem 1–50. What is the magnitude of this vector product?

1–53 Find the angle between the two vectors $A = 3i + 4j + 5k$ and $B = 3i + 4j - 5k$.

CHALLENGE PROBLEMS

1–54 Obtain a *unit vector* perpendicular to the two vectors given in Problem 1–50.

1–55 Prove that for any three vectors A, B, and C,

$$A \cdot (B \times C) = (A \times B) \cdot C.$$

1–56

a) Prove that for any three vectors A, B, and C,

$$A \times (B \times C) = B(A \cdot C) - C(A \cdot B).$$

b) Consider the two repeated vector products $A \times (B \times C)$ and $(A \times B) \times C$. Are these two products equal in either magnitude *or* direction? Prove your answer.

1–57 The dot and cross product will appear later in many physical applications; for now, consider the following applications.

a) Let A and B be vectors along the sides of a parallelogram, where the sides have lengths A and B.

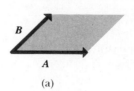
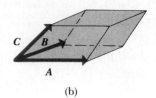

(a)   (b)

FIGURE 1–16

(See Fig. 1–16a.) Show that C is the area of the parallelogram, where $C = A \times B$.

b) The volume of a parallelepiped with sides of length A, B, and C can be expressed in an elegant way as a product of three vectors A, B, and C, where A is along the edge of length A, and so on. (See Fig. 1–16b.) Express the volume of the parallelepiped as the magnitude of the appropriate product of the three vectors A, B, and C.

2

MOTION ALONG
A STRAIGHT LINE

OUR STUDY OF PHYSICS BEGINS IN THE AREA OF *MECHANICS,* WHICH DEALS
with the relations among force, matter, and motion. In this chapter and the
next we discuss mathematical methods for describing motion, and in subsequent chapters we begin to explore the relationship of motion to force. The
part of mechanics concerned with describing motion is called **kinematics.**

Motion is a continuous change of position. We can think of a moving body
as a particle when it is small and when there is no rotation or change of shape.
We say that the particle is a **model** for a moving body; it provides a simplified,
idealized description of the position and motion of the body.

The simplest case is motion of a particle along a straight line, and we will
always take that line to be a **coordinate axis.** Later we will consider more
general motions in space, but these can always be represented by means of
their projections onto three coordinate axes. Displacement is in general a vector quantity, as discussed in Chapter 1, but we first consider situations in
which only one component of displacement is different from zero. Then the
particle moves along one coordinate axis, and its position is described by a
single **coordinate.**

Coordinates: describing the position of a
point

2–1 AVERAGE VELOCITY

Let us consider a hockey player skating the puck down the center line of the
ice toward the opposition's net. The puck moves along a straight line, which
we will use as the *x*-axis of our coordinate system, as shown in Fig. 2–1a. The
puck's distance from the origin O at center court is given by the coordinate x,
which varies with time. At time t_1 the puck is at point P, with coordinate x_1,
and at time t_2 it is at point Q, where its coordinate is x_2. The displacement
during the time interval from t_1 to t_2 is the vector from P to Q: the *x*-compo-
nent of this vector is $(x_2 - x_1)$, and all other components are zero.

It is convenient to represent the quantity $(x_2 - x_1)$, the *change* in x, by
means of a notation using the Greek letter Δ (capital delta) to designate a
change in any quantity. Thus we write

$$\Delta x = x_2 - x_1 \qquad (2\text{–}1)$$

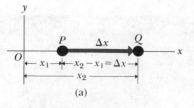

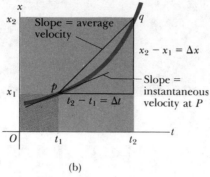

(b)

2–1 (a) Hockey puck moving on the x-axis. (b) Coordinate–time graph of the motion. The average velocity between t_1 and t_2 equals the slope of the line pq. The instantaneous velocity at P equals the slope of the tangent at p.

in which Δx is not a product but is a single symbol representing the change in the quantity. Similarly, we denote the time interval from t_1 to t_2 as $\Delta t = t_2 - t_1$.

Average velocity: distance divided by time

The **average velocity** of the puck is defined as the ratio of the displacement Δx to the time interval Δt. We represent average velocity by the letter v with a subscript "av" to signify average value. Thus

$$v_{\mathrm{av}} = \frac{x_2 - x_1}{t_2 - t_1} = \frac{\Delta x}{\Delta t}. \qquad (2\text{–}2)$$

Strictly speaking, average velocity is a vector quantity, and this defines the x-component of the average velocity. However, in this chapter all vectors have *only* x-components, so we do not need to distinguish between a vector and its x-component. In later chapters we will return to this distinction in two-dimensional motion.

In Fig. 2–1b the coordinate x is graphed as a function of time. The curve in this figure does not represent the path of the puck in space; as Fig. 2–1a shows, the path is a straight line. Rather, the graph is a representation of the change of position with time. The points on the coordinate–time graph corresponding to points P and Q are labeled p and q.

In Fig. 2–1b, the average velocity v_{av} of the puck is represented by the *slope* of the line pq, that is, the ratio of vertical to horizontal intervals of the triangle. Thus v_{av} is the ratio of $x_2 - x_1$ or Δx to $t_2 - t_1$ or Δt.

We have not specified whether the speed of the hockey puck is or is not constant during the time interval $\Delta t = t_2 - t_1$. It may have started from rest, reached a maximum speed, and then slowed down. But that does not matter: To calculate the average velocity we need only the total displacement $\Delta x = x_2 - x_1$ and the total time interval $\Delta t = t_2 - t_1$.

2–2 INSTANTANEOUS VELOCITY

Instantaneous velocity: What do we mean by "velocity at a point"?

Even when the velocity of a moving particle varies, we can still define a velocity at any one specific instant of time or at one specific point in the path. Such a velocity is called **instantaneous velocity,** and it needs to be defined carefully.

Suppose we want to find the instantaneous velocity of the hockey puck in Fig. 2–1 at the point P. We associate the average velocity between points P and Q with the entire displacement Δx and with the entire time interval Δt. But now we can imagine the second point Q as taken closer and closer to the first point P, and we can compute the average velocity over these shorter and shorter displacements and time intervals. We can then define the instantaneous velocity at P as the limiting value approached by this series of average velocities for shorter and shorter intervals as Q becomes closer and closer to P.

In the notation of calculus, the limit of $\Delta x/\Delta t$ as Δt approaches zero is written dx/dt and is called the **derivative** of x with respect to t. Thus instantaneous velocity is defined as

The derivative: calculus as the language of physics

$$v = \lim_{\Delta t \to 0} \frac{\Delta x}{\Delta t} = \frac{dx}{dt}. \qquad (2-3)$$

Since Δt is assumed positive, v has the same algebraic sign as Δx. Hence if the positive x-axis points to the right, a positive velocity indicates motion toward the right.

In more general motion, velocity must be treated as a vector quantity. In that case Eq. (2–3) becomes the x-component of the instantaneous velocity. However, in one-dimensional motion along a straight line, it is customary to call v the instantaneous velocity. When we use the term *velocity*, we always mean instantaneous rather than average velocity, unless we state otherwise.

As point Q approaches point P in Fig. 2–1a, point q approaches point p in Fig. 2–1b. In the limit, the slope of the line pq equals the slope of the tangent to the curve at point p, obtained by dividing a vertical interval (with distance units) by a horizontal interval (with time units). *The instantaneous velocity at any point of a coordinate–time graph therefore equals the slope of the tangent to the graph at that point.* If the tangent slopes upward to the right, its slope is positive, the velocity is positive, and the motion is toward the right. If the tangent slopes downward to the right, the velocity is negative. At a point where the tangent is horizontal, its slope is zero and the velocity is zero.

If distance is expressed in meters and time in seconds, velocity is expressed in meters per second ($\text{m}\cdot\text{s}^{-1}$). Other common units of velocity are feet per second ($\text{ft}\cdot\text{s}^{-1}$), centimeters per second ($\text{cm}\cdot\text{s}^{-1}$), miles per hour ($\text{mi}\cdot\text{hr}^{-1}$), and knots (1 knot = 1 nautical mile or 6080 ft per hour).

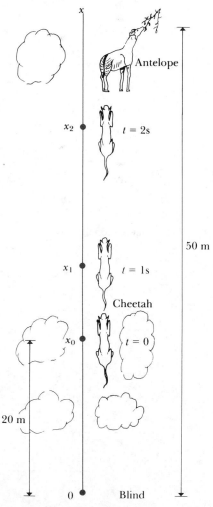

2–2 A cheetah attacking an antelope from ambush.

EXAMPLE 2–1 A cheetah is crouched in ambush 20 m to the north of an observer's blind (see Fig. 2–2). At time $t = 0$, the cheetah charges an antelope in a clearing 50 m north of the observer. The observer estimates that during the first 2 s of the attack, the cheetah's motion is described by the equation $x = a + bt^2$, where $a = 20$ m and $b = 5$ $\text{m}\cdot\text{s}^{-2}$.

a) Find the displacement of the cheetah in the time interval between $t_1 = 1$ s and $t_2 = 2$ s.

At time $t_1 = 1$ s, the cheetah's position is

$$x_1 = 20 \text{ m} + (5 \text{ m}\cdot\text{s}^{-2})(1 \text{ s})^2 = 25 \text{ m}.$$

At time $t_2 = 2$ s,

$$x_2 = 20 \text{ m} + (5 \text{ m}\cdot\text{s}^{-2})(2 \text{ s})^2 = 40 \text{ m}.$$

The displacement is therefore

$$x_2 - x_1 = 40 \text{ m} - 25 \text{ m} = 15 \text{ m}.$$

b) Find the average velocity in this time interval:

$$v_{av} = \frac{x_2 - x_1}{t_2 - t_1} = \frac{15 \text{ m}}{1 \text{ s}} = 15 \text{ m·s}^{-1}.$$

c) Find the instantaneous velocity at time $t_1 = 1$ s.
    The position at time $t = 1 \text{ s} + \Delta t$ is

$$x = 20 \text{ m} + (5 \text{ m·s}^{-2})(1 \text{ s} + \Delta t)^2$$
$$= 25 \text{ m} + (10 \text{ m·s}^{-1})\Delta t + (5 \text{ m·s}^{-2})(\Delta t)^2.$$

The displacement during the interval Δt is

$$\Delta x = 25 \text{ m} + (10 \text{ m·s}^{-1})\Delta t + (5 \text{ m·s}^{-2})(\Delta t)^2 - 25 \text{ m}$$
$$= (10 \text{ m·s}^{-1})\Delta t + (5 \text{ m·s}^{-2})(\Delta t)^2.$$

The average velocity during Δt is

$$v_{av} = \frac{\Delta x}{\Delta t} = 10 \text{ m·s}^{-1} + (5 \text{ m·s}^{-2}) \Delta t.$$

For the instantaneous velocity at $t = 1$ s, we let Δt approach zero in the expression for v_{av}: $v = 10 \text{ m·s}^{-1}$. This corresponds to the slope of the tangent at point p in Fig. 2–1b.

d) Derive a general expression for the instantaneous velocity as a function of time, and from it find v at $t = 1$ s and $t = 2$ s.
    We take the derivative of $x = a + bt^2$, obtaining

$$v = \frac{dx}{dt} = 2bt = (10 \text{ m·s}^{-1})t.$$

At time $t = 1$ s, $v = 10 \text{ m·s}^{-1}$, in agreement with part (c). At time $t = 2$ s, $v = 20 \text{ m·s}^{-1}$.

Speed: a somewhat ambiguous term

The term **speed** has two different meanings. It may mean the *magnitude* of the instantaneous velocity. For example, when two cars travel at 50 km·hr^{-1}, one north and the other south, both have speeds of 50 km·hr^{-1}. In a different sense, referring to an *average* quantity, the speed of a body is the total length of path, divided by the elapsed time. Thus if a car travels 90 km in 3 hr, its average speed is 30 km·hr^{-1}, even if the trip starts and ends at the same point. In the latter case the average velocity would be zero, because the total displacement is zero.

2–3 AVERAGE AND INSTANTANEOUS ACCELERATION

When the velocity of a moving body changes with time, we say that the body has an *acceleration*. Just as velocity is a quantitative description of the rate of change of position with time, so acceleration is a quantitative description of the rate of change of velocity with time.

Average acceleration: change in velocity divided by time

Considering again the motion of a particle along the *x*-axis, suppose that at time t_1 the particle is at point P and has velocity v_1, and that at a later time t_2 it is at point Q and has velocity v_2.

The **average acceleration** a_{av} of the particle as it moves from P to Q is defined as the ratio of the change in velocity to the elapsed time.

$$a_{av} = \frac{v_2 - v_1}{t_2 - t_1} = \frac{\Delta v}{\Delta t}. \qquad (2\text{--}4)$$

Again, strictly speaking, v_1 and v_2 are values of the x-component of instantaneous velocity, and Eq. (2–4) defines the x-component of average acceleration.

EXAMPLE 2–2 An astronaut has left the space shuttle on a tether line to test a new personal maneuvering device. Her on-board partner measures her velocity before and after certain maneuvers, as follows:

a) $v_1 = 0.8$ m·s^{-1}, $\qquad v_2 = 1.2$ m·s^{-1};

b) $v_1 = 1.6$ m·s^{-1}, $\qquad v_2 = 1.2$ m·s^{-1};

c) $v_1 = -0.4$ m·s^{-1}, $\qquad v_2 = -1.0$ m·s^{-1};

d) $v_1 = -1.6$ m·s^{-1}, $\qquad v_2 = -0.8$ m·s^{-1}.

If $t_1 = 2$ s and $t_2 = 4$ s, find the average acceleration for each set of data.

SOLUTION

a) $a_{av} = \dfrac{1.2 \text{ m·s}^{-1} - 0.8 \text{ m·s}^{-1}}{4 \text{ s} - 2 \text{ s}} = 0.2$ m·s^{-2}.

b) $a_{av} = \dfrac{1.2 \text{ m·s}^{-1} - 1.6 \text{ m·s}^{-1}}{4 \text{ s} - 2 \text{ s}} = -0.2$ m·s^{-2}.

c) $a_{av} = \dfrac{-1.0 \text{ m·s}^{-1} - (-0.4 \text{ m·s}^{-1})}{4 \text{ s} - 2 \text{ s}} = -0.3$ m·s^{-2}.

d) $a_{av} = \dfrac{-0.8 \text{ m·s}^{-1} - (-1.6 \text{ m·s}^{-1})}{4 \text{ s} - 2 \text{ s}} = +0.4$ m · s^{-2}.

When the acceleration has the *same* direction as the initial velocity, the body goes faster; when it is in the opposite direction, the particle slows down. When the body moves in the negative direction with increasing speed (c), the acceleration is negative; when it moves in the negative direction with decreasing speed (d), the acceleration is positive.

We can now define **instantaneous acceleration,** following the same procedure used for instantaneous velocity. Consider this situation: A sports car driver has just entered the final straightaway at the Grand Prix. He reaches point P at time t_1, moving with velocity v_1, and crosses the finish line at point Q at time t_2 with velocity v_2, as shown in Fig. 2–3a. Figure 2–3b is a graph of instantaneous velocity v plotted as a function of time, points p and q corresponding to positions P and Q in Fig. 2–3a. The average acceleration is represented by the slope of the line pq, computed by using the appropriate scales and units of the graph.

The instantaneous acceleration of a body—that is, its acceleration at some one instant of time or at some one point of its path—is defined in the same way as instantaneous velocity. We take the second point Q in Fig. 2–3a to be closer and closer to the first point P, so the average acceleration is computed over shorter and shorter intervals of time. The instantaneous acceleration at the first point is the limit approached by the average acceleration when the

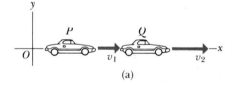

(a)

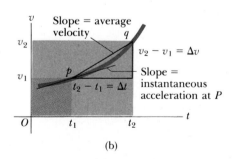

(b)

2–3 (a) Car moving on the x-axis. (b) Velocity–time graph of the motion. The average acceleration between t_1 and t_2 equals the slope of the line pq. The instantaneous acceleration at P equals the slope of the tangent at p.

Instantaneous acceleration: instantaneous rate of change of velocity second point is taken closer and closer to the first:

$$a = \lim_{\Delta t \to 0} \frac{\Delta v}{\Delta t} = \frac{dv}{dt}. \qquad (2\text{--}5)$$

Instantaneous acceleration plays an essential role in the laws of mechanics, while average acceleration is less frequently used. From now on when we use the term *acceleration,* we always mean instantaneous acceleration.

As point Q in Fig. 2–3a approaches point P, point q in the graph (Fig. 2–3b) approaches point p, and the slope of the line pq approaches the slope of the line tangent to the curve at point p. Thus *the instantaneous acceleration at any point on the graph equals the slope of the line tangent to the curve at that point.* As with velocity, we have to use appropriate scales and units for the graph.

The acceleration $a = dv/dt$ can be expressed in various ways. Since $v = dx/dt$, it follows that

$$a = \frac{dv}{dt} = \frac{d}{dt}\left(\frac{dx}{dt}\right) = \frac{d^2x}{dt^2}. \qquad (2\text{--}6)$$

Acceleration as a second derivative

The acceleration is therefore the *second* derivative of the coordinate with respect to time.

If we express velocity in meters per second and time in seconds, then acceleration is in meters per second, per second ($m \cdot s^{-1} \cdot s^{-1}$). This is usually written as $m \cdot s^{-2}$, and is read "meters per second squared." Other common units of acceleration are feet per second squared ($ft \cdot s^{-2}$) and centimeters per second squared ($cm \cdot s^{-2}$).

EXAMPLE 2–3 Suppose the velocity of the car in Fig. 2–3 is given by the equation

$$v = m + nt^2,$$

where $m = 10\ m \cdot s^{-1}$ and $n = 2\ m \cdot s^{-3}$.

a) Find the change in velocity of the car in the time interval between $t_1 = 2$ s and $t_2 = 5$ s.

At time $t_1 = 2$ s,

$$v_1 = 10\ m \cdot s^{-1} + (2\ m \cdot s^{-3})\,(2\ s)^2$$
$$= 18\ m \cdot s^{-1}.$$

At time $t_2 = 5$ s,

$$v_2 = 10\ m \cdot s^{-1} + (2\ m \cdot s^{-3})\,(5\ s)^2$$
$$= 60\ m \cdot s^{-1}.$$

The change in velocity is therefore

$$v_2 - v_1 = 60\ m \cdot s^{-1} - 18\ m \cdot s^{-1}$$
$$= 42\ m \cdot s^{-1}.$$

b) Find the average acceleration in this time interval:

$$a_{av} = \frac{v_2 - v_1}{t_2 - t_1} = \frac{42\ m \cdot s^{-1}}{3\ s} = 14\ m \cdot s^{-2}.$$

This corresponds to the slope of the line pq in Fig. 2–3b.

c) Find the instantaneous acceleration at time $t_1 = 2$ s.

At time $t = 2$ s $+$ Δt,

$$v = 10 \text{ m·s}^{-1} + (2 \text{ m·s}^{-3}) (2 \text{ s} + \Delta t)^2$$
$$= 18 \text{ m·s}^{-1} + (8 \text{ m·s}^{-2}) \Delta t + (2 \text{ m·s}^{-3}) (\Delta t)^2.$$

The change in velocity during Δt is

$$\Delta v = 18 \text{m·s}^{-1} + (8 \text{ m·s}^{-2}) \Delta t + (2 \text{ m·s}^{-3})(\Delta t)^2 - 18 \text{ m·s}^{-1}$$
$$= (8 \text{ m·s}^{-2}) \Delta t + (2 \text{ m·s}^{-3}) (\Delta t)^2.$$

The average acceleration during Δt is

$$a_{\text{av}} = \frac{\Delta v}{\Delta t} = 8 \text{ m·s}^{-2} + (2 \text{ m·s}^{-3}) \Delta t.$$

The instantaneous acceleration at $t = 2$ s, obtained by letting Δt approach zero, is $a = 8$ m·s^{-2}. This corresponds to the slope of the tangent at point p in Fig. 2–3b.

d) Derive an expression for the instantaneous acceleration at any time, and use it to find the acceleration at $t = 2$ s and at $t = 5$ s.

Starting with $v = 10 \text{m·s}^{-1} + (2 \text{ m·s}^{-1})t^2$ and using Eq. (2–6), we find

$$a = \frac{dv}{dt}$$

$$= \frac{d}{dt}[10 \text{ m·s}^{-1} + (2 \text{ m·s}^{-3})t^2] = 2(2 \text{ m·s}^{-3})t.$$

When $t = 2$ s,

$$a = (2) (2 \text{m·s}^{-3}) (2 \text{ s}) = 8 \text{ m·s}^{-2}.$$

When $t = 5$ s,

$$a = (2) (2 \text{ m·s}^{-3}) (5 \text{ s}) = 20 \text{ m·s}^{-2}.$$

Note that neither of these values is equal to the average acceleration found in (b); the instantaneous acceleration varies with time. Automotive engineers sometimes call the time rate of change of acceleration the "jerk."

A few remarks about the *sign* of acceleration may be helpful. When the acceleration and velocity of a body have the same sign, the body is speeding up. If both are positive, the body moves in the positive direction with increasing speed. If both are negative, the body moves in the negative direction with a velocity that becomes more and more negative with time, and again the body's speed increases.

When v and a have opposite signs, the body is slowing down. If v is positive and a negative, the body moves in the positive direction with decreasing speed. If v is negative and a positive, the body moves in the negative direction with a velocity that is becoming less negative, and again the body slows down.

The term *deceleration* is sometimes used, either for a negative value of a or for a decrease in speed. Because of the ambiguity, it is best to avoid this term. We do not use it in this book; instead we recommend careful attention to the interpretation of the algebraic sign of a in relation to that of v.

What does the *sign* of acceleration mean?

Deceleration is a bad word.

Constant acceleration: an important special case.

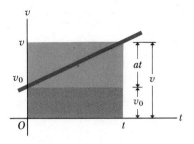

2-4 Velocity–time graph for straight-line motion with constant acceleration.

How to find the velocity at any time

2-4 MOTION WITH CONSTANT ACCELERATION

The simplest accelerated motion is straight-line motion with *constant* acceleration, when the velocity changes at the same rate throughout the motion. The graph of velocity as a function of time is then a straight line, as in Fig. 2–4; the velocity increases by equal amounts in equal time intervals. Hence in Eq. (2–4) we can replace the average acceleration a_{av} by the constant (instantaneous) acceleration a. We then have

$$a = \frac{v_2 - v_1}{t_2 - t_1}. \qquad (2\text{–}7)$$

Now let $t_1 = 0$ and let t_2 be any arbitrary later time t. Let v_0 represent the velocity when $t = 0$ (called the *initial* velocity), and let v be the velocity at the later time t. Then Eq. (2–7) becomes

$$a = \frac{v - v_0}{t - 0}, \qquad \text{or} \qquad v = v_0 + at. \qquad (2\text{–}8)$$

We can interpret this equation as follows: The acceleration a is the constant rate of change of velocity, or the change per unit time. The term at is the product of the change in velocity per unit time, a, and the time interval t. Therefore it equals the *total* change in velocity. The velocity v at any time t then equals the initial velocity v_0 (at the time $t = 0$) plus the change in velocity at. Graphically, we can consider the ordinate v at time t in Fig. 2–4 as the sum of two segments: one with length v_0 equal to the initial velocity, the other with length at equal to the change in velocity during time t.

Here is an alternative route to Eq. (2–8). We know that v is some function of t. If the derivative of that function with respect to t is the constant a, what must the function itself be? (Such a function might be called an *antiderivative*, but the more usual term is *indefinite integral*.) One possibility is the function at; its derivative with respect to t is the constant a. But the derivative of the function $at + C$, where C is any constant, is also equal to a, because the derivative of any constant is zero. Thus we conclude that the function for v must have the form

$$v = at + C. \qquad (2\text{–}9)$$

Now this function must also satisfy the additional requirement that at time $t = 0$ it yields the value v_0. Substituting $t = 0$ and equating the result to v_0, we find

$$v_0 = a(0) + C, \qquad \text{or} \quad C = v_0. \qquad (2\text{–}10)$$

Putting all this together, we again obtain

$$v = v_0 + at.$$

How to find the position at any time

We may use a similar procedure to find an expression for the position x of the particle as a function of time, if the position at time $t = 0$ is denoted by x_0. We know that $v = dx/dt$; what function of t must x be, in order for its derivative with respect to t to equal $v_0 + at$? That is, what is the antiderivative of $v_0 + at$? The function

$$x = v_0 t + \tfrac{1}{2} a t^2 + D, \qquad (2\text{–}11)$$

where D is any constant, satisfies this requirement; this may be checked easily by taking the derivative of Eq. (2–11).

We also know that at time $t = 0$, Eq. (2–11) must give the value x_0. Substituting the value $t = 0$ and equating the result to x_0, we find $x_0 = D$. Combining this with Eq. (2–11), we obtain

$$x = x_0 + v_0 t + \tfrac{1}{2}at^2. \qquad (2\text{–}12)$$

This equation states that if at time $t = 0$ a particle is at position x_0 and has velocity v_0, its new position x at any later time t is the sum of three terms: its initial position x_0, plus the distance $v_0 t$ it would move if its velocity were constant, plus an additional distance $at^2/2$ caused by the changing velocity.

We may also combine Eqs. (2–8) and (2–12) to obtain a relation for x, v, and a that does not contain t. We first solve Eq. (2–8) for t and then substitute the resulting expression into Eq. (2–12) and simplify:

Getting rid of the time variable: relation of position to velocity

$$t = \frac{v - v_0}{a},$$

$$x = x_0 + v_0\left(\frac{v - v_0}{a}\right) + \frac{1}{2}a\left(\frac{v - v_0}{a}\right)^2.$$

Transfer the term x_0 to the left side and multiply through by $2a$:

$$2a(x - x_0) = 2v_0 v - 2v_0^2 + v^2 - 2v_0 v + v_0^2,$$

or, finally,

$$v^2 = v_0^2 + 2a(x - x_0). \qquad (2\text{–}13)$$

Equations (2–8), (2–12), and (2–13) are the *equations of motion with constant acceleration.* Any kinematic problem involving motion of a particle on a straight line with constant acceleration can be solved by using these equations.

The curve in Fig. 2–5 is a graph of the coordinate x as a function of time for motion with constant acceleration, for a case where v_0 and a are positive. That is, it is a graph of Eq. (2–12); the curve is a *parabola*. The slope of the tangent at $t = 0$ equals the initial velocity v_0, and the slope of the tangent at any time t equals the velocity v at that time. The slope continuously increases with t, and measurements would show that the *rate* of increase with t is constant.

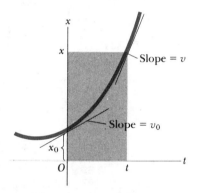

2–5 Coordinate–time graph for motion with constant acceleration.

PROBLEM-SOLVING STRATEGY: *Motion with constant acceleration*

1. It is essential to decide at the beginning of a problem where the origin of coordinates is and which axis direction is positive. The choices are usually made on the basis of convenience; often it is easiest to place the particle at the origin at time $t = 0$. A diagram showing these choices is always helpful.

2. Once you have chosen the positive axis direction, the positive directions for velocity and acceleration are also determined. It would be inconsistent to define x as positive to the right of the origin and velocities toward the left as positive.

3. In constant-acceleration problems it is always useful to make a list of unknown quantities such as x, x_0, v, v_0, and a. In general, some of these will be known, some unknown. When possible, select an equation from Eqs. (2–8), (2–12), and (2–13) that contains only one

of the unknowns; solve for the unknown, then substitute the known values and compute the value of the unknown.

4. It often helps to restate the problem in prose first, then translate that description into symbols and equations. *When* does the particle arrive at a certain point? (I.e., at what value of t?) *Where* is the particle when its velocity has a certain value? (I.e., what is the value of x when v has the specified value?) In the following example we ask: "Where is the motorcyclist when his velocity is 5 m·s⁻¹?" Translated into symbols, this becomes: "What is the value of x when $v = 5$ m·s⁻¹?"

5. Take a hard look at your results to see whether they make sense. Are they within the general range of magnitudes you expected?

How fast can a motorcyclist get out of town?

EXAMPLE 2-4 A motorcyclist heading east through a small Iowa town accelerates as he passes the signpost at $x = 0$ marking the city limits. His acceleration is constant, $a = 4 \text{ m·s}^{-2}$. At time $t = 0$ he is at $x = 5$ m and has velocity $v = 3 \text{ m·s}^{-1}$.

a) Find his position and velocity at time $t = 2$ s.

b) Where is he when his velocity is 5 m·s^{-1}?

SOLUTION We take the signpost as the origin of coordinates, and the positive x-axis points east. At the initial time $t = 0$, the initial position is $x_0 = 5$ m, and the initial velocity is $v_0 = 3 \text{ m·s}^{-1}$. We want to know the position and velocity (i.e., the values of x and v) at the later time $t = 2$ s. With reference to Eqs. (2–8), (2–12), and (2–13), we have

$$x_0 = 5 \text{ m}, \qquad v_0 = 3 \text{ m·s}^{-1}, \qquad a = 4 \text{ m·s}^{-2}.$$

a) From Eq. (2–12), which gives position x as a function of time t,

$$\begin{aligned} x &= x_0 + v_0 t + \tfrac{1}{2}at^2 \\ &= 5 \text{ m} + (3 \text{ m·s}^{-1})(2 \text{ s}) + \tfrac{1}{2}(4 \text{ m·s}^{-2})(2 \text{ s})^2 \\ &= 19 \text{ m}. \end{aligned}$$

From Eq. (2–8), which gives velocity v as a function of time t,

$$\begin{aligned} v &= v_0 + at \\ &= 3 \text{ m·s}^{-1} + (4 \text{ m·s}^{-2})(2 \text{ s}) \\ &= 11 \text{ m·s}^{-1}. \end{aligned}$$

b) From Eq. (2–13)

$$v^2 = v_0{}^2 + 2a(x - x_0),$$

$$(5 \text{ m·s}^{-1})^2 = (3 \text{ m·s}^{-1})^2 + 2(4 \text{ m·s}^{-2})(x - 5 \text{ m}),$$

$$x = 7 \text{ m}.$$

Alternatively, we may use Eq. (2–8) to find first the *time* when $v = 5 \text{ m·s}^{-1}$:

$$5 \text{ m·s}^{-1} = 3 \text{ m·s}^{-1} + (4 \text{ m·s}^{-2})(t),$$

$$t = \tfrac{1}{2} \text{ s}.$$

Then use Eq. (2–12),

$$\begin{aligned} x &= 5 \text{ m} + (3 \text{ m·s}^{-1})(\tfrac{1}{2} \text{ s}) + \tfrac{1}{2}(4 \text{ m·s}^{-2})(\tfrac{1}{2} \text{ s})^2 \\ &= 7 \text{ m}. \end{aligned}$$

2–5 VELOCITY AND COORDINATE BY INTEGRATION

In deriving the equations of Section 2–4 we assumed that the acceleration is constant; when it is not, we cannot use them. When a varies with time, we can still use the relation $v = dx/dt$ to find the velocity v as a function of time if the position x is a given function of time. Similarly, we can use $a = dv/dt$ to find the acceleration a as a function of time if the velocity v is a given function of time.

Integration: Reversing the process of differentiation

We can also reverse this process. Suppose v is known as a function of time; how can we find x as a function of time? To answer this question, we first consider a graphical approach. Figure 2–6 shows a velocity-versus-time curve for a situation where the acceleration (the slope of the curve) is not constant but increases with time. Considering the motion during the interval between times t_1 and t_2, we divide this total interval into many smaller intervals, calling a typical one Δt. Let the velocity during that interval be v. Of course, the velocity changes during Δt, but if the interval is very small, the change will also be very small. This displacement during that interval, neglecting the variation of v, is given by

$$\Delta x = v\,\Delta t.$$

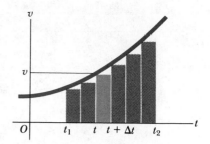

2–6 The area under a velocity–time graph equals the displacement.

This corresponds graphically to the area of the shaded strip with height v and width Δt, that is, the area under the curve corresponding to the interval Δt. Since the total displacement in any interval (say, t_1 to t_2) is the sum of the displacements in the small subintervals, the total displacement is given graphically by the *total* area under the curve between the vertical lines t_1 and t_2. In the limit, when all the Δt's become very small and their number very large, this is simply the **integral** of v (which is in general a function of t) from t_1 to t_2. Thus x_1 is the position at time t_1 and x_2 the position at time t_2:

$$x_2 - x_1 = \int_{x_1}^{x_2} dx = \int_{t_1}^{t_2} v\,dt. \qquad (2\text{–}14)$$

A similar analysis with the acceleration-versus-time curve, where a is in general a function of t, shows that if v_1 is the velocity at time t_1 and v_2 the velocity at time t_2, the change in velocity Δv during a small time interval Δt is approximately equal to $a\,\Delta t$, and the total change in velocity $(v_2 - v_1)$ during the interval $t_2 - t_1$ is given by

$$v_2 - v_1 = \int_{v_1}^{v_2} dv = \int_{t_1}^{t_2} a\,dt. \qquad (2\text{–}15)$$

EXAMPLE 2–5 An automobile travels along a straight highway. Its acceleration is given as a function of time by

$$a = 2.0 \text{ m·s}^{-2} - (0.1 \text{ m·s}^{-3})t.$$

The position at time $t = 0$ is given by $x_0 = 0$, and the velocity at time $t = 0$ is $v_0 = 10 \text{ m·s}^{-1}$.

a) Derive expressions for the velocity and position as functions of time.

b) At what time is the velocity of the car greatest?

c) What is the maximum velocity?

SOLUTION

a) We let t_1 be the initial time $t = 0$ and t_2 be any later time t. Similarly, v_1 is the initial velocity (at time $t = 0$), $v_0 = 10 \text{ m·s}^{-1}$. We use Eq. (2–15) to find the velocity v at time t:

$$v - 10 \text{ m·s}^{-1} = \int_0^t [2.0 \text{ m·s}^{-2} - (0.1 \text{ m·s}^{-3})t]\,dt$$
$$= (2.0 \text{ m·s}^{-2})t - \tfrac{1}{2}(0.1 \text{ m·s}^{-2})t^2.$$

Then we use Eq. (2–14) to find x as a function of t, taking x_1 as the initial position $x_0 = 0$:

$$x - 0 = \int_0^t [10 \text{ m·s}^{-1} + (2.0 \text{ m·s}^{-2})t - \tfrac{1}{2}(0.1 \text{ m·s}^{-2})t^2]\, dt$$
$$= (10 \text{ m·s}^{-1})t + \tfrac{1}{2}(2.0 \text{ m·s}^{-1})t^2 - \tfrac{1}{6}(0.1 \text{ m·s}^{-2})t^3.$$

b) The maximum value of v occurs at the time when v stops increasing and begins to decrease. At this instant, $dv/dt = 0$. Taking the derivative of the expression above for v and setting it equal to zero, we obtain

$$0 = (2.0 \text{ m·s}^{-2}) - (0.1 \text{ m·s}^{-3})t,$$
$$t = \frac{2.0 \text{ m·s}^{-2}}{0.1 \text{ m·s}^{-3}} = 20 \text{ s}.$$

This result can also be obtained immediately from the expression for a by noting that a is positive between $t = 0$ and $t = 20$ s, and negative after that. It is zero at $t = 20$ s, corresponding to the fact that $dv/dt = 0$ at this time. Thus the automobile speeds up until $t = 20$ s and then begins to slow down again.

c) The maximum velocity is obtained by substituting $t = 20$ s (the time when the maximum velocity is attained) into the general velocity equation:

$$v_{\text{max}} = (2.0 \text{ m·s}^{-2})(20 \text{ s}) - \tfrac{1}{2}(0.1 \text{ m·s}^{-3})(20 \text{ s})^2 + 10 \text{ m·s}^{-1}$$
$$= 30 \text{ m·s}^{-1}$$

We can also use Eq. (2–14) for an alternative derivation of Eq. (2–12), the constant-acceleration case. Using Eq. (2–8) in Eq. (2–14) and carrying out the integration, we obtain:

$$x - x_0 = \int_0^t (v_0 + at)\, dt$$
$$= v_0 t + \tfrac{1}{2}at^2.$$

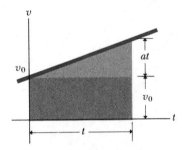

2–7 The area under a velocity–time graph equals the displacement $x - x_0$.

Finally, we can use the graphical approach that led us to Eq. (2–14) for still another derivation of Eq. (2–12). The graph of Eq. (2–8) is shown in Fig. 2–7. The total displacement $x - x_0$ from time $t = 0$ to any later time t is represented by the area under the graph. This can be obtained by finding the rectangular and triangular areas and adding. Carrying out this program, we find

$$x - x_0 = v_0 t + \tfrac{1}{2}(t)(at)$$
$$= v_0 t + \tfrac{1}{2}at^2.$$

Thus the various methods we have described are all consistent, and they lead to the same results in the special case of constant acceleration.

2–6 FREELY FALLING BODIES

Free fall: Do heavy bodies fall faster than light ones?

The most familiar example of motion with (nearly) constant acceleration is that of a body falling toward the earth. The motion of a falling body under the influence of the earth's gravity alone is often called **free fall;** it has held the attention of philosophers and scientists since ancient times. Aristotle thought (erroneously) that heavy objects fall faster than light objects, in proportion to their weight. Galileo argued that the motion of a falling body should be nearly independent of its weight and should have constant acceleration. According to legend, Galileo tested his theory by dropping cannonballs and bullets from the Leaning Tower of Pisa, although there is no reference to such experiments in his own writings.

In more recent times the motion of falling bodies has been studied with great precision. When air resistance can be neglected, we find that all bodies at a particular location fall with the same acceleration, regardless of their size or weight. If the distance of fall is small compared to the radius of the earth, the acceleration is constant throughout the fall. In the following discussion we neglect the effects of air resistance and the decrease of acceleration with increasing altitude. We call this idealized motion *free fall*, although it includes rising as well as falling motion.

The constant acceleration of a freely falling body is called the **acceleration due to gravity,** the *acceleration of gravity*, or (most accurately) the *acceleration of free fall*, and its magnitude is denoted by the letter g. At or near the earth's surface, the value of g is approximately 9.8 m·s^{-2}, 980 cm·s^{-2}, or 32 ft·s^{-2}. Later we will discuss the variation of g with latitude and elevation and the effect of the earth's rotation. On the surface of the moon, the acceleration of gravity is caused by the attractive force of the moon rather than the earth, and there $g = 1.67 \text{ m·s}^{-2}$. Near the surface of the sun, $g = 274 \text{ m·s}^{-2}$!

In the following examples we use the constant-acceleration equations developed in Section 2–4. We suggest you review the problem-solving strategies discussed in that section before you study these examples.

A body falls with constant acceleration, independent of its mass (almost).

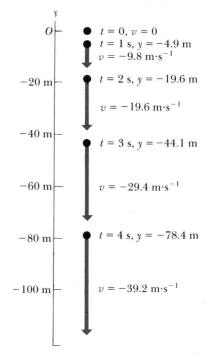

2–8 Position and velocity of a freely falling body.

EXAMPLE 2–6 A bullet is dropped from the Leaning Tower of Pisa; it starts from rest and falls freely. Compute its position and velocity after 1, 2, 3, and 4 s. Take the origin O at the elevation of the starting point, the y-axis as vertical, and the upward direction as positive.

SOLUTION The initial coordinate y_0 and the initial velocity v_0 are both zero. The acceleration is downward, in the negative y-direction, so $a = -g = -9.8 \text{ m·s}^{-2}$. From Eqs. (2–12) and (2–8),

$$y = v_0 t + \tfrac{1}{2}at^2 = 0 - \tfrac{1}{2}gt^2 = (-4.9 \text{ m·s}^{-2})t^2,$$

$$v = v_0 + at = 0 - gt = (-9.8 \text{ m·s}^{-2})t.$$

When $t = 1$ s, $y = (-4.9 \text{ m·s}^{-2})(1 \text{ s})^2 = -4.9 \text{ m}$, and $v = (-9.8 \text{ m·s}^{-2})(1 \text{ s}) = -9.8 \text{ m·s}^{-1}$. The body is therefore 4.9 m below the origin (y is negative) and has a downward velocity (v is negative) of magnitude 9.8 m·s^{-1}.

The position and velocity at 2, 3, and 4 s are found in the same way. The results are shown in Fig. 2–8: check the numerical values for yourself.

EXAMPLE 2–7 Suppose you throw a ball vertically upward from the cornice of a tall building, and it leaves your hand with an upward speed of 15 m·s^{-1}. On its way back down it just misses the cornice. In Fig. 2–9, the downward path is displaced a little to the right from its actual position, for clarity. Find (a) the position and velocity of the ball 1 s and 4 s after it leaves your hand; (b) the velocity when the ball is 5 m above the cornice; (c) the maximum height reached and the time at which it is reached. Take the origin at the elevation at which the ball leaves your hand, the y-axis vertical and positive upward.

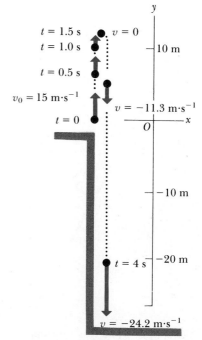

2–9 Position and velocity of a body thrown vertically upward.

SOLUTION The initial position y_0 is zero. The initial velocity v_0 is $+15$ m·s^{-1}, and the acceleration is $a = -9.8$ m·s^{-2}. The velocity at any time is

$$v = v_0 + at = 15 \text{ m·s}^{-1} + (-9.8 \text{ m·s}^{-2})t. \qquad (2\text{–}16)$$

The position at any time is

$$y = v_0 t + \tfrac{1}{2}at^2 = (15 \text{ m·s}^{-1})t + \tfrac{1}{2}(-9.8 \text{ m·s}^{-2})t^2. \qquad (2\text{–}17)$$

The velocity at any position is

$$v^2 = v_0^2 + 2ay = (15 \text{ m·s}^{-1})^2 + 2(-9.8 \text{ m·s}^{-2})y. \qquad (2\text{–}18)$$

a) When $t = 1$ s, Eqs. (2–16) and (2–17) give

$$y = +10.1 \text{ m}, \qquad v = +5.2 \text{ m·s}^{-1}.$$

The ball is 10.1 m above the origin (y is positive) and it has an upward velocity (v is positive) of 5.2 m·s^{-1} (less than the initial velocity, as expected). When $t = 4$ s, again from Eqs. (2–16) and (2–17),

$$y = -18.4 \text{ m}, \qquad v = -24.2 \text{ m·s}^{-1}.$$

The ball has passed its highest point and is 18.4 m *below* the origin (y is negative). It has a *downward* velocity (v is negative) of magnitude 24.4 m·s^{-1}. Note that it is not necessary to find the highest point reached or the time at which it was reached. The equations of motion give the position and velocity at *any* time, whether the ball is on the way up or the way down.

b) When the ball is 5 m above the origin,

$$y = +5 \text{ m}$$

and, from Eq. (2–18),

$$v^2 = 127 \text{ m}^2\text{·s}^{-2}, \qquad v = \pm 11.3 \text{ m·s}^{-1}.$$

The ball passes this point *twice*, once on the way up and again on the way down. The velocity on the way up is $+11.3$ m·s^{-1}, and on the way down it is -11.3 m·s^{-1}.

c) At the highest point, $v = 0$. Hence, from Eq. (2–18),

$$0 = (15 \text{ m·s}^{-1})^2 - (19.6 \text{ m·s}^{-2})y$$

and

$$y = 11.5 \text{ m}.$$

The time can now be found from Eq. (2–16), setting $v = 0$:

$$0 = 15 \text{ m·s}^{-1} + (-9.8 \text{ m·s}^{-2})t.$$

$$t = 1.53 \text{ s}.$$

Alternatively, to find the maximum height we may ask first *when* the maximum height is reached. That is, at what t is $v = 0$? As just shown, $v = 0$ when $t = 1.53$ s; substituting this value of t back into Eq. (2–17), we find

$$y = (15 \text{ m·s}^{-1})(1.53 \text{ s}) + \tfrac{1}{2}(-9.8 \text{ m·s}^{-2})(1.53 \text{ s})^2$$
$$= 11.5 \text{ m}.$$

Although at the highest point the velocity is instantaneously zero, the *accelera-*

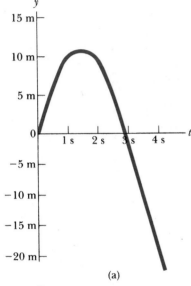

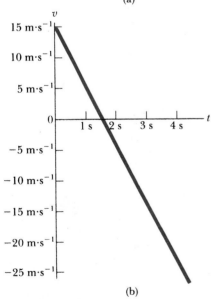

2–10 Graphs of y and v as functions of t for Example 2–7.

tion at this point is still −9.8 m·s⁻². The ball stops for an instant, but its velocity is continuously changing; the acceleration is constant throughout.

Figure 2–10 shows graphs of position and velocity as functions of time for this problem.

Figure 2–11 is a multiflash photograph of a falling golf ball, taken with a stroboscopic light source. This light source produces a series of intense flashes, each so short (a few millionths of a second) that there is no blur in the image of even a rapidly moving body. The time interval between flashes can be adjusted; the camera shutter is left open during the entire motion, and as each flash occurs, the position of the ball at that instant is recorded on the film.

Strobe photography: a useful tool to study motion

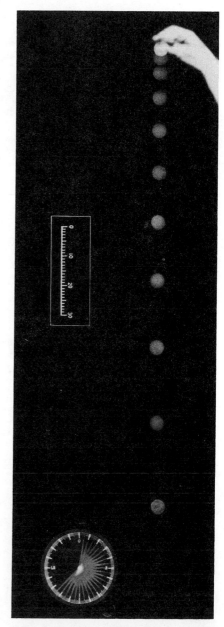

The equally spaced light flashes subdivide the motion into equal time intervals Δt, so the velocity of the ball between any two flashes is directly proportional to the separation of its corresponding images in the photograph. If the velocity were constant, the images would be equally spaced. The increasing separation of the images during the fall shows that the velocity is continually increasing; the ball accelerates. By comparing two successive displacements of the ball, we can find the *change* in velocity in the corresponding time interval. Careful measurement shows that this change is the same in each time interval, indicating that the motion is one of *constant* acceleration.

2–7 RELATIVE VELOCITY

The velocity of a body, like its position, is always described with reference to a coordinate system or **frame of reference.** Often this system is taken to be fixed in some other body, but this second body may also be in motion relative to a third, and so on. Thus one frame of reference may be moving with respect to another. When we speak of "the velocity of a car," we usually mean its velocity relative to the earth. But the earth is in motion relative to the sun, the sun is in motion relative to some other star, and so on. We use the term **relative velocity** to describe such relationships.

Suppose a long train of flatcars is moving to the right along a straight, level track, as in Fig. 2–12. Such trains are used in the Alps to transport automobiles through long tunnels. Suppose a daredevil is driving a car to the right along the flatcars. In Fig. 2–12, v_{FE} represents the velocity of the flatcars F relative to the earth E, and v_{AF} the velocity of the automobile A relative to the flatcars. In any time interval, the total displacement of the automobile relative to the earth is the sum of its displacement relative to the flatcars and their displacement relative to the earth. Thus the automobile's velocity relative to the earth, v_{AE}, is equal to the *sum* of the relative velocities v_{AF} and v_{FE}:

$$v_{AE} = v_{AF} + v_{FE}. \tag{2–19}$$

Thus if the flatcars are traveling relative to the earth at 30 km·hr⁻¹ (= v_{FE}), and the automobile is traveling relative to the flatcars at 40 km·hr⁻¹ (= v_{AF}), the velocity of the automobile relative to the earth (v_{AE}) is 70 km·hr⁻¹.

2–11 Multiflash photograph (retouched) of freely falling golf ball.

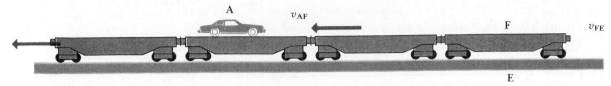

2–12 Vector v_{FE} is the velocity of the flatcars F relative to the earth E, and v_{AF} is the velocity of automobile A relative to the flatcars.

In straight-line motion the velocity v_{AE} is the *algebraic* sum of v_{AF} and v_{FE}. Thus if the automobile were traveling to the *left* with a velocity of 40 km·hr^{-1} relative to the flatcars, $v_{AF} = -40$ km·hr^{-1} and the velocity of the automobile relative to the earth would be -10 km·hr^{-1}; that is, it would be traveling to the left, relative to the earth.

PROBLEM-SOLVING STRATEGY: *Relative velocity*

Note the order of the double subscripts on the velocities: v_{AB} always means "velocity of A relative to B." In writing equations such as Eq. (2–19), make sure that the first subscript on the left side of the equation is the same as the first subscript in the *first* term on the right, and the second on the left is the same as the second in the *last* term on the right. Also, adjacent subscripts in adjacent terms on the right must match. Thus

$$v_{AE} = v_{AB} + v_{BC} + v_{CD} + v_{DE}.$$

Catching a speeding driver: an example of relative velocity

EXAMPLE 2–8   A man is driving a car at 65 km·hr^{-1} on a straight, level road where the speed limit is 40 km·hr^{-1}. He is spotted by a motorcycle officer, who accelerates in pursuit. By the time the driver sees the motorcycle's flashing blue light, the motorcycle is traveling at 80 km·hr^{-1}. What is the motorcycle's velocity relative to the car?

SOLUTION   Let the car be C, the motorcycle M, and the earth E. Then

$$v_{CE} = 65 \text{ km·hr}^{-1} \quad \text{and} \quad v_{ME} = 80 \text{ km·hr}^{-1},$$

and we wish to find v_{MC}.

From the rule for combining velocities,

$$v_{ME} = v_{MC} + v_{CE}.$$

Thus

$$80 \text{ km·hr}^{-1} = v_{MC} + 65 \text{ km·hr}^{-1},$$

$$v_{MC} = 15 \text{ km·hr}^{-1},$$

and the officer is overtaking the driver at 15 km·hr^{-1}.

EXAMPLE 2–9   How would the relative velocities be altered if the motorcycle were ahead of the car?

SOLUTION   Not at all. The relative *positions* of the bodies do not matter. The velocity of M relative to C is still $+15$ km·hr^{-1}, but he is now pulling ahead of C at this rate.

In more general motion, where the velocities do not all lie along a straight line, vector addition must be used to combine velocities. We will return to these more general problems at the end of Chapter 3.

SUMMARY

When a particle moves along a straight line, we describe its position with respect to an origin O by means of a coordinate such as x.

The particle's average velocity during a time interval Δt is defined as

$$v_{av} = \frac{\Delta x}{\Delta t}. \tag{2-2}$$

The instantaneous velocity at any time t is defined as

$$v = \lim_{\Delta t \to 0} \frac{\Delta x}{\Delta t} = \frac{dx}{dt}. \tag{2-3}$$

The average acceleration during a time interval Δt is defined as

$$a_{av} = \frac{\Delta v}{\Delta t}. \tag{2-4}$$

The instantaneous acceleration at any time t is defined as

$$a = \lim_{\Delta t \to 0} \frac{\Delta v}{\Delta t} = \frac{dv}{dt}. \tag{2-5}$$

Positive and negative directions for velocity are determined by the choice of positive direction for the coordinate.

When the acceleration is constant, the position x and velocity v at any time t are related to the acceleration a, the initial position x_0, and the initial velocity v_0 (both at time $t = 0$) by the following equations:

$$x = x_0 + v_0 t + \tfrac{1}{2}at^2, \tag{2-12}$$

$$v = v_0 + at, \tag{2-8}$$

$$v^2 - v_0{}^2 = 2a(x - x_0). \tag{2-13}$$

When the acceleration is not constant but is a known function of time, we can find the velocity and coordinate as functions of time by integrating the acceleration function.

When a body A moves relative to a body F, and F moves relative to the earth E or other frame of reference, we denote the velocity of A relative to F by v_{AF}, the velocity of F relative to E by v_{FE}, and the velocity of A relative to E by v_{AE}. These velocities are related by

$$v_{AE} = v_{AF} + v_{FE}. \tag{2-19}$$

KEY TERMS

kinematics

model

coordinate axis

coordinate

average velocity

instantaneous velocity

derivative

speed

average acceleration

instantaneous acceleration

integral

free fall

acceleration due to gravity

frame of reference

relative velocity

QUESTIONS

2-1 When a particle moves in a circular path, how many coordinates are required to describe a position, assuming that the radius of the circle is given?

2-2 What is meant by the statement that space is three-dimensional?

2-3 Does the speedometer of a car measure speed, or velocity?

2-4 Some European countries have highway speed limits of $100 \text{ km} \cdot \text{hr}^{-1}$. What is the equivalent number of miles per hour?

2–5 A student claims that a speed of 60 mi·hr^{-1} is the same as 88 ft·s^{-1}. Is this relationship exact, or only approximate?

2–6 In a given time interval, is the total displacement of a particle equal to the product of the average velocity and the time interval, even when the velocity is not constant?

2–7 Under what conditions is average velocity equal to instantaneous velocity?

2–8 When one flies in an airplane at night in smooth air, there is no sensation of motion even though the plane may be moving at 800 km·hr^{-1} (500 mi·hr^{-1}). Why is this?

2–9 An automobile is traveling north. Can it have a velocity toward the north and at the same time have an acceleration toward the south? Under what circumstances?

2–10 A ball is thrown straight up in the air. What is its acceleration at the instant it reaches it highest point?

2–11 Is the acceleration of a car greater when the accelerator is pushed to the floor, or when the brake pedal is pushed hard?

2–12 Under constant acceleration, the average velocity of a particle is half the sum of its initial and final velocities. Is this still true if the acceleration is *not* constant?

2–13 A baseball is thrown straight up in the air. Is the acceleration greater while it is being thrown, or after it is thrown?

2–14 How could you measure the acceleration of an automobile by using only instruments located within the automobile?

2–15 If the initial position and initial velocity of a vehicle are known, and a record is kept of the acceleration at each instant, can its position after a certain time be computed from these data? Explain how this might be done.

EXERCISES

Section 2–1 Average Velocity

2–1 In 1954 Roger Bannister became the first human to run a mile in less than 4 min. Suppose that a runner on a straight track covers a distance of 1 mi in exactly 4 min. What was the magnitude of his average velocity in

a) mi·hr^{-1} b) ft·s^{-1}

2–2 A hiker travels in a straight line for 40 min with an average velocity of 1.6 m·s^{-1}. After the 40 min what distance has he covered?

2–3 A mouse moves along a straight line; its distance from the origin at any instant is given by the equation $x = at + bt^2$, where $a = 8$ cm·s^{-1} and $b = -3$ cm·s^{-2}. Find the average velocity of the mouse in the interval from $t = 0$ to $t = 1$ s, and in the interval from $t = 0$ to $t = 4$ s.

Section 2–2 Instantaneous Velocity

2–4 A physics professor leaves his house and walks along the sidewalk toward campus. After 5 min he realizes it is raining and returns home. His distance from his house as a function of time is shown in Fig. 2–13. At which of the labeled points is his velocity

a) zero?   b) constant and positive?
c) constant and negative?   d) increasing in magnitude?
e) decreasing in magnitude?

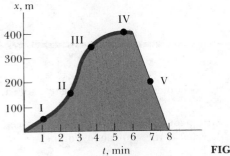

FIGURE 2–13

2–5 A hobbyist is testing a new model rocket engine by using it to propel a cart along a model railroad track. He determines that its motion along the x-axis is described by the equation $x = bt^2$, where $b = 10$ cm·s^{-2}. Compute the instantaneous velocity of the cart at time $t = 3$ s.

2–6 A car is initially stopped at a traffic light. It then travels along a straight road such that its distance from the light is given by $x = bt + ct^2$, where $b = 4.0$ m·s^{-1} and $c = 0.5$ m·s^{-2}.

a) Calculate the average velocity of the car for the time interval $t = 0$ to $t = 10$ s.

b) Calculate the instantaneous velocity of the car at
   i) $t = 0$;
   ii) $t = 5$ s;
   iii) $t = 10$ s.

Section 2–3 Average and Instantaneous Acceleration

2–7 A test driver at Incredible Motors, Inc., is testing a new model car whose speedometer is calibrated to read m·s^{-1} rather than mi·hr^{-1}. The following series of speedometer readings was obtained during a test run:

Time (s)	0	2	4	6	8	10	12	14	16
Velocity (m·s^{-1})	0	0	2	5	10	15	20	22	22

a) Compute the average acceleration during each 2-s interval. Is the acceleration constant? Is it constant during any part of the time?

b) Make a velocity–time graph of the data above, using scales of 1 cm = 1 s horizontally, and 1 cm = 2 m·s^{-1} vertically. Draw a smooth curve through the plotted points. By measuring the slope of your curve, find the instantaneous acceleration at $t = 8$ s, 13 s, and 15 s.

2–8 An astronaut has left Spacelab V to test a new space scooter for use in constructing Space Habitat I. His partner measures the following velocity changes, each taking place in a 10-s interval. What are the magnitude, the algebraic sign, and the direction of the average acceleration in each interval?

a) At the beginning of the interval the astronaut is moving toward the right along the x-axis at 5 m·s⁻¹, and at the end of the interval he is moving toward the right at 20 m·s⁻¹.

b) At the beginning he is moving toward the right at 20 m·s⁻¹, and at the end he is moving toward the right at 5 m·s⁻¹.

c) At the beginning he is moving toward the left at 5 m·s⁻¹, and at the end he is moving toward the left at 20 m·s⁻¹.

d) At the beginning he is moving toward the left at 20 m·s⁻¹, and at the end he is moving toward the left at 5 m·s⁻¹.

e) At the beginning he is moving toward the right at 20 m·s⁻¹, and at the end he is moving toward the left at 20 m·s⁻¹.

f) At the beginning he is moving toward the left at 20 m·s⁻¹, and at the end he is moving toward the right at 20 m·s⁻¹.

g) In which of these cases is the scooter speeding up? In which cases is it slowing down?

2–9 Figure 2–14 is a graph of the coordinate of a spider crawling along the x-axis. Sketch the graphs of its velocity and acceleration as functions of time.

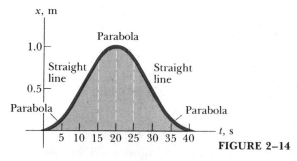

FIGURE 2–14

2–10 An automobile has a velocity as a function of time given by $v(t) = \alpha + \beta t^2$, where $\alpha = 5.0$ m·s⁻¹ and $\beta = 0.3$ m·s⁻³. Calculate

a) the average acceleration for the time interval $t = 0$ to $t = 5$ s;

b) the instantaneous acceleration for
  i) $t = 0$,
  ii) $t = 5$ s.

Section 2–4 Motion with Constant Acceleration

2–11 The makers of a certain automobile advertise that it will accelerate from 15 to 50 mi·hr⁻¹ in 13 s. Compute

a) the acceleration in ft·s⁻²;

b) the distance the car travels in this time, assuming the acceleration to be constant.

2–12 An airplane travels 500 m down the runway before taking off. If it starts from rest, moves with constant acceleration, and becomes airborne in 20 s, with what velocity in m·s⁻¹ does it take off?

2–13 A car moving with constant acceleration covers the distance between two points 60 m apart in 6 s. Its velocity as it passes the second point is 15 m·s⁻¹.

a) What is its velocity at the first point?

b) What is the acceleration?

2–14 The fastest time for a 440 yd race from a standing start is 7.08 s according to the 1984 *Guinness Book of World Records*. This was done on a 1200 cc motorcycle in 1980. Assuming constant acceleration:

a) What was the acceleration of the cycle (in ft·s⁻²)?

b) What was the final velocity of the cycle (in ft·s⁻¹)?

c) With this acceleration how long would it take to go from 0 to 55 mph?

2–15 The graph in Fig. 2–15 shows the velocity of a motorcycle police officer plotted as a function of time.

a) Find the instantaneous acceleration at $t = 3$ s, at $t = 7$ s, and at $t = 11$ s.

b) How far does the officer go in the first 5 s? The first 9 s? The first 13 s?

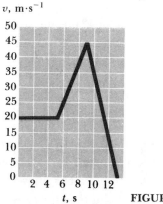

FIGURE 2–15

2–16 Figure 2–16 is a graph of the acceleration of a model railroad locomotive moving on the x-axis. Sketch the graphs of its velocity and coordinate as functions of time, if $x = v = 0$ when $t = 0$.

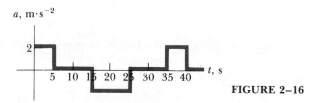

FIGURE 2–16

2–17 The "reaction" time of the average automobile driver is about 0.7 s. (The reaction time is the interval between the perception of a signal to stop and the application of the brakes.) If an automobile can slow down with an acceleration −16 ft·s⁻², compute the total distance covered in coming to a stop after a signal is observed

a) from an initial velocity of 15 mi·hr⁻¹ (in a school zone);

b) from an initial velocity of 55 mi·hr⁻¹.

2–18 A subway train starts from rest at a station and accelerates at a rate of 2 m·s^{-2} for 10 s. It runs at constant speed for 30 s, and slows down at -4 m·s^{-2} until it stops at the next station. Find the *total* distance covered.

2–19 A hypothetical spaceship takes a straight-line path from the earth to the moon, a distance of about 400,000 km. Suppose it accelerates at 10 m·s^{-2} for the first 10 min of the trip, then travels at constant speed until the last 10 min, when it accelerates at -10 m·s^{-2}, just coming to rest as it reaches the moon.

a) What is the maximum speed attained?

b) What fraction of the total distance is traveled at constant speed?

c) What total time is required for the trip?

Section 2–5 Velocity and Coordinate by Integration

2–20 The motion of a particle along a straight line is described by the function

$$x = (6 \text{ m}) + (5 \text{ m·s}^{-2})t^2 - (1 \text{ m·s}^{-4})t^4.$$

Assume that t is positive.

a) Find the position, velocity, and acceleration at time $t = 2$ s.

b) During what time interval is the velocity positive?

c) During what time interval is x positive?

d) What is the maximum positive velocity attained by the particle?

2–21 The acceleration of a motorcycle is given by $a = (1.2 \text{ m·s}^{-3})t - (0.12 \text{ m·s}^{-4})t^2$. It is at rest at the origin at time $t = 0$.

a) Find its position and velocity as functions of time.

b) Calculate the maximum velocity it attains.

2–22 The acceleration of a bus is given by $a = (2 \text{ m·s}^{-3})t$.

a) If the bus's velocity at time $t = 1$ s is 5 m·s^{-1}, what is its velocity at time $t = 2$ s?

b) If the bus's position at time $t = 1$ s is 6 m, what is its position at time $t = 2$ s?

Section 2–6 Freely Falling Bodies

2–23

a) With what velocity must a ball be thrown vertically upward in order to rise to a height of 20 m?

b) How long will it be in the air?

2–24 A brick is dropped from the roof of a building. The brick strikes the ground in 4 s.

a) How tall, in meters, is the building?

b) What is the velocity of the brick just before it reaches the ground?

2–25 A ball is thrown vertically downward from the top of a building, leaving the thrower's hand with a velocity of 10 m·s^{-1}.

a) What will be its velocity after falling for 2 s?

b) How far will it fall in 2 s?

c) What will be its velocity after falling 10 m?

d) If it moved a distance of 1 m while in the thrower's hand, find its acceleration while in his hand (assuming it is constant).

e) If the point where the ball leaves the thrower's hand is 40 m above the ground, in how many seconds will the ball strike the ground?

f) What will the velocity of the ball be just before it strikes the ground?

2–26 A hot-air balloonist, rising vertically with a velocity of 5 m·s^{-1}, releases a sandbag at an instant when the balloon is 20 m above the ground.

a) Compute the position and velocity of the sandbag at the following times after its release: $\frac{1}{4}$ s, $\frac{1}{2}$ s, 1 s, 2 s.

b) How many seconds after its release will the bag strike the ground?

c) With what velocity will it strike?

2–27 A ball is thrown nearly vertically upward from a point near the cornice of a tall building. It just misses the cornice on the way down, and passes a point 160 ft below its starting point 5 s after it leaves the thrower's hand.

a) What was the initial velocity of the ball?

b) How high did it rise above its starting point?

c) What were the magnitude and direction of its velocity at the highest point?

d) What were the magnitude and direction of its acceleration at the highest point?

e) What was the magnitude of its velocity as it passed a point 64 ft below the starting point?

2–28 A ball is thrown vertically upward from the ground with a velocity of 30 m·s^{-1}.

a) How long will it take to rise to its highest point?

b) How high does the ball rise?

c) How long after being thrown will the ball have a velocity of 10 m·s^{-1} upward?

d) Of 10 m·s^{-1} downward?

e) When is the displacement of the ball zero?

f) When is the magnitude of the ball's velocity equal to half its initial velocity?

g) When is the magnitude of the ball's displacement equal to half the greatest height to which it rises?

h) What are the magnitude and direction of the acceleration while the ball is moving upward?

i) While moving downward?

j) When at the highest point?

2–29 The rocket-driven sled Sonic Wind No. 2, used for investigating the physiological effects of large accelerations, runs on a straight, level track 3500 ft long. Starting from rest, it can reach a speed of 1000 mi·hr^{-1} in 1.8 s.

a) Compute the acceleration in ft·s^{-2}, assuming it to be constant.

b) What is the ratio of this acceleration to that of a free-falling body, g?

c) What is the distance covered in 1.8 s?

d) A magazine article states that at the end of a certain run the speed of the sled was decreased from 632 mi·hr^{-1} to zero in 1.4 s, and that during this time its passenger was subjected to more than 40 times the pull of gravity (that is, the acceleration was greater than 40 g). Are these figures consistent?

2–30 Suppose the acceleration of gravity were only 1.0 m·s^{-2}, instead of 9.8 m·s^{-2}.

a) Estimate the height to which you could jump vertically from a standing start.

b) How high could you throw a baseball?

c) Estimate the maximum height of a window from which you would care to jump to a concrete sidewalk below. (Each story of an average building is about 4 m high.)

d) With what speed, in kilometers per hour, would you strike the sidewalk?

e) How many seconds would be required for you to reach the sidewalk?

Section 2–7 Relative Velocity

2–31 A "moving sidewalk" in an airport terminal building moves 1 m·s^{-1} and is 150 m long. If a man steps on at one end and walks 2 m·s^{-1} relative to the moving sidewalk, how much time does he require to reach the opposite end if he walks

a) in the same direction the sidewalk is moving?

b) in the opposite direction?

2–32 Two piers A and B are located on a river, 1610 m apart. Two men must make round trips from pier A to pier B and return. One man is to row a boat at a velocity of 6.4 km·hr^{-1} relative to the water, and the other man is to walk on the shore at a velocity of 6.4 km·hr^{-1}. The velocity of the river is 3.2 km·hr^{-1} in the direction from A to B. How long does it take each man to make the round trip?

PROBLEMS

2–33 A motorcycle's position is described by $x = A + Bt + Ct^3$, where A, B, and C are numerical constants.

a) Calculate the velocity of the cycle, as a function of time.

b) From your answer to (a), calculate the acceleration of the motorcycle as a function of time.

2–34 A jet-propelled motorcycle starts from rest, moves in a straight line with constant acceleration, and covers a distance of 64 m in 4 s.

a) What is the final velocity?

b) How much time was required to cover half the total distance?

c) What is the distance covered in one-half the total time?

d) What is the velocity when half the total distance has been covered?

e) What is the velocity after one-half the total time?

f) When will the instantaneous velocity equal the average velocity for the 0-to-4 s time interval?

2–35 A sled starts from rest at the top of a hill and slides down with constant acceleration. At some later time it is 32.0 m from the top; two seconds after that it is 50.0 m from the top, two seconds later 72.0 m from the top, and two seconds later 98.0 m.

a) What is the average velocity of the sled during each of the 2-s intervals after passing the 32.0-m point?

b) What is the acceleration of the sled?

c) What was the velocity of the sled when it passed the 32.0-m point?

d) How long did it take to go from the top to the 32.0-m point?

e) How far did the sled go during the first second after passing the 32.0-m point?

f) How long does it take the sled to go from the 32.0-m point to the midpoint between the 32.0-m and the 50.0-m points?

g) What is the velocity of the sled as it passes the midpoint in part (f)?

2–36 The engineer of a passenger train traveling at 30 m·s^{-1} sights a freight train whose caboose is 200 m ahead on the same track. The freight train is traveling in the same direction as the passenger train with a velocity of 10 m·s^{-1}. The engineer of the passenger train immediately applies the brakes, causing a constant acceleration of -1 m·s^{-2}, while the freight train continues with constant speed.

a) Will there be a collision?

b) If so, where will it take place?

2–37 At the instant the traffic light turns green, an automobile that has been waiting at an intersection starts ahead with a constant acceleration of 2 m·s^{-2}. At the same instant a truck, traveling with a constant velocity of 10 m·s^{-1}, overtakes and passes the automobile.

a) How far beyond its starting point will the automobile overtake the truck?

b) How fast will it be traveling?

2–38 An automobile and a truck start from rest at the same instant, with the automobile initially at some distance behind the truck. The truck has a constant acceleration of 2 m·s^{-2} and the automobile an acceleration of 3 m·s^{-2}. The automobile overtakes the truck after the truck has moved 75 m.

a) How long does it take the automobile to overtake the truck?

b) How far was the automobile behind the truck initially?

c) What is the velocity of each when they are abreast?

2–39 A delivery truck's velocity is given by

$$v(t) = (4 \text{ m·s}^{-2})t + (3 \text{ m·s}^{-4})t^3.$$

It is at $x = 0$ when $t = 0$. Calculate its position and acceleration as functions of time.

2–40 An object's velocity is measured to be

$$v(t) = 4 \text{ m·s}^{-1} - (5 \text{ m·s}^{-3})t^2.$$

At $t = 0$ the object is at $x = 0$.

a) Calculate the object's position and acceleration as functions of time.

b) What is the object's maximum *positive* displacement from the origin?

2–41 The position of a particle is given by $x = A \sin \omega t$, where A and ω are constants.

a) Find the velocity of the particle as a function of time.

b) Find the acceleration as a function of time.

c) Find the velocity as a function of x. At what point is the magnitude of the velocity greatest? Least?

d) Find the acceleration as a function of x. At what point is the magnitude of the acceleration greatest? Least?

e) What is the maximum distance of the particle from the origin ($x = 0$)?

f) What is the maximum velocity?

g) What is the maximum acceleration?

h) Sketch graphs of x, v, and a as functions of time.

2–42 A student determined to test the law of gravity for himself walks off a skyscraper 300 m high, stopwatch in hand, and starts his free fall (zero initial velocity). Five seconds later, Superman arrives at the scene and dives off the roof to save the student.

a) What must Superman's initial velocity be for him to catch the student just before the ground is reached? (Assume that Superman's acceleration is that of any free-falling body.)

b) If the height of the skyscraper is less than some minimum value, even Superman cannot save the student. What is this minimum height?

2–43 A football is kicked vertically upward from the ground, and a student gazing out of the window sees it moving upward past her at 5 m·s^{-1}. The window is 10 m above the ground.

a) How high does the ball go above ground?

b) How long does it take to go from a height of 10 m to its highest point?

c) Find its velocity and acceleration $\frac{1}{2}$ s after it leaves the ground.

2–44 A flower pot falls off a window sill and past the window below. It takes 0.25 s to pass this window, which is 3 m high. How far is the top of the window below the window sill above?

2–45 A ball is released from rest at the top of an inclined plane 18 m long and reaches the bottom 3 s later. At the same instant that the first ball is released, a second ball is projected upward along the plane from its bottom with a certain initial velocity. The second ball is to travel part way up the plane, stop, and return to the bottom so that it arrives simultaneously with the first ball. Both balls have the same constant acceleration.

a) Find the acceleration of the balls.

b) What must be the initial velocity of the second ball?

c) How far up the plane will the second ball travel?

2–46 A stone is dropped from the top of a tall cliff, and 1 s later a second stone is thrown vertically downward with a velocity of 20 m·s^{-1}. How far below the top of the cliff will the second stone overtake the first?

2–47 A marble is dropped, and then 1 s later and from a point 5.0 m lower a second marble is dropped. When will the two marbles be 15.0 m apart?

2–48 A juggler performs in a room whose ceiling is 3 m above the level of his hands. He throws a ball vertically upward so that it just reaches the ceiling.

a) With what initial velocity does he throw the ball?

b) How much time is required for the ball to reach the ceiling?

He throws the second ball upward with the same initial velocity, at the instant that the first ball is at the ceiling.

c) How long after the second ball is thrown do the two balls pass each other?

d) When the balls pass each other, how far are they above the juggler's hands?

2–49 The driver of a car wishes to pass a truck that is traveling at a constant speed of 20 m·s^{-1} (about 45 mi·hr^{-1}). The car initially is also traveling at 20 m·s^{-1}. The car's maximum acceleration in this speed range is 0.5 m·s^{-2}. Initially the vehicles are separated by 25 m, and the car pulls back into the truck's lane after it is 25 m ahead of the truck. The car is 5 m long and the truck 20 m.

a) How much time is required for the car to pass the truck?

b) What distance does the car travel during this time?

c) What is the final speed of the car, assuming its acceleration is constant while passing the truck?

2–50 Two cars, A and B, travel in a straight line. The distance of A from the starting point is given as a function of time by $x_A = (4 \text{ m·s}^{-1})t + (1 \text{ m·s}^{-2})t^2$; the distance of B from the starting point is $x_B = (2 \text{ m·s}^{-2})t^2 + (2 \text{ m·s}^{-3})t^3$.

a) Which car is ahead just after they leave the starting point?

b) At what values of t are the cars at the same point?

c) At what values of t is the velocity of B relative to A zero?

d) At what values of t is the distance from A to B neither increasing nor decreasing?

CHALLENGE PROBLEMS

2–51 As implied by Eqs. 2–3 and 2–5, the definition of the derivative df/dt of the function $f(t)$ is

$$\frac{df}{dt} = \lim_{\Delta t \to 0} \frac{\Delta f}{\Delta t},$$

where $\Delta f = f(t + \Delta t) - f(t)$. Use this definition to prove the following:

a) $(d/dt)t^n = nt^{n-1}$. (*Hint:* Use the binomial theorem, which may be found in Appendix B.)

b) $(d/dt)[af(t)] = a(df/dt)$.

c) $(d/dt)[f(t) + g(t)] = df/dt + dg/dt$.

2–52 An alert hiker sees a boulder fall from the top of a distant cliff, and notes that it takes 2 s for the boulder to fall the last third of the way to the ground.

a) What is the height of the cliff, in meters?

b) If in part (a) you get two roots to a quadratic equation and use one for your answer, what does the other root represent?

2–53 A student is running to catch the campus shuttle bus, which is stopped at the bus stop. The student is running at a constant velocity of 6 m·s^{-1}; she cannot run any faster. When the student is still 80 m from the bus, it starts to pull away. The bus moves with a constant acceleration of 0.2 m·s^{-2}.

a) For how much time and how far will the student have to run before she overtakes the bus?

b) When she reaches the bus, how fast will the bus be traveling?

c) Sketch a graph showing $x(t)$ for both the student and the bus. Take $x = 0$ as the initial position of the student.

d) The equations you used in (a) to find the time have a second solution, corresponding to a later time for which the student and bus will again be at the same place if they continue their specified motions. Explain the significance of this second solution. How fast will the bus be traveling at this point?

e) If the student's constant velocity is 4 m·s^{-1}, will she catch the bus?

f) What is the *minimum* speed the student must have just to catch up with the bus? How long and how far will she have to run in that case?

2–54 A ball is thrown straight up from the edge of the roof of a building. A second ball is dropped from the roof 2 s later.

a) If the height of the building is 40 m, what must be the initial velocity of the first ball if both are to hit the ground at the same time?

Consider the same situation as above, but now let the initial velocity v_0 of the first ball be given and treat the height h of the building as an unknown.

b) What must the height of the building be for both balls to reach the ground at the same time for each of the

following values of v_0:
  i) 13.0 m·s^{-1};
  ii) 19.5 m·s^{-1}?

c) If v_0 is greater than some value v_{max}, a value of h does not exist that allows both balls to hit the ground at the same time. Solve for v_{max}. The value v_{max} has a simple physical interpretation; what is it?

d) If v_0 is less than some value v_{min}, a value of h does not exist that allows both balls to hit the ground at the same time. Solve for v_{min}. The value v_{min} also has a simple physical interpretation; what is it?

2–55

a) Suppose cars are lined up bumper to bumper at a red light. Assume each car is 4.6 m in length. When the light turns green, the first car accelerates at 1.22 m·s^{-2}. When this car is 6 m in front of the second car, the second car begins to accelerate at 1.22 m·s^{-2}. When the second car is 6 m in front of the third car, it begins to accelerate at 1.22 m·s^{-2}. This continues in similar fashion for the fourth, fifth, etc., car. If the light stays green for a minute and a half, how many cars have made it to the beginning of the intersection before the light turns red again?

b) Now assume instead that the cars stopped at the red light are spaced such that there is 9 m between each pair of cars. If, when the light turns green, each car immediately accelerates at 1.22 m·s^{-2} for 12 s and then goes at constant velocity, how many cars make it to the beginning of the intersection before the light turns red again? Assume as in (a) that the light is green for 90 s and that each car is 4.6 m long. Compare your answer to that of (a); you may find the results surprising.

2–56 The acceleration of an object suspended from a spring and oscillating vertically is $a = -Ky$, where K is a constant and y is the coordinate measured from the equilibrium position. Suppose that an object moving in this way is given an initial velocity v_0 at the coordinate y_0. Find the expression for the velocity v of the object as a function of its coordinate y. (*Hint:* Use the expression $a = v(dv/dy)$)

2–57 The motion of an object falling from rest in a resisting medium is described by the equation

$$\frac{dv}{dt} = A - Bv,$$

where A and B are constants. In terms of A and B, find

a) the initial acceleration;

b) the velocity at which the acceleration becomes zero (the terminal velocity).

c) Show that the velocity at any given t is given by

$$v = \frac{A}{B}(1 - e^{-Bt}).$$

2–58 After the engine of a moving motorboat is cut off, the boat has an acceleration in the opposite direction to its velocity and directly proportional to the square of its velocity. That is, $dv/dt = -kv^2$, where k is constant.

a) Show that the magnitude v of the velocity at a time t after the engine is cut off is given by

$$\frac{1}{v} = \frac{1}{v_0} + kt.$$

b) Show that the distance x traveled in a time t is

$$x = \frac{1}{k} \ln(v_0 kt + 1).$$

c) Show that the velocity after traveling a distance x is

$$v = v_0 e^{-kx}.$$

As a numerical example, suppose the engine is cut off when the velocity is 6 m·s^{-1}, and that the velocity decreases to 3 m·s^{-1} in a time of 15 s.

d) Find the numerical value of the constant k, and the unit in which it is expressed.

e) Find the acceleration at the instant the engine is cut off.

f) Calculate x, v, and a at 10-s intervals for the first 60 s after the engine is cut off. Sketch graphs of x, v, and a as functions of t, for the first 60 s of the motion.

3

MOTION IN A PLANE

IN CHAPTER 2 WE STUDIED MOTION ALONG A STRAIGHT LINE. NOW WE broaden our discussion to include motion along a path that lies in a plane. A few familiar examples are the flight of a thrown or batted baseball, a projectile shot from a gun, a ball whirled in a circle at the end of a cord, the motion of the moon or of a satellite around the earth, and the motions of the planets around the sun. To describe these motions we need to generalize the kinematic language introduced in Chapter 2. Velocity and acceleration are *vector* quantities, and so we need to use the methods of vector algebra introduced in Chapter 1.

Things don't always move in straight lines.

The content of this chapter, like that of the preceding one, is classified as *kinematics;* we are concerned only with describing motion, not with relating motion to its causes. The language introduced here, however, will be an essential tool in later chapters when we use Newton's laws of motion to study the relation between force and motion.

3–1 AVERAGE AND INSTANTANEOUS VELOCITY

To begin our study of motion along a curved path in a plane, we consider the situation of Fig. 3–1a. A particle moves along a curved path in the xy-plane; points P and Q represent the positions of the particle at two different times. We can describe the position of the particle at P by means of the displacement vector r from the origin 0 to P; we call r the **position vector** of the particle. The components of r are the coordinates x and y, and we may write

How to describe the position of a point in a plane

$$r = xi + yj, \qquad (3-1)$$

where i and j are the unit vectors introduced in Section 1–8.

As the particle moves from P to Q, its displacement is the change Δr in the position vector r, as shown. The x- and y-components of Δr are the quantities Δx and Δy, as shown.

Let Δt be the time interval during which the particle moves from P to Q. The **average velocity** v_{av} during this interval is defined to be a vector quantity

Average velocity: displacement divided by time

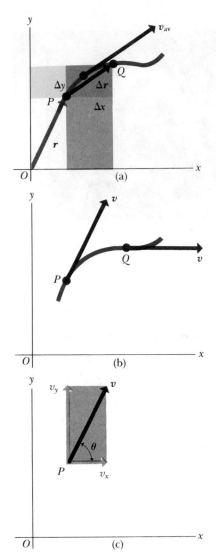

3–1 (a) The displacement $\Delta \boldsymbol{r}$ between points P and Q has components Δx and Δy. The average velocity $\boldsymbol{v}_{av}$ has the same direction as $\Delta \boldsymbol{r}$. (b) The instantaneous velocity $\boldsymbol{v}$ at each point is always tangent to the path at that point. (c) Components of the instantaneous velocity at P are shown.

equal to the displacement divided by the time interval:

$$v_{av} = \frac{\Delta r}{\Delta t}. \tag{3–2}$$

That is, the average velocity is a vector quantity having the same direction as $\Delta \boldsymbol{r}$ and a magnitude equal to the magnitude of $\Delta \boldsymbol{r}$ divided by Δt. The magnitude of $\Delta \boldsymbol{r}$ is always the straight-line distance from P to Q, regardless of the actual shape of the path taken by the particle. Thus the average velocity would be the same for any path that would take the particle from P to Q in the time interval Δt.

The **instantaneous velocity** $\boldsymbol{v}$ at point P is defined in magnitude and direction as the *limit approached by the average velocity* when point Q is taken closer and closer to point P (as $\Delta t \to 0$).

$$\text{Instantaneous velocity } v = \lim_{\Delta t \to 0} \frac{\Delta r}{\Delta t} = \frac{dr}{dt}. \tag{3–3}$$

As point Q approaches point P, the direction of the vector $\Delta \boldsymbol{r}$ approaches that of the tangent to the path at P, so that the instantaneous velocity vector at any point is tangent to the path at that point. The instantaneous velocities at points P and Q are shown in Fig. 3–1b. Just as in Chapter 2, we define average velocity with reference to a finite time interval, while instantaneous velocity describes the motion at a specific point and at a specific instant of time.

It follows from Eq. (3–1) that the components $(v_x)_{av}$ and $(v_y)_{av}$ of average velocity are the corresponding components of displacement, divided by Δt. That is,

$$(v_x)_{av} = \frac{\Delta x}{\Delta t}, \qquad (v_y)_{av} = \frac{\Delta y}{\Delta t}.$$

Similarly, the instantaneous rates of change of x and y, denoted by v_x and v_y, are equal respectively to the x- and y-components of the instantaneous velocity $\boldsymbol{v}$, and so we may write

$$v_x = \lim_{\Delta t \to 0} \frac{\Delta x}{\Delta t} = \frac{dx}{dt}, \qquad v_y = \lim_{\Delta t \to 0} \frac{\Delta y}{\Delta t} = \frac{dy}{dt}. \tag{3–4}$$

The magnitude of the instantaneous velocity is given by

$$|v| = \sqrt{v_x^2 + v_y^2},$$

and the angle θ in Fig. 3–1c by

$$\tan \theta = \frac{v_y}{v_x}.$$

In terms of unit vectors,

$$v = \frac{dr}{dt} = \frac{d}{dt}(x\boldsymbol{i} + y\boldsymbol{j}) = \frac{dx}{dt}\boldsymbol{i} + \frac{dy}{dt}\boldsymbol{j}. \tag{3–5}$$

If we know x and y as functions of time, we can compute the derivatives in Eqs. (3–4) and (3–5) and determine the velocity at any time. In many of the problems of this chapter our objective will be to obtain x and y as functions of time; these functions then provide a complete description of the motion.

Thus we may represent velocity, a vector quantity, in terms of its components or in terms of its magnitude and direction, just as with other vector quantities such as displacement and force. The *direction* of the instantaneous velocity of a particle at any point is *always* tangent to the path at that point.

3–2 AVERAGE AND INSTANTANEOUS ACCELERATION

In Fig. 3–2a the vectors v_1 and v_2 represent the instantaneous velocities, at points P and Q, of a particle moving in a curved path. The velocity v_2 may differ in both magnitude and direction from the velocity v_1.

The **average acceleration** a_{av} of the particle as it moves from P to Q is defined as the *vector change in velocity*, Δv, divided by the time interval Δt:

$$\text{Average acceleration} = a_{av} = \frac{\Delta v}{\Delta t}. \tag{3–6}$$

Average acceleration is a vector quantity, in the same direction as the vector Δv.

The vector change in velocity, Δv, means the vector difference $v_2 - v_1$:

$$\Delta v = v_2 = v_1,$$

or

$$v_2 = v_1 + \Delta v.$$

Thus v_2 is the vector sum of the original velocity v_1 and the change Δv. This relationship is shown in Fig. 3–2b. The average acceleration vector, $a_{av} = \Delta v/\Delta t$, is shown in Fig. 3–2a.

The **instantaneous acceleration** a at point P is defined in magnitude and direction as the limit approached by the average acceleration when point Q approaches point P and Δv and Δt both approach zero:

$$\text{Instantaneous acceleration} = a = \lim_{\Delta t \to 0} \frac{\Delta v}{\Delta t} = \frac{dv}{dt}. \tag{3–7}$$

Just as with velocity, average acceleration refers to a finite time interval during which the velocity changes, while instantaneous acceleration is the rate of change of velocity at a specific point and at a specific instant of time. The instantaneous acceleration vector at point P is shown in Fig. 3–2c. Note that it does not have the same direction as the velocity vector; in general there is no reason it should. Reference to the construction in Fig. 3–2c shows that the acceleration vector must always lie on the *concave* side of the curved path.

3–3 COMPONENTS OF ACCELERATION

We often represent the acceleration of a particle in terms of the components of this vector quantity. Figure 3–3 again shows the motion of a particle as described in a rectangular coordinate system. The x- and y-components, a_x and a_y, of the instantaneous acceleration a of the particle are

$$a_x = \lim_{\Delta t \to 0} \frac{\Delta v_x}{\Delta t} = \frac{dv_x}{dt}, \qquad a_y = \lim_{\Delta t > 0} \frac{\Delta v_y}{\Delta t} = \frac{dv_y}{dt}. \tag{3–8}$$

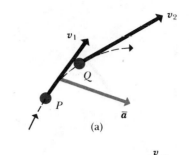

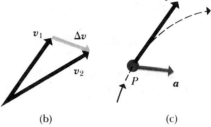

3–2　(a) The vector $a_{av} = \Delta v/\Delta t$ represents the average acceleration between P and Q. (b) Construction for obtaining $\Delta v = v - v_1$. (c) Instantaneous acceleration a at point P. Vector v is tangent to the path; vector a points toward the concave side of the path.

Instantaneous acceleration: acceleration at a point

Acceleration is a vector quantity; it has components.

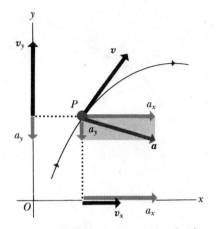

3–3 The acceleration **a** is resolved into its components a_x and a_y.

In terms of unit vectors,

$$a = \frac{dv_x}{dt}i + \frac{dv_y}{dt}j. \tag{3–9}$$

By using Eqs. (3–4) and (3–5), we may also express the acceleration as

$$a_x = \frac{d^2x}{dt^2}, \qquad a_y = \frac{d^2y}{dt^2}, \tag{3–10}$$

or

$$a = \frac{d^2x}{dt^2}i + \frac{d^2y}{dt^2}j. \tag{3–11}$$

If we know the components a_x and a_y, we can find the magnitude and direction of the acceleration **a,** just as with velocity.

EXAMPLE 3–1　The coordinates of a particle moving in the *xy*-plane are given as functions of time by

$$x = 1 \text{ m} + (2 \text{ m·s}^{-2})t^2,$$
$$y = (2 \text{ m·s}^{-1})t + (1 \text{ m·s}^{-3})t^3.$$

Find the particle's position, velocity, and acceleration at time $t = 2$ s.

SOLUTION　At time $t = 2$ s,

$$x = 1 \text{ m} + (2 \text{ m·s}^{-2})(2 \text{ s})^2 = 9 \text{ m},$$
$$y = (2 \text{ m·s}^{-1})(2 \text{ s}) + (1 \text{ m·s}^{-3})(2 \text{ s})^3 = 12 \text{ m},$$

or

$$r = (9 \text{ m})i + (12 \text{ m})j.$$

The particle's distance from the origin at this time is

$$r = \sqrt{x^2 + y^2} = \sqrt{(9 \text{ m})^2 + (12 \text{ m})^2} = 15 \text{ m}.$$

The velocity at any given time t is given by

$$v_x = \frac{dx}{dt} = 2(2 \text{ m·s}^{-2})t, \qquad v_y = \frac{dy}{dt} = (2 \text{ m·s}^{-1}) + 3(1 \text{ m·s}^{-3})t^2.$$

The velocity at time $t = 2$ s is given by

$$v_x = 2(2 \text{ m·s}^{-2})(2 \text{ s}) = 8 \text{ m·s}^{-1},$$
$$v_y = (2 \text{ m·s}^{-1}) + 3(1 \text{ m·s}^{-3})(2 \text{ s})^2 = 14 \text{ m·s}^{-1},$$

or

$$v = (8 \text{ m·s}^{-1})i + (14 \text{ m·s}^{-1})j.$$

The magnitude of the velocity at this time is

$$v = \sqrt{v_x{}^2 + v_y{}^2} = \sqrt{(8 \text{ m·s}^{-1})^2 + (14 \text{ m·s}^{-1})^2} = 16.1 \text{ m·s}^{-1},$$

and its direction with respect to the positive *x*-axis is

$$\theta = \arctan \frac{14 \text{ m·s}^{-1}}{8 \text{ m·s}^{-1}} = 60.3°.$$

The acceleration at any time is given by

$$a_x = \frac{dv_x}{dt} = \frac{d^2x}{dt^2} = 2(2 \text{ m·s}^{-2}), \qquad a_y = \frac{dv_y}{dt} = \frac{d^2y}{dt^2} = 6(1 \text{ m·s}^{-3})t.$$

The acceleration at time $t = 2$ s is

$$a_x = 4 \text{ m·s}^{-2}, \qquad a_y = 6(1 \text{ m·s}^{-3})(2 \text{ s}) = 12 \text{ m·s}^{-2},$$

or

$$\boldsymbol{a} = (4 \text{ m·s}^{-2})\boldsymbol{i} + (12 \text{ m·s}^{-2})\boldsymbol{j}.$$

The magnitude of the acceleration at this time is

$$a = \sqrt{a_x^2 + a_y^2} = \sqrt{(4 \text{ m·s}^{-2})^2 + (12 \text{ m·s}^{-2})^2} = 12.6 \text{ m·s}^{-2},$$

and its direction with respect to the positive x-axis is

$$\theta = \arctan \frac{a_y}{a_x} = \arctan \frac{12 \text{ m·s}^{-2}}{4 \text{ m·s}^{-2}} = 71.6°.$$

Note that the direction of the acceleration is different from that of the velocity, and that it is *not* tangent to the particle's path.

The acceleration of a particle moving in a curved path can also be represented in terms of rectangular components $a_\perp$ and $a_\parallel$, in directions **normal** (perpendicular) and **tangential** (parallel) to the path, as shown in Fig. 3–4a. Unlike the rectangular components referred to a set of fixed axes, the normal and tangential components do not have fixed directions in space. They do, however, have a direct physical significance. The parallel component $a_\parallel$ corresponds to a change in the *magnitude* of the velocity vector v, while the normal component $a_\perp$ is associated with a change in the *direction* of the velocity.

Figure 3–4 shows two special cases. In Fig. 3–4b, the acceleration is *parallel* to the velocity v_1. Then because $\boldsymbol{a}$ gives the rate of change of velocity, the change in v during a small time interval Δt is a vector Δv having the same direction as $\boldsymbol{a}$ and hence the same direction as v_1. Thus the velocity v_2 at the end of Δt, given by $v_2 = v_1 + \Delta v$, is a vector having the same direction as v_1 but somewhat greater magnitude.

In Fig. 3–4c, the acceleration is *perpendicular* to the velocity. In an interval Δt, the change Δv is a vector perpendicular to v_1, as shown. Again $v_2 = v_1 + \Delta v$, but in this case v_1 and v_2 have different directions. As the time interval Δt approaches zero, the angle θ in the figure also approaches zero, Δv becomes perpendicular to both v_1 and v_2, and v_1 and v_2 have the same magnitude.

Thus when $\boldsymbol{a}$ is parallel to v, its effect is to change the magnitude of v but not its direction; when $\boldsymbol{a}$ is perpendicular to v, its effect is to change the direction of v but not its magnitude. In general $\boldsymbol{a}$ may have components both parallel and perpendicular to v, but the above statements are still valid for the individual components. In particular, when a particle travels along a curved path with constant speed, its acceleration is not zero, even though the magnitude of v does not change. In this case the acceleration is always perpendicular to v at each point. When a particle moves in a circle with constant speed, the acceleration is at each instant directed toward the center of the circle. We will consider this special case in detail in Section 3–5.

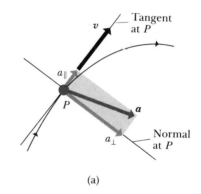

(a)

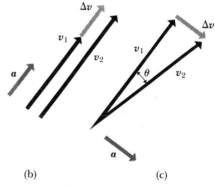

(b)   (c)

3–4 (a) The acceleration is resolved into a component $a_\parallel$ parallel or *tangent* to the path and a component $a_\perp$ perpendicular or *normal* to the path. (b) When $\boldsymbol{a}$ is parallel to v, the magnitude of v increases but its direction does not change. (c) When $\boldsymbol{a}$ is perpendicular to v, the direction of v changes but its magnitude is constant.

3–4 PROJECTILE MOTION

Projectile motion: What do a kicked football and a fireworks rocket have in common?

Λ **projectile** is any body that is given an initial velocity and then follows a path determined by the effect of the gravitational acceleration and by air resistance. A batted baseball, a thrown football, an object dropped from an airplane, and a bullet shot from a rifle are all examples of projectiles. The path followed by a projectile is called its *trajectory*. The motion of a freely falling body, discussed in Chapter 2, is a special case of projectile motion; in that case the trajectory is a vertical straight line.

To simplify the analysis we will consider only trajectories that have a range short enough that the acceleration of gravity may be considered constant in both magnitude and direction. We will also omit effects associated with air resistance and the rotation of the earth. These simplifications form the basis of an idealized *model* of the physical problem; we neglect minor details in order to focus attention on the most important features of the problem.

For our analysis of projectile motion we will use a set of rectangular coordinate axes, taking the *x*-axis horizontal and the *y*-axis vertically upward.

After the diver's feet leave the board, his center of mass follows a parabolic path. (Dr. Harold Edgerton, M.I.T., Cambridge, Massachusetts)

There is no acceleration in the horizontal direction, and the acceleration in the vertical direction is the acceleration due to gravity, which we studied in Section 2–6. Thus the components of a are

$$a_x = 0, \qquad a_y = -g. \tag{3–12}$$

The negative sign for a_y arises because we have chosen the positive y-axis to be upward, while the acceleration due to gravity is downward.

Thus the horizontal component of acceleration is zero and the vertical component is downward and equal to that of a freely falling body. Since zero acceleration means constant velocity, the motion can be described as a combination of *horizontal motion with constant velocity* and *vertical motion with constant acceleration*.

The key to analysis of projectile motion is the fact that we can express all the needed vector relationships in terms of separate equations for the x- and y-components of these vector quantities. Each component of velocity is the rate of change of the corresponding coordinate, and each component of acceleration is the rate of change of the corresponding velocity component. In this sense the x and y motions are independent and may be analyzed separately. The actual motion is then the superposition of these separate motions.

The x- and y-motions can be analyzed separately.

Suppose that at time $t = 0$ our particle is at the point (x_0, y_0) and has velocity components v_{0x} and v_{0y}. The components of acceleration are $a_x = 0$, $a_y = -g$. The time variation of each coordinate is an example of motion with constant acceleration, and we can use Eqs. (2–8) and (2–12) directly. Substituting v_{0x} for v_0 and 0 for a, we find for x

$$v_x = v_{0x}, \tag{3–13}$$
$$x = x_0 + v_{0x}t. \tag{3–14}$$

Similarly, substituting v_{0y} for v_0 and $-g$ for a,

$$v_y = v_{0y} - gt, \tag{3–15}$$
$$y = y_0 + v_{0y}t - \tfrac{1}{2}gt^2. \tag{3–16}$$

Equations of projectile motion: all you ever wanted to know about projectiles

The content of Eqs. (3–13) through (3–16) can also be represented by the vector equations

$$v = v_0 - gt\mathbf{j}, \tag{3–17}$$
$$r = r_0 + v_0t - \tfrac{1}{2}gt^2\mathbf{j}, \tag{3–18}$$

where r_0 is the position vector at time $t = 0$.

Usually it is convenient to take the initial position as the origin; in this case, $x_0 = y_0 = 0$ or $r_0 = 0$. This might be, for example, the position of a ball at the instant it leaves the thrower's hand, or the position of a bullet at the instant it leaves the gun barrel.

Figure 3–5 shows the path of a projectile that passes through the origin at time $t = 0$. The position, velocity, and velocity components of the projectile are shown at a series of times separated by equal intervals. As the figure shows, v_x does not change, but v_y changes by equal amounts in successive intervals, corresponding to constant y-acceleration.

The initial velocity v_0 may be represented by its magnitude v_0 (the initial speed) and the angle θ_0 it makes with the positive x-axis. In terms of these quantities, the *components* v_{0x} and v_{0y} of initial velocity are

$$v_{0x} = v_0 \cos \theta_0,$$
$$v_{0y} = v_0 \sin \theta_0. \tag{3–19}$$

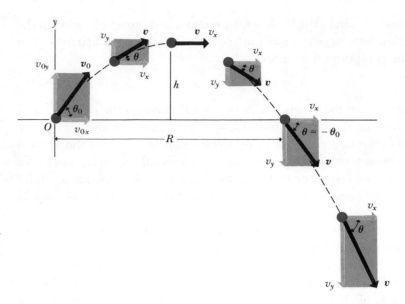

3–5 Trajectory of a body projected with an initial velocity v_0 at an angle of departure θ_0. *The distance R is the horizontal range, and h is the maximum height.*

Using these relations in Eqs. (3–13) through (3–16) and setting $x_0 = y_0 = 0$, we obtain

$$x = (v_0 \cos \theta_0)t, \tag{3-20}$$

$$y = (v_0 \sin \theta_0)t - \tfrac{1}{2}gt^2, \tag{3-21}$$

$$v_x = v_0 \cos \theta_0, \tag{3-22}$$

$$v_y = v_0 \sin \theta_0 - gt. \tag{3-23}$$

These equations describe the position and velocity of the projectile in Fig. 3–5 at any time t.

We can obtain a variety of additional information from Eqs. (3–20) through (3–23). For example, the distance r of the projectile from the origin at any time (the magnitude of the position vector r) is given by

$$r = \sqrt{x^2 + y^2}. \tag{3-24}$$

The projectile's speed (the magnitude of its resultant velocity) is

$$v = \sqrt{v_x{}^2 + v_y{}^2}. \tag{3-25}$$

The *direction* of the velocity, in terms of the angle θ it makes with the positive x-axis, is given by

$$\tan \theta = \frac{v_y}{v_x}. \tag{3-26}$$

The velocity vector v is tangent to the trajectory at each point.

Equations (3–20) and (3–21) give the position of the particle in terms of the parameter t. We can also obtain an equation for the shape of the trajectory, in terms of x and y, by eliminating t. We find $t = x/v_0 \cos \theta_0$ and

$$y = (\tan \theta_0)x - \frac{g}{2v_0{}^2 \cos^2 \theta_0}x^2. \tag{3-27}$$

The quantities v_0, $\tan \theta_0$, $\cos \theta_0$, and g are constants, so the equation has the form

$$y = ax - bx^2,$$

where a and b are constants. This is the equation of a *parabola*.

Figure 3–6 shows parabolic trajectories of a bouncing golf ball.

3–6 Stroboscopic photograph of a bouncing golf ball, showing parabolic trajectories after each bounce. Successive images are separated by equal time intervals, as in Fig. 3–5. Each peak in the trajectories is lower than the preceding one because of energy loss during the "bounce" or collision with the horizontal surface. (Dr. Harold Edgerton, M.I.T., Cambridge, Massachusetts.)

PROBLEM-SOLVING STRATEGY: *Projectile problems*

The same strategies used in Section 2–4 for solving problems with constant acceleration along a straight line are also useful here.

1. Define your coordinate system. Make a sketch showing your axes; label the positive direction for each and show the location of the origin.

2. Make lists of known and unknown quantities. In some problems the components (or magnitude and direction) of initial velocity will be given, and you can use Eqs. (3–20) and (3–21) to find the coordinates and velocity components at some later time. In other problems you may know two points on the trajectory and be asked to find the initial velocity. Be sure you know which quantities are given and which are to be found.

3. It often helps to state the problem in prose and then translate into symbols. *When* does the particle arrive at a certain point (i.e., at what value of t)? *Where* is the particle when its velocity has a certain value (i.e., what are the values of x and y when v_x or v_y has the specified value)? And so on.

4. At the highest point in a trajectory, $v_y = 0$. So the question "When does the projectile reach its highest point?" translates into "What is the value of t when $v_y = 0$?" Similarly, if $y_0 = 0$, then "When does the projectile return to its initial elevation?" translates into "What is the value of t when $y = 0$?" And so on.

EXAMPLE 3–2 A motorcycle stunt rider rides off the edge of a cliff with a horizontal velocity of magnitude 5 m·s^{-1}. Find the rider's position and velocity after $\frac{1}{4}$ s (see Fig. 3–7).

An adventurous motorcyclist rides off a cliff.

SOLUTION The coordinate system is shown in Fig. 3–7. The initial angle θ_0 is zero, so $v_{0x} = 5 \text{ m·s}^{-1}$ and $v_{0y} = 0$. The horizontal velocity component equals the initial velocity and is constant.

Where is the motorcycle at $t = \frac{1}{4}$ s? The x- and y-coordinates, when $t = \frac{1}{4}$ s, are

$$x = v_x t = (5 \text{ m·s}^{-1})(\tfrac{1}{4} \text{ s}) = 1.25 \text{ m},$$
$$y = -\tfrac{1}{2} g t^2 = -\tfrac{1}{2}(9.8 \text{ m·s}^{-2})(\tfrac{1}{4} \text{ s})^2 = -0.306 \text{ m}.$$

The distance from the origin at this time is

$$r = \sqrt{x^2 + y^2} = \sqrt{(1.25 \text{ m})^2 + (-0.306 \text{ m})^2} = 1.29 \text{ m}.$$

What is the velocity at time $t = \frac{1}{4}$ s? The components of velocity at time $t = \frac{1}{4}$ s are

$$v_x = v_0 = 5 \text{ m·s}^{-1},$$
$$v_y = -gt = (-9.8 \text{ m·s}^{-2})(\tfrac{1}{4} \text{ s}) = -2.45 \text{ m·s}^{-1}.$$

The resultant velocity has magnitude

$$v = \sqrt{v_x{}^2 + v_y{}^2} = 5.57 \text{ m·s}^{-1}.$$

The angle θ is

$$\theta = \arctan \frac{v_y}{v_x} = \arctan \frac{-2.45 \text{ m·s}^{-1}}{5.00 \text{ m·s}^{-1}} = -26.1°.$$

That is, at this time the velocity is 26.1° *below* the horizontal.

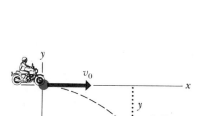

3–7 Trajectory of a body projected horizontally.

EXAMPLE 3–3 In Fig. 3–5, suppose the projectile is a baseball hit by Superman with an initial speed $v_0 = 50 \text{ m·s}^{-1}$ at an initial angle $\theta_0 = 53.1°$. The ball is probably struck a meter or so above ground level; we neglect this distance and assume it

How far can Superman hit a major-league fastball?

starts at ground level, where $y = 0$. Then

$$v_{0x} = v_0 \cos \theta_0 = (50 \text{ m·s}^{-1})(0.60) = 30 \text{ m·s}^{-1},$$
$$v_{0y} = v_0 \sin \theta_0 = (50 \text{ m·s}^{-1})(0.80) = 40 \text{ m·s}^{-1}.$$

a) Find the position of the ball, and the magnitude and direction of its velocity, when $t = 2.0$ s.

Using the coordinate system shown in Fig. 3–5, we want to find x, y, v_x, and v_y at time $t = 2.0$ s. From Eqs. (3–20) through (3–23),

$$x = (30 \text{ m·s}^{-1})(2.0 \text{ s}) = 60 \text{ m},$$
$$y = (40 \text{ m·s}^{-1})(2.0 \text{ s}) - \tfrac{1}{2}(9.8 \text{ m·s}^{-2})(2.0 \text{ s})^2 = 60.4 \text{ m},$$
$$v_x = 30 \text{ m·s}^{-1},$$
$$v_y = 40 \text{ m·s}^{-1} - (9.8 \text{ m·s}^{-2})(2.0 \text{ s}) = 20.4 \text{ m·s}^{-1},$$
$$v = \sqrt{v_x{}^2 + v_y{}^2} = 36.3 \text{ m·s}^{-1},$$
$$\theta = \arctan \frac{20.4 \text{ m·s}^{-1}}{30 \text{ m·s}^{-1}} = \arctan 0.680 = 34.2°.$$

b) Find the time at which the ball reaches the highest point of its flight, and find the height of this point.

At the highest point, the vertical velocity v_y is zero. When does this happen? If t_1 is the time at which this point is reached,

$$v_y = 0 = 40 \text{ m·s}^{-1} - (9.8 \text{ m·s}^{-2})t_1, \qquad t_1 = 4.08 \text{ s}.$$

The height h of the point is the value of y when $t = 4.08$ s:

$$h = (40 \text{ m·s}^{-1})(4.08 \text{ s}) - \tfrac{1}{2}(9.8 \text{ m·s}^{-2})(4.08 \text{ s})^2 = 81.6 \text{ m}.$$

c) Find the *horizontal range R*, that is, the horizontal distance from the starting point to the point where the ball returns to earth, where $y = 0$.

When does the ball return to earth? Let t_2 be the time when y becomes zero. Then

$$y = 0 = (40 \text{ m·s}^{-1})t_2 - \tfrac{1}{2}(9.8 \text{ m·s}^{-2})t_2{}^2.$$

This is a quadratic equation for t_2; it has two roots,

$$t_2 = 0 \quad \text{and} \quad t_2 = 8.16 \text{ s},$$

corresponding to the two times at which $y = 0$. The value $t_2 = 0$ is, of course, the time the ball left the ground; $t_2 = 8.16$ s is the time of its return. Note that this is just twice the time to reach the highest point. The time of descent therefore equals the time of rise. (This is *always* true if the starting point and end point are at the same elevation; can you prove this?)

The horizontal range R is the value of x when the ball returns to the ground, that is, when $t = 8.16$ s:

$$R = v_x t_2 = (30 \text{ m·s}^{-1})(8.16 \text{ s}) = 245 \text{ m}.$$

Thus, the ball is a home run. The vertical component of velocity at this point is

$$v_y = 40 \text{ m·s}^{-1} - (9.8 \text{ m·s}^{-2})(8.16 \text{ s}) = -40 \text{ m·s}^{-1}.$$

That is, the vertical velocity has the same magnitude as the initial vertical velocity, but the opposite direction. Since v_x is constant, the angle *below* the horizontal at this point equals the initial angle θ_0.

d) If the ball did not hit the ground, it would continue to travel on below its original level. The playing field might be located atop a flat-topped hill that drops off steeply on one side. Then negative values of y, corresponding to times greater than 8.16 s, are possible. We challenge the reader to compute the position and velocity at a time 10 s after the start, corresponding to the last position shown in Fig. 3–5. The results are

$$x = 300 \text{ m}, \qquad y = -90 \text{ m},$$
$$v_x = 30 \text{ m·s}^{-1}, \qquad v_y = -58 \text{ m·s}^{-1}.$$

EXAMPLE 3–4 In Fig. 3–8, a boy shoots an arrow from ground level at an apple hanging in a tree. At the same instant he releases the arrow, the apple falls from the tree and drops straight down, starting from rest. Show that the arrow's path curves just enough for it to hit the apple, regardless of its initial velocity.

A very talented archer hits the apple.

SOLUTION We have to prove that the apple and the arrow both arrive at the same time at some point directly below the apple's initial location. The initial elevation of the apple is $x \tan \theta_0$, and in time t it falls a distance $\frac{1}{2}gt^2$. Its elevation at the instant of collision is therefore

$$y = x \tan \theta_0 - \tfrac{1}{2}gt^2.$$

In this same time the arrow travels the distance x with constant x-component of velocity $v_0 \cos \theta_0$, so $x = v_0 \cos \theta_0 t$. Solving this for t and substituting in the preceding equation, we obtain

$$y = x \tan \theta_0 - \frac{1}{2}g\left(\frac{x}{v_0 \cos \theta_0}\right)^2.$$

But this is the same expression as Eq. (3–27) for the path of the arrow. Thus at the instant the arrow reaches the line along which the apple is falling, the two heights are the same, and the two meet at this point.

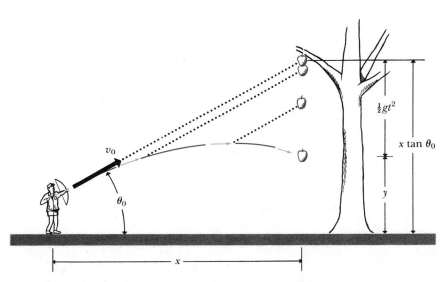

3–8 Trajectory of an arrow shot directly at a freely falling apple.

EXAMPLE 3–5 For a projectile launched with speed v_0 at initial angle θ_0, derive general expressions for the maximum height h and range R shown in Fig. 3–5. For a given v_0, what value of θ_0 gives maximum range?

How to find the range and maximum height of a projectile

SOLUTION We follow the same pattern as in Example 3–3, part (b). First, when does the projectile reach its maximum height? At this point $v_y = 0$, so the time t_1

at maximum height is given by

$$v_y = v_0 \sin \theta_0 - gt_1 = 0, \qquad t_1 = \frac{v_0 \sin \theta_0}{g}.$$

Next, what is the value of y at this time? From Eq. (3–21),

$$h = v_0 \sin\theta_0 \left(\frac{v_0 \sin \theta_0}{g}\right) - \frac{1}{2} g \left(\frac{v_0 \sin \theta_0}{g}\right)^2$$

$$= \frac{v_0^2 \sin^2 \theta_0}{2g}. \tag{3–28}$$

As expected, the maximum value of h occurs when the projectile is launched straight up, for then $\theta_0 = 90°$, $\sin \theta_0 = 1$, and $h = v_0^2/2g$. If it is launched horizontally, $\theta_0 = 0$ and the maximum height is zero!

To find the range, we first find the time t_2 when the projectile returns to the ground. At that time $y = 0$ and, from Eq. (3–21),

$$t_2(v_0 \sin \theta_0 - \tfrac{1}{2}gt_2) = 0.$$

The two roots of this quadratic equation for t_2 are $t_2 = 0$ and $t_2 = 2v_0 \sin \theta_0/g$. The first is obviously the time the projectile *leaves* the ground; the range R is the value of x at the second time. From Eq. (3–20),

$$R = (v_0 \cos \theta_0)\left(\frac{2v_0 \sin \theta_0}{g}\right).$$

According to a familiar trigonometric identity, $2 \sin \theta_0 \cos \theta_0 = \sin 2\theta_0$, so

$$R = \frac{v_0^2 \sin 2\theta_0}{g}. \tag{3–29}$$

The maximum value of $\sin 2\theta_0$, namely unity, occurs when $2\theta_0 = 90°$, or $\theta_0 = 45°$, and this angle gives the maximum range for a given initial speed. Finally, Eqs. (3–28) and (3–29) can be used only when the initial and final values of y are equal. Be careful; there are many end-of-chapter problems where they are *not* applicable.

How to be a Super Bowl quarterback.

EXAMPLE 3–6 In some problems we want to know what the departure angle θ_0 should be for a given v_0 to produce a certain range R. Suppose a football player wants to throw a football at 20 m·s^{-1} to a receiver 30 m away. At what angle should he throw it?

SOLUTION From Eq. (3–29),

$$\theta_0 = \frac{1}{2} \arcsin \frac{Rg}{v_0^2}$$

$$= \frac{1}{2} \arcsin \frac{(30 \text{ m})(9.80 \text{ m·s}^{-2})}{(20 \text{ m·s}^{-1})^2}$$

$$= \frac{1}{2} \arcsin 0.735.$$

There are *two* values of θ_0 between 0 and 90° satisfying this equation: arcsin $0.735 = 47.3°$ or $132.7°$, giving $\theta_0 = 23.7°$ or $66.3°$. Both these angles give the same range; the time of flight and the maximum height are greater for the higher-angle trajectory. Incidentally, the sum of these two values of θ_0 is exactly 90°. This is not a coincidence. Can you prove this?

Figure 3–9 is copied from a multiflash photograph of the trajectory of a ball; x- and y-axes and the initial velocity vector have been added. The horizontal distances between consecutive positions are all equal, showing that the horizontal velocity component is constant. The vertical distances first decrease and then increase, indicating that the vertical motion is accelerated.

Figure 3–10 is made from a composite photograph of three trajectories of a ball projected from a spring gun with angles of 30°, 45°, and 60°. The initial speed v_0 is approximately the same in all three cases. The horizontal ranges are nearly the same for the 30° and 60° angles, and the range for 45° is greater than either.

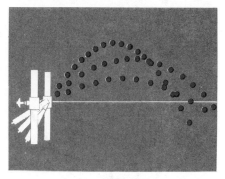

3–9 Trajectory of a body projected at an angle with the horizontal. (Reproduction of a multiflash photograph.)

3–5 CIRCULAR MOTION

We discussed components of acceleration in Section 3–3. When a particle moves along a curved path, it must have a component of acceleration perpendicular (normal) to the path, even if its speed is constant. When the path is a *circle,* there is a simple relation between the normal component of acceleration, the speed of the particle, and the radius of the circle. We now derive this relation for the special case when a particle moves in a circle with *constant speed.* This motion is called **uniform circular motion.** Note that this case is different from that of Section 3–4 because here the acceleration is *not* constant.

Figure 3–11 shows a particle moving in a circular path of radius R with center at O. The vector change in velocity, Δv, is shown in Fig. 3–11b. The particle moves from P to Q in a time Δt.

The triangles OPQ and opq in Fig. 3–11 are similar, since both are isosceles triangles and the angles labeled $\Delta \theta$ are the same. Hence

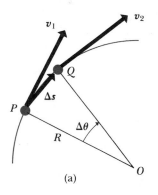

$$\frac{\Delta v}{v_1} = \frac{\Delta s}{R} \quad \text{or} \quad \Delta v = \frac{v_1}{R} \Delta s.$$

The magnitude of the average normal acceleration $(a_\perp)_{av}$ during Δt is therefore

$$(a_\perp)_{av} = \frac{\Delta v}{\Delta t} = \frac{v_1}{R} \frac{\Delta s}{\Delta t}.$$

(a)

The *instantaneous* acceleration $a_\perp$ at point P is the limit of this expression, as point Q is taken closer and closer to point P:

$$a_\perp = \lim_{\Delta t \to 0} \frac{v_1}{R} \frac{\Delta s}{\Delta t} = \frac{v_1}{R} \lim_{\Delta t \to 0} \frac{\Delta s}{\Delta t}.$$

However, the limit of $\Delta s/\Delta t$ is the speed v_1 at point P, and since P can be any point of the path, we can drop the subscript from v_1 and let v represent the speed at any point. Then

$$a_\perp = \frac{v^2}{R}. \tag{3–30}$$

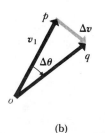

(b)

3–11 Construction for finding change in velocity, Δv, of a particle moving in a circle.

The magnitude of the instantaneous normal acceleration is equal to the square of the speed divided by the radius of the circle. The direction is perpendicular to v and inward along the radius. Because the acceleration is always directed toward the center of the circle, it is sometimes called **centripetal acceleration;** the word *centripetal* is derived from two Greek words meaning "seeking the center."

3–10 An angle of departure of 45° gives the maximum horizontal range. (Reproduction of a multiflash photograph.)

A particle moving in a circle with constant speed *must* have $a_\perp = v^2/R$. When a particle travels with a constant speed of 4 m·s⁻¹ in a circle of radius 2 m, its acceleration has magnitude

$$a = \frac{(4 \text{ m·s}^{-1})^2}{2 \text{ m}} = 8 \text{ m·s}^{-2}.$$

Figure 3–12 shows the directions of the velocity and acceleration vectors at several points for a particle moving with uniform circular motion.

The magnitude of the acceleration can also be expressed in terms of the **period** τ of the motion, the time for one revolution. If a particle travels once around the circle, a distance of $2\pi R$, in a time τ, its speed v is given by

$$v = \frac{2\pi R}{\tau}. \tag{3-31}$$

Thus Eq. (3–30) can also be written as

$$a_\perp = \frac{4\pi^2 R}{\tau^2}. \tag{3-32}$$

We have assumed that the particle's speed is constant. If the speed varies, Eq. (3–30) still gives the normal component of acceleration, but in that case there is also a *tangential* component of acceleration. From the discussion at the end of Section 3–3, we see that the tangential component of acceleration is equal to the rate of change of speed:

$$a_\parallel = \lim_{\Delta t \to 0} \frac{\Delta v_\parallel}{\Delta t} = \frac{dv_\parallel}{dt}. \tag{3-33}$$

If the speed is constant, there is no tangential component of acceleration and the acceleration is purely normal, resulting from the continuous change in direction of the velocity.

The period is the time for one revolution.

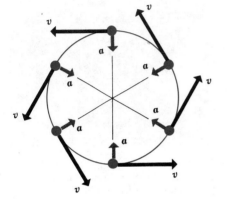

3–12 Velocity and acceleration vectors of a particle in uniform circular motion.

EXAMPLE 3–7 A car traveling at a constant speed of 20 m·s⁻¹ rounds a curve of radius 100 m. What is its acceleration?

SOLUTION The magnitude of the acceleration is given by Eq. (3–30):

$$a_\perp = \frac{v^2}{R} = \frac{(20 \text{ m·s}^{-1})^2}{100 \text{ m}} = 4.0 \text{ m·s}^{-2}.$$

The direction of $\boldsymbol{a}$ at each instant is perpendicular to the velocity and directed toward the center of the circle.

EXAMPLE 3–8 In a carnival ride, the passengers travel in a circle of radius 5.0 m, making one complete circle in 4.0 s. What is the acceleration?

SOLUTION The speed is the circumference of the circle divided by the period τ (the time for one revolution):

$$v = \frac{2\pi R}{\tau} = \frac{2\pi (5.0 \text{ m})}{4.0 \text{ s}} = 7.85 \text{ m·s}^{-1}.$$

The centripetal acceleration is

$$a_\perp = \frac{v^2}{R} = \frac{(7.85 \text{ m·s}^{-1})^2}{5.0 \text{ m}} = 12.3 \text{ m·s}^{-2}.$$

Or, from Eq. (3–32),

$$a_\perp = \frac{4\pi^2(5.0 \text{ m})}{(4.0 \text{ s})^2} = 12.3 \text{ m·s}^{-2}.$$

As in the preceding example, the direction of $\boldsymbol{a}$ is always toward the center of the circle. The magnitude of $\boldsymbol{a}$ is greater than g, the acceleration due to gravity, so this is quite a wild ride!

Relative velocity again: What you see depends on how you're moving. Vector addition of velocities.

3–6 RELATIVE VELOCITY

In Section 2–7 we introduced the concept of **relative velocity** for motion along a straight line. We can easily extend this concept to include motion in a plane or in space. Suppose that in Section 2–7 (Fig. 2–12) the automobile is not traveling in the same direction as the train, but perpendicular to this direction, across the flatcars. In any time interval the displacement of the automobile relative to the earth is the vector sum of its displacement relative to the flatcars and their displacement relative to the earth. Thus the velocity v_{AE} of the automobile relative to the earth is the vector sum of its velocity v_{AF} relative to the flatcars and their velocity v_{FE} relative to the earth. That is, Eq. (2–19) is a special case of the more general *vector equation:*

$$v_{AE} = v_{AF} + v_{FE}. \tag{3–34}$$

Thus if the automobile is moving across the flatcars at 40 km·hr^{-1}, and they are moving relative to the earth at 30 km·hr^{-1}, the appropriate vector diagram is shown in Fig. 3–13. The automobile's velocity relative to the earth is 50 km·hr^{-1} at an angle of 53° to the direction of the train's motion.

We can extend Eq. (3–34) to include any number of relative velocities. For example, if a bug B crawls along the floor of the automobile with a velocity v_{BA} relative to the automobile, the bug's velocity relative to the earth is the vector sum of its velocity relative to the automobile and that of the velocity of the automobile relative to the earth:

$$v_{BE} = v_{BA} + v_{AE}.$$

Combining this with Eq. (3–34), we find

$$v_{BE} = v_{BA} + v_{AF} + v_{FE}. \tag{3–35}$$

This equation illustrates the general rule for combining relative velocities.

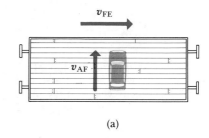

(a)

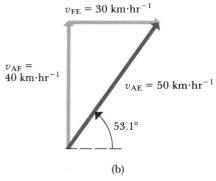

(b)

3–13 (a) An automobile being driven across a moving railroad flatcar. (b) Velocity vector diagram showing how the automobile's velocity v_{AE} relative to the earth is related to its velocity v_{AF} relative to the flatcar and the flatcar's velocity v_{FE} relative to the earth.

PROBLEM-SOLVING STRATEGY: *Relative velocity*

The same strategy introduced in Section 2–7 is also useful here.

1. Write each velocity with a double subscript, *in the proper order,* as "velocity of (first subscript) relative to (second subscript)."

2. In adding relative velocities, make sure that the first subscript of each velocity is the same as the last subscript of the preceding velocity.

3. The *first* subscript of the *first* velocity in the sum is the same as the *first* subscript of the velocity representing the sum, and the *last* subscript of the *last* velocity in the sum is the same as the *last* subscript of the velocity representing the sum. This sounds complicated, but referring to Eq. (3–35) will help you see how it works.

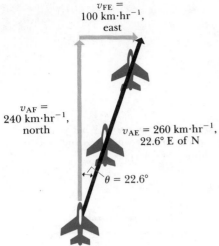

3–14 Vector diagram for aircraft flying north, wind blowing east, and resultant velocity vector of the plane.

Any of the relative velocities in an equation such as Eq. (3–34) can be transferred from one side of the equation to the other, with the sign reversed. Thus Eq. (3–34) can be written

$$v_{AF} = v_{AE} - v_{FE}.$$

The velocity of the automobile relative to the flatcar equals the *vector difference* between the velocities of automobile and flatcar, each relative to the earth.

One more point should be noted. The velocity of body A relative to body B, v_{AB}, is the negative of the velocity of B relative to A, v_{BA}:

$$v_{AB} = -v_{BA}.$$

That is, v_{AB} is equal in magnitude and opposite in direction to v_{BA}. If the automobile is traveling to the *left* at 40 mi·hr^{-1}, relative to the flatcars, the flatcars are traveling to the *right* at 40 mi·hr^{-1}, relative to the automobile.

EXAMPLE 3–9 The compass of an aircraft indicates that it is headed due north, and its airspeed indicator shows that it is moving through the air at 240 km·hr^{-1}. If there is a wind of 100 km·hr^{-1} from west to east, what is the velocity of the aircraft relative to the earth?

SOLUTION Let subscript A refer to the aircraft, and subscript F to the moving air (which now corresponds to the flatcar in Fig. 2–12). Subscript E refers to the earth. We have given

$$v_{AF} = 240 \text{ km·hr}^{-1}, \text{ due north,}$$
$$v_{FE} = 100 \text{ km·hr}^{-1}, \text{ due east,}$$

and we wish to find the magnitude and direction of v_{AE}:

$$v_{AE} = v_{AF} + v_{FE}.$$

The three relative velocities and their relationship are shown in Fig. 3–14. It follows from this diagram that

$$v_{AE} = 260 \text{ km·hr}^{-1}, \qquad \theta = \arctan \frac{100 \text{ km·hr}^{-1}}{240 \text{ km·hr}^{-1}} = 22.6° \text{ E of N.}$$

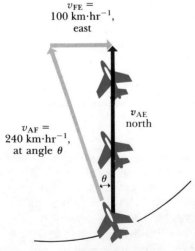

3–15 The resultant vector v_{AF} shows the direction in which the pilot should point the plane in order to have the plane travel due north.

EXAMPLE 3–10 In Example 3–9, what direction should the pilot head in order to travel due north? What will then be the plane's velocity relative to the earth? (The magnitude of the airspeed and the wind velocity are the same as in the preceding example.)

SOLUTION We now have given

$$v_{AF} = 240 \text{ km·hr}^{-1}, \text{ direction unknown,}$$
$$v_{FE} = 100 \text{ km·hr}^{-1}, \text{ due east.}$$

We want to find v_{AE}; its magnitude is unknown, but we know that its direction is due north. (Note that both this and the preceding example require us to determine two unknown quantities. In Example 3–9 these were the magnitude and direction of v_{AE}; in this example the unknowns are the direction of v_{AF} and the magnitude of v_{AE}.)

The three relative velocities must still satisfy the vector equation

$$v_{AE} = v_{AF} + v_{FE}.$$

The appropriate vector diagram is shown in Fig. 3–15. We find

$$v_{AE} = \sqrt{(240 \text{ km·hr}^{-1})^2 - (100 \text{ km·hr}^{-1})^2} = 218 \text{ km·hr}^{-1},$$

$$\theta = \arcsin \frac{100 \text{ km·hr}^{-1}}{240 \text{ km·hr}^{-1}} = 24.6°.$$

That is, the pilot should head 24.6° W of N, and his ground speed is then 218 km·hr^{-1}.

SUMMARY

The position vector r of a point P in a plane is the displacement vector from the origin to P. Its components are the coordinates x and y.

The average velocity v_{av} during the time interval Δt is the displacement Δr (the change in the position vector r) divided by Δt, that is, $v_{av} = \Delta r/\Delta t$. Its components are $(v_x)_{av} = \Delta x/\Delta t$ and $(v_y)_{av} = \Delta y/\Delta t$.

Instantaneous velocity is $v = dr/dt$; its components are $v_x = dx/dt$ and $v_y = dy/dt$.

The average acceleration a_{av} during the time interval Δt is the velocity change Δv divided by Δt; that is, $a_{av} = \Delta v/\Delta t$; its components are $(a_x)_{av} = \Delta v_x/\Delta t$ and $(a_y)_{av} = \Delta v_y/\Delta t$.

Instantaneous acceleration is $a = dv/dt$. Its components are $a_x = dv_x/dt$ and $a_y = dv_y/dt$.

Acceleration can also be represented in terms of its components parallel and normal to the path, $a_{\parallel} = dv/dt$ and $a_{\perp}$, respectively.

In projectile motion, $a_x = 0$ and $a_y = -g$. The coordinates and velocity components are given as functions of time by

$$x = (v_0 \cos \theta_0)t, \quad y = (v_0 \sin \theta_0)t - \tfrac{1}{2}gt^2, \quad v_x = v_0 \cos \theta_0, \quad v_y = v_0 \sin \theta_0 - gt.$$

The shape of the path in projectile motion is always a parabola.

When a particle moves in a circular path of radius R with speed v, it has an acceleration with magnitude $a_{\perp} = v^2/R$, directed always toward the center of the circle and perpendicular to v. The period τ of a circular motion is the time for one revolution. If the speed is constant, then $v = 2\pi R/\tau$ and $a_{\perp} = 4\pi^2 R/\tau^2$. If the speed is not constant, there is also a component of a parallel to the path, equal to $a_{\parallel} = dv/dt$.

When a body A moves relative to a body F, and F moves relative to the earth E or other reference frame, we denote the velocity of A relative to F by v_{AF}, the velocity of F relative to E by v_{FE}, and the velocity of A relative to E by v_{AE}. These velocities are related by

$$v_{AE} = v_{AF} + v_{FE}.$$

KEY TERMS

position vector
average velocity
instantaneous velocity
average acceleration
instantaneous acceleration
normal component
tangential component
projectile
uniform circular motion
centripetal acceleration
period
relative velocity

QUESTIONS

3–1 A simple pendulum (a mass swinging at the end of a string) swings back and forth in a circular arc. What is the direction of its acceleration at the ends of the swing? At the midpoint?

3–2 A football is thrown in a parabolic path. Is there any point at which the acceleration is parallel to the velocity? Perpendicular to the velocity?

3–3 When a rifle is fired at a distant target, the barrel is not lined up exactly on the target. Why not? Does the angle of correction depend on the distance of the target?

3–4 The acceleration of a falling body is measured in an elevator traveling upward at a constant speed of 9.8 m·s^{-1}. What result is obtained?

3–5 One can play catch with a softball in an airplane in level flight just as though the plane were at rest. Is this still possible when the plane is making a turn?

3–6 A package is dropped out of an airplane in level flight. If air resistance could be neglected, how would the motion of the package look to the pilot? To an observer on the ground?

3–7 No matter what the initial velocity, the motion of a projectile (neglecting air resistance) is *always* confined to a single plane. Why?

3–8 If a jumper can give himself the same initial speed regardless of the direction he jumps (forward or straight up), how is his maximum vertical jump (high jump) related to his maximum horizontal jump (broad jump)?

3–9 A manned space-flight projectile is launched in a parabolic trajectory. A man in the capsule feels weightless. Why? In what sense is he weightless?

3–10 The maximum range of a projectile occurs when it is aimed at 45°, if air resistance is neglected. Is this still true if air resistance is included? If not, is the optimum angle greater or less than 45°? Why?

3–11 A passenger in a car rounding a sharp curve is thrown toward the outside of the curve. What force throws him in this direction?

3–12 In uniform circular motion, what is the *average* velocity during one revolution? The average acceleration?

3–13 In uniform circular motion the acceleration is perpendicular to the velocity at every instant, even though both change continuously in direction. Is there any other motion having this property, or is uniform circular motion unique?

3–14 If an artificial earth satellite is in an orbit with a period of exactly one day, how does its motion look to an observer on the rotating earth? (Such an orbit is said to be *geosynchronous;* most communications satellites are placed in such orbits.)

3–15 Raindrops hitting the side windows of a car in motion often leave diagonal streaks. Why? What about diagonal streaks on the windshield? Is the explanation the same or different?

3–16 In a rainstorm with a strong wind, what determines the best position in which to hold an umbrella?

EXERCISES

Section 3–3 Components of Acceleration

3–1 The coordinates of a particle moving in the xy-plane are given as functions of time by $x = 2 \text{ m} - (4 \text{ m·s}^{-1})t$ and $y = (3 \text{ m·s}^{-2})t^2$.

a) Calculate the velocity and acceleration vectors of the particle as functions of time.

b) Calculate the magnitude and direction of the particle's velocity and acceleration at $t = 3$ s.

3–2 A particle moves in the xy-plane with acceleration

$$\boldsymbol{a} = (2 \text{ m·s}^{-4})t^2\boldsymbol{i} + (4 \text{ m·s}^{-3})t\boldsymbol{j}.$$

a) Assuming that the particle is at rest at the origin at time $t = 0$, derive expressions for the velocity and position vectors as functions of time.

b) Sketch the path of the particle.

c) Find the magnitude and direction of the velocity at $t = 3$ s.

Section 3–4 Projectile Motion

3–3 A golf ball is hit horizontally from a tee at the edge of a cliff. Its x- and y-coordinates are given as functions of time by

$$x = (20 \text{ m·s}^{-1})t, \qquad y = -(4.9 \text{ m·s}^{-2})t^2.$$

a) Compute the x- and y-coordinates at times $t = 0$ s, 1 s, 2 s, 3 s, and 4 s. Plot these positions on graph paper and sketch the trajectory.

b) Determine the ball's initial velocity vector and its acceleration vector.

c) Find the x- and y-components of the velocity at time $t = 2$ s. Plot the velocity vector at the appropriate point on the trajectory obtained in (a). Is the velocity tangent to the trajectory?

3–4 A ball rolls off the edge of a table top 1 m above the floor and strikes the floor at a point 1.5 m horizontally from the edge of the table.

a) Find the time of flight.

b) Find the initial velocity.

c) Find the magnitude and direction of the velocity of the ball just before it strikes the floor. Draw a diagram to scale.

3–5 A book slides off a horizontal table top with a speed of 4 m·s^{-1}. It is observed to strike the floor in 0.5 s. Find

a) the height of the table top above the floor;

b) the horizontal distance from the edge of the table to the point where the book strikes the floor;

c) the horizontal and vertical components of its velocity when it reaches the floor.

3–6 An airplane flying horizontally at a speed of 100 m·s^{-1} drops a box at an elevation of 2000 m.

a) How much time is required for the box to reach the earth?

b) How far does it travel horizontally while falling?

c) Find the horizontal and vertical components of its velocity when it strikes the earth.

d) Where is the airplane when the box strikes the earth?

3–7 A marksman fires a .22-caliber rifle horizontally at a target; the bullet has a muzzle velocity of 900 ft·s⁻¹. How much does the bullet drop in flight if the target is

a) 50 yd away?

b) 150 yd away?

3–8 A baseball is thrown with an initial upward velocity component of 20 m·s⁻¹ and a horizontal velocity component of 25 m·s⁻¹.

a) Find the horizontal and vertical components of the displacement and velocity of the ball after 1 s, 2 s, 3 s, and 4 s.

b) How much time is required to reach the highest point of the trajectory?

c) How high is this point?

d) How much time (after being thrown) is required for the ball to return to its original level? How does this compare with the time calculated in part (b)?

e) How far has it traveled horizontally during this time? Show your results in a neat sketch large enough to show all features clearly.

3–9 A baseball is thrown at an angle of 53° above the horizontal with an initial speed of 40 m·s⁻¹.

a) What is the maximum height above the thrower's hand reached by the baseball?

b) How long does it take to reach the maximum height?

c) At what *two* times is the baseball at a height of 25 m above the point from which it was thrown?

d) Calculate the horizontal and vertical components of the baseball's velocity at each of the two times calculated in (c).

e) What are the magnitude and direction of the baseball's velocity when it returns to the level from which it was thrown?

3–10 A batted baseball leaves the bat at an angle of 30° above the horizontal and is caught by an outfielder 400 ft from the plate. Assume that the height of the point where it was struck by the bat equals the height of the point where it was caught.

a) What was the initial speed of the ball?

b) How high did it rise above the point where it struck the bat?

3–11 A man stands on the roof of a building that is 50 m tall and throws a rock with a velocity of 60 m·s⁻¹ at an angle of 37° above the horizontal. Calculate:

a) the maximum height, above the roof, reached by the rock;

b) the magnitude of the resultant velocity of the rock just before it strikes the ground;

c) the horizontal distance from the base of the building to the point where the rock strikes the ground.

3–12 Suppose the departure angle θ_0 in Fig. 3–8 is 14° and the distance x is 5 m. Where will the arrow and apple meet

if the initial speed of the arrow is

a) 20 m·s⁻¹?

b) 12 m·s⁻¹?

Sketch both trajectories.

c) What will happen if the initial speed of the arrow is 8 m·s⁻¹?

Section 3–5 Circular Motion

3–13 The earth has a radius of 6.38×10^6 m and turns around on its axis once in 24 hr. What is the radial acceleration of an object at the equator of the earth, in units of m·s⁻²?

3–14 The radius of the earth's orbit around the sun (assumed circular) is 1.49×10^{11} m, and the earth travels around this orbit in 365 days.

a) What is the magnitude of the orbital velocity of the earth, in meters per second?

b) What is the radial acceleration of the earth toward the sun, in meters per second squared?

3–15 A Ferris wheel of radius 12 m is turning about a horizontal axis through its center, such that the linear speed of a passenger on the rim is constant and equal to 9 m·s⁻¹.

a) What are the magnitude and direction of the acceleration of the passenger as he passes through the lowest point in his circular motion?

b) How long does it take the Ferris wheel to make one revolution?

3–16 A model of a helicopter rotor has four blades, each 2 m in length from the central shaft to the blade tip, and is rotated in a wind tunnel at 1500 rev·min⁻¹.

a) What is the linear speed of the blade tip, in meters per second?

b) What is the radial acceleration of the blade tip, expressed as a multiple of the acceleration of gravity, g?

Section 3–6 Relative Velocity

3–17 An airplane pilot wishes to fly due north. A wind of 96 km·hr⁻¹ (about 60 mi·hr⁻¹) is blowing toward the west.

a) If the flying speed of the plane (its speed in still air) is 290 km·hr⁻¹ (about 180 mi·hr⁻¹), in what direction should the pilot head?

b) What is the speed of the plane over the ground? Illustrate with a vector diagram.

3–18 A passenger on a ship traveling due east with a speed of 18 knots observes that the stream of smoke from the ship's funnels makes an angle of 20° with the ship's wake. The wind is blowing from south to north. Assume that the smoke acquires a velocity (with respect to the earth) equal to the velocity of the wind, as soon as it leaves the funnels. Find the magnitude of the velocity of the wind, in knots. (A knot is a unit of speed used by sailors; 1 knot = 1.852 km·hr⁻¹.)

3–19 A river flows due north with a velocity of 2 m·s^{-1}. A man rows a boat across the river; his velocity relative to the water is 3 m·s^{-1} due east; the river is 1000 m wide.

a) What is his velocity relative to the earth?

b) How much time is required to cross the river?

c) How far north of his starting point will he reach the opposite bank?

3–20

a) In what direction should the rowboat in Exercise 3–19 be headed in order to reach a point on the opposite bank directly east from the starting point?

b) What will be the velocity of the boat relative to the earth?

c) How much time is required to cross the river?

PROBLEMS

3–21 The coordinates of a particle moving in the xy-plane are given as functions of time by

$$x = (2 \text{ m·s}^{-1})t, \qquad y = 19 \text{ m} - (2 \text{ m·s}^{-2})t^2.$$

a) What is the particle's distance from the origin at time $t = 2$ s?

b) What is the particle's velocity (magnitude and direction) at time $t = 2$ s?

c) What is the particle's acceleration (magnitude and direction) at time $t = 2$ s?

d) At what times is the particle's velocity perpendicular to its acceleration?

e) At what times is the particle's velocity perpendicular to its position vector? What are the locations of the particle at these times?

f) What is the particle's minimum distance from the origin? At what time does this minimum occur?

g) Sketch the path of the particle.

3–22 A faulty model rocket moves in the xy-plane with an acceleration whose components in a coordinate system in which the positive y-direction is vertically upward are given by $a_x = (3 \text{ m·s}^{-4})t^2$ and $a_y = 10 \text{ m·s}^{-2} - (2 \text{ m·s}^{-3})t$. At $t = 0$ the rocket is at the origin and has an initial velocity $v_0 = 2 \text{ m·s}^{-1}\boldsymbol{i} + 6 \text{ m·s}^{-1}\boldsymbol{j}$.

a) Calculate the velocity and position vectors as functions of time.

b) What is the maximum height reached by the rocket?

c) What is the horizontal displacement of the rocket when it returns to $y = 0$?

3–23 A bird flies in the xy-plane with a velocity vector given by

$$v = (2 \text{ m·s}^{-1} - 3 \text{ m·s}^{-3}t^2)\boldsymbol{i} + (5 \text{ m·s}^{-2}t)\boldsymbol{j},$$

where the positive y-direction is vertically upward. At $t = 0$ the bird is at the origin.

a) Calculate the position and acceleration vectors of the bird as functions of time.

b) What is the bird's altitude (y-coordinate) as it flies over $x = 0$ for the first time after $t = 0$?

3–24 A player kicks a football at an angle of 37° with the horizontal and with an initial speed of 15 m·s^{-1}. A second player standing at a distance of 30 m from the first in the direction of the kick starts running to meet the ball at the instant it is kicked. How fast must he run in order to catch the ball just before it hits the ground?

3–25 According to the *Guiness Book of World Records* the longest home run ever measured was hit by Roy "Dizzy" Carlyle in a minor league game and traveled 618 ft before landing on the ground outside the ballpark.

a) Assuming the ball's initial velocity was 45° above the horizontal and neglecting air resistance, what would the initial speed of the ball need to be to produce such a home run if hit at a point 3.0 ft above ground level? Assume that the ground is perfectly flat.

b) How far above a fence 10 ft in height would the ball be if the fence were 380 ft from home plate?

3–26 A projectile shot at an angle of 60° above the horizontal strikes a building 30 m away at a point 15 m above the point of projection.

a) Find the initial velocity of the projectile, the velocity with which it is projected.

b) Find the magnitude and direction of the velocity of the projectile just before it strikes the building.

3–27 An airplane diving at an angle of 36.9° with the horizontal drops a bag of sand from an altitude of 800 m. The bag is observed to strike the ground 5 s after its release.

a) What is the speed of the plane?

b) How far does the bag travel horizontally during its fall?

c) What are the horizontal and vertical components of its velocity just before it strikes the ground?

3–28 A snowball rolls off a roof that slopes downward at an angle of 40°, as shown in Fig. 3–16. The edge of the roof is

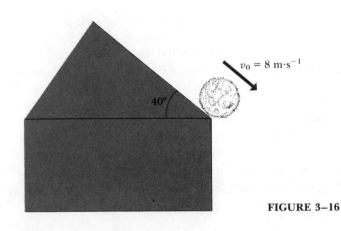

$v_0 = 8$ m·s^{-1}

40°

FIGURE 3–16

12 m above the ground, and the snowball has a speed of 8 m·s^{-1} as it rolls off the roof.

a) How far from the edge of the house will the snowball strike the ground, if it does not strike anything else while falling?

b) A man 2 m tall is standing 4 m from the edge of the house. Will he be hit by the snowball?

3–29 Prove that a projectile launched at angle θ_0 has the same range as one launched with the same speed at angle $(90° - \theta_0)$.

3–30 A rock is thrown from the roof of a building with a velocity v_0 at an angle of θ relative to the horizontal. The building has height h. Calculate the magnitude of the resultant velocity of the rock just before it strikes the ground, and show that this velocity is independent of θ.

3–31 A particle moves in the xy-plane; its coordinates are given as functions of time by

$$x = R \cos \omega t, \qquad y = R \sin \omega t,$$

where R and ω are constants.

a) Show that the particle's distance from the origin is constant and equal to R, i.e., that its path is a circle of radius R.

b) Show that at every point the particle's velocity is perpendicular to its position vector.

c) Show that the particle's acceleration is always opposite in direction to its position vector and has magnitude $\omega^2 R$.

d) Show that the magnitude of the particle's velocity is constant and equal to ωR.

e) Combine the results of parts (c) and (d) to show that the particle's acceleration has constant magnitude v^2/R.

3–32 In an action-adventure film the hero is supposed to throw a grenade from his car, which is going 80.5 km·hr^{-1}, to his enemy's car, which is going 133 km·hr^{-1}. The enemy's car is 14.6 m in front of the hero's when he lets go of the grenade. If the hero throws the grenade so its initial velocity relative to the hero is at an angle of 45° with the horizontal, what should be the magnitude of the velocity? The cars are both traveling in the same direction on a level road. Find the magnitude of the velocity both relative to the hero and relative to the earth.

3–33 An airplane pilot sets a compass course due west and maintains an air speed of 240 km·hr^{-1}. After flying for $\frac{1}{2}$ hr, he finds himself over a town that is 150 km west and 40 km south of his starting point.

a) Find the wind velocity, in magnitude and direction.

b) If the wind velocity were 120 km·hr^{-1} due south, in what direction should the pilot set his course in order to travel due west? Take the same air speed of 240 km·hr^{-1}.

3–34 When a train's speed is 10 m·s^{-1} eastward, raindrops that are falling vertically with respect to the earth make traces that are inclined 30° to the vertical on the windows of the train.

a) What is the horizontal component of a drop's velocity with respect to the earth? With respect to the train?

b) What is the magnitude of the velocity of the raindrop with respect to the earth? With respect to the train?

3–35 A motorboat is observed to travel 16 km·hr^{-1} relative to the earth in the direction 37° north of east. If the velocity of the boat due to the wind is 3.2 km·hr^{-1} eastward and that due to the current is 6.4 km·hr^{-1} southward, what are the magnitude and direction of the velocity of the boat due to its own power?

CHALLENGE PROBLEMS

3–36 A man is riding on a flatcar traveling with a constant speed of 9.1 m·s^{-1} (Fig. 3–17). He wishes to throw a ball though a stationary hoop 4.9 m above the height of his hands in such a manner that the ball will move horizontally as it passes through the hoop. He throws the ball with a speed of 12.2 m·s^{-1} with respect to himself.

a) What must be the vertical component of the initial velocity of the ball?

b) How many seconds after he releases the ball will it pass through the hoop?

c) At what horizontal distance in front of the hoop must he release the ball?

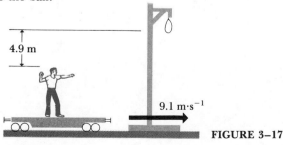

4.9 m

9.1 m·s⁻¹

FIGURE 3–17

3–37 A physics professor did daredevil stunts in his spare time. His last stunt was to attempt to jump across a river on a motorcycle (Fig. 3–18). The take-off ramp was inclined at 53°, the river was 40 m wide, and the far bank was 15 m

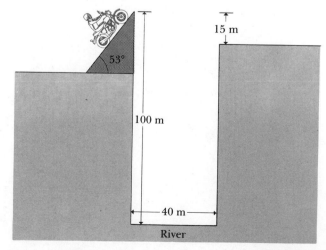

53°

15 m

100 m

40 m

River

FIGURE 3–18

lower than the top of the ramp. The river itself was 100 m below ramp. What should his velocity have been at the top of the ramp to have just made it to the edge of the far bank?

3–38 For a basketball free throw, at what angle and with what initial velocity should you shoot a basketball so that the ball goes through the basket at the smallest angle of entry (smallest angle relative to the horizontal) possible without touching the rim? For a free throw you stand at the foul line 13.8 ft from the center of the basket which is 10.0 ft above the floor. Assume that the ball leaves your hands at a point 5.0 ft above the floor. The basketball is 30 in. in circumference and the basket is 18 in. in diameter according to the official rules. (For more on the physics involved in basketball, see the article by P. J. Brancazio in *Amer. Jour. of Phys.*, Vol. 49 (1981), pp. 356–365.)

3–39 A projectile is given an initial velocity v_0 at an angle ϕ above the surface of an incline, which is in turn inclined at an angle θ above the horizontal; see Fig. 3–19.

a) Calculate the distance, measured along the incline, from the launch point to where the object strikes the incline. Your answer will be in terms of v_0, g, θ, and ϕ.

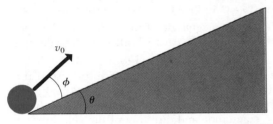

FIGURE 3–19

b) What angle ϕ gives the maximum range, measured along the incline?

(*Note.* You may be interested in the three different methods of solution presented by I. R. Lapidus in *Amer. Jour. of Phys.*, Vol. 51 (1983), pp. 806 and 847. See also H. A. Buckmaster in *Amer. Jour. of Phys.*, Vol. 53 (1985), pp. 638–641, for a thorough study of this and some similar problems.)

3–40 Refer to the previous problem.

a) An archer on ground of constant upward slope of 30° aims at a target 50 m farther up the incline. The arrow in the bow and the bull's-eye at the center of the target are each 1.5 m above the ground. Let the initial velocity of the arrow just after it leaves the bow be 28 m·s⁻¹. At what angle above the *horizontal* should the archer aim to hit the bull's-eye? If there are two such angles, calculate the smaller of the two. You may have to solve the equation for the angle by iteration, that is, by trial and error. How does the angle compare to that required when the ground is level, with zero slope?

b) Repeat the above, for ground of constant *downward* slope of 30°.

3–41 An object is traveling in a circle of radius $R = 2$ m with a constant speed of $v = 5$ m·s⁻¹. Let v_1 be the velocity vector at time t_1, and v_2 be the velocity vector at time t_2. Consider $\Delta v = v_2 - v_1$ and $\Delta t = t_2 - t_1$. Recall that $a_{av} = \Delta v / \Delta t$. For $\Delta t = 0.5$ s, 0.1 s, and 0.05 s, calculate the magnitude (to four significant figures) and direction (relative to v_1) of the average acceleration a_{av}. Compare your results to the general expression for the instantaneous acceleration a for uniform circular motion that is derived in the text.

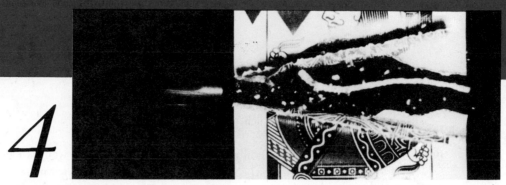

4 NEWTON'S LAWS OF MOTION

IN THE TWO PRECEDING CHAPTERS WE STUDIED *KINEMATICS,* THAT IS, THE methods for *describing* motion. With this chapter we begin a study of more general problems involving the relation of motion to its causes; these problems form the area called **dynamics.** In addition to the kinematic quantities introduced in Chapters 2 and 3, we need two additional concepts: *force* and *mass. Statics* is a division of dynamics dealing with special cases where the acceleration is zero, that is, with bodies in *equilibrium.*

All of dynamics is based on three principles called *Newton's laws of motion.* The first law states that when the vector sum of forces on a body is zero, the acceleration of the body is also zero. The second law relates force to acceleration when the vector sum of forces is *not* zero. And the third law relates the pairs of forces that interacting bodies exert on each other.

Newton's laws are *empirical* laws, deduced from experiment; they cannot be derived from anything more fundamental. They were clearly stated for the first time by Sir Isaac Newton (1642–1727) and were published in 1686 in his *Philosophiae Naturalis Principia Mathematica* (The Mathematical Principles of Natural Science). Many other scientists before Newton contributed to the foundations of mechanics, especially Galileo Galilei (1564–1642), who died the same year Newton was born. Indeed Newton himself said, "If I have been able to see a little farther than other men, it is because I have stood on the shoulders of giants."

4–1 FORCE

Force is a central concept in all of physics. A force on a body resulting from direct contact with another body is called a *contact force.* For example, when we push or pull on a body, we exert a force on it. A stretched spring exerts forces on the bodies attached to its ends; compressed air exerts a force on the wall of its container; and a locomotive exerts a force on the train it is pulling or pushing. Viewed on an atomic scale, contact forces result principally from the electrical attractions and repulsions of the electrons and nuclei in the atoms of the materials.

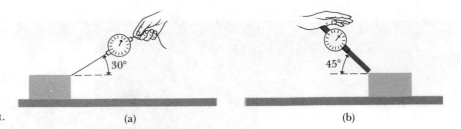

(a)   (b)

4–1 Force may be exerted on the box by either (a) pulling it, or (b) pushing it.

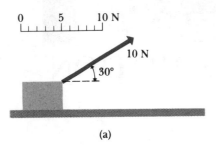

0   5   10 N

10 N

30°

(a)

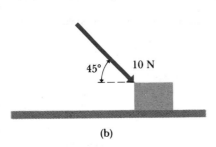

45°   10 N

(b)

4–2 Force diagram for the forces acting on the box in Fig. 4–1.

Measuring force

Forces can be combined by vector addition.

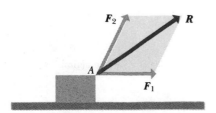

F_2 R

A

F_1

4–3 A force represented by the vector R, equal to the vector sum of F_1 and F_2, produces the same effect as the forces F_1 and F_2 acting simultaneously.

In contrast, gravitational forces, as well as electric and magnetic forces, can act through empty space; bodies do not have to be in contact for these forces to act. The force of gravitational attraction exerted on a body by the earth is called the *weight* of the body.

As we mentioned in Section 1–6, force is a *vector* quantity. This means that to describe a force, we need to describe the *direction* in which it acts as well as its *magnitude*, the quantity that describes "how much" or "how hard" the force pushes. In Section 4–4 we will define units of force in terms of the units of mass, length, and time. As mentioned in Section 1–6, the SI unit of force is the *newton*, abbreviated N. In the cgs version of the metric system (discussed in Section 4–4) the unit of force is the *dyne*, equal to 10^{-5} N, and in the British system the unit of force is the *pound*.

A familiar instrument for measuring forces is the spring balance. It consists of a coil spring, enclosed in a case for protection, with a pointer attached to one end. When forces are applied to the ends of the spring, it stretches; the amount of stretch depends on the force. We can calibrate such an instrument by using a number of identical bodies, each having a weight of exactly 1 N. Then when two, three, or more of these are suspended simultaneously from the balance, the total force stretching the spring is 2 N, 3 N, and so on, and the corresponding positions of the pointer can be labeled 2 N, 3 N, and so on. Then we can use the instrument to measure the magnitude of an unknown force.

Suppose we slide a box along the floor by pulling it with a string or pushing it with a stick, as in Fig. 4–1. The interaction of the box with the other bodies that push or pull it is described in terms of the *forces* they exert on the box. Thus our point of view is that the motion of the box is caused by the *forces* these bodies exert on it. The forces in the two cases can be represented as in Fig. 4–2. The labels indicate the magnitude and direction of the force; the length of the arrow, to some chosen scale, also shows the magnitude.

What happens when *two* forces, represented by the vectors F_1 and F_2 in Fig. 4–3, are applied simultaneously at the same point A of a body? Is it possible to produce the same effect on the body by applying a *single* force at A? If so, what should be its magnitude and direction? These questions can be answered only by experiment. Investigation shows that a single force equal to the *vector sum* $F_1 + F_2$ of the original forces produces the same effect as the two forces together. This single force is often called the **resultant** of the two forces. Hence the mathematical process of *vector addition* of two force vectors corresponds to the physical operation of finding the *resultant of two forces*, simultaneously applied at a given point. The same statement can be extended to combining any number of forces.

The fact that forces can be combined by vector addition is of the utmost importance, as we will see in the following chapters. Furthermore, this fact also allows us to represent a force by means of *components*, as we have done with displacements in Section 1–7. In Fig. 4–4a, force F acts on a body at point O. The components of F in the directions Ox and Oy are F_x and F_y; we find that simultaneous application of the forces F_x and F_y, as in Fig. 4–4b, is equivalent in all respects to the effect of the original force. *Any force can be replaced by its components, acting at the same point.*

As a numerical example, let

$$F = 10.0 \text{ N}, \qquad \theta = 30°.$$

Then,

$$F_x = F \cos \theta = (10.0 \text{ N})(0.866) = 8.66 \text{ N},$$
$$F_y = F \sin \theta = (10.0 \text{ N})(0.500) = 5.00 \text{ N},$$

and the effect of the original 10.0-N force is equivalent to the simultaneous application of a horizontal force, to the right, with magnitude 8.66 N, and a lifting force with magnitude 5.00 N.

The axes we use to obtain components of a vector need not be vertical and horizontal. For example, Fig. 4–5 shows a block being pulled up an inclined plane by a force F, represented by its components F_x and F_y, parallel and perpendicular to the sloping surface of the plane.

We will often need to find the vector sum of several forces acting on a body, which again may be called their *resultant*. The Greek letter Σ (sigma) is often used in a shorthand notation for this sum. If the forces are labeled F_1, F_2, F_3, and so on, and their resultant is R, then the operation

$$R = F_1 + F_2 + F_3 + \cdots$$

is often abbreviated

$$R = \Sigma F, \qquad (4\text{–}1)$$

where ΣF is read "sum of the forces." (The Greek letter Σ, equivalent to the Roman S, is an abbreviation for "sum.") This means, of course, the *vector* sum. In terms of components, we may write

$$R_x = \Sigma F_x, \qquad R_y = \Sigma F_y, \qquad (4\text{–}2)$$

in which ΣF_x is read "the sum of the x-components of the forces." We also note that a boldface Σ is used in Eq. (4–1) as a reminder that the sum is a vector sum, and a lightface Σ is used in Eqs. (4–2) because components are ordinary numbers.

Finally, we can combine these components to form the resultant R. Its magnitude is

$$R = \sqrt{R_x^2 + R_y^2},$$

since R_x and R_y are perpendicular to each other.

The angle α between R and the x-axis can now be found from any one of its trigonometric functions. For example, $\tan \alpha = R_y/R_x$. As with the individual vectors, the components R_x and R_y may be positive or negative, and the angle α may be in any of the four quadrants.

Forces can be represented by components.

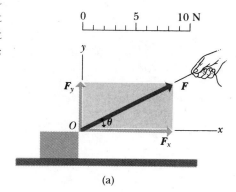

(a)

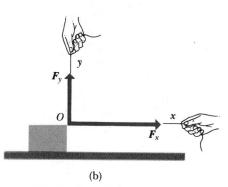

(b)

4–4 The inclined force F may be replaced by its rectangular components F_x and F_y. $F_x = F \cos \theta$, $F_y = F \sin \theta$.

Resultant force: the vector sum of forces

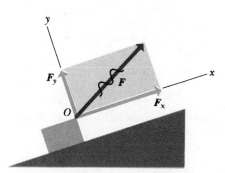

4–5 F_x and F_y are the rectangular components of F, parallel and perpendicular to the sloping surface of the inclined plane.

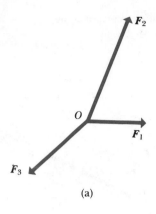

(a)

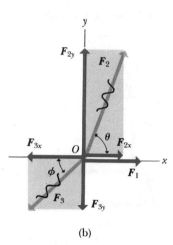

(b)

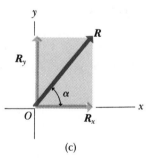

(c)

4–6 Vector **R**, the resultant of F_1, F_2, and F_3, is obtained by the method of rectangular resolution. The rectangular components of **R** are $R_x = \Sigma F_x$ and $R_y = \Sigma F_y$.

Equilibrium: the simplest state of motion

EXAMPLE 4–1 In Fig. 4–6a, three forces F_1, F_2, F_3 lying in the *xy*-plane act at point O. Let $F_1 = 120$ N, $F_2 = 200$ N, $F_3 = 150$ N, $\theta = 60°$, and $\phi = 45°$. Find the *x*- and *y*-components of the resultant or vector sum **R,** and find its magnitude and direction.

SOLUTION The computations can be arranged systematically as in Table 4–1.

TABLE 4–1

Force	Angle	x-component	y-component
$F_1 = 120$ N	0	+120 N	0
$F_2 = 200$ N	60°	+100 N	−173 N
$F_3 = 150$ N	45°	−106 N	−106 N
		$R_x = \Sigma F_x = +114$ N	$R_y = \Sigma F_y = +67$ N

$$R = \sqrt{(114 \text{ N})^2 + (67 \text{ N})^2} = 132 \text{ N},$$
$$\alpha = \arctan \frac{67 \text{ N}}{114 \text{ N}} = \arctan 0.588 = 30.4°.$$

4–2 EQUILIBRIUM AND NEWTON'S FIRST LAW

We have mentioned that one possible effect of a force is to alter the state of motion of the body on which it acts. In general this motion can consist of motion of the body as a whole, called *translational* motion, and of *rotational* motion of the body about its center. A familiar example is a thrown football, which spins as it moves through the air. In contrast, the motion of a *particle,* represented as a single point, is described in terms of translational motion only. In this chapter and the next we assume the body can be represented as a point, so we need not be concerned yet with the possibility of rotation.

When several forces act on a body at the same time, their effects can compensate or cancel one another. When this happens, Newton's first law predicts that there is *no* change in the motion, and we say that the body is in **equilibrium.** This means that the body either remains at rest or moves in a straight line with constant velocity.

In Fig. 4–7 a small body such as a hockey puck rests on a horizontal surface having negligible friction, such as an air-hockey table or a slab of wet ice. If the body is initially at rest and a single force F_1 acts on it, as in Fig. 4–7a, the body immediately starts to move. If it is in motion at the start, the effect of the force is to change the motion, either in direction or speed, or both. In either case, the body is *not* in equilibrium.

Now suppose we apply a second force F_2, as in Fig. 4–7b, equal in magnitude to F_1 but opposite in direction. Experiments show that the body is then in equilibrium; if it is initially at rest, it remains at rest, and if it is initially moving, it continues to move in the same direction with constant speed. In this case the two forces are negatives of each other, $F_2 = -F_1$, and so their vector sum is zero:

$$R = F_1 + F_2 = 0.$$

For brevity, we speak of two forces as being "equal and opposite," meaning that their magnitudes are equal and that one is the negative of the other.

We can generalize this discussion to a body with any number of forces acting on it. When a body is in equilibrium, the vector sum, or resultant, of all the forces acting on it must be zero. Each component of the resultant must therefore be zero. Hence, for a body in equilibrium,

$$R = \Sigma F = 0,$$

or

$$\Sigma F_x = 0, \qquad \Sigma F_y = 0. \qquad (4-3)$$

When Eqs. (4–3) are satisfied, the body is in equilibrium, provided that it can be represented as a point or that all the forces acting on it are applied at the same point.

The condition for equilibrium represented by Eqs. (4–3) is called *Newton's first law of motion*. Newton did not state his first law in exactly these words. His original statement (translated from the Latin in which the *Principia* was written) reads:

> Every body continues in its state of rest, or of uniform motion in a straight line, unless it is compelled to change that state by forces impressed on it.

Newton's first law of motion is not as self-evident as it may seem. This law asserts that in the absence of any applied force a body either remains at rest or moves uniformly in a straight line. It follows that *once a body has been set in motion, no force is needed to keep it moving.*

This assertion may seem to be contradicted by everyday experience. Suppose you exert a force with your hand to push a book along a horizontal table top. After the book has left your hand, and you are no longer exerting a force on it, it does *not* continue to move indefinitely, but slows down and eventually comes to rest. To keep it moving uniformly you have to continue to push it. But the reason this force is needed is because a frictional force is exerted on the sliding book by the table top, in a direction *opposite* to the book's motion. The smoother the surfaces of the book and table, the smaller the frictional force and the smaller the force needed to keep the book moving. The first law asserts that if the frictional force could be eliminated completely, no forward force at all would be required to keep the book moving once it is set in motion. The law also states that if the *resultant* force on the book is zero, as it is when the frictional force is balanced by an equal forward force, the book also continues to move uniformly. In other words, to keep the book moving uniformly, *zero resultant force is equivalent to no force at all.*

4–3 MASS AND NEWTON'S SECOND LAW

We know from experience that a body at rest never starts to move by itself; some other body has to apply a push or pull to it. Similarly, when a body is already in motion, a force is required to slow it down or stop it. To make a moving body deviate from straight-line motion, we must apply a sideways force. All these processes (speeding up, slowing down, or changing direction) involve a *change* in either the magnitude or the direction of the velocity. In each case the body has an *acceleration,* and a force must act on it to cause this acceleration.

Let us consider several fundamental experiments. A small body, which we may model as a particle, rests on a level, frictionless surface and moves to the

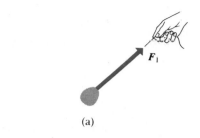
(a)

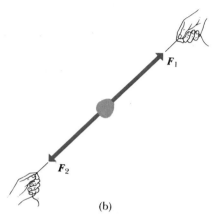
(b)

4–7 A small body acted on by two forces is in equilibrium if the forces are equal in magnitude and opposite in direction.

A moving body keeps moving unless some force makes it stop.

No force at all is equivalent to a set of forces with a vector sum of zero.

Bodies resist *changes* in their motion.

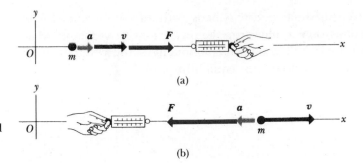

4–8 The acceleration **a** is proportional to the force **F** and is in the same direction as the force.

right along the x-axis of a reference system, as in Fig. 4–8a. We apply a constant horizontal force **F** to the body. This force might be supplied by a spring balance, as described in Section 4–1, with the spring stretched a constant amount. We find that during the time the force is acting, the velocity of the body increases at a constant rate; that is, the body moves with *constant acceleration*. If the magnitude of the force is changed, the acceleration changes in the same proportion. Doubling the force doubles the acceleration, halving the force halves the acceleration, and so on. When the force is removed, the acceleration becomes zero: The body then moves with constant velocity, as discussed in Section 4–2.

Before the time of Galileo and Newton, it was generally believed that a force was necessary just to keep a body moving, even on a level, frictionless surface or in outer space. Galileo and Newton realized that *no* net force is necessary to keep a body moving, once it has been set in motion, and that the effect of a force is not to *maintain* the velocity of a body, but to *change* its velocity. The *rate of change* of velocity for a given body is directly proportional to the force acting on it.

In Fig. 4–8b, the velocity of the body is also toward the right, but the *force* is toward the left. Under these conditions the body moves more and more slowly to the right until it stops. It then reverses its direction of motion and begins to move more and more rapidly toward the left. During this entire process the direction of the acceleration is toward the *left,* in the same direction as the force **F.** Hence the *magnitude* of the acceleration is proportional to that of the force, and the *direction* of the acceleration is the same as that of the force, regardless of the direction of the velocity.

Mass: the ratio of force to acceleration

To say that the acceleration of a body is directly proportional to the force exerted on it is to say that the *ratio* of the force to the acceleration is a constant, regardless of the magnitude of the force. This ratio is called the **mass** m of the body. Thus

$$m = \frac{F}{a},$$

or

$$F = ma. \tag{4–4}$$

We can think of the mass of a body as the *force per unit of acceleration.* For example, if the acceleration of a certain body is $5 \ \text{m·s}^{-2}$ when the force is 20 N, the mass of the body is

$$m = \frac{20 \ \text{N}}{5 \ \text{m·s}^{-2}} = 4 \ \text{N·m}^{-1}\text{·s}^2,$$

and a force of 4 N must be exerted on the body for each m·s^{-2} of acceleration.

This relationship can also be used to compare masses quantitatively. Suppose we apply a certain force F to a body having mass m_1 and observe an acceleration of a_1. We then apply the *same* force to another body having mass m_2, observing an acceleration a_2. Then, according to Eq. (4–4),

$$m_1 a_1 = m_2 a_2,$$

or

$$\frac{m_2}{m_1} = \frac{a_1}{a_2}. \qquad (4\text{–}5)$$

We can use this relation to compare any mass with a standard mass. If m_1 is a standard mass and m_2 an unknown mass, we can apply the same force to each and measure the accelerations; the ratio of the masses is the inverse of the ratio of the accelerations. When a large force is needed to give a body a certain acceleration (i.e., speed it up, slow it down, or deviate it if it is in motion), the mass of the body is large; if only a small force is needed for the same acceleration, the mass is small. Thus the mass of a body is a quantitative measure of the property described in everyday language as *inertia*.

To identify another important property of mass, we measure the masses of two bodies, using the procedure just described, and then fasten them together and measure the mass of the composite body. If m_1 and m_2 are the individual masses, the mass of the composite body is always found to be $m_1 + m_2$. This very important result shows that mass is an *additive* quantity, and that it is directly correlated with quantity of matter. Indeed, the concept of mass is one way to give the term *quantity of matter* a precise meaning.

In the discussion above, the particle moves along a straight line (the *x*-axis), and the force also lies along this direction. This is of course a special case. More generally, the force may also have a component in the *y*-direction, and the particle's motion need not be confined to a straight line. Furthermore, more than one force may act on the particle. Thus this formulation needs to be generalized to include motion in a plane or in space and the possibility of several forces acting simultaneously.

Experiments show that *when several forces act on a particle at the same time, the acceleration is the same as would be produced by a single force equal to the vector sum of these forces.* This sum is usually most conveniently handled by using the method of components. When several forces act on a particle moving along the *x*-axis,

$$\sum F_x = ma. \qquad (4\text{–}6)$$

When a particle moves in a plane, with position described by coordinates (x,y), the velocity is a vector quantity with components v_x and v_y equal to the time rates of change of x and y, respectively, and the acceleration is a vector quantity with components a_x and a_y equal to the rates of change of v_x and v_y, respectively. Then a more general formulation of the relation of force to acceleration is

$$\sum F_x = ma_x, \qquad \sum F_y = ma_y. \qquad (4\text{–}7)$$

This pair of equations is equivalent to the single vector equation

$$\sum \mathbf{F} = m\mathbf{a}, \qquad (4\text{–}8)$$

Comparing masses by comparing accelerations

Mass: a quantitative description of inertia

Mass is an *additive* property of matter.

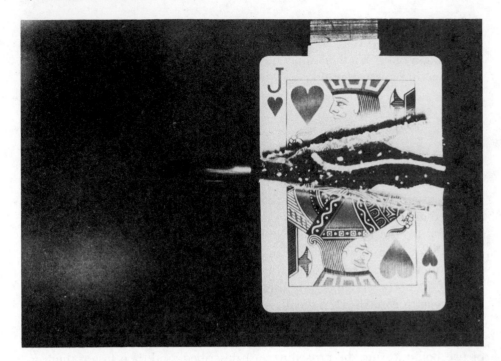

A playing card is cut in two by a .30-caliber bullet striking it edge-on. Although the card accelerates under the action of the force applied by the bullet, it does not have time to move any appreciable distance during the time of the bullet's travel through it (roughly 1/10,000 s). The actual exposure time of the photograph was less than a millionth of a second. (Dr. Harold Edgerton, M.I.T., Cambridge, Massachusetts.)

Newton's second law: the most important relation in mechanics

where we write the left-hand side explicitly as ΣF to emphasize that the acceleration is determined by the resultant of all the forces acting on the particle. If the particle moves in three dimensions, then of course Eqs. (4–7) include a third equation for the z-components, $\Sigma F_z = ma_z$.

Equation (4–8), or the equivalent pair of equations (4–7), is the mathematical statement of Newton's *second law of motion*. The acceleration of a body (the rate of change of its velocity) is equal to the resultant (vector sum) of all forces acting on the particle, divided by its mass, and has the same direction as the resultant force.

In this chapter we consider only straight-line motion, and we usually take the line of motion to be the x-axis. In these problems v_y and a_y are always zero. Individual forces may have y-components, but the *sum* of the y-components will always be zero. In Chapter 5 we consider more general motion in which v_y, a_y, and ΣF_y need not be zero.

4–4 SYSTEMS OF UNITS

In the preceding section we did not discuss the *units* to be used to measure force, mass, and acceleration. From the equation $\Sigma F = ma$ it is clear that a force of one unit gives a mass of one unit an acceleration of one unit, no matter what system we use. Thus the three units cannot be chosen independently; if we choose units for mass and acceleration, the unit of force is determined.

Several systems of units are in common use in the United States; one is the meter-kilogram-second (mks) system, which forms the basis of the Système International (SI) introduced in Chapter 1. Other systems are the British or foot-pound-second system and the now-obsolete centimeter-gram-second (cgs) version of the metric system. This book uses principally SI units, not only because this system is convenient and widely used in scientific work, but also because there seems little doubt that SI will eventually be adopted worldwide for commerce and industry as well as scientific work. The United States and Great Britain are among the few major countries that do not use the metric system exclusively, and Great Britain is now in the process of converting.

In SI units, the meter, the kilogram, and the second are defined as described in Section 1–2. The unit of force in this system is then defined as *the magnitude of force that gives an acceleration of 1 m·s⁻² to a body of mass one kilogram.* This force is called one **newton** (1 N). Thus in SI units

Definitions of units of force

$$F \text{ (N)} = m \text{ (kg)} \times a(\text{m·s}^{-2}).$$

That is,

$$1 \text{ N} = 1 \text{ kg·m·s}^{-2}.$$

This is a very useful relationship to remember; we will use it often in the next several chapters.

In the *centimeter-gram-second* (cgs) system, the unit of mass is one gram (1 g), equal to 1/1000 kg, and the unit of acceleration is 1 cm·s⁻². The unit of force is called one *dyne;* one dyne gives one gram an acceleration of 1 cm·s⁻². Since 1 kg = 10^3 g and 1 m·s⁻² = 10^2 cm·s⁻², it follows that 1 N = 10^5 dyn. In the cgs system,

$$F \text{ (dyn)} = m \text{ (g)} \times a(\text{cm·s}^{-2}).$$

In the SI and cgs systems, we first selected units of mass and acceleration, and defined the unit of force in terms of these. In the British system, we first select a unit of force (1 lb) and a unit of acceleration (1 ft·s⁻²), and then define the unit of mass as *the mass of a body whose acceleration is* 1 ft·s⁻² *when the resultant force on the body is* 1 lb. This unit of mass is called one *slug.* In the British system,

$$F \text{ (lb)} = m \text{ (slugs)} \times a \text{ (ft·s}^{-2}).$$

The pound is used in everyday life as a unit of quantity of matter (e.g., a pound of butter), but properly speaking it is a unit of *force* or *weight.* Thus a pound of butter is an amount having a weight of 1 lb. A useful fact in converting between mks and British units is that an object with a *mass* of 1 kg has a *weight* of about 2.2 lb (more precisely, 2.2046 lb).

Confusion between mass and weight: We don't always say what we mean.

The units of force, mass, and acceleration in the three systems are summarized in Table 4–2.

TABLE 4–2

System of units	Force	Mass	Acceleration
SI	newton (N)	kilogram (kg)	m·s⁻²
cgs	dyne (dyn)	gram (g)	cm·s⁻²
Engineering	pound (lb)	slug	ft·s⁻²

We emphasize again that because of the definitions given above, there is an equivalence between force units and those of mass times acceleration. For example, 1 N = 1 kg·m·s⁻²; in checking equations for consistency of units it is always appropriate to replace "N" wherever it appears by "kg·m·s⁻²." Similarly, 1 lb can be replaced by 1 slug·ft·s⁻²; this conversion can be used in unit checks.

4–5 MASS AND WEIGHT

In the introductory discussion of force in Section 4–1 we mentioned that the **weight** of a body on earth, a familiar kind of force, is the result of the gravitational interaction of the body with the earth. We will study gravitational interactions in detail in Chapter 6, but some preliminary analysis is appropriate at this point. The terms *mass* and *weight* are often misused and interchanged in everyday conversation, and it is absolutely essential for us to keep clearly in mind the distinctions between these two physical quantities.

Relation between mass and weight

Mass, on the one hand, characterizes the *inertial* properties of a body. The greater the mass, the greater the force needed to produce a given acceleration; this meaning is reflected in Newton's second law, $\Sigma F = ma$. Weight, on the other hand, is a *force,* exerted on a body by the earth or some other large body. Of course, everyday experience shows us that bodies having large mass also have large weight; a cart loaded with bricks is hard to get rolling because of its large mass, and it is also hard to lift off the ground because of its large weight. Thus we are led to ask what the relationship is between mass and weight.

Weight and free fall: Newton's apple tree

The answer to this question, according to legend, came to Newton as he sat under an apple tree watching the apples fall. A falling body has an acceleration, and according to Newton's second law this requires a force. If a 1-kg body falls with an acceleration of 9.80 m·s^{-2}, the force required to cause this acceleration is

$$F = ma = (1 \text{ kg})(9.8 \text{ m·s}^{-2}) = 9.8 \text{ kg·m·s}^{-2}$$
$$= 9.8 \text{ N}.$$

But the force that makes the body accelerate downward is its *weight.* Hence any body having a mass of 1 kg *must* have a weight of 9.8 N. We can generalize this result: Denoting the weight of the body by w, we say that a body having a mass m must have a weight w given by

$$w = mg. \tag{4–9}$$

Variation of g with location

The value of g varies somewhat from point to point on the earth's surface. Part of this variation results from the fact that the earth's density is not uniform because of local deposits of ore, oil, or other materials of anomalous density. In addition, the earth is not perfectly spherical but is flattened at the poles and is slightly egg shaped. There are also complications associated with the rotation of the earth. For now we ignore these, but we will consider some aspects of these effects in Chapter 6.

Weight depends on location.

As mentioned above, the weight of a standard kilogram at a point where $g = 9.80$ m·s^{-2} is $w = 9.80$ N. At a second point where $g = 9.78$ m·s^{-2}, the weight is $w = 9.78$ N. Thus, unlike the mass of a body, which is constant, the weight of a body varies from one location to another. If we take a standard kilogram to the moon, where the acceleration of free fall is 1.67 m·s^{-2}, its weight is 1.67 N but its mass is still 1 kg.

Because acceleration and force are both vector quantities, Eq. (4–9) may be written as a vector equation:

$$w = mg. \tag{4–10}$$

It is important to understand that the weight of a body, as given by Eq. (4–10), acts on the body all the time, whether it is actually in free fall or not. When a 1-kg body hangs suspended from a string, it is in equilibrium, and its acceleration is zero. But its weight is still acting on it and is given by Eq. (4–10). In this case there is also an upward force on the body provided by the string; the *vector sum* of the forces is zero, and the body is in equilibrium.

The mass of a body can be measured in several different ways. One way is to use the relationship $m = F/a$. We apply a known force to the body, measure its acceleration, and compute the mass as the ratio of force to acceleration. This method, or some variation of it, is often used to measure the masses of atomic and subatomic particles.

Another method is to use a comparison technique, finding by trial some other body whose mass is equal to that of the given body but is already known. Often it is easier to compare *weights* than masses. The weight of a body equals the product of its mass and the acceleration of gravity, so if the weights of two bodies are equal at a given location, their masses are also equal. Balances can be used to determine with great precision (up to 1 part in 10^6) when the weights of two bodies are equal, and hence when their masses are equal.

Measuring masses by comparing weights

Thus mass plays two rather different roles in mechanics. On the one hand, the gravitational force acting on a body is proportional to its mass, so mass is *the property of matter that causes bodies to exert attractive gravitational forces on each other*. We may call this property *gravitational mass*. On the other hand, Newton's second law tells us that the force (which need not be gravitational) required to cause an acceleration of a body is proportional to its mass. This inertial property of the body may be called its *inertial mass*.

It is not obvious that the *gravitational* mass of a particle has to be the same as its *inertial* mass, but very precise experiments have established (within about 1 part in 10^{12}) that in fact the two are the same. That is, if we have to push twice as hard on body A as on body B to cause a given acceleration, then the weight of A at a given location is *precisely* twice that of body B at the same location. Thus inertial and gravitational masses really are identical, and we do not have to distinguish between them. This equivalence is also the fundamental reason why the acceleration of free fall is independent of mass.

Inertial and gravitational masses are identical.

Finally, we remark that the SI units for mass and weight are frequently misused in everyday speech. Expressions such as "This box weighs 6 kg" are nearly universal. What is meant, of course, is that the *mass* of the box (probably determined indirectly by *weighing*) is 6 kg. This usage is so common that there is probably no hope of straightening things out, but it is essential to recognize that the term *weight* is often used when *mass* is meant.

4–6 NEWTON'S THIRD LAW

Any individual force on a body is only one aspect of a mutual interaction between *two* bodies; I cannot push on you unless you push back on me at the same time. Experiment shows that *whenever one body exerts a force on another, the second body exerts simultaneously on the first a force that is equal in magnitude and opposite in direction to the force of the first on the second.* Thus it is impossible to have a single isolated force. A force on a particular body must be exerted by some other body or bodies, and it in turn exerts forces back on the other body or bodies. This property of forces is Newton's *third law of motion*. In his words,

Newton's third law: Action and reaction are equal.

> To every action there is always opposed an equal reaction; or, the mutual actions of two bodies upon each other are always equal, and directed to contrary parts.

An astronaut outside the cabin of the Space Shuttle Challenger uses a jet of nitrogen to maneuver. As the nitrogen escapes through the jet orifice, it exerts a reaction force on its container. This force causes the acceleration of the astronaut needed for him to maneuver in space. (Courtesy of NASA.)

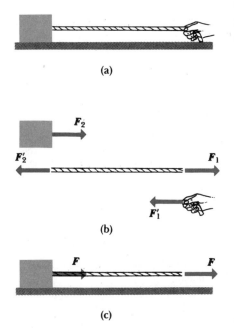

(a)

(b)

(c)

4–9 (a) A man pulls on a rope attached to a block. The forces that the rope exerts on the block and on the man are shown. (b) Separate diagrams showing the forces acting on the block, on the rope, and on the man. (c) If the rope is in equilibrium so that $F'_2 = F_1$, the rope can be considered to transmit a force from the man to the block, and vice versa.

The two forces involved in a mutual interaction between two bodies are often called an **action** and a **reaction.** This does not imply any basic difference in their nature or that one is cause and the other result. Either force may be considered the action and the other the reaction.

As an example, suppose that a man pulls on one end of a rope attached to a box, as in Fig. 4–9. The box may or may not be in equilibrium. The resulting action–reaction pairs of horizontal forces are shown in the figure. (The lines of action of all the forces lie along the rope; the force vectors have been offset from this line to show them more clearly.) Vector F_1 represents the force exerted *on* the rope *by* the man. Its reaction is the equal and opposite force F'_1 exerted *on* the man *by* the rope. Vector F_2 represents the force exerted on the box by the rope. The reaction to it is the equal and opposite force F'_2, exerted on the rope by the box:

$$F'_1 = -F_1, \qquad F'_2 = -F_2. \qquad (4–11)$$

It is very important to realize that the forces F_1 and F'_2, although they are opposite in direction and have the same line of action, are *not* an action–reaction pair. These forces act on the *same* body (the rope), whereas an action and its reaction must act on *different* bodies. Furthermore, the forces F_1 and F'_2 are not necessarily equal in magnitude. If the box and rope are moving to the right with increasing speed, the rope is not in equilibrium because it is accelerating. In that case F_1 is greater in magnitude than F'_2. Only in the special case when the rope remains at rest or moves with constant speed are the forces F_1 and F'_2 equal in magnitude—but this is an example of Newton's *first* law, not his *third*. Even when the speed of the rope is changing, however, the action–reaction forces F_1 and F'_1 are equal in magnitude to each other, and the action–reaction forces F_2 and F'_2 are equal in magnitude to each other, although then F_1 is not equal to F'_2.

In the special case when the rope is in equilibrium, and when no forces act on it except those at its end, F_2' equals $-F_1$ because of Newton's *first* law. Since F_2 *always* equals $-F_2'$ by Newton's *third* law, then in this special case F_2 also equals F_1, and the force exerted on the box by the rope is equal to the force exerted on the rope by the man. The rope can therefore be considered to "transmit" to the box, without change, the force exerted on it by the man, as in Fig. 4–9c. This point of view is often useful, but we must remember that it applies only under the restricted conditions as stated.

A body such as the rope in Fig. 4–9, which is subjected to pulls at its ends, is said to be in **tension.** The tension at any point equals the force exerted at that point. Thus in Fig. 4–9b the tension at the right-hand end of the rope equals the magnitude of F_1 (or of F_1'), and the tension at the left-hand end equals the magnitude of F_2 (or of F_2'). If the rope is in equilibrium and if no forces act except at its ends, the tension is the same at both ends. If, for example, the magnitudes of F_1 and F_2 are each 50 N, the tension in the rope is 50 N (*not* 100 N).

One statement made above needs emphasis: The two forces in an action–reaction pair can *never* act on the same body. Remembering this general principle can often help clear up confusion regarding action–reaction pairs.

Tension in a rope: Forces act at both ends.

4–7 APPLICATIONS OF NEWTON'S LAWS

The examples in this section illustrate applications of Newton's laws and the usefulness of the following strategy.

Free-body diagrams: an indispensable problem-solving aid

PROBLEM-SOLVING STRATEGY: *Newton's laws of motion*

1. Always define your coordinate system; a diagram showing the location of the origin and positive axis directions is always helpful. It is often convenient to take the positive direction to be the direction of the acceleration, if this is known.

2. Be consistent with signs. If the positive end of the *x*-axis is defined as positive, then velocities, accelerations, and forces to the right are also positive.

3. In applying Newton's laws, always decide on a specific body to which the laws will be applied. Draw a diagram showing all the forces (magnitudes and directions) acting on the body, but *do not* include forces that the body exerts on any other body. Such a

diagram is called a **free-body diagram.** The acceleration of the body is determined by the forces acting *on it*, not by the forces it exerts on something else.

4. Identify the known and unknown quantities, and give each unknown an algebraic symbol.

5. Always check for unit consistency; use the conversion $1\text{ N} = 1\text{ kg·m·s}^{-2}$ when appropriate.

6. We will elaborate this strategy further in the following chapters in the context of more complex problems. It is important to use it consistently from the start, however, to develop good habits in the systematic analysis of problems.

EXAMPLE 4–2 A constant horizontal force with magnitude 2 N is applied to a box of mass 4 kg resting on a level, frictionless surface. What is the acceleration of the box?

SOLUTION We take the +*x*-axis in the direction of the horizontal force. The free-body diagram (Fig. 4–10) includes this force and the two vertical forces: the weight of the box and the upward supporting force exerted on it by the surface. The acceleration is given by Newton's second law, Eq. (4–5). There is only one

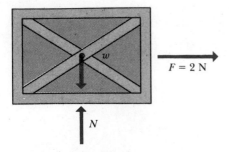

4–10 Free-body diagram for Example 4–2, showing side view of box.

horizontal force, and we have

$$a = \frac{F}{m} = \frac{2\ \text{N}}{4\ \text{kg}} = 0.5\ \text{N} \cdot \text{kg}^{-1} = 0.5\ (\text{kg} \cdot \text{m} \cdot \text{s}^{-2}) \cdot \text{kg}^{-1}$$
$$= 0.5\ \text{m} \cdot \text{s}^{-2}.$$

The force is constant, so the acceleration is also constant. Hence if we know the initial position and velocity of the box, we can find the position and velocity at any later time from the equations of motion with constant acceleration.

EXAMPLE 4–3 A student shoves a lab notebook of mass 0.2 kg toward the right along a level lab bench. As the book leaves contact with the student's hand, it has an initial velocity of 0.4 m·s^{-1}. As it slides, it slows down because of the horizontal friction force exerted on it by the bench top. It slides a distance of 1.0 m before coming to rest. What are the magnitude and direction of the friction force F acting on it?

SOLUTION Suppose the book slides along the x-axis in the +x-direction, starting at the point x_0 with the given initial velocity. A free-body diagram is shown in Fig. 4–11. We assume that the friction force is constant. The acceleration is then constant also. From the equations of motion with constant acceleration, we have

$$v^2 = v_0{}^2 + 2ax, \qquad 0 = (0.4\ \text{m} \cdot \text{s}^{-1})^2 + (2a)(1.0\ \text{m});$$
$$a = -0.08\ \text{m} \cdot \text{s}^{-2}.$$

The negative sign means that the acceleration is toward the *left* (although the velocity is toward the right). The friction force on the book is

$$F = ma = (0.2\ \text{kg})(-0.08\ \text{m} \cdot \text{s}^{-2})$$
$$= -0.016\ \text{N},$$

and is toward the left also. (A force of equal magnitude, but directed toward the right, is exerted *on* the table *by* the sliding book.)

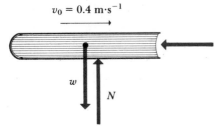

$v_0 = 0.4\ \text{m} \cdot \text{s}^{-1}$

4–11 Free-body diagram for Example 4–3, showing side view of lab notebook.

EXAMPLE 4–4 A machine part having a mass of 5 kg slides along a rail aligned with the x-axis. Its position is given as a function of time by

$$x = (18\ \text{m} \cdot \text{s}^{-2})t^2 - (3\ \text{m} \cdot \text{s}^{-3})t^3.$$

Find the force acting on the body, as a function of time. What is the force at time $t = 5$ s? For what times is the force positive? Negative? Zero?

SOLUTION The free-body diagram is just like that of Example 4–2. We first find the acceleration a by taking the second derivative of x:

$$a = d^2x/dt^2 = 36\ \text{m} \cdot \text{s}^{-2} - (18\ \text{m} \cdot \text{s}^{-3})t.$$

Then, from Newton's second law,

$$F = ma = (5\ \text{kg})[36\ \text{m} \cdot \text{s}^{-2} - (18\ \text{m} \cdot \text{s}^{-3})t]$$
$$= [180 - (90\ \text{s}^{-1})t]\ \text{N}.$$

At time $t = 5$ s, the force is 180 N $- (90\ \text{s}^{-1})(5\ \text{s})$ N $= -270$ N. The force is zero when 36 m·s^{-2} $- (18\ \text{m} \cdot \text{s}^{-3})t = 0$, that is, when $t = 2$ s. When $t < 2$ s, F is positive, and when $t > 2$ s, F is negative.

EXAMPLE 4–5 A flowerpot having a mass of 10 kg is suspended by a chain from the ceiling. What is its weight? What force (magnitude and direction) does the chain exert on it? What is the tension in the chain? Assume the weight of the chain itself is negligible.

SOLUTION Free-body diagrams for the flowerpot and the chain are shown in Fig. 4–12. The weight of the pot is given by Eq. (4–9):

$$w = (10 \text{ kg})(9.8 \text{ m·s}^{-2}) = 98 \text{ N}.$$

The pot is in equilibrium, so the vector sum of forces acting on it must be zero. Thus the chain must pull *up* with a force T of magnitude 98 N. By Newton's third law, the pot pulls *down* on the chain with a force of 98 N. Because the chain is also in equilibrium, there must be an upward force of 98 N exerted on the chain at its top end to make the vector sum of forces on the chain equal zero.

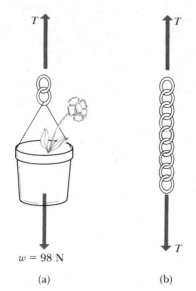

4–12 Free-body diagrams for Example 4–5. (a) Forces acting on the flowerpot; (b) forces acting on the chain. Each body is in equilibrium, and the sum of the forces on each must be zero. The weight of the chain is assumed to be negligible.

4–8 INERTIAL FRAMES OF REFERENCE

A set of coordinate axes attached to (or moving with) some specified body or bodies is called a **frame of reference** (or a *reference system*). We always describe the motion of a given body with reference to, or relative to, some other body. Its motion relative to one body may be very different from that relative to another. Thus a passenger in an airplane making its takeoff run is at rest relative to the airplane but is moving faster and faster relative to the earth.

Newton's first law defines by implication a special class of frames of reference. Suppose we consider a frame of reference attached to the airplane mentioned above. During the takeoff run, while the plane is going faster and faster, a passenger feels the back of his seat pushing him forward, although he remains at rest relative to the airplane. Therefore Newton's first law does not describe the situation correctly; a forward force does act on the passenger, but nevertheless (relative to the airplane) he remains at rest.

But now suppose that the passenger is standing in the aisle on roller skates. He then starts to move *backward,* relative to the plane, when the takeoff run begins, even though no backward force acts on him. Again Newton's first law is not obeyed.

Thus Newton's first law is obeyed in some frames of reference and not in others. We now define an **inertial frame of reference** as a frame of reference in which Newton's first law is obeyed. Relative to an inertial frame, a body does remain at rest or move uniformly in a straight line when no force (or no resultant force) acts on it.

An airplane gaining speed during takeoff is evidently *not* an inertial frame. For many purposes a reference system attached to the earth can be considered an inertial system, although it is not precisely so because of effects related to the earth's rotation and other motions. But if a frame of reference *is* inertial, then a second frame moving uniformly (i.e., with constant speed in a straight line, without rotation) relative to it is also inertial, since any body in uniform motion as seen by an observer in the first frame will also appear to an observer in the second frame to be in uniform motion. The speed and direction appear different to the two observers, but both observe that Newton's first law is obeyed.

Inertial frames of reference: the coordinate systems where Newton's laws work

There is no single, unique inertial frame of reference. There are infinitely many, but the motion of any one relative to any other is uniform in the sense noted above. It follows that the concepts of "absolute rest" and "absolute motion" have no physical meaning. An airplane in steady flight over the earth is just as suitable an inertial frame of reference as the earth itself.

In most of this book we assume that the earth is an adequate approximation to an inertial frame of reference and that errors resulting from its being not precisely inertial are negligible. When we observe a body at rest relative to the earth begin to move, or when a moving body speeds up, slows down, or changes direction, we know that a force must be acting on it, or perhaps several forces with a vector sum or resultant that is different from zero. The quantitative relationship between the body's acceleration and the vector sum of the forces is Newton's second law. Like the first law, it is obeyed only in inertial frames of reference.

SUMMARY

KEY TERMS

dynamics

force

resultant

equilibrium

mass

newton

weight

action and reaction

tension

free-body diagram

frame of reference

inertial frame of reference

Force is a quantitative measure of the mechanical interaction between two bodies. It is a *vector* quantity, having a magnitude and a direction. When several forces act on a body at the same time, the effect is the same as when a single force, equal to the *vector sum* or *resultant* of the forces, acts on the body.

Newton's first law states that when no force acts on a body, or when the vector sum of all forces acting on it is zero, the body is in *equilibrium*. If the body is initially at rest, it remains at rest; if initially in motion, it continues to move with constant velocity.

The inertial properties of a body are characterized by its *mass*. The acceleration of a body under the action of a given force has a magnitude that is directly proportional to the magnitude of the force and inversely proportional to the mass of the body. This relationship is Newton's second law, $\Sigma F = ma$.

Units of force are defined in terms of the fundamental units of mass, length, and time. In SI the unit of force is 1 newton (1 N), equal to 1 kg·m·s^{-2}. In the cgs system it is 1 dyne (1 dyn), equal to 1 g·cm·s^{-2}.

The weight of a body is the gravitational force exerted on it by the earth (or whatever other body exerts the gravitational force). Weight is a force and is therefore a vector quantity. The magnitude of the weight of a body at any specific location is equal to the product of its mass and the magnitude of the acceleration of free fall (acceleration due to gravity) at that location; that is, $w = mg$. The weight of a body depends on its location, but the mass is independent of location.

Newton's third law states that when two bodies interact, they exert forces on each other simultaneously; these forces are equal in magnitude and opposite in direction. This principle is often stated as "action equals reaction." The two forces in an action-reaction pair always act on two *different* bodies; they never act on the same body.

Newton's laws of motion are valid only in *inertial frames of reference*. If one frame of reference is inertial, any other frame moving relative to it with constant velocity is also inertial, but a frame that accelerates relative to the first is *not* inertial. The earth is approximately an inertial frame, but not precisely so because of its rotation about its axis and orbital motion around the sun.

QUESTIONS

4–1 When a heavy weight is lifted by a string that is barely strong enough, the weight can be lifted by a steady pull, but if the string is jerked it will break. Why?

4–2 When a car accelerates, starting from rest, where is the force applied to the car to cause its acceleration? By what other body is the force exerted?

4–3 When a car stops suddenly, the passengers are thrown forward, away from their seats. What force causes this motion?

4–4 For medical reasons it is important for astronauts in outer space to determine their body mass at regular intervals. Devise a scheme for measuring body mass in a zero-gravity environment.

4–5 In SI units, suppose the fundamental units had been chosen to be force, length, and time instead of mass, length, and time. What would be the units of mass, in terms of the fundamental units?

4–6 When a bullet is fired from a gun, what is the origin of the force that accelerates the bullet?

4–7 Why can a person dive into water from a height of 10 m without injury, while a person who jumps off the roof of a 10-m building and lands on a concrete street is likely to be seriously injured?

4–8 A passenger in a bus notices that a ball that has been at rest in the aisle suddenly starts to move toward the rear of the bus. Think of two different possible explanations, and devise a way to decide which is correct.

4–9 Two bodies on the two sides of an equal-arm balance exactly balance each other. If the balance is placed in an elevator and given an upward acceleration, do they still balance?

4–10 If a man in an elevator drops his briefcase but it does not fall to the floor, what can he conclude about the elevator's motion?

4–11 Suppose you are sealed inside a box on the planet Vulcan. After some time you notice that objects do not fall to the floor when released. How can you determine whether you and the box are in free fall or whether the Vulcans have somehow found a way to turn off the force of gravity?

4–12 Automotive engineers, in discussing the vertical motion of an automobile driving over a bump, call the time derivative of the acceleration the "jerk." Why is this a useful quantity in characterizing the riding qualities of an automobile?

EXERCISES

Section 4–1 Force

4–1 A raccoon pushes a tin box along the floor as in Fig. 4–1b with a force of 40 N that makes an angle of 30° with the horizontal. Find the horizontal and vertical components of the force.

4–2 A man is dragging a trunk up the loading ramp of a mover's truck. The ramp has a slope angle of 20°, and the man pulls upward with force F whose direction makes an angle of 30° with the ramp.

a) How large a force F is necessary for the component F_x parallel to the plane to be 16 N?

b) How large will the component F_y then be?

4–3 Two men pull horizontally on ropes attached to a post; the angle between the ropes is 45°. If man A exerts a force of 68 lb and man B a force of 50 lb, find the magnitude of the resultant force and the angle it makes with A's pull.

4–4 Two forces, F_1 and F_2, act at a point. The magnitude of F_1 is 8 N and its direction is 60° above the x-axis in the first quadrant. The magnitude of F_2 is 5 N and its direction is 53° below the x-axis in the fourth quadrant.

a) What are the horizontal and vertical components of the resultant force?

b) What is the magnitude of the resultant?

c) What is the magnitude of the vector difference $F_1 - F_2$?

Section 4–5 Mass and Weight

4–5

a) What is the mass of a book that weighs 1 N at a point where $g = 9.80$ m·s^{-2}?

b) At the same point, what is the weight of a dog whose mass is 8 kg?

4–6

a) What is the mass of a radio that weighs 8 lb at a point where $g = 32$ ft·s^{-2}?

b) At the same point, what is the weight of a television set whose mass is 0.8 slug?

4–7 At the surface of Mars the acceleration due to gravity is $g = 3.7$ m·s^{-2}. A watermelon weighs 52 N at the surface of the earth. What would be its mass and weight on the surface of Mars?

Section 4–6 Newton's Third Law

4–8 Imagine that you are holding a book weighing 4 N at rest on the palm of your hand. Complete the following sentences:

a) A downward force of magnitude 4 N is exerted on the book by _____.

b) An upward force of magnitude _____ is exerted on _____ by the hand.

c) Is the upward force (b) the reaction to the downward force (a)?

d) The reaction to force (a) is a force of magnitude _____, exerted on _____ by _____. Its direction is _____.

e) The reaction to force (b) is a force of magnitude _____, exerted on _____ by _____. Its direction is _____.

f) That the forces (a) and (b) are equal and opposite is an example of Newton's _____ law.

g) That forces (b) and (e) are equal and opposite is an example of Newton's _____ law.

Suppose now that you exert an upward force of magnitude 5 N on the book.

h) Does the book remain in equilibrium?

i) Is the force exerted on the book by the hand equal and opposite to the force exerted on the book by the earth?

j) Is the force exerted on the book by the earth equal and opposite to the force exerted on the earth by the book?

k) Is the force exerted on the book by the hand equal and opposite to the force exerted on the hand by the book?

Finally, suppose that you snatch your hand away while the book is moving upward.

l) How many forces then act on the book?

m) Is the book in equilibrium?

n) What balances the downward force exerted on the book by the earth?

4–9 A bottle is given a push along a table top and slides off the edge of the table. Neglect air resistance.

a) What forces are exerted on it while it is falling from the table to the floor?

b) What is the reaction to each force; that is, on what body and by what body is the reaction exerted?

Section 4–7 Applications of Newton's Laws

4–10 A crate of mass 1.2 slugs initially at rest on a frictionless horizontal plane is acted on by a horizontal force of 36 lb.

a) What acceleration is produced?

b) How far will the crate travel in 10 s?

c) What will be its velocity at the end of 10 s?

4–11 A constant horizontal force of 40 N acts on a block of ice on a smooth horizontal plane. The block starts from rest and is observed to move 100 m in 5 s.

a) What is the mass of the block of ice?

b) If the force ceases to act at the end of 5 s, how far will the block move in the next 5 s?

4–12 A .22 rifle bullet, traveling at 360 m·s^{-1}, strikes a block of soft wood, which it penetrates to a depth of 0.1 m. The mass of the bullet is 1.8 g. Assume a constant retarding force.

a) How much time was required for the bullet to stop?

b) What was the accelerating force, in newtons?

4–13 A hockey puck of mass 0.5 kg is at rest at the origin $x = 0$ on a frictionless horizontal surface. At time $t = 0$ a force of 0.001 N is applied to the puck parallel to the x-axis, and 5 s later this force is removed

a) What are the position and velocity of the puck at $t = 5$ s?

b) If the same force is again applied at $t = 15$ s, what are the position and velocity of the puck at $t = 20$ s?

4–14 An electron (mass = 9.11×10^{-31} kg) leaves the cathode of a vacuum tube with zero initial velocity and travels in a straight line to the anode, which is 1 cm away. It reaches the anode with a velocity of 6.00×10^6 m·s^{-1}. If the accelerating force is constant, compute

a) the acceleration;

b) the time to reach the anode;

c) the accelerating force, in newtons.

(The gravitational force on the electron may be neglected.)

4–15 An object of mass m moves along the x-axis. Its position as a function of time is given by $x(t) = At + Bt^3$, where A and B are constants. Calculate the resultant force on the object, as a function of time.

PROBLEMS

4–16 The three forces shown in Fig. 4–13 act on an object located at the origin.

a) Find the x- and y-components of each of the three forces.

b) Find the resultant of these forces.

c) Find the magnitude and direction of a fourth force that must be added to make the resultant force zero. Indicate the fourth force by a diagram.

4–17 Two forces F_1 and F_2 act on an object such that the resultant force R has a magnitude equal to that of F_1 and makes an angle of 90° with F_1. Let $F_1 = R = 10$ N. Find the magnitude of the second force, and its direction (relative to F_1).

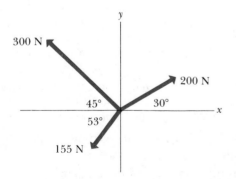

FIGURE 4–13

4–18 The resultant of four forces is 1000 N in the direction 30° west of north. Three of the forces are 400 N, 60° north of east; 200 N, south; and 400 N, 53° west of south. Find the magnitude and direction of the fourth force.

4–19 Two men and a boy want to push a crate in the direction marked x in Fig. 4–14. The two men push with forces F_1 and F_2, whose magnitudes and directions are indicated in the figure. Find the magnitude and direction of the *smallest* force that the boy should exert.

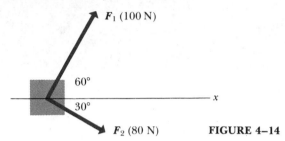

F_1 (100 N)

60°

30°

x

F_2 (80 N)   **FIGURE 4–14**

4–20 A short commuter train consists of a locomotive and two cars. The mass of the locomotive is 6000 kg and that of each car 2000 kg. The train pulls away from a station with an acceleration of 0.5 m·s⁻².

a) Find the tension in the coupler joining the locomotive to the first car, and in the coupler joining the two cars.

b) What total horizontal force must the locomotive wheels exert on the track?

4–21 A loaded elevator with very worn cables has a total mass of 2000 kg, and the cables can withstand a maximum tension of 24,000 N.

a) What is the maximum upward acceleration for the elevator if the cables are not to break?

b) What is the answer to (a) if the elevator is taken to the moon, where $g = 1.67$ m·s⁻²?

4–22 An 80-kg man steps off a platform 2.0 m above the ground. He keeps his legs straight as he falls, but at the moment his feet touch the ground his knees begin to bend, and his torso moves an additional 0.6 m before coming to rest.

a) What is his speed at the instant his feet touch the ground?

b) What is the acceleration of his torso as he slows down, assuming the acceleration is constant?

c) What force do his feet exert on the ground while he slows down? Express this force in newtons and also as a multiple of his weight.

4–23 An object of mass 2 kg moves in the x–y plane such that its x- and y-coordinates vary in time according to

$$x(t) = 2 \text{ m} - (5 \text{ m·s}^{-3})t^3, \qquad y(t) = (16 \text{ m·s}^{-2})t^2.$$

a) Calculate the x- and y-components of the resultant force on the object, as functions of time.

b) What are the magnitude and direction of the resultant force at $t = 3$ s?

4–24 An object of mass m initially at rest is acted on by a force $F = k_1 i + k_2 t^2 j$, where k_1 and k_2 are constants. Calculate the velocity $v(t)$ of the object as a function of time.

CHALLENGE PROBLEM

4–25 An object of mass m is at rest at the origin at time $t = 0$. A force $F(t)$ is then applied that has components

$$F_x(t) = k_1 + k_2 y, \qquad F_y(t) = k_3 t,$$

where k_1, k_2, and k_3 are constants. Calculate the position $r(t)$ and velocity $v(t)$ vectors as functions of time.

5

APPLICATIONS OF NEWTON'S LAWS—I

NEWTON'S LAWS OF MOTION FORM THE FOUNDATION FOR ALL OF CLASSICAL mechanics, including statics (the study of bodies in equilibrium) and dynamics (the general relationships between force and motion). In this chapter we encounter no new fundamental principles; instead, we concentrate on learning to apply Newton's laws to problems of fundamental and practical importance. We begin with a discussion of the various classes of forces found in nature. Then we study in detail the contact forces between two bodies, including friction forces. Next we consider examples of bodies in equilibrium, using Newton's first law, the concept of an inertial frame of reference, and the vector language introduced in Chapter 1. Finally, we study several examples of Newton's second law, analyzing in detail the relationships between force and motion for bodies that are *not* in equilibrium. We find that much of the same problem-solving strategy can be used for both statics and dynamics problems.

5–1 FORCES IN NATURE

Our present-day understanding of the physical world includes four distinct classes of forces. Two are familiar in everyday experience. The other two are concerned with interactions of fundamental particles. We cannot observe the latter with the unaided senses; studying them requires sophisticated and elaborate experiments.

Forces found in nature: two familiar kinds

Of the two familiar classes of force, **gravitational interactions** were the first to be studied in detail. The *weight* of a body results from the earth's gravitational attraction acting on it. The sun's gravitational force on the earth is responsible for making the earth move in a nearly circular orbit instead of following a straight-line path as it would do if there were no force. Indeed, one of Newton's great achievements was to recognize that both the motions of the planets around the sun and the free fall of objects on earth are manifestations of gravitational forces. We will study gravitational interactions in greater detail in Chapter 6, and will analyze their vital role in the motions of planets and satellites.

The second familiar class of forces, **electromagnetic interactions,** includes electric and magnetic forces. When you run a comb through your hair, you can then use the comb to pick up bits of paper or fluff; this interaction is the result of electric charge on the comb. We encounter magnetic forces in interactions between magnets or between a magnet and a piece of iron. These may seem to fall in a different category, but detailed study shows that magnetic interactions are actually the result of electric charges in motion. An electromagnet causes magnetic interactions as a result of an electric current in a coil of wire. We will study electric and magnetic interactions in detail in the second half of this book.

These two familiar kinds of interactions differ enormously in their strength. For example, the electrical repulsion between two protons at a given

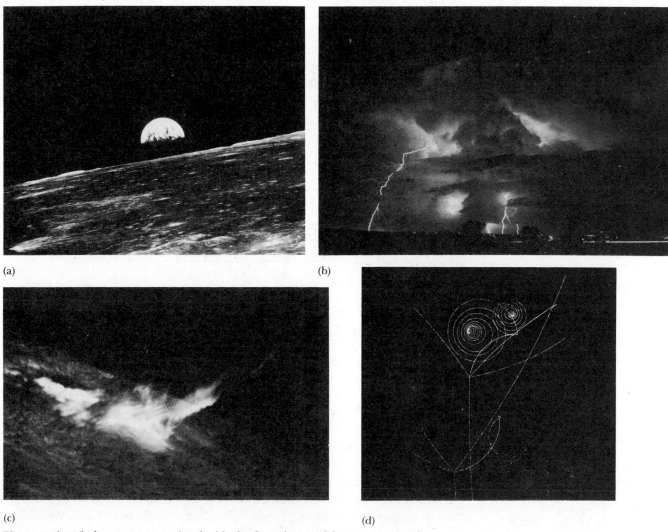

(a) (b)

(c) (d)

Photographs of phenomena associated with the four classes of forces occurring in nature. (a) The orbital motions of the earth around the sun and of the moon around the earth result from *gravitational* forces. (NASA.) (b) Lightning is a flow of electric charge resulting from the electrical interactions of charged particles, one aspect of the *electromagnetic* interaction. (National Center for Atmospheric Research/National Science Foundation.) (c) A solar flare, the result of a thermonuclear reaction associated with *strong* interactions between nuclear particles. (Sacramento Peak Observatory Association of Universities for Research in Astronomy, Inc.) (d) An experiment to detect the presence of neutrinos, which have only *weak* interactions. (Lawrence Berkeley Laboratory.)

distance apart is stronger than the gravitational attraction by a factor of the order of 10^{35}. Thus gravitational forces play no significant role in determining the microscopic structure of atoms, molecules, and materials. But in bodies of astronomical size, protons and electrons are ordinarily present in nearly equal numbers; their charges are opposite and nearly cancel out. Gravitational interactions are the dominant influence in the motion of planets and also in the internal structure of stars.

The strong interaction holds nuclei together.

The other two classes of interactions are less familiar. One, the **strong interaction,** is responsible for holding the nuclei of atoms together. Nuclei contain electrically neutral and positively charged particles. The charged particles repel each other; a nucleus could not be stable it it were not for the presence of an attractive force of a different kind that counteracts the repulsive electrical interactions. In this context the strong interaction is also called the *nuclear force.* It has shorter range than electrical interactions, but within its range it is much stronger.

The weak interaction is responsible for the decay of some unstable particles.

Finally, there are the **weak interactions;** they play no direct role in the behavior of ordinary matter but are of vital importance in interactions among fundamental particles. The weak interaction is responsible for the emission of electrons (beta particles) from radioactive nuclei. In a beta-emitting nucleus a neutron is converted into a proton, an electron, and a third particle called a neutrino. The weak interaction is also responsible for the decay of many unstable particles produced in high-energy collisions of fundamental particles.

Grand unification: Are the four kinds of forces really all the same?

Since about 1970, attempts have been made to understand all four classes of interactions on the basis of a single unified theory called a *grand unified theory.* Such theories are still very speculative; as yet none has been completely successful. In 1983, however, evidence was found that supports a unified theory of the electromagnetic and weak interactions. The entire area is a very active field of present-day theoretical and experimental research.

5–2 CONTACT FORCES AND FRICTION

Describing the interaction of bodies in contact

When two bodies interact by direct contact (touching) of their surfaces, the interaction forces are called **contact forces.** We have already remarked that on a microscopic level these forces arise mainly from the electrical interactions of the electrons and positively charged nuclei of the atoms in the bodies. But in this chapter we simply describe the behavior of the contact forces without going into their electrical basis.

Normal and frictional components of the contact force

To discuss the nature of the contact force, let us first think about a body sliding across a surface. It might be, for example, the lab notebook in Example 4–3 (Section 4–7). In studying the motion of the body, we need to know the forces acting on it, including the force applied to it by the surface. We call this force the contact force; the force acting on the body can always be represented in terms of a component perpendicular to the surface and a component parallel to the surface. We call the perpendicular component the **normal force,** denoted by $\mathbf{n}$. (*Normal* is a synonym for *perpendicular*.) The component parallel to the surface is the **friction force,** denoted by $\mathcal{F}$.

The friction force $\mathcal{F}$ is always the component of the contact force *parallel* to the surface, and the normal force $\mathbf{n}$ is always the component *normal* (perpendicular) to the surface. We use special script symbols for these quantities to emphasize their special role in representing the contact force. In addition, use of the script letter n prevents our confusing this symbol for normal force with the abbreviation for newton, N. The normal force is not necessarily equal in magnitude to the weight of the body, because there may be additional components of force normal to the surface. Nevertheless, n always denotes the normal component of contact force exerted by the surface on the body. By definition, n and $\mathcal{F}$ are always perpendicular.

If the surface is perfectly smooth and perfectly lubricated so the body can slide without any resistance, then there is *no* friction force. In the real world this is an unattainable idealization, but we sometimes talk about an idealized "frictionless surface," where the friction force is negligibly small. In such cases the contact force has *only* a normal component and is perpendicular to the surface.

The general behavior of friction forces is rather complex, but some aspects are quite familiar. The friction force on each interacting body is opposite in direction to the motion of that body relative to the other. Thus when a book slides from left to right along a table top, the friction force on it acts to the left. According to Newton's third law, the book exerts on the table a force of equal magnitude, directed to the right. It tries to drag the table along with it.

The *magnitude* of the friction force depends on the magnitude of the normal force, usually increasing with the normal force. It is harder to push a full oil barrel across a floor than to push the same barrel when it is empty. Often the friction force $\mathcal{F}_k$ is found to be approximately *proportional* to the normal force; in such cases we represent the relation by the equation

> The frictional force is proportional to the normal force (sometimes).

$$\mathcal{F}_k = \mu_k n, \qquad (5\text{–}1)$$

where μ_k is a constant called the **coefficient of kinetic friction.** The adjective *kinetic* and the corresponding subscript k refer to the fact that the two surfaces are moving relative to each other. Note that because the friction force and the normal force are always perpendicular, Eq. (5–1) is not a vector equation but a relation between the *magnitudes* of the two forces.

The coefficient of kinetic friction depends on the nature of the materials and the textures of the surfaces. For teflon on steel, μ_k is about 0.04; for rubber tires on dry concrete, μ_k can be greater than unity. The more slippery the surface, the smaller the coefficient of friction. Table 5–1 shows a few representative values of μ_k. Friction forces can also depend on the relative *velocity* of the interacting surfaces; we will ignore that complication here in order to concentrate on the simplest cases.

TABLE 5–1   Coefficients of Friction

Materials	Static, μ_s	Kinetic, μ_k
Steel on steel	0.74	0.57
Aluminum on steel	0.61	0.47
Copper on steel	0.53	0.36
Brass on steel	0.51	0.44
Zinc on cast iron	0.85	0.21
Copper on cast iron	1.05	0.29
Glass on glass	0.94	0.40
Copper on glass	0.68	0.53
Teflon on Teflon	0.04	0.04
Teflon on steel	0.04	0.04
Rubber on concrete (dry)	1.0	0.80
Rubber on concrete (wet)	0.30	0.25

Friction forces may also act when there is *no* relative motion. For example, when we push horizontally on a heavy packing case resting on the floor, the case may not move because the floor exerts an equal and opposite friction force on the case. This is called a **static friction force** $\mathcal{F}_s$, exerted on the box by the surface.

Static friction: when there is no relative motion of the surfaces in contact

Figure 5–1 shows a box at rest on a horizontal surface. Its weight w and the upward normal force n exerted on it by the surface have equal magnitude, and it is in equilibrium. Now we attach a rope to the box as in Fig. 5–1b and gradually increase the tension T in the rope. At first the box remains at rest because the force of static friction $\mathcal{F}_s$ is equal in magnitude and opposite in direction to the force T. As T increases further, a critical value is reached at which the box suddenly "breaks loose" and starts to slide. In other words, there is a certain *maximum* value that the force of static friction $\mathcal{F}_s$ can have. When that value is not sufficient to prevent motion, the box starts to move. Figure 5–1c is the force diagram when T is just below its critical value. If T exceeds this value, the box is no longer in equilibrium.

For a given pair of surfaces, the magnitude of the maximum value of $\mathcal{F}_s$ depends on the magnitude of the normal force n; when n increases, $\mathcal{F}_s$ also increases, and a greater force is needed to make the box start to slide. In some cases the maximum value of $\mathcal{F}_s$ is approximately *proportional* to n; we call the proportionally factor μ_s the **coefficient of static friction.** The actual force of static friction can have any magnitude between zero (when there is no other force parallel to the surface) and a maximum value given by $\mu_s n$. Thus,

$$\mathcal{F}_s \leq \mu_s n. \tag{5–2}$$

The equality sign holds only when the applied force T, parallel to the surface, has such a value that motion is about to start (Fig. 5–1c). When T is less than this value (Fig. 5–1b), the inequality sign holds and the magnitude of the friction force must be computed from the conditions of equilibrium.

As soon as sliding begins, the friction force usually (but not always) *decreases.* Thus for any given pair of surfaces, the coefficient of kinetic friction is usually less than the coefficient of static friction. A few representative values of both μ_s and μ_k are given in Table 5–1. This table lists values only for solids. Liquids and gases also show frictional effects, but the simple equation $\mathcal{F} = \mu n$ does not hold. The friction force between two surfaces sliding over each other with a layer of liquid or gas between them is determined by a property of the fluid called *viscosity*. Gases have much smaller viscosities than

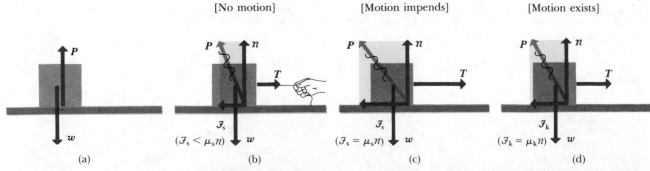

5–1 The magnitude of the friction force $\mathcal{F}_s$ is less than or equal to $\mu_s n$ when there is no relative motion and is equal to $\mu_k n$ when there is motion.

(a)

(b)

5–2 (a) The Ealing–Stull linear air track. Inverted Y-shaped sliders ride on a layer of air streaming through many fine holes in the inverted V-shaped surface. (b) The Ealing–Daw two-dimensional air table. Plastic pucks slide on a cushion of air issuing from more than a thousand minute holes in the tabletop. (Courtesy of the Ealing Corporation.)

liquids at normal temperatures, and therefore friction can be made very small if a body can be made to slide on a layer of gas.

This is the principle of the hovercraft, a type of vehicle containing powerful blowers that maintain a cushion of air between the vehicle and the ground so it literally floats on a layer of air. A familiar laboratory example of this principle is the linear air track, shown in Fig. 5–2a. The frictional force is velocity dependent, but at typical speeds the effective coefficient of friction is of the order of 0.001.

A similar device is the frictionless air table shown in Fig. 5–2b. The pucks are supported by an array of small air jets about 2 cm apart. Two-dimensional collisions can be demonstrated on this table. The same principle is used in "air-hockey" games.

EXAMPLE 5–1 Suppose the body shown in Fig. 5–1 is a 500-N crate full of home exercise equipment that a delivery company has just unloaded in your driveway. You find that to get it moving toward your garage you have to push with a horizontal force of magnitude 200 N; but once it is moving, you can keep it moving with constant speed by pushing with 100-N force. What are the coefficients of static and kinetic friction?

The role of friction in moving heavy boxes

SOLUTION When you are getting the crate moving, the appropriate force diagram is Fig. 5–1c; from this and the above data, we have

$$\Sigma F_y = n - w = n - 500 \text{ N} = 0, \qquad n = 500 \text{ N},$$
$$\Sigma F_x = T - \mathcal{F}_s = 200 \text{ N} - \mathcal{F}_s = 0, \qquad \mathcal{F}_s = 200 \text{ N}.$$
$$\mathcal{F}_s = \mu_s n \qquad \text{(motion impends)}.$$

Hence

$$\mu_s = \frac{\mathcal{F}_s}{n} = \frac{200 \text{ N}}{500 \text{ N}} = 0.40.$$

After the crate is moving, the forces are as in Fig. 5–1d, and we have

$$\Sigma F_y = n - w = n - 500 \text{ N} = 0, \qquad n = 500 \text{ N},$$
$$\Sigma F_x = T - \mathcal{F}_k = 100 \text{ N} - \mathcal{F}_k = 0, \qquad \mathcal{F}_k = 100 \text{ N}.$$
$$\mathcal{F}_k = \mu_k n \qquad \text{(motion exists)}.$$

Hence

$$\mu_k = \frac{\mathcal{F}_k}{n} = \frac{100 \text{ N}}{500 \text{ N}} = 0.20.$$

EXAMPLE 5–2 In Example 5–1, what is the friction force if the crate is at rest on the surface and a horizontal force of 50 N is exerted on it?

SOLUTION We have

$$\Sigma F_x = T - \mathcal{F}_s = 50 \text{ N} - \mathcal{F}_s = 0 \quad \text{(first law).}$$

$$\mathcal{F}_s = 50 \text{ N.}$$

Note that in this case $\mathcal{F}_s < \mu_s n$

Rolling friction: why the wheel is a useful invention

Ever since the invention of the wheel, we have known that it is a lot easier to move a heavy load across a horizontal surface by using a cart with wheels than it is to drag the same load across the surface. How much easier? We can define a coefficient of rolling friction μ_r for a wheeled vehicle as the ratio of the horizontal force needed to make it move with constant speed, to the upward normal force exerted by the surface (equal in this case to the weight of the vehicle and load). Transportation engineers call this ratio the *tractive resistance*. Typical values of tractive resistance are 0.002 to 0.003 for steel wheels on steel rails, and 0.01 to 0.02 for rubber tires on concrete. These values show why railroad trains are in general much more fuel-efficient than highway trucks.

EXAMPLE 5–3 A 1200-kg car weighs about 12,000 N (about 2700 lb). If the tractive resistance is $\mu_r = 0.01$, what horizontal force must be provided to make the car move with constant speed?

SOLUTION From the definition of tractive resistance, a horizontal force of $(0.01)(12,000 \text{ N}) = 120 \text{ N}$ (about 50 lb) is required.

The actual behavior of tractive resistance forces is a lot more complicated than this. An important factor is speed: Air-resistance forces usually increase proportionally to the square of the speed. The numerical values cited above are for relatively slow speeds.

5–3 EQUILIBRIUM OF A PARTICLE

Equilibrium: The forces on a body must add up to zero.

In this section we consider several problems involving the **equilibrium** of particles. It is surprising how many situations of interest and importance in engineering, in the life and earth sciences, and in everyday life involve equilibrium of particles. The essential physical principle is Newton's first law: When a particle is at rest in an inertial frame of reference, the vector sum of all the forces acting on it must be zero. In symbols,

$$\Sigma F = 0. \tag{5–3}$$

We will usually use this in component form:

$$\Sigma F_x = 0,$$
$$\Sigma F_y = 0. \qquad\qquad (5\text{--}4)$$

In the analysis of equilibrium problems, we strongly recommend strict adherence to the following strategy.

PROBLEM-SOLVING STRATEGY: *Equilibrium of a particle*

1. Make a simple sketch of the apparatus or structure, showing dimensions and angles.

2. Choose some object (a knot in a rope, for example) as the particle in equilibrium. Draw a separate diagram of this object and show with arrows (use a colored pencil or pen) *all* of the forces exerted *on* it by other bodies. This is called the *force diagram* or **free-body diagram.** When a system is composed of several particles, you may need to draw a separate free-body diagram for each one. Do *not* show, in the free-body diagram of a chosen particle, any of the forces exerted *by* it on other bodies.

3. Draw a set of coordinate axes and resolve into components all forces acting on the particle. Cross out lightly those forces that have been resolved. Often one particular choice of axes can simplify the problem considerably. For example, when friction is present, it is usually simplest to take the axes in the directions of the normal and frictional forces, even when these are not vertical and horizontal.

4. Set the algebraic sum of all *x*-forces (or force components) equal to zero, and the algebraic sum of all *y*-forces (or components) equal to zero. This provides two independent equations, which can be solved simultaneously for two unknown quantities; these may be forces, angles, distances, and so on.

A remark about notation is in order. We have consistently denoted vector quantities by boldface italic symbols, and we will continue this usage throughout this book. However, it is sometimes convenient in diagrams to label a vector by using a lightface italic symbol to represent the *magnitude* of the quantity, and to describe its direction (if that is not obvious in the diagram) separately, perhaps by use of an angle. When both light and bold symbols appear on a diagram, you can avoid confusion by recalling that a bold symbol always represents the vector quantity itself, while a light symbol represents either the magnitude of a vector quantity or a component.

How to label forces on free-body diagrams

In many of the problems we will encounter, one of the forces is the *weight* of a body, that is, the force of gravitational attraction that the earth exerts on the body. In such a case, the body attracts the earth with a force equal in magnitude and opposite in direction to the earth's force on the body. Thus if a body weighs 10 N, the earth pulls *down* on it with a force of 10 N, and the body pulls *up* on the earth with a force of 10 N. This is an example of Newton's third law and an action–reaction pair.

The weight of a body

EXAMPLE 5–4 A gymnast has just begun climbing up a rope hanging from a gymnasium ceiling, as in Fig. 5–3a. She stops, suspended from the lower end of the rope by her hands. Her weight is 600 N, and the weight of the rope is 100 N. Analyze the forces on the gymnast and on the rope.

SOLUTION Figure 5–3b is a free-body diagram for the gymnast. The forces acting on her are her weight w_1 and the upward force T_1 exerted on her by the rope. If

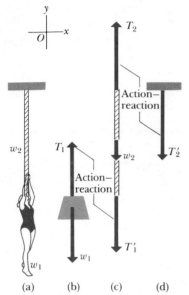

5–3 (a) Gymnast hanging at rest from vertical rope. (b) The gymnast is isolated and all forces acting on her are shown. (c) Forces on the rope. (d) Downward force on the ceiling. Lines connect action–reaction pairs.

we take the x-axis horizontal and the y-axis vertical, there are no x-components of force. The y-components are those associated with the forces T_1 and w_1. Force T_1 acts in the positive y-direction, and its y-component is just the magnitude T_1, a positive (scalar) quantity. But w_1 acts in the negative y-direction, and its y-component is the *negative* of the magnitude w_1. Thus the algebraic sum of y-components is $T_1 - w_1$. Then from the equilibrium condition, $\Sigma F_y = 0$, we have

$$\Sigma F_y = T_1 - w_1 = 0,$$
$$T_1 = w_1.$$

The forces w_1 and T_1 are *not* an action–reaction pair, although they are equal in magnitude, opposite in direction, and have the same line of action. The weight w_1 is a force of attraction exerted on the body by the earth. Its reaction is an equal and opposite force of attraction exerted on the earth by the body. This reaction is one of the set of forces acting *on* the earth, and therefore it does not appear in the free-body diagram of the suspended body. The reaction to the force T_1 is a downward force, T_1', of equal magnitude:

$$T_1 = T_1' \qquad \text{(third law)}.$$

The force T_1' is shown in part (c), which is the free-body diagram for the rope. The other forces on the rope are its own weight w_2 and the upward force T_2 exerted on its upper end by the ceiling. The y-component of T_2 is positive, while those of w_2 and T_2' are negative. Since the rope is also in equilibrium,

$$\Sigma F_y = T_2 - w_2 - T_1' = 0,$$
$$T_2 = w_2 + T_1' \qquad \text{(first law)}.$$

The reaction to T_2 is the downward force T_2' in part (d), exerted on the ceiling by the rope:

$$T_2 = T_2' \qquad \text{(third law)}.$$

With the numbers given, the gymnast's weight w_1 is 600 N, and the weight of the rope w_2 is 100 N. Then

$$T_1 = w_1 = 600 \text{ N},$$
$$T_1' = T_1 = 600 \text{ N},$$
$$T_2 = w_2 + T_1' = 100 \text{ N} + 600 \text{ N} = 700 \text{ N},$$
$$T_2' = T_2 = 700 \text{ N}.$$

EXAMPLE 5–5 In Fig. 5–4a, a hanging lamp of weight w hangs from a cord, which is knotted at point O to two other cords, one fastened to the ceiling, the other to the wall. We wish to find the tensions in these three cords, assuming the weights of the cords to be negligible.

SOLUTION To use the conditions of equilibrium to determine unknown forces (in this case, the tensions in the cords), we must consider some body that is in equilibrium and on which the desired forces act. The suspended lamp is one such body. As shown in the preceding examples, the tension T_1 in the vertical cord supporting the lamp is equal in magnitude to the weight of the lamp. The other cords do not exert forces on the lamp (because they are not attached directly to it), but they do act on the knot at O. Hence we consider the *knot* as a particle in equilibrium, considering the weight of the knot itself as negligible.

Free-body diagrams for the lamp and the knot are shown in Fig. 5–4b, where T_1, T_2, and T_3 are the *magnitudes* of the forces shown; the directions of these forces are indicated by the vectors on the diagram. An *xy*-coordinate axis system is also shown, and the force of magnitude T_3 has been resolved into its *x*- and *y*-components.

Considering the lamp first, we note that there are no *x*-components of force. The *y*-component of force exerted by the cord is in the positive *y*-direction and is just T_1. The *y*-component of the weight is, however, in the negative *y*-direction and is $-w$. The equilibrium condition for the lamp, that the algebraic sum of *y*-components of force must be zero, is

$$T_1 + (-w) = 0.$$

Hence,
$$T_1 = w.$$

The tension in the vertical cord equals the weight of the lamp, as already noted.

We now consider the knot at O. Both *x*- and *y*-components of force are acting on it, so there are two separate equilibrium conditions. The algebraic sum of the *x*-components must be zero, and the algebraic sum of *y*-components must separately be zero. (Note that *x*- and *y*-components are *never* added together in a single equation.) We find

$$\Sigma F_x = 0: \qquad T_3 \cos 60° - T_2 = 0.$$
$$\Sigma F_y = 0: \qquad T_3 \sin 60° - T_1 = 0.$$

Because $T_1 = w$, the second equation can be rewritten as

$$T_3 = \frac{T_1}{\sin 60°} = \frac{w}{\sin 60°} = 1.155\,w.$$

This result can now be used in the first equation:

$$T_2 = T_3 \cos 60° = (1.155w) \cos 60° = 0.577w.$$

Thus all three tensions can be expressed as multiples of the weight w of the lamp, which is assumed to be known. To summarize,

$$T_1 = w,$$
$$T_2 = 0.577w,$$
$$T_3 = 1.155w.$$

If the lamp's weight is $w = 50.0$ N, then

$$T_1 = 50.0 \text{ N},$$
$$T_2 = (0.577)(50.0 \text{ N}) = 28.9 \text{ N},$$
$$T_3 = (1.155)(50.0 \text{ N}) = 57.7 \text{ N}.$$

We note that T_3 is greater than the weight of the lamp. If this seems strange, note that T_3 must be large enough so that its vertical component is equal to w in magnitude; thus T_3 itself must have somewhat *larger* magnitude than w.

EXAMPLE 5-6 A repairman of weight w_1 is trying to find a leak in an icy, perfectly frictionless roof with slope angle θ. He is secured by a rope passing over the top and down the vertical side of the building. The other end of the rope is held by his helper (weight w_2) on the ground. The helper finds that he must put his full weight on the rope to hold the man on the roof. If we know the weight w_1 of the repairman, what is the weight w_2 of the helper?

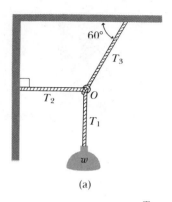

(a)

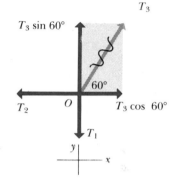

(b)

5-4 (a) A lamp of weight w is suspended from a cord knotted at O to two other cords. (b) Free-body diagrams for the lamp and for the knot, showing the components of force acting on each.

Fixing a slippery roof: how physics can help

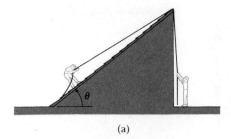

(a)

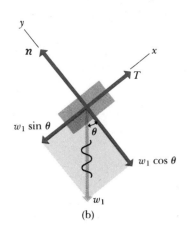

(b)

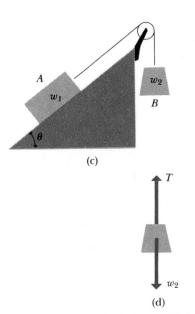

(c)

T

w_2

(d)

5–5 Forces on a body in equilibrium on a frictionless inclined plane. (a) The physical situation; (b) free-body diagram for body A; (c) idealized model of situation; (d) free-body diagram for body B.

Moving heavy boxes again, this time with a rope

SOLUTION We represent the situation by the idealized model shown in Fig. 5–5, where A is the repairman and B the helper. Free-body diagrams for the two bodies are shown in parts (b) and (d) of the figure. Here, as in preceding examples, we label and refer to forces by using light italic symbols, which in equations represent the *magnitudes* of these forces. The forces on body B are its weight w_2 and the force T exerted on it by the cord. Because it is in equilibrium,

$$T = w_2.$$

Body A is acted on by its weight w_1, the force T exerted on it by the rope, and the force n exerted on it by the plane. We use the same symbol T for the force exerted on each body because, as discussed above, these forces are equivalent to an action–reaction pair and have the same magnitude. The pulley, which corresponds to the frictionless top edge of the roof, changes the directions of the forces exerted by the rope but not their magnitudes; and the two forces at the ends of the rope have the same magnitude even though their directions are different. Because there is no friction, the force n can have no component along the plane that would resist the motion of the body on the plane; hence it must be perpendicular, or *normal*, to the surface of the plane.

It is simplest to choose x- and y-axes parallel and perpendicular to the surface of the plane, because then only the weight w_1 needs to be resolved into components. The conditions of equilibrium for body A give

$$\sum F_x = T - w_1 \sin \theta = 0,$$
$$\sum F_y = n - w_1 \cos \theta = 0.$$

Thus if $w_1 = 800$ N and $\theta = 30°$, we find

$$w_2 = T = w_1 \sin \theta = (800 \text{ N})(0.500) = 400 \text{ N},$$

and

$$n = w_1 \cos \theta = (800 \text{ N})(0.866) = 693 \text{ N}.$$

Note carefully that, *if there is no friction*, the same weight w_2 of 400 N is required for equilibrium whether the system remains at rest or moves with constant speed in *either* direction. This is not the case when friction is present.

EXAMPLE 5–7 In Example 5–1 (Section 5–2) suppose you try to move the crate by tying a rope around it and pulling upward on the rope at an angle of 30° to the horizontal. What tension in the rope is required to make the crate move with constant speed?

SOLUTION Figure 5–6 is a free-body diagram showing the forces on the crate. We note that the normal force n is *not* equal in magnitude to the weight of the crate because the force exerted by the rope has an additional vertical component. From the equilibrium conditions,

$$\sum F_x = T \cos 30° - 0.2n = 0,$$
$$\sum F_y = T \sin 30° + n - 500 \text{ N} = 0.$$

These are two simultaneous equations for the two unknown quantities T and n. To solve them we can eliminate one unknown and solve for the other, as follows. Rearrange the second equation to the form

$$n = 500 \text{ N} - T \sin 30°.$$

Substitute this expression for n back into the first equation, obtaining

$$T \cos 30° - 0.2(500 \text{ N} - T \sin 30°) = 0.$$

Finally, solve this equation for T, then substitute the result back into either of the original equations to obtain n. The results are

$$T = 103 \text{ N}, \qquad n = 448 \text{ N}.$$

EXAMPLE 5–8 A toboggan loaded with vacationing students slides down a long, snow-covered slope having a coefficient of kinetic friction μ_k. The slope has just the right angle to make the toboggan slide with constant speed. Find the slope angle.

SOLUTION We represent the situation by a block sliding down a plane inclined at an angle θ, as shown in Fig. 5–7. The forces on the block (toboggan) are its weight w and the normal and frictional components of the force exerted on it by the plane. We take axes perpendicular and parallel to the surface of the plane and represent the weight in terms of its components in these two directions, as shown. The equilibrium conditions in component form are then

$$\Sigma F_x = w \sin \theta - \mathcal{F}_k = w \sin \theta - \mu_k n = 0,$$
$$\Sigma F_y = n - w \cos \theta = 0.$$

Hence

$$\mu_k n = w \sin \theta, \qquad n = w \cos \theta.$$

Dividing the first of these equations by the second, we find

$$\mu_k = \frac{\sin \theta}{\cos \theta} = \tan \theta.$$

It follows that a body, regardless of its weight, slides down an inclined plane with constant speed if the tangent of the slope angle of the plane equals the coefficient of kinetic friction. Measurement of this angle then provides a simple experimental method for determining the coefficient of kinetic friction.

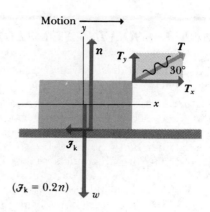

5–6 Forces on a box being dragged to the right on a level surface at constant speed.

A not-too-fast toboggan ride down a not-too-steep slope

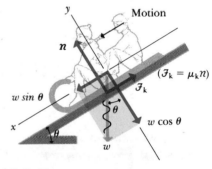

5–7 Forces on a toboggan sliding down a hill (with friction) at constant speed.

5–4 APPLICATIONS OF NEWTON'S SECOND LAW

We now discuss several **dynamics** problems, showing applications of Newton's second law to systems that are *not* in equilibrium. In this case,

$$\Sigma F = ma. \tag{5–5}$$

We will usually use this relation in component form:

$$\Sigma F_x = ma_x,$$
$$\Sigma F_y = ma_y. \tag{5–6}$$

The problem-solving strategies appropriate for these problems are very similar to those used for equilibrium problems in Section 5–3. We urge you to study these carefully, watch for their applications in the example problems, and use them when you solve end-of-chapter problems.

Dynamics: the relation of force to acceleration

PROBLEM-SOLVING STRATEGY: *Newton's second law*

1. Select a body for analysis. Newton's second law will be applied to this body.

2. Draw a free-body diagram. Be sure to include all the forces acting *on* the chosen body, but be equally careful *not* to include any force exerted *by* the body on some other body. Some of the forces may be unknown; label them with algebraic symbols. Usually one of the forces will be the body's weight. Often it is useful to label this immediately as *mg* rather than *w*. If a numerical value of mass is given, the corresponding numerical value of weight can be computed.

3. Show your coordinate axes explicitly in the free-body diagram, and then determine components of forces with reference to these axes. When a force is represented in terms of its components, cross out the original force so as not to include it twice. When the direction of the acceleration is known in advance, it is usually best to take its direction as that of the +*x*-axis. When there are two or more bodies, it is often useful to use a separate axis system for each body; there is no reason you have to use the same axis system for all the bodies.

4. If more than one body is involved, steps 1–3 must be carried out for each body. In addition there may be *geometrical* relationships between the motions of two or more bodies; express these in algebraic form.

5. Write the Newton's-second-law equations for each body, usually Eqs. (4–3), and solve to find the unknown quantities.

6. Check special cases or extreme values of quantities, where possible, and compare the results for these particular cases with your intuitive expectations. Ask: "Does this result make sense?"

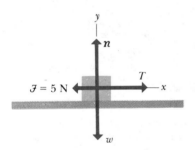

5–8 Force diagram for block in Example 5–9.

EXAMPLE 5–9 A 10.0-kg block of ice rests on a horizontal surface. What constant horizontal force T is required to give it a velocity of 4.0 m·s^{-1} in 2.0 s? Assume that the block starts from rest, that the friction force between the block and the surface is constant and equal to 5.0 N, and that all forces act at the center of the block (see Fig. 5–8).

SOLUTION The mass of the block is given. The *y*-component of its acceleration is zero. The *x*-component of acceleration can be found from the data on the velocity acquired in a given time. Since the forces are constant, the *x* acceleration is constant, and, from the equations of motion with constant acceleration,

$$a_x = \frac{v - v_0}{t} = \frac{4 \text{ m·s}^{-1} - 0}{2 \text{ s}} = 2 \text{ m·s}^{-2}.$$

The resultant of the *x*-forces is

$$\Sigma F_x = T - \mathcal{F}.$$

and that of the *y*-forces is

$$\Sigma F_y = n - w.$$

(Note that the *magnitude* of the friction force, $\mathcal{F} = 5.0$ N, is a positive quantity, but the *component* of this force in the *x*-direction is negative, equal to $-\mathcal{F}$ or -5.0 N. Similarly, the magnitude of the weight is the positive quantity *w*, but the *y*-component of this force is $-w$.) Hence, from Newton's second law in component form,

$$T - \mathcal{F} = ma_x, \qquad n - w = ma_y = 0.$$

From the second equation, we find that

$$n = w = mg = (10.0 \text{ kg})(9.80 \text{ m·s}^{-2}) = 98.0 \text{ N},$$

and from the first,

$$T = \mathcal{F} + ma_x = 5.0 \text{ N} + (10.0 \text{ kg})(2.0 \text{ m·s}^{-2}) = 25 \text{ N}.$$

EXAMPLE 5–10   An elevator and its load, shown in Fig. 5–9, have a total mass of 800 kg. Find the tension T in the supporting cable when the elevator, originally moving downward at 10 m·s^{-1}, is brought to rest with constant acceleration in a distance of 25 m.

SOLUTION   The weight of the elevator is

$$w = mg = (800 \text{ kg})(9.8 \text{ m·s}^{-2}) = 7840 \text{ N}.$$

From the equations of motion with constant acceleration,

$$v^2 = v_0{}^2 + 2a(y - y_0), \qquad a = \frac{v^2 - v_0{}^2}{2(y - y_0)}.$$

The initial velocity v_0 is -10 m·s^{-1}; the final velocity v is zero. If we take the origin at the point where the acceleration begins, then $y_0 = 0$ and $y = -25$ m. Hence

$$a = \frac{0 - (-10 \text{ m·s}^{-1})^2}{2(-25 \text{ m})} = 2 \text{ m·s}^{-2}.$$

The acceleration is therefore positive (upward). From the free-body diagram (Fig. 5–9) the resultant force is

$$\Sigma F = T - w = T - 7840 \text{ N}.$$

Since $\Sigma F = ma$,

$$T - 7840 \text{ N} = (800 \text{ kg})(2 \text{ m·s}^{-2}) = 1600 \text{ N},$$

$$T = 9440 \text{ N}.$$

The tension must be *greater* than the weight by 1600 N to cause the upward acceleration while the elevator is stopping.

EXAMPLE 5–11   In Example 5–10, with what force do the feet of a passenger press downward on the floor, if the passenger's mass is 80.0 kg?

SOLUTION   We first find the force the floor exerts *on* the passenger's feet, which is the reaction force to the force required. The forces on the passenger are his weight $(80.0 \text{ kg})(9.80 \text{ m·s}^{-2}) = 784$ N and the upward force F caused by the floor. Thus Newton's second law applied to the passenger, whose acceleration is the same as the elevator's, gives

$$F - 784 \text{ N} = (80.0 \text{ kg})(2.00 \text{ m·s}^{-2}) = 160 \text{ N},$$

and

$$F = 944 \text{ N}.$$

Thus the passenger pushes down on the floor with a force of 944 N while the elevator is stopping, and he *feels* a greater strain in his legs and feet than while standing at rest. If he stands on a bathroom scale calibrated in newtons, what will the scale read?

EXAMPLE 5–12   A child slides down a frictionless playground slide inclined at an angle θ with the horizontal (Fig. 5–10a). What is the child's acceleration?

Is the elevator cable strong enough?

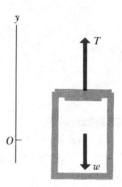

5–9 The resultant vertical force is $T - w$.

Does a heavy child slide faster than a light child?

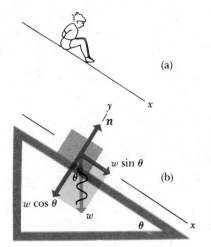

5-10 (a) A body on a frictionless inclined plane. (b) Free-body diagram.

SOLUTION The only forces acting on the child are the weight w and the normal force n exerted by the plane (Fig. 5–10b).

Take axes parallel and perpendicular to the surface of the plane and resolve the weight into x- and y-components. Then

$$\Sigma F_y = n - w \cos \theta,$$
$$\Sigma F_x = w \sin \theta.$$

But we know that $a_y = 0$, so from the equation $\Sigma F_y = ma_y$, we find that $n = w \cos \theta$. From the equation $\Sigma F_x = ma_x$, we have

$$w \sin \theta = ma_x,$$

and since $w = mg$,

$$a_x = g \sin \theta.$$

The mass does not appear in the final result, which means that any body, regardless of its mass, will slide down a frictionless inclined plane with an acceleration of $g \sin \theta$. In particular, when $\theta = 0$, $a_x = 0$, and when the plane is vertical, $\theta = 90°$ and $a_x = g$, as we should expect.

EXAMPLE 5–13 Consider the situation of Fig. 5–11 (which is the same figure as Fig. 4–9). Let the mass of the block be 4 kg and that of the rope be 0.5 kg. If the force F_1 is 9 N, what are the forces F_1', F_2, and F_2'? The surface on which the block moves is level and frictionless.

SOLUTION We know from Newton's third law that $F_1 = F_1'$ and that $F_2 = F_2'$. Hence $F_1' = 9$ N. The force F_2 could be computed by applying Newton's second law to the block, if its acceleration were known, or the force F_2' could be computed by applying this law to the rope if its acceleration were known. The acceleration is not given, but it can be found by considering the block and rope together as a single system. We know there is no vertical motion, so the vertical forces need not be considered. Since there is no friction, the resultant force acting *on* the system is the force F_1. (The forces F_2 and F_2' are not forces acting on the system but rather are *internal* forces exerted by one part of the system on another. The force F_1' does not act on the system, but *on the man*.) Then, from Newton's second law,

$$\Sigma F = ma,$$
$$9 \text{ N} = (4 \text{ kg} + 0.5 \text{ kg})a,$$
$$a = 2 \text{ m·s}^{-2}.$$

We can now apply Newton's second law to the block:

$$\Sigma F = ma,$$
$$F_2 = (4 \text{ kg})(2 \text{ m·s}^{-2}) = 8 \text{ N}.$$

If we consider the rope alone, the resultant force on it is

$$\Sigma F = F_1 - F_2' = 9 \text{ N} - F_2',$$

and from the second law,

$$9 \text{ N} - F_2' = (0.5 \text{ kg})(2 \text{ m·s}^{-2}) = 1 \text{ N},$$
$$F_2' = 8 \text{ N}.$$

We find that F_2 and F_2' are equal in magnitude. This result agrees with Newton's third law, which we used tacitly when we omitted the forces F_2 and F_2' in

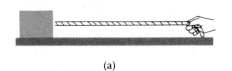

(a)

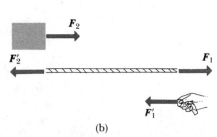

(b)

5-11 (a) A man pulls on a rope attached to a block. (b) Forces acting on the block, the rope, and the man's hand.

considering the system as a whole. Note, however, that the forces F_1 and F_2' are *not* equal and opposite (the rope is not in equilibrium) and that these forces are *not* an action–reaction pair.

EXAMPLE 5–14 In Fig. 5–12, a block of mass m_1 and weight $w_1 = m_1g$ moves on a level, frictionless surface. It is connected by a light, flexible cord passing over a small, frictionless pulley to a second hanging block of mass m_2 and weight $w_2 = m_2g$. What is the acceleration of the system, and what is the tension in the cord connecting the two blocks?

SOLUTION The diagram shows the forces acting on each block. Because there is no friction in the pulley and the cord is considered massless, the tension is the same throughout the string. Thus the forces applied to the blocks by the ends of the cord have equal magnitude; we have used the same symbol T for each. For the block on the surface,

$$\Sigma F_x = T = m_1 a_x,$$
$$\Sigma F_y = n - m_1g = m_1 a_y = 0.$$

Applying Newton's second law to the hanging block, we obtain

$$\Sigma F_y = m_2g - T = m_2 a_y.$$

Note that there is no necessity to use the same coordinate axes for the two bodies. It is convenient to take the $+y$ direction as downward for m_2, so both bodies move in positive axis directions.

Now assuming that the string does not stretch, the speeds of the two bodies at any instant are equal, and so their accelerations have the same magnitude. That is, a_x for m_1 equals a_y for m_2. We can thus denote this common magnitude simply as a. Then the two equations containing accelerations are

$$T = m_1 a.$$
$$m_2g - T = m_2 a.$$

These are two simultaneous equations for the unknowns T and a. Adding the two equations eliminates T, giving

$$m_2g = m_1 a + m_2 a = (m_1 + m_2)a$$

and

$$a = \frac{m_2}{m_1 + m_2} g.$$

Substituting this back into the first equation yields

$$T = \frac{m_1 m_2}{m_1 + m_2} g.$$

We see that the tension T is not equal to the weight m_2g of mass m_2, but is *less* by a factor of $m_1/(m_1 + m_2)$. Further thought shows that T *cannot* be equal to m_2g; if it were, m_2 would be in equilibrium, and it is not.

As an example of step 6 in the problem strategy, we can check some special cases. If $m_1 = 0$, we would expect that m_2 would fall freely and there would be no tension in the string. The equations do give $T = 0$ and $a = g$ for this case. Also, if $m_2 = 0$, we expect no tension and no acceleration; for this case the equations give $T = 0$ and $a = 0$. Thus in these two cases the general analytical results agree with intuitive expectations.

How helpful would a frictionless pulley be?

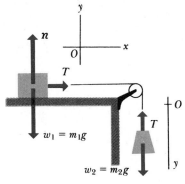

5–12 Force diagram for block on a frictionless horizontal surface and for the hanging block.

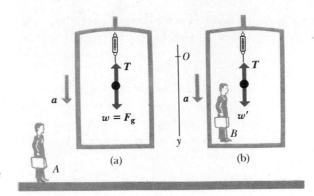

5–13 (a) To observer A, the body has a downward acceleration a, and he writes $w - T = ma$. (b) To observer B, the acceleration of the body is zero. He writes $w' = T$.

Can you weigh things in an accelerating elevator?

EXAMPLE 5–15 A body of mass m is suspended from a spring balance attached to the roof of an elevator, as shown in Fig. 5–13. What is the reading of the balance if the elevator has an acceleration a relative to the earth? Consider the earth's surface to be an inertial reference system.

SOLUTION The forces on the body are its weight w (the gravitational force F_g exerted on it by the earth) and the upward force T exerted on it by the balance. The reading on the scale of the balance is T. The body is at rest relative to the elevator and hence has an acceleration a relative to the earth. (We take the downward direction as positive.) The resultant force on the body is $w - T$, so, from Newton's second law,

$$w - T = ma, \qquad T = w - ma = m(g - a).$$

By Newton's third law, the body pulls down on the balance with a force equal and opposite to T, or equal to $(w - ma)$, so the balance reading T equals $(w - ma)$. The same analysis can be applied to a person standing on a bathroom scale in an accelerating elevator, and the result is the same; in that case T is the scale reading.

If the same body were suspended in equilibrium from a balance attached to the earth, the balance reading would equal the weight w. To an observer riding in the accelerating elevator, the body *appears* (in this noninertial frame of reference) to be acted on by a downward force w' equal in magnitude to the balance reading, as in Fig. 5–13b. This apparent force w' can be called the *apparent weight* of the body. Then

$$w' = w - ma = m(g - a). \tag{5–7}$$

In a noninertial frame of reference, apparent weight is different from true weight.

If the elevator is at rest, or moving vertically (either up or down) with constant velocity, $a = 0$ and the apparent weight equals the true weight. If the acceleration is *downward*, as in Fig. 5–13, so that a is positive, the apparent weight is less than the true weight: the body appears "lighter." If the acceleration is *upward*, a is negative, the apparent weight is greater than the true weight, and the body appears "heavier." These effects can be felt by taking a few steps in an elevator that is starting to move up or down after a stop. If the elevator falls freely, $a = g$, and since the true weight w also equals mg, the apparent weight w' is zero and the body appears "weightless." It is in this sense that an astronaut orbiting the earth in a space capsule is said to be "weightless."

The effect of this condition is exactly the same as though the body were in outer space with no gravitational force at all. The physiological effect of prolonged weightlessness is an interesting medical problem that is being actively explored. Gravity plays a role in blood distribution in the body, and one reaction to weightlessness is a decrease in volume of blood through increased excretion of water. In some cases, astronauts returning to earth have experienced temporary impairment of their sense of balance and a greater tendency toward motion sickness.

EXAMPLE 5–16 Figure 5–14a represents a simple *accelerometer*. A small body is fastened at one end of a light rod pivoted freely at point P. When the system has an acceleration a toward the right, the rod makes an angle θ with the vertical. In a practical instrument, some form of damping must be provided to keep the rod from swinging when the acceleration changes. For example, the rod might hang in a tank of oil. You can make a primitive version of this device by tying a thread to the ceiling light in an automobile, tying a key or nearly any other small object to the other end, and using a protractor. The problem: Given m and θ, what is a?

How to make an accelerometer

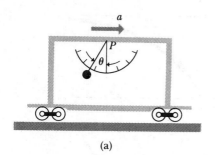

(a)

SOLUTION As shown in the free-body diagram, Fig. 5–14b, two forces act on the body: its weight $w = mg$ and the tension T in the rod. (We neglect the weight of the rod.) The resultant horizontal force is

$$\Sigma F_x = T \sin \theta,$$

and the resultant vertical force is

$$\Sigma F_y = T \cos \theta - mg.$$

The x-acceleration is the acceleration a of the system, and the y-acceleration is zero. Hence

$$T \sin \theta = ma, \qquad T \cos \theta = mg.$$

When the first equation is divided by the second, we get

$$a = g \tan \theta,$$

and the acceleration a is proportional to the tangent of the angle θ. When $\theta = 0$, the acceleration is zero; when $\theta = 45°$, $a = g$, and so on. We note that θ can never be $90°$ because that would require an infinite acceleration.

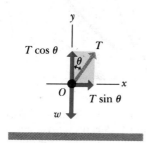

(b)

5–14 (a) A simple accelerometer. (b) The forces on the body are w and T.

EXAMPLE 5–17 Sometimes a frictional force opposing the motion of a body is not constant but increases with the body's speed. Two familiar examples are air resistance on a falling body and the viscous resisting force on a body falling through a liquid. Suppose we drop a penny into a deep pond. For small speeds the resisting force f of the liquid is approximately proportional to the penny's speed v:

$$f = kv, \tag{5–8}$$

Free fall with air resistance

where k is a proportionality constant that depends on the dimensions of the penny and the properties of the fluid. Taking the positive direction to be downward and neglecting any force associated with buoyancy in the fluid, we find that the net vertical component of force is $mg - kv$. Newton's second law gives

$$mg - kv = ma.$$

When the penny first starts to move, $v = 0$, the resisting force is zero, and the initial acceleration is $a = g$. As its speed increases, the resisting force also increases, until finally it equals the weight in magnitude. At this time $mg - kv = 0$, the acceleration becomes 0, and there is no further increase in speed. The final speed, also

Terminal speed

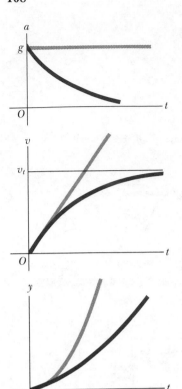

5–15 Graphs of acceleration, velocity, and position versus time for a body falling in a viscous fluid, shown as solid color curves. The light color curves show the corresponding relations if there is *no* viscous friction.

called the *terminal* speed, is given by this equation. Solving for v, we find

$$v_t = \frac{mg}{k}. \tag{5–9}$$

Figure 5–15 shows how the acceleration, velocity, and position vary with time.

To find the relation between speed and time during the interval before the terminal speed is reached, we go back to Newton's second law, which we rewrite as

$$m\frac{dv}{dt} = mg - kv.$$

After rearranging terms and replacing mg/k by v_t, we find

$$\frac{dv}{v - v_t} = -\frac{k}{m}\,dt.$$

We now integrate both sides, noting that $v = 0$ when $t = 0$:

$$\int_0^v \frac{dv}{v - v_t} = -\frac{k}{m}\int_0^t dt,$$

which integrates to

$$\ln\frac{v_t - v}{v_t} = -\frac{k}{m}\,t, \qquad \text{or} \qquad 1 - \frac{v}{v_t} = e^{-(k/m)t},$$

and finally

$$v = v_t(1 - e^{-(k/m)t}). \tag{5–10}$$

From this result we can derive expressions for the acceleration and position as functions of time. The derivations are left as an exercise. The results are

$$a = ge^{-(k/m)t},$$

$$y = v_t[t - \frac{m}{k}(1 - e^{-(k/m)t})]. \tag{5–11}$$

Now look again at Fig. 5–15, which shows graphs of these three relations.

In high-speed motion through air, the resisting force is approximately proportional to v^2 rather than to v; it is then called *air drag* or simply *drag*. Falling raindrops, airplanes, and cars moving at high speed all experience air drag. In this case Eq. (5–8) is replaced by $f = Kv^2$. We challenge the reader to show that the terminal speed is then given by

$$v_t = \sqrt{\frac{mg}{K}}, \tag{5–12}$$

and that the units of the constant K are $\text{N}\cdot\text{s}^2\cdot\text{m}^{-2}$ or $\text{kg}\cdot\text{m}^{-1}$.

The physics of sky-diving

EXAMPLE 5–18 For a human body falling through air, the numerical value of the constant K in Eq. (5–12) is found to be about $0.25\ \text{kg}\cdot\text{m}^{-1}$. If an 80-kg sky diver jumps out of an airplane and reaches terminal velocity before he opens his parachute, the terminal velocity is, from Eq. (5–12),

$$v_t = \sqrt{\frac{(80\ \text{kg})(9.8\ \text{m·s}^{-2})}{0.25\ \text{kg·m}^{-1}}}$$
$$= 56\ \text{m·s}^{-1}.$$

SUMMARY

Four classes of forces are found in nature. Two—the gravitational and electromagnetic interactions—are familiar in everyday life; the other two—the strong interaction and the weak interaction—are seen directly only in experiments with fundamental particle interactions.

The contact force between two bodies can always be represented in terms of a normal component n perpendicular to the interaction surface and a frictional component $\mathcal{F}$ parallel to the surface. When sliding occurs between the surfaces, the friction force $\mathcal{F}_k$ is approximately proportional to the normal force, and the proportionality constant is μ_k, the coefficient of kinetic friction:

$$\mathcal{F}_k = \mu_k n. \tag{5–1}$$

When there is no relative motion, the maximum possible friction force is approximately proportional to the normal force, and the proportionality constant is μ_s, the coefficient of static friction:

$$\mathcal{F}_s \le \mu_s n. \tag{5–2}$$

The actual static friction force may be anything from zero to this maximum value, and it is determined by the details of the specific problem at hand. Usually μ_k is less than μ_s for a given pair of surfaces.

When a particle is in equilibrium in an inertial frame of reference, the vector sum of all forces acting on it must be zero:

$$\sum \mathbf{F} = \mathbf{0}. \tag{5–3}$$

The vector sum is often calculated most easily by using components:

$$\sum F_x = 0,$$
$$\sum F_y = 0. \tag{5–4}$$

Free-body diagrams are useful in identifying the forces acting on the body being considered. Newton's third law is also frequently needed in equilibrium problems. The two forces in an action–reaction pair never act on the same body.

When the vector sum of forces on a body is not zero, the body has an acceleration determined by Newton's second law:

$$\sum \mathbf{F} = m\mathbf{a}. \tag{5–5}$$

In problems this is usually used in component form:

$$\sum F_x = ma_x,$$
$$\sum F_y = ma_y. \tag{5–6}$$

If you apply the problem-solving strategies outlined in this chapter, you will be rewarded with an improved percentage of correct problem solutions!

KEY TERMS

gravitational interactions
electromagnetic interactions
strong interaction
weak interactions
contact forces
normal force
friction force
coefficient of kinetic friction
static friction force
coefficient of static friction
equilibrium
free-body diagram

QUESTIONS

5–1 Can a body be in equilibrium when only one force acts on it?

5–2 A helium balloon hovers in midair, neither ascending nor descending. Is it in equilibrium? What forces act on it?

5–3 If the two ends of a rope in equilibrium are pulled with forces of equal magnitude and opposite direction, why is the total tension in the rope not zero?

5–4 A horse is hitched to a wagon. Since the wagon pulls back on the horse just as hard as the horse pulls on the wagon, why doesn't the wagon remain in equilibrium, no matter how hard the horse pulls?

5–5 A clothesline is hung between two poles, and then a shirt is hung near the center. No matter how tightly the line is stretched, it always sags a little at the center. Explain why.

5–6 A man sits in a chair that is suspended from a rope. The rope passes over a pulley suspended from the ceiling, and the man holds the other end of the rope in his hands. What is the tension in the rope, and what force does the chair exert on the man?

5–7 How can pushing *down* on a bicycle pedal make the bicycle move *forward*?

5–8 A car is driven up a steep hill at constant speed. Discuss all the forces acting on the car; in particular, what pushes it up the hill?

5–9 Can the coefficient of friction ever be greater than unity? If so, give an example; if not, explain why not.

5–10 A block rests on an inclined plane with enough friction to prevent sliding down. To start the block moving, is it easier to push it up the plane, down the plane, or sideways? Why?

5–11 In pushing a box up a ramp, is it better to push horizontally or to push parallel to the ramp?

5–12 In stopping a car on an icy road, it is better to push the brake pedal hard enough to "lock" the wheels and make them slide, or to push gently so the wheels continue to roll? Why?

5–13 When one stands with bare feet in a wet bathtub, the grip feels fairly secure, and yet a catastrophic slip is quite possible. Discuss this situation with respect to the two coefficients of friction.

5–14 The horrible squeak made by a piece of chalk held at the wrong angle against a blackboard results from alternate sticking and slipping of the chalk against the blackboard. Interpret this phenomenon in terms of the two coefficients of friction. Can you think of other examples of "slip-stick" behavior?

EXERCISES

Section 5–2 Contact Forces and Friction

5–1

a) A large rock rests on a rough horizontal surface. A bulldozer pushes on the rock with a horizontal force T that is slowly increased, started from zero. Draw a graph with T along the x-axis and the friction force $\mathcal{F}$ along the y-axis, starting at $T = 0$ and showing the region of no motion, the point where motion impends, and the region where motion exists.

b) A block of weight w rests on a rough horizontal plank. The slope angle of the plank θ is gradually increased until the block starts to slip. Draw two graphs, both with θ along the x-axis. In one graph show the ratio of the normal force to the weight, n/w as a function of θ. In the second graph show the ratio of the friction force to the weight, $\mathcal{F}/w$. Indicate the region of no motion, the point where motion impends, and the region where motion exists.

5–2 A box of bananas weighing 20 N rests on a horizontal surface. The coefficient of static friction between the block and the surface is 0.40, and the coefficient of sliding friction is 0.20.

a) How large is the friction force exerted on the box?

b) How great will the friction force be if a horizontal force of 5 N is exerted on the box?

c) What is the minimum force that will start the box in motion?

d) What is the minimum force that will keep the box in motion once it has been started?

e) If the horizontal force is 10 N, how great is the friction force?

5–3 A box of mass 0.6 slug is moving with a constant velocity of 15 ft·s^{-1} on a horizontal surface. The coefficient of sliding friction between the box and the surface is 0.20.

a) What horizontal force is required to maintain the motion?

b) If the force is removed, how soon will the box come to rest?

5–4 A hockey puck leaves a player's stick with a velocity of 10 m·s^{-1} and slides 40 m before coming to rest. Find the coefficient of friction between the puck and the ice.

5–5

a) If the coefficient of friction between tires and dry pavement is 0.8, what is the shortest distance in which an automobile can be stopped by locking the brakes when traveling at 27 m·s^{-1} (about 60 mi·hr^{-1})?

b) On wet pavement the coefficient of friction may be only 0.25. How fast should you drive on wet pavement in order to be able to stop in the same distance as in part (a)?

(*Note.* Locking the brakes is *not* the safest way to stop.)

5–6 Find the ratio of the stopping distance of an automobile on wet concrete with the wheels locked ($\mu_k = 0.25$) to the stopping distance of the same automobile stopped by tractive friction only (where the tractive resistance is 0.015). Assume the same initial velocity for both cases.

5–7 Air pressure greatly affects the tractive resistance of bicycle tires. To study this effect, two tires are set rolling with the same initial speed of 5 m·s^{-1} along a long, straight road, and the distance each travels before its speed is reduced by half is measured. One tire is inflated to a pressure of 40 psi and the other is at 105 psi. The low-pressure tire goes 45.6 m, and the high-pressure tire goes 213 m. What is the tractive resistance for each? Assume the net horizontal force is due to tractive friction only.

Section 5–3 Equilibrium of a Particle

5–8 In each of the situations in Fig. 5–16 the objects suspended from the rope have weight w. The pulleys are frictionless. Calculate in each case the tension T in the rope in terms of the weight w.

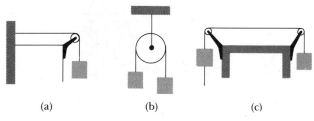

(a)   (b)   (c)

FIGURE 5–16

5–9 Two 10-N weights are suspended at opposite ends of a rope that passes over a light, frictionless pulley. The pulley is attached to a chain that goes to the ceiling.

a) What is the tension in the rope?

b) What is the tension in the chain?

5–10 A picture frame hung against a wall is suspended by two wires attached to its upper corners. If the two wires make the same angle with the vertical, what must this angle be for the tension in each wire to equal the weight of the frame?

5–11 Figure 5–17 illustrates a mountaineering technique called a Tyrolean traverse. A rope is stretched tightly

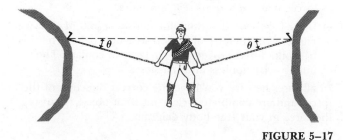

FIGURE 5–17

between two points, and the climber slides across the rope. The climber's weight is 800 N, and the breaking strength of the rope (typically nylon, 11 mm diameter) is 20,000 N.

a) If the angle θ is 15°, find the tension in the rope.

b) What is the smallest value the angle θ can have if the rope is not to break?

5–12 Find the tension in each cord in Fig. 5–18 if the weight of the suspended object is w.

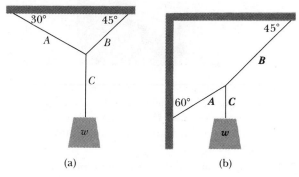

(a)   (b)

FIGURE 5–18

5–13 In Fig. 5–19 the tension in the diagonal string is 20 N.

a) Find the magnitudes of the horizontal forces F_1 and F_2 that must be applied to hold the system in the position shown.

b) What is the weight of the suspended block?

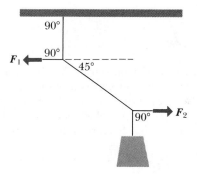

FIGURE 5–19

5–14 A tether ball leans against the post it is attached to, as shown in Fig. 5–20. If the string the ball is attached to is 1.8 m long, the ball has a radius of 0.20 m, and the ball has a mass of 0.20 kg, what are the tension in the rope and the force the pole exerts on the ball? Assume there is so little friction between the ball and the pole that it can be neglected.

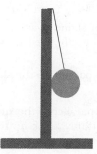

FIGURE 5–20

5–15 Two blocks, each of weight w, are held in place on a frictionless incline as shown in Fig. 5–21. In terms of w and the angle θ of the incline, calculate the tension in

a) the rope connecting the blocks;

b) the rope that connects block A to the wall.

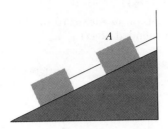

FIGURE 5–21

5–16 A man is pushing a piano weighing 315 lb at constant velocity up a ramp that is inclined at 37° above the horizontal. Neglect friction. If the force applied by the man is parallel to the incline, calculate the magnitude of this force.

5–17 Two crates connected by a rope lie on a horizontal surface as shown in Fig. 5–22. Crate A has weight w_A, and crate B weight w_B. The coefficient of kinetic friction between the crates and the surface is μ_k. The crates are pulled to the right at constant velocity by a horizontal force P. In terms of w_A, w_B, and μ_k calculate

a) the magnitude of the force P;

b) the tension in the rope connecting the blocks.

FIGURE 5–22

5–18 Consider the system shown in Fig. 5–23. The coefficient of kinetic friction between block A, of weight w_A, and the table top is μ_k. Calculate the weight w_B of the hanging block required if this block is to descend at constant speed once it has been set into motion.

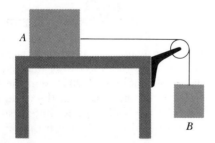

FIGURE 5–23

5–19 A safe weighing 2000 N is to be lowered at constant speed down skids 4 m long from a truck 2 m high.

a) If the coefficient of sliding friction between safe and skids is 0.30, will the safe need to be pulled down or held back?

b) How great a force parallel to the skids is needed?

5–20

a) If a force of 86 N parallel to the surface of a 20° inclined plane will push a 120-N block up the plane at constant

speed, what force parallel to the plane will push it down at constant speed?

b) What is the coefficient of sliding friction?

5–21 A large crate of weight w rests on a horizontal floor. The coefficients of friction between the crate and the floor are μ_s and μ_k. A man pushes on the crate with a force P that is directed at an angle θ below the horizontal.

a) What magnitude of force P is required to keep the crate moving at constant velocity?

b) If μ_s is larger than some critical value, the man cannot start the crate moving no matter how hard he pushes. Calculate this critical value of μ_s.

5–22 Two blocks A and B, are placed as in Fig. 5–24 and connected by ropes to block C. Both A and B weigh 20 N each, and the coefficient of sliding friction between each block and the surface is 0.5. Block C descends with constant velocity.

a) Draw two separate force diagrams showing the forces acting on A and on B.

b) Find the tension in the rope connecting blocks A and B.

c) What is the weight of block C?

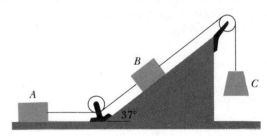

FIGURE 5–24

Section 5–4 Applications of Newton's Second Law

5–23 The first two steps in the solution of Newton's-second-law problems are to select an object for analysis and then to draw free-body diagrams for that object. Do exactly that for the following situations:

a) a mass M sliding down a frictionless inclined plane of angle θ;

b) a mass M sliding up a frictionless inclined plane of angle θ;

c) a mass M sliding up an inclined plane of angle θ with kinetic friction present;

d) masses M and m sliding down an inclined plane of angle θ with friction present, as shown in Fig. 5–25a. Here draw free-body diagrams for both m and M. Identify the forces that are action–reaction pairs.

e) Draw free-body diagrams for masses m and M shown in Fig. 5–25b. Identify all action–reaction pairs. There is a frictional force between all surfaces in contact. The pulley is frictionless and massless.

In all cases be sure you have the correct direction of the forces and are completely clear on what object is causing the force in your free-body diagram.

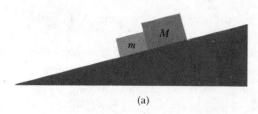

(a)

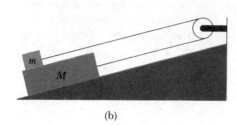

(b)   FIGURE 5–25

5–24 An elevator with mass 2000 kg rises with an acceleration of 1 m·s^{-2}. What is the tension in the supporting cable?

5–25 A 4-kg bucket of water is accelerated upward by a cord whose breaking strength is 60 N. Find the maximum upward acceleration that can be given to the bucket without breaking the cord.

5–26 A large fish hangs from a spring balance supported from the roof of an elevator.

a) If the elevator has an upward acceleration of 2.45 m·s^{-2} and the balance reads 50 N, what is the true weight of the fish?

b) Under what circumstances will the balance read 30 N?

c) What will the balance read if the elevator cable breaks?

5–27 A train (an engine plus four cars) is accelerating at 1.5 ft·s^{-2}. If each car has a mass of 2700 slugs and if each car has negligible frictional forces acting on it, what is

a) the force of the engine on the first car?

b) the force of the first car on the second car?

c) the force of the second car on the third car?

d) the force of the fourth car on the third car?

e) What would be these same four forces if the train were slowing down with an acceleration of -1.5 ft·s^{-2}?

When solving, note the importance of selecting the correct set of cars as your object.

5–28 A transport plane is to take off from a level landing field with two gliders in tow, one behind the other. Each glider has a mass of 1200 kg, and the friction force or drag on each may be assumed constant and equal to 2000 N. The tension in the towrope between the transport plane and the first glider is not to exceed 10,000 N.

a) If a velocity of 40 m·s^{-1} is required for takeoff, what minimum length of runway is needed?

b) What is the tension in the towrope between the two gliders while they are accelerating for the takeoff?

5–29 A 5-kg block slides down a plane inclined at 30° to the horizontal. Find the acceleration of the block

a) if the plane is frictionless;

b) if the coefficient of kinetic friction is 0.2.

5–30 A physics student playing with an air hockey table (just a frictionless surface) finds that if she gives the puck a velocity of 15.0 ft·s^{-1} along the length (5.5 ft) of the table at one end, by the time it has reached the other end the puck has drifted 1.1 in. to the right but still has a velocity along the length of 15.0 ft·s^{-1}. She concludes correctly that the table is not level, and correctly calculates its inclination from the above information. What does she get for the angle of inclination?

5–31 A 600-N man stands on a bathroom scale in an elevator. As the elevator starts moving, the scale reads 800 N.

a) Find the acceleration of the elevator (magnitude and direction).

b) What is the acceleration if the scale reads 450 N?

c) If the scale reads zero, should the man worry? Explain.

5–32 A packing crate rests on an inclined plane that makes an angle θ with the horizontal. The coefficient of sliding friction is 0.50, and the coefficient of static friction is 0.75.

a) As the angle θ is increased, find the minimum angle at which the crate starts to slip.

b) At this angle, find the acceleration (in m·s^{-2}) once the crate has begun to move.

c) How much time is required for the crate to slip 8 m along the inclined plane?

5–33 A block of mass 6 kg resting on a horizontal surface is connected by a cord passing over a light, frictionless pulley to a hanging block of mass 4 kg. The coefficient of kinetic friction between the block and the horizontal surface is 0.5. After the blocks are released find

a) the acceleration of each block;

b) the tension in the cord.

5–34 A box of mass m is dragged across a rough, level floor having a coefficient of kinetic friction μ_k by a rope that is pulled upward at an angle θ to the horizontal with a force of magnitude F. In terms of m, μ_k, θ, F, and g, obtain expressions for

a) the normal force;

b) the frictional force;

c) the acceleration of the box;

d) the magnitude of force required to move the box with constant speed.

5–35 Two 0.2-kg blocks hang at the ends of a light, flexible cord passing over a small, frictionless pulley, as in Fig. 5–26. A 0.1-kg block is placed on the block on the right, and then removed after 2 s.

a) How far will each block move in the first second after the 0.1-kg block is removed?

b) What was the tension in the cord before the 0.1-kg block was removed? After it was removed?

c) What was the tension in the cord supporting the pulley before the 0.1-kg block was removed? Neglect the weight of the pulley.

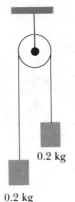

0.2 kg

0.2 kg   **FIGURE 5–26**

5–36 Which way will the accelerometer in Fig. 5–14 deflect under the following conditions?

a) The cart is moving toward the right and traveling faster.

b) The cart is moving toward the right and traveling slower.

c) The cart is moving toward the left and traveling faster.

d) The cart is moving toward the left and traveling slower.

e) The cart is at rest on a sloping surface.

f) The cart is given an upward velocity on a frictionless inclined plane. It first moves up, then stops, and then moves down. What is the deflection in each stage of the motion?

PROBLEMS

5–37 In Fig. 5–27, a man lifts a weight w by pulling down on a rope with a force F. The upper pulley is attached to the ceiling by a chain, and the lower pulley is attached to the weight by another chain. In terms of w, find the tension in each chain and the force F if the weight is lifted at constant speed. Assume the weights of the rope, pulleys, and chains to be negligible.

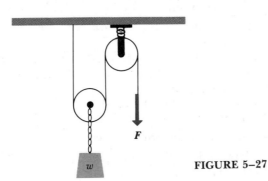

F

w **FIGURE 5–27**

5–38 Figure 5–28 shows a technique called rappelling, used by mountaineers for descending vertical rock faces. The

15°

FIGURE 5–28

climber sits in a rope seat, and the rappel rope passes through a friction device attached to this seat. Suppose the rock is perfectly smooth (no friction), and that the climber's feet push horizontally onto the rock. If the climber's weight is 800 N, find the tension in the rope and the force his feet exert on the rock face.

5–39 A refrigerator of weight w is being pushed up an incline at constant velocity by a man who applies a force $\textbf{\textit{P}}$. The surface of the incline is at an angle θ above the horizontal. *Neglect friction.* If the force $\textbf{\textit{P}}$ is *horizontal*, calculate its magnitude in terms of w and θ.

5–40 Most problems in this book do not give the mass of ropes, cords, or cables, and therefore we do not consider their mass or weight. In some problems the cords are said to be massless. It is understood that the ropes, cords, and cables, compared to other objects in the problem, have so little mass that their mass can be neglected. But if the rope is the *only* object in the problem, then clearly its mass cannot be neglected. For example, suppose we have a clothesline attached to two poles as in Fig. 5–29. The clothesline has a mass M, and each end makes an angle θ with the horizontal. What is

a) the tension at the ends of the clothesline?

b) the tension at the lowest point?

The curve of the clothesline, or of any flexible cable hanging under its own weight, is called a catenary curve. For a more advanced treatment of this curve, see J. L. Synge and B. A. Griffith, *Principles of Mechanics* (New York: McGraw-Hill, 1959), pp. 94–97.

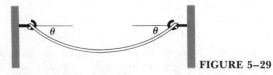

θ θ

FIGURE 5–29

5–41 If the coefficient of static friction between a table and a rope is μ_s, what fraction of the rope can hang over the edge of a table without the rope sliding?

5–42 A dictionary is pulled to the right at constant velocity by a 10-N force acting 30° above the horizontal. The coefficient of sliding friction between the dictionary and the surface is 0.5. What is the weight of the dictionary?

5–43

a) Block A in Fig. 5–30 weighs 100 N. The coefficient of static friction between the block and the surface on which it rests is 0.30. The weight w is 20 N and the system is in equilibrium. Find the friction force exerted on block A.

b) Find the maximum weight w for which the system will remain in equilibrium.

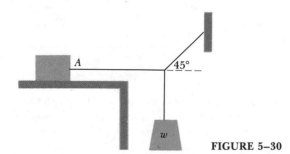

FIGURE 5–30

5–44 A block weighing 100 N is placed on an inclined plane of slope angle 30° and is connected to a second hanging block of weight w by a cord passing over a small, frictionless pulley, as in Fig. 5–31. The coefficient of static friction is 0.40, and the coefficient of sliding friction is 0.30.

a) Find the weight w for which the 100-N block moves up the plane at constant speed, once it has been set in motion.

b) Find the weight w for which it moves down the plane at constant speed.

c) For what range of values of w will the block remain at rest, if it is released from rest?

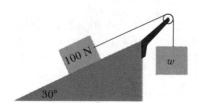

FIGURE 5–31

5–45 A window washer pushes his scrub brush up a vertical window by applying a force P as shown in Fig. 5–32. The brush weighs 6 N, and the coefficient of kinetic friction is $\mu_k = 0.4$. Calculate

a) the magnitude of the force P;

b) the normal force exerted by the window on the brush.

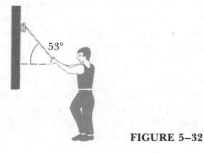

FIGURE 5–32

5–46 Block A in Fig. 5–33 weighs 4 N and block B weighs 8 N. The coefficient of sliding friction between all surfaces is 0.25. Find the force P necessary to drag block B to the left at constant speed

a) if A rests on B and moves with it;

b) if A is held at rest.

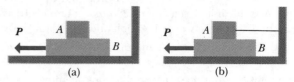

(a)   (b)

FIGURE 5–33

5–47 Block A, of weight w, slides down an inclined plane S of slope angle 37° at a constant velocity while plank B, also of weight w, rests on top of A. The plank is attached by a cord to the top of the plane (Fig. 5–34).

a) Draw a diagram of all the forces acting on block A.

b) If the coefficient of friction is the same between surfaces A and B and between S and A, determine its value.

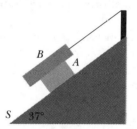

FIGURE 5–34

5–48 A man attempts to push a box of books of weight w up a ramp inclined at an angle θ above the horizontal. The coefficients of friction between the ramp and the box are μ_s and μ_k. The force P applied by the man is *horizontal*.

a) If μ_s is greater than some critical value, the man cannot start the box moving up the ramp no matter how hard he pushes. Calculate this critical value of μ_s.

b) Assume that μ_s is less than this critical value. What magnitude of force must the man apply to keep the box moving up the ramp at constant velocity?

5–49 A 40-kg packing case is initially at rest on the floor of a truck. The coefficient of static friction between the case and the truck floor is 0.30, and the coefficient of sliding friction is 0.20. Find the magnitude and direction of the friction force acting on the case

a) when the truck accelerates at 2.5 m·s^{-2};

b) when it accelerates at -3.0 m·s^{-2}.

5–50 A 20-kg box rests on the flat floor of a truck. The coefficients of friction between box and floor are $\mu_s = 0.15$ and $\mu_k = 0.10$. The truck stops at a stop sign and then starts to move, with an acceleration of 2.0 m·s^{-2}. If the box is 5.0 m from the rear of the truck when it starts, how much time elapses before it falls off the rear of the truck? How far does the truck travel in this time?

5–51 A truck traveling down a 10° slope is slowing down with an acceleration of magnitude 2.0 m·s^{-2}. A box weighing 900 N is resting on the back of the truck.

a) Will the box slide forward on the truck bed if the coefficient of static friction between the box and truck bed is 0.30?

b) If your answer in (a) is yes, what will be the acceleration of the box relative to the truck bed if the coefficient of kinetic friction is 0.22?

5–52 A book whose mass is 2 kg is projected up a long 30° incline with an initial velocity of 22 m·s^{-1}. The coefficient of kinetic friction between the book and the plane is 0.3.

a) Find the friction force acting on the book as it moves up the plane.

b) How much time does it take the book to move up the plane?

c) How *far* does the book move up the plane?

d) How much time does it take the book to slide from its position in part (c) back to its starting point?

e) With what velocity does it arrive at this point?

f) If the mass of the book had been 5 kg instead of 2 kg, would the answers in the preceding parts be different?

5–53 Block A in Fig. 5–35 has a mass of 2 kg and block B 20 kg. The coefficient of kinetic friction between B and the horizontal surface is 0.1.

a) What is the mass of block C if the acceleration of B is 2 m·s^{-2} toward the right?

b) What is the tension in each cord when B has the acceleration stated above?

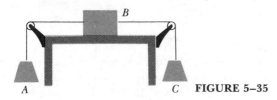

FIGURE 5–35

5–54 Two blocks connected by a cord passing over a small, frictionless pulley rest on frictionless planes, as shown in Fig. 5–36.

a) Which way will the system move?

b) What is the acceleration of the blocks?

c) What is the tension in the cord?

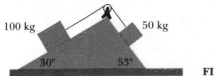

FIGURE 5–36

5–55 In terms of m_1, m_2, and g, find the accelerations of both blocks in Fig. 5–37. There is no friction anywhere in the system.

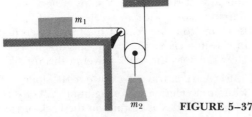

FIGURE 5–37

5–56 The masses of objects A and B in Fig. 5–38 are 20 kg and 10 kg, respectively. They are initially at rest on the floor and are connected by a weightless string passing over a weightless and frictionless pulley. An upward force F is applied to the pulley. Find the accelerations a_1 of object A and a_2 of object B when F is

a) 98 N,   b) 196 N,   c) 394 N,   d) 788 N.

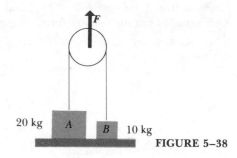

FIGURE 5–38

5–57 The two blocks in Fig. 5–39 are connected by a heavy, uniform rope of mass 4 kg. An upward force of 200 N is applied as shown.

a) What is the acceleration of the system?

b) What is the tension at the top of the heavy rope?

c) What is the tension at the midpoint of the rope?

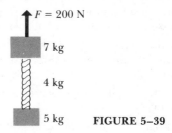

FIGURE 5–39

5–58 Two blocks with masses of 4 kg and 8 kg, respectively, are connected by a string and slide down a 30° inclined plane, as in Fig. 5–40. The coefficient of sliding friction between the 4-kg block and the plane is 0.25; between the 8-kg block and the plane it is 0.50.

a) Calculate the acceleration of each block.

b) Calculate the tension in the string.

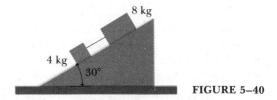

FIGURE 5–40

5–59 Two objects with masses of 5 kg and 2 kg, respectively, hang 1 m above the floor from the ends of a cord 3 m long passing over a frictionless pulley. Both objects start from rest. Find the maximum height reached by the 2-kg object.

5–60 A man of mass 80 kg stands on a platform of mass 40 kg. He pulls on a rope that is fastened to the platform and runs over a pulley on the ceiling. With what force does he have to pull in order to give himself and the platform an upward acceleration of 1 m·s^{-2}?

5–61 What acceleration must the cart in Fig. 5–41 have in order that the block A will not fall? The coefficient of static friction between the block and the cart is μ_s. How would the behavior of the block be described by an observer on the cart?

FIGURE 5–41

5–62 A 20-kg monkey has a firm hold on a light rope that passes over a frictionless pulley and is attached to a 20-kg bunch of bananas, as shown in Fig. 5–42. The monkey looks upward, sees the bananas, and starts to climb the rope to get at the bananas.

a) As he climbs, do the bananas move up, down, or remain at rest?

b) As he climbs, does the distance between him and the bananas decrease, increase, or remain constant?

c) The monkey releases his hold on the rope. What about the distance between him and the bananas while he is falling?

d) Before he reaches the ground, he grabs the rope to stop his fall. What do the bananas do?

FIGURE 5–42

CHALLENGE PROBLEMS

5–63 A box of weight w is pulled at constant speed along a level floor by a force P that is at an angle ϕ above the horizontal. The coefficient of kinetic friction between the floor and box is μ_k.

a) In terms of ϕ, μ_k, and w, calculate P.

b) For $w = 400$ N and $\mu_k = 0.4$, calculate P for ϕ ranging from 0° to 90° in increments of 10°. Sketch a graph of P versus ϕ.

c) From the general expression in (a), calculate the value of ϕ for which the P required to maintain constant speed is a minimum. For the special case of $w = 400$ N and $\mu_k = 0.4$, evaluate this optimal ϕ and compare your result to the graph you constructed in part (b).

(*Note*. At the value of ϕ where the function $P(\phi)$ has a minimum, $dP(\phi)/d\phi = 0$.)

5–64 A rock of mass $m = 5$ kg falls from rest in a viscous medium. The rock is acted on by a net constant downward force of 20 N and by a viscous retarding force proportional to its speed and equal to $(5 \text{ N·s·m}^{-1})v$, where v is the speed in meters per second. (See Example 5–17.)

a) Find the initial acceleration, a_0.

b) Find the acceleration when the speed is 3 m·s^{-1}.

c) Find the speed when the acceleration equals $0.1a_0$.

d) Find the terminal velocity, v_t.

e) Find the coordinate, velocity, and acceleration 2 s after the start of the motion.

f) Find the time required to reach a speed $0.9v_t$.

g) Construct a graph of v versus t, for the first 3 s of the motion.

5–65 The assumption that the coefficient of kinetic friction is independent of speed when an object is sliding is true only over a suitably small range of speeds. For example, in some cases for a speed of 130 km·hr^{-1} (about 80 mi·hr^{-1}), μ_k may be only $\frac{1}{2}$ what it is for speeds of 8 to 16 km·hr^{-1}. For a car sliding with its brakes locked, take $\mu_k - 1.0$ at 8 km·hr^{-1} and $\mu_k = 0.5$ at 130 km·hr^{-1}, and assume that μ_k varies linearly with speed. Then how far would a car slide after the brakes are applied at 130 km·hr^{-1} before coming to rest? (For a still more realistic characterization of the friction forces involved in sliding, see J. D. Edwards, Jr., *Amer. Jour. Phys.*, Vol. 48 (1980), pp. 253–254.)

5–66 A marble falls from rest through a medium that exerts a resisting force that varies directly with the square of the velocity ($R = -Kv^2$).

a) Draw a diagram showing the direction of motion and indicate with the aid of vectors all of the forces acting on the marble.

b) Apply Newton's second law and infer from the resulting equation the general properties of the motion.

c) Show that the marble acquires a terminal velocity that is as given in Eq. (5–8).

d) Derive the equation for the velocity at any time. $\Big($*Note*.

$$\int \frac{dx}{a^2 - x^2} = \frac{1}{\arctanh(x/a)},$$

where the hyperbolic tangent is defined by

$$\tanh(x) = \frac{e^x - e^{-x}}{e^x + e^{-x}} = \frac{e^{2x} - 1}{e^{2x} + 1}.\Big)$$

5–67 A particle of mass m, originally at rest, is subjected to a force whose direction is constant but whose magnitude varies with time according to the relation

$$F = F_0\left[1 - \left(\frac{t - T}{T}\right)^2\right],$$

where F_0 and T are constants. The force acts only for the time interval $2T$.

a) Make a rough graph of F versus t.

b) Prove that the speed v of the particle after time $2T$ has elapsed is equal to $4F_0T/3m$.

c) Choose numbers for v, T, and m that might be appropriate to a batted baseball, and calculate the force F_0. Judge whether the answer is sensible.

5–68

a) A wedge of mass M rests on a frictionless horizontal table top. A block of mass m is placed on the wedge, as shown in Fig. 5–43a. There is no friction between the block and the wedge. The system is released from rest.

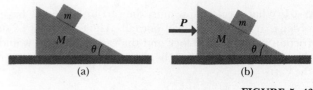

FIGURE 5–43

Calculate

 i) the acceleration of the wedge;

 ii) the horizontal and vertical components of the acceleration of the block.

Do your answers reduce to the correct results when M is very large?

b) The wedge and block are as in part (a). Now a horizontal force P is applied to the wedge, as in Fig. 5–43b. What magnitude must P have if the block is to remain at constant height above the table top?

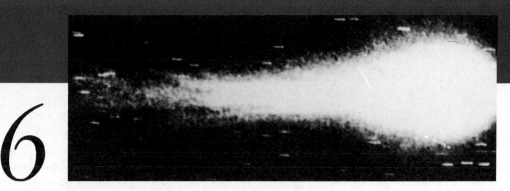

6

APPLICATIONS OF NEWTON'S LAWS—II

IN CHAPTER 5 WE STUDIED SEVERAL SIMPLE DYNAMICS PROBLEMS SHOWING applications of Newton's laws. In this chapter we continue our study of dynamics, concentrating especially on *circular motion*. Examples of motion of a particle in a circular path turn up in a wide variety of situations, such as racetracks, children's swings, and the motions of planets and satellites; thus it is worthwhile to study this special class of motions in detail. We also return to *gravitational* interactions for a detailed study of Newton's law of gravitation; this law plays a central role in planet and satellite motion and also provides a fuller understanding of the concept of *weight*. We suggest you review the discussion of the kinematics of circular motion in Section 3–5 to help you understand what is to come in this chapter.

6–1 FORCE IN CIRCULAR MOTION

When a particle moves in a circular path of radius R with constant speed v, the motion is called **uniform circular motion.** In Section 3–5 we showed that in this motion the acceleration of the particle is always directed toward the center of the circle, perpendicular to the instantaneous velocity. This is called **centripetal acceleration.** Its magnitude, as derived in Section 3–5, is

Uniform circular motion is accelerated motion.

$$a_\perp = \frac{v^2}{R}. \qquad (6–1)$$

The symbol $\perp$ is a reminder that at each point the acceleration is perpendicular to the instantaneous velocity. The **period** τ is the time for one revolution. It is given by

$$\tau = \frac{2\pi R}{v}. \qquad (6–2)$$

In terms of the period, the centripetal acceleration can also be expressed as

$$a_\perp = \frac{4\pi^2 R}{\tau^2}. \qquad (6–3)$$

Circular motion, like all other motion of a particle, is governed by Newton's second law. The particle's acceleration toward the center of the circle must be caused by a *force* also directed radially toward the center. The magnitude of the radial acceleration is given by v^2/R, so the magnitude of the net radial force F on a particle of mass m must be

$$F = ma_\perp = m\frac{v^2}{R}. \tag{6–4}$$

A familiar example of such a radial force occurs when you tie an object to a string and whirl it in a circle. The string must constantly pull in toward the center; if it breaks, then the inward force no longer acts, and the object flies off along a tangent to the circle.

What force causes centripetal acceleration?

More generally, several forces may be acting on a body in uniform circular motion. In this case the *vector sum* of all forces on the body must have the magnitude given by Eq. (6–4) and must be directed toward the center of the circle.

The force in Eq. (6–4) is sometimes called *centripetal force*. This usage is unfortunate because it implies that this force is somehow different from ordinary forces, or that the fact of circular motion somehow generates an additional force. Neither of these assumptions is correct. *Centripetal* refers to the *effect* of the force—specifically, to the fact that it causes circular motion in which the *direction* of the velocity changes but not its *magnitude*. In the equation $\Sigma F = ma$, the sum of forces must include only the real physical forces: pushes or pulls exerted by strings, rods, or other agencies. The quantity mv^2/R does *not* appear in ΣF but belongs on the ma side of the equation.

PROBLEM-SOLVING STRATEGY: *Circular motion*

The strategies for dynamics problems outlined at the beginning of Section 5–4 are equally applicable here, and we suggest you reread them before beginning your study of the examples below.

A serious peril in circular-motion problems is the temptation to regard mv^2/R as a *force*, as though the body's circular motion somehow generates an extra force in addition to the real physical forces exerted by strings, contact with other bodies, gravitational interactions, or whatever. Resist this temptation! Draw the free-body diagram with *green* arrows for the forces and a *red* arrow for the acceleration. Then in the $\Sigma F = ma$ equations, write the force terms in green and the mv^2/R term in red. This may sound silly, but if it helps, do it!

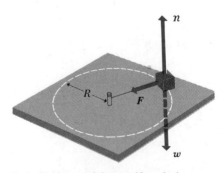

6–1 Body revolving uniformly in a circle on a horizontal frictionless surface.

EXAMPLE 6–1   A small plastic box having mass 0.200 kg revolves uniformly in a circle on a frictionless horizontal surface such as an air-hockey table. It is attached by a cord 0.200 m long to a pin set in the surface. If the body makes two complete revolutions per second, find the force F exerted on it by the cord. (See Fig. 6–1.)

SOLUTION   The circumference of the circle is

$$2\pi(0.200 \text{ m}) = 1.26 \text{ m},$$

so the speed is $v = 2.51 \text{ m·s}^{-1}$. The magnitude of the centripetal acceleration is

$$a = \frac{v^2}{R} = \frac{(2.51 \text{ m·s}^{-1})^2}{0.200 \text{ m}} = 31.6 \text{ m·s}^{-2}.$$

The body has no vertical acceleration, so the forces n and w are equal and opposite. Thus the force F is the resultant force, and the tension F in the cord is

$$F = ma = (0.200 \text{ kg})(31.6 \text{ m·s}^{-2})$$
$$= 6.32 \text{ N}.$$

EXAMPLE 6–2 A child's indoor swing consists of a rope of length L anchored to the ceiling, with a seat at the lower end. The total mass of child and seat is m. They swing in a horizontal circle with constant speed v, as shown in Fig. 6–2; as they swing around, the rope makes a constant angle θ with the vertical. Assuming the time τ for one revolution (i.e., the period) is known, find the tension T in the rope and the angle θ.

Dynamics of a child's swing

SOLUTION First, note that the center of the circular path is at the center of the shaded area, *not* at the top end of the rope. A free-body diagram for the system is shown in the figure. The forces on the system in the position shown are the weight mg and the tension T in the rope. The tension has a horizontal component $T \sin \theta$ and a vertical component $T \cos \theta$. (Note that the force diagram in Fig. 6–2 is exactly like that in Fig. 5–14b. The only difference is that in this case the acceleration a is the *radial* acceleration, v^2/R.) The system has no vertical acceleration, so the sum of the vertical components of force is zero. The horizontal force must equal the mass m times the radial (centripetal) acceleration. Thus the $\Sigma F = ma$ equations are

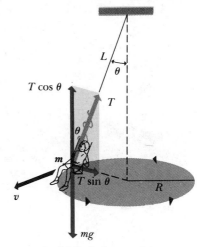

6–2 A child's indoor swing.

$$T \sin \theta = m \frac{v^2}{R}, \qquad T \cos \theta = mg.$$

When the first of these equations is divided by the second, the result is

$$\tan \theta = \frac{v^2}{Rg}. \tag{6–5}$$

Also, the radius R of the circle is given by $R = L \sin \theta$, and the speed is the circumference $2\pi L \sin \theta$ divided by the period τ:

$$v = \frac{2\pi L \sin \theta}{\tau}.$$

When these relations are used to eliminate R and v from Eq. (6–5), we obtain

$$\cos \theta = \frac{g\tau^2}{4\pi^2 L}, \tag{6–6}$$

or

$$\tau = 2\pi\sqrt{L(\cos \theta)/g}. \tag{6–7}$$

Once θ is determined, the tension T can be found from $T = mg/\cos \theta$. For a given length L, $\cos \theta$ decreases as the time is made shorter, and the angle θ increases. The angle never becomes 90°, however, since this requires that $\tau = 0$ or $v = \infty$. Equation (6–7) is similar in form to the expression for the time of swing of a simple pendulum, which will be derived in Chapter 11. Because of this similarity, and because the moving rope traces out a cone, the present system is called a *conical pendulum*.

You may feel tempted to add to the forces in Fig. 6–2 an extra outward force, to "keep the body out there" at angle θ, or to "keep it in equilibrium." Perhaps you have heard the term *centrifugal force; centrifugal* means "fleeing from the center."

Don't make up imaginary forces that aren't there.

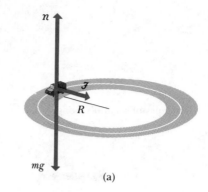

(a)

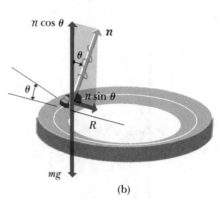

(b)

6–3 (a) Forces on a vehicle rounding a curve on a level track. (b) Forces when the track is banked.

The temptation to add an extra force must be resisted with all possible strength. First, the body does not stay "out there"; it is in constant motion around its circular path. Its velocity is constantly changing in direction, and the body is *not* in equilibrium. Second, if there were an additional outward ("centrifugal") force to balance the inward force, there would then be *no* resultant inward force to cause the circular motion. In that case the body would move in a straight line, not a circle. What is sometimes called centrifugal force is, at least in an inertial frame of reference, not a force at all but a reflection of Newton's first law. In the absence of force, a body moves in a straight line with constant speed, and for circular motion to occur there must be a net *inward* force.

EXAMPLE 6–3 The driver of a sports car is rounding a flat, unbanked curve of radius R. If the coefficient of friction between tires and road is μ_s, what is the maximum speed v at which he can take the curve without sliding?

SOLUTION Figure 6–3a is a free-body diagram for the situation. The acceleration v^2/R toward the center of the curve must be caused by the friction force $\mathcal{F}$, and there is no vertical acceleration. Thus we have

$$\mathcal{F} = m\frac{v^2}{R}, \qquad n = mg.$$

The maximum friction force available is $\mathcal{F} = \mu_s n = \mu_s mg$. Combining these relations to eliminate $\mathcal{F}$, we find

$$\mu_s mg = m\frac{v^2}{R},$$

Friction keeps the car on the road.

or

$$v = \sqrt{\mu_s gR}.$$

If $\mu_s = 0.50$ and $R = 25$ m, then

$$v = \sqrt{(0.50)(9.8 \text{ m·s}^{-2})(25 \text{ m})} = 11.1 \text{ m·s}^{-1},$$

or about 25 mi·hr^{-1}. Why did we use the coefficient of *static* friction for a moving car?

Banking a curve to help the car make the turn

EXAMPLE 6–4 An engineer proposes to rebuild the curve in Example 6–3, banking it so that at a certain speed v no friction at all is needed for the car to make the curve. At what angle θ should it be banked?

SOLUTION The free-body diagram is shown in Fig. 6–3b. The normal force is no longer vertical but is perpendicular to the roadway, at an angle θ with the vertical. Thus it has a vertical component $n \cos \theta$ and a horizontal component $n \sin \theta$, as shown. The horizontal component of n must now cause the acceleration v^2/R, and again there is no vertical acceleration. Thus

$$n \sin \theta = \frac{mv^2}{R}, \qquad n \cos \theta = mg.$$

Dividing the first of these equations by the second, we find

$$\tan \theta = \frac{v^2}{Rg}. \tag{6–8}$$

If $R = 25$ m and $v = 11.1$ m·s^{-1}, as in Example 6–3, then

$$\theta = \arctan \frac{(11.1 \text{ m·s}^{-1})^2}{(25 \text{ m})(9.8 \text{ m·s}^{-2})} = 26.6°.$$

Equation (6–8) shows that the tangent of the banking angle is proportional to the square of the speed and inversely proportional to the radius. For a given radius, no one angle is correct for all speeds. Hence in the design of highways and railroads, curves are banked for the *average speed* of the traffic over them. The same considerations apply to the correct banking angle of a plane when it makes a turn in level flight. We also note that the banking angle is given by the same expression as that for the angle of a conical pendulum, Example 6–2, Eq. (6–5). In fact, the free-body diagrams of Figs. 6–2 and 6–3b are identical.

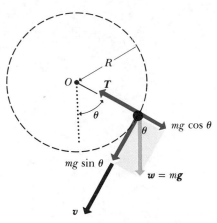

6–4 Forces on a body whirling in a vertical circle with center at O.

6–2 MOTION IN A VERTICAL CIRCLE

Figure 6–4 shows a small ball attached to a string of length R and whirling in a *vertical* circle about a fixed point O; the other end of the string is attached to this point. The motion is circular, but it is *not* uniform (unlike the examples in Section 6–1), because the speed increases on the way down and decreases on the way up. The normal component of acceleration is still given by $a_\perp = v^2/R$, as discussed above, but v changes from point to point along the path. Because the magnitude of the velocity is changing, there must also be a component of acceleration $a_\parallel$ *parallel* or *tangent* to the path, as discussed in Section 3–3.

The forces on the body at any point are its weight $w = mg$ and the tension T in the cord. We resolve the weight into a normal component, of magnitude $mg \cos \theta$, and a tangential component of magnitude $mg \sin \theta$, as in Fig. 6–4. The resultant tangential and normal forces are then

$$F_\parallel = mg \sin \theta \quad \text{and} \quad F_\perp = T - mg \cos \theta.$$

The tangential acceleration, from Newton's second law, is

$$a_\parallel = \frac{F_\parallel}{m} = g \sin \theta;$$

this is the same as the acceleration of a body sliding on a frictionless inclined plane with slope angle θ. The normal (radial) acceleration $a_\perp = v^2/R$ is

$$a_\perp = \frac{F_\perp}{m} = \frac{T - mg \cos \theta}{m} = \frac{v^2}{R};$$

To find the tension in the cord, we solve this for T:

$$T = m\left(\frac{v^2}{R} + g \cos \theta\right). \tag{6–9}$$

At the lowest point of the path, $\theta = 0$, $\sin \theta = 0$, and $\cos \theta = 1$. Hence at this point $F_\parallel = 0$, $a_\parallel = 0$, and the acceleration is purely radial (upward). The magnitude of the tension, from Eq. (6–9), is

$$T = m\left(\frac{v^2}{R} + g\right).$$

As the passengers in this amusement-park ride travel in a nearly circular path, they experience accelerations toward the center of the circle. (Tidal Wave, Great America, California)

At the highest point, $\theta = 180°$, $\sin \theta = 0$, $\cos \theta = -1$, and the acceleration is again purely radial (downward). The tension is

$$T = m\left(\frac{v^2}{R} - g\right). \tag{6-10}$$

With motion of this sort, it is a familiar fact that when the speed at the highest point is less than some critical value v_c, the cord becomes slack and the path is no longer circular. To find this critical speed, we set $T = 0$ in Eq. (6-10):

$$0 = m\left(\frac{v_c^2}{R} - g\right), \qquad v_c = \sqrt{Rg}.$$

If $R = 1$ m, then

$$v_c = \sqrt{(1 \text{ m})(9.8 \text{ m·s}^{-2})} = 3.13 \text{ m·s}^{-1}.$$

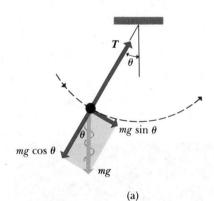

EXAMPLE 6–5 In Fig. 6–5, a small body of mass $m = 0.10$ kg swings in a vertical circle at the end of a cord of length $R = 1.0$ m. If its speed is 2.0 m·s^{-1} when the cord makes an angle $\theta = 30°$ with the vertical, find

a) the radial and tangential components of its acceleration at this instant;

b) the magnitude and direction of the resultant acceleration;

c) the tension T in the cord.

SOLUTION

a) The radial component of acceleration is

$$a_\perp = \frac{v^2}{R} = \frac{(2.0 \text{ m·s}^{-1})^2}{1.0 \text{ m}} = 4.0 \text{ m·s}^{-2}.$$

The tangential component of acceleration, due to the tangential force $mg \sin \theta$, is

$$a_\parallel = g \sin \theta = (9.8 \text{ m·s}^{-2})(0.50) = 4.9 \text{ m·s}^{-2}.$$

b) The magnitude of the resultant acceleration (see Fig. 6–5b) is

$$a = \sqrt{a_\parallel^2 + a_\perp^2} = 6.33 \text{ m·s}^{-2}.$$

The angle ϕ is

$$\phi = \arctan \frac{a_\parallel}{a_\perp} = 50.8°.$$

c) The tension in the cord is given by $F_\perp = ma_\perp$: $T - mg \cos \theta = mv^2/R$, so

$$T = m\left(\frac{v^2}{R} + g \cos \theta\right) = 1.25 \text{ N}.$$

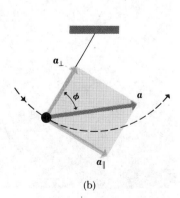

6–5 (a) Forces on a body swinging in a vertical circle. (b) The radial and tangential components of acceleration are combined to obtain the resultant acceleration a.

Note that the magnitude of the tangential acceleration is not constant but is proportional to the sine of the angle θ. Hence the equations of motion with constant acceleration *cannot* be used to find the speed at other points of the path. We will show in the next chapter, however, how the speed at any point can be found from *energy* considerations.

6–3 NEWTON'S LAW OF GRAVITATION

We have already encountered gravitational interactions several times. The **weight** of a body, the force that attracts it toward the earth (or whatever celestial body it is near), is an example of gravitational attraction. In this section we study this phenomenon in more detail.

Newton discovered the **law of gravitation** during his study of the motions of planets around the sun, and he published it in 1686. It may be stated as follows:

A general formulation of gravitational interactions

> Every particle of matter in the universe attracts every other particle with a force that is directly proportional to the product of the masses of the particles and inversely proportional to the square of the distance between them.

Thus

$$F_g = G\frac{m_1 m_2}{r^2}, \qquad (6\text{–}11)$$

where F_g is the magnitude of the gravitational force on either particle, m_1 and m_2 are their masses, r is the distance between them, and G is a fundamental physical constant called the **gravitational constant.** The numerical value of G depends on the system of units used.

The directions of the gravitational forces on the interacting bodies are along the straight line joining them. These forces form an action–reaction pair. Even when the masses of the particles are different, the interaction forces have equal magnitude.

Newton's law of gravitation describes the interaction forces between two *particles*. But these forces also obey the **principle of superposition:** If two masses each exert forces on a third, the *total* force on the third mass is the vector sum of the individual forces of the first two. It can also be shown that *the gravitational force exerted on or by any homogeneous sphere is the same as though the entire mass of the sphere were concentrated in a point at its center.* Thus if the earth were a homogeneous sphere, of mass m_E, the force exerted by it on a small body of mass m, at a distance r from its center, would be

Superposition: how gravitational forces combine

$$F_g = G\frac{m m_E}{r^2}, \qquad (6\text{–}12)$$

provided that the body lies outside the earth, i.e., that r is greater than the radius of the earth. A force of the same magnitude would be exerted *on* the earth by the whole body. We will learn how to prove these statements in Chapter 25 in connection with analogous problems with electric charge. (See also Challenge Problem 6–42.)

At points *inside* the earth, these statements need to be modified. If we could drill a hole to the center of the earth and measure the force of gravity on a body at various distances from the center, the force would be found to *decrease* as the center is approached, rather than increasing as $1/r^2$. Qualitatively, it is easy to see why this should be so: As the body enters the interior of the earth (or other spherical body), some of the earth's mass is on the side of the body opposite from the center of the earth and pulls the body in the opposite direction. Exactly at the center of the earth, the gravitational force on the body is zero!

The Cavendish balance: a way to measure very small forces

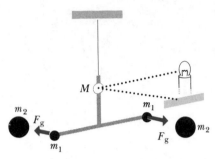

6–6 Principle of the Cavendish balance.

The magnitude of the gravitational constant G can be found experimentally by measuring the force of gravitational attraction between two bodies of known masses m_1 and m_2, at a known separation. For bodies of moderate size the force is extremely small, but it can be measured with an instrument invented by the Rev. John Mitchell and first used for this purpose by Sir Henry Cavendish in 1798. The same type of instrument was also used by Coulomb for studying forces of electrical and magnetic attraction and repulsion.

The Cavendish balance consists of a light, rigid T-shaped rod (Fig. 6–6) supported by a fine vertical fiber such as a quartz thread or a thin metallic ribbon. Two small spheres, each of mass m_1, are mounted at the ends of the horizontal portion of the T, and a small mirror M, fastened to the vertical portion, reflects a beam of light onto a scale. To use the balance, two large spheres, each of mass m_2, are brought up to the positions shown. The forces of gravitational attraction between the large and small spheres twist the system through a small angle, thereby moving the reflected light beam along the scale.

By use of an extremely thin fiber, the deflection of the mirror may be made sufficiently large so that the gravitational force can be measured quite accurately. The gravitational constant obtained in this way is found to be (in SI units)

$$G = 6.670 \times 10^{-11} \text{ N·m}^2\text{·kg}^{-2}.$$

Gravitational forces on small bodies are tiny!

EXAMPLE 6–6 The mass m_1 of one of the small spheres of a Cavendish balance is 0.00100 kg; the mass m_2 of one of the large spheres is 0.500 kg; and the center-to-center distance between the spheres is 0.0500 m. Find the gravitational force on each sphere.

SOLUTION The force is

$$F_g = 6.67 \times 10^{-11} \text{ N·m}^2\text{·kg}^{-2} \frac{(0.00100 \text{ kg})(0.500 \text{ kg})}{(0.0500 \text{ m})^2}$$
$$= 1.33 \times 10^{-11} \text{ N},$$

or about one hundred-billionth of a newton!

EXAMPLE 6–7 Suppose the spheres in Example 6–6 are placed 0.0500 m from each other at a point in space far removed from all other bodies. What is the acceleration of each, relative to an inertial system?

SOLUTION The acceleration a_1 of the smaller sphere is

$$a_1 = \frac{F_g}{m_1} = \frac{1.33 \times 10^{-11} \text{ N}}{1.0 \times 10^{-3} \text{ kg}} = 1.33 \times 10^{-8} \text{ m·s}^{-2}.$$

The acceleration a_2 of the larger sphere is

$$a_2 = \frac{F_g}{m_2} = \frac{1.33 \times 10^{-11} \text{ N}}{0.5 \text{ kg}} = 2.67 \times 10^{-11} \text{ m·s}^{-2}.$$

In this case, the accelerations are *not* constant because the gravitational force increases as the spheres approach each other.

EXAMPLE 6–8 Some of the masses from Example 6–6 are arranged as shown in Fig. 6–7. Find the magnitude and direction of the total force exerted on the small sphere by both large ones.

SOLUTION We use the principle of superposition: The total force on the small sphere is the vector sum of the two forces due to the two individual large spheres. These forces have different directions, so we have to compute a vector sum. We first find the magnitudes of the forces. The force F_1 on the small mass due to the upper large one is

$$F_1 = \frac{(6.67 \times 10^{-11} \text{ N·m}^2\text{·kg}^{-2})(0.5 \text{ kg})(0.001 \text{ kg})}{(0.2 \text{ m})^2 + (0.2 \text{ m})^2}$$
$$= 4.17 \times 10^{-13} \text{ N},$$

and the force F_2 due to the lower large mass is

$$F_2 = \frac{(6.67 \times 10^{-11} \text{ N·m}^2\text{·kg}^{-2})(0.5 \text{ kg})(0.001 \text{ kg})}{(0.2 \text{ m})^2}$$
$$= 8.34 \times 10^{-13} \text{ N}.$$

The x- and y-components of these forces are

$$F_{1x} = (4.17 \times 10^{-13} \text{ N}) \cos 45° = 2.95 \times 10^{-13} \text{ N},$$
$$F_{1y} = (4.17 \times 10^{-13} \text{ N}) \sin 45° = 2.95 \times 10^{-13} \text{ N},$$
$$F_{2x} = 8.34 \times 10^{-13} \text{ N},$$
$$F_{2y} = 0.$$

Or, in terms of unit vectors,

$$\boldsymbol{F}_1 = (2.95 \times 10^{-13} \text{ N})\boldsymbol{i} + (2.95 \times 10^{-13} \text{ N})\boldsymbol{j},$$
$$\boldsymbol{F}_2 = (8.34 \times 10^{-13} \text{ N})\boldsymbol{i}.$$

Thus the total force on the small mass is the vector sum

$$\boldsymbol{F} = [(2.95 + 8.34) \times 10^{-13} \text{ N}]\boldsymbol{i} + (2.95 \times 10^{-13} \text{ N})\boldsymbol{j},$$

with components

$$F_x = 11.29 \times 10^{-13} \text{ N},$$
$$F_y = 2.95 \times 10^{-13} \text{ N}.$$

The magnitude of this force is

$$F = \sqrt{(11.29 \times 10^{-13} \text{ N})^2 + (2.95 \times 10^{-13} \text{ N})^2}$$
$$= 11.7 \times 10^{-13} \text{ N},$$

and its direction is

$$\theta = \arctan \frac{2.95}{11.29} = 14.6°.$$

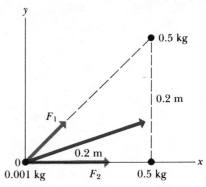

6–7 The total gravitational force on the 0.001-kg mass is the vector sum of the forces exerted by the two 0.5-kg masses.

We can now define the *weight* of a body more generally as *the resultant gravitational force exerted on the body by all other bodies in the universe.* When the body is near the surface of the earth, the earth's attraction is much greater than that of any other body; in that case we can ignore all other gravitational forces and consider the weight as due entirely to the earth's attraction. Similarly, at the surface of the moon, or of another planet, the weight of a body

A more general definition of *weight*

results almost entirely from the gravitational attraction of the moon or the planet. Thus if the earth were a homogeneous sphere of radius R and mass m_E, the weight w of a small body of mass m at or near its surface would be

$$w = F_g = \frac{Gmm_E}{R^2}. \tag{6-13}$$

Equation (6–13) shows that the weight of a given body decreases inversely with the square of its distance from the earth's center; at a radial distance of two earth radii, for example, it has decreased to one quarter of its value at the earth's surface.

The *apparent* weight of a body at the surface of the earth differs slightly in magnitude and direction from the earth's force of gravitational attraction because the rotation of the earth about its axis makes it not precisely an inertial frame of reference. For the present, we ignore this effect and assume that the earth *is* an inertial system. Then when a body is allowed to fall freely, the force accelerating it is its weight *w*, and the acceleration produced by this force is the acceleration due to gravity, *g*. The general relation

$$\Sigma F = ma$$

therefore becomes, for the special case of a freely falling body,

$$w = mg. \tag{6-14}$$

Since, according to Eq. (6–13)

$$w = mg = G\frac{mm_E}{R^2},$$

it follows that

$$g = \frac{Gm_E}{R^2}, \tag{6-15}$$

showing that the acceleration due to gravity is the same for *all* bodies (since m cancels out) and is very nearly constant (since G and m_E are constants and R varies only slightly from point to point on the earth).

Determining the mass of the earth

When Eq. (6–15) is solved for m_E, we find

$$m_E = \frac{R^2 g}{G}, \tag{6-16}$$

where R is the earth's radius. All of the quantities on the right are known, so the mass of the earth, m_E, can be calculated. Taking $R = 6370$ km $= 6.37 \times 10^6$ m, and $g = 9.80$ m·s^{-2}, we find

$$m_E = 5.96 \times 10^{24} \text{ kg.}$$

The volume of the earth is

$$V = \frac{4}{3}\pi R^3 = 1.08 \times 10^{21} \text{ m}^3.$$

Density of the earth

The mass of a body divided by its volume is known as its average *density*. (The density of water is 1 g·cm^{-3} = 1000 kg·m^{-3}.) The average density of the earth is therefore

$$\frac{m_E}{V} = 5520 \text{ kg·m}^{-3} = 5.52 \text{ g·cm}^{-3}.$$

The density of most rock near the earth's surface, such as granites and gneisses, is about 3 g·cm^{-3} = 3000 kg·m^{-3}, so the interior of the earth must have much higher density. Some basaltic rock found on the surface has a density of about 5 g·cm^{-3}.

6–4 GRAVITATIONAL FIELD

Newton's law of gravitation can be restated in a useful way with the concept of **gravitational field.** Instead of calculating the interaction forces between two masses by using Eq. (6–11) directly, we consider a two-stage process. First, we think of one mass as creating a change at each point in the space surrounding it. This change is called a *gravitational field.* To describe it, we associate a vector quantity *g* with each point in space. The significance of *g* is that a particle of mass *m* at a point where the gravitational field is *g* experiences a gravitational *force* given by *F = mg.* To measure a gravitational field we may place a small mass, called a *test mass*, at the point. If the mass experiences a gravitational force, then there is a gravitational field at that point.

Gravitational field: a useful reformulation of gravitational interactions

In Example 6–8, instead of calculating directly the forces on the 0.001-kg body by the large bodies, we may take the point of view that the large bodies create a *gravitational field* at the location of the small body, and that the field exerts a force on any mass located at the point.

Because force is a vector quantity, gravitational field is also a vector quantity. We define the gravitational field *g* at a point as the quotient of the force *F* experienced by a test mass *m* and the mass. Thus

$$g = \frac{F}{m}. \qquad (6-17)$$

To say it another way: At a point where the gravitational field is *g*, the force *F* on a mass *m* is given by

$$F = mg. \qquad (6-18)$$

The gravitational field due to a point mass *m*, at a distance *r* away from it, has magnitude

$$g = \frac{Gm}{r^2}, \qquad (6-19)$$

and is directed away from the point mass. Gravitational fields also obey the principle of superposition: The *total* gravitational field at a point, due to several point masses, is the *vector sum* of the gravitational fields of the separate masses. Gravitational field is a very useful concept; when we know the field, we can calculate quickly the gravitational force on any body.

The gravitational field in the vicinity of any collection of masses varies from one point to another in the region. Thus it is not a single vector quantity but rather a whole collection of vector quantities, one vector associated with each point in space. Indeed, this is what is meant by the general term *vector field.* Another familiar example of a vector field is the velocity in a flowing fluid: different parts of the fluid have different velocities, and so we speak of the *velocity field* in the fluid. During our study of electricity and magnetism, we will discuss in detail the properties of the electric and magnetic fields; both of these are vector fields.

Vector field: a vector quantity associated with each point in a region of space

We assert that when a particle of mass m is acted on *only* by gravitational force, its acceleration at each instant is equal in magnitude and direction to the gravitational field g at its instantaneous location. Can you prove this?

EXAMPLE 6–9 In Example 6–8, find the magnitude of the gravitational field at the location of the 0.001-kg mass.

SOLUTION From Eq. (6–17), the gravitational field at this point has magnitude

$$g = \frac{11.7 \times 10^{-13}\ \text{N}}{0.001\ \text{kg}} = 11.7 \times 10^{-10}\ \text{m·s}^{-2}.$$

The direction of the field is the same as that of the force on the small body.

EXAMPLE 6–10 A ring-shaped body of radius a has total mass M. Find the gravitational field at point P, at a distance x from the center, along the line perpendicular to the plane of the ring, through its center.

SOLUTION The situation is shown in Fig. 6–8. We consider first a small segment Δs of the ring; we call the mass of this segment ΔM. At point P, this segment produces a gravitational field of magnitude

$$g = \frac{G\,\Delta M}{r^2} = \frac{G\,\Delta M}{x^2 + a^2}.$$

The component of this field along the x-axis is given by

$$g_x = -g \cos \theta = -\frac{G\,\Delta M}{x^2 + a^2}\,\frac{x}{\sqrt{x^2 + a^2}} = -\frac{G\,\Delta M x}{(x^2 + a^2)^{3/2}}. \qquad (6\text{–}20)$$

Now x is the same for every segment around the ring, so to find the *total* x-component of gravitational field from all the mass elements ΔM, we simply replace ΔM by the *total* mass M. The result is

$$g_x = -\frac{GMx}{(x^2 + a^2)^{3/2}}. \qquad (6\text{–}21)$$

The y-component of gravitational field at point P is zero because the y-components of field produced by two segments at opposite sides of the ring cancel each other. Similarly, the contributions of all such pairs added over the entire ring sum to zero.

Equation (6–21) shows that at the center of the ring ($x = 0$) the total gravitational field is zero. We should expect this. Mass elements on opposite sides pull in opposite directions, and their fields cancel. Also, when x is much larger than a, Eq. (6–21) becomes approximately equal to GM/x^2, corresponding to the fact that at distances much greater than the dimensions of the ring, it appears as a point mass.

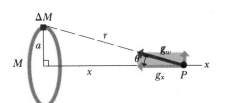

6–8 Gravitational field of a ring of mass M and radius a.

Calculations with gravitational fields

6–5 SATELLITE MOTION

In our discussion of projectile motion in Section 3–4, we assumed that the acceleration of the projectile was constant in magnitude and direction. Because this acceleration is caused by the projectile's weight w, an equivalent assumption is that the weight is the same at all points in the trajectory, corresponding to motion in a uniform gravitational field **g.** These conditions are approximately satisfied if the projectile remains near the earth's surface and if the length of flight is small compared to the earth's radius. The trajectory is then a *parabola* (provided that air resistance can be neglected).

When the size of the trajectory becomes comparable to the radius of the earth, we can no longer ignore the variation of weight with position. At any point outside the earth, the gravitational force on the projectile is directed toward the center of the earth, and its magnitude is inversely proportional to the square of the distance from the center; thus the force is *not* constant in either magnitude or direction. It turns out that under such an attractive "inverse-square" force, the projectile's path may be a circle, an ellipse, a parabola, or a hyperbola. Circular paths will be of special interest to us, and they are also the easiest paths to analyze.

To gain intuitive understanding of satellite motion, let us imagine a tall tower, as in Fig. 6–9a. We launch a projectile from point A in the direction AB tangent to the earth's surface. If the initial speed is rather small, the trajectory will be like that labeled (1) in the figure; this is a portion of an ellipse with the

Orbits of satellites and planets

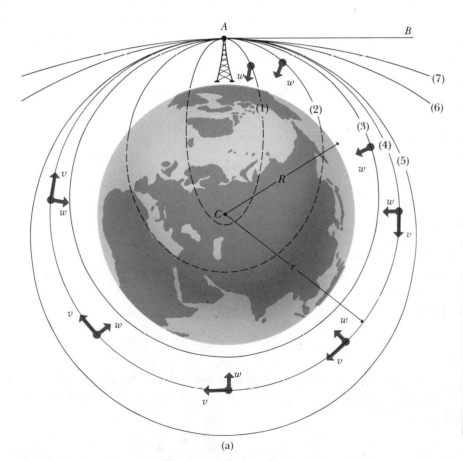

(a)

(b)

6–9 (a) Trajectories of a body projected from point A in the direction AB with different initial velocities. Orbits (1) and (2) would be completed as shown if the earth were a point mass at C. (b) A communications satellite before launch. The main body is 2.16 m in diameter and 2.82 m in height. Solar-energy collector panels on the sides and a parabolic antenna at the end can be seen. (Courtesy Hughes Aircraft Co.)

Halley's comet moves in an elliptical path around the sun under the action of the sun's gravitational attraction. A complete trip around the very flat ellipse requires about 76 years; the comet was in its closest proximity to the sun and the planets of the solar system during 1985 and 1986. (AP/Wide World Photos)

earth's center C at one focus. If the trajectory is so short that changes in the magnitude and direction of w can be neglected, the ellipse is approximately the same shape as a parabola.

Trajectories (2) to (7) illustrate the effect of increasing the initial velocity. Trajectory (2) is again a portion of an ellipse. Trajectory (3), which just misses the earth, is a *complete* ellipse, and the projectile has become an earth satellite. Its velocity when it returns to point A is the same as its initial velocity, and if there is no retarding force (such as air resistance), it repeats its motion indefinitely. The earth's rotation will have moved the tower to a different point by the time the satellite returns to point A, but this does not affect the orbit.

Trajectory (4) is a special case in which the orbit is a circle. Trajectory (5) is again an ellipse, (6) is a parabola, and (7) is a hyperbola. Trajectories (6) and (7) are not closed orbits; for these the projectile never returns to its starting point, but "escapes" permanently from the earth's gravitational field.

We will consider only the simplest case, *circular* orbits. Many manmade satellites have nearly circular orbits, and the orbits of the planets of our solar system around the sun are also nearly circular. We have learned that a particle in uniform circular motion with speed v and radius r experiences a centripetal acceleration $a_\perp = v^2/r$ directed always toward the center of the circle. The *force* that provides this acceleration is the gravitational attraction of the earth (mass m_E) for the satellite (mass m). If the radius of the circular orbit, measured from the *center* of the earth, is r, then the gravitational force is given by $F = Gm_Em/r^2$, as discussed in Section 6–3. Using Newton's second law, we equate this to the product of the satellite's mass m and its acceleration v^2/r:

$$\frac{Gm_Em}{r^2} = \frac{mv^2}{r}.$$

When this equation is solved for v, the result is

$$v = \sqrt{\frac{Gm_E}{r}}. \qquad (6\text{–}22)$$

This relation shows that the orbit radius r and the satellite's speed v cannot be chosen independently but must be related by Eq. (6–22).

We can also derive a relation between the orbit radius r and the *period* τ, the time for one revolution. As pointed out in Section 3–5, the speed v is the distance $2\pi r$ traveled in one revolution, divided by the time τ for one revolution. Thus

$$v = \frac{2\pi r}{\tau} = 2\pi r f, \qquad (6\text{–}23)$$

The period of a satellite depends on its orbit radius but not on its mass.

where f is the *frequency* (number of revolutions per unit time), equal to $1/\tau$. We can obtain an expression for τ by solving this equation for τ and combining it with Eq. (6–22):

$$\tau = \frac{2\pi r}{v} = 2\pi r \sqrt{\frac{r}{Gm_E}} = \frac{2\pi r^{3/2}}{\sqrt{Gm_E}}. \qquad (6\text{–}24)$$

Equations (6–22) and (6–24) show that larger orbits correspond to longer periods and slower speeds.

EXAMPLE 6–11 Suppose you want to place the communications satellite shown in Fig. 6–9b into a circular orbit 300 km above the earth's surface. What must be its speed, its period, and its radial acceleration? The earth's radius is 6380 km = 6.38×10^6 m, and its mass is 5.98×10^{24} kg. (See Appendix F.)

SOLUTION The radius of the orbit is

$$r = 6380 \text{ km} + 300 \text{ km} = 6680 \text{ km} = 6.68 \times 10^6 \text{ m}.$$

From Eq. (6–22),

$$v = \sqrt{\frac{(6.67 \times 10^{-11} \text{ N} \cdot \text{m}^2 \cdot \text{kg}^{-2})(5.98 \times 10^{24} \text{ kg})}{6.68 \times 10^6 \text{ m}}}$$
$$= 7728 \text{ m} \cdot \text{s}^{-1}.$$

From Eq. (6–24),

$$\tau = \frac{2\pi(6.68 \times 10^6 \text{ m})}{7728 \text{ m} \cdot \text{s}^{-1}} = 5431 \text{ s} = 90.5 \text{ min.}$$

The radial acceleration is

$$a_\perp = \frac{v^2}{r} = \frac{(7728 \text{ m} \cdot \text{s}^{-1})^2}{6.68 \times 10^6 \text{ m}} = 8.94 \text{ m} \cdot \text{s}^{-2}.$$

This is somewhat less than the value of the free-fall acceleration g at the earth's surface and is, in fact, equal to the value of g at a height of 300 km above the earth's surface.

Our discussion has centered around the motion of manmade earth satellites, but it should be clear that these principles are also applicable to the circular motion of *any* body under its gravitational attraction to a stationary body. Other familiar examples are the motion of our moon and that of the moons of Jupiter and the motions of the planets around the sun in nearly circular orbits. The rings of Saturn, shown in Fig. 6–10, are composed of small particles in circular orbits around the planet.

Historically the study of planetary motion played a pivotal role in the development of physics. Johannes Kepler (1571–1630) spent several painstaking years analyzing the motions of the planets, basing his work on remarkably precise measurements made by the Danish astronomer Tycho Brahe (1546–1601) *without the aid of a telescope.* (The telescope was invented in 1608.) Kepler discovered that the orbits of the planets are (nearly circular) ellipses, and that the period of a planet in its orbit is proportional to the three-halves power of the orbit radius, as Eq. (6–24) shows.

It remained for Newton (1642–1727) to show, with his laws of motion and law of gravitation, that this behavior of the planets could be understood on the basis of the very same physical principles he had developed to analyze *terrestrial* motion. According to legend the law of gravitation came to Newton as he watched an apple fall out of a tree and asked himself what force accelerated it toward the earth. But to arrive at the $1/r^2$ form of the law of gravitation he had to do something much subtler: He had to take Brahe's observations of planetary motions, as systematized by Kepler, and show that Kepler's rules demanded a $1/r^2$ force law.

6–10 The rings of Saturn, photographed from Voyager 1 on November 1, 1980, from a range of 700,000 km. Individual particles move in circular orbits, with periods given by Eq. (6–24), using Saturn's mass. The outer portions of the rings take more time for one revolution than the inner portions. (Courtesy of Jet Propulsion Laboratory/NASA.)

The Newtonian synthesis: getting it all together

From our historical perspective three hundred years later, there is absolutely no doubt that this **Newtonian synthesis,** as it has come to be called, is one of the greatest achievements in the entire history of science, certainly comparable in significance to the development of quantum mechanics, the theory of relativity, and the understanding DNA in our own century.

6–6 EFFECT OF THE EARTH'S ROTATION ON g

On a rotating earth, apparent weight is different from real weight.

Because the earth rotates on its axis, it is not precisely an inertial frame of reference. For this reason the apparent weight of a body on earth is not precisely equal to the earth's gravitational attraction w_0, which we define as the **true weight** of the body. Figure 6–11 is a cutaway view of the earth, showing three observers, each holding a body of mass m hanging from a spring balance. Each balance applies a certain tension force T to the body hanging from it, and the reading on each balance is the magnitude T of this force. Each observer *thinks* that his scale reading equals the weight of the body. Also, each observer, if he is unaware of the earth's rotation, thinks the body on his spring balance is in equilibrium; if so, he thinks, the tension T must be opposed by an equal and opposite force w, which we define to be the **apparent weight.** As we will see, the apparent weight is different at different points on the earth.

If we assume that the earth is spherically symmetric, then the true weight w_0 has magnitude Gmm_E/R^2, where m_E and R are the mass and radius of the earth. This value is the same for all points on the earth's surface. If the center of the earth can be taken as the origin of an inertial coordinate system, then the body at the north pole really *is* in equilibrium in an inertial system, and the reading on that observer's spring balance is equal to w_0. But the body at the equator is moving in a circle of radius R with speed v, and there must be a net inward force equal to the mass times the centripetal acceleration:

$$w_0 - T = \frac{mv^2}{R}. \tag{6–25}$$

Thus the magnitude of the apparent weight (equal to T at this location) is

$$w = w_0 - \frac{mv^2}{R}. \tag{6–26}$$

We also note that if the earth were not rotating, the body when released would have a free-fall acceleration g_0 given by $g_0 = w_0/m$, but that its actual acceleration relative to the observer at the equator is given by $g = w/m$. Dividing Eq. (6–26) by m and using these relations, we find

$$g = g_0 - \frac{v^2}{R}. \tag{6–27}$$

To evaluate v^2/R, we note that in one day (86,400 s) a point on the equator moves a distance equal to the earth's circumference, $2\pi R = 2\pi(6.38 \times 10^6 \text{ m})$. Thus

$$v = \frac{2\pi(6.38 \times 10^6 \text{ m})}{86,400 \text{ s}} = 464 \text{ m·s}^{-1},$$

and

$$\frac{v^2}{R} = \frac{(464 \text{ m·s}^{-1})^2}{6.38 \times 10^6 \text{ m}} = 0.0337 \text{ m·s}^{-2}.$$

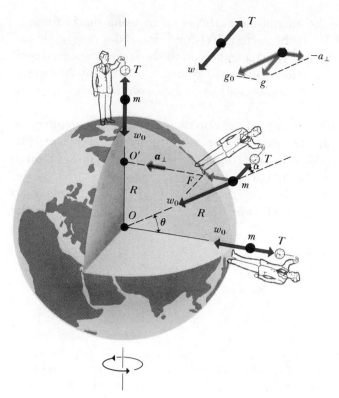

6-11 The resultant of the forces T and w_0 is equal to m times the centripetal acceleration $a_\perp = v^2/r$.

So for a spherically symmetric earth, the acceleration due to gravity should be about 0.03 m·s^{-2} less at the equator than at the poles.

At locations intermediate between the equator and the poles, we can write a vector equation corresponding to Eq. (6–26). From Fig. 6–11 we see that the appropriate equation is

$$w = w_0 - ma_\perp = mg_0 - ma_\perp. \qquad (6-28)$$

The difference in the magnitudes of g and g_0 is intermediate between zero and 0.0337 m·s^{-2}, and the *direction* of the apparent weight differs from the radial direction by an angle α, as shown.

Table 6–1 gives the values of g at several locations, showing variations with latitude of approximately this magnitude. There are also small additional variations due to the lack of perfect spherical symmetry of the earth, local variations in density, and differences in elevation.

In this discussion we have ignored the *orbital* motion of the earth around the sun; this motion makes the center of the earth not precisely an inertial

The acceleration of free fall on earth depends on location.

TABLE 6–1 Variations of g with Latitude and Elevation

Station	North latitude	Elevation, m	g, m·s^{-2}	g, ft·s^{-2}
Canal Zone	9°	0	9.78243	32.0944
Jamaica	18°	0	9.78591	32.1059
Bermuda	32°	0	9.79806	32.1548
Denver	40°	1638	9.79609	32.1393
Cambridge, Mass.	42°	0	9.80398	32.1652
Pittsburgh, Pa.	40.5°	235	9.80118	32.1561
Greenland	70°	0	9.82534	32.2353

origin. We can estimate the magnitude of this error by using the fact that in one year (about 3.15×10^7 s) the earth travels once around a nearly circular orbit of radius 1.49×10^{11} m. We challenge you to show that the centripetal acceleration of the orbital motion is about 0.006 m·s^{-2}. This is only 20% of the acceleration due to the earth's rotation about its axis, so it is a less important effect.

Weightlessness: Is a body in an orbiting satellite *really* weightless?

Our discussion of apparent weight can also be applied to the phenomenon of "weightlessness" in satellites. Bodies in an orbiting satellite are *not* weightless; the earth's gravitational attraction continues to act on them just as though they were at rest relative to the earth. But a space vehicle in orbit has an acceleration $a_\perp$ toward the earth's center equal to the value of the acceleration of gravity g_0 at its orbit radius. The apparent weight w is given as before by

$$w = w_0 - m\mathbf{a}_\perp = mg_0 - m\mathbf{a}_\perp.$$

But in this case,

$$g_0 = a_\perp,$$

so

$$w = 0.$$

It is in this sense that an astronaut or other body in the spacecraft is said to be "weightless" or in a state of "zero g." (See Fig. 6–12.) The earth's gravity acts on both the spacecraft and the bodies in it, giving each the same acceleration. Thus a body released inside the spacecraft does not fall relative to it, and it appears to be weightless. The situation is similar to that of the freely falling elevator in Example 5–15 (Section 5–4), except that the spacecraft also has a large constant tangential speed along its orbit.

6–12 Astronauts conducting experiments in zero-gravity environment. (Courtesy of NASA, Johnson Space Center.)

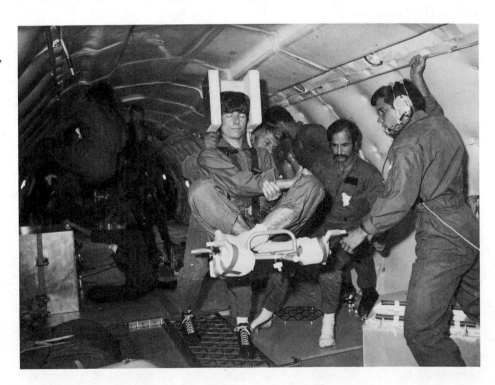

SUMMARY

A body in uniform circular motion with speed v and radius R has an acceleration $a_\perp$ directed toward the center of the circle, with magnitude

$$a_\perp = v^2/R. \tag{6–1}$$

The period τ is the time for one revolution:

$$\tau = 2\pi R/v. \tag{6–2}$$

According to Newton's second law, this acceleration requires a resultant force on the body, directed toward the center, with magnitude mv^2/R. Care must be taken not to regard mv^2/R as an additional force; it is the $m\boldsymbol{a}$ side of $\Sigma \mathbf{F} = m\boldsymbol{a}$.

A particle moving in a *vertical* circle has a centripetal acceleration $a_\perp = v^2/R$, and in addition a tangential component of acceleration $a_\parallel$ determined by the tangential components of force.

Newton's law of gravitation states that two bodies having masses m_1 and m_2, a distance r apart, attract each other with forces of magnitudes

$$F = \frac{Gm_1 m_2}{r^2}. \tag{6–11}$$

These forces form an action–reaction pair and obey Newton's third law. When two or more bodies exert gravitational forces on a particular body, the total gravitational force on it is the vector sum of the forces exerted by the individual bodies.

The weight of a body is defined as the total gravitational force exerted on it by all other bodies in the universe. Near the surface of the earth, where this is essentially equal to the earth's gravitational force alone, w is given by

$$w = \frac{Gmm_\mathrm{E}}{R^2}, \tag{6–12}$$

and the acceleration due to gravity is given by

$$g = \frac{Gm_\mathrm{E}}{R^2}. \tag{6–15}$$

The gravitational field $\boldsymbol{g}$ at a point in space is defined as the gravitational force exerted on a mass m at that point, divided by m. The gravitational field caused by a mass m at a distance r from it has magnitude

$$g = \frac{Gm}{r^2}. \tag{6–19}$$

The total gravitational field at any point due to several masses is the vector sum of gravitational fields due to the individual masses.

When a satellite moves in a circular orbit, the centripetal acceleration is provided by the gravitational attraction of the earth. The speed and period of a satellite in an orbit of radius r are given by

$$v = \sqrt{\frac{Gm_\mathrm{E}}{r}}, \tag{6–22}$$

$$\tau = \frac{2\pi r^{3/2}}{\sqrt{Gm_\mathrm{E}}}. \tag{6–24}$$

KEY TERMS

uniform circular motion
centripetal acceleration
period
weight
Newton's law of gravitation
gravitational constant
principle of superposition
gravitational field
Newtonian synthesis
true weight
apparent weight

Because of the earth's rotation, the apparent weight of a body on earth differs from its true weight by about 0.3% at the equator, and the free-fall acceleration g differs from the value it would have without rotation by the same amount.

QUESTIONS

6–1 "It's not the fall that hurts you; it's the sudden stop at the bottom." Translate this saying into the language of Newton's laws of motion.

6–2 If action and reaction are always equal in magnitude and opposite in direction, why don't they always cancel each other and leave no net force for acceleration of the body?

6–3 Scales for weighing objects can be classified as those that use springs and those using standard weights to balance unknown weights. Which group would have greater accuracy when used in an accelerating elevator? When used on the moon? Does it matter whether you are trying to determine mass or weight?

6–4 Because of air resistance, two objects of unequal mass do *not* fall at precisely the same rate. If two bodies of identical shape but unequal mass are dropped simultaneously from the same height, which one reaches the ground first?

6–5 In discussing the Cavendish balance, a student claimed that the reason the small pivoted masses move toward the larger stationary masses rather than the larger toward the smaller is that the larger masses exert a stronger gravitational pull than the smaller. Please comment.

6–6 Since the earth is constantly attracted toward the sun by the gravitational interaction, why does it not fall into the sun and burn up?

6–7 A student wrote: "The only reason an apple falls downward to meet the earth instead of the earth falling upward to meet the apple is that the earth is much more massive and so exerts a much greater pull." Please comment.

6–8 Cavendish described his measurement of the gravitational constant G as "weighing the earth." This is an apt description not of the experiment itself but of the significance of the result. Why?

6–9 In uninformed discussions of satellites, one hears questions such as: "What keeps the satellite moving in its orbit?" and "What keeps the satellite up?" How do you answer these questions? Are your answers applicable to the moon?

6–10 A certain centrifuge is claimed to operate at 100,000 g. What does this mean?

EXERCISES

Section 6–1 Forces in Circular Motion

6–1 A stone of mass 1 kg is attached to one end of a string 1 m long, of breaking strength 500 N, and is whirled in a horizontal circle on a frictionless table top. The other end of the string is kept fixed. Find the maximum velocity the stone can attain without breaking the string.

6–2 A flat (unbanked) curve on a highway has a radius of 240 m, and a car rounds the curve at a speed of 22 m·s^{-1}. What must be the minimum coefficient of friction to prevent sliding?

6–3 A highway curve of radius 1200 ft is to be banked so that a car traveling 50 mi·hr^{-1} will not skid sideways even in the absence of friction. At what angle should it be banked?

6–4 A coin placed on a record of diameter 12 in. will revolve with the record when it is brought up to a speed of $33\frac{1}{3}$ rev·min^{-1}, provided that the coin is not more than 4 in. from the axis.

a) What is the coefficient of static friction between the coin and the record?

b) How far from the axis can the coin be placed, without slipping, if the turntable rotates at 45 rev·min^{-1}?

6–5 One of the problems for humans living in outer space is the fact that they would be weightless. One way around this problem is to design space stations as shown in Fig. 6–13 and have them spin about their center at a constant rate. This would create a sort of artificial gravity. If the diameter of the space station were 1000 m, how many revolutions per minute would be needed for the artificial gravity acceleration to be 9.8 m·s^{-2}?

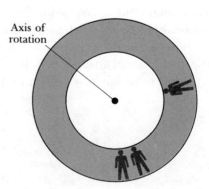

Axis of rotation

FIGURE 6–13

6–6 The "Giant Swing" at a county fair consists of a vertical central shaft with a number of horizontal arms attached at its upper end. Each arm supports a seat suspended from a cable 5 m long, with the upper end of the cable fastened to the arm at a point 4 m from the central shaft.

a) Find the time of one revolution of the swing if the cable supporting a seat makes an angle of 30° with the vertical.

b) Does the angle depend on the weight of the passenger for a given rate of revolution?

Section 6–2 Motion in a Vertical Circle

6–7 A cord is tied to a pail of water, and the pail is swung in a vertical circle of radius 1 m. What must be the minimum velocity of the pail at the highest point of the circle if no water is to spill from the pail?

6–8 A bowling ball weighing 16 lb is attached to the ceiling by a 15-ft rope. The ball is pulled to one side and released; it then swings back and forth as a pendulum. As it swings through the vertical, its velocity is 12 ft·s^{-1}.

a) What is the centripetal acceleration, in magnitude and direction, at this instant?

b) What is the tension in the rope at this instant?

6–9 A stunt pilot who has been diving vertically at a velocity of 180 m·s^{-1} pulls out of the dive by changing his course to a circle in a vertical plane.

a) What is the minimum radius of the circle for the acceleration at the lowest point not to exceed "$7g$"?

b) What is the apparent weight of a 80-kg pilot at the lowest point of the pullout?

Section 6–3 Newton's Law of Gravitation

6–10 A communications satellite of mass 200 kg is in a circular orbit of radius 40,000 km measured from the center of the earth. What is the gravitational force on the satellite, and what fraction is this of its weight at the surface of the earth?

6–11 What is the ratio of the gravitational pull of the sun on the moon to that of the earth on the moon? Use the data in Appendix F. What is the significance of the result to the motion of the moon?

6–12 The mass of the moon is about one eighty-first, and its radius one fourth, that of the earth. What is the acceleration due to gravity on the surface of the moon?

6–13 In an experiment using the Cavendish balance to measure the gravitational constant G, it is found that a mass of 0.8 kg attracts another sphere of mass 0.004 kg with a force of 13×10^{-11} N when the distance between the centers of the spheres if 0.04 m. The acceleration of gravity at the earth's surface is 9.80 m·s^{-2}, and the radius of the earth is 6400 km. Compute the mass of the earth from these data.

6–14 A 0.1-kg point mass is located on the line between a 5-kg and a 10-kg point mass. It is 4 m to the right of the 5-kg mass and 6 m to the left of the 10-kg mass. What are the magnitude and direction of the force on the 0.1-kg mass?

Section 6–4 Gravitational Field

6–15 What is the magnitude of the gravitational field 3 m from a 5-kg point mass?

6–16 What are the magnitude and direction of the gravitational force on a 0.1-kg point mass placed at a point where the gravitational field has components $g_x = 4 \text{ m·s}^{-2}$ and $g_y = -8 \text{ m·s}^{-2}$?

6–17 The gravitational force F on a 0.01-kg test mass is found to be $F = (-0.1 \text{ N})i + (0.4 \text{ N})j$ at a certain point. What are the components of the gravitational field vector at that point?

6–18 What are the magnitude and direction of the gravitational field at a point midway between two point masses, one of mass m_1 and the other of mass m_2 where $m_2 > m_1$? The two point masses are a distance d apart.

Section 6–5 Satellite Motion

6–19 To launch a satellite in a circular orbit 1000 km above the surface of the earth, what orbital speed must be imparted to the satellite?

6–20 What is the period of revolution of a manmade satellite of mass m that is orbiting the earth in a circular path of radius 8000 km (about 1620 km above the surface of the earth)?

6–21 An earth satellite rotates in a circular orbit of radius 7000 km (about 620 km above the earth's surface) with an orbital speed of 7550 m·s^{-1}.

a) Find the time of one revolution.

b) Find the acceleration of gravity at the orbit.

Section 6–6 Effect of the Earth's Rotation on g

6–22 The weight of a man as determined by a spring balance at the equator is 800 N. By how much does this differ from the true force of gravitational attraction at the same point?

6–23 The acceleration due to gravity at the north pole of Jupiter is approximately 25 m·s^{-2}. Jupiter has mass 1.9×10^{27} kg, radius 7.1×10^7 m, and makes one revolution about its axis in about 10 hr.

a) What is the gravitational force on a 2-kg object at the north pole of Jupiter?

b) What is the apparent weight of this same object at the equator of Jupiter?

PROBLEMS

6–24 You are riding in a bus. As the bus rounds a flat curve of radius 60 m, a horseshoe of mass 0.5 kg suspended from the ceiling of the bus by a 2-m-long string is found to be in equilibrium when the string makes an angle of 37° with respect to the vertical. What is the speed v of the bus?

6–25 Consider a roadway banked as in Example 6–4, where there is a coefficient of static friction of 0.35 and a coefficient of kinetic friction of 0.25 between the tires and the roadway. The radius of the curve is $R = 25$ m.

a) If the banking angle is $\theta = 25°$, what is the *maximum* velocity the automobile can have before sliding *up* the banking?

b) What is the *minimum* velocity the automobile can have before sliding *down* the banking?

6–26 A curve of 200-m radius on a level road is banked at the correct angle for a velocity of 15 m·s⁻¹. If an automobile rounds this curve at 30 m·s⁻¹, what is the minimum coefficient of friction between tires and road so that the automobile will not skid? Assume all forces to act at the center of gravity.

6–27 In the ride "Spindletop" at the Six Flags Over Texas amusement park, people stand against the inner wall of a hollow vertical cylinder of radius 3 m. The cylinder starts to rotate, and when it reaches a constant rotation rate of 0.5 rev·s⁻¹, the floor on which people are standing drops about 0.5 m. The people remain pinned against the wall.

a) Draw a force diagram for a person in this ride, after the floor has dropped.

b) What minimum coefficient of static friction is required if the person in the ride is not to slide downward to the new position of the floor?

c) Does your answer in (b) depend on the mass of the passenger?

(*Note.* When the ride is over, the cylinder is slowly brought to rest; as it slows down, people slide down the walls to the floor.)

6–28 The 4-kg block in Fig. 6–14 is attached to a vertical rod by means of two strings. When the system rotates about the axis of the rod, the strings are extended as shown in the diagram.

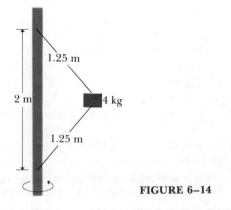

FIGURE 6–14

a) How many revolutions per minute must the system make for the tension in the upper string to be 60 N?

b) What is then the tension in the lower string?

6–29 A bead can slide without friction on a circular hoop of radius 0.1 m in a vertical plane. The hoop rotates at a constant rate of 2 rev·s⁻¹ about a vertical diameter, as in Fig. 6–15.

a) Find the angle θ at which the bead is in vertical equilibrium. (Of course, it has a radial acceleration toward the axis.)

b) Is it possible for the bead to "ride" at the same elevation as the center of the hoop?

c) What will happen if the hoop rotates at 1 rev·s⁻¹?

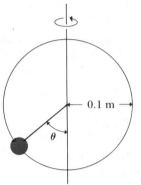

FIGURE 6–15

6–30 A physics major is working to pay his college tuition by performing in a traveling carnival. He rides a motorcycle inside a hollow steel sphere of radius 15 m. The surface of the sphere is full of holes so the audience can see in. After gaining sufficient speed, he travels in a vertical circle. The physics major has mass 60 kg, and his motorcycle has mass 40 kg.

a) What minimum velocity must he have at the top of the circle if the tires of the motorcycle are not to lose contact with the steel sphere?

b) At the bottom of the circle his velocity is twice the value calculated in (a). What then is the magnitude of the normal force exerted on the motorcycle by the steel sphere at this point?

6–31 The radius of a Ferris wheel is 5 m, and it makes one revolution in 10 s.

a) Find the difference between the apparent weight of a passenger at the highest and lowest points, expressed as a fraction of his weight. (That is, find the difference between the upward force exerted on the passenger by the seat at these two points.)

b) What would be the time for one revolution if his apparent weight at the highest point were zero?

c) What would then be his apparent weight at the lowest point?

d) What would happen, at this rate of revolution, if his seat belt broke at the highest point, and if he did not hang onto his seat?

6–32 A ball is held at rest at position *A* in Fig. 6–16 by two light cords. The horizontal cord is cut, and the ball starts swinging as a pendulum. Point *B* is the farthest to the right that the ball goes as it swings back and forth. What is the ratio of the tension in the supporting cord in position *B* to what it was at *A* before the horizontal cord was cut?

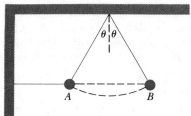

FIGURE 6–16

6–33 A spaceship travels from Earth directly toward the sun. At what distance from the center of the earth do the gravitational forces of the sun and the earth on the ship exactly cancel? Use the data in Appendix F.

6–34 Two spheres, each of mass 6.4 kg, are fixed at points *A* and *B* (Fig. 6–17). Find the magnitude and direction of the initial acceleration of a sphere of mass 0.010 kg if released from rest at point *P* and acted on only by forces of gravitational attraction of the spheres at *A* and *B*.

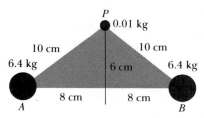

FIGURE 6–17

6–35 Three point masses are fixed at positions shown in Fig. 6–18.

a) Calculate the *x*- and *y*-components of the gravitational field at point *P*, which is at the origin of coordinates, due to these three point masses.

b) What would be the magnitude and direction of the force on a 0.01-kg mass placed at *P*?

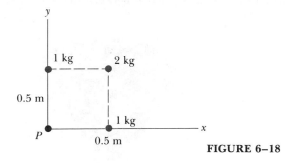

FIGURE 6–18

6–36 Two point masses are located at (*x*,*y*) coordinates as follows: 4 kg at (1 m, 0) and 3 kg at (0, −0.5 m).

a) What are the components of the gravitational field at the origin, due to these two point masses?

b) What will be the magnitude and direction of the gravitational force on a 0.01-kg test mass placed at the origin?

6–37 The asteroid Toro was discovered in 1964. Its radius is about 5 km.

a) Assuming the density (mass per unit volume) of Toro is the same as that of the earth, find its total mass, and find the acceleration due to gravity at its surface.

b) Suppose an object is to be placed in a circular orbit around Toro, with a radius just slightly larger than the asteroid's radius. What is the speed of the object? Could you launch yourself into orbit around Toro by running?

6–38 Several satellites are moving in a circle in the earth's equatorial plane. They are at such a height above the earth's surface that they always remain above the same point. Find the height of these satellites. (Such an orbit is said to be geosynchronous.)

6–39 What would be the length of a day (that is, the time required for one revolution of the earth on its axis) if the rate of revolution of the earth were such that $g = 0$ at the equator?

CHALLENGE PROBLEMS

6–40 Mass *M* is distributed uniformly along a line of length 2*L*. Calculate the gravitational field components perpendicular and parallel to the line at a point that is at a distance *a* from the line, on its perpendicular bisector. Does your result reduce to the correct expression as *a* becomes very large?

6–41 Mass *M* is distributed uniformly over a disk of radius *a*. Find the gravitational force (magnitude and direction) between this disk-shaped mass and a point mass *m* located a distance *x* above the center of the disk. Does your result reduce to the correct expression as *x* becomes very large? (*Hint:* Divide the disk into infinitesimally thin concentric rings, use the expression derived in Section 6–4 for the

gravitational field due to each ring, and integrate to find the total field.)

6–42 Mass *M* is distributed uniformly over a thin spherical shell of radius *R*. Calculate the gravitational field at a distance *r* from the center of the shell for

a) $r > R$;   b) $r < R$.

(*Hint:* Divide the spherical shell into infinitesimally thin concentric rings, use the expression derived in Section 6–4 for the gravitational field due to each ring, and integrate to find the total field.)

6–43 A small block of mass *m* is placed inside an inverted cone that is rotating about a vertical axis such that the time for one revolution of the cone is τ (see Fig. 6–19). The walls of the cone make an angle θ with the vertical. The coefficient of static friction between the block and the cone is μ_s. If the block is to remain at a constant height h above the apex of the cone, what are the maximum and minimum values of τ?

6–44 A small block of mass m on a frictionless horizontal table is attached to a spring of unstretched length l_0. The other end of the spring is looped around a vertical rod at the center of the table, so that the block and spring can rotate freely about the rod. The spring has negligible mass and force constant k, defined by $F_{\text{spring}} = -k(l - l_0)$ where l is the length of the stretched spring. The block is set into uniform circular motion with velocity of magnitude v.

a) Calculate the length l of the spring as a function of the speed v of the block.

b) Express your result in (a) as the length l of the spring in terms of the frequency f of revolution of the block, where f is the number of revolutions it makes each second. Make a sketch of l as a function of f.

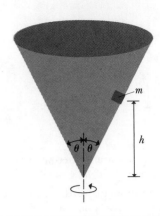

FIGURE 6–19

PART TWO

MECHANICS—FURTHER DEVELOPMENTS

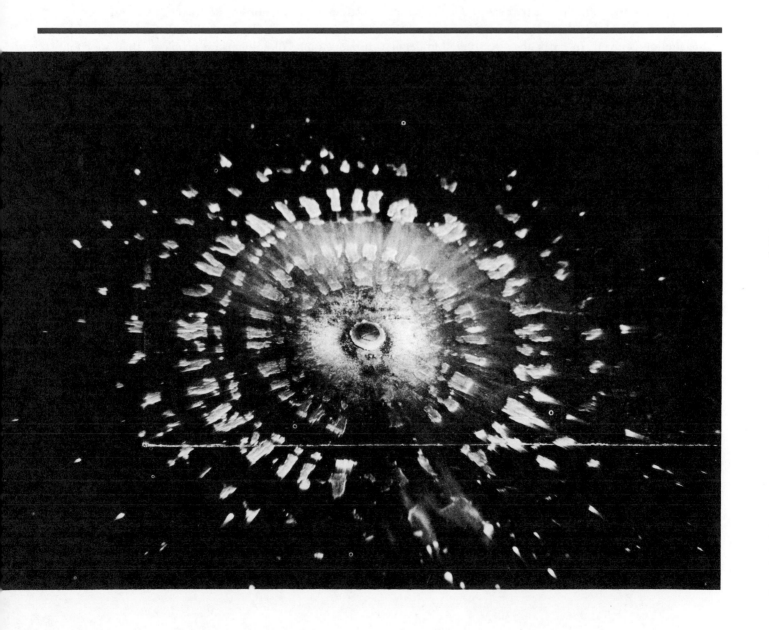

PERSPECTIVE

In the opening chapters of this book we developed mathematical language for describing the position and motion of a particle, using the concepts of displacement, velocity, and acceleration. We used the particle as a simplified *model* to represent any body small enough to be considered as a point, or whose shape is irrelevant in the problem at hand. With the aid of two additional concepts, *force* and *mass,* we have also been able to state general principles relating the motion of a particle to the forces acting on it. We learned how to determine the motion of a particle under the action of given forces and how to determine the forces when the motion is given. In particular, we studied in detail the motion of a body in a uniform gravitational field, including free fall and parabolic trajectories, and we applied Newton's laws to the motion of a variety of other simple mechanical systems. With the aid of one additional principle, the *law of gravitation,* we can also analyze in detail the motions of planets around the sun and of satellites around the earth.

However, there are many problems where it is not feasible to apply Newton's laws directly, either because the motion is too complicated to describe in detail or because the forces are not known. An example is a collision between two bodies, where large, variable, and unknown forces act on the bodies during their interaction. To discuss this and similar situations, we need additional principles that will enable us to relate the motion of a system at one time to its motion at a later time, without having to know all the details of what happened in between. The concepts of *energy* and *momentum,* which we study next, enable us to formulate such principles and the related *conservation laws.* These concepts have the added significance that their relevance extends to other areas of physics, such as thermodynamics, electromagnetism, and optics;

thus they provide unifying threads that help to integrate the whole fabric of physics.

Next we broaden the scope of our mechanical principles to include bodies that cannot be represented adequately as particles. In particular, we consider in detail the rotational motion of an extended body. We introduce the idealized model of a *rigid body:* a body that does not deform or change shape when forces are applied. We develop kinematic language for describing this motion and the dynamical principles that relate angular motion to the forces acting on the body. These principles also allow us to deal with a broader range of equilibrium problems than we could handle using only particle models. After this we undertake a detailed study of *periodic motion,* a class of motions in which a physical system repeats the same motion over and over again in a definite cycle. Periodic motion has far-reaching applications that go beyond mechanics and extend into nearly all other areas of physics. The study of periodic motion prepares us for later work in analyzing mechanical waves, electrical circuits, optics, and many other areas of physics. Once again, a concept developed in a mechanical context serves as a unifying theme, tying together several areas of physics.

In the opening chapters, we relied heavily on experimental evidence as the basis for the general principles we call Newton's laws. Our approach there was largely *inductive,* generalizing from specific observations to general principles. In the next few chapters our process is more *deductive,* deriving additional results from Newton's laws. Thus these chapters have a somewhat different flavor from that of the opening chapters. Taken together, these first eleven chapters show the vital interplay between inductive and deductive reasoning in the development of physical science.

7

WORK AND ENERGY

ENERGY IS ONE OF THE MOST IMPORTANT UNIFYING CONCEPTS IN ALL OF physical science. Its importance stems from the principle of **conservation of energy,** which states that in any isolated system the total energy of all forms is constant. In this chapter we concentrate on mechanical energy—energy associated with the motion and position of mechanical systems. We begin by introducing the concepts of work and kinetic energy and the relationship between them, and we use these in a variety of problems. Potential energy provides a convenient way to calculate work associated with certain kinds of forces. Finally, we introduce the term *power* to characterize the rate of doing work or transferring energy in or out of a system.

7–1 CONSERVATION OF ENERGY

The concept of **energy** appears throughout every area of physics, and yet it is difficult to define in a general way just what energy is. Energy plays a central role in one of a class of fundamental laws of nature called **conservation laws,** and looking at this role is as good a way as any to approach the question of what energy is. The various conservation laws of physics, including conservation of energy, momentum, angular momentum, electric charge, and others, play vital roles in every area of physics and help to unify the subject.

A conservation law always concerns a transformation or an interaction that occurs within some physical system or between a system and its surroundings. The state or condition of the system at any time is described by physical quantities such as position and velocity. These quantities may change during the transformation or interaction, but one or more quantities may remain constant, or be *conserved*. A familiar example is conservation of *mass* in chemical reactions. A large amount of experimental evidence has established that the total mass of the reactants in a chemical reaction is always equal to the total mass of all the products of the reaction. That is, the total mass is always the same after the reaction occurs as before. This generalization is called the principle of *conservation of mass,* and it is obeyed in all chemical reactions.

Conservation laws play a vital role in all of physical science.

Conservation of mass: Matter cannot be created or destroyed.

145

Kinetic energy is energy associated with mass in motion.

Something similar happens in collisions between bodies. For a body of mass m moving with speed v, we can define a quantity $\frac{1}{2}mv^2$ associated with the motion and called the *kinetic energy* of the body. When two highly elastic or "springy" bodies (such as two hard steel ball bearings) collide, we find that the individual speeds change, but the total kinetic energy (the sum of the $\frac{1}{2}mv^2$ quantities for all the colliding bodies) is the same after the collision as before. We say that kinetic energy is *conserved* in such collisions. This result does not tell us what kinetic energy *is*, but only that it is useful in representing a conservation principle in certain kinds of interactions.

In contrast, when two soft, deformable bodies, such as two balls of putty or chewing gum, collide, experiment shows that kinetic energy is *not* conserved. However, something else happens: The bodies become *warmer*. Furthermore, it turns out that there is a definite relationship between the temperature rise of the material and the loss of kinetic energy. We can define a new quantity, *internal energy*, that increases with temperature in a definite way, so that the *sum* of kinetic energy and internal energy is conserved in these collisions.

Conservation of energy can be generalized to include energy associated with every kind of physical phenomenon.

The significant discovery here is that it is *possible* to extend the principle of conservation of energy to a broader class of phenomena by defining a new form of energy. This is precisely how the principle has developed. In any situation where it seems that the total energy in all known forms is *not* conserved, it has been found possible to define a new form of energy so that the total energy, including the new form, *is* conserved. Thus we find energy associated with heat, with elastic deformations, with electric and magnetic fields, and, in relativity theory, even with mass itself. Conservation of energy has the status, along with a small number of partners, of a *universal* conservation principle: No exception to its validity has ever been found.

In this chapter we are concerned primarily with *mechanical* energy, that is, with energy associated with motion, position, and deformation of material bodies. In some interactions mechanical energy is conserved, and in others there is a conversion from mechanical energy to other forms of energy, or the reverse, so that mechanical energy by itself is not conserved. In Chapters 15 and 18 we will study in greater detail the relation of mechanical energy to heat, and in later chapters still other forms of energy will be considered.

7–2 WORK

In everyday life *work* is any activity that requires muscular or mental exertion. Physicists use the term in a much more specific sense, involving a *force* acting on a body while the body undergoes a *displacement*. When a body moves a distance s along a straight line while a constant force of magnitude F, directed along the line, acts on it, the **work** W done by the force is defined as

$$W = Fs. \tag{7–1}$$

In this chapter we will develop a very useful relationship between work and quantities that describe the *motion* of the body.

The force need not have the same direction as the displacement. In Fig. 7–1, the force $\boldsymbol{F}$, assumed constant, makes an angle θ with the displacement. The work W done by this force when its point of application undergoes a displacement s is defined as the product of the magnitude of the displacement and the *component* of the force in the direction of the displacement.

A force acting on a body does work on the body only when the body moves.

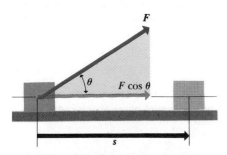

7–1 The work done by the force $\boldsymbol{F}$ during a displacement s is $(F \cos \theta)s$.

The component of F in the direction of s is $F \cos \theta$. Then

$$W = (F \cos \theta)s. \qquad (7\text{–}2)$$

An alternative interpretation of Eq. (7–2) is that $s \cos \theta$ is the component of displacement in the direction of F. Thus the work is also the component of displacement in the direction of F multiplied by the magnitude of F.

Comparing Eq. (7–2) with the definition of the *scalar product* of two vectors, Eq. (1–19), we see that work can also be expressed compactly as

$$W = \boldsymbol{F} \cdot \boldsymbol{s}. \qquad (7\text{–}3)$$

Although it is calculated from two vector quantities, work itself is a *scalar* quantity. A 5-N force toward the east acting on a body that is displaced 6 m to the east does exactly the same amount of work as a 5-N force toward the north acting on a body displaced 6 m to the north. Work is an *algebraic* quantity: It can be positive or negative. When the component of the force is in the *same* direction as the displacement, the work W is *positive*. When it is *opposite* to the displacement, the work is *negative*. If the force is at *right angles* to the displacement, it has no component in the direction of the displacement, and the work is *zero*.

> Work is a scalar quantity, even though force and displacement are vectors.

Thus, when a body is lifted, the work done by the lifting force is positive; when a spring is stretched, the work done by the stretching force is positive; when a gas is compressed in a cylinder, the work done by the compressing force is positive. On the other hand, the work done by the *gravitational* force on a body being lifted is negative, since the (downward) gravitational force is opposite to the (upward) displacement. It is considered "hard work" to hold a heavy object stationary at arm's length, but no work is done in the technical sense because there is no motion! Even if you walk along a level floor while carrying the object, no work would be done, because the (vertical) supporting force has no component in the direction of the horizontal motion. Similarly, when a body slides along a surface, the work done by the normal force acting on the body is zero, and when a body moves in a circle, the work done by the centripetal force on the body is also zero.

> When force and displacement are perpendicular, the force does no work.

The unit of work in any system is the unit of force multiplied by the unit of distance. In SI units the unit of force is the newton and the unit of distance is the meter; thus in this system the unit of work is 1 *newton meter* (1 N·m). This combination of units appears so frequently in mechanics that it is given a special name, the **joule** (abbreviated J).

> The joule is the SI unit of work and of energy of all kinds.

$$1 \text{ joule} = (1 \text{ newton})(1 \text{ meter}) \quad \text{or} \quad 1 \text{ J} = 1 \text{ N·m}.$$

In the cgs system, the unit of work is 1 dyne-centimeter, also called 1 *erg*. Because 1 dyn $= 10^{-5}$ N and 1 cm $= 10^{-2}$ m,

$$1 \text{ erg} = 10^{-7} \text{ J}.$$

In the British system the unit of work is 1 *foot-pound* (1 ft-lb); no special name is given this unit. The following conversions are useful:

$$1 \text{ J} = 0.7376 \text{ ft·lb}, \quad 1 \text{ ft·lb} = 1.356 \text{ J}.$$

When several external forces act on a body, it is useful to consider the work done by each separate force. Each of these may be computed from the definition of work in Eq. (7–2). Then, since work is a scalar quantity, the total work is the algebraic sum of the individual works.

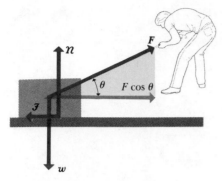

7–2 A box on a rough horizontal surface moving to the right under the action of a force **F** inclined at an angle θ.

When several forces act on a moving body, the total work can be calculated in several ways.

EXAMPLE 7–1 Figure 7–2 shows a box being dragged along a horizontal surface by a worker applying a constant force **F** that makes a constant angle θ with the direction of motion. The other forces on the box are its weight w, the upward normal force n exerted by the surface, and the friction force $\mathcal{F}$. What is the work done by each force when the box moves a distance s along the surface to the right?

SOLUTION The component of **F** in the direction of motion is $F \cos \theta$. The work done by the force **F** is therefore

$$W_F = (F \cos \theta)s.$$

The forces w and n are both at right angles to the displacement. Hence

$$W_w = 0, \qquad W_n = 0.$$

The friction force $\mathcal{F}$ is opposite to the displacement, so the work done by the friction force is

$$W_{\mathcal{F}} = -\mathcal{F}s.$$

Since work is a scalar quantity, the total work ΣW done by all forces on the box is the *algebraic* sum (not the vector sum) of the individual works:

$$\Sigma W = W_F + W_w + W_n + W_{\mathcal{F}}$$
$$= (F \cos \theta)s + 0 + 0 - \mathcal{F}s$$
$$= (F \cos \theta - \mathcal{F})s.$$

But $(F \cos \theta - \mathcal{F})$ is the *resultant* force on the box. Hence *the total work done by all forces is equal to the work done by the resultant force.*

Suppose that $F = 50$ N, $\mathcal{F} = 15$ N, $\theta = 36.9°$, and $s = 20$ m. Then

$$W_F = (F \cos \theta)s$$
$$= (50 \text{ N})(0.800)(20 \text{ m}) = 800 \text{ N·m},$$

$$W_{\mathcal{F}} = -\mathcal{F}s$$
$$= (-15 \text{ N})(20 \text{ m}) = -300 \text{ N·m},$$

$$\Sigma W = W_F + W_{\mathcal{F}} = 500 \text{ N·m}.$$

As a check, the total work may be expressed as

$$\Sigma W = (F \cos \theta - \mathcal{F})s$$
$$= (40 \text{ N} - 15 \text{ N})(20 \text{ m}) = 500 \text{ N·m}.$$

When several forces act on a body, there are always two equivalent ways to calculate the total work. We may calculate the work done by each force separately and take the algebraic sum of these works, or we may compute the vector sum or resultant of the forces and compute the work done by the resultant.

7–3 WORK DONE BY A VARYING FORCE

In the preceding section we defined the work done by a *constant* force. Often, however, work is done by a force that varies in magnitude or direction during the displacement of the body on which it acts. Thus when a spring is stretched slowly, the force required to stretch it increases steadily as the spring elongates; when a body is projected vertically upward, the gravitational force exerted on it by the earth decreases inversely with the square of its distance from the earth's center.

Suppose a particle moves along a line under the action of a force directed along the line but varying with the particle's position. In Fig. 7–3a the force magnitude is shown as a function of the particle's coordinate x. To find the work done by this force, we divide the displacement into short segments Δx_1, Δx_2, and so on, as in Fig. 7–3b. We approximate the varying force by one that is constant within each segment. The force then has approximately the value F_1 in segment Δx_1, F_2 in segment Δx_2, and so on. The work done in the first segment is then $F_1 \Delta x_1$, that in the second is $F_2 \Delta x_2$, and so on. The *total* work is

$$W = F_1 \Delta x_1 + F_2 \Delta x_2 + \cdots.$$

As the number of segments becomes very large and the size of each very small, this sum becomes (in the limit) the *integral* of F from x_1 to x_2:

$$W = \int_{x_1}^{x_2} F \, dx. \tag{7–4}$$

Note that $F_1 \Delta x_1$ represents the *area* of the first vertical strip in Fig. 7–3b, and so on, and that the integral in Eq. (7–4) represents the area under the curve in Fig. 7–3a.

If the force also varies in *direction* during the displacement, then F in Eq. (7–4) must be replaced by the *component* of force in the direction of displacement. This is given by $F \cos \theta$, where θ is the angle between F and the x-axis at each point. The angle θ may vary from point to point. Then we have

$$W = \int_{x_1}^{x_2} F \cos \theta \, dx. \tag{7–5}$$

As an example of this method, let us compute the work done when a spring is stretched. To keep a spring stretched at an elongation x beyond its unstretched length, we must apply a force F at each end, as shown in Fig. 7–4. If the elongation is not too great, F is directly proportional to x:

$$F = kx, \tag{7–6}$$

where k is a constant called the **force constant,** or the *stiffness,* of the spring. This direct proportion between force and elongation, for elongations that are not too great, was discovered by Robert Hooke in 1678 and is known as **Hooke's law.** We will discuss it more fully in Chapter 12.

Suppose that equal and opposite forces are applied to the ends of the spring, and that the forces are gradually increased from zero. One end of the spring is held stationary; no work is done by the force at this end, but work is

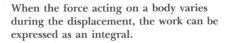

When the force acting on a body varies during the displacement, the work can be expressed as an integral.

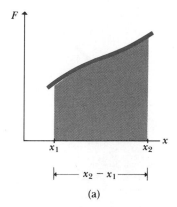

$x_2 - x_1$

(a)

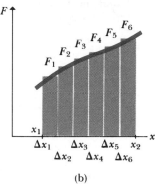

(b)

7–3 (a) Curve showing how F varies with x. (b) If the area is partitioned into small rectangles, their sum approximates the total work done during the displacement; the greater the number of rectangles used, the closer is the approximation.

Hooke's law: The force required to stretch a spring is proportional to the amount of stretch.

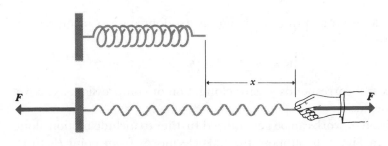

7–4 The force to stretch a spring is proportional to its elongation: $F = kx$.

The work required to stretch a spring is proportional to the square of the amount of stretch.

done by the varying force F at the moving end. In Fig. 7–5, F is plotted vertically, and the displacement x of the moving end, representing the elongation of the spring, is plotted horizontally.

The work required to stretch the spring from $x = 0$ (no elongation) to $x = X$ is

$$W = \int_0^X F\, dx = \int_0^X kx\, dx = \tfrac{1}{2}kX^2. \qquad (7\text{–}7)$$

This result can also be obtained graphically: The area of the shaded triangle in Fig. 7–5, representing the total work, is equal to half the product of base and altitude, or

$$W = \tfrac{1}{2}(X)(kX) = \tfrac{1}{2}kX^2,$$

in agreement with the above result.

The work is therefore proportional to the *square* of the final elongation X. When the elongation is doubled, the work increases by a factor of four. Another view of this relationship is that the total work W is equal to the *average* force $\tfrac{1}{2}kX$ multiplied by the total displacement X.

The spring also exerts a force on the hand, which moves during the stretching process, and we may ask what work the spring does on the hand. The displacement of the hand is the same as that of the moving end of the spring, but the force on it is the negative of the force on the spring because the two forces form an action–reaction pair. Thus the work done on the hand is the negative of the work done on the spring, namely, $-\tfrac{1}{2}kX^2$. In work calculations, it is always essential to specify on which body the work is being done!

In some cases, such as a spring with spaces between the coils, Hooke's law holds for compression as well as stretching. In this case, the work done on the spring in compressing it a distance X is also given by Eq. (7–7).

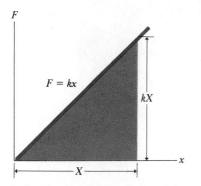

7–5 The work done in stretching a spring is equal to the area of the shaded triangle.

Bathroom scales often use springs to measure weight.

EXAMPLE 7–2 A woman weighing 600 N steps on a bathroom scale containing a heavy spring. The spring is compressed 1.0 cm under her weight. Find the force constant of the spring and the total work done on it during the compression.

SOLUTION From Eq. (7–6),

$$k = \frac{F}{x} = \frac{600\ \text{N}}{0.01\ \text{m}} = 60{,}000\ \text{N·m}^{-1}.$$

Then from Eq. (7–7),

$$W = \tfrac{1}{2}kX^2 = \tfrac{1}{2}(60{,}000\ \text{N·m}^{-1})(0.01\ \text{m})^2 = 3\ \text{N·m} = 3\ \text{J}.$$

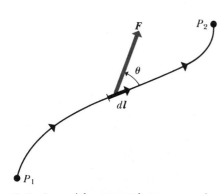

7–6 A particle moves along a curved path from point P_1 to P_2, acted on by a force F that varies in magnitude and direction. During an infinitesimal displacement dl (a small segment of the path), the work dW done by the force is given by $dW = F \cdot dl$.

The work done on a spring in changing its elongation or compression from x_1 to x_2 is

$$W = \tfrac{1}{2}kx_2{}^2 - \tfrac{1}{2}kx_1{}^2,$$

provided that $x = 0$ corresponds to zero elongation or compression, as above. (Can you prove this?)

The definition of work can be generalized further to include motion along a *curved* path. In Fig. 7–6, suppose the particle moves from point P_1 to P_2

along the curve. We imagine dividing the portion of the curve between these points into many infinitesimal vector displacements, and we call a typical one of these $d\boldsymbol{l}$. Each $d\boldsymbol{l}$ is tangent to the path at its position. Let $\boldsymbol{F}$ be the force at a typical point along the path, and θ the angle between $\boldsymbol{F}$ and $d\boldsymbol{l}$ at this point. Then the small element of work dW done on the particle during the displacement $d\boldsymbol{l}$ may be written as

$$dW = F \cos \theta \, dl = F_{\parallel} \, dl = \boldsymbol{F} \cdot d\boldsymbol{l},$$

where as before $F_{\parallel} = F \cos \theta$ is the component of $\boldsymbol{F}$ in the direction of $d\boldsymbol{l}$. The total work done on the particle by $\boldsymbol{F}$ as the particle moves from P_1 to P_2 is then represented symbolically as

> When a particle moves along a curved path, the work done on it can be expressed as a line integral.

$$W = \int_{P_1}^{P_2} F \cos \theta \, dl = \int_{P_1}^{P_2} F_{\parallel} \, dl = \int_{P_1}^{P_2} \boldsymbol{F} \cdot d\boldsymbol{l}. \qquad (7\text{--}8)$$

This integral is called a *line integral*. Actually evaluating Eq. (7–8) in a specific problem requires that we have some sort of detailed description of the path and of the variation of $\boldsymbol{F}$ along the path.

EXAMPLE 7–3 A child of weight w sits on a swing of length R, as shown in Fig. 7–7. A *variable* horizontal force $\boldsymbol{P}$, which starts at zero and gradually increases, is used to pull the child very slowly (so that equilibrium exists at all times) until the ropes make an angle θ_0 with the vertical. Calculate the work of the force $\boldsymbol{P}$.

> Work done while pushing a child on a swing: an example of motion along a curved path

SOLUTION The body is in equilibrium, so the sum of the horizontal forces equals zero:

$$P = T \sin \theta.$$

The sum of the vertical forces is also zero:

$$w = T \cos \theta.$$

Dividing these two equations, we find

$$P = w \tan \theta.$$

The point of application of $\boldsymbol{P}$ swings through the arc s. Since $s = R \theta$, $dl = R \, d\theta$, and

$$W = \int \boldsymbol{P} \cdot d\boldsymbol{l} = \int P \cos \theta \, dl$$

$$= \int_{0}^{\theta_0} w \tan \theta \cos \theta \, R \, d\theta$$

$$= wR \int_{0}^{\theta_0} \sin \theta \, d\theta = wR(1 - \cos \theta_0). \qquad (7\text{--}9)$$

If $\theta_0 = 0$, there is no displacement; in that case $\cos \theta_0 = 1$ and $W = 0$, as expected. If $\theta_0 = 90°$, then $\cos \theta_0 = 0$ and $W = wR$. In that case the work is the same as though the body had been lifted straight up a distance R by a force equal to its weight w. In fact, as you should verify, the quantity $R(1 - \cos \theta_0)$ is the increase in the height of the body during the displacement; so for any value of θ_0, the work done by force $\boldsymbol{P}$ is the change in height multiplied by the weight. We will prove this result more generally in Section 7–5.

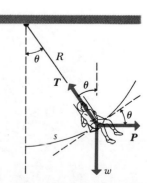

7–7 A variable horizontal force $\boldsymbol{P}$ acts on a body while the displacement varies from zero to s.

7–4 WORK AND KINETIC ENERGY

The work done on a body by a force is related very directly to the resulting change in the body's motion. To develop this relationship, we consider first a body of mass m moving along a straight line under the action of a constant resultant force of magnitude F directed along the line. The body's acceleration is given by Newton's second law, $F = ma$. Suppose the speed increases from v_1 to v_2 while the body undergoes a displacement $s = x_2 - x_1$. Then, from the analysis of motion with constant acceleration, and from Eq. (2–13), we have

$$v_2^2 = v_1^2 + 2as,$$

$$a = \frac{v_2^2 - v_1^2}{2s}.$$

Hence

$$F = m\,\frac{v_2^2 - v_1^2}{2s},$$

and

$$Fs = \tfrac{1}{2}mv_2^2 - \tfrac{1}{2}mv_1^2. \tag{7–10}$$

The product Fs is the work W done by the resultant force F. The quantity $\tfrac{1}{2}mv^2$, one-half the product of the mass of the body and the square of its speed, is called its **kinetic energy, K**:

$$K = \tfrac{1}{2}mv^2. \tag{7–11}$$

The work done on a body equals its change in kinetic energy.

The first term on the right side of Eq. (7–10) is the final kinetic energy of the body, $K_2 = \tfrac{1}{2}mv_2^2$, and the second term is the initial kinetic energy, $K_1 = \tfrac{1}{2}mv_1^2$. The difference between these terms is the *change* in kinetic energy, and we have the important result that *the work done by the resultant external force on a body is equal to the change in kinetic energy of the body:*

$$W = K_2 - K_1 = \Delta K. \tag{7–12}$$

Kinetic energy is a scalar quantity, even though the particle's velocity is a vector quantity.

Kinetic energy, like work, is a *scalar* quantity. The kinetic energy of a moving body depends only on its speed (the *magnitude* of its velocity) but not on the *direction* in which it is moving. The *change* in kinetic energy depends only on the work $W = Fs$ and not on the individual values of F and s. That is, the force F could have been large and the displacement s small, or the reverse. If the mass m and the speeds v_1 and v_2 are known, the work done by the resultant force can be found without any knowledge of the force F and the displacement s.

If the work W is *positive*, the final kinetic energy is greater than the initial kinetic energy and the kinetic energy *increases*. If the work is *negative*, the kinetic energy *decreases*. In the special case in which the work is *zero*, the kinetic energy remains *constant*.

In Eq. (7–12), W is the work done by the *resultant* force, that is, the *vector sum* of all the forces acting on the body. Alternatively, one may calculate the work done by each separate force. W is then the *algebraic sum* of all these quantities of work. Example 7–1 (Section 7–2) illustrates these two alternatives.

Although we derived Eq. (7–12) for the special case of a constant resultant force, it is true even when the force varies in an arbitrary way. The work done on a particle by any resultant force equals the change in kinetic energy of the body. To prove this statement, we divide the total displacement x into a large number of small segments Δx, just as in the calculation of work done by a varying force. The change of kinetic energy in segment Δx_1 is equal to the work $F_1 \Delta x_1$, and so on; the total kinetic energy change is the sum of the changes in the individual segments and is thus equal to the total work done.

In SI units, m is measured in kilograms and v in meters per second. From Eq. (7–12), kinetic energy must have the same units as work. To verify this we recall that $1 \text{ N} = 1 \text{ kg·m·s}^{-2}$. Hence

The units of work and kinetic energy are the same.

$$1 \text{ J} = 1 \text{ N·m} = 1 \text{ kg·m·s}^{-2} \cdot \text{m} = 1 \text{ kg·m}^2 \cdot \text{s}^{-2}.$$

The joule is thus the SI unit of *both* work and kinetic energy and, as we will see later, of all kinds of energy. Similarly, in the cgs system, with m in grams and v in centimeters per second, we have

$$1 \text{ erg} = 1 \text{ dyn·cm} = 1 \text{ g·cm·s}^{-2} \cdot \text{cm} = 1 \text{ g·cm}^2 \cdot \text{s}^{-2}.$$

In the British system.

$$1 \text{ ft·lb} = 1 \text{ ft·slug·ft·s}^{-2} = 1 \text{ slug·ft}^2 \cdot \text{s}^{-2}.$$

PROBLEM-SOLVING STRATEGY: *Work and kinetic energy*

1. Make a list of all the forces acting on the body, and calculate the work done by each force. In some cases one or more forces may be unknown; represent the unknowns by algebraic symbols. Be sure to check signs: When a force has a component in the *same* direction as the displacement, its work is positive; when it has a component in the direction opposite to the displacement, its work is negative.

2. Add the works done by the separate forces to find the total work. Again be careful with signs. Sometimes, though not often, it may be easier to calculate the resultant (vector sum) of the forces first, and then find the work done by the resultant force.

3. List the initial and final kinetic energies K_1 and K_2. If a quantity, such as v_1 or v_2 is unknown, leave it in terms of the corresponding algebraic symbol.

4. Use the relationship $W = K_2 - K_1$; insert the results from the above steps and solve for whatever unknown is required.

EXAMPLE 7–4 Consider again the box in Fig. 7–2 and the numbers in Example 7–1. We found that the total work done by the forces was 500 N·m = 500 J. Hence the kinetic energy of the box must increase by 500 J. Suppose the initial speed v_1 is 4 m·s^{-1} and the mass of the box is 10 kg. What is the final speed?

Pushing a box with increasing speed: an application of the work-energy relationship

SOLUTION The initial kinetic energy is

$$K_1 = \tfrac{1}{2}mv_1{}^2 = \tfrac{1}{2}(10 \text{ kg})(4 \text{ m·s}^{-1})^2 = 80 \text{ J}.$$

The final kinetic energy is given by Eq. (7–12):

$$K_2 = K_1 + W = 80 \text{ J} + 500 \text{ J} = 580 \text{ J}.$$

Thus

$$K_2 = \tfrac{1}{2}(10 \text{ kg})v_2{}^2 = 580 \text{ J}$$

and

$$v_2 = 10.8 \text{ m·s}^{-1}.$$

To verify this we may find the acceleration from $F = ma$ and then use the equations of motion with constant acceleration to find v_2:

$$a = \frac{F}{m} = \frac{40 \text{ N} - 15 \text{ N}}{10 \text{ kg}} = 2.5 \text{ m·s}^{-2}.$$

Then

$$\begin{aligned}v_2{}^2 &= v_1{}^2 + 2as = (4 \text{ m·s}^{-1})^2 + 2(2.5 \text{ m·s}^{-2})(20 \text{ m}) \\ &= 116 \text{ m}^2\text{·s}^{-2}, \\ v_2 &= 10.8 \text{ m·s}^{-1}.\end{aligned}$$

This is the same result we obtained with the work-energy approach, but there we avoided the intermediate step of finding the acceleration. Several examples and problems in this chapter can be done without using energy considerations but are easier when energy methods are used. Example 7–4 illustrates this point.

7–5 GRAVITATIONAL POTENTIAL ENERGY

When a gravitational force acts on a body while the body undergoes a vertical displacement, the force does work on the body. This work can be expressed conveniently in terms of the initial and final *positions* of the body. In Fig. 7–8a, a body having mass m and weight $w = mg$ moves vertically from a height y_1 above some reference level to a height y_2. The positive direction for y is upward. We assume that the body remains close enough to the earth's surface so that g and w can be considered constant. In the figure, P represents the result of all other forces on the body. The direction of w is opposite to the upward displacement, and the work done by this force is

$$W_{\text{grav}} = Fs = -w(y_2 - y_1) = -(mgy_2 - mgy_1). \tag{7–13}$$

Gravitational potential energy: a convenient way to calculate the work done by a gravitational force

You should verify that this expression also has the correct sign when the body moves *downward* (so y_2 is less than y_1) and also that it gives the correct resultant when y_1, or y_2, or both, are negative, corresponding to positions *below* the reference plane.

Thus we can express W_{grav} in terms of the values of the quantity mgy at the beginning and end of the displacement. This quantity, the product of the

7–8 Work done by the gravitational force w during the motion of an object from one point in a gravitational field to another. (a) A vertical displacement from height y_1 to y_2. (b) A displacement along a curved path. (c) The work done by the gravitational force w depends only on the vertical component of displacement Δy.

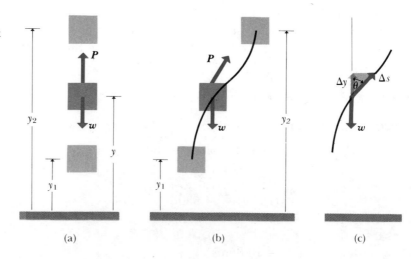

weight mg and the height y above the reference level (the origin of coordinates), is called the **gravitational potential energy, U**:

$$U = mgy. \tag{7–14}$$

Later, when we also consider other kinds of potential energy, we will sometimes denote it as $U_{grav} = mgy$. The initial value of gravitational potential energy is $U_1 = mgy_1$ and the final value is $U_2 = mgy_2$. We can then express the work W_{grav} done by the gravitational force during the displacement from y_1 to y_2 as

$$W_{grav} = U_1 - U_2 = -\Delta U. \tag{7–15}$$

Thus when the body moves *downward,* y decreases, the gravitational force does *positive* work, and the potential energy *decreases.* When the body moves *upward,* the work done by the gravitational force is *negative* and the potential energy *increases.*

The gravitational potential energy of a body depends on its position.

Note that if we shift the origin for y, then y_1 and y_2 change, but the difference $(y_1 - y_2)$ does not. Similarly, U_1 and U_2 change, but the difference $(U_1 - U_2)$ is the same as before. The choice of origin is arbitrary; the physically significant quantity is not the value of U at a particular point, but only the *difference* in U between two points.

Now let W_{other} represent the total work done by P, that is, by all forces other than the gravitational force. The total work done by *all* forces is then $W = W_{grav} + W_{other}$. Since the total work equals the change in kinetic energy,

$$W_{other} + W_{grav} = K_2 - K_1 = \Delta K,$$
$$W_{other} - (mgy_2 - mgy_1) = (\tfrac{1}{2}mv_2^2 - \tfrac{1}{2}mv_1^2). \tag{7–16}$$

The quantities $\tfrac{1}{2}mv_2^2$ and $\tfrac{1}{2}mv_1^2$ depend only on the final and initial *speeds;* the quantities mgy_2 and mgy_1 depend only on the initial and final *elevations.* Let us therefore rearrange this equation, transferring the quantities mgy_2 and mgy_1 from the "work" side of the equation to the "energy" side:

The work-energy theorem can be expressed in terms of gravitational potential energy and other work.

$$W_{other} = (\tfrac{1}{2}mv_2^2 - \tfrac{1}{2}mv_1^2) + (mgy_2 - mgy_1)$$
$$= \Delta K + \Delta U. \tag{7–17}$$

The left side of Eq. (7–17) contains only the work done by the force P. The terms on the right depend only on the final and initial states of the body (its speed and position). The first expression in parentheses on the right of Eq. (7–17) is the change in kinetic energy of the body, and the second is the change in its gravitational potential energy.

The sum of kinetic and potential energies is called the **total mechanical energy, $E = K + U$**. Equation (7–17) can also be written

Total mechanical energy is the sum of kinetic and potential energies.

$$W_{other} = (\tfrac{1}{2}mv_2^2 + mgy_2) - (\tfrac{1}{2}mv_1^2 + mgy_1)$$
$$= (K_2 + U_2) - (K_1 + U_1) = E_2 - E_1 = \Delta E. \tag{7–18}$$

Hence *the work done by all forces acting on the body, **with the exception of the gravitational force,** equals the change in the total mechanical energy of the body.* If the work W_{other} is positive, the mechanical energy increases. If W_{other} is negative, the mechanical energy decreases.

In the special case where the *only* force on the body is the gravitational force, the work W_{other} is zero. Equation (7–18) can then be written as $E_1 = E_2$, or

$$K_1 + U_1 = K_2 + U_2,$$

or

$$\tfrac{1}{2}mv_1^2 + mgy_1 = \tfrac{1}{2}mv_2^2 + mgy_2. \qquad (7\text{-}19)$$

Total mechanical energy is constant (conserved) when the only work on the body is the work done by the gravitational force.

Under these conditions *the total mechanical energy is constant;* that is, it is *conserved.* This is a particular case of the principle of **conservation of mechanical energy.**

PROBLEM-SOLVING STRATEGY: *Conservation of mechanical energy I*

1. Decide what the initial and final states of the system are; use the subscript 1 for the initial state, 2 for the final state. Use algebraic symbols for any unknown coordinates or velocities.

2. Define your coordinate system, particularly the level at which $y = 0$. This will be used to compute potential energies. Equation (7–14) assumes that the positive direction for y is upward; we suggest you use this choice consistently.

3. Make a list of the initial and final kinetic and potential energies, that is, K_1, K_2, U_1, and U_2. In general, some of these will be known, some unknown.

4. Identify all nongravitational forces that do work. A free-body diagram is often helpful. Calculate the work done by all these forces. If some of the needed quantities are unknown, represent them by algebraic symbols.

5. Use Eq. (7–16) or (7–17) to relate these quantities. If there is no nongravitational work, Eq. (7–19) may be used. Then solve to find whatever unknown quantity is required.

Throwing a ball in the air: a problem made simpler by use of conservation of energy

EXAMPLE 7–5   A man holds a ball of mass $m = 0.2$ kg at rest in his hand. He then throws the ball vertically upward. In this process, his hand moves up 0.5 m before the ball leaves his hand with an upward velocity of 20 m·s^{-1}. Discuss the motion of the ball from the work-energy standpoint, assuming $g = 10$ m·s^{-2}.

SOLUTION   First, consider the throwing process. Take the reference level ($y = 0$) at the initial position of the ball. Then $K_1 = 0$, $U_1 = 0$. Take point 2 at the point where the ball leaves the thrower's hand. Then

$$U_2 = mgy_2 = (0.2 \text{ kg})(10 \text{ m·s}^{-2})(0.5 \text{ m}) = 1.0 \text{ J},$$
$$K_2 = \tfrac{1}{2}mv_2^2 = \tfrac{1}{2}(0.2 \text{ kg})(20 \text{ m·s}^{-1})^2 = 40 \text{ J}.$$

Let P represent the upward force exerted on the ball by the man in the throwing process. The work W_{other} is then the work done by this force and is equal to the sum of the changes in kinetic and potential energy of the ball. The kinetic energy of the ball increases by 40 J and its potential energy by 1 J. The work W_{other} done by the upward force P is therefore 41 J.

If the force P is constant, the work done by this force is given by

$$W_{\text{other}} = P(y_2 - y_1),$$

and the force P is then

$$P = \frac{W_{\text{other}}}{y_2 - y_1} = \frac{41 \text{ J}}{0.5 \text{ m}} = 82 \text{ N}.$$

However, the *work* done by the force P is 41 J whether the force is constant or not.

Now consider the flight of the ball after it leaves the thrower's hand. In the absence of air resistance, the only force of the ball is then its weight $w = mg$. Hence the total mechanical energy of the ball remains constant. The calculations are simpler if we take a new reference level at the point where the ball leaves the thrower's hand. Calling this point 1, we have

$$K_1 = 40 \text{ J}, \qquad U_1 = 0,$$
$$K_1 + U_1 = K_2 + U_2 = 40 \text{ J} + 0,$$

and the total mechanical energy at *any* point in the path is 40 J.

Suppose we want to find the speed of the ball at a height of 15 m above the reference level. Its potential energy at this elevation is

$$U_2 = mgy = (0.2 \text{ kg})(10 \text{ m·s}^{-2})(15 \text{ m}) = 30 \text{ J}.$$

Therefore the kinetic energy at this point is $K_2 = 10$ J, and the speed v_2 at this point is given by

$$\tfrac{1}{2}mv^2 = K_2, \qquad v = \pm \sqrt{2K_2/m} = \pm 10 \text{ m·s}^{-1}.$$

The significance of the $\pm$ sign is that the ball passes this point *twice*, once on the way up and again on the way down. Its *potential* energy at this point is the same whether it is moving up or down. Hence its kinetic energy is the same and its speed is the same. The algebraic sign of the velocity is $+$ when the ball is moving up and $-$ when it is moving down.

Next, let us find the highest point the ball reaches. At this point $v = 0$ and $K = 0$. Therefore at this point $U = 40$ J, and the maximum height h of the ball above the thrower's hand is given by

$$mgh = (0.2 \text{ kg})(10 \text{ m·s}^{-2})h = 40 \text{ J}$$

and

$$h = 20 \text{ m}.$$

Finally, suppose we are asked to find the ball's speed at a point 30 m above the reference level. The potential energy at this point would be 60 J. But the *total* energy is only 40 J, so the ball can never reach a height of 30 m.

What happens when a body travels from an initial elevation y_1 to a final elevation y_2 along a slanted or curved path, as shown in Fig. 7–8b? We assert that the work done by the gravitational force during this displacement is the same as when the body travels straight up, as in Fig. 7–8a. To prove this we divide the path into a large number of small segments Δs; a typical one is shown in Fig. 7–8c. The work done during this displacement is the component of displacement in the direction of the force, multiplied by the magnitude of the force. As shown in Fig. 7–8c, the vertical component of displacement has magnitude $\Delta s \cos \theta = \Delta y$, but its direction is opposite to that of the force w. Thus the work done by w is

$$-mg \, \Delta s \cos \theta = -mg \, \Delta y.$$

This is the same as though the body had been displaced straight up a distance Δy. Similarly, the *total* work done by the gravitational force depends only on the *total* vertical displacement $(y_2 - y_1)$. It is given by $-mg(y_2 - y_1)$ and is independent of any horizontal motion that may occur.

In a uniform gravitational field, the change of potential energy of a body depends only on the change in elevation.

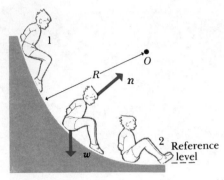

7–9 A child sliding down a frictionless curved slide.

A child on a playground slide: another problem simplified by energy considerations

EXAMPLE 7–6 A child slides down a curved playground slide that is one quadrant of a circle of radius R, as in Fig. 7–9. If he starts from rest and there is no friction, find his speed at the bottom of the track. (The motion of this child is exactly the same as that of a child on a swing of length R, with the other end held at point O.)

SOLUTION We cannot use the equations of motion with constant acceleration because the acceleration decreases during the motion. (The slope angle of the slide becomes smaller and smaller as the body descends.) If there is no friction, however, the only force on the child other than his weight is the normal force n exerted on him by the slide. The work done by this force is zero because at each point it is perpendicular to the small element of displacement near that point. Thus $W_{\text{other}} = 0$ and mechanical energy is conserved. Take point 1 at the starting point and point 2 at the bottom of the slide. Take the reference level at point 2. Then $y_1 = R$, $y_2 = 0$, and

$$K_2 + U_2 = K_1 + U_1.$$
$$\tfrac{1}{2}mv_2{}^2 + 0 = 0 + mgR,$$
$$v_2 = \pm\sqrt{2gR}.$$

The speed is therefore the same as if the child had fallen *vertically* through a height R. (What is now the significance of the $\pm$ sign?)

As a numerical example, let $R = 3.00$ m. Then

$$v = \pm\sqrt{2(9.80 \text{ m·s}^{-2})(3.00 \text{ m})} = \pm 7.67 \text{ m·s}^{-2}.$$

EXAMPLE 7–7 Suppose a child of mass 25.0 kg slides down a slide of radius $R = 3.00$ m, like that in Fig. 7–9, but his speed at the bottom is only 3.00 m·s^{-1}. What work was done by the frictional force acting on the child?

SOLUTION In this case, $W_{\text{other}} = W_{\mathcal{F}}$, and

$$\begin{aligned} W_{\mathcal{F}} &= (\tfrac{1}{2}mv_2{}^2 - \tfrac{1}{2}mv_1{}^2) + (mgy_2 - mgy_1) \\ &= \tfrac{1}{2}(25.0 \text{ kg})(3.00 \text{ m·s}^{-1})^2 - 0 + 0 - (25.0 \text{ kg})(9.80 \text{ m·s}^{-2})(3.00 \text{ m}) \\ &= 112 \text{ J} - 735 \text{ J} = -623 \text{ J}. \end{aligned}$$

The frictional work was therefore -623 J, and the total mechanical energy decreased by 623 J. The mechanical energy of a body is *not* conserved when friction forces act on it.

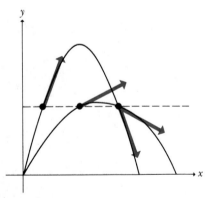

7–10 For the same initial speed, the speed is the same at all points at the same elevation.

EXAMPLE 7–8 In the absence of air resistance, the only force on a ball after it is thrown is its weight, and the mechanical energy of the ball is constant. Figure 7–10 shown two trajectories of a ball with the same initial speed (hence the same total energy) but with different angles of departure. At all points at the same elevation the potential energy is the same; hence the kinetic energy is the same and the speed is the same.

EXAMPLE 7–9 A child of weight w sits on a swing of length l, as shown in Fig. 7–11. A *variable* horizontal force P that starts at zero and gradually increases is used to pull the child very slowly (so the kinetic energy is negligibly small) until the swing makes an angle θ with the vertical. Calculate the work done by the force P.

Child on a swing again: an easy way to calculate work

SOLUTION The sum of the works W_{other} done by all the forces other than the gravitational force must equal the change of total energy, that is, the change of kinetic energy plus the change of gravitational potential energy. Hence

$$W_{\text{other}} = W_P + W_T = \Delta K + \Delta U = \Delta E.$$

Since T is perpendicular to the path at its point of application, $W_T = 0$; and since the swing was pulled very slowly at all times, the change of kinetic energy is also zero. Hence

$$W_P = \Delta U = w \, \Delta y,$$

where Δy is the distance that the child has been raised. From Fig. 7–11, $\Delta y = l(1 - \cos \theta)$. Therefore

$$W_P = wl(1 - \cos \theta).$$

Note that this result agrees with that obtained in Example 7–3, where W_P was calculated directly.

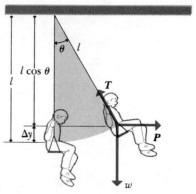

7–11 $\Delta y = l(1 - \cos \theta).$

Thus far in this section we have assumed that the gravitational force on a body is constant in magnitude and direction. But we learned in Section 6–3 that the earth's gravitational force on a body of mass m is given more generally by

$$F_g = \frac{Gmm_{\text{E}}}{r^2}, \qquad (7\text{–}20)$$

Potential energy in a nonuniform gravitational field, obtained by integration

where m_{E} is the mass of the earth and r is the distance of the body from the earth's center. When the change in r is sufficiently large, we may no longer consider the gravitational force constant; it decreases as $1/r^2$. To find the corresponding generalization of gravitational potential energy, we must compute the work W_{grav} done by the gravitational force when r increases from r_1 to r_2. This is given by

$$W_{\text{grav}} = \int_{r_1}^{r_2} F_r \, dr,$$

where F_r is the *component* of the gravitational force in the direction of increasing r, that is, the direction outward from the center of the earth. Because this force actually points in the direction of *decreasing* r, F_r differs from Eq. (7–20) by a minus sign. That is, the magnitude of the gravitational force is positive, but its component along the direction of increasing r (the radial direction) is negative.

Thus W_{grav} is given by

$$W_{\text{grav}} = -Gmm_{\text{E}} \int_{r_1}^{r_2} \frac{dr}{r^2} = \frac{Gmm_{\text{E}}}{r_2} - \frac{Gmm_{\text{E}}}{r_1}. \qquad (7\text{–}21)$$

The quantity $-Gmm_{\text{E}}/r$ is therefore the general expression for the gravitational potential energy of a body attracted by the earth:

$$U(\text{gravitational}) = -G\frac{mm_{\text{E}}}{r}. \qquad (7\text{–}22)$$

If the gravitational force is the only force on the body, then W_{grav} is equal to the change in kinetic energy from r_1 to r_2:

$$Gmm_{\text{E}}\left(\frac{1}{r_2} - \frac{1}{r_1}\right) = \frac{1}{2}mv_2{}^2 - \frac{1}{2}mv_1{}^2$$

$$\frac{1}{2}mv_1{}^2 - \frac{Gmm_{\text{E}}}{r_1} = \frac{1}{2}mv_2{}^2 - \frac{Gmm_{\text{E}}}{r_2}. \qquad (7\text{–}23)$$

Conservation of energy in a nonuniform gravitational field

The *total mechanical energy* of the body, the sum of its kinetic energy and potential energy, is

$$E = K + U = \frac{1}{2}mv^2 - G\frac{mm_E}{r}. \qquad (7\text{--}24)$$

If the only force on the body is the gravitational force, the total mechanical energy remains constant, or is *conserved*.

How to fly to the moon: an application of gravitational potential energy

EXAMPLE 7–10 In Jules Verne's story "From the Earth to the Moon" (written in 1865) three men were shot to the moon in a shell fired from a giant cannon sunk in the earth in Florida. What muzzle velocity would be needed (a) to raise a total mass m to a height above the earth equal to the earth's radius; (b) to escape from the earth completely? To simplify the calculation, neglect the gravitational pull of the moon.

SOLUTION

a) In Eq. (7–23), let m be the total projectile mass and let v_1 be the initial velocity. Then

$$r_1 = R, \qquad r_2 = 2R, \qquad v_2 = 0,$$

and

$$\frac{1}{2}mv_1{}^2 - G\frac{mm_E}{R} = 0 - G\frac{mm_E}{2R},$$

or

$$v_1{}^2 = \frac{Gm_E}{R}.$$

$$v_1 = \sqrt{\frac{(6.67 \times 10^{-11}\ \text{N·m}^2\text{·kg}^{-2})(5.98 \times 10^{24}\ \text{kg})}{6.38 \times 10^6\ \text{m}}}$$

$$= 7920\ \text{m·s}^{-1}\ (= 17{,}715\ \text{mi·hr}^{-1}).$$

b) When v_1 is the escape velocity, $r_1 = R$, $r_2 = \infty$, $v_2 = 0$. Then

$$\frac{1}{2}mv_1{}^2 - G\frac{mm_E}{R} = 0 \qquad \text{or} \qquad v_1{}^2 = \frac{2Gm_E}{R}.$$

$$v_1 = \sqrt{\frac{2(6.67 \times 10^{-11}\ \text{N·m}^2\text{·kg}^{-2})(5.98 \times 10^{24}\ \text{kg})}{6.38 \times 10^6\ \text{m}}}$$

$$= 1.12 \times 10^4\ \text{m·s}^{-1}\ (= 25{,}050\ \text{mi·hr}^{-1}).$$

Note that the speed of an earth satellite in a circular orbit of radius just slightly greater than R is, from Eq. (6–22), the same as the result of part (a), and that the escape velocity is larger than this by exactly a factor of $\sqrt{2}$.

It may seem strange that the general expression for gravitational potential energy, Eq. (7–22), should contain a minus sign. The reason lies in the choice of reference level at which the potential energy is considered zero. If we set $U = 0$ in Eq. (7–22) and solve for r, we find $r = \infty$. That is, *the gravitational potential energy of a body is zero when the body is at an infinite distance from the earth*.

Since the potential energy decreases as the body approaches the earth, it must be negative at any finite distance from the earth. The *change* in potential energy of a body as it moves from one point to another is the same, whatever the choice of reference level, and it is only changes in potential energy that are significant.

Finally, note that Eq. (7–21) can be rewritten as

$$W_{\text{grav}} = Gmm_{\text{E}}\left(\frac{r_1 - r_2}{r_1 r_2}\right).$$

If the particle stays close to the earth, then in the denominator we may replace r_1 and r_2 by R, the earth's radius, obtaining

$$W_{\text{grav}} = Gmm_{\text{E}}\left(\frac{r_1 - r_2}{R^2}\right).$$

But according to Eq. (6–15), $g = Gm_{\text{E}}/R^2$, so we finally obtain

$$W_{\text{grav}} = mg(r_1 - r_2),$$

which agrees with Eq. (7–13) with the y's replaced by r's. Thus Eq. (7–13) may be considered a special case of the more general Eq. (7–21).

The relationship between the gravitational force on a body, given by Eq. (7–20), and its gravitational potential energy, given by Eq. (7–22), can be expressed in a different way. When the distance r changes by a small amount dr, the work done by the gravitational force is $dW = F_r\, dr$, and the corresponding change in potential energy dU can be represented as $dU = (dU/dr)\, dr$. Because of the general relation of potential energy to work, we also know that $dW = -dU$. Thus we have the general relation

$$F_r = -\frac{dU}{dr}. \qquad (7\text{–}25)$$

For the gravitational force of the earth on a body of mass m, we have

$$F_r = -\frac{Gmm_{\text{E}}}{r^2},$$

$$\frac{dU}{dr} = \frac{Gmm_{\text{E}}}{r^2},$$

which is consistent with Eq. (7–25). We will find analogous relationships when we study electric fields and their associated potential energies in Chapters 25 and 26.

Relation between potential energy and force: Force is the negative of the derivative of potential energy.

7–6 ELASTIC POTENTIAL ENERGY

The concept of potential energy is useful in calculations of work done by *elastic* forces, such as the spring discussed in Section 7–3. Figure 7–12 shows a body of mass m on a level surface. One end of a spring is attached to the body, and the other end is held stationary. We take our origin of coordinates ($x = 0$) as the position of the body when the spring is neither stretched nor compressed. We now apply a force P, causing the body to accelerate. As soon as the spring begins to stretch, it begins to exert a force F on the body; we may call this an *elastic force*. If the force P is removed, the elastic force pulls the body back

Elastic potential energy: how to calculate work involved in stretching a spring

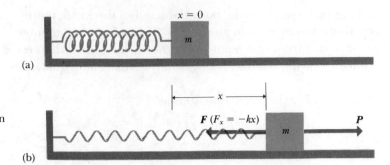

7-12 When an applied force P produces an extension x of a spring, an elastic restoring force F acts on the mass m. The x-component of F is $F_x = -kx$.

toward the original position, where the spring is unstretched. Hence F may be called a *restoring force*.

In Section 7-3 we spoke of the work done on a spring. Now, in the light of the relation between work done on a body and its change in kinetic energy, we shift our attention to the work that the spring does on the body of mass m; this is the negative of the work done on the spring. Thus in an elongation from $x = 0$ to a final value x, the work done *on* the mass *by* the spring is $-\frac{1}{2}kx^2$. Similarly, in a displacement from an initial elongation x_1 to a final elongation x_2, the elastic restoring force does an amount of work W_{el} given by

$$W_{el} = -\tfrac{1}{2}kx_2{}^2 - (-\tfrac{1}{2}kx_1{}^2).$$

The quantity $\frac{1}{2}kx^2$, one-half the product of the force constant and the square of the displacement from the unstretched position, is called the **elastic potential energy** of the system, denoted by U. That is,

$$U = \tfrac{1}{2}kx^2. \tag{7-26}$$

We use the symbol U for any form of potential energy. When there is more than one kind of potential energy in a problem, we will use $U_{el} = \frac{1}{2}kx^2$ for elastic potential energy. Note that $U = 0$ when $x = 0$. We could add any arbitrary constant to Eq. (7-26) if we chose, because the difference between the values of U at any two points would not change.

Thus we can express the work done on the body by the elastic force in terms of a change in potential energy, just as we did for the work done by a gravitational force:

$$W_{el} = \tfrac{1}{2}kx_1{}^2 - \tfrac{1}{2}kx_2{}^2 = U_1 - U_2. \tag{7-27}$$

When x increases, the work done on the mass by the elastic force is negative and the potential energy increases; when x decreases, the elastic force does positive work and the potential energy decreases. If the spring can be compressed as well as stretched, then x is negative when it is compressed. In this case U is still positive, and Eq. (7-26) is still valid.

As before, we let W_{other} be the work done by forces other than the elastic force, in this case the work done by the force P. Setting the total work equal to the change in kinetic energy of the body, we have

$$W_{other} + W_{el} = \Delta K,$$
$$W_{other} - (\tfrac{1}{2}kx_2{}^2 - \tfrac{1}{2}kx_1{}^2) = (\tfrac{1}{2}mv_2{}^2 - \tfrac{1}{2}mv_1{}^2).$$

The quantities $\frac{1}{2}kx_2{}^2$ and $\frac{1}{2}kx_1{}^2$ depend only on the initial and final positions of the body. When we transfer them from the "work" side of the equa-

Elastic potential energy is proportional to the square of the amount of stretch or compression.

tion to the "energy" side, we obtain:

$$W_{\text{other}} = (\tfrac{1}{2}mv_2^2 - \tfrac{1}{2}mv_1^2) + (\tfrac{1}{2}kx_2^2 - \tfrac{1}{2}kx_1^2). \qquad (7\text{--}28)$$

Hence the work W_{other} of the force $\boldsymbol{P}$ equals the sum of the change in the kinetic energy of the body and the change in its elastic potential energy.

Equation (7–28) can also be written

$$\begin{aligned} W_{\text{other}} &= (\tfrac{1}{2}mv_2^2 + \tfrac{1}{2}kx_2^2) - (\tfrac{1}{2}mv_1^2 + \tfrac{1}{2}kx_1^2) \\ &= (K_2 + U_2) - (K_1 + U_1) \\ &= E_2 - E_1 = \Delta E. \qquad (7\text{--}29) \end{aligned}$$

The sum of the kinetic and potential energies of the body is its total mechanical energy; and *the work done by all forces acting on the body,* **with the exception of the elastic force,** *equals the change in the total mechanical energy of the body.*

If the work W_{other} is positive, the mechanical energy increases. If W_{other} is negative, it decreases. In the special case where $W_{\text{other}} = 0$, the mechanical energy is constant, or is conserved. In that special case we can rewrite Eq. (7–29) as

$$\tfrac{1}{2}mv_1^2 + \tfrac{1}{2}kx_1^2 = \tfrac{1}{2}mv_2^2 + \tfrac{1}{2}kx_2^2. \qquad (7\text{--}30)$$

We can easily generalize these results to cases where we have *both* gravitational and elastic forces and their associated potential energies. We may still use the relationship

$$W_{\text{other}} = (K_2 + U_2) - (K_1 + U_1) = E_2 - E_1, \qquad (7\text{--}31)$$

where now U_1 and U_2 are the initial and final values of the *total* potential energy, including both gravitational and elastic. If the gravitational and elastic forces are the *only* forces that do work on the body, then $W_{\text{other}} = 0$; in that special case we have

$$K_1 + U_1 = K_2 + U_2, \qquad (7\text{--}32)$$

where again U_1 and U_2 are the total potential energies.

PROBLEM-SOLVING STRATEGY: *Conservation of energy II*

The strategy outlined in Section 7–5 is equally useful here. In the list of kinetic and potential energies in item 3, include both gravitational and elastic potential energies where appropriate. Keep in mind that when gravitational, elastic, and other forces are present, the work done by the gravitational and elastic forces is accounted for by the respective potential energies, and that the work of the other forces, W_{other}, has to be included separately. In some cases, though, there may be "other" forces that do no work. This was the case in Example 7–6 (Section 7–5): The normal force exerted by the slide on the child does no work.

EXAMPLE 7–11 In Fig. 7–12, let the force constant k of the spring be 24.0 N·m^{-1}, and let the mass of the body be 4.00 kg. The body is initially at rest, and the spring is initially stretched 0.500 m. Then the body is released and moves

Motion of a body pulled by a spring: a problem where energy methods are easier than Newton's laws back toward its equilibrium position. What is its speed when the spring is stretched 0.300 m?

SOLUTION The speed can be found most easily by using energy considerations. The spring force is the only force that does work on the body. Thus $W_{\text{other}} = 0$, and we may use Eq. (7–32). The various energy quantities can be listed as follows:

$$K_1 = \tfrac{1}{2}(4.00 \text{ kg})(0)^2 = 0,$$
$$U_1 = \tfrac{1}{2}(24.0 \text{ N·m}^{-1})(0.500 \text{ m})^2 = 3.00 \text{ J},$$
$$K_2 = \tfrac{1}{2}(4.00 \text{ kg})(v_2)^2,$$
$$U_2 = \tfrac{1}{2}(24.0 \text{ N·m}^{-1})(0.300 \text{ m})^2 = 1.08 \text{ J}.$$

Then from Eq. (7–32),

$$0 + 3.00 \text{ J} = (2.00 \text{ kg})v_2{}^2 + 1.08 \text{ J},$$

from which we obtain $v_2 = \pm 0.98 \text{ m·s}^{-1}$. (What is the physical significance of the $\pm$ sign?)

Alternatively, we may use Eq. (7–30). We have $v_1 = 0, x_1 = 0.5 \text{ m}, x_2 = 0.3 \text{ m}$, and v_2 is to be found. From Eq. (7–30),

$$\tfrac{1}{2}(4.00 \text{ kg})(0)^2 + \tfrac{1}{2}(24.0 \text{ N·m}^{-1})(0.500 \text{ m})^2$$
$$= \tfrac{1}{2}(4.00 \text{ kg})(v_2)^2 + \tfrac{1}{2}(24.0 \text{ N·m}^{-1})(0.300 \text{ m})^2.$$

Evaluating the left side, we find that the total mechanical energy is 3 J, and solving for v_2 again yields the result

$$v_2 = \pm 0.98 \text{ m·s}^{-1}.$$

Note that this problem *cannot* be done by using the equations of motion with constant acceleration because the spring force varies with position. The energy method, by contrast, offers a simple and elegant solution.

EXAMPLE 7–12 For the system of Example 7–11, suppose the body is initially at rest and the spring is initially unstretched. Then a constant force P of magnitude 10.0 N is applied to the body. There is no friction. What is the speed of the body when it has moved 0.500 m?

The equations of motion with constant acceleration cannot be used with spring problems because the force isn't constant.

SOLUTION The equations of motion with constant acceleration cannot be used, since the resultant force on the body varies as the spring is stretched. However, the speed can be found from energy considerations:

$$W_{\text{other}} = \Delta K + \Delta U,$$
$$(10.0 \text{ N})(0.500 \text{ m}) = \tfrac{1}{2}(4.00 \text{ kg})v_2{}^2 - 0$$
$$+ \tfrac{1}{2}(24.0 \text{ N·m}^{-1})(0.500 \text{ m})^2 - 0,$$
$$v_2 = 1.00 \text{ m·s}^{-1}.$$

EXAMPLE 7–13 In Example 7–12, suppose the force P is removed when the body has moved 0.500 m. How much *farther* does the body move before coming to rest?

SOLUTION The elastic force is now the only force, and total mechanical energy is conserved. The initial kinetic energy is $\tfrac{1}{2}mv^2 = 2 \text{ J}$, and the potential energy is $\tfrac{1}{2}kx^2 = 3 \text{ J}$. The total energy is therefore 5 J (equal to the work of the force P). When the body comes to rest, its kinetic energy is zero and its potential energy is therefore 5 J. Hence

$$\tfrac{1}{2}kx_{max}{}^2 = \tfrac{1}{2}(24.0 \text{ N·m}^{-1})x_{max}{}^2 = 5.00 \text{ J},$$
$$x_{max} = 0.645 \text{ m}.$$

EXAMPLE 7–14 A brick of mass m, initially at rest, is dropped from a height h onto a spring whose force constant is k. Find the maximum distance y that the spring will be compressed (see Fig. 7–13).

SOLUTION The gravitational and elastic forces are the only forces acting on the brick. Their work can be represented in terms of potential energies, and the total energy $K + U$ is conserved. At the instant of release, the brick is at rest; hence $v_1 = 0$ and $K_1 = 0$. At the instant when maximum compression occurs, the brick is again at rest, so $K_2 = 0$. Hence the total potential energy is the same at the beginning and at the end. The gravitational potential energy *decreases* by an amount $mg(h + y)$, and the elastic potential energy *increases* by an amount $\tfrac{1}{2}ky^2$. Equating these two quantities, we find

$$mg(h + y) = \tfrac{1}{2}ky^2, \quad \text{or} \quad y^2 - \frac{2mg}{k}y - \frac{2mgh}{k} = 0.$$

Therefore, from the quadratic formula,

$$y = \frac{1}{2}\left[\frac{2mg}{k} \pm \sqrt{\left(\frac{2mg}{k}\right)^2 + \frac{8mgh}{k}}\right].$$

The positive root is the desired result; the negative root corresponds to the height to which the brick and spring would rebound if they were fastened together after contact.

Dropping a brick onto a spring-mounted platform: using elastic and gravitational potential energies together

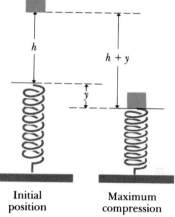

7–13 The total fall of the block is $h + y$.

7–7 CONSERVATIVE AND DISSIPATIVE FORCES

We have seen that when a body acted on by a gravitational force moves from one position to another, the work done by the gravitational force is independent of the body's path. It can be represented as the difference between the initial and final values of a function called the gravitational potential energy, $W = U_1 - U_2$. If the gravitational force is the *only* force acting on the body, the total mechanical energy (the sum of kinetic and potential energies) is constant, or conserved. For this reason the gravitational force is called a **conservative force.** When a body ascends in the earth's gravitational force, the work done by this force is negative, the kinetic energy decreases, and the potential energy increases. When it descends to its original level, the gravitational force does positive work and the kinetic energy increases to its original value. Thus the kinetic energy lost during the ascent is completely recovered during descent.

A similar situation occurs when a body is attached to a spring and moved from one position to another, changing the extension or compression of the spring. When the extension increases, the spring does negative work, the kinetic energy decreases, and the potential energy increases. When the body returns to its original position, the spring does positive work and the kinetic energy returns to its original value. In each case the work done by the spring force can again be represented as $W = U_1 - U_2$.

When kinetic energy decreases because a force does negative work, can it be recovered? Maybe!

The work done by a conservative force is reversible; when the direction of motion reverses, so does the work.

In both these situations, the work is *reversible:* On the return trip the work is always exactly the negative of that on the first part of the trip. Thus the work done by a conservative force always has these properties:

1. It is independent of the path of the body and depends only on the starting point and endpoint.
2. It is equal to the difference between the initial and final values of a *potential energy* function.
3. It is completely reversible.
4. When the starting point and endpoint are the same—that is, the path forms a closed loop—the total work is zero.

Work done by a friction force is not reversible, and the force is not conservative.

For comparison, we may consider the friction force acting on a body sliding on a stationary surface. There is no potential-energy function here. When the body slides in one direction and then back to its original position, the total work done on it by the frictional force is *not* zero. The reason is that when the direction of motion reverses, so does the friction force. Thus it does negative work in both directions. There is no such change in the direction of a gravitational force. When a body slides across a surface with decreasing speed (and therefore decreasing kinetic energy), there is no way to recover this lost kinetic energy. When friction forces act in such situations, the total mechanical energy is *not* conserved. Such a force is therefore called a *nonconservative force* or a **dissipative force.** We can then describe the energy relations in terms of additional kinds of energy, and a more general energy-conservation principle includes these additional energies. For example, when a body slides on a rough surface, the surface becomes hotter, and energy is associated with this change in the condition of the material.

EXAMPLE 7–15   In Example 7–7 (Section 7–5) a body is acted on by a dissipative friction force. The initial mechanical energy of the body is its initial potential energy of 735 J. Its final mechanical energy is its final kinetic energy of 112 J. The work $W_{\mathscr{I}}$ done by the friction force is −623 J. As the body slides down the track, the surfaces become warmer, and the same temperature changes could have been produced by adding 623 J of heat to the bodies. Thus their *internal* energy increases by 623 J; the sum of this energy and the final mechanical energy equals the initial mechanical energy, and the total energy of the system is conserved.

7–8 INTERNAL WORK AND ENERGY

We have discussed mechanical energy in the context of bodies that can be represented as particles. For more complex systems that have to be represented in terms of many particles, new subtleties in energy considerations appear. We cannot discuss these in detail, but here are three examples.

First consider a man standing on frictionless roller skates on a level surface, facing a rigid wall, as shown in Fig. 7–14. He pushes against the wall, setting himself in motion backward (to the right). The forces acting on him are his weight w, the upward normal forces n_1 and n_2 exerted by the ground on his skates, and the horizontal force P exerted on him by the wall. There is no

7–14 External forces acting on a man who is pushing against a wall. The work done by these forces is zero.

vertical displacement, so w, n_1, and n_2 do no work. The force P is the horizontal force that gives the system its acceleration to the right, but the point where that force is applied (i.e., the man's hands) does not move, so the force P also does no work. Where does the man's kinetic energy come from?

The difficulty is that representing the man as a single point is not an adequate model. For the motion to occur as we have described it, different parts of the man's body must have different motions. One part of the body may exert forces and do work on another part; therefore there may be changes in the *total* kinetic energy of this composite system, even though no work is done by forces applied by bodies outside the system. This would not be possible with a system that can be represented as a single point. In Chapter 8 we will consider the motion of a collection of moving particles interacting with each other. We will discover that the total kinetic energy of such a system can change even when no work is done on any part of the system by anything outside it.

As a second example, we consider a body sliding across a rough horizontal surface. The friction force on the body acts in the direction opposite to its motion relative to the surface, so it does negative work on the body. If that is the only horizontal force on the body, the body's kinetic energy decreases. At the same time, the body exerts an equal and opposite reaction force on the surface. But because the surface does not move, that force does no work. The work done on the surface is *not* the negative of the work done on the body. Hence there is a decrease in the body's kinetic energy but no corresponding increase in kinetic energy of the surface. So where does the energy go?

We have already pointed out in Section 7–7 that in this situation, mechanical energy is not conserved. The surfaces become warmer; the corresponding energy transferred to the materials is called *internal energy*. In later chapters we will study internal energy and its relation to temperature changes, heat, and work in considerable detail. This is the heart of the area of physics called *thermodynamics*.

As a final example, suppose we drop a box onto a moving conveyor belt. If the box is dropped from a very small initial height, it is essentially at rest the instant it contacts the belt. It slips at first, but eventually acquires the same velocity as the belt. In doing so, it acquires kinetic energy, and some force has to act on the box, doing positive work on it, to give it this kinetic energy. What is the force?

Clearly, the answer has to be "the friction force exerted on the box by the belt"; there is no other horizontal force on the box. So here is a case where a friction force on a moving body does *positive* work on the body and *increases* its kinetic energy. The same thing happens when a truck starts from rest and accelerates. If there is a box sitting in the truck, the force that gives the box its forward acceleration and adds to its kinetic energy is again the friction force exerted on the box, this time by the moving floor of the truck. Although we are accustomed to think of moving bodies slowing down and stopping as a result of friction forces, in these two examples friction forces do positive work and increase the kinetic energy of the bodies on which they act.

Still, we are left with the nagging feeling that somehow friction forces ought to dissipate or waste mechanical energy. After all, don't we oil bearings and sliding surfaces to decrease friction and make them more efficient? True enough! Going back to the conveyor belt, let us consider the work done *on the belt* by the friction force exerted on the belt by the box. This is negative,

Some systems cannot be represented by a single point. One part of a system can do work on another part of the system and change the kinetic energy even when no work is done by external forces.

Internal energy and energy associated with heat: a preview of things to come

Box on a conveyer belt: a case where friction does positive work

7–15 Stroboscopic photograph of a pole-vaulter. The athlete's initial kinetic energy is partly converted to elastic potential energy in the flexed pole, then to gravitational potential energy as he rises and clears the bar, then back to kinetic energy as he drops on the other side. The elastic and gravitational forces are conservative. (Dr. Harold Edgerton, M.I.T., Cambridge, Massachusetts.)

because the force on the belt has opposite direction to its motion. By Newton's third law, the two forces have equal magnitude. Furthermore, during the sliding of the box the belt moves farther than the box; thus the negative work done on the belt has greater magnitude than the positive work done on the box. The *total* work done by these two forces is negative, and in this energy-exchange process there is indeed a net loss of mechanical energy. As mentioned above, there is a corresponding increase of internal energy of the materials, showing itself as a rise in temperature.

Despite these complications, it remains true that for any system that can be adequately represented as a moving *point* mass, the total change in kinetic energy in any process is *always* equal to the total work done by all the forces acting on the system. Only when a system consists of several interacting masses do complications develop. A dramatic example of these complications is shown in Fig. 7–15.

7–9 POWER

Power: a description of how quickly work is done

Time considerations are not involved in the definition of work. When you lift a body weighing 100 N through a vertical distance of 0.5 m, you do 50 J of work, whether it takes you 1 second, 1 hour, or 1 year to do it. There are many situations, though, where it is important for us to know how quickly work is done, that is, the *rate* of doing work. The time rate at which work is done or energy transferred is called **power.**

When a quantity of work ΔW is done during a time interval Δt, the **average power** P_{av} is defined as

$$\text{Average power} = \frac{\text{work done}}{\text{time interval}}, \qquad P_{av} = \frac{\Delta W}{\Delta t}.$$

If the rate at which work is done is not constant, this ratio may vary; in this case we may define an *instantaneous power* P as the limit of this quotient as $\Delta t \to 0$:

$$P = \lim_{\Delta t \to 0} \frac{\Delta W}{\Delta t} = \frac{dW}{dt}. \tag{7–33}$$

The SI unit of power, 1 joule per second (1 J·s^{-1}), is called 1 *watt* (1 W). The kilowatt (1 kW = 10^3 W) and the megawatt (1 MW = 10^6 W) are also commonly used. The cgs power unit is 1 erg per second (1 erg·s^{-1}).

In the British system, where work is expressed in foot-pounds and time in seconds, the unit of power is 1 foot-pound per second. A larger unit called the *horsepower* (hp) is also used:

$$1 \text{ hp} = 550 \text{ ft·lb·s}^{-1}$$
$$= 33,000 \text{ ft·lb·min}^{-1}.$$

That is, a 1-hp motor running at full load does 33,000 ft·lb of work every minute.

The watt is a familiar unit of *electrical* power; a 100-W light bulb converts electrical energy into light and heat at the rate of 100 joules per second. But there is nothing inherently electrical about the watt; the electric power consumption of a light bulb could be expressed in horsepower, and many automobile manufacturers now rate their engines in kilowatts rather than horsepower.

The watt, the kilowatt, and the horsepower: all units of power

From the relations between the newton, pound, meter, and foot, we can show that

$$1 \text{ hp} = 746 \text{ W} = 0.746 \text{ kW},$$

or about $\frac{3}{4}$ of a kilowatt, a useful figure to remember.

Because power is energy or work per unit time, the units of power may be used to define new units of work or energy. The kilowatt-hour (kWh) is commonly used as a unit of electrical energy. One kilowatt-hour is the work done in 1 hour by an agent working at a constant power of 1 kilowatt (10^3 J·s^{-1}). Such an agent does 1000 J of work each second; the work done in 1 hr is $3600 \times 1000 = 3,600,000$ J:

$$1 \text{ kWh} = 3.6 \times 10^6 \text{ J} = 3.6 \text{ MJ}.$$

The kilowatt-hour is a unit of *work* or *energy*, not power.

The kilowatt-hour: a familiar unit of energy, not power

Although energy is an abstract physical quantity, it nevertheless has a monetary value. A newton of force or a meter per second of velocity is not a thing that is bought and sold as such, but a kilowatt-hour of energy is a quantity offered for sale at a definite market rate. In the form of electrical energy, a kilowatt-hour can be purchased at a price varying from a few tenths of a cent to around 10 cents, depending on the locality and the quantity purchased.

When a force acts on a moving body, doing work on it, the corresponding power can be expressed in terms of the force and the velocity. Suppose a force $\mathbf{F}$ acts on a body while it undergoes a displacement of magnitude Δs. If $F_\parallel$ is the component of $\mathbf{F}$ tangent to the path, then work is given by $\Delta W = F_\parallel \Delta s$, and the average power is

$$P_{av} = \frac{\Delta W}{\Delta t} = F_\parallel \frac{\Delta s}{\Delta t} = F_\parallel v_{av}.$$

In the limit as $\Delta t \to 0$ we obtain the instantaneous power:

$$P = F_{\parallel}v, \tag{7-34}$$

where v is the magnitude of the instantaneous velocity. We can also express this relation in terms of the scalar product:

$$P = \boldsymbol{F} \cdot \boldsymbol{v}. \tag{7-35}$$

Power output of a jet engine.

EXAMPLE 7–16 A jet airplane engine develops a thrust (a forward force on the plane) of 15,000 N (roughly 3000 lb). When the plane is flying at 300 m·s^{-1} (roughly 600 mph), what horsepower does the engine develop?

SOLUTION

$$P = Fv = (1.50 \times 10^4 \text{ N})(300 \text{ m·s}^{-1}) = 4.50 \times 10^6 \text{ W}$$

$$= (4.50 \times 10^6 \text{ W})\left(\frac{1 \text{ hp}}{746 \text{ W}}\right) = 6030 \text{ hp}.$$

EXAMPLE 7–17 As part of a charity fund-raising drive, a Chicago marathon runner of mass 50.0 kg runs up the stairs to the top of the Sears tower, the tallest building in the United States (443 m), in 15.0 minutes. What is her average power output, in watts? In kilowatts? In horsepower?

SOLUTION The total work is

$$W = mgh = (50.0 \text{ kg})(9.80 \text{ m·s}^{-2})(443 \text{ m}) = 2.17 \times 10^5 \text{ J}.$$

The time is 15 min = 900 s, so the average power is

$$P_{\text{av}} = \frac{2.17 \times 10^5 \text{ J}}{900 \text{ s}} = 241 \text{ W} = 0.241 \text{ kW} = 0.323 \text{ hp}.$$

Alternatively, the average vertical component of velocity is (443 m)/(900 s) = 0.492 m·s^{-1}, so the average power is

$$P_{\text{av}} = Fv_{\text{av}} = (mg)v_{\text{av}}$$
$$= (50.0 \text{ kg})(9.80 \text{ m·s}^{-2})(0.492 \text{ m·s}^{-1}) = 241 \text{ W}.$$

7–10 AUTOMOTIVE POWER

Energy relations in a gasoline-powered automobile provide a familiar and interesting example of some of the energy and power concepts in this chapter. First, burning 1 liter of gasoline provides about 3.5×10^7 J of energy. Not all of this is converted to *mechanical* energy; the laws of thermodynamics, which we will encounter in Chapters 18 and 19, impose fundamental limitations on the efficiency of converting heat to mechanical energy or work. In a typical car engine, two-thirds of the heat from gasoline combustion is wasted in the cooling system and the exhaust. Another 20% or so is converted to mechanical energy but is lost in friction in the drive train or is used by accessories such as electric-power generators, air-conditioners, and power steering. This leaves about 15% of the energy to propel the car.

Rolling friction: It's easier to roll things than to drag them.

Most of this energy is used to overcome rolling friction and air resistance. Each of these can be described in terms of a force that resists the motion of the car. We described rolling friction in Section 5–2 in terms of an effective coefficient of rolling friction called *tractive resistance* μ_r. A typical value of μ_r for rubber tires on hard pavement is 0.015. For a 1000-kg compact car weighing about 10,000 N, the resisting force of rolling friction is thus

$$F_{rol} = (0.015)(10,000\text{ N}) = 150\text{ N}.$$

This force is nearly independent of car speed.

The air resistance force increases rapidly with speed, and in fact the relationship can be expressed approximately by the equation

$$F_{air} = \tfrac{1}{2}CA\rho v^2, \tag{7–36}$$

The drag coefficient: a way to calculate air resistance of a moving car

where A is the silhouette area of the car (seen from the front), ρ is the density of air (about 1.2 kg·m^{-3} at ordinary temperatures), v is the car's speed, and C is a dimensionless constant, called the *drag coefficient*, that depends on the shape of the moving body. A typical value of C for a car designed with reasonable attention to streamlining is 0.5. Assuming $A = 2$ m^2 for our car, we find an air-resistance force of

$$F_{air} = \tfrac{1}{2}(0.5)(2\text{ m}^2)(1.2\text{ kg·m}^{-3})v^2$$
$$= (0.6\text{ N·s}^2\text{·m}^{-2})v^2.$$

Thus in a residential speed zone, where $v = 10$ m·s^{-1} (about 22 mi·hr^{-1}), the air-resistance force is about

$$F_{air} = (0.6\text{ N·s}^2\text{·m}^{-2})(10\text{ m·s}^{-1})^2 = 60\text{ N}.$$

At a moderate speed of 15 m·s^{-1} (34 mi·hr^{-1}), F_{air} is 135 N, and at a highway speed of 30 m·s^{-1} (67 mi·hr^{-1}), it is 540 N. Thus at slow speeds air resistance is less important than rolling friction. At moderate speeds they are comparable, and at highway speeds air resistance is the dominant effect.

What does this mean in terms of the *power* needed from the engine? In constant-speed driving, the forward force supplied by the driven wheels must just balance the sum of these two resisting forces, and the power is this force multiplied by the velocity. Thus for our hypothetical car, the power needed for constant speed v is

$$P = (F_{rol} + F_{air})v$$
$$= [150\text{ N} + (0.6\text{ N·s}^2\text{·m}^{-2})v^2]v.$$

You can do the arithmetic yourself. For the three speeds mentioned above, you will find the following results:

v (m·s^{-1})	F_{rol} (N)	F_{air} (N)	F_{tot} (N)	P (kW)	P (hp)
10	150	60	210	2.10	2.81
15	150	135	285	4.28	5.73
30	150	540	690	20.70	27.70

Now, how is all this related to fuel consumption? Let's look at the 15 m·s^{-1} case. The power required is 4.28 kW = 4280 J·s^{-1}. In one hour (3600 s) the total energy required is

$$(4280\text{ J·s}^{-1})(3600\text{ s}) = 1.54 \times 10^7\text{ J}.$$

During that hour the car travels a distance of

$$(15 \text{ m}\cdot\text{s}^{-1})(3600 \text{ s}) = 5.4 \times 10^4 \text{ m} = 54 \text{ km}.$$

Fuel consumption: The faster you go, the more it costs.

If the entire 3.5×10^7 J of energy obtained by burning one liter of gasoline were available to propel the car, the amount consumed in an hour would be

$$(1.54 \times 10^7 \text{ J})/(3.5 \times 10^7 \text{ J}\cdot\text{liter}^{-1}) = 0.44 \text{ liter}.$$

But because only 15% of the energy is actually available for propulsion, as explained above, the total amount consumed is $(0.44 \text{ liter})/0.15 = 2.93$ liter. That amount of gasoline gets us 54 km, so the rate of consumption per unit distance is $(54 \text{ km})/(2.93 \text{ liter}) = 18.4 \text{ km}\cdot\text{liter}^{-1}$. You may want to work out the conversion factor between kilometers per liter and miles per gallon; this comes out to about 43.3 miles per gallon.

The power required for a steady $15 \text{ m}\cdot\text{s}^{-1}$ is 4.28 kW, but the power required for acceleration and hill climbing may be much greater. Suppose the car accelerates from zero to $30 \text{ m}\cdot\text{s}^{-1}$ in 20 s. The final kinetic energy is

$$K = \tfrac{1}{2}mv^2 = \tfrac{1}{2}(1000 \text{ kg})(30 \text{ m}\cdot\text{s}^{-1})^2 = 4.5 \times 10^5 \text{ J}.$$

The average power required is

$$P_{\text{av}} = (4.5 \times 10^5 \text{ J})/(20 \text{ s}) = 2.25 \times 10^4 \text{ W} = 22.5 \text{ kW} = 30.2 \text{ hp}.$$

Thus this relatively rapid acceleration requires five times as much power as cruising at a steady, moderate speed.

What about hill climbing? A 5% grade, which is about the maximum found on most interstate highways, rises 5 meters for every 100 meters of horizontal distance. A car driving at $15 \text{ m}\cdot\text{s}^{-1}$ up a 5% grade is gaining elevation at the rate of $(0.05)(15 \text{ m}\cdot\text{s}^{-1}) = 0.75 \text{ m}\cdot\text{s}^{-1}$. To lift a car weighing 10,000 N at this rate requires a power of

$$\begin{aligned} P = Fv &= (10,000 \text{ N})(0.75 \text{ m}\cdot\text{s}^{-1}) \\ &= 7500 \text{ J}\cdot\text{s}^{-1} = 7.50 \text{ kW} = 10.0 \text{ hp}. \end{aligned}$$

The *total* power required is this amount plus the 4.28 kW needed to maintain $15 \text{ m}\cdot\text{s}^{-1}$ on a level road, that is,

$$P_{\text{tot}} = 7.50 \text{ kW} + 4.28 \text{ kW} = 11.78 \text{ kW}.$$

Finally, let us compare these energy and power quantities with some purely thermal considerations. This is a little premature; we will study the relation of heat to mechanical energy in detail in later chapters, but we can anticipate that discussion here. We will learn that to heat one kilogram of water from 0°C to 100°C requires an energy input to the water of 4.18×10^5 J. Thus the 3.5×10^7 J obtained from one liter of gasoline is enough to heat $(3.5 \times 10^7)/(4.18 \times 10^5) = 83.7$ kg of water from freezing to boiling. That is not much water, only about 23 gallons. So the amount of energy needed to heat 23 gallons of water from freezing to boiling is enough to push our car over 18 km! Does that surprise you?

SUMMARY

The concept of energy plays a central role in one of the universal conservation laws of physics. Mechanical energy includes kinetic energy, associated with motion, and potential energy, associated with position.

When a force F acts on a particle that undergoes a displacement s, the work W done by the force is defined as

$$W = Fs \cos \theta = F \cdot s,$$

where θ is the angle between the directions of F and s. The unit of work in SI units is 1 newton·meter = 1 joule (1 N·m = 1 J). Work is a scalar quantity; it has an algebraic sign (positive or negative) but no direction in space.

When the force varies during the displacement, the work done by the force is given by

$$W = \int_{x_1}^{x_2} F \, dx \tag{7-4}$$

if the force has the same direction as the displacement, or

$$W = \int_{x_1}^{x_2} F \cos \theta \, dx \tag{7-5}$$

if it makes an angle θ with the displacement. The angle θ may vary during the displacement.

The kinetic energy K of a particle having mass m and speed v is given by $K = \frac{1}{2}mv^2$. Kinetic energy is a scalar quantity; it has a magnitude (always positive) but no direction in space. When forces act on a particle while it undergoes a displacement, the total work W_{tot} done on it by all the forces equals the change in the particle's kinetic energy:

$$W_{\text{tot}} = K_2 - K_1.$$

The work done on a particle in a uniform gravitational field can be represented in terms of a potential energy $U = mgy$, as

$$W_{\text{grav}} = U_1 - U_2 = mgy_1 - mgy_2. \tag{7-15}$$

More generally, the potential energy of a particle of mass m in the gravitational field of a particle of mass m_{E} is

$$U = -\frac{Gmm_{\text{E}}}{r}. \tag{7-22}$$

If gravitational and other forces act on a body, the work W_{other} done by the forces other than gravitational potential energy:

$$W_{\text{other}} = K_2 - K_1 + U_2 - U_1 = E_2 - E_1,$$

where $E = K + U$ is the total energy (kinetic plus potential).

The work done by a stretched or compressed spring that exerts a force $F = -kx$ on a particle, where x is the amount of stretch or compression, can be represented in terms of a potential-energy function $U = \frac{1}{2}kx^2$:

$$W_{\text{el}} = U_1 - U_2.$$

If other forces also act on the body, then the work of the other forces equals the total change in kinetic energy plus elastic potential energy:

$$\begin{aligned} W_{\text{other}} &= K_2 - K_1 + U_2 - U_1 = E_2 - E_1 \\ &= (\tfrac{1}{2}mv_2^2 + \tfrac{1}{2}kx_2^2) - (\tfrac{1}{2}mv_1^2 + \tfrac{1}{2}kx_1^2). \end{aligned} \tag{7-31}$$

conservation laws
work
joule
force constant
Hooke's law
kinetic energy
gravitational potential energy
total mechanical energy
conservation of mechanical energy
elastic potential energy
conservative force
dissipative force
power
average power

In the special cases where no forces other than elastic or gravitational do work on the body, the total energy $E = K + U$ is conserved or constant, and

$$K_1 + U_1 = K_2 + U_2,$$

where U may in general include both gravitational and elastic potential energies.

A conservative force is one for which the work–kinetic energy relation is completely reversible. The work of a conservative force can always be represented in terms of a potential energy, but the work of a nonconservative force cannot.

Power is time rate of doing work. If an amount of work ΔW is done in a time Δt, the average power P_{av} is defined as

$$P_{av} = \frac{\Delta W}{\Delta t},$$

and the instantaneous power P is defined as

$$P = \frac{dW}{dt}. \tag{7-33}$$

When a force $\boldsymbol{F}$ acts on a particle moving with velocity $\boldsymbol{v}$, the instantaneous power, or rate at which the force does work, is

$$P = \boldsymbol{F} \cdot \boldsymbol{v}.$$

Power, like work and kinetic energy, is a scalar quantity.

QUESTIONS

7–1 An elevator is hoisted by its cables at constant speed. Is the total work done on the elevator positive, negative, or zero?

7–2 A rope tied to a body is pulled, causing the body to accelerate. But according to Newton's third law the body pulls back on the rope with an equal and opposite force. Is the total work done then zero? If so, how can the body's kinetic energy change?

7–3 In Fig. 7–10, the projectile has the same initial kinetic energy in each case. Why does it not then rise to the same maximum height in each case?

7–4 Are there any cases where a frictional force can *increase* the mechanical energy of a system? If so, give examples.

7–5 An automobile jack is used to lift a heavy weight by exerting a force much smaller in magnitude than the weight. Does this mean that less work is done than if the weight had been lifted directly?

7–6 A compressed spring is clamped in its compressed position and is then dissolved in acid. What becomes of its potential energy?

7–7 A child standing on a playground swing can increase the amplitude of his motion by "pumping up," pulling back on the swing ropes at the appropriate points during the motion. Where does the added energy come from?

7–8 In a siphon, water is lifted above its original level during its flow from one container to another. Where does it get the needed potential energy?

7–9 A man bounces on a trampoline, going a little higher with each bounce. Explain how he increases his total mechanical energy.

7–10 Is it possible for the second hill on a roller-coaster track to be higher than the first? What would happen if it were higher?

7–11 Does the kinetic energy of a car change more when it speeds up from 10 to 15 $\mathrm{m \cdot s^{-1}}$ or from 15 to 20 $\mathrm{m \cdot s^{-1}}$?

7–12 A car accelerates from an initial speed to a greater final speed while the engine develops constant power. Is the acceleration greater at the beginning of this process or at the end?

7–13 Time yourself while running up a flight of steps, and compute your maximum power, in horsepower. Are you stronger than a horse?

7–14 When a constant force is applied to a body moving with constant acceleration, is the power of the force constant? If not, how would the force have to vary with speed for the power to be constant?

7–15 An advertisement for a power saw states: "This power tool uses the energy of the motor to stop the blade rotation within 5 seconds after the switch is turned off." Is this an accurate statement? Please explain.

EXERCISES

Section 7–2 Work

7–1 A block is pushed 2 m along a horizontal table top by a horizontal force of 2 N. The opposing force of friction is 0.4 N.

a) How much work is done by the 2-N force?

b) What is the work done by the friction force?

7–2 A 40-kg crate is pushed a distance of 5 m along a level floor at a constant speed by a force that is at an angle of 30° below the horizontal. The coefficient of kinetic friction between the crate and floor is 0.25.

a) How much work is done by this force?

b) How much work is done by friction?

c) How much work is done by the normal force? By gravity?

7–3 A 22-lb block is pulled up a frictionless plane inclined at 30° to the horizontal, by a force P of 18 lb acting parallel to the plane. If the block travels 15 ft along the incline, calculate

a) the work done by the force P,

b) the work done by the gravity force,

c) the work done by the normal force,

d) the total work done on the block.

7–4 A 5-kg package slides 2 m down a surface that is inclined at 53° below the horizontal. The coefficient of kinetic friction between the package and the surface is $\mu_k = 0.4$. Calculate

a) the work done by friction on the package,

b) the work done by gravity,

c) the work done by the normal force,

d) the total work done on the package.

Section 7–3 Work Done by a Varying Force

7–5 A force of 100 N is observed to stretch a certain spring a distance of 0.4 m.

a) What force is required to stretch the spring 0.1 m? To compress the spring 0.2 m?

b) How much work must be done to stretch the spring 0.1 m beyond its unstretched length? How much work to compress the spring 0.2 m from its equilibrium position?

7–6 A force F that is parallel to the x-axis acts on an object. The magnitude of the force varies with the x-coordinate of the object as shown in Fig. 7–16. Calculate the work done by the force F when the object moves from $x = 0$ to $x = 7$ m.

7–7 An object is attracted toward the origin with a force given by $F = -(6\,\text{N·m}^{-3})x^3$, where F is in newtons and x is in meters.

a) What is the force F when the object is at point a, 1 m from the origin?

b) At point b, 2 m from the origin?

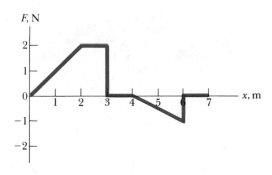

FIGURE 7–16

c) How much work is done by the force F when the object moves from a to b? Is this work positive or negative?

7–8 An object is attracted toward the origin with a force given by $F = -k/x^2$. Calculate the work done by the force F when the object moves in the x-direction from x_1 to x_2. If $x_2 > x_1$, is the work done by F positive or negative?

Section 7–4 Work and Kinetic Energy

7–9 Compute the kinetic energy, in joules, of a 2-g rifle bullet traveling at 500 m·s^{-1}.

7–10 An object of mass 2 slugs is initially at rest on a frictionless horizontal surface. It is then pulled 4 ft by a horizontal force of magnitude 25 lb. Use the work-energy relation (Eq. 7–12) to find its final speed.

7–11 An object of mass 2 kg moves in a straight line on a frictionless horizontal surface. It has an initial speed of 10 m·s^{-1} and then is pulled 4 m by a force of magnitude 25 N and in the direction of the initial velocity. What is the object's final speed? Use the work-energy relation.

7–12 A block of mass 8 kg moves in a straight line on a frictionless horizontal surface. At one point in its path its speed is 4 m·s^{-1}, and after it has traveled 3 m its speed is 5 m·s^{-1} in the same direction. Use the work-energy relation to find the force acting on the block, assuming that this force is constant.

7–13 A car is traveling on a level road with velocity v_0 at the instant the brakes are applied.

a) Use the work-energy relation (Eq. 7–12) to calculate the minimum stopping distance of the car in terms of v_0, the coefficient of friction μ between the tires and the road, and the acceleration g due to gravity.

b) It is observed that the car stops in a distance of 140 ft if $v_0 = 40$ mi·hr^{-1}. What is the stopping distance if $v_0 = 60$ mi·hr^{-1}? Assume that μ and hence the friction force remain the same.

7–14 A small block of mass 0.05 kg is attached to a cord passing through a hole in a frictionless horizontal surface, as in Fig. 7–17. The block is originally revolving at a distance of 0.2 m from the hole with a speed of 0.6 m·s^{-1}. The cord is then pulled from below, shortening the radius of the circle in which the block revolves to 0.1 m. At this new distance the speed of the block is observed to be 1.2 m·s^{-1}. How much work was done by the person who pulled the cord?

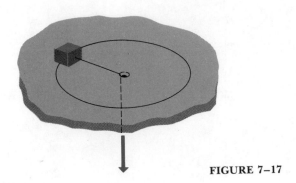

FIGURE 7-17

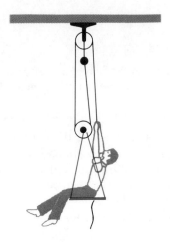

FIGURE 7-18

7-15 A proton of mass 1.67×10^{-27} kg is propelled at an initial speed of $2 \times 10^4 \text{m·s}^{-1}$ directly toward a gold nucleus that is 4 m away. The proton is repelled by the gold nucleus with a force of magnitude $F = (1.82 \times 10^{-26} \text{ N·m}^2)/x^2$, where x is the separation between the two objects. Assume that the gold nucleus remains at rest.

a) What is the speed of the proton when it is 1.0×10^{-7} m from the gold nucleus?

b) How close to the gold nucleus does the proton get? (Calculate the position of the proton for which the work done by the repulsive force has reduced the velocity of the proton momentarily to zero.)

7-16 An object of mass 5 kg is initially at rest on a frictionless horizontal surface. A force F is then applied to the object, and as a result the object moves along the x-axis such that its position as a function of time is given by $x(t) = (5\text{m·s}^{-2})t^2 + (2\text{m·s}^{-3})t^3$, where t is in seconds and x is in meters.

a) Calculate the velocity of the object when $t = 4$ s.

b) Calculate the work done by the force F during the first 4 s of the motion. (*Hint:* Use Eq. 7-12.)

Section 7-5 Gravitational Potential Energy

7-17 What is the potential energy of an 800-kg elevator at the top of the Empire State Building, 380 m above street level? Assume the potential energy at street level to be zero.

7-18 A 5-kg block is lifted vertically at a constant velocity of 4 m·s^{-1} through a height of 12 m.

a) How great a force is required?

b) How much work is done? What becomes of this work?

7-19 A man of mass 80 kg sits on a platform suspended from a movable pulley as shown in Fig. 7-18 and raises himself at constant speed by a rope passing over a fixed pulley. The platform and the pulleys have negligible mass. Assuming no friction losses, find

a) the force he must exert;

b) the increase in his energy when he raises himself 1 m.

Answer part (b) by calculating his increase in potential energy, and also by computing the product of the force on the rope and the length of the rope passing through his hands.

7-20 A barrel of mass 120 kg is suspended by a rope 10 m long.

a) What horizontal force is necessary to hold the barrel in a position displaced sideways 2 m from the vertical?

b) How much work is done in moving it to this position?

7-21 A ball is thrown from the roof of a 90-ft-tall building with an initial velocity of 60 ft·s^{-1} at an angle of 37° above the horizontal.

a) What is the speed of the ball just before it strikes the ground? Use energy methods.

b) What would be the answer in (a) if the initial velocity was at an angle of 37° *below* the horizontal?

7-22 A small sphere of mass m is fastened to a weightless string of length 0.5 m to form a pendulum. The pendulum is swinging so as to make a maximum angle of 60° with the vertical. What is the speed of the sphere when it passes through the vertical position?

7-23 An object of mass 0.1 kg is released from rest at point A, which is at the top edge of a hemispherical bowl of radius $R = 0.4$ m (Fig. 7-19). When it reaches point B at the bottom of the bowl, the object is observed to have speed $v = 1.8$ m·s^{-1}. Calculate the *work* done by friction on the block when it moves from A to B. (*Note.* The friction force is not constant. You can easily solve for the friction *work* but not for the friction *force*.)

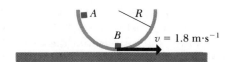

FIGURE 7-19

7-24 A 5-kg block is pushed up a frictionless plane inclined at 30° to the horizontal. It is pushed 2.0 m along the incline by a constant 100 N force parallel to the plane. If its speed at the bottom is 2 m·s^{-1}, what is its speed at the top? Use energy methods.

7–25 A 12-kg block is pushed 20 m up the sloping surface of a plane inclined at an angle of 37° to the horizontal, by a constant force F of 120 N acting parallel to the plane. The coefficient of kinetic friction between the block and plane is 0.25.

a) What is the work of the force F?

b) What is the work done by the friction force?

c) Compute the increase in potential energy of the block.

d) Use your answers to (a), (b), and (c) to calculate the increase in kinetic energy of the block.

e) Use $\Sigma F = ma$ to calculate the acceleration of the block. Assuming that the block is initially at rest, use the acceleration to calculate the block's velocity after traveling 20 m. Compute from this the increase in kinetic energy of the block, and compare to the answer you got in (d).

7–26 Use the results of Example 7–10 to calculate the escape velocity for an object

a) from the surface of the moon,

b) from the surface of Saturn.

c) Why is the escape velocity for an object independent of the object's mass?

7–27 The asteroid Toro, discovered in 1964, has a radius of about 5 km and a mass of about 2×10^{15} kg. Calculate the escape velocity for an object at the surface of Toro. Could a person *jump* off Toro, i.e., jump with a velocity greater than the escape velocity?

7–28 An object of mass m is shot vertically upward from the surface of the earth. If the object's initial velocity is 6.0×10^3 m·s^{-1}, to what height above the surface of the earth will it rise?

Section 7–6 Elastic Potential Energy

7–29 A force of 1200 N stretches a certain spring a distance of 0.1 m.

a) What is the potential energy of the spring when it is stretched 0.1 m? Compressed 0.05 m?

b) What is the potential energy of the spring when a 60-kg mass hangs vertically from it?

7–30 The spring of a spring gun has a force constant of $k = 500$ N·m^{-1}. The spring is compressed 0.05 m and a ball of mass 0.01 kg is placed in the barrel against the compressed spring. The spring is then released and the ball is propelled out the barrel.

a) Compute the speed with which the ball leaves the gun, if friction forces are negligible.

b) Determine the speed of the ball as it leaves the end of the barrel if a constant resisting force of 10 N acts on the ball while it travels the 0.05 m along the barrel. Note that in this case the maximum speed does not occur at the end of the barrel.

7–31 A 2-kg block is dropped from a height of 0.4 m onto a spring whose force constant k is 1960 N·m^{-1}. Find the maximum distance the spring will be compressed.

7–32 A block of mass 5 kg is moving at 8 m·s^{-1} along a frictionless horizontal surface toward a spring of force constant $k = 500$ N·m^{-1} that is attached to a wall (Fig. 7–20). Find the maximum distance the spring will be compressed. The spring has negligible mass.

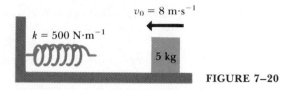

FIGURE 7–20

7–33 A block of mass 0.5 kg is pressed against a vertical spring of force constant $k = 500$ N·m^{-1} such that the spring is compressed 0.2 m. When the block is released, how high will the block rise from this position? (The block and the spring are *not* attached. The spring has negligible mass.)

7–34 A 2-kg block is pushed against a spring of force constant $k = 400$ N·m^{-1}, compressing it 0.3 m. The block is then released and moves along a frictionless horizontal surface and then up a frictionless incline of slope 37° (Fig. 7–21).

a) What is the speed of the block as it slides along the horizontal surface after having left the spring?

b) How far does the object travel up the incline before starting to slide back down?

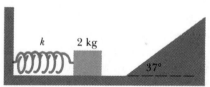

FIGURE 7–21

7–35 A block of mass 1 kg is forced against a horizontal spring of negligible mass, compressing the spring a distance $x_1 = 0.2$ m. When released, the block moves on a horizontal table top a distance $x_2 = 1.0$ m before coming to rest. The spring constant k is 100 N·m^{-1} (Fig. 7–22). What is the coefficient of kinetic friction, μ_k, between the block and the table?

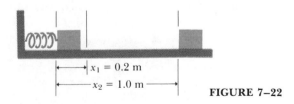

FIGURE 7–22

Section 7–9 Power

7–36 At 5 cents per kWh, what does it cost to operate a 10-hp motor for 8 hr?

7–37 A tandem (two-person) bicycle team must overcome a force of 34.0 lb to maintain a speed of 30.0 ft·s^{-1}. Find the power required per rider, assuming they each contribute equally. Express your answer in horsepower.

7–38 The total consumption of electrical energy in the United States is of the order of 10^{19} joules per year.

a) What is the average rate of electrical energy consumption, in watts?

b) If the population of the United States is 200 million, what is the average rate of electrical energy consumption per person?

c) The sun transfers energy to the earth by radiation, at a rate of approximately 1.4 kW per square meter of surface. If this energy could be collected and converted to electrical energy with 100% efficiency, how great an area (in square kilometers) would be required to collect the electrical energy used by the United States?

7–39 A man whose mass is 70 kg walks up to the third floor of a building. This is a vertical height of 12 m above the street level.

a) By how much has he increased his potential energy?

b) If he climbs the stairs in 20 s, what is his rate of working, in watts?

7–40 The hammer of a pile driver has a weight of 1100 lb and must be lifted a vertical distance of 6 ft in 3 s. What horsepower engine is required?

7–41 The Grand Coulee Dam is 1270 m long and 170 m high. The electrical power output from generators at its base is approximately 2000 megawatts. How much water in $m^3 \cdot s^{-1}$ must be flowing over the dam to produce this amount of power? (The density, or mass per unit volume, of water is 1.0×10^3 kg$\cdot$m^{-3}.)

7–42 A ski tow is to be operated on a 37° slope 300 m long. The rope is to move at 12 km$\cdot$hr^{-1}, and power must be provided for 80 riders at one time, with an average mass

per rider of 70 kg. Estimate the power required to operate the tow.

7–43 The engine of a motor boat delivers 30 kW to the propeller while the boat is moving at 10 m$\cdot$s^{-1}. What would be the tension in the towline if the boat were being towed at the same speed?

7–44 The engine of an automobile develops 15 kW (20 hp) when the automobile is traveling at 50 km$\cdot$hr^{-1}.

a) What is the resisting force acting on the automobile, in newtons?

b) If the resisting force is proportional to the velocity, what power will drive the car at 25 km$\cdot$hr^{-1}? At 100 km$\cdot$hr^{-1}? Give your answers in kilowatts and in horsepower.

Section 7–10 Automotive Power

7–45 For a touring bicyclist, the drag coefficient is 1.0, the frontal area is 0.463 m^2, and the tractive resistance is 0.0045. The rider weighs 712 N and his bike 111 N.

a) To maintain a speed of 14 m$\cdot$s^{-1} (about 31 mi$\cdot$hr^{-1}) on a level road, what must the rider's power output be?

b) For racing, the same rider uses a different bike, one with tractive resistance 0.003 and weight 89 N. He also crouches down, reducing his drag coefficient to 0.88 and reducing his frontal area to 0.366 m^2. What then must his power output be to maintain a speed of 14 m$\cdot$s^{-1}?

c) For the situation in part (b), what power output is required to maintain a speed of 7 m$\cdot$s^{-1}? Note the great drop in power requirement when the speed is only halved. (See "The Aerodynamics of Human-Powered Land Vehicles," in *Scientific American*, December 1983. The article discusses aerodynamic speed limitations for a wide variety of human-powered vehicles.)

PROBLEMS

7–46 A projectile of mass m is fired from a gun with a muzzle velocity of v_0 at an angle of θ above the horizontal. Neglecting air resistance and using energy methods, find the maximum height attained by the projectile.

7–47 The system in Fig. 7–23 is released from rest with the 12-kg block 3 m above the floor. Use the principle of conservation of energy to find the speed with which the

block strikes the floor. Neglect friction and the inertia of the pulley.

7–48 A small object of mass m slides without friction around the loop-the-loop apparatus shown in Fig. 7–24. It starts from rest at point A at a height h above the bottom of the loop.

a) What is the minimum value of h (in terms of R) such that the object moves around the loop without falling off at the top (point B)?

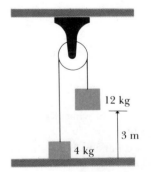

FIGURE 7–23

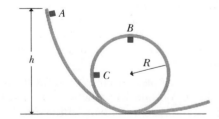

FIGURE 7–24

b) If $h = 3R$, compute the velocity, radial acceleration, and tangential acceleration of the object at point C, which is at the end of a horizontal diameter.

7–49 A skier starts at the top of a very large frictionless spherical snowball, with a very small initial velocity, and skis straight down the side. At what point does he lose contact with the snowball and fly off at a tangent? That is, at the instant he loses contact with the snowball, what angle does a radial line from the center to the skier make with the vertical?

7–50 In an ore-processing plant a conveyor belt carrying small pieces of the refined metal moves at a constant velocity v_0. The conveyor belt goes around a cylinder of radius R as shown in Fig. 7–25. At some angle θ the pieces of metal no longer contact the conveyor belt. In order to design the next phase of the plant the engineer needs to know this angle θ. Find θ in terms of g, R, and v_0.

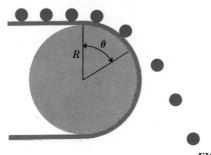

FIGURE 7–25

7–51 A ball of mass 0.5 kg is tied to a string of length 1.0 m, and the other end of the string is tied to a rigid support. The ball is held straight out horizontally from the point of support, with the string pulled taut, and is then released.

a) What is the speed of the ball at the lowest point of its motion?

b) What is the tension in the string at this point?

7–52 A ball is tied to a cord and set in rotation in a vertical circle. Prove that the tension in the cord at the lowest point exceeds that at the highest point by six times the weight of the ball.

7–53 A block of mass 2 kg is released from rest at point A on a track that is one quadrant of a circle of radius 1 m (Fig. 7–26). It slides down the track and reaches point B with a speed of 4 m·s⁻¹. From point B it slides on a level surface a distance of 3 m to point C, where it comes to rest.

a) What was the coefficient of sliding friction on the horizontal surface?

b) How much work was done by friction as the block slid down the circular arc from A to B?

FIGURE 7–26

7–54 Consider the system shown in Fig. 7–27. The coefficient of kinetic friction between the 8 kg block and the table top is $\mu_k = 0.3$. Neglect the mass of the rope and of the pulley, and assume that the pulley is frictionless. Use

FIGURE 7–27

energy methods to calculate the speed of the 6-kg block after it has descended 5 m, starting from rest.

7–55 A pump is required to lift 800 kg (about 200 gal) of water per minute from a well 10 m deep and eject it with a speed of 20 m·s⁻¹.

a) How much work is done per minute in lifting the water?

b) How much in giving it kinetic energy?

c) What power engine is needed?

7–56 An automobile of mass m accelerates, starting from rest, while the engine supplies constant power P.

a) Show that the speed is given as a function of time by $v = (2Pt/m)^{1/2}$.

b) Show that the displacement is given as a function of time by $x - x_0 = (8P/9m)^{1/2}t^{3/2}$.

7–57 The force exerted by a gas in a cylinder on a piston whose area is A is given by $F = pA$, where p is the force per unit area, or *pressure*. The work W in a displacement of the piston from x_1 to x_2 is

$$W = \int_{x_1}^{x_2} F \, dx = \int_{x_1}^{x_2} pA \, dx = \int_{V_1}^{V_2} p \, dV,$$

where dV is the accompanying infinitesimal change of volume of the gas.

a) During an expansion of a gas at constant temperature (isothermal), the pressure depends on the volume according to the relation

$$p = \frac{nRT}{V},$$

where n and R are constants and T is the constant temperature. Calculate the work done by the force F exerted by the gas when the gas expands isothermally from volume V_1 to volume V_2.

b) During an expansion of a gas at constant entropy (adiabatic), the pressure depends on the volume according to the relation

$$p = \frac{k}{V^{\gamma}},$$

where k and γ are constants. Calculate the work done by the gas when it expands adiabatically from V_1 to V_2.

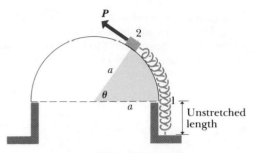

FIGURE 7–28

7–58 A variable force **P** is maintained tangent to a frictionless surface of radius a, as shown in Fig. 7–28. By slowly varying the force, a block of weight w is moved and the spring to which it is attached is stretched from position 1 to position 2. Calculate the work done by the force **P**.

7–59 An object has several forces acting on it. One of these forces is $F = 3xy\mathbf{i}$, a force in the x-direction whose magnitude depends on the position of the object. If x and y are in meters, then F is in newtons. Calculate the work done by this force for each of the following displacements of the object:

a) The object starts at the point $x = 0$, $y = 4$ m and moves parallel to the x-axis to the point $x = 2$ m, $y = 4$ m.

b) The object starts at this point $x = 2$ m, $y = 0$ and moves in the y-direction to the point $x = 2$ m, $y = 4$ m.

c) The object starts at the origin and moves on the line $y = 2x$ to the point $x = 2$ m, $y = 4$ m.

7–60 An object has several forces acting on it. One of these forces is $F = -2xy^2\mathbf{j}$, a force in the negative y-direction whose magnitude depends on the position of the object. For x and y in meters, F is newtons. Consider the displacement of the object from the origin to the point $x = 2$ m, $y = 2$ m.

a) Calculate the work done by F if this displacement is along the straight line $y = x$ that connects these two points.

b) Calculate the work done by F if this displacement is carried out by the object first moving out along the x-axis to the point $x = 2$ m, $y = 0$, and then moving parallel to the y-axis to $x = 2$ m, $y = 2$ m.

c) Compare the work done by F along these two paths. Is the force F conservative or nonconservative?

7–61 There are two equations from which a change in the gravitational potential energy U of a mass m can be calculated. One is $U = mgy$ (Eq. 7–14). The other is $U = -G(mm_E/r)$ (Eq. 7–22). As shown in Section 7–5, the first equation is correct only if g is a constant over the change in height Δy. The second is always correct. Actually g is never exactly constant over any change in height, but the variation may be so small that we ignore it. Consider the difference in U between a mass at the earth's surface and a distance h above it by using both equations, and find for what height h Eq. (7–14) is in error by 1%. Express this h as a fraction of the earth's radius, and also obtain a numerical value for it.

7–62 An object is thrown straight up from the surface of the asteroid Toro (see Exercise 7–27) with a velocity of 15 m·s^{-1}, which is greater than the escape velocity. What is the speed of the object when it is far from Toro? Neglect the gravitational forces due to all other astronomical objects.

7–63 An object of mass m is dropped from a height h above the earth's surface. This height is not necessarily small compared with the radius R_E of the earth. If air resistance is neglected, derive an expression for the speed v of the object when it reaches the surface of the earth. Your expression should involve h, R_E, and g.

7–64 A shaft is drilled from the surface to the center of the earth. The gravitational force on an object of mass m that is inside the earth at a distance r from the center has magnitude $F = mgr/R_E$, where R_E is the radius of the earth, and points toward the center of the earth. If an object is released in the shaft at the surface of the earth, what speed will it have when it reaches the center of the earth? (See Challenge Problem 6–42 and Problem 11–34.)

7–65 A certain spring is found *not* to obey Hooke's law but rather exerts a restoring force $F(x) = -80x - 15x^2$ if it is stretched or compressed a distance x. The units of the numerical factors are such that if x is in meters, then F will be in newtons.

a) Calculate the potential energy function $U(x)$ for this spring. Let $U = 0$ when $x = 0$.

b) An object of mass 2 kg is attached to this spring, pulled a distance 1.0 m to the right on a frictionless horizontal surface, and released. What is the velocity of the object when it is 0.5 m to the right of the $x = 0$ equilibrium position?

7–66 An 80 kg man jumps from a height of 2 m onto a platform mounted on springs. As the springs compress, the platform is pushed down a maximum distance of 0.2 m below its initial position, and then it rebounds. The platform and springs have negligible mass.

a) What is the man's speed at the instant the platform is depressed 0.1 m?

b) If the man had just stepped gently onto the platform, how far would it have been pushed down?

7–67 A block of mass 5 kg is placed against a compressed spring at the bottom of an incline of slope 37° (point A). When the spring is released, it projects the block up the incline. At point B, a distance of 6 m up the incline from A, the block is observed to have a velocity of 4 m·s^{-1}, directed up the incline. The coefficient of kinetic friction between the block and incline is $\mu_k = 0.5$. Calculate the amount of potential energy that was initially stored in the spring.

7–68 If an object is attached to a vertical spring and slowly lowered to its equilibrium position, it is found to stretch the spring by an amount d. If the same object is attached to the unstretched spring and then allowed to fall from rest, through what maximum distance does it stretch the spring? (*Hint:* Calculate the force constant of the spring in terms of the distance d and the mass m of the object.)

CHALLENGE PROBLEMS

7–69 A 0.5-kg block *attached* to a spring of length 0.6 m and force constant $k = 40$ N·m^{-1} is at rest at point A on a frictionless horizontal surface (Fig. 7–29). A constant horizontal force $P = 20$ N is applied to the block and moves the block to the right along the surface.

a) What is the velocity of the block when it reaches point B, which is 0.25 m to the right of point A?

b) When the block reaches point B, the force P is suddenly removed. In the subsequent motion, how close will the block get to the wall?

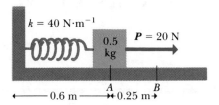

FIGURE 7–29

7–70 A 2-kg block is released on an incline 5 m from a long spring of force constant $k = 80$ N·m^{-1} that is attached at the bottom of the incline. The incline makes an angle of 53° with the horizontal, and the coefficients of friction between the block and the incline are $\mu_s = 0.4$ and $\mu_k = 0.2$.

a) What is the velocity of the block just before it reaches the spring?

b) What will be the maximum compression of the spring?

c) The spring will propel the block back up the incline. How close will the block get to its initial position?

7–71 A particle of mass m moves with a one-dimensional potential-energy function

$$U(x) = \frac{\alpha}{x^2} - \frac{\beta}{x}.$$

The particle is released from rest at $x_0 = \alpha/\beta$.

a) Show that $U(x)$ can be written as

$$U(x) = \frac{\alpha}{x_0{}^2}\left[\left(\frac{x_0}{x}\right)^2 - \left(\frac{x_0}{x}\right)\right].$$

Sketch $U(x)$. Calculate $U(x_0)$ and thereby locate the point x_0 on the sketch.

b) Calculate $v(x)$, the velocity of the particle as a function of position. Draw a sketch of v as a function of x, and give a qualitative description of the motion.

c) For what value of x is the velocity of the particle a maximum, and what is the value of that maximum velocity?

d) What is the force on the particle at the point where the velocity is maximum?

e) Now let the particle be released instead at $x_1 = 3(\alpha/\beta)$. Locate the point x_1 on the sketch of $U(x)$. Calculate $v(x)$ and give a qualitative description of the motion.

f) In each case, when the particle is released at $x = x_0$ and at $x = x_1$, what are the maximum and minimum values of x reached during the motion?

8

IMPULSE AND MOMENTUM

IN THE PRECEDING CHAPTER WE STUDIED THE CONCEPT OF ENERGY AND THE principle of conservation of energy, one of the great conservation principles of physical science. In the course of this study, we developed the concepts of work and mechanical energy from Newton's laws of motion. In this chapter we develop an equally important and fundamental conservation law, conservation of *momentum*. We introduce two new concepts, impulse and momentum; we show how they also arise from Newton's laws and how they are related. This information is especially useful in situations such as impacts and collisions, where very large forces act for short times and cause sudden changes in motion of a body. The concept of momentum and the principle of conservation of momentum appear in many other areas of physics as well as in mechanics.

8–1 IMPULSE AND MOMENTUM

To introduce the concepts of impulse and momentum, let us first consider a particle of mass m moving along a straight line. For the moment we assume that the force F on this particle is constant and directed along the line of motion. If the particle's velocity at some initial time $t = 0$ is v_0, the velocity v at a later time t is given by

$$v = v_0 + at,$$

where the constant acceleration a is determined from the relation $F = ma$. When we multiply the above equation by m and replace ma by F, the result is

$$mv = mv_0 + Ft,$$

or

$$mv - mv_0 = Ft. \tag{8–1}$$

The right side of this equation, the product of the force and the time during which it acts, is called the **impulse** of the force, denoted by J. More generally, when a constant force F acts during a time interval from t_1 to t_2, the impulse of the force is defined to be

Impulse depends on the force and time during which it acts.

$$\text{Impulse} = J = F(t_2 - t_1). \tag{8–2}$$

The left side of Eq. (8–1) contains the product of mass and velocity of the particle at two different times. This product is also given a special name: **momentum.** It is also sometimes called *linear momentum* to distinguish it from a similar quantity called *angular momentum,* to be discussed later. We use the symbol p for momentum:

Momentum is mass times velocity.

$$\text{Momentum} = p = mv. \tag{8–3}$$

In terms of these newly defined quantities, the meaning of Eq. (8–1) is that the impulse of the force from time zero to time t is equal to the change of momentum during that interval. There is nothing special about these two times; we may also say that if the particle's velocity at time t_1 is v_1, and its velocity at time t_2 is v_2, then

Impulse equals change in momentum.

$$F(t_2 - t_1) = mv_2 - mv_1. \tag{8–4}$$

Momentum depends on the motion of the particle; impulse depends on the force acting on it. Thus Eq. (8–4) is a relation between force and motion. Indeed, it is closely related to Newton's second law, which can be restated in the form

$$F = ma = m\frac{dv}{dt} = \frac{d(mv)}{dt}$$
$$= \frac{dp}{dt}.$$

Straight-line motion is, of course, a special case. Strictly speaking, the quantities F, v, and a should have been called the *components* along the straight line of the vector quantities $\boldsymbol{F}$, $\boldsymbol{v}$, and $\boldsymbol{a}$, respectively. The definitions of momentum and impulse and the relationship expressed by Eq. (8–4) are easily generalized for motion not confined to a straight line. We define impulse and momentum to be *vector quantities,* as follows:

Impulse and momentum are vector quantities.

$$\text{Impulse} = \boldsymbol{J} = \boldsymbol{F}(t_2 - t_1), \tag{8–5}$$

$$\text{Momentum} = \boldsymbol{p} = m\boldsymbol{v}. \tag{8–6}$$

The corresponding generalization of Eq. (8–4) is the vector equation

$$\boldsymbol{F}(t_2 - t_1) = m\boldsymbol{v}_2 - m\boldsymbol{v}_1. \tag{8–7}$$

This relationship between impulse and momentum change has the same form as the relation between work and change of kinetic energy developed in Chapter 7. Both are relations between force and change of motion. There are important differences, however. First, impulse is a product of a force and a *time* interval, while work is a product of force and a *distance* and depends on the angle between force and displacement. Furthermore, both force and velocity are *vector* quantities, so impulse and momentum are also vector quantities, unlike work and kinetic energy, which are scalars. Even in straight-line motion, where only one component of a vector is involved, force and velocity may have components along this line that are either positive or negative.

EXAMPLE 8–1 A particle of mass 2 kg moves along the x-axis with an initial velocity of 3 m·s^{-1}. A force $F = -6$ N (i.e., in the negative x-direction) is applied for a period of 3 s. Find the final velocity.

SOLUTION From Eq. (8–4),

$$(-6 \text{ N})(3 \text{ s}) = (2 \text{ kg})v_2 - (2 \text{ kg})(3 \text{ m·s}^{-1}),$$

or

$$v_2 = -6 \text{ m·s}^{-1}.$$

The particle's final velocity is in the negative x-direction.

Units of impulse and momentum

The unit of impulse, in any system, equals the product of the units of force and time in that system. Thus, in SI units, the unit is one newton-second (1 N·s); in the cgs system it is one dyne-second (1 dyn·s); and in the British system it is one pound-second (1 lb·s).

The unit of momentum in SI units is one kilogram meter per second (1 kg·m·s^{-1}); in the cgs system it is one gram centimeter per second (1 g·cm·s^{-1}), and in the British system it is 1 slug foot per second (1 slug·ft·s^{-1}). Because

$$1 \text{ kg·m·s}^{-1} = (1 \text{ kg·m·s}^{-2})\text{s} = 1 \text{ N·s},$$

the units of momentum and impulse are equivalent.

We have defined impulse only for a constant force, but the concept can be generalized to include forces that vary with time; both the magnitude and the direction of the force may vary. Consider a particle of mass m moving in the xy-plane, or in space, acted on by a varying resultant force $\boldsymbol{F}$. Newton's second law states that

$$\boldsymbol{F} = m\boldsymbol{a} = m\frac{d\boldsymbol{v}}{dt},$$

or

$$\boldsymbol{F} \, dt = m \, d\boldsymbol{v}.$$

When the force varies with time, the impulse can be expressed as an integral.

If $\boldsymbol{v}_1$ is the particle's velocity when $t = t_1$, and $\boldsymbol{v}_2$ is its velocity when $t = t_2$, then

$$\int_{t_1}^{t_2} \boldsymbol{F} \, dt = \int_{v_1}^{v_2} m \, d\boldsymbol{v}. \tag{8–8}$$

The integral on the left is the impulse $\boldsymbol{J}$ of the force $\boldsymbol{F}$ in the time interval $t_2 - t_1$, and is a vector quantity:

$$\text{Impulse} = \boldsymbol{J} = \int_{t_1}^{t_2} \boldsymbol{F} \, dt.$$

This integral can be evaluated, of course, only when the force is known as a function of time. The integral on the right, however, is just m multiplied by the total change in $\boldsymbol{v}$:

$$\int_{v_1}^{v_2} m \, d\boldsymbol{v} = m \int_{v_1}^{v_2} d\boldsymbol{v} = m(\boldsymbol{v}_2 - \boldsymbol{v}_1).$$

Thus Eq. (8–8) may be written as

$$\int_{t_1}^{t_2} \mathbf{F} \, dt = m\mathbf{v}_2 - m\mathbf{v}_1. \tag{8–9}$$

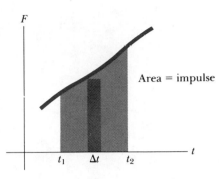

Area = impulse

In the special case when $\mathbf{F}$ is constant, this reduces to Eq. (8–7). When $\mathbf{F}$ varies with time, we can define an *average* force $\mathbf{F}_{av}$ such that the product $\mathbf{F}_{av}(t_2 - t_1)$ represents the same area under the curve in Fig. 8–1. This can be done even when the force varies in direction, so the statement

$$\mathbf{F}_{av}(t_2 - t_1) = \mathbf{J} = m\mathbf{v}_2 - m\mathbf{v}_1 \tag{8–10}$$

is perfectly general, whether the actual force $\mathbf{F}$ is constant or not.

Equation (8–9) is a vector equation; it is equivalent (for forces and velocities in the *xy*-plane) to the two scalar equations

$$\int_{t_1}^{t_2} F_x \, dt = mv_{2x} - mv_{1x},$$

$$\int_{t_1}^{t_2} F_y \, dt = mv_{2y} - mv_{1y}. \tag{8–11}$$

8–1 The area of the small rectangle is approximately equal to the total change in momentum in the interval Δt.

Impulse and momentum can be expressed in terms of their components.

The impulse of any force component, or any force whose direction is constant, can be represented graphically by plotting the force vertically and time horizontally, as in Fig. 8–1. The *area* under the curve, between vertical lines at t_1 and t_2, is equal to the impulse of the force in this time interval.

When the component of the impulse of a force, in a given direction, is *positive,* the corresponding component of momentum of the body on which the impulse acts *increases* algebraically; when the component of impulse is negative, the momentum component decreases. If the impulse is zero, there is no change in momentum.

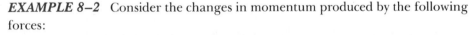

EXAMPLE 8–2 Consider the changes in momentum produced by the following forces:

a) A hockey puck on a frictionless horizontal ice surface moves along the *x*-axis; it is acted on for 2 s by a constant force of 10 N toward the right.

b) The puck is acted on for 2 s by a constant force of 10 N toward the right and then for 2 s by a constant force of 20 N toward the left.

c) The puck is acted on for 2 s by a constant force of 10 N toward the right, then for 1 s by a constant force of 20 N toward the left. The forces for each case are shown graphically in Fig. 8–2.

SOLUTION

a) The impulse of the force is (+ 10 N)(2 s) = +20 N·s. Hence the momentum of *any* body on which this force acts increases by 20 kg·m·s^{-1}. This change is the same whatever the mass of the body and whatever the magnitude and direction of its initial velocity.

   Suppose the mass of the puck is 2 kg and that it is initially at rest. Its *final* momentum then equals its *change* in momentum, and its final *velocity* is 10 m·s^{-1} toward the right. You may verify this by computing the acceleration.

Hitting a hockey puck: analysis in terms of impulse and momentum

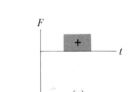

(a)

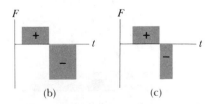

(b)   (c)

8–2 Graphic representation of the forces in Example 8–2.

If the puck had been moving initially toward the *right* at 5 m·s⁻¹, its initial momentum would have been 10 kg·m·s⁻¹ toward the right, its final momentum 30 kg·m·s⁻¹, and its final velocity 15 m·s⁻¹ toward the right.

If the puck has been moving initially toward the *left* at 5 m·s⁻¹, its initial momentum would have been −10 kg·m·s⁻¹, its final momentum +10 kg·m·s⁻¹, and its final velocity 5 m·s⁻¹ toward the right. That is, the constant force of 10 N toward the right would first have brought the body to rest and then given it a velocity in the direction opposite to its initial velocity.

b) The impulse of this force is

$$(+10 \text{ N})(2 \text{ s}) - (20 \text{ N})(2 \text{ s}) = -20 \text{ N·s}.$$

The momentum of *any* body on which this impulse acts is changed by −20 kg·m·s⁻¹. We invite you to check various possibilities.

c) The impulse of this force is

$$(+10 \text{ N})(2 \text{ s}) - (20 \text{ N})(1 \text{ s}) = 0.$$

Hence the momentum of any body on which it acts is not changed. Of course, the momentum of the body is increased during the first 2 s, but it is *decreased* by an equal amount in the next second. As an exercise, describe the motion of a body of mass 2 kg, moving initially to the left at 5 m·s⁻¹, and acted on by this force. It helps to construct a graph of velocity versus time.

A ball bouncing off a wall: using impulse and momentum to find the impact force

EXAMPLE 8–3 A ball of mass 0.40 kg is thrown against a brick wall. When it strikes the wall, the ball is moving horizontally to the left at 30 m·s⁻¹, and it rebounds horizontally to the right at 20 m·s⁻¹. Find the impulse of the force exerted on the ball by the wall. If the ball is in contact with the wall for 0.010 s, find the average force on the ball during the impact.

SOLUTION The initial momentum of the ball is

$$(0.40 \text{ kg})(-30 \text{ m·s}^{-1}) = -12 \text{ kg·m·s}^{-1}.$$

The final momentum is +8.0 kg·m·s⁻¹. The *change* in momentum is

$$mv_2 - mv_1 = 8.0 \text{ kg·m·s}^{-1} - (-12 \text{ kg·m·s}^{-1})$$
$$= 20 \text{ kg·m·s}^{-1} = J.$$

Hence the impulse of the force exerted on the ball was 20 kg·m·s⁻¹ = 20 N·s. Since the impulse is *positive,* the force must be toward the right.

The general nature of the force–time graph is shown by one of the curves in Fig. 8–3. The force is zero before impact, rises to a maximum, and decreases to zero when the ball leaves the wall. If the ball is relatively rigid, like a baseball, the time of collision is small and the maximum force is large, as in curve (a). If the ball is more yielding, like a tennis ball, the collision time is larger and the maximum force is less, as in curve (b). In any event, the *area* under the curve represents the impulse $J = 20$ N·s.

If the collision time is 0.01 s, then from Eq. (8–10),

$$F_{av}(0.01 \text{ s}) = 20 \text{ N·s}, \qquad F_{av} = 2000 \text{ N}.$$

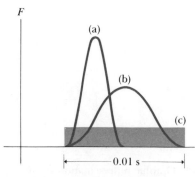

8–3 Force–time graph for Example 8–3.

This average force is represented by the horizontal line (c) in Fig. 8–3. The actual force might vary with time as shown by curve (b) in the figure.

Figure 8–4 is a stroboscopic photograph showing the impact of a tennis ball and racket during a serve.

EXAMPLE 8–4 Suppose the ball in Example 8–3 is a soccer ball with mass 0.4 kg. Initially it is moving to the left at 30 m·s^{-1}, but then it is kicked and given a velocity at 45° upward and to the right, with magnitude 30 m·s^{-1}. Find the impulse of the force and the average force, assuming a collision time of 0.010 s.

SOLUTION The velocities are not along the same line, and we must consider the vector nature of momentum and impulse explicitly. Taking the x-axis horizontal to the right and the y-axis vertically upward, we find the following velocity components:

$$v_{1x} = -30 \text{ m·s}^{-1}, \qquad v_{1y} = 0,$$

$$v_{2x} = v_{2y} = (0.707)(30 \text{ m·s}^{-1}) = 21.2 \text{ m·s}^{-1}.$$

Alternatively,

$$\boldsymbol{v}_1 = (-30 \text{ m·s}^{-1})\boldsymbol{i},$$

$$\boldsymbol{v}_2 = (21.2 \text{ m·s}^{-1})\boldsymbol{i} + (21.2 \text{ m·s}^{-1})\boldsymbol{j}.$$

The x-component of impulse is equal to the x-component of momentum change, and similarly for the y-components:

$$\begin{aligned} J_x &= m(v_{2x} - v_{1x}) \\ &= (0.4 \text{ kg})[21.2 \text{ m·s}^{-1} - (-30 \text{ m·s}^{-1})] \\ &= 20.5 \text{ kg·m·s}^{-1}, \end{aligned}$$

$$\begin{aligned} J_y &= m(v_{2y} - v_{1y}) \\ &= (0.4 \text{ kg})(21.2 \text{ m·s}^{-1} - 0) \\ &= 8.48 \text{ kg·m·s}^{-1}. \end{aligned}$$

Or:

$$\boldsymbol{J} = (20.5 \text{ kg·m·s}^{-1})\boldsymbol{i} + (8.48 \text{ kg·m·s}^{-1})\boldsymbol{j}.$$

The components of average force on the ball are

$$F_x = J_x/\Delta t = 2050 \text{ N}, \qquad F_y = J_y/\Delta t = 848 \text{ N},$$

and

$$\boldsymbol{F}_{av} = (2050 \text{ N})\boldsymbol{i} + (848 \text{ N})\boldsymbol{j}.$$

The magnitude and direction of the average force are

$$F_{av} = \sqrt{(2050 \text{ N})^2 + (848 \text{ N})^2} = 2218 \text{ N},$$

$$\theta = \arctan \frac{848 \text{ N}}{2050 \text{ N}} = 22.5°.$$

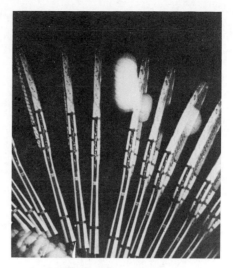

8–4 Multiflash stroboscopic photograph of a tennis racket hitting a ball during a serve. The exposure rate was 300 pictures per second. The ball is in contact with the racket for approximately 0.01 s. The ball flattens noticeably on both sides, and the frame of the racket bends and vibrates during and after the impact. (Dr. Harold Edgerton, M.I.T., Cambridge, Massachusetts.)

8–2 CONSERVATION OF MOMENTUM

The concept of momentum is most useful in situations involving several interacting bodies. To explore this topic, we consider first a system consisting of two bodies that interact with each other but not with anything else. Each body exerts a force on the other, so the momentum of each body changes. According to Newton's third law, the forces the bodies exert on each other are always

The total momentum of a system is the vector sum of the momenta of its parts.

equal in magnitude and opposite in direction. Thus the *impulses* given to the two bodies in any time interval are also equal and opposite, and therefore the *momentum changes* of the two bodies are equal and opposite.

To state this principle more compactly, we define the **total momentum** of the system, **P**, as the vector sum of momenta of the bodies in the system:

$$P = mv_1 + mv_2 + \cdots = p_1 + p_2 + \cdots. \tag{8-12}$$

If the change in momentum of one body is exactly the negative of that of the other, then the change in the *total* momentum must be zero. Thus *when two bodies interact only with each other, their total momentum is constant.*

Internal and external forces: Is a force on one part of a system due to another part or to something outside the system?

A force that one part of a system exerts on another is called an **internal force,** and a force exerted on a part of the system by some agency outside the system is an **external force.** When no external forces act on a system, we call it an **isolated system.** Thus we may state the principle of **conservation of momentum** as follows:

> *The total momentum of an isolated system is constant, or conserved.*

When no external forces act, the total momentum is constant.

More generally, if there are external forces but the vector sum of all external forces is zero, the total momentum is constant. The *internal* forces can change the momentum of an individual body within the system, but they cannot change the *total* momentum of the system.

The principle of conservation of momentum is one of the most fundamental and important principles in mechanics. In some respects it is more general than the principle of conservation of mechanical energy. For example, it is valid even if the internal forces are not conservative, while mechanical energy is conserved only when they *are* conservative. In this chapter we will see situations in which both momentum and energy are conserved, and others in which only momentum is conserved.

PROBLEM-SOLVING STRATEGY: *Conservation of momentum*

Momentum is a vector quantity, and the rules of vector addition *must* be used in computing the total momentum of a system. Usually the use of components offers the simplest method.

1. Define your coordinate system. Make a sketch showing the coordinate axes, including the positive direction for each. Often the analysis is simplified if you choose the *x*-axis to have the same direction as one of the initial velocities.

2. Draw a sketch, including the coordinate axes and vectors representing all known velocity vectors. Label the vectors with magnitudes, angles, components, or whatever information is given, and give algebraic symbols to all unknown magnitudes, angles, or components.

3. Compute the *x*- and *y*-components of momentum of each body, both before and after the collision, impact, or other interaction, by using the relations $p_x = mv_x$ and $p_y = mv_y$. Even when all the velocities lie along a line (such as the *x*-axis), the components of velocity

along this line can be positive or negative, and careful attention to signs is essential.

4. Write an equation equating the total *initial* *x*-component of momentum to the total *final* *x*-component of momentum. Write another equation for the *y*-components. These two equations express conservation of momentum in component form. Some of the components will be expressed in terms of symbols representing unknown quantities.

5. Solve these equations to determine whatever results are required. In some problems you will have to convert from the *x*- and *y*-components of a velocity to its magnitude and direction, or the reverse.

6. Remember that the *x*- and *y*-components of velocity or momentum are *never* added together in the same equation.

7. In some problems energy considerations give additional relationships among the various velocities, as we will see later in this chapter.

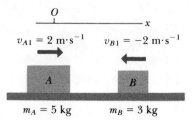

EXAMPLE 8–5 Figure 8–5 shows two gliders moving toward each other on a linear air track. The resultant vertical force on each body is zero; the resultant force on each glider is the force exerted on it by the other glider, and the total momentum of the system is constant in magnitude and direction.

Suppose that, after the gliders collide, B moves away with a final velocity of $+2$ m·s^{-1}. What is the final velocity of A?

8–5 Momentum diagram for Example 8–5.

SOLUTION Let the final velocity of A be v_{A2}. Then write an equation showing the equality of the total momentum before and after the collision:

$$(0.5 \text{ kg})(2 \text{ m·s}^{-1}) + (0.3 \text{ kg})(-2 \text{ m·s}^{-1}) = (0.5 \text{ kg})v_{A2} + (0.3 \text{ kg})(+2 \text{ m·s}^{-1}).$$

Solving this equation for v_{A2}, we find $v_{A2} = -0.4$ m·s^{-1}.

Collision of two gliders on an air track: Which one moves away faster?

EXAMPLE 8–6 Figure 8–6 shows two chunks of ice sliding on a frictionless frozen pond. Chunk A, having mass $m_A = 5$ kg, moves with initial velocity $v_{A1} = 2$ m·s^{-1} parallel to the x-axis. It collides with chunk B, which has mass $m_B = 3$ kg and is initially at rest. After the collision, the velocity of A is found to be $v_{A2} = 1$ m·s^{-1} in a direction making an angle $\theta = 30°$ with the initial direction. What is the final velocity of B?

Colliding ice chunks on a skating pond: use of components in a conservation-of-momentum calculation

SOLUTION The total momentum of the system is the same before and after the collision. The velocities are not all along a single line, and we have to recognize the *vector* nature of momentum. The simplest procedure is to find the components of each momentum in the x- and y-directions. Then momentum conservation requires that the sum of the x-components before the collision must equal the sum after the collision, and similarly for the y-components. Just as with force equilibrium problems, we write a separate equation for each component. For the x-components, we have

$$(5 \text{ kg})(2 \text{ m·s}^{-1}) + (3 \text{ kg})(0) = (5 \text{ kg})(1 \text{ m·s}^{-1})(\cos 30°) + (3 \text{ kg})v_{B2x}.$$

From this we find $v_{B2x} = 1.9$ m·s^{-1}.

Similarly, conservation of the y-component of total momentum gives

$$(5 \text{ kg})(0) + (3 \text{ kg})(0) = (5 \text{ kg})(1 \text{ m·s}^{-1})(\sin 30°) + (3 \text{ kg})v_{B2y},$$

from which $v_{B2y} = -0.83$ m·s^{-1}.

The magnitude of v_{B2} is

$$v_{B2} = \sqrt{(1.9 \text{ m·s}^{-1})^2 + (-0.83 \text{ m·s}^{-1})^2}$$
$$= 2.06 \text{ m·s}^{-1},$$

and the angle of its direction from the positive x-axis is

$$\phi = \arctan \frac{-0.83 \text{ m·s}^{-1}}{1.9 \text{ m·s}^{-1}}$$
$$= -24°.$$

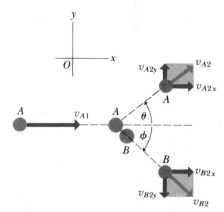

8–6 Velocity vector diagram for Example 8–6.

8–3 COLLISIONS

Momentum considerations alone are not always sufficient to predict all the details of the outcome of a collision. For example, suppose we are given the masses m_A and m_B and initial velocities v_{A1} and v_{B1} of two colliding bodies, and we want to find their velocities v_{A2} and v_{B2} after the collision. If the collision is

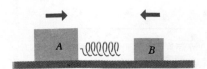

8–7 Body A (with spring) and body B approach each other on a frictionless surface.

In an elastic collision, the total kinetic energy is conserved, and so is the total momentum.

In an inelastic collision, the total momentum is conserved, but the total kinetic energy is not.

When two colliding bodies stick together, the collision is inelastic.

"head-on," as in Example 8–5, all the velocities are along the same line. Momentum conservation gives the equation

$$m_A v_{A1} + m_B v_{B1} = m_A v_{A2} + m_B v_{B2} \qquad (8\text{–}13)$$

for the two unknowns v_{A2} and v_{B2}. This equation by itself does not provide enough information to determine two unknowns. If the collision is of the more general type of Example 8–6, then we have two equations, one each for the x- and y-components of momentum; but then we have four unknowns, the x- and y-components of each final velocity. Again we need more information to determine these unknowns.

If the interaction forces between the bodies are *conservative,* the total *kinetic energy* of the system is the same after the collision as before, and this provides an additional relation among the velocities. Such a collision is called an **elastic collision,** or sometimes *completely elastic* or *perfectly elastic collision.* A collision between two glass balls or two hard steel balls or two ivory billiard balls is almost completely elastic. As a model for elastic collisions, we may consider the situation shown in Fig. 8–7. When the bodies collide, the spring is momentarily compressed and some of the original kinetic energy is momentarily converted to elastic potential energy. The spring then expands, and when the bodies separate, this potential energy is reconverted to kinetic energy.

A collision in which the total kinetic energy after the collision is *less* than before is called an **inelastic collision.** In one kind of inelastic collision, the colliding bodies stick together and move *as a unit* after the collision. In Fig. 8–7, we could replace the spring with a ball of putty or chewing gum that squashes and sticks the two bodies together. A gum ball striking a window shade, a bullet embedding itself in a block of wood, or two cars colliding and bending their fenders are examples of inelastic collisions.

8–4 INELASTIC COLLISIONS

In this section we consider inelastic collisions of the particular type where the two bodies stick together after the collision. Their two final velocities are then equal:

$$v_{A2} = v_{B2} = v_2.$$

When this is combined with the principle of conservation of momentum, we obtain

$$m_A v_{A1} + m_B v_{B1} = (m_A + m_B)v_2, \qquad (8\text{–}14)$$

and the final velocity can be computed if the initial velocities and the masses are known.

The total kinetic energy of a system after an inelastic collision is always less than before the collision. Suppose, for example, a body having mass m_A and initial velocity v_1 collides inelastically with a body of mass m_B initially at rest. After the collision the two bodies have a common velocity v_2 given by Eq. (8–14):

$$v_2 = \frac{m_A}{m_A + m_B} v_1. \qquad (8\text{–}15)$$

The kinetic energies K_1 and K_2 before and after the collision, respectively, are

$$K_1 = \tfrac{1}{2} m_A v_1^{\,2},$$

$$K_2 = \tfrac{1}{2}(m_A + m_B)\left(\frac{m_A}{m_A + m_B}\right)^2 v_1^{\,2},$$

so the ratio of final to initial kinetic energy is

$$\frac{K_2}{K_1} = \frac{m_A}{m_A + m_B}. \qquad (8-16)$$

The right side is always less than unity because the denominator of the fraction is always greater than the numerator. Thus in such a collision the final kinetic energy is always less than the initial value. Even when the initial velocity of m_B is not zero, it is not difficult to prove that the kinetic energy after a collision where the bodies stick together is always less than before.

EXAMPLE 8–7 Suppose that in the collision in Fig. 8–5 the two bodies stick together after the collision; the masses and initial velocities are as shown. From momentum conservation,

$$(5 \text{ kg})(2 \text{ m·s}^{-1}) + (3 \text{ kg})(-2 \text{ m·s}^{-1}) = (5 \text{ kg} + 3 \text{ kg})v_2,$$

and

$$v_2 = 0.5 \text{ m·s}^{-1}.$$

Since v_2 is positive, the system moves to the right after the collision. The kinetic energy of body A before the collision is

$$\tfrac{1}{2}m_A v_{A1}{}^2 = \tfrac{1}{2}(5 \text{ kg})(2 \text{ m·s}^{-1})^2 = 10 \text{ J},$$

and that of body B is

$$\tfrac{1}{2}m_B v_{B1}{}^2 = \tfrac{1}{2}(3 \text{ kg})(-2 \text{ m·s}^{-1})^2 = 6 \text{ J}.$$

The impact of this bullet on a hard surface is an inelastic collision; most of the bullet's kinetic energy is lost, and it becomes appreciably hotter as its kinetic energy is converted to internal energy. (Dr. Harold Edgerton, M.I.T., Cambridge, Massachusetts.)

The total kinetic energy before collision is therefore 16 J. Note that the kinetic energy of body B is positive, although its velocity v_{B1} and its momentum $m v_{B1}$ are both negative.

The kinetic energy after the collision is

$$\tfrac{1}{2}(m_A + m_B)v_2{}^2 = \tfrac{1}{2}(5 \text{ kg} + 3 \text{ kg})(0.5 \text{ m·s}^{-1})^2 = 1 \text{ J}.$$

In an inelastic collision, the kinetic energy is always less after the collision than before.

Hence, far from remaining constant, the final kinetic energy is only $\tfrac{1}{16}$ of the original, and $\tfrac{15}{16}$ is "lost" in the collision.

The energy is not really lost, of course; it is converted from mechanical energy to various other forms. If there is a ball of putty between the masses, it squashes irreversibly and becomes warmer. If the masses couple together like two freight cars, the energy goes into elastic waves that are eventually dissipated. If there is a spring between the bodies that is compressed when they are locked together, then the energy is stored as potential energy of the spring. In all of these cases the *total* energy of the system is conserved, although the *kinetic* energy is not. In an isolated system, however, momentum is *always* conserved, whether the collision is elastic or not.

EXAMPLE 8–8 Figure 8–8 shows a *ballistic pendulum*, a device for measuring the speed of a bullet. The bullet is allowed to make a completely inelastic collision with a body of much greater mass. The momentum of the system immediately after the collision equals the original momentum of the bullet, but since the *velocity* is very much smaller, it can be determined more easily. Although the ballistic pendulum has now been superseded by other devices, it is still an important laboratory experiment for illustrating the concepts of momentum and energy.

Ballistic pendulum: using conservation of momentum to measure the speed of a bullet

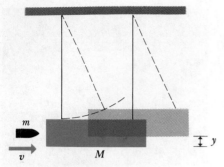

8–8 The ballistic pendulum.

In Fig. 8–8, the pendulum, consisting perhaps of a large wooden block of mass M, hangs vertically by two cords. A bullet of mass m, traveling with a velocity v, strikes the pendulum and remains embedded in it. If the collision time is very small compared with the time of swing of the pendulum, the supporting cords remain practically vertical during the collision. Hence no external horizontal forces act on the system during the collision, and the horizontal momentum is conserved. Then if V represents the velocity of bullet and block immediately after the collision,

$$mv = (m + M)V, \qquad v = \frac{m + M}{m}V.$$

The kinetic energy of the system immediately after the collision is $K = \frac{1}{2}(m + M)V^2$.

The pendulum now swings to the right and upward until its kinetic energy is converted to gravitational potential energy. Hence

$$\tfrac{1}{2}(m + M)V^2 = (m + M)gy,$$

$$V = \sqrt{2gy},$$

and

$$v = \frac{m + M}{m}(\sqrt{2gy}).$$

By measuring m, M, and y, we can compute the original velocity v of the bullet.

Suppose $m = 5.00 \text{ g} = 0.00500 \text{ kg}$, $M = 2.00 \text{ kg}$, and $h = 3.00 \text{ cm} = 0.0300 \text{ m}$. Working backwards, we find the velocity V of the block just after impact:

$$V = \sqrt{2(9.80 \text{ m·s}^{-2})(0.0300 \text{ m})} = 0.767 \text{ m·s}^{-1}.$$

Then we use momentum conservation to find the bullet's velocity v just before impact:

$$(0.00500 \text{ kg})v = (2.00 \text{ kg} + 0.00500 \text{ kg})(0.767 \text{ m·s}^{-1}),$$

$$v = 307 \text{ m·s}^{-1}.$$

Note that the kinetic energy just before impact is $\frac{1}{2}(0.00500 \text{ kg})(307 \text{ m·s}^{-1})^2 = 236 \text{ J}$, while just after impact it is $\frac{1}{2}(2.005 \text{ kg})(0.767 \text{ m·s}^{-1})^2 = 0.589 \text{ J}$. Nearly all the kinetic energy disappears as the wood splinters and the bullet becomes hotter.

8–5 ELASTIC COLLISIONS

Next let us consider an *elastic* collision between two bodies A and B. We first discuss a head-on collision in which all the initial and final velocities lie along the same line. Later in this section we will consider more general collisions.

Both kinetic energy and momentum are conserved. From conservation of kinetic energy we have

Conservation of kinetic energy in an elastic collision

$$\tfrac{1}{2}m_A v_{A1}{}^2 + \tfrac{1}{2}m_B v_{B1}{}^2 = \tfrac{1}{2}m_A v_{A2}{}^2 + \tfrac{1}{2}m_B v_{B2}{}^2,$$

and from conservation of momentum,

$$m_A v_{A1} + m_B v_{B1} = m_A v_{A2} + m_B v_{B2}.$$

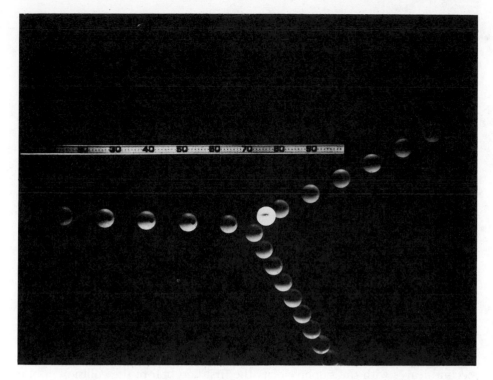

In this multiflash photograph, a billiard ball approaching from the left collides elastically with a second ball having equal mass, initially at rest. Both momentum and kinetic energy are conserved, and the final velocities of the two balls are perpendicular. The speeds can be determined from the distances between successive photographic images (representing equal time intervals), and the conservation principles can be verified directly from this photograph. (*PSSC Physics*, 2e, 1965; D.C. Heath and Company, Inc., with Education Development Center, Inc., Newton, Mass.)

If the masses and initial velocities are known, these are two independent equations that can be solved simultaneously to find the two final velocities. The general solution of these equations is somewhat involved, so we will concentrate on a few particular cases, such as having one body at rest before the collision.

Suppose mass m_B is initially at rest. We can then simplify the velocity notation by letting v be the initial velocity of A, and v_A and v_B the final velocities of A and B, respectively. Then the kinetic energy and momentum conservation equations are, respectively,

$$\tfrac{1}{2}m_A v^2 = \tfrac{1}{2}m_A v_A{}^2 + \tfrac{1}{2}m_B v_B{}^2, \qquad (8-17)$$

and

$$m_A v = m_A v_A + m_B v_B. \qquad (8-18)$$

Assuming the masses and v are known, we may solve for v_A and v_B. The simplest approach is somewhat indirect and uncovers an additional interesting feature of elastic collisions.

We first rearrange Eqs. (8–17) and (8–18) as follows:

$$m_B v_B{}^2 = m_A(v^2 - v_A{}^2) = m_A(v - v_A)(v + v_A), \qquad (8-19)$$

and

$$m_B v_B = m_A(v - v_A). \qquad (8-20)$$

We now divide Eq. (8–19) by Eq. (8–20) to obtain

$$v_B = v + v_A. \qquad (8-21)$$

We now substitute this back into Eq. (8–20) to eliminate v_B and then solve for v_A:

$$m_B(v + v_A) = m_A(v - v_A),$$

$$v_A = \frac{m_A - m_B}{m_A + m_B}v. \qquad (8\text{–}22)$$

Finally, we substitute this result back into Eq. (8–21) to obtain

$$v_B = \frac{2m_A}{m_A + m_B}v. \qquad (8\text{–}23)$$

A somewhat tedious calculation, but there is no substitute for persistence!

Now we can interpret the results. Suppose first that A is a ping-pong ball and B is a bowling ball. Then we expect A to bounce off with a velocity nearly equal to the original value but in the opposite direction, and we expect B's velocity to be much smaller. Indeed, when m_A is much smaller than m_B, the fraction in Eq. (8–22) is approximately equal to (-1), and the fraction in Eq. (8–23) is much smaller than unity. We challenge you to invent a similar check for the opposite case where A is the bowling ball and B the ping-pong ball.

Another interesting case is that of *equal* masses. If $m_A = m_B$, then Eqs. (8–22) and (8–23) give $v_A = 0$ and $v_B = v$. That is, the first body stops and the second leaves with the same velocity the first had before the collision, a phenomenon familiar to pool players.

Now comes the dividend. Equation (8–21) may be rewritten as

$$v = v_B - v_A. \qquad (8\text{–}24)$$

In an elastic collision, the relative velocity has the same magnitude before and after: an alternative definition of an elastic collision.

Now $v_B - v_A$ is just the velocity of B relative to A after the collision, and v, apart from sign, is the velocity of B relative to A before the collision. Hence *the relative velocity has the same magnitude before and after the collision*. Although we have proved this only for one special case, it turns out to be a general property of *all* elastic collisions, even when both bodies are moving initially and the velocities do not all lie along the same line. The constancy of the magnitude of relative velocity provides an alternative definition of an elastic collision; in any collision where this condition is satisfied, the total kinetic energy is also conserved.

EXAMPLE 8–9 Suppose the collision shown in Fig. 8–5 is completely elastic. What are the velocities of A and B after the collision?

SOLUTION From the principle of conservation of momentum,

$$(5 \text{ kg})(2 \text{ m·s}^{-1}) + (3 \text{ kg})(-2 \text{ m·s}^{-1}) = (5 \text{ kg})v_{A2} + (3 \text{ kg})v_{B2},$$

$$5v_{A2} + 3v_{B2} = 4 \text{ m·s}^{-1}.$$

Because the collision is completely elastic,

$$v_{B2} - v_{A2} = -(v_{B1} - v_{A1})$$
$$= -(-2 \text{ m·s}^{-1} - 2 \text{ m·s}^{-1}) = 4 \text{ m·s}^{-1}.$$

Solving these equations simultaneously, we obtain

$$v_{A2} = -1 \text{ m·s}^{-1}, \qquad v_{B2} = 3 \text{ m·s}^{-1}.$$

Both bodies therefore reverse their directions of motion, A traveling to the left at 1 m·s^{-1} and B to the right at 3 m·s^{-1}.

The total kinetic energy after the collision is

$$\tfrac{1}{2}(5 \text{ kg})(-1 \text{ m·s}^{-1})^2 + \tfrac{1}{2}(3 \text{ kg})(3 \text{ m·s}^{-1})^2 = 16 \text{ J},$$

which equals the total kinetic energy before the collision.

If an elastic collision is not head-on, the velocities do not all lie on a single line. Each final velocity then has two unknown components, and we have four unknowns in all. Conservation of energy and of the x- and y-components of momentum gives only three equations, so additional information is needed. This may take the form of the direction of one of the final velocities, perhaps obtained from geometrical considerations, or of one of the final velocity magnitudes.

EXAMPLE 8–10   In Fig. 8–6 suppose the 5-kg mass m_A has an initial velocity of 4 m·s^{-1} in the positive direction and a final velocity of 2 m·s^{-1} in an unknown direction. Find the final speed v_{B2} of the 3-kg mass, and the angles θ and ϕ in the figure, assuming the collision is perfectly elastic.

An elastic collision: using momentum components to find unknown directions

SOLUTION   Because the collision is elastic, the total final kinetic energy is the same as the initial kinetic energy. Thus

$$\tfrac{1}{2}(5 \text{ kg})(4 \text{ m·s}^{-1})^2 = \tfrac{1}{2}(5 \text{ kg})(2 \text{ m·s}^{-1})^2 + \tfrac{1}{2}(3 \text{ kg})v_{B2}{}^2,$$

from which

$$v_{B2} = 4.47 \text{ m·s}^{-1}.$$

Conservation of the x- and y-components of total momentum gives, respectively,

$$(5 \text{ kg})(4 \text{ m·s}^{-1}) = (5 \text{ kg})(2 \text{ m·s}^{-1}) \cos \theta + (3 \text{ kg})(4.47 \text{ m·s}^{-1}) \cos \phi,$$

$$0 = (5 \text{ kg})(2 \text{ m·s}^{-1}) \sin \theta - (3 \text{ kg})(4.47 \text{ m·s}^{-1}) \sin \phi.$$

These are two simultaneous equations for θ and ϕ; the simplest solution is to eliminate ϕ as follows. We solve the first equation for $\cos \phi$ and the second for $\sin \phi$; we then square each equation and add; since $\sin^2 \phi + \cos^2 \phi = 1$, this eliminates ϕ and leaves an equation that may be solved for $\cos \theta$ and hence for θ. This value may then be substituted back into either of the two equations and the result solved for ϕ. The details are left as an exercise; the results are

$$\theta = 36.9°, \qquad \phi = 26.6°.$$

The examples in this section and the preceding one show that collisions can be classified according to energy considerations. A collision in which kinetic energy is conserved is called an *elastic collision*. A collision in which the total kinetic energy decreases is called an *inelastic* or *completely inelastic collision*. We have discussed inelastic collisions where the two bodies stick together after the collision. There are also inelastic collisions where some kinetic energy is

Can the kinetic energy ever be greater after the collision than before?

lost, but not enough for the two bodies to reach a common final velocity. In some cases the final kinetic energy is *greater* than the initial value. This occurs, for example, when an explosive between the two bodies is detonated during the collision, blowing them apart with great speed. Such a collision might be called a *superelastic collision*, although this term is not in common use. A special case of a superelastic collision is **recoil,** discussed in the next section.

Finally, momentum conservation can be applied even to systems that are not really isolated. Some external forces may be acting on the colliding bodies; but if the internal forces during the collision are much larger than the external, then the external forces during the actual collision can be neglected. This is certainly the case, for example, when two cars collide at an icy intersection.

8–6 RECOIL

Recoil: a case where the total kinetic energy increases

Figure 8–9 shows two toy cars A and B, with a compressed spring between them. We hold both cars, and then release them without giving them any initial motion. The spring begins to expand, pushing the cars apart. As it does so, it exerts equal and opposite forces on the cars until it has expanded back to its original uncompressed length. It then drops to the surface, while the cars roll on. The initial momentum of the system was zero, and if frictional forces can be neglected, the resultant external force on the system is zero. The total momentum of the system therefore remains constant and equal to zero during this process. Then if v_A and v_B are the velocities acquired by A and B, we have

$$m_A v_A + m_B v_B = 0, \qquad \frac{v_A}{v_B} = -\frac{m_B}{m_A}. \qquad (8\text{--}25)$$

The velocities are of opposite sign, and their magnitudes are inversely proportional to the corresponding masses.

The original kinetic energy of the system is also zero. The final kinetic energy is $K = \frac{1}{2}m_A v_A{}^2 + \frac{1}{2}m_B v_B{}^2$. The source of this energy is the original elastic potential energy of the spring. The ratio of kinetic energies is

$$\frac{\frac{1}{2}m_A v_A{}^2}{\frac{1}{2}m_B v_B{}^2} = \frac{m_A}{m_B}\left(\frac{v_A}{v_B}\right)^2 = \frac{m_B}{m_A}. \qquad (8\text{--}26)$$

Thus although the final momenta are equal in magnitude, the final kinetic energies are inversely proportional to the corresponding masses; the body of smaller mass receives the larger share of the original potential energy. The reason is that the change in *momentum* of a body equals the *impulse* of the force acting on it, while the change in *kinetic energy* equals the *work* of the force. The forces on the two bodies are equal in magnitude and act for equal *times*, so they produce equal and opposite changes in momentum. The points of application of the forces, however, *do not* move through equal *distances* (except when $m_A = m_B$), since the acceleration, velocity, and displacement of the smaller body are greater than those of the larger. Hence more *work* is done on the body of smaller mass.

8–9 Conservation of momentum in recoil.

We can apply these considerations to the firing of a rifle. The initial momentum of the system is zero. When the rifle is fired, the bullet and the powder gases acquire a forward momentum, and the rifle (together with the person holding it) acquires a rearward momentum of the same magnitude. Because of the relatively large mass of the rifle compared with that of the bullet and powder charge, the velocity and kinetic energy of the rifle are much *smaller* than those of the bullet and powder gases.

Firing a rifle: an application of momentum conservation

EXAMPLE 8–11   A hunter holds a 3.00-kg rifle loosely in his hands and fires a bullet of mass 5.00 g with a muzzle velocity of 300 m·s^{-1}. What is the recoil velocity of the rifle? What is the final kinetic energy of the bullet? Of the rifle?

SOLUTION   The total momentum is zero before and after firing. Thus

$$0 = (0.00500 \text{ kg})(300 \text{ m·s}^{-1}) + (3.00 \text{ kg})v,$$

$$v = -0.500 \text{ m·s}^{-1}.$$

The negative sign means that the recoil is in the direction opposite to that of the bullet.

The kinetic energy of the bullet is

$$K_B = \tfrac{1}{2}(0.00500 \text{ kg})(300 \text{ m·s}^{-1})^2 = 225 \text{ J},$$

and the kinetic energy of the rifle is

$$K_R = \tfrac{1}{2}(3.00 \text{ kg})(0.500 \text{ m·s}^{-1})^2 = 0.375 \text{ J}.$$

The bullet acquires much more kinetic energy than the rifle because the interaction force on it acts over a much longer distance, and hence does more work, than on the rifle. Finally, we note that

$$\frac{K_B}{K_R} = \frac{225 \text{ J}}{0.375 \text{ J}} = 600 = \frac{3.00 \text{ kg}}{0.00500 \text{ kg}}.$$

in agreement with Eq. (8–26).

8–7 CENTER OF MASS

The principle of conservation of momentum for an isolated system can be restated in a useful way by referring to the concept of **center of mass.** Suppose we have a collection of particles having masses m_1, m_2, and so on. Let the coordinates of m_1 be (x_1, y_1), those of m_2 be (x_2, y_2), and so on. The center of mass of the system is defined as the point having coordinates (X, Y) given by

$$X = \frac{m_1 x_1 + m_2 x_2 + m_3 x_3 + \cdots}{m_1 + m_2 + m_3 + \cdots},$$

$$Y = \frac{m_1 y_1 + m_2 y_2 + m_3 y_3 + \cdots}{m_1 + m_2 + m_3 + \cdots}.$$

(8–27)

Thus the center of mass represents a weighted average position of the particles.

Center of mass: the average position of a set of particles

The center of the mass of this wrench is marked with an X. The total external force acting on the wrench is zero. As it spins on a smooth horizontal surface, the center of mass moves in a straight line with constant velocity. (*PSSC Physics*, 2e, 1965; D. C. Heath and Company, Inc., with Education Development Center, Inc., Newton, Mass.)

The total momentum of a system depends on the velocity of its center of mass.

Newton's second law: The center of mass moves as though all the mass and force were concentrated at that point.

The *x*- and *y*-components of velocity of the center of mass are dX/dt and dY/dt. Taking the time derivatives of Eqs. (8–27), we see that the *velocity* of the center of mass is given by

$$V = \frac{m_1 v_1 + m_2 v_2 + m_3 v_3 + \cdots}{m_1 + m_2 + m_3 + \cdots}. \qquad (8\text{–}28)$$

We denote the *total* mass $m_1 + m_2 + \cdots$ by M; we can then rewrite Eq. (8–28) as

$$MV = m_1 v_1 + m_2 v_2 + m_3 v_3 + \cdots = P. \qquad (8\text{–}29)$$

The right side is simply the total momentum P of the system; hence we have proved that *the total momentum is equal to the total mass times the velocity of the center of mass*. It follows that for an isolated system, where the total momentum is constant, the velocity of the center of mass is also constant.

Proceeding one additional step, we note that the *rates of change* of the various velocities, that is, the accelerations, are related in the same way:

$$MA = m_1 a_1 + m a_2 + m_3 a_3 + \cdots. \qquad (8\text{–}30)$$

Now $m_1 a_1$ is equal to the vector sum of forces on the first particle, and so on, and the right side of Eq. (8–30) is equal to the vector sum of *all* the forces on all the particles. Just as in Section 8–2, we may classify each force as internal or external; because of Newton's third law, the internal forces all cancel in pairs. That is, $\Sigma F = \Sigma F_\text{ext} + \Sigma F_\text{int}$, and $\Sigma F_\text{int} = 0$, from Newton's third law. Thus what survives on the right side is the sum of the *external* forces only, and we have

$$\Sigma F_\text{ext} = MA. \qquad (8\text{–}31)$$

That is, *when a body or a collection of particles is acted on by external forces, the center of mass moves just as though all the mass were concentrated at that point and it were acted on by a resultant force equal to the sum of the external forces on the system*. For example, suppose we mark the center of mass of a hammer, which will be at some point partway down the handle. We throw the hammer spinning through the air. The motion appears complicated, but the point representing the position of the center of mass follows a parabolic path just as though all the mass were concentrated at the center of mass. Momentum is not conserved because there is an external force (the hammer's weight), but the effect of the force is to cause the parabolic motion of the center of mass. Or suppose a shell traveling in a parabolic trajectory explodes in flight, splitting into two fragments of equal mass. The fragments follow new parabolic paths, but the center of mass continues on the original trajectory just as though all the mass were still concentrated at that point. This phenomenon is shown in Fig. 8–10. A

8–10 A shell explodes in flight. The fragments follow individual parabolic paths, but the center of mass continues on the same trajectory as the shell's path before exploding.

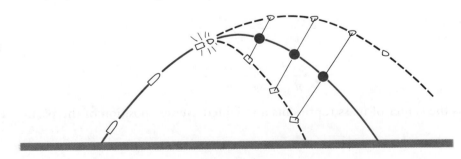

Fourth-of-July skyrocket exploding in air is a more spectacular example of the same idea.

This property of the center of mass is important in the analysis of the motion of rigid bodies, where the motion is described as a combination of motion of the center of mass and rotational motion about an axis through the center of mass. We return to this topic in Chapter 9.

Finally, we note that $MA = M\,dV/dt = dP/dt$, so the rate of change of total momentum of the system equals the sum of the external forces:

$$\Sigma F_{ext} = \frac{dP}{dt}. \tag{8–32}$$

Thus when $\Sigma F = 0$, $A = 0$, V is constant, and P is constant; conservation of momentum corresponds to constant velocity of the center of mass and to zero total external force.

In principle the position of the center of mass of a body can always be calculated by regarding it as a collection of particles and using Eqs. (8–27). In practice these calculations can become quite complex. Usually we will deal with symmetric bodies in which the center of mass lies at the geometric center. We will discuss other calculation techniques in Chapter 10 in connection with the related concept of *center of gravity*.

EXAMPLE 8–12 A 2-kg body and a 3-kg body are moving along the *x*-axis. At a particular instant the 2-kg body is 1.0 m from the origin and has a velocity of 3 m·s^{-1}, and the 3-kg body is 2 m from the origin and has a velocity of -1 m·s^{-1}. Find the position and velocity of the center of mass, and also find the total momentum.

SOLUTION From Eq. (8–27),

$$X = \frac{(2 \text{ kg})(1 \text{ m}) + (3 \text{ kg})(2 \text{ m})}{2 \text{ kg} + 3 \text{ kg}} = 1.6 \text{ m},$$

and from Eq. (8–28),

$$V = \frac{(2 \text{ kg})(3 \text{ m·s}^{-1}) + (3 \text{ kg})(-1 \text{ m·s}^{-1})}{2 \text{ kg} + 3 \text{ kg}} = 0.60 \text{ m·s}^{-1}.$$

Strictly speaking, this is the *x*-component of velocity of the center of mass. The total momentum (actually, its *x*-component) is

$$P = (2 \text{ kg})(3 \text{ m·s}^{-1}) + (3 \text{ kg})(-1 \text{ m·s}^{-1}) = 3 \text{ kg·m·s}^{-1}.$$

Alternatively, from Eq. (8–29),

$$P = (5 \text{ kg})(0.6 \text{ m·s}^{-1}) = 3 \text{ kg·m·s}^{-1}.$$

8–8 ROCKET PROPULSION

We can use momentum and impulse considerations as a basis for analyzing rocket propulsion. A rocket is propelled by rearward ejection of part of its mass. The forward force on the rocket is the reaction to the backward force on the ejected material, and as more material is ejected, the mass of the rocket

In a rocket, burned fuel is ejected backward; the reaction force pushes the rocket forward.

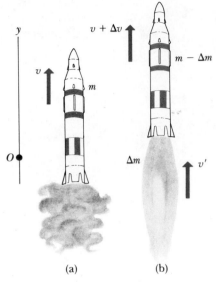

(a)   (b)

8–11 (a) Rocket at time t after takeoff, with mass m and upward velocity v. Its momentum is mv. (b) At time $t + \Delta t$, the mass of the rocket (and unburned fuel) is $m - \Delta m$, its velocity is $v + \Delta v$, and its momentum is $(m - \Delta m)(v + \Delta v)$. The ejected gas has momentum $\Delta m(v - v_r)$.

The mass of a rocket decreases during flight.

Calculating the thrust of a rocket: an application of momentum and impulse

decreases. As a simple example, consider a rocket fired vertically upward in a uniform gravitational field; neglect air resistance.

Figure 8–11a represents the rocket at a time t after takeoff, when its mass is m and its upward velocity is v. The total momentum at this instant is thus mv. In a short time interval dt, a mass dm of gas is ejected from the rocket. Let v_r represent the downward speed of this gas *relative to the rocket*. The velocity v' of the gas relative to the earth is then

$$v' = v - v_r,$$

and its momentum is

$$dm\, v' = dm(v - v_r).$$

At the end of the time interval dt, the mass of rocket and unburned fuel has decreased to $m - dm$, and its velocity has increased to $v + dv$. Its momentum is therefore

$$(m - dm)(v + dv).$$

Thus the *total* momentum at time $t + dt$ is

$$(m - dm)(v + dv) + dm(v - v_r).$$

Figure 8–11b represents rocket and ejected gas at this time.

We now use the impulse–momentum relation. The product of the resultant external force F on a system and the time interval dt during which it acts is equal to the change in momentum of this system. If air resistance is neglected, the external force F on the rocket is its weight, $-mg$. (We take the upward direction as positive.) The change in momentum, in time dt, is the difference between the momentum of the system at the end and at the beginning of the time interval. Hence

$$-mg\, dt = (m - dm)(v + dv) + dm(v - v_r) - mv$$

$$= m\, dv - dm\, v_r - dm\, dv.$$

The term $dm\, dv$ may be dropped because it is a product of two small quantities and thus is much smaller than the other terms. Dropping this term, dividing by dt, and rearranging, we obtain

$$m\left(\frac{dv}{dt}\right) = v_r\left(\frac{dm}{dt}\right) - mg. \tag{8–33}$$

Now dv/dt is the acceleration of the rocket, so the left side of this equation (mass times acceleration) equals the resultant force on the rocket. The first term on the right equals the upward thrust on the rocket, and the resultant force equals the difference between this thrust and the weight of the rocket mg. We see that the upward thrust is proportional both to the relative velocity, v_r, of the ejected gas and to the mass of gas ejected per unit time, dm/dt.

The acceleration of the rocket is

$$\frac{dv}{dt} = \frac{v_r}{m}\left(\frac{dm}{dt}\right) - g. \tag{8–34}$$

As the rocket rises, the value of g decreases, according to Newton's law of gravitation. (In "outer space," far from all other bodies, g becomes negligibly small.) If the values of v_r and dm/dt remain approximately constant while the fuel is being consumed and m continually decreases, the acceleration *increases* until all the fuel is burned.

Equation (8–34) can be integrated to find a relation between the velocity at any time and the remaining mass. From Eq. (8–34),

$$dv = v_r \frac{dm}{m} - g \, dt.$$

Now dm is a positive quantity, representing the mass ejected in time dt, so the change in mass of the rocket in that time is $-dm$. Thus in computing the total mass change *in the rocket* we must change the sign of the term containing dm.

Let the mass and velocity at time $t = 0$ be m_0 and v_0, respectively; then

$$\int_{v_0}^{v} dv = -\int_{m_0}^{m} v_r \frac{dm}{m} - \int_{0}^{t} g \, dt,$$

and

$$v - v_0 = -v_r \ln \frac{m}{m_0} - gt,$$

$$v = v_0 + v_r \ln \frac{m_0}{m} - gt. \tag{8–35}$$

EXAMPLE 8–13 In the first second of its flight, a rocket ejects $\frac{1}{60}$ of its mass with a relative velocity of 2400 m·s⁻¹. What is the acceleration of the rocket?

SOLUTION We have $dm = m/60$, $dt = 1$ s. From Eq. (8–34),

$$\frac{dv}{dt} = \frac{2400 \text{ m·s}^{-1}}{(60)(1 \text{ s})} - 9.8 \text{ m·s}^{-2}$$
$$= 30.2 \text{ m·s}^{-2}.$$

EXAMPLE 8–14 Suppose the ratio of initial mass m_0 to final mass m for the rocket in Example 8–13 is 4, and that the fuel is consumed in a time $t = 60$ s. Find the rocket's velocity at the end of this time.

SOLUTION From Eq. (8–35),

$$v = (2400 \text{ m·s}^{-1})(\ln 4) - (9.8 \text{ m·s}^{-2})(60 \text{ s})$$
$$= 2740 \text{ m·s}^{-1}.$$

At the start of the flight, when the velocity of the rocket is zero, the ejected gases are moving downward, relative to the earth, with a velocity equal to the relative velocity v_r. When the velocity of the rocket has increased to v_r, the ejected gases have a velocity zero relative to the earth. When the rocket velocity becomes greater than v_r, the velocity of the ejected gases is in the same direction as that of the rocket. Thus the velocity acquired by the rocket can be greater (and is often much greater) than the relative velocity v_r. In the example above, where the final velocity of the rocket was 2740 m·s⁻¹ and the relative velocity was 2400 m·s⁻¹, the last portion of the ejected fuel had an upward velocity (relative to the earth) of $(2740 - 2400)$ m·s⁻¹ = 340 m·s⁻¹.

Figure 8–12 shows a dramatic example of rocket propulsion.

8–12 Launch of a space shuttle, a dramatic example of rocket propulsion. Successful launch and landing of the Space Shuttle are the first steps toward feasibility of a manned space station in orbit around the earth. Despite a catastrophic set-back, research and progress continue. (Courtesy NASA, Kennedy Space Center.)

SUMMARY

The impulse J of a force F acting for a time interval from t_1 to t_2 is defined for a constant force as

$$J = F(t_2 - t_1) \tag{8-5}$$

and for a force that varies with time as

$$J = \int_{t_1}^{t_2} F \, dt. \tag{8-8}$$

The momentum p of a particle of mass m moving with velocity v is defined as

$$p = mv. \tag{8-6}$$

In any time interval, the total impulse of all forces acting on a body equals the change in its momentum:

$$J = mv_2 - mv_1 = p_2 - p_1. \tag{8-9}$$

When the force is not constant, the impulse still equals the average force multiplied by the time interval:

$$F_{av}(t_2 - t_1) = J = m(v_2 - v_1). \tag{8-10}$$

The total momentum P of a system is the vector sum of momenta of all bodies in the system:

$$P = p_1 + p_2 + \cdots = m_1 v_1 + m_2 v_2 + \cdots. \tag{8-12}$$

The total momentum of an isolated system is constant, or conserved. When momenta are represented in terms of their components, each component of momentum is separately conserved.

Collisions can be classified according to energy relations and final velocities:

Elastic collision: Energy is conserved, and the initial and final relative velocities have the same magnitude.

Inelastic collision: The final energy is less than the initial; the two bodies may or may not have the same final velocity.

Superelastic collision: The final energy is greater than the initial.

In recoil phenomena the total momentum is often zero both before and after the interaction.

The coordinates of the center of mass of a system of particles are defined as

$$X = \frac{m_1 x_1 + m_2 x_2 + m_3 x_3 + \cdots}{m_1 + m_2 + m_3 + \cdots},$$

$$Y = \frac{m_1 y_1 + m_2 y_2 + m_3 y_3 + \cdots}{m_1 + m_2 + m_3 + \cdots} \tag{8-27}$$

The total momentum of the system equals the total mass M multiplied by the velocity V of the center of mass:

$$MV = m_1 v_1 + m_2 v_2 + \cdots = P. \tag{8-29}$$

In rocket propulsion the mass of the rocket changes as the fuel is burned. Analysis of momentum relations must include the momentum carried away by the fuel as well as the momentum of the rocket itself.

QUESTIONS

8–1 Suppose you catch a baseball, and then someone invites you to catch a bullet with the same momentum or with the same kinetic energy. Which would you choose?

8–2 In splitting logs with a hammer and wedge, is a heavy hammer more effective than a lighter hammer? Why?

8–3 When a large, heavy truck collides with a passenger car, the occupants of the car are much more likely to be hurt than the truck driver. Why?

8–4 A glass dropped on the floor is more likely to break if the floor is concrete than if it is wood. Why?

8–5 "It ain't the fall that hurts you; it's the sudden stop at the bottom." Discuss.

8–6 When a person fires a rifle or shotgun, it is advisable to hold the butt firmly against the shoulder rather than a bit away from it, to minimize the impact on the shoulder. Why?

8–7 When a catcher in a baseball game catches a fast ball, he does not hold his arms rigid, but relaxes them so the mitt moves several inches while the ball is being caught. Why is this important?

8–8 When rain falls from the sky, what becomes of its momentum as it hits the ground? Is your answer also valid for Newton's famous apple?

8–9 A man stands in the middle of a perfectly smooth, frictionless frozen lake. He can set himself in motion by throwing things, but suppose he has nothing to throw. Can he propel himself to shore *without* throwing anything?

8–10 A machine gun is fired at a steel plate. Is the average force on the plate from the bullet impact greater if the bullets bounce off, or if they are squashed and stick to the plate?

8–11 How do Mexican jumping beans work? Do they violate conservation of momentum? Of energy?

8–12 Early critics of Robert Goddard, a pioneer in the use of rocket propulsion, claimed that rocket engines could not be used in outer space where there is no air for the rocket to push against. How would you answer such criticism? Would the criticism be valid for a jet engine in an ordinary airplane?

8–13 In a zero-gravity environment, can a rocket-propelled spaceship ever attain a speed greater than the relative speed with which the burnt fuel is exhausted?

EXERCISES

Section 8–1 Impulse and Momentum

8–1

a) What is the momentum of a 10,000-kg truck whose speed is 20 m·s^{-1}? What speed must a 5000-kg truck attain in order to have

b) the same momentum?

c) the same kinetic energy?

8–2 A bullet having a mass of 0.05 kg, moving with a speed of 400 m·s^{-1}, penetrates a distance of 0.1 m into a wooden block firmly attached to the earth. Assume the force that stops it is constant. Compute

a) the acceleration of the bullet,

b) the accelerating force,

c) the time of the acceleration,

d) the impulse of the force.

Compare the answer to part (d) with the initial momentum of the bullet.

8–3 A block of ice of mass 2 kg is moving on a frictionless horizontal surface. At $t = 0$ the block is moving to the right with a velocity of 6 m·s^{-1}. Calculate the velocity of the block after each of the following forces has been applied for 5 s:

a) a force of 5 N, directed to the right;

b) a force of 7 N, directed to the left.

8–4 A baseball has a mass of about 0.2 kg.

a) If the velocity of a pitched ball is 30 m·s^{-1}, and after being batted it is 50 m·s^{-1} in the opposite direction, find the change in momentum of the ball and the impulse applied to it by the bat.

b) If the ball remains in contact with the bat for 0.002 s, find the average force applied by the bat.

8–5 A golf ball of mass 0.10 kg initially at rest is given a speed of 50 m·s^{-1} when it is struck by a club. If the club and ball are in contact for 0.002 s, what average force acted on the ball? Is the effect of the ball's weight during the time of contact significant?

8–6 A baseball of mass 0.25 kg is struck by a bat. Just before impact, the ball is traveling horizontally at 40 m·s^{-1}, and it leaves the bat at an angle of 30° above horizontal with a speed of 60 m·s^{-1}. If the ball and bat were in contact for 0.005 s, find the horizontal and vertical components of the average force on the ball.

8–7 A force of magnitude $F(t) = A + Bt^2$ and directed to the right is applied to an object of mass m starting at $t = 0$ and continuing until $t = t_2$.

a) What is the impulse J of the force?

b) If the object was initially at rest, what is its velocity at time t_2?

Section 8–2 Conservation of Momentum

8–8 On a frictionless horizontal surface block A, of mass 3 kg, is moving toward block B, of mass 5 kg, which is initially at rest. After the collision block A has a velocity of 1.2 m·s^{-1} to the left and block B has a velocity of 7.9 m·s^{-1} to the right.

a) What was the velocity of block A before the collision?

b) Calculate the decrease in kinetic energy that occurs during the collision.

8–9 Ice hockey star Wayne Gretzky is skating at 13 m·s^{-1} toward a defender, who is in turn skating at 4 m·s^{-1} toward Gretzky. Gretzky's weight is 756 N, that of the defender is 900 N. Immediately after the collision Gretzky is moving at 8 m·s^{-1} in his original direction. Neglect external horizontal forces applied by the ice to the skaters during the collision.

a) What is the velocity of the defender immediately after the collision?

b) Calculate the decrease in total kinetic energy of the two players.

8–10 A railroad handcar is moving along straight, frictionless tracks. In each of the following cases, the car initially has a total mass (car and contents) of 200 kg and is traveling with a velocity of 4 m·s^{-1}. Find the *final velocity* of the car in each of the three cases.

a) A 20-kg mass is thrown sideways out of the car with a velocity of 2 m·s^{-1} relative to the car.

b) A 20-kg mass is thrown backwards out of the car with a velocity of 4 m·s^{-1} relative to the *initial* motion of the car.

c) A 20-kg mass is thrown into the car with a velocity of 6 m·s^{-1} relative to the ground and opposite in direction to the velocity of the car.

8–11 A hockey puck B rests on a smooth ice surface and is struck by a second puck A, which was originally traveling at 30 m·s^{-1} and which is deflected 30° from its original direction (Fig. 8–13). Puck B acquires a velocity at 45° with the original velocity of A.

a) Compute the speed of each puck after the collision.

b) What fraction of the original kinetic energy of puck A is "lost"?

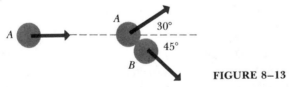

FIGURE 8–13

8–12 A block of mass 0.60 kg is initially at rest. It is struck by a second block of mass 0.40 kg initially moving with a velocity of 0.125 m·s^{-1} toward the right along the x-axis. After the collision the 0.40-kg block has a velocity of 0.10 m·s^{-1} at an angle of 37° above the x-axis in the first quadrant. Both blocks move on a frictionless horizontal plane.

a) What are the magnitude and direction of the velocity of the 0.60-kg block after the collision?

b) What is the loss of kinetic energy during the collision?

Section 8–4 Inelastic Collisions

8–13

a) An empty freight car of mass 10,000 kg rolls at 2 m·s^{-1} along a level track and collides with a loaded car of mass 20,000 kg, standing at rest with brakes released. If the cars couple together, find their speed after the collision.

b) Find the decrease in kinetic energy as a result of the collision.

c) With what speed should the loaded car be rolling toward the empty car for both to be brought to rest by the collision?

8–14 On a frictionless table, a 3-kg block moving 4 m·s^{-1} to the right collides with an 8-kg block moving 1.5 m·s^{-1} to the left.

a) If the two blocks stick together, what is the final velocity?

b) How much mechanical energy is dissipated in the collision?

8–15 A 2000-kg automobile going eastward on Chestnut Street at 60 km·hr^{-1} collides with a 4000-kg truck that is going southward *across* Chestnut Street at 20 km·hr^{-1}. If they become coupled on collision, what are the magnitude and direction of their velocity immediately after colliding?

8–16 A rifle bullet of mass 10 g is fired with a velocity of 800 m·s^{-1} into a ballistic pendulum of mass 5 kg, suspended from a cord 1 m long. Compute

a) the vertical height through which the pendulum rises,

b) the initial kinetic energy of the bullet,

c) the kinetic energy of the bullet and pendulum immediately after the bullet becomes embedded in the pendulum.

8–17 A 5-g bullet is fired horizontally into a 3-kg wooden block resting on a horizontal surface. The coefficient of sliding friction between block and surface is 0.20. The bullet remains embedded in the block, which is observed to slide 0.25 m along the surface before stopping. What was the velocity of the bullet?

Section 8–5 Elastic Collisions

8–18 Two blocks of mass 0.30 kg and 0.20 kg are moving toward each other along a frictionless horizontal surface with velocities of 0.50 m·s^{-1} and 1.00 m·s^{-1}, respectively.

a) If the blocks collide and stick together, find their final velocity.

b) Find the loss of kinetic energy during the collision.

c) Find the final velocity of each block if the collision is completely elastic, so the blocks do not stick together.

8–19 Supply the details of the calculation of θ and ϕ in Example 8–10.

8–20 A small steel ball moving with speed v_0 in the positive x-direction makes a perfectly elastic noncentral collision with an incident ball originally at rest. After impact, the first ball moves with speed v_1 in the first quadrant at an angle θ_1 with the x-axis and the second with speed v_2 in the fourth quadrant at an angle θ_2 with the x-axis.

a) Write the equations expressing conservation of linear momentum in the x-direction and in the y-direction.

b) Square these equations and add them.

c) At this point, introduce the fact that the collision is perfectly elastic.

d) Prove that $\theta_1 + \theta_2 = \pi/2$.

8–21 In Exercise 8–11, suppose the collision is perfectly elastic, and A is deflected 30° from its initial direction. Find the final speed of each puck and the direction of B's velocity.

Section 8–6 Recoil

8–22 An 80-kg man standing on ice throws a 0.2-kg ball horizontally with a speed of 30 m·s^{-1}. With what speed and in what direction will the man begin to move?

8–23 Block A in Fig. 8–14 has a mass of 1 kg, and block B has a mass of 2 kg. The blocks are forced together, compressing a spring S between them, and the system is released from rest on a level, frictionless surface. The spring is not fastened to either block and drops to the surface after it has expanded. Block B acquires a speed of 0.5 m·s^{-1}.

a) What is the final velocity acquired by block A?

b) How much potential energy was stored in the compressed spring?

FIGURE 8–14

8–24 An open-topped freight car of mass 10,000 kg is coasting without friction along a level track. It is raining very hard, with the rain falling vertically. The car is originally empty and moving with a velocity of 1 m·s^{-1}. What is the velocity of the car after it has traveled long enough to collect 1000 kg of rain water?

8–25 One of James Bond's adversaries is standing on a frozen lake. He throws his steel-lined hat with a velocity of 20 m·s^{-1} at 37° above the horizontal, hoping to hit James. If his mass is 160 kg and that of his hat is 9 kg, what is his horizontal recoil velocity?

Section 8–7 Center of Mass

8–26 A 1000-kg automobile is moving along a straight highway at 10 m·s^{-1}. Another car, with mass 2000 kg and speed 20 m·s^{-1} is 30 m ahead of the first.

a) Find the position of the center of mass of the two automobiles.

b) Find the total momentum from the above data.

c) Find the velocity of the center of mass.

d) Find the total momentum, using the velocity of the center of mass. Compare your result with that of part (b).

8–27 Find the position of the center of mass of the earth–moon system. Use the data in Appendix F.

8–28 Three particles have the following masses and coordinates: (1) 2 kg, (3 m, 2 m); (2) 3 kg, (1 m, −4 m); (3) 4 kg, (−3 m, 5 m). Find the coordinates of the center of mass of the system.

Section 8–8 Rocket Propulsion

8–29 A small rocket burns 0.05 kg of fuel per second, ejecting it as a gas with a velocity relative to the rocket of 5000 m·s^{-1}.

a) What force does this gas exert on the rocket?

b) Would the rocket operate in free space?

c) If so, how would you steer it? Could you brake it?

8–30 If a single-stage rocket, fired vertically from rest at the earth's surface, burns its fuel in a time of 30 s, and the relative velocity $v_r = 3000$ m·s^{-1}, what must be the mass ratio m_0/m for a final velocity v of 8 km·s^{-1} (about equal to the orbital velocity of an earth satellite)?

8–31 Obviously rockets can be made to go very fast, but what is a reasonable top speed? Assume that a rocket is fired from rest at a space station where there is no gravitational force.

a) If the rocket ejects gas at a relative velocity of 2400 m·s^{-1} and you want the rocket's velocity eventually to be $0.001c$, where c is the velocity of light, what fraction of the initial mass of the rocket and fuel is *not* fuel?

b) What is this fraction if the final velocity is to be 3000 m·s^{-1}?

PROBLEMS

8–32 A steel ball of mass 0.5 kg is dropped from a height of 4 m onto a horizontal steel slab. The collision is elastic, and the ball rebounds to its original height.

a) Calculate the impulse delivered to the ball during impact.

b) If the ball is in contact with the slab for 0.002 s, find the average force on the ball during impact.

8–33 A bullet emerges from the muzzle of a gun with a velocity of 300 m·s^{-1}. The resultant force on the bullet, while it is in the gun barrel, is given by

$$F = 400 \text{ N} - \frac{(4 \times 10^5 \text{ N·s}^{-1})t}{3}.$$

a) Construct a graph of F versus t.

b) Compute the time required for the bullet to travel the length of the barrel, assuming the force becomes zero just at the end of the barrel.

c) Find the impulse of the force.

d) Find the mass of the bullet.

8–34 A tennis ball weighing 0.5 N has $v_1 = (22\ \text{m·s}^{-1})i - (4\ \text{m·s}^{-1})j$ before being struck by a racket. The racket applies a force $F = -(600\ \text{N})i + (80\ \text{N})j$ that we will assume to be constant during the 0.004 s that the racket and ball are in contact.

a) What are the x- and y-components of the impulse of the force applied to the ball?

b) What are the x- and y-components of the final velocity of the ball?

8–35 A force $F = (20\ \text{N·s}^{-2}t^2)i - (10\ \text{N} - 30\ \text{N·s}^{-1}t)j$ is applied to an object of mass 2 kg. If the object was originally at rest, what is its velocity after the force has acted for 0.5 s?

8–36 Objects A, of mass 2 kg; B, of mass 3 kg; and C, of mass 5 kg, are each approaching the origin as shown in Fig. 8–15. The initial velocities of A and B are given in the figure. All three objects arrive at the origin at the same time. What must be the x- and y-components of the initial velocity of C if all three objects are to end up at rest after the collision?

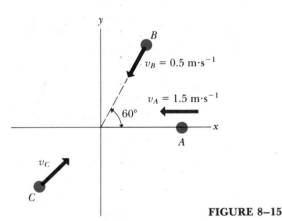

FIGURE 8–15

8–37 A station wagon traveling west collides with a pickup truck traveling north. They stick together as a result of the collision. The station wagon has mass 1400 kg and the pickup truck has mass 900 kg. A police officer estimates from the skid marks that immediately after the collision the combined object was traveling at 9 m·s⁻¹ in a direction 37° west of north. Calculate the velocities of the station wagon and pickup just before the collision.

8–38 A bullet of mass 5 g is shot *through* a 1-kg wood block suspended on a string 2 m long. The center of gravity of the block is observed to rise a distance of 0.50 cm. Find the speed of the bullet as it emerges from the block if the initial speed is 300 m·s⁻¹.

8–39 A bullet of mass 2 g, traveling in a horizontal direction with a velocity of 500 m·s⁻¹, is fired into a wooden block of mass 1 kg, initially at rest on a level surface. The bullet passes through the block and emerges with its velocity reduced to 100 m·s⁻¹. The block slides a distance of 0.20 m along the surface from its initial position.

a) What was the coefficient of sliding friction between block and surface?

b) What was the decrease in kinetic energy of the bullet?

c) What was the kinetic energy of the block at the instant after the bullet passed through it?

8–40 A rifle bullet of mass 0.01 kg strikes and embeds itself in a block of mass 0.99 kg, which rests on a frictionless horizontal surface and is attached to a coil spring, as shown in Fig. 8–16. The impact compresses the spring 10 cm. Calibration of the spring shows that a force of 1.0 N is required to compress the spring 1 cm.

a) Find the maximum potential energy of the spring.

b) Find the velocity of the block just after impact.

c) What was the initial velocity of the bullet?

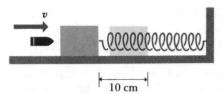

FIGURE 8–16

8–41 A frame of mass 0.20 kg, when suspended from a certain coil spring, is found to stretch the spring 0.10 m. A lump of putty of mass 0.20 kg is dropped from rest onto the frame from a height of 0.30 m (Fig. 8–17). Find the maximum distance the frame moves downward.

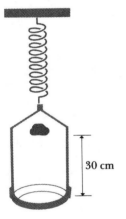

FIGURE 8–17

8–42

a) Prove that when a moving object makes a perfectly inelastic collision with a second object of equal mass, initially at rest, one-half the original kinetic energy is "lost."

b) Prove that when a very heavy particle makes a perfectly elastic head-on collision with a very light particle that is at rest, the light one goes off with twice the velocity of the heavy one.

8–43 Two railroad cars roll along and couple with a third car, which is initially at rest. These three roll along and couple to a fourth. This process continues until the speed of the final collection of railroad cars is one-tenth the speed of the initial two railroad cars. Ignoring friction, how many cars are in the final collection of cars? All the cars are identical.

8–44 A block of mass 0.20 kg, sliding with a velocity of 0.12 m·s⁻¹ on a smooth, level surface, makes a perfectly elastic head-on collision with a block of mass m, initially at rest. After the collision, the velocity of the 0.20 kg block is 0.04 m·s⁻¹ in the same direction as its initial velocity. Find

a) the mass m;   b) its velocity after the collision.

8–45 A stone whose mass is 0.10 kg rests on a frictionless horizontal surface. A bullet of mass 2.5 g, traveling horizontally at 400 m·s⁻¹, strikes the stone and rebounds horizontally at right angles to its original direction with a speed of 300 m·s⁻¹.

a) Compute the magnitude and direction of the velocity of the stone after it is struck.

b) Is the collision perfectly elastic?

8–46 A neutron of mass m collides elastically with a nucleus of mass M, which is initially at rest. Show that if the neutron's initial kinetic energy is K_0, the maximum kinetic energy that it can *lose* during the collision is

$$4mMK_0/(M + m)^2.$$

(*Hint:* The maximum energy loss occurs in a head-on collision.)

8–47 A neutron of mass 1.67×10^{-27} kg, moving with a velocity of 2.0×10^4 m·s⁻¹, makes a head-on collision with a boron nucleus of mass 17.0×10^{-27} kg, originally at rest.

a) If the collision is completely inelastic, what is the final kinetic energy of the system, expressed as a fraction of the original kinetic energy?

b) If the collision is perfectly elastic, what fraction of its original kinetic energy does the neutron transfer to the boron nucleus?

8–48 A man and a woman are sitting in a sleigh that is at rest on frictionless ice. The weight of the man is 800 N, the weight of the woman is 600 N, and that of the sleigh is 1200 N. The people suddenly see a poisonous spider on the floor of the sleigh and jump out. The man jumps to the left with a velocity of 5 m·s⁻¹ at 30° above the horizontal, and the woman to the right at 9 m·s⁻¹ at 37° above the horizontal. Calculate the horizontal velocity (magnitude and direction) that the sleigh has after they jump out.

8–49 Fission, the process that supplies energy in nuclear power plants, occurs when a heavy nucleus is split into two medium-sized nuclei. One such reaction would occur if a neutron colliding with a ²³⁵U nucleus split that nucleus into a ¹⁴¹Ba nucleus and a ⁹²Kr nucleus. In this reaction two neutrons also would be split off from the original ²³⁵U. Before the collision we have the arrangement in Fig. 8–18a. After the collision we have the Ba nucleus moving in the +z-direction and the Kr nucleus in the −z-direction. The three neutrons are moving in the xy-plane as shown in Fig. 8–18b. If the incoming neutron has an initial velocity of 3.0×10^6 m·s⁻¹ and a final velocity of 1.5×10^6 m·s⁻¹ in the directions shown, what are the velocities of the other two neutrons and what can you say about the velocities of the Ba and Kr nuclei? (The mass of the Ba nucleus is approximately 2.3×10^{-25} kg and that of Kr is about 1.5×10^{-25} kg.)

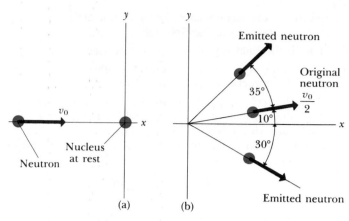

(a)   (b)

FIGURE 8–18

8–50 A uniform steel rod 1 m in length is bent in a 90° angle at its midpoint. Determine the position of its center of mass (*Hint:* The mass of each side of the angle may be assumed to be concentrated at its center.)

8–51 Two asteroids with masses m_A and m_B are moving with velocities v_A and v_B with respect to an astronomer in a space vehicle.

a) Show that the total kinetic energy as measured by the astronomer is

$$K = \tfrac{1}{2}MV^2 + \tfrac{1}{2}(m_Av_A'^2 + m_Bv_B'^2),$$

with V and M defined as in Section 8–7, $v_A' = v_A - V$, and $v_B' = v_B - V$. In this expression the total kinetic energy of the two asteroids is the energy associated with their center of mass plus that associated with the internal motion relative to the center of mass.

b) If the asteroids collide, what is the *minimum* possible kinetic energy they can have after the collision, as measured by the astronomer?

8–52 A man of mass 80 kg stands up in a 30-kg canoe of length 5 m. He walks from a point 1 m from one end to a point 1 m from the other end. If resistance to motion of the canoe in the water can be neglected, how far does the canoe move during this process?

8–53 This problem illustrates the advantage of using a multistage rather than a single-stage rocket. Suppose that the first stage of a two-stage rocket has a total mass of 12,000 kg, of which 9000 kg is fuel. The total mass of the second stage is 1000 kg, of which 750 kg is fuel. Assume that the relative velocity v_r of ejected material is constant, and neglect any effect of gravity. (The latter effect is small during the firing period if the rate of fuel consumption is large.)

a) Suppose that the entire fuel supply carried by the two-stage rocket was utilized in a single-stage rocket of the same total mass of 13,000 kg. What would be the velocity of the rocket, starting from rest, when its fuel was exhausted?

b) What is the velocity when the fuel of the first stage is exhausted, if the first stage carries the second stage with it to this point? This velocity then becomes the initial velocity of the second stage. At this point the second stage separates from the first stage.

c) What is the final velocity of the second stage?

d) What value of v_r would be required to give the second stage of the rocket a velocity of 8 km·s^{-1}?

8–54 In the rocket-propulsion problem the mass is variable. Another such problem is a raindrop falling through a cloud of small water droplets. Some of these droplets adhere to the raindrop, thereby *increasing* its mass as it falls. The force on the raindrop is

$$F_{ext} = \frac{dp}{dt} = m\frac{dv}{dt} + v\frac{dm}{dt}.$$

Suppose the mass of the raindrop depends on how far it has fallen. Then $m = kx$, where k is a constant, and $dm/dt = kv$. Since $F_{ext} = mg$, this gives

$$mg = m\frac{dv}{dt} + v(kv).$$

Or, dividing by k,

$$xg = x\frac{dv}{dt} + v^2.$$

This is a differential equation that has a solution of the form $v = at$. Take the initial velocity of the raindrop to be zero.

a) Using the proposed solution for v, find the acceleration a.

b) Find the distance the raindrop has fallen in $t = 0.4$ s.

c) Given that $k = 2.0$ g·m^{-1}, find the mass of the raindrop at $t = 0.4$ s.

For many more intriguing aspects of this problem, see K. S. Krane, *Amer. Jour. Phys.*, Vol. 49, pp. 113–117 (1981).

CHALLENGE PROBLEMS

8–55 This problem illustrates the usefulness of the concept of center of mass. Suppose one-third of a rope of length l is hanging down over the edge of a frictionless table. The rope has a linear density (mass per unit length) λ, and the end already on the table is held by a person. How much work is done by the person when he pulls on the rope to raise the rest of the rope slowly onto the table? Do the problem in two ways.

a) Find the force that the person must exert to raise the rope, and from this the work done. Note this is a variable force because at different times different amounts of rope are hanging over the edge.

b) Suppose the segment of the rope initially hanging over the edge has all of its mass concentrated at the center of mass. Find the work necessary to raise this to table height. You will probably find this approach simpler than that of part (a). How do the answers compare?

8–56 A 20-kg projectile is fired at an angle of 60° above the horizontal and with a muzzle velocity of 250 m·s^{-1}. At the highest point of its trajectory the projectile explodes into two fragments of equal mass, one of which falls vertically with zero initial speed.

a) How far from the point of firing does the other fragment strike if the terrain is level?

b) How much energy was released during the explosion?

8–57 Block A of mass m_A is moving on a frictionless horizontal surface in the $+x$-direction with velocity v_{A1} and makes an elastic collision with block B of mass m_B that is initially at rest.

a) Calculate the velocity of the center of mass (cm) of the two-block system before the collision.

b) Consider a coordinate system whose origin is at the cm and moves with it. Is this an inertial reference frame?

c) What are the initial velocities u_{A1} and u_{B1} of the two blocks in this cm reference frame, and what is the total momentum in this frame?

d) Use conservation of momentum and energy, applied in the cm frame, to relate the final momentum of each block to its initial momentum, and hence the final velocity of each block to its initial velocity. Your results should show that a one-dimensional elastic collision has a very simple description in cm coordinates.

e) Let $m_A = 2$ kg, $m_B = 4$ kg, and $v_{A1} = 2$ m·s^{-1}. Find the cm velocities u_{A1} and u_{B1}, apply the simple result found in (d), and transform back to velocities in a stationary frame to find the final velocities of the blocks. Does your result agree with Eqs. (8–22) and (8–23)?

8–58 A jet of liquid of cross-sectional area A and density ρ moves with speed v_J in the positive x-direction and impinges against a perfectly smooth blade B, which deflects the stream at right angles but does not slow it down, as shown in Fig. 8–19.

a) If the blade is *stationary*, prove that the rate of arrival of mass at the blade is $dm/dt = \rho A v_J$.

b) If the impulse–momentum theorem is applied to a small mass dm, prove that the x-component of the force acting on this mass for the time interval dt is given by

$$F_x = -\frac{dm}{dt}v_J.$$

c) Prove that the *steady* force exerted *on* the blade in the x-direction is

$$F_x = \rho A v_J^2.$$

Consider now that the blade moves to the right with a speed v_B ($v_B < v_J$). The stream is deflected at right angles to

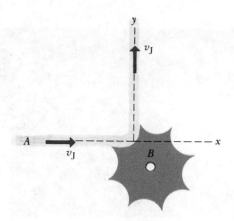

FIGURE 8–19

the *moving* blade. Derive the equations for

d) the rate of arrival of mass at the moving blade,

e) the force F_x on the blade,

f) the power delivered to the blade.

8–59

a) Show that the acceleration of a rocket fired vertically upward is given by

$$a = -\frac{v_r}{m}\frac{dm}{dt} - g,$$

where dm/dt is the rate of change of the mass of the rocket.

b) Suppose that the rate of ejection of mass by the rocket is constant; that is, the rate of decrease of mass is $dm/dt = -km_0$, where k is a positive constant and m_0 is the initial mass. What is the numerical value of k and in what units is it expressed for the rocket in Exercise 8–30?

c) Show that the mass at any time t is given by the equation

$$m = m_0(1 - kt).$$

d) Show that the acceleration at any time is equal to

$$\frac{v_r k}{1 - kt} - g.$$

e) Find the initial acceleration of the rocket in Exercise 8–30, in terms of the acceleration of gravity, g.

f) Find the acceleration 15 s after the motion starts.

g) Sketch the acceleration–time graph.

9 ROTATIONAL MOTION

THE MOTIONS OF REAL-WORLD BODIES CAN BE VERY COMPLEX. A BODY CAN have rotational as well as translational motions, and it can deform as the forces acting on it stretch, twist, and squeeze it. Our discussion of motion thus far has centered around a pointlike particle, an idealized model that is adequate when rotation and deformation can be ignored. Now we go on to a more sophisticated idealized model, a body that has finite size and definite shape and that can have rotational as well as translational motion. We continue to neglect deformations and assume that the body has a perfectly definite and unchanging shape and size. This idealized model is called a **rigid body,** and the study of rotational motion of a rigid body is the principal subject of this chapter. We introduce language for *describing* rotational motion and then develop the dynamic principles relating the forces on the body to its motion. During this development we introduce several new physical quantities, including torque, moment of inertia, and angular momentum. As we will see, several aspects of rotational motion have direct analogs in translational motion. We postpone until Chapter 12 a detailed study of the deformations of real-world bodies when forces act on them.

A rigid body: an idealized body that doesn't change shape at all

9–1 ANGULAR VELOCITY

Let us first consider a rigid body that rotates about a stationary axis; it might be a motor shaft, a chunk of roast beef on a barbeque skewer, or possibly a merry-go-round. In Fig. 9–1, a rigid body rotates about a stationary line passing through point O, perpendicular to the plane of the diagram. Line OP is fixed in the body and rotates with it. The angle between this line and the horizontal line in the figure is θ. Once we know the position of the axis of rotation, θ describes the position of the body completely. Thus θ serves as a *coordinate* to describe the rotational position of the body.

It is convenient to measure the angle θ in **radians.** As shown in Fig. 9–2a, one radian (1 rad) is the angle subtended at the center of a circle by an arc of length equal to the radius of the circle. The circumference is 2π times the radius, so there are 2π or about 6.283 radians in one complete revolution or

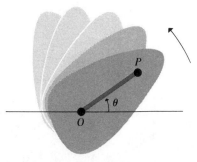

9–1 Body rotating about a fixed axis through point O.

An angle can be measured in degrees or radians; in rotational motion, radians are more convenient.

360°. Hence

$$1 \text{ rad} = \frac{360°}{2\pi} = 57.3°,$$

$$360° = 2\pi \text{ rad} = 6.28 \text{ rad},$$

$$180° = \pi \text{ rad} = 3.14 \text{ rad},$$

$$90° = \pi/2 \text{ rad} = 1.57 \text{ rad},$$

$$60° = \pi/3 \text{ rad} = 1.05 \text{ rad},$$

$$45° = \pi/4 \text{ rad} = 0.79 \text{ rad, and so on.}$$

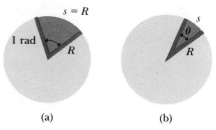

(a)   (b)

9–2 An angle θ in radians is defined as the ratio of the arc s to the radius R.

Figure 9–2b shows an arbitrary angle θ subtended by an arc of length s of the circumference of a circle of radius R; θ (in radians) is equal to s divided by R:

$$\theta = \frac{s}{R}, \qquad s = R\theta. \qquad (9\text{–}1)$$

For example, if $\theta = 2\pi$ rad, then s is the circumference of the circle.

An angle in radians is the quotient of a length and a length, so it is a pure number, without units. If $s = 1.5$ m and $R = 1$ m, the angle is usually described as $\theta = 1.5$ rad, but it would be equally correct to say simply $\theta = 1.5$.

Rotational *motion* of a body can be described in terms of the rate of change of θ. In Fig. 9–3, a reference line OP in a rotating body makes an angle θ_1 with the reference line Ox, at a time t_1. At a later time t_2 the angle has changed to θ_2. We define the **average angular velocity** of the body, ω_{av}, in the time interval $\Delta t = t_2 - t_1$, as the ratio of the angular displacement $\theta_2 - \theta_1$, or $\Delta\theta$, to Δt:

Angular velocity: describing how fast a body is rotating

$$\omega_{\text{av}} = \frac{\Delta\theta}{\Delta t}.$$

The **instantaneous angular velocity** ω is defined as the limit of this ratio as Δt approaches zero, that is, the derivative of θ with respect to t:

$$\omega = \lim_{\Delta t \to 0} \frac{\Delta\theta}{\Delta t} = \frac{d\theta}{dt}. \qquad (9\text{–}2)$$

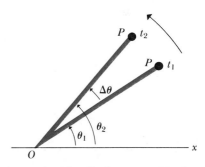

Because the body is rigid, *all* lines in it rotate through the same angle in the same time, and the angular velocity is characteristic of the body as a whole. If the angle θ is in radians, the unit of angular velocity is one radian per second ($1 \text{ rad}\cdot\text{s}^{-1}$ or simply 1 s^{-1}). Other units, such as the revolution per minute ($\text{rev}\cdot\text{min}^{-1}$), are in common use. We note that $1 \text{ rev}\cdot\text{s}^{-1} = 2\pi \text{ rad}\cdot\text{s}^{-1}$, and $1 \text{ rev}\cdot\text{min}^{-1} = 1 \text{ rpm} = (2\pi/60)\text{rad}\cdot\text{s}^{-1}$.

9–3 Angular displacement $\Delta\theta$ of a rotating body.

9–2 ANGULAR ACCELERATION

When the angular velocity of a body changes, it has an angular acceleration. If ω_1 and ω_2 are the instantaneous angular velocities at times t_1 and t_2, we define the **average angular acceleration** α_{av} as

Angular acceleration: describing the rate of change of rotational motion

$$\alpha_{\text{av}} = \frac{\omega_2 - \omega_1}{t_2 - t_1} = \frac{\Delta\omega}{\Delta t},$$

and the **instantaneous angular acceleration** α as the limit of this ratio as

$\Delta t \to 0$:

$$\alpha = \lim_{\Delta t \to 0} \frac{\Delta \omega}{\Delta t} = \frac{d\omega}{dt}. \tag{9–3}$$

The unit of angular acceleration is 1 rad·s^{-2} or 1 s^{-2}. Angular velocity and angular acceleration are exactly analogous to linear velocity and acceleration. In each case, velocity is the time derivative of position, and acceleration is the time derivative of velocity.

Because $\omega = d\theta/dt$, the angular acceleration can also be written:

$$\alpha = \frac{d}{dt} \frac{d\theta}{dt} = \frac{d^2\theta}{dt^2}. \tag{9–4}$$

Notation: using Greek letters for angular quantities

Note that we are using Greek letters for angular kinematic quantities: θ for angular position, ω for angular velocity, and α for angular acceleration. These are analogous to x for position, v for velocity, and a for acceleration, respectively, in straight-line motion.

9–3 ROTATION WITH CONSTANT ANGULAR ACCELERATION

Rotation with constant angular acceleration is completely analogous to straight-line motion with constant acceleration.

In Chapter 2 we found that analyzing straight-line motion is particularly simple when the acceleration is *constant*. This is also true of rotational motion; when the angular acceleration is constant, it is easy to derive equations for angular velocity and angular position as functions of time by integration. We begin with

$$\frac{d\omega}{dt} = \alpha = \text{constant}.$$

We integrate this with respect to t:

$$\int d\omega = \int \alpha \, dt, \qquad \omega = \alpha t + C_1,$$

where C_1 is an integration constant. If ω_0 is the angular velocity when $t = 0$, C_1 is equal to ω_0 and

$$\omega = \omega_0 + \alpha t. \tag{9–5}$$

Also, $\omega = d\theta/dt$; integrating again, we find

$$\int d\theta = \int \omega_0 \, dt + \int \alpha t \, dt, \qquad \theta = \omega_0 t + \frac{1}{2}\alpha t^2 + C_2.$$

The integration constant C_2 is the value of θ when $t = 0$ (the initial position), which we denote as θ_0. Thus

$$\theta = \theta_0 + \omega_0 t + \tfrac{1}{2}\alpha t^2. \tag{9–6}$$

We can also derive an equation relating ω and θ, by the same procedure we used to derive Eq. (2–13): Solve Eq. (9–5) for t, substitute the result into Eq. (9–6) to eliminate t, and simplify the resulting equation. We leave the details for you to work out; the final result is

$$\omega^2 = \omega_0{}^2 + 2\alpha(\theta - \theta_0). \tag{9–7}$$

Table 9–1 shows the similarity between Eqs. (9–5), (9–6), and (9–7) for motion with constant angular acceleration and the equations for motion with constant linear acceleration.

TABLE 9–1

Motion with constant linear acceleration	Motion with constant angular acceleration
a = constant	α = constant
$v = v_0 + at$	$\omega = \omega_0 + \alpha t$
$x = x_0 + v_0 t + \frac{1}{2}at^2$	$\theta = \theta_0 + \omega_0 t + \frac{1}{2}\alpha t^2$
$v^2 = v_0{}^2 + 2a(x - x_0)$	$\omega^2 = \omega_0{}^2 + 2\alpha(\theta - \theta_0)$

EXAMPLE 9–1   The angular velocity of a bicycle wheel is 4.0 rad·s^{-1} at time $t = 0$, and its angular acceleration is constant and equal to 2.0 rad·s^{-2}. A spoke OP on the wheel is horizontal at time $t = 0$.

a)   What angle does this spoke make with the horizontal at time $t = 3.0$ s?

b)   What is the wheel's angular velocity at this time?

SOLUTION   We can use Eqs. (9–5) and (9–6) to find θ and ω at any time, in terms of the given initial conditions.

Motion of a speeding-up bicycle wheel: an example of rotational kinematics

a)   The angle θ is given as a function of time by

$$\theta = \theta_0 + \omega_0 t + \tfrac{1}{2}\alpha t^2.$$

At time $t = 3.0$ s,

$$\theta = 0 + (4.0 \text{ rad·s}^{-1})(3.0 \text{ s}) + \tfrac{1}{2}(2.0 \text{ rad·s}^{-2})(3.0 \text{ s})^2$$

$$= 21 \text{ rad} = \frac{21}{2\pi} \text{ rev} = 3.34 \text{ rev}.$$

The body turns through three complete revolutions plus an additional 0.34 rev or (0.34 rev) $(2\pi \text{ rad·rev}^{-1}) = 2.15$ rad $= 123°$. The line OP thus turns through $123°$ and makes an angle of $57°$ with the horizontal.

b)   In general, $\omega = \omega_0 + \alpha t$. At time $t = 3.0$ s,

$$\omega = 4.0 \text{ rad·s}^{-1} + (2.0 \text{ rad·s}^{-2})(3.0 \text{ s}) = 10 \text{ rad·s}^{-1}.$$

Alternatively, from Eq. (9–7),

$$\omega^2 = \omega_0{}^2 + 2\alpha(\theta - \theta_0) = (4 \text{ rad·s}^{-1})^2 + 2(2 \text{ rad·s}^{-2})(21 \text{ rad})$$

$$= 100 \text{ rad}^2\text{·s}^{-2},$$

$$\omega = 10 \text{ rad·s}^{-1}.$$

9–4 RELATION BETWEEN ANGULAR AND LINEAR VELOCITY AND ACCELERATION

When a rigid body rotates about a stationary axis, every particle in the body moves in a circle lying in a plane perpendicular to this axis, with the center of the circle on the axis. In Section 3–5 we worked out a relation for the acceleration of a particle moving in a circular path, in terms of its speed and the radius; this relation is still valid when the particle is part of a rotating rigid body.

A spiral galaxy. As material moves away from the rapidly rotating center of the galaxy, its angular momentum is approximately constant. Thus the angular velocity must decrease as the distance from the center increases, forming the spiral pattern. Can you deduce which way the galaxy is rotating? (Photo by R. V. Willstrop; © 1975 Anglo-Australian Telescope Board.)

The speed of a particle in a rotating rigid body depends on its distance from the axis and on the angular velocity.

The speed of a particle in a rigid body is directly proportional to the body's angular velocity. In Fig. 9–4, point P is a distance r away from the axis of rotation, and it moves in a circle of radius r. When the angle θ increases by a small amount $\Delta\theta$ in a time interval Δt, the particle moves through an arc length $\Delta s = r\,\Delta\theta$. If $\Delta\theta$ is very small, the arc is nearly a straight line, and the average speed of the particle is given by

$$v_{av} = \frac{\Delta s}{\Delta t} = r\frac{\Delta\theta}{\Delta t}. \tag{9–8}$$

In the limit, as $\Delta t \to 0$, this becomes

$$v = r\frac{d\theta}{dt} = r\omega. \tag{9–9}$$

The *direction* of the particle's velocity is tangent to its circular path at each point.

If the angular velocity changes by $\Delta\omega$, the particle's speed changes by an amount Δv given by

$$\Delta v = r\,\Delta\omega.$$

The acceleration of a particle in a rotating rigid body can be expressed in terms of tangential and radial components.

This corresponds to a component of acceleration $a_\parallel$ tangent to the circle. If these changes take place in a small time interval Δt, then

$$\frac{\Delta v}{\Delta t} = r\frac{\Delta\omega}{\Delta t}.$$

and, in the limit, as $\Delta t \to 0$,

$$a_\parallel = r\frac{d\omega}{dt} = r\alpha, \tag{9–10}$$

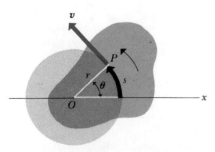

9–4 The distance s moved through by point P equals $r\theta$.

where $a_\parallel$ is the **tangential component of acceleration** of a point at a distance r from the axis.

The **radial component of acceleration** of the point, as worked out in Section 3–5, is $a_\perp = v^2/r$. We can also express this in terms of ω by using Eq. (9–9):

$$a_\perp = \frac{v^2}{r} = \omega^2 r. \qquad (9\text{–}11)$$

This is true at each instant *even when ω and v are not constant.*

The tangential and radial components of acceleration of a point P in a rotating body are shown in Fig. 9–5. Their vector sum is the acceleration $\boldsymbol{a}$, as shown.

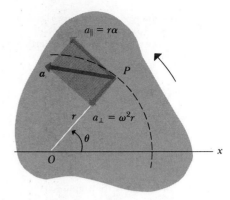

9–5 Nonuniform rotation about a fixed axis through point O. The tangential component of acceleration of point P equals $r\alpha$; the radial component equals $\omega^2 r$.

EXAMPLE 9–2 A discus thrower turns with angular acceleration $\alpha = 50$ rad·s^{-2}. What are the radial and tangential components of acceleration of the discus at the instant when the angular velocity is 10 rad·s^{-1}? The thrower's arm is 0.80 m long.

SOLUTION The acceleration components are given by Eqs. (9–10) and (9–11):

$$a_\perp = \omega^2 r = (10\ \text{s}^{-1})^2(0.80\ \text{m}) = 80\ \text{m·s}^{-2},$$

$$a_\parallel = r\alpha = (0.80\ \text{m})\,(50\ \text{s}^{-2}) = 40\ \text{m·s}^{-2}.$$

The magnitude of the acceleration vector is

$$a = \sqrt{a_\perp{}^2 + a_\parallel{}^2} = 89\ \text{m·s}^{-2},$$

or about nine times the acceleration due to gravity. Note that in using Eqs. (9–10) and (9–11) we *must* express the angular quantities in radians. Also note that in unit cancellations, we may drop out "radians" when it is convenient because an angle expressed in radians is a unitless number.

9–5 KINETIC ENERGY OF ROTATION

Because a rotating rigid body consists of particles in motion, it has kinetic energy. It turns out that we can express the kinetic energy simply in terms of the body's angular velocity and a quantity called the *moment of inertia* of the body. To develop this relationship, we use Eq. (9–9) to find the speed v of a particle in a rigid body that rotates with angular velocity ω about a stationary axis. That is, $v = r\omega$, where r is the particle's distance from the axis. If the particle has mass m, its kinetic energy is

$$\tfrac{1}{2}mv^2 = \tfrac{1}{2}mr^2\omega^2.$$

The *total* kinetic energy of the body is the *sum* of the kinetic energies of all particles in the body,

$$K = \tfrac{1}{2}m_1 r_1{}^2\omega^2 + \tfrac{1}{2}m_2 r_2{}^2\omega^2 + \cdots$$
$$= \Sigma\tfrac{1}{2}mr^2\omega^2.$$

Because ω is the same for all particles in a rigid body, we can rewrite this as

$$K = \tfrac{1}{2}(m_1 r_1{}^2 + m_2 r_2{}^2 + \cdots)\omega^2$$
$$= \tfrac{1}{2}[\Sigma mr^2]\omega^2.$$

The kinetic energy of a rotating rigid body can be expressed in terms of angular velocity and moment of inertia.

To obtain the sum Σmr^2, we subdivide the body (in our imagination) into a large number of particles, multiply the mass of each particle by the square of its distance from the axis, and add these products for all particles. The result is called the **moment of inertia** I of the body, about the axis of rotation:

$$I = \Sigma mr^2. \tag{9–12}$$

In SI units the unit of moment of inertia is 1 kilogram-meter² (1 kg·m²).

We can now express the rotational kinetic energy of a rigid body as

$$K = \tfrac{1}{2}I\omega^2. \tag{9–13}$$

Moment of inertia is the rotational analog of mass.

The form of this expression is analogous to that for translational kinetic energy:

$$K = \tfrac{1}{2}mv^2.$$

That is, for rotation about a stationary axis, moment of inertia I is analogous to mass m, and angular velocity ω is analogous to velocity v.

Calculating the moment of inertia of a set of particles by adding their separate moments of inertia

EXAMPLE 9–3 An engineer is designing a one-piece machine part consisting of three heavy connectors linked by light molded struts, as in Fig. 9–6. The connectors can be considered as massive particles connected by massless rods.

What is the moment of inertia of this machine part

a)   about an axis through point A, perpendicular to the plane of the diagram?

b)   about an axis coinciding with rod BC?

c)   If the body rotates about an axis through A perpendicular to the plane of the diagram, with angular velocity $\omega = 4.0$ rad·s⁻¹, what is the rotational kinetic energy?

SOLUTION

a)   The particle at point A lies on the axis. Its distance *from* the axis is zero, and it contributes nothing to the moment of inertia. Therefore, from Eq. (9–12),

$$I = \Sigma mr^2 = (0.10 \text{ kg})(0.50 \text{ m})^2 + (0.20 \text{ kg})(0.40 \text{ m})^2$$
$$= 0.057 \text{ kg·m}^2$$

b)   The particles at B and C both lie on the axis, so neither contributes to the moment of inertia; only A contributes, and we have

$$I = \Sigma mr^2 = (0.30 \text{ kg})(0.40 \text{ m})^2 = 0.048 \text{ kg·m}^2.$$

This illustrates the important fact that the moment of inertia of a body, unlike its mass, is *not* a unique property of the body; it depends on the position of the axis about which it is computed.

c)   From Eq. (9–13),

$$K = \tfrac{1}{2}I\omega^2 = \tfrac{1}{2}(0.057 \text{ kg·m}^2)(4.0 \text{ rad·s}^{-1})^2 = 0.456 \text{ J}.$$

9–6 A strangely shaped machine part.

In Example 9–3 the body could be represented as several point masses, and the sum in Eq. (9–12) could be evaluated directly. When the body is a *continuous* distribution of matter, such as a solid cylinder or plate, this sum must be evaluated by integration. Several examples of calculations of mo-

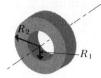
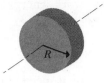
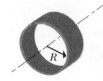

$\frac{1}{12}ml^2$　　　$\frac{1}{12}m(a^2 + b^2)$　　　$\frac{1}{2}m(R_1{}^2 + R_2{}^2)$　　　$\frac{1}{2}mR^2$　　　mR^2　　　$\frac{2}{5}mR^2$

(a) Slender rod, axis through center　　(b) Rectangular plate, axis through center　　(c) Hollow cylinder　　(d) Solid cylinder　　(e) Thin-walled hollow cylinder　　(f) Solid sphere

9–7 Moments of inertia. For each body, the axis is shown as a broken line.

ments of inertia are given in Section 9–6; meanwhile, Fig. 9–7 gives moments of inertia for several simple shapes.

You may be tempted to try to compute the moment of inertia of a body by assuming that all the mass is concentrated at the center of mass, and then multiplying the total mass by the square of the distance from the center of mass to the axis. Resist that temptation; it doesn't work! For example, when a uniform, thin rod of length L and mass M is pivoted about an axis through one end, perpendicular to the rod, the moment of inertia is $I = ML^2/3$. If we took the mass as concentrated at the center, a distance $L/2$ from the axis, we would obtain the *incorrect* result $I = M(L/2)^2 = ML^2/4$.

Now that we have learned how to calculate the kinetic energy of a rotating rigid body, we can apply the energy principles of Chapter 7 to rotational motion. The following examples illustrate this technique.

PROBLEM-SOLVING STRATEGY: *Rotational energy*

1. We suggest you review the strategy outlined in Section 7–5; it is equally useful here. The only new idea is that the kinetic energy K is expressed in terms of the moment of inertia I and angular velocity ω of the body instead of its mass M and speed v. We can then use work–energy relations and conservation of energy, where appropriate, to find relations involving position and motion of a rotating body.

2. The kinematic relations of Section 9–4, especially Eqs. (9–9) and (9–10), are often useful, especially when a rotating cylindrical body functions as any sort of pulley. Example 9–4 illustrates this point.

EXAMPLE 9–4 A light, flexible rope is wrapped several times around a solid cylinder of mass 50 kg and diameter 0.12 m, which rotates on frictionless bearings about a stationary horizontal axis. The free end of the rope is pulled with a constant force of magnitude 9.0 N for a distance of 2.0 m. If the cylinder is initially at rest, find its final angular velocity and the final speed of the rope.

An example of the work–energy theorem with rotational kinetic energy

SOLUTION Because no energy is lost in friction, the final kinetic energy $\frac{1}{2}I\omega^2$ of the cylinder is equal to the work Fd done by the force, which is (9.0 N) (2.0 m) = 18 J. From Fig. 9–7, the moment of inertia is

$$I = \tfrac{1}{2}MR^2 = \tfrac{1}{2}(50 \text{ kg})(0.060 \text{ m})^2 = 0.090 \text{ kg·m}^2.$$

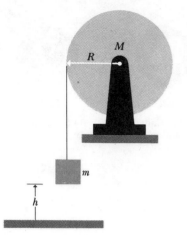

9–8 As the cylinder rotates, the rope unwinds and mass m drops.

An example of conservation of energy with rotational kinetic energy

The work–energy relation then gives

$$\tfrac{1}{2}(0.090 \text{ kg·m}^2)\omega^2 = 18 \text{ J},$$

$$\omega = 20 \text{ rad·s}^{-1}.$$

The final speed of the rope is equal to the tangential speed of the cylinder, which is given by Eq. (9–9):

$$v = r\omega = (0.060 \text{ m})(20 \text{ rad·s}^{-1}) = 1.2 \text{ m·s}^{-1}.$$

EXAMPLE 9–5 We wrap a light, flexible rope around a solid cylinder of mass M and radius R, which rotates with no friction about a stationary horizontal axis, as in Fig. 9–8. We tie the free end of the rope to a mass m and release the mass with no initial velocity, at a distance h above the floor. Find its speed and the angular velocity of the cylinder just as mass m strikes the floor.

SOLUTION Initially the system has no kinetic energy ($K_1 = 0$), but it has potential energy $U_1 = mgh$. Just as mass m strikes the floor, the potential energy is zero ($U_2 = 0$), but both this mass and the cylinder have kinetic energy. The total kinetic energy K_2 is

$$K_2 = \tfrac{1}{2}mv^2 + \tfrac{1}{2}I\omega^2. \qquad (9\text{–}14)$$

Now, according to Fig. 9–7, the moment of inertia of the cylinder is $I = \tfrac{1}{2}MR^2$. Furthermore, v and ω are related by $v = R\omega$, since the speed of mass m must be equal to the tangential speed at the outer surface of the cylinder. Using these relations and the energy relation $K_1 + U_1 = K_2 + U_2$, we obtain

$$mgh = \tfrac{1}{2}mv^2 + \tfrac{1}{2}(\tfrac{1}{2}MR^2)\left(\frac{v}{R}\right)^2 = \tfrac{1}{2}(m + \tfrac{1}{2}M)v^2,$$

$$v = \sqrt{\frac{2gh}{1 + M/2m}}.$$

When M is much larger than m, v is very small, as might be expected. When M is much smaller than m, v is nearly equal to the speed of a body in free fall with initial height h, namely, $\sqrt{2gh}$.

9–6 MOMENT-OF-INERTIA CALCULATIONS

Although the moment of inertia of a solid body is defined in principle by Eq. (9–12), this equation can be applied directly only in cases where the body consists of a few point masses, as was the case in Example 9–3. When the body consists of a *continuous* distribution of matter, we can express the sum in terms of an integral.

Imagine dividing the entire volume of the body into small volume elements dV so that all points in a particular element are very nearly the same distance from the axis of rotation; we call this distance r, as before. Let dm be the mass in a volume element dV. The moment of inertia may then be expressed as

$$I = \int r^2 \, dm. \qquad (9\text{–}15)$$

Density ρ is mass per unit volume, $\rho = dm/dV$, so we may also write

Finding the moment of inertia of a solid body by integration

$$I = \int r^2 \, \rho \, dV.$$

If the body is homogeneous (uniform in density), then ρ may be taken outside the integral:

$$I = \rho \int r^2 \, dV. \tag{9-16}$$

In using this equation, we express the volume element dV in terms of the differentials of the integration variables, usually the coordinates of the volume element. The element dV must always be chosen so that all points within it are at very nearly the same distance from the axis of rotation. For regularly shaped bodies this integration can often be carried out quite easily. Three examples are given below.

EXAMPLE 9–6 *Uniform, slender rod; axis perpendicular to length.* Figure 9–9 shows a uniform, slender rod of mass M and length l. We wish to compute its moment of inertia about an axis through O, at an arbitrary distance h from one end. Using Eq. (9–15), we choose as an element of mass a short section having length dx at a distance x from point O. The ratio of the mass dm of this element to the total mass M is equal to the ratio of its length dx to the total length l. Thus

$$\frac{dm}{M} = \frac{dx}{l}.$$

We solve this for dm, substitute into Eq. (9–15), and add the appropriate integration limits on x, to obtain

$$I_0 = \int x^2 \, dm = \frac{M}{l} \int_{-h}^{l-h} x^2 \, dx$$

$$= \frac{M}{l} \left. \frac{x^3}{3} \right]_{-h}^{l-h} = \frac{1}{3}M(l^2 - 3lh + 3h^2).$$

From this general expression we can find the moment of inertia about an axis through any point on the rod. For example, if the axis is at the left end, $h = 0$ and

$$I = \frac{1}{3}Ml^2. \tag{9-17}$$

If the axis is at the right end, $h = l$ and

$$I = \frac{1}{3}Ml^2,$$

as would be expected. If the axis passes through the center,

$$h = \frac{l}{2} \quad \text{and} \quad I = \frac{1}{12}Ml^2, \tag{9-18}$$

as shown also in Fig. 9–7.

EXAMPLE 9–7 *Hollow or solid cylinder; axis of symmetry.* Figure 9–10 shows a hollow cylinder of length l and inner and outer radii R_1 and R_2. We choose as the most convenient volume element a thin cylindrical shell of radius r, thickness dr, and length l. The volume of this shell is very nearly equal to that of a flat sheet of thickness dr, length l, and width $2\pi r$ (the circumference of the shell). Then

$$dm = \rho \, dV = 2\pi \rho l r \, dr.$$

Dividing a thin rod into short segments to compute its moment of inertia

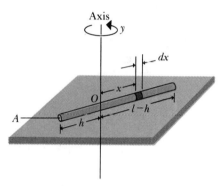

9–9 Moment of inertia of a thin rod. The mass element is a segment of length dx.

Dividing a cylinder into cylindrical shells to compute its moment of inertia

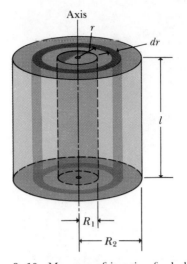

9–10 Moment of inertia of a hollow cylinder. The mass element is a cylindrical shell of radius r and thickness dr.

The moment of inertia is given by

$$I = \rho \int r^2 \, dV$$

$$= 2\pi\rho l \int_{R_1}^{R_2} r^3 \, dr$$

$$= \frac{\pi\rho l}{2}(R_2{}^4 - R_1{}^4)$$

$$= \frac{\pi\rho l}{2}(R_2{}^2 - R_1{}^2)(R_2{}^2 + R_1{}^2). \tag{9–19}$$

It is usually more convenient to express the moment of inertia in terms of the total mass M of the body, which is its density multiplied by the total volume. The volume is given by

$$\pi l(R_2{}^2 - R_1{}^2).$$

Hence

$$M = \pi l\rho(R_2{}^2 - R_1{}^2),$$

and the moment of inertia is

$$I = \tfrac{1}{2}M(R_1{}^2 + R_2{}^2), \tag{9–20}$$

as shown also in Fig. 9–7.

If the cylinder is solid, $R_1 = 0$; letting the outer radius be R, we find that the moment of inertia of a solid cylinder of radius R is

$$I = \tfrac{1}{2}MR^2. \tag{9–21}$$

If the cylinder is very thin walled (like a stovepipe), R_1 and R_2 are very nearly equal; if R represents this common radius,

$$I = MR^2.$$

Note that the moment of inertia of a cylinder about an axis coinciding with its axis of symmetry does not depend on the length l. Two hollow cylinders of the same inner and outer radii, one of wood and one of brass, but having the same mass M, have equal moments of inertia even though the length of the wood cylinder is much greater. Moment of inertia depends only on the *radial* distribution of mass, not on its distribution along the axis. Thus Eq. (9–20) holds also for a very short cylinder, such as a washer, and Eq. (9–21) for a thin disk.

EXAMPLE 9–8 *Uniform sphere of radius R; axis through center.* Divide the sphere into thin disks, as indicated in Fig. 9–11. The radius r of the disk shown is

$$r = \sqrt{R^2 - x^2}.$$

Its volume is

$$dV = \pi r^2 \, dx = \pi(R^2 - x^2) \, dx;$$

and its mass is

$$dm = \rho \, dV.$$

Hence from Eq. (9–21) its moment of inertia is

$$dI = \frac{\pi\rho}{2}(R^2 - x^2)^2 \, dx.$$

Slicing a sphere into disks to compute its moment of inertia

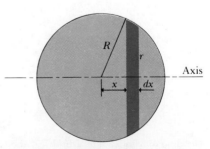

9–11 Moment of inertia of a sphere. The mass element is a disk of thickness dx.

Integrating this expression from 0 to R gives the moment of inertia of the right hemisphere; from symmetry, the total I for the entire sphere is just twice this:

$$I = (2)\frac{\pi\rho}{2}\int_0^R (R^2 - x^2)^2 \, dx.$$

Carrying out the integration, we obtain

$$I = \frac{8\pi\rho}{15}R^5.$$

The mass M of the sphere is

$$M = \rho V = \frac{4\pi\rho R^3}{3}.$$

Hence

$$I = \tfrac{2}{5}MR^2.$$

9–7 PARALLEL-AXIS THEOREM

Here is a theorem that is often useful in finding moments of inertia with respect to various axes. If the moment of inertia I_{cm} of a body about an axis through its center of mass is known, then the moment of inertia I_P about any other axis parallel to the original one but displaced from it by a distance d is easily obtained by means of a relation called the **parallel-axis theorem,** which states that

$$I_P = I_{cm} + Md^2. \tag{9–22}$$

To prove this theorem we consider the body shown in Fig. 9–12. The origin of coordinates has been chosen to coincide with the center of mass. We wish to compute the moment of inertia about an axis through point P, perpendicular to the plane of the figure. Point P has coordinates (a, b), and its distance from the origin is d. We note that $d^2 = a^2 + b^2$.

How to relate the moments of inertia of a body about two different axes

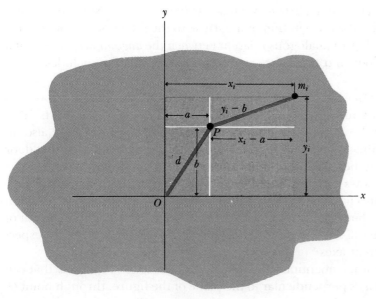

9–12 The mass element m_i has coordinates (x_i, y_i) with respect to the center of mass at O, and coordinates $(x_i - a, y_i - b)$ with respect to point P.

Let m_i be a typical mass element, with coordinates (x_i, y_i). Then the moment of inertia about an axis through O perpendicular to the figure is

$$I_{cm} = \Sigma m_i(x_i{}^2 + y_i{}^2),$$

and the moment of inertia about the axis through P is

$$I_P = \Sigma m_i\left[(x_i - a)^2 + (y_i - b)^2\right].$$

We expand the squared terms and regroup, obtaining

$$I_P = \Sigma m_i(x_i{}^2 + y_i{}^2) - 2a\Sigma m_i x_i - 2b\Sigma m_i y_i + (a^2 + b^2)\Sigma m_i.$$

The first sum is I_{cm}. The second and third sums are zero because they represent the x- and y-coordinates of the center of mass, which are zero because we have taken the origin to be the center of mass. The final term is d^2 multiplied by the total mass, so the theorem is proved.

EXAMPLE 9–9 The moment of inertia of a thin rod about an axis through its midpoint, perpendicular to the rod, is $I = ML^2/12$ (as shown in Fig. 9–7). Find the moment of inertia about an axis perpendicular to the rod at one end.

SOLUTION We have $I_{cm} = ML^2/12$ and $d = L/2$; Eq. (9–22) yields

$$I_P = \frac{ML^2}{12} + M\left(\frac{L^2}{4}\right) = \frac{ML^2}{3}.$$

EXAMPLE 9–10 Find the moment of inertia of a thin, uniform disk about an axis perpendicular to its plane at the edge.

SOLUTION From Fig. 9–7 we have $I_{cm} = MR^2/2$, and in this case $d = R$. Thus

$$I_P = \frac{MR^2}{2} + MR^2 = \frac{3MR^2}{2}.$$

Torque is the tendency of a force to cause a rotation.

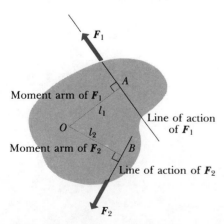

9–13 The moment or torque of a force about an axis is the product of the force and its moment arm.

9–8 TORQUE

In studying dynamics of a particle, we made extensive use of Newton's second law, which relates the acceleration of a particle to the forces acting on it. Now we need to develop an analogous relation between the *angular acceleration* of a rotating rigid body and the forces acting on it. This relation includes a new concept, *torque*.

A torque is always associated with a force. Qualitatively speaking, torque is the tendency of a force to cause a rotation of the body on which it acts. This tendency depends on the magnitude and direction of the force, and also on the location of the point where it acts. For example, it is easier to push a door open by pushing near the doorknob side than near the hinge side.

Torque is always defined with reference to a specific *axis* of rotation. In the problems in this chapter, the axis will usually be an actual axis of rotation about which the body can turn. In Chapter 10, when we study equilibrium of a rigid body, we will see that it is often useful to compute torques with respect to several different axes.

To define torque quantitatively, we consider in Fig. 9–13 a body that can rotate about an axis perpendicular to the plane of the figure, through point O.

Forces F_1 and F_2 act on the body; both forces act along lines that lie in a plane perpendicular to the axis. The tendency of force F_1 in Fig. 9–13 to cause a rotation about the axis through O depends on both the magnitude F_1 of the force and the perpendicular distance l_1 between the line of action of the force and the axis. If $l_1 = 0$, there is *no* tendency to cause rotation. The role of distance l_1 is analogous to that of a wrench handle; we can turn a tight bolt more easily by using a long-handled wrench than with a short-handled one. The distance l_1 is called the **moment arm** of force F_1 about the axis through O, and the product $F_1 l_1$ is called the **torque, or moment,** of the force about point O, denoted by the Greek letter Γ (capital "gamma"). The terms *torque* and *moment* are synonymous; we will usually use *torque*, but *moment arm* is the more usual term for the distance l_1:

$$\Gamma = Fl. \qquad (9\text{--}23)$$

> The torque of a force depends on the point where the force is applied, with reference to the axis of rotation.

The moment arm of F_1 is the perpendicular distance OA or l_1, and the moment arm of F_2 is the perpendicular distance OB or l_2.

Force F_1 tends to cause *counterclockwise* rotation about the axis, while F_2 tends to cause *clockwise* rotation. To distinguish between these directions of rotation, we will usually use the convention that *counterclockwise torques are positive and clockwise torques are negative*. Hence the torque Γ_1 of the force F_1 about the axis through O is

$$\Gamma_1 = +F_1 l_1,$$

and the torque Γ_2 of F_2 is

$$\Gamma_2 = -F_2 l_2.$$

When the line along which a force acts passes through the axis of rotation, the moment arm for that force is zero and its torque with respect to that axis is zero. In SI units, where the unit of force is the newton and the unit of length the meter, the unit of torque is the newton-meter.

In problems involving torques in this and the following chapter, we use the symbol

to indicate the choice of positive direction of rotation.

Often one of the important forces acting on a body is its *weight*. This force is not concentrated at a point but is distributed over the entire body. Nevertheless, it is always possible to calculate the corresponding torque by assuming all the weight to be concentrated at the center of mass of the body. We postpone proof of this statement until Chapter 10, but meanwhile we can use it in some of the problems in this chapter.

Figure 9–14 shows that there are several alternative ways to calculate torque. We can find the moment arm l and use $\Gamma = Fl$. Or we can determine the angle θ and use $\Gamma = rF \sin \theta$. Finally, we can represent F in terms of components F_1 parallel to r and F_2 perpendicular to r, as shown by the light-colored vectors. Then $F_2 = F \sin \theta$, and $\Gamma = F_2 r = rF \sin \theta$. The component F_1 has no torque with respect to O because its moment arm with respect to that point is zero. Whenever the line of action of a force goes through the point we are using to calculate torques, the torque of that force is zero.

In more advanced work, where rotations about axes in various directions have to be considered, we generalize the definition of torque as follows. When

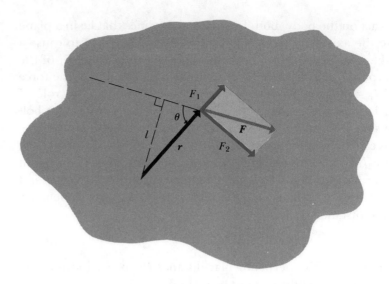

9–14 The vector torque $\boldsymbol{\Gamma}$ of force $\boldsymbol{F}$ with respect to point O is defined as $\boldsymbol{\Gamma} = \boldsymbol{r} \times \boldsymbol{F}$. In this example $\boldsymbol{\Gamma}$ points into the plane of the figure.

The torque of a force can be expressed as a vector product.

a force $\boldsymbol{F}$ acts at a point having a position vector $\boldsymbol{r}$ with respect to an origin O, as in Fig. 9–14, the torque $\boldsymbol{\Gamma}$ of the force with respect to O is defined to be the vector quantity

$$\boldsymbol{\Gamma} = \boldsymbol{r} \times \boldsymbol{F}. \qquad (9\text{–}24)$$

Recalling the definition of the vector product from Section 1–9, we note that the magnitude of $\boldsymbol{\Gamma}$ is $rF \sin \theta$. From the figure we see that $r \sin \theta = l$, so the magnitude of $\boldsymbol{\Gamma}$ is equal to Fl, in agreement with our previous definition of torque. The *direction* of $\boldsymbol{\Gamma}$ is perpendicular to the plane of the figure, in this case *into* the plane, as determined by the right-hand rule. If the sense of rotation were opposite, the direction of $\boldsymbol{\Gamma}$ would be *out of* the plane. This would be the case, for instance, if $\boldsymbol{F}$ were applied at the bottom edge of the body.

9–9 TORQUE AND ANGULAR ACCELERATION

We are now ready to consider the *dynamics* of rotational motion of a rigid body about a stationary axis. As we will see, the angular acceleration of a rotating body is directly proportional to the sum of the torques with respect to the axis of rotation. The proportionality factor is the moment of inertia.

To develop this relationship, we again imagine the body as made up of a large number of particles. A typical particle has mass m and is at distance r from the axis. We represent the *total force* acting on this particle in terms of a component $F_\perp$ that acts along the radial direction and a component $F_\parallel$ that is tangent to the circle of radius r in which the particle moves during rotation. Applying Newton's second law to this particle gives

$$F_\parallel = ma_\parallel. \qquad (9\text{–}25)$$

Now $a_\parallel$ may be expressed in terms of the angular acceleration α, according to Eq. (9–10): $a_\parallel = r\alpha$. Using this relation and multiplying both sides of Eq. (9–25) by r, we obtain

$$F_\parallel r = mr^2\alpha. \qquad (9\text{–}26)$$

Note that $F_\parallel r$ is just the torque of the force, and that mr^2 is the moment of inertia of the particle. (The component $F_\perp$ acts along a line passing through the axis and so has no torque with respect to the axis.) Thus Eq. (9–26) may be rewritten as

$$\Gamma = I\alpha.$$

Torque and angular acceleration: the rotational analog of Newton's second law

We can write an equation like this for every particle in the body and add all these equations. The left side of the resulting equation is the sum of all the torques acting on all the particles, and the right side is the total moment of inertia multiplied by the angular acceleration, which is the same for every particle. Thus for the entire body

$$\sum\Gamma = I\alpha. \qquad (9\text{–}27)$$

Finally, the sum $\sum\Gamma$ includes only the torques of the *external* forces, that is, the forces exerted on the body by agencies outside it. Torques corresponding to internal forces that the particles of the body exert on each other cancel out in pairs. The two forces that a given pair of particles exert on each other are equal and opposite, and if they act along the line joining the particles, their moment arms with respect to any axis are equal. Hence the torques of these two forces add to zero. Similarly, all other internal torques cancel out in pairs. (The requirement that the interaction forces for each pair of particles are not only equal and opposite but also act along the same line is called the *strong form* of Newton's third law; in previous developments we have needed only the fact that the forces are equal and opposite.)

Equation (9–27) is the rotational analog of Newton's second law, $\sum\mathbf{F} = m\mathbf{a}$. It provides the basis for relating the rotational motion of a rigid body to the forces acting on it.

PROBLEM-SOLVING STRATEGY: *Rotational dynamics*

The strategy we recommend for problems in rotational dynamics is very similar to that used in Section 5–4 for applications of Newton's laws:

1. Select a body for analysis. $\sum\mathbf{F} = m\mathbf{a}$ or $\sum\Gamma = I\alpha$ will be applied to this body.

2. Draw a free-body diagram. Be sure to include all the forces acting *on* the chosen body, but be equally careful *not* to include any force exerted *by* the body on some other body. Some of the forces may be unknown; label them with algebraic symbols. One of the forces may be the body's weight; it is often useful to label it immediately as mg rather than w. If a numerical value of mass is given, the corresponding numerical value of weight may be computed.

3. Choose coordinate axes for each body and also indicate a positive sense of rotation for each rotating body. If you know the direction of α in advance, it is usually easiest to pick that as the positive sense of rotation. When appropriate, determine components of

force with reference to the chosen axes. When a force is represented in terms of its components, cross out the original force so as not to include it twice.

4. If more than one body is involved, carry out steps 1–3 for each body. Some problems will include one or more bodies having translational motion and one or more others having rotational motion. There may also be geometrical relations between the motions of two or more bodies. Express these in algebraic form, usually as relations between two accelerations or an acceleration and an angular acceleration.

5. Write the appropriate dynamical equations, cited in step 1, and solve them to find the unknown quantities. Often this involves solving a set of simultaneous equations.

6. Check special cases or extreme values of quantities, where possible, and compare the results for these particular cases with your intuitive expectations. Ask: "Does this result make sense?"

Spinning a cylinder with a rope: a simple example of rotational dynamics

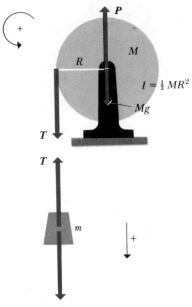

9–15 Free-body diagrams for Example 9–12.

Torque and angular acceleration in a two-body problem

The tension in the string does not equal the weight of the hanging body!

EXAMPLE 9–11 A rope is wrapped several times around a uniform solid cylinder of radius 0.1 m and mass 50 kg, pivoted so it can rotate about its axis. What is the angular acceleration when the rope is pulled with a force of 20 N?

SOLUTION The torque is $\Gamma = (20 \text{ N})(0.1 \text{ m}) = 2.0$ N·m, and the angular acceleration is

$$\alpha = \frac{\Gamma}{I} = \frac{2.0 \text{ N·m}}{\frac{1}{2}(50 \text{ kg})(0.1 \text{ m})^2} = 8 \text{ rad·s}^{-2}.$$

EXAMPLE 9–12 In Example 9–5 (Section 9–5), find the acceleration of mass m and the angular acceleration of the cylinder.

SOLUTION We treat the two bodies separately. Figure 9–15 shows the forces acting on each body. We take the positive sense of rotation for the cylinder to be counterclockwise and the positive direction for m to be downward. Applying Newton's second law to m yields the relation

$$mg - T = ma.$$

Applying Eq. (9–27) to the cylinder gives

$$RT = I\alpha = \frac{1}{2}MR^2\,\alpha.$$

Now the velocity of the mass and string at any instant is equal to the tangential velocity of the point on the cylinder where the string is tangent. Thus the acceleration a of mass m must equal the tangential acceleration of a point on the surface of the cylinder, which, according to Eq. (9–10), is given by $a_\parallel = R\alpha$. We replace $(R\alpha)$ with a in the cylinder equation, divide by R, and substitute the resulting expression for T into the equation for m, obtaining

$$mg - \frac{1}{2}Ma = ma, \qquad a = \frac{mg}{m + M/2} = \frac{g}{1 + M/2m}.$$

Note that the tension in the rope is *not* equal to the weight mg of mass m; if it were, m could not accelerate. From the above relations,

$$T = mg - ma = \frac{mg}{1 + 2m/M}.$$

When M is much larger than m, the tension is nearly equal to mg, and the acceleration is correspondingly much less than g. When M is zero, $T = 0$ and $a = g$; the mass then falls freely.

If mass m starts from rest at a height h above the floor, its velocity v when it strikes the ground is given by $v^2 - v_0{}^2 = 2ah$. In this case $v_0 = 0$ and

$$v = \sqrt{2ah} = \sqrt{\frac{2gh}{1 + M/2m}},$$

in agreement with the result obtained from energy considerations in Section 9–5.

EXAMPLE 9–13 In Fig. 9–16a, mass m_1 slides without friction on the horizontal surface, the pulley is in the form of a thin cylindrical shell of mass M and radius R, and the string turns the pulley without slipping. Find the acceleration of each mass, the angular acceleration of the pulley, and the tension in each part of the string.

SOLUTION Figure 9–16b shows free-body diagrams for the three bodies involved, and also shows a choice of positive directions for the various coordinates. Note that the two tensions T_1 and T_2 *cannot* be equal; if they were, the pulley could not have an angular acceleration. Hence to label the tension in both parts of the string as simply T would be a serious error.

Three interacting masses: linear and rotational dynamics

The equations of motion for masses m_1 and m_2 are

$$T_1 = m_1 a_1 \qquad (9–28)$$

and

$$m_2 g - T_2 = m_2 a_2. \qquad (9–29)$$

The unknown normal force N_2 acting on the axis of the pulley has no torque with respect to the axis of rotation, and the equation of motion of the pulley is

$$T_2 R - T_1 R = I\alpha = (MR^2)\alpha. \qquad (9–30)$$

Assuming the string does not stretch or slip, we have the additional *kinematic* relations

$$a_1 = a_2 = R\alpha. \qquad (9–31)$$

(The accelerations of m_1 and m_2 have different directions but the same magnitude.)

Equation (9–31) can be used to eliminate a_2 and α from Eqs. (9–28) through (9–30). The result is the following set of equations for the three unknowns T_1, T_2, and a_1:

$$T_1 = m_1 a_1,$$
$$m_2 g - T_2 = m_2 a_1,$$
$$T_2 - T_1 = M a_1.$$

These may be solved simultaneously; the simplest procedure is to add the three equations to eliminate T_1 and T_2, and then solve for a_1. The result is

$$a_1 = \frac{m_2 g}{m_1 + m_2 + M}.$$

This result may then be substituted back into Eqs. (9–28) and (9–29) to find the tensions. The results are

$$T_1 = \frac{m_1 m_2 g}{m_1 + m_2 + M}, \qquad T_2 = \frac{(m_1 + M)m_2 g}{m_1 + m_2 + M}.$$

Note that if either m_1 or M is much larger than m_2, the accelerations are very small and T_2 is approximately $m_2 g$, while if m_2 is much larger than either m_1 or M, the acceleration is approximately g, as should be expected.

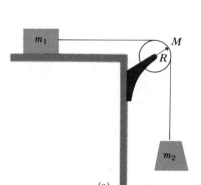

(a)

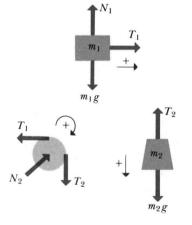

(b)

9–16 (a) System for Example 9–13. (b) Free-body diagrams.

9–10 WORK AND POWER IN ROTATIONAL MOTION

A force applied to a rotating body does *work* on the body. In problems involving energy changes and power it is important to know how to calculate this work, which may be expressed in terms of the torque of the force and the angular displacement.

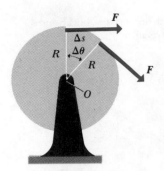

9–17 A force applied to a rotating body does work on the body.

Suppose a force **F** acts as shown in Fig. 9–17 at the rim of a pivoted wheel of radius R while the wheel rotates through a small angle $d\theta$. If this angle is small enough, the force may be regarded as constant during the correspondingly small time interval. By definition, the work done by the force **F** is

$$dW = F\, ds.$$

But $ds = R\, d\theta$, so that

$$dW = FR\, d\theta.$$

Now FR is the *torque*, Γ, due to the force **F**, so we have finally

$$dW = \Gamma\, d\theta. \tag{9–32}$$

If the torque is constant while the angle changes by a finite amount from θ_1 to θ_2,

$$W = \Gamma(\theta_2 - \theta_1) = \Gamma\, \Delta\theta. \tag{9–33}$$

That is, *the work done by a constant torque equals the product of the torque and the angular displacement.* If Γ is expressed in newton·meters, the work is in joules.

The force in Fig. 9–17 has no component along the *radial* direction. Such a component, if it existed, would do no work, since the displacement of the point of application has no radial component. Similarly, a radial component of force would make no contribution to the torque. Hence Eqs. (9–32) and (9–33) are still correct even when **F** does have a radial component.

When both sides of Eq. (9–32) are divided by the time interval dt during which the displacement occurs, we obtain

How to calculate the power transmitted by a rotating motor shaft

$$\frac{dW}{dt} = \Gamma\left(\frac{d\theta}{dt}\right).$$

But dW/dt is the rate of doing work, or the *power*, and $d\theta/dt$ is the angular velocity. Hence

$$P = \Gamma\omega. \tag{9–34}$$

That is, *the instantaneous power developed by an agent exerting a torque equals the product of the torque and the instantaneous angular velocity.* This is the analog of $P = Fv$ for linear motion.

EXAMPLE 9–14 The driveshaft of an automobile rotates at 3600 rpm and transmits 80 hp from the engine to the rear wheels. Compute the torque developed by the engine.

SOLUTION We first find ω (in rad·s^{-1}) and then use Eq. (9–34):

$$\omega = \frac{(3600 \text{ rev·min}^{-1})(2\pi \text{ rad·rev}^{-1})}{(60 \text{ s·min}^{-1})} = 120\,\pi \text{ rad·s}^{-1},$$

$$80 \text{ hp} = (80 \text{ hp})(746 \text{ W·hp}^{-1}) = 59{,}700 \text{ W},$$

$$\Gamma = \frac{P}{\omega} = \frac{59{,}700 \text{ W}}{120\pi \text{ rad·s}^{-1}} = 158 \text{ N·m}.$$

EXAMPLE 9–15 Suppose an electric motor exerts a constant torque of $\Gamma = 10$ N·m on a grindstone mounted on its shaft; the moment of inertia of the grindstone is $I = 2$ kg·m^2. The system starts from rest. Find the work done by the motor in 8 s and the kinetic energy at the end of this time. What was the average power of the motor?

SOLUTION From $\Gamma = I\alpha$, the angular acceleration is $5\ \text{s}^{-2}$. The angular velocity after 8 s is

$$\omega = \alpha t = (5\ \text{s}^{-1})(8\ \text{s}) = 40\ \text{s}^{-1}.$$

The kinetic energy at this time is

$$K = \tfrac{1}{2}I\omega^2 = \tfrac{1}{2}(2\ \text{kg·m}^2)(40\ \text{s}^{-1})^2 = 1600\ \text{J}.$$

The total angle through which the system turns in 8 s is

$$\theta = \tfrac{1}{2}\alpha t^2 = \tfrac{1}{2}(5\ \text{s}^{-1})(8\ \text{s})^2 = 160\ \text{rad},$$

and the total work done by the torque is

$$W = \Gamma\theta = (160\ \text{rad})(10\ \text{N·m}) = 1600\ \text{J}.$$

This equals the total kinetic energy, as of course it must.

The average power is the total work divided by the time interval:

$$P_{\text{av}} = \frac{1600\ \text{J}}{8\ \text{s}} = 200\ \text{J·s}^{-1} = 200\ \text{W}.$$

The instantaneous power, given by $P = \Gamma\omega$, is not constant, since ω increases continuously. But we can compute the total work by taking the time integral of P, as follows:

$$W = \int P\,dt = \int \Gamma\omega\,dt = \int \Gamma(\alpha t)\,dt$$
$$= \int_0^{8\ \text{s}} (10\ \text{N·m})(5\ \text{s}^{-2})t\,dt$$
$$= 1600\ \text{J},$$

as we found previously. The instantaneous power increases from zero at the start to $(10\ \text{N·m})(40\ \text{s}^{-1}) = 400\ \text{W}$ at time $t = 8$ s. The angular velocity and the power increase uniformly with time, so the *average* power is just half this maximum value, or 200 W.

A motor-driven grindstone: The work done by the rotating motor increases the grindstone's kinetic energy.

9–11 ROTATION ABOUT A MOVING AXIS

Our analysis of the dynamics of rotational motion of a rigid body (Section 9–9) can be extended to some cases where the axis of rotation moves, that is, where both translational and rotational motion occur at once. Familiar examples of such motion include a ball rolling down a hill or a yo-yo unwinding at the end of a string. The key to this more general analysis, which we shall not derive in detail, is that $\Sigma\Gamma = I\alpha$ remains valid when the axis of rotation moves, *if the axis passes through the center of mass of the body and does not change its direction.*

Another useful relationship, which we also state without proof, is an expression for the total kinetic energy of a rigid body having both translational and rotational motion. For a body of mass M, moving with a center-of-mass velocity V and rotating with angular velocity ω about an axis through the center of mass, the *total kinetic energy* of the body is

$$K = \tfrac{1}{2}MV^2 + \tfrac{1}{2}I_c\omega^2,$$

where I_c is the moment of inertia about the axis through the center of mass.

A rigid body can have translational and rotational motion at the same time.

PROBLEM-SOLVING STRATEGY: Rotation about a moving axis

The strategy outlined in Section 9–9 is equally useful here. There is one new wrinkle: When a body undergoes translational and rotational motion at the same time, we need two separate equations of motion *for the same body.* One of these is based on $\Sigma F = ma$ for the translational motion of the center of mass. The discussion of Section 8–7 shows that the acceleration of the center of mass is the same as that of a point mass equal to the total mass of the body, acted on by all the forces on the actual body. The other equation of motion is based on $\Sigma \Gamma = I\alpha$ for the rotational motion about the axis through the center of mass. In addition, there is often a kinematic relation between the two motions, such as a wheel that rolls without slipping or a string that unwinds from a pulley while turning it. Such relations are needed to relate the various accelerations.

Analyzing a yo-yo: dynamics of combined translational and rotational motion

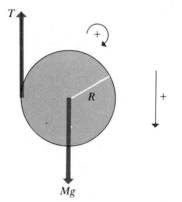

9–18 A cylinder rotates and drops as string unwinds.

EXAMPLE 9–16 A string is wrapped several times around a solid cylinder. The end of the string is held stationary while the cylinder is released with no initial motion. Find the downward acceleration of the cylinder and the tension in the string.

SOLUTION Figure 9–18 shows a free-body diagram for the situation, including the choice of positive coordinate directions. The equation for the translational motion of the center of mass is

$$Mg - T = Ma, \tag{9–35}$$

and the equation for rotational motion about the axis through the center of mass is

$$TR = I\alpha = \tfrac{1}{2}(MR^2)\alpha. \tag{9–36}$$

In addition, if the string unwinds without slipping, we have the kinematic relation

$$a = R\alpha. \tag{9–37}$$

This may be used to eliminate α from Eq. (9–36), and then Eqs. (9–35) and (9–36) may be solved simultaneously for T and a. The results are

$$a = \tfrac{2}{3}g, \qquad T = \tfrac{1}{3}Mg.$$

EXAMPLE 9–17 A solid bowling ball rolls without slipping down a ramp inclined at angle θ to the horizontal. What is the acceleration of its center?

A ball rolling down a hill: another example of combined translational and rotational dynamics.

SOLUTION Figure 9–19 shows a free-body diagram, with positive coordinate directions indicated. The equations of motion for translational and rotational motion, respectively, are

$$mg \sin \theta - \mathcal{F} = ma,$$
$$\mathcal{F}R = I\alpha = (\tfrac{2}{5}mR^2)\alpha.$$

If the ball rolls without slipping, then $a = R\alpha$. We use this to eliminate α and then solve for a and $\mathcal{F}$ to obtain

$$a = \tfrac{5}{7}g \sin \theta, \qquad \mathcal{F} = \tfrac{2}{7}mg \sin \theta.$$

Note that the acceleration is just $\tfrac{5}{7}$ as large as it would be if the ball could *slide* without friction down the slope. Also, the friction force is essential to prevent slipping and thus to cause the angular acceleration of the ball. An expression for the minimum coefficient of friction can be obtained by noting that the normal force is $n = mg \cos \theta$. To prevent slipping, the coefficient of (static) friction must be at least as great as

$$\mu_s = \frac{\mathcal{F}}{n} = \frac{\frac{2}{7}mg \sin \theta}{mg \cos \theta} = \tfrac{2}{7} \tan \theta.$$

If the plane is tilted only slightly, θ is small and only a small value of μ_s is needed to prevent slipping; but as the angle increases, the required value of μ_s increases, as we might expect intuitively.

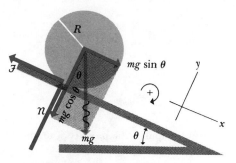

9–19 Free-body diagram for a ball rolling down a plane.

9–12 ANGULAR MOMENTUM AND ANGULAR IMPULSE

The concept of **angular momentum** plays a role in rotational motion that is closely analogous to that of momentum in particle motion, and there is a corresponding conservation principle. To introduce this concept, we begin by defining angular momentum of a particle.

Figure 9–20a shows a particle with mass m moving in the plane of the figure with velocity v and momentum mv. We define the angular momentum L of the particle, about an axis that passes through O and is perpendicular to the plane of the figure, as the product of the magnitude of its momentum and the perpendicular distance from the axis to its instantaneous line of motion:

Angular momentum of a particle depends on its momentum and its position with reference to the axis of rotation.

$$\text{Angular momentum} = L = mvr. \tag{9–38}$$

This definition is analogous to the definition of torque of a force.

Now we define the *total* angular momentum of a body of finite size as the sum of the angular momenta of the particles of the body. For a rotating rigid body, this can be expressed easily in terms of the body's moment of inertia and angular velocity. Figure 9–20b shows a rigid body rotating about an axis through O. The speed v of a small element of the body is related to the angular velocity ω of the body by $v = \omega r$. The angular momentum of the element is therefore

The angular momentum of a rigid body can be expressed in terms of its moment of inertia and angular velocity.

$$L = mvr = \omega m r^2,$$

and the total angular momentum of the body is

$$\Sigma \omega m r^2 = \omega \Sigma m r^2.$$

But $\Sigma m r^2$ is the moment of inertia of the body about its axis of rotation. Hence the angular momentum can be written as $I\omega$. Again denoting angular momentum by the symbol L, we have

$$L = I\omega. \tag{9–39}$$

This is analogous to the definition of linear momentum mv for a particle.

We can use the same sign convention for angular momentum as for angular velocity. I is always positive, so the sign of L is the same as that of ω. In Fig. 9–20b, where the positive direction for ω is defined to be clockwise, L is positive for clockwise rotations, negative for counterclockwise.

When a constant torque Γ acts on a body having moment of inertia I, for a time interval from t_1 to t_2, the angular velocity changes from ω_1 to ω_2, accord-

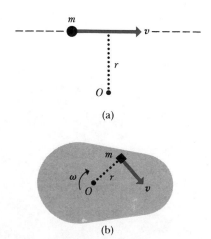

9–20 Angular momentum.

ing to the relation

$$\Gamma = I\alpha = I\left(\frac{\omega_2 - \omega_1}{t_2 - t_1}\right).$$

Rearranging this equation, we obtain

$$\Gamma(t_2 - t_1) = I\omega_2 - I\omega_1 = L_2 - L_1 = \Delta L. \tag{9-40}$$

Angular impulse: the product of torque and the time during which it acts

The product of the torque and the time interval during which it acts is called the **angular impulse** of the torque; we denote this quantity by J_θ.

$$\text{Angular impulse} = J_\theta = \Gamma(t_2 - t_1). \tag{9-41}$$

This quantity is the rotational analog of the impulse of a force, defined in Section 7–1. Equation (9–40) states that *the angular impulse acting on a body equals the change of angular momentum of the body about the same axis.*

For a torque that varies with time, we generalize the definition of angular impulse to

$$J_\theta = \int_{t_1}^{t_2} \Gamma \, dt. \tag{9-42}$$

Thus the general relationship between angular impulse and angular momentum is

$$J_\theta = I\omega_2 - I\omega_1 = L_2 - L_1, \tag{9-43}$$

with J_θ given by Eq. (9–41) if Γ is constant, and by Eq. (9–42) otherwise. A torque that is large but acts only during a short time interval is called an *impulsive torque.*

The basic dynamic relation for rigid-body rotation, given by Eq. (9–27), may be restated in terms of angular momentum. Taking the time derivative of Eq. (9–39), we find

$$\frac{dL}{dt} = I\frac{d\omega}{dt} = I\alpha.$$

Combining this with Eq. (9–27) yields the relation

$$\Sigma\Gamma = \frac{dL}{dt}. \tag{9-44}$$

Vector definition of angular momentum of a particle

In more complex rotational motion, where the direction of the axis of rotation may change, we generalize the definition of angular momentum of a particle to a vector quantity L defined as

$$L = r \times mv, \tag{9-45}$$

where r is the position vector of the particle with respect to the origin O. Comparing this definition with Fig. 9–20a, we see that in this figure L is a vector perpendicular to the plane of the figure, pointing *into* the plane, with magnitude mvr, in agreement with Eq. (9–38). The basic dynamic relationship for rigid body motion can then be stated as

$$\Sigma\Gamma = \frac{dL}{dt}, \tag{9-46}$$

with the vector torque Γ defined as in Eq. (9–24). In Section 9–14 we will explore the usefulness of this formulation.

9–13 CONSERVATION OF ANGULAR MOMENTUM

Here is an example of the usefulness of the concepts of angular impulse and angular momentum. Figure 9–21 shows two disks with moments of inertia I and I'; initially they are rotating with constant angular velocities ω_0 and ω_0', respectively. The disks are then pushed together by a force directed along the axis, which does not exert any torque on either disk. After a short time the disks reach a common final angular velocity ω.

During this time the larger disk exerts a torque Γ' on the smaller, and the smaller exerts a torque Γ on the larger. Both Γ and Γ' vary during the contact, and both become zero after the common final angular velocity is reached. At each instant the two torques are equal in magnitude and opposite in direction, because of Newton's third law. To see why this must be so, consider a point of contact between the two bodies. The two forces that the bodies exert on each other at this common point have equal magnitude and opposite direction, according to Newton's third law. Because they act at the same point, they also have the same moment arm with respect to any axis. Thus the torque of one force is the negative of the torque of the other. This is true of all the pairs of forces at all the other contact points. Thus at any instant $\Gamma = -\Gamma'$, and the total impulses on the two bodies are related by $J_\theta = -J_\theta'$.

Now, according to Eq. (9–43), the impulse on each disk equals the change of angular momentum of that disk:

$$J_\theta = I\omega - I\omega_0,$$

$$J_\theta' = I'\omega - I'\omega_0'.$$

But because $J_\theta = -J_\theta'$, these changes are equal and opposite:

$$I\omega - I\omega_0 = -(I'\omega - I'\omega_0')$$

or

$$I\omega_0 + I'\omega_0' = (I + I')\omega. \qquad (9\text{--}47)$$

The left side of Eq. (9–47) is the total angular momentum of the system before contact, and the right side is the total angular momentum after contact. We have therefore derived the important conclusion that the total angular momentum of the whole system is unaltered. When both disks are regarded as one system, then the torques Γ and Γ' are *internal* torques; as the disks are pushed together, no *external* torque acts. *When the resultant external torque on a system is zero, the angular momentum of the system remains constant;* hence any internal interaction between the parts of a system cannot alter its total angular momentum. This is the principle of **conservation of angular momentum,** and it ranks with the principles of conservation of linear momentum and conservation of energy as one of the most fundamental of physical laws.

A circus acrobat, a diver, or a skater performing a pirouette on the toe of one skate, all take advantage of this principle. Suppose an acrobat has just left a swing as in Fig. 9–22, with arms and legs extended and with a counterclockwise angular momentum. When he pulls his arms and legs in, his moment of inertia I becomes much smaller. His angular momentum $I\omega$ remains constant and I decreases, so his angular velocity ω increases. (Note that the change of ω cannot be determined from $\Sigma\Gamma = I\alpha$ because I is not constant.)

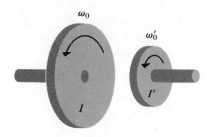

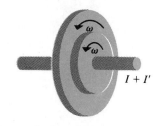

9–21 An impulsive torque acts when two rotating disks engage.

When no external torques act on a system, its total angular momentum is constant.

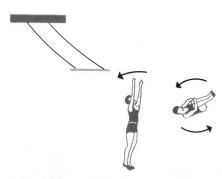

9–22 Conservation of angular momentum.

Acrobats, divers, and skaters: examples of conservation of angular momentum

EXAMPLE 9–18 In the situation of Fig. 9–21, suppose the large disk has mass 2 kg, radius 0.2 m, and initial angular velocity 50 rad·s^{-1}, and the small disk has mass 4 kg, radius 0.1 m, and initial angular velocity 200 rad·s^{-1} (about 1900 rpm). Find the common final angular velocity after the disks are pushed into contact. Is kinetic energy conserved during this process?

SOLUTION The moments of inertia of the two disks are

$$I = \tfrac{1}{2}(2 \text{ kg})(0.2 \text{ m})^2 = 0.04 \text{ kg·m}^2,$$

and

$$I' = \tfrac{1}{2}(4 \text{ kg})(0.1 \text{ m})^2 = 0.02 \text{ kg·m}^2.$$

From conservation of angular momentum, we have

$$(0.04 \text{ kg·m}^2)(50 \text{ rad·s}^{-1}) + (0.02 \text{ kg·m}^2)(200 \text{ rad·s}^{-1})$$
$$= (0.04 \text{ kg·m}^2 + 0.02 \text{ kg·m}^2)\omega,$$
$$\omega = 100 \text{ rad·s}^{-1}.$$

The initial kinetic energy is

$$K_0 = \tfrac{1}{2}(0.04 \text{ kg·m}^2)(50 \text{ rad·s}^{-1})^2 + \tfrac{1}{2}(0.02 \text{ kg·m}^2)(200 \text{ rad·s}^{-1})^2$$
$$= 450 \text{ J}.$$

The final kinetic energy is

$$K = \tfrac{1}{2}(0.04 \text{ kg·m}^2 + 0.02 \text{ kg·m}^2)(100 \text{ rad·s}^{-1})^2 = 300 \text{ J}.$$

One-third of the kinetic energy was lost during this "angular collision," the rotational analog of an inelastic collision. We should not expect kinetic energy to be conserved, even though the resultant external force and torque are zero, since nonconservative (frictional) internal forces act during the contact.

EXAMPLE 9–19 In Fig. 9–23, a man stands at the center of a turntable, holding his arms extended horizontally, with a 5-kg mass in each hand. He is set rotating about a vertical axis with an angular velocity of one revolution in 2 s. Find his new angular velocity if he drops his hands to his sides. The man's moment of inertia may be assumed constant and equal to 6 kg·m^2. The original distance of the weights from the axis is 1 m, and their final distance is 0.2 m.

SOLUTION If friction in the turntable is neglected, no external torques act about the vertical axis, and the angular momentum about this axis is constant. That is,

$$I_i\omega_i = I_f\omega_f,$$

where I_i and ω_i are the initial moment of inertia and angular velocity, and I_f and ω_f are the final values of these quantities. In each case, $I = I_{man} + I_{weights}$.

$$I_i = 6 \text{ kg·m}^2 + 2(5 \text{ kg})(1.0 \text{ m})^2 = 16 \text{ kg·m}^2,$$
$$I_f = 6 \text{ kg·m}^2 + 2(5 \text{ kg})(0.2 \text{ m})^2 = 6.4 \text{ kg·m}^2,$$
$$\omega_i = 2\pi(\tfrac{1}{2})\text{rad·s}^{-1}$$
$$\omega_f = \omega_i\frac{I_i}{I_f} = \pi \text{ rad·s}^{-1}\frac{16 \text{ kg·m}^2}{6.4 \text{ kg·m}^2}$$
$$= 2.5\pi \text{ rad·s}^{-1} = 1.25 \text{ rev·s}^{-1}.$$

That is, the angular velocity is more than doubled.

9–23 Conservation of angular momentum about a fixed axis.

The initial kinetic energy is

$$K_0 = \tfrac{1}{2}(16 \text{ kg·m}^2)(\pi \text{ rad·s}^{-1})^2 = 79 \text{ J}.$$

The final kinetic energy is

$$K = \tfrac{1}{2}(6.4 \text{ kg·m}^2)(2.5\pi \text{ rad·s}^{-1})^2 = 197 \text{ J}.$$

Where did the extra energy come from?

EXAMPLE 9–20   A door 1.0 m wide, having a mass of 15 kg, is hinged at one side so it can rotate without friction about a vertical axis. A bullet having mass 10 g and speed 400 m·s^{-1} is fired into the door, in a direction perpendicular to the plane of the door, and embeds itself at the exact center of the door. Find the angular velocity of the door just after the bullet embeds itself. Is kinetic energy conserved?

Firing a bullet into a door: a dangerous way to open a door, but a good way to illustrate conservation of angular momentum

SOLUTION   There is no external torque about the axis defined by the hinges, so angular momentum about this axis is conserved. The initial angular momentum of the bullet is given by Eq. (9–38):

$$L = mvr = (0.01 \text{ kg})(400 \text{ m·s}^{-1})(0.5 \text{ m}) = 2.0 \text{ kg·m}^2\text{·s}^{-1}.$$

This is equal to the final angular momentum $I\omega$, where $I = I_{\text{door}} + I_{\text{bullet}}$ is the total moment of inertia of door and bullet. To find I for the door alone, note that dimensions parallel to the axis are irrelevant, and that its moment of inertia is the same as that of a rod 1 m long, pivoted at one end. As discussed in Example 9–9 (Section 9–7), this is given by

$$I_{\text{door}} = \frac{ML^2}{3} = \frac{(15 \text{ kg})(1.0 \text{ m})^2}{3} = 5.0 \text{ kg·m}^2.$$

The moment of inertia of the bullet is

$$I_{\text{bullet}} = mr^2 = (0.01 \text{ kg})(0.5 \text{ m})^2 = 0.0025 \text{ kg·m}^2.$$

Conservation of angular momentum requires that $mvr = I\omega$, or

$$2.0 \text{ kg·m}^2\text{·s}^{-1} = (5.0 \text{ kg·m}^2 + 0.0025 \text{ kg·m}^2)\omega,$$

$$\omega = 0.4 \text{ rad·s}^{-1}.$$

The collision of bullet and door is inelastic, so we do not expect energy to be conserved. To check, we calculate initial and final kinetic energies:

$$K_i = \tfrac{1}{2}mv^2 = \tfrac{1}{2}(0.01 \text{ kg})(400 \text{ m·s}^{-1})^2 = 800 \text{ J};$$

$$K_f = \tfrac{1}{2}I\omega^2 = \tfrac{1}{2}(5.0025 \text{ kg·m}^2)(0.4 \text{ rad·s}^{-1})^2 = 0.4 \text{ J}.$$

The final kinetic energy is only $\frac{1}{2000}$ of the initial value.

9–14 VECTOR REPRESENTATION OF ANGULAR QUANTITIES

A rotational quantity associated with an axis of rotation, such as angular velocity, angular momentum, or torque, can be represented by a *vector* lying along that axis. Thus we can define angular velocity as a vector quantity $\boldsymbol{\omega}$ having a

Representing rotational quantities as
vectors: the relation of the axis of
rotation to directions of angular velocity
and other quantities

magnitude equal to the number of radians through which the body turns per
unit time, and a direction along the axis of rotation.

A vector lying along a given line can have either of two opposite directions
on that line. In defining the direction of the angular momentum vector, we
use the same right-hand rule used for the vector product in Section 1–9. This
rule is shown in Fig. 9–24; the direction of **ω** is the direction in which a screw
with a right-hand thread would advance if the screw turned with the body.
Alternatively, wrap the fingers of your right hand around the axis, with your
fingers pointing in the direction of rotation; your thumb then points in the
direction of the vector **ω.**

We mentioned in Section 9–12 the vector definition of angular momentum of a particle. Using this definition, we can prove that for a body rotating
about an axis of symmetry, the total angular momentum **L** is given by

$$L = I\boldsymbol{\omega}, \tag{9–48}$$

where I is the moment of inertia about the axis of rotation. Deriving this
equation would be beyond our scope, but note that it is valid *only* when the axis
of rotation is a symmetry axis of the body, such as a motor shaft or the axle of
a turning wheel. When a body rotates about an axis that is *not* a symmetry axis,
such as a disk with an axis through its center but not perpendicular to its
plane, the angular-momentum vector does not have the same direction as the
angular velocity. As a result, a body rotating with constant angular velocity
does not have constant angular momentum. This makes the body tend to wob-
ble, and torques must be supplied by the bearings supporting the body to
prevent this. A general analysis of rigid body motion is a problem of considerable complexity.

The direction of the torque vector, as defined by Eq. (9–24), can be understood as follows. A torque tends to cause a rotation about a certain axis,
namely, the axis about which the body would begin to rotate if it were initially
at rest with only the torque under consideration acting on it. As mentioned in
Section 9–12, the generalized relation between torque and angular momentum is

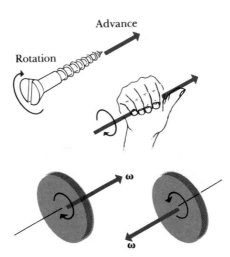

9–24 Vector angular velocity of a
rotating body.

$$\boldsymbol{\Sigma\Gamma} = \frac{d\boldsymbol{L}}{dt}. \tag{9–49}$$

We have not needed this generalized equation thus far in this chapter, because
when the axis of rotation always keeps the same direction, only one component of angular velocity is different from zero. Such rotational motion is analogous to motion of a particle along a straight line. But Eq. (9–49) also includes
the possibility that **Γ** and **L** may have different directions; in that case, **L** and
$d\boldsymbol{L}/dt$ have different directions, and the direction of the axis of rotation may
change. In such situations it is essential to consider the vector nature of the
various angular quantities.

We cannot discuss the general formulation of the dynamics of rotation in
detail here, but here is an example of its application. Figure 9–25 shows a
familiar toy gyroscope. We set the flywheel spinning by wrapping a string
around its shaft and pulling. When the shaft is supported at only one end, as
shown in the figure, one possible motion is a steady circular motion of the axis
in a horizontal plane. This is an interesting phenomenon, and quite unexpected if you have not seen it before. Intuition suggests that the free end of
the axis should simply drop if it is not supported. Vector angular momentum
considerations provide the key to understanding this behavior.

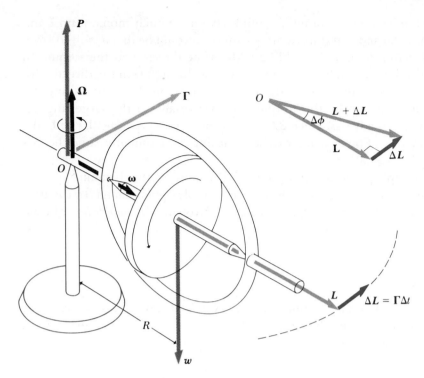

9–25 Vector **ΔL** is the change in angular momentum produced in time Δt by the moment **Γ** of the force **w**. Vectors **ΔL** and **Γ** are in the same direction.

The forces acting on the gyroscope are its weight **w**, acting downward at the center of mass, and the upward force **P** at the pivot point O. These forces acting on a body *at rest* would tend to cause a rotation about a horizontal axis perpendicular to the gyroscope's axis; the associated torque **Γ** is in the direction shown. However, the body is *not* at rest, and the effect of this torque is not to *initiate* a rotational motion but to *change* a motion that already exists. At the instant shown, the body has a definite, nonzero angular momentum described by the vector **L,** and the *change* **ΔL** in angular momentum in a short time interval following this is given by Eq. (9–49). Thus **ΔL** has the same direction as **Γ**. After time Δt, the angular momentum is **L** + **ΔL** and, as the vector diagram shows, this means that the gyroscope axis has turned through a small angle $\Delta\phi$ given by $\Delta\phi = |\Delta L|/|L|$. Thus the motion of the axis is consistent with the torque–angular-momentum relationship. This motion of the axis is called **precession.**

The rate at which the axis moves, $\Delta\phi/\Delta t$, is called the *precession angular velocity*. Denoting this quantity by Ω, we find

$$\Omega = \frac{\Delta\phi}{\Delta t} = \frac{|\Delta L|/|L|}{\Delta t} = \frac{\Gamma}{L} = \frac{wR}{I\omega}. \tag{9–50}$$

Thus the precession angular velocity Ω is *inversely* proportional to the angular velocity ω of spin about the axis. A rapidly spinning gyroscope precesses slowly; and as it slows down, the precession angular velocity *increases*!

We may now return to the question of why the unsupported end of the axis does not fall, the reason is that the upward force **P** exerted on the gyroscope by the pivot is just equal in magnitude to its weight **w**. The resultant vertical *force* is zero, and the vertical acceleration of the center of gravity is zero. However, the resultant *torque* of these forces is *not* zero, and the angular momentum changes.

Precession of a gyroscope: obeying the laws of motion by defying common sense

If the top were not rotating, it would have no angular momentum L initially. Its angular momentum ΔL after a time Δt would be that acquired from the torque acting on it and would have the same direction as the torque. In other words, the top would rotate about an axis through O in the direction of the vector $\boldsymbol{\Gamma}$. But if the top is originally rotating, the *change* in its angular momentum produced by the torque adds vectorially to the large angular momentum it already has. Since ΔL is horizontal and perpendicular to L, the result is a motion of precession, with both the angular momentum and the axis remaining horizontal.

This example gives us a glimpse of the richness of the dynamics of rotational motion, which can involve some very complex phenomena. In the next chapter we consider a *simpler* set of problems involving the *equilibrium* of rigid bodies.

SUMMARY

KEY TERMS

rigid body

radians

average angular velocity

instantaneous angular velocity

average angular acceleration

instantaneous angular acceleration

tangential component of acceleration

radial component of acceleration

moment of inertia

parallel-axis theorem

moment arm

torque

moment

angular momentum

angular impulse

conservation of angular momentum

precession

When a rigid body rotates about a stationary axis, its position is described by an angular coordinate θ. The angular velocity ω is defined as the derivative of the angular coordinate.

$$\omega = \frac{d\theta}{dt}, \tag{9-2}$$

and the angular acceleration α is defined as the derivative of the angular velocity, or the second derivative of the angular coordinate:

$$\alpha = \frac{d\omega}{dt} = \frac{d^2\theta}{dt^2}. \tag{9-3}$$

When a body rotates with constant angular acceleration, the angular position, velocity, and acceleration are related by

$$\omega = \omega_0 + \alpha t, \tag{9-5}$$

$$\theta = \theta_0 + \omega_0 t + \tfrac{1}{2}\alpha t^2, \tag{9-6}$$

$$\omega^2 = \omega_0{}^2 + 2\alpha(\theta - \theta_0), \tag{9-7}$$

where θ_0 and ω_0 are the initial values of angular position and angular velocity, respectively.

A particle in a rotating rigid body has a speed v given by

$$v = r\omega \tag{9-9}$$

and an acceleration $\boldsymbol{a}$ with tangential component

$$a_{\parallel} = r\alpha \tag{9-10}$$

and radial component

$$a_{\perp} = \frac{v^2}{r} = \omega^2 r. \tag{9-11}$$

The moment of inertia I of a body with respect to a given axis is defined as

$$I = \Sigma m r^2, \tag{9-12}$$

where r is the distance of mass m from the axis of rotation. The kinetic energy of a rotating rigid body is given by

$$K = \tfrac{1}{2}I\omega^2. \tag{9–13}$$

When the moment of inertia I_{cm} of a body about an axis through the center of mass is known, the moment of inertia I_P about a parallel axis at a distance d from the first axis is given by

$$I_P = I_{cm} + Md^2. \tag{9–22}$$

When a force $\boldsymbol{F}$ acts on a body, the torque Γ of that force with respect to a point O is given by

$$\Gamma = Fl. \tag{9–23}$$

A sign convention is needed, such as "counterclockwise positive, clockwise negative" or the reverse. A generalized definition of torque as a vector quantity $\boldsymbol{\Gamma}$ is

$$\boldsymbol{\Gamma} = \boldsymbol{r} \times \boldsymbol{F}, \tag{9–24}$$

where $\boldsymbol{r}$ is the position vector of the point at which the force acts.

The angular acceleration α of a rigid body rotating about a stationary axis is related to the total torque $\Sigma\Gamma$ and the moment of inertia I of the body by

$$\Sigma\Gamma = I\alpha. \tag{9–27}$$

This relation is also valid for a moving axis, provided the axis passes through the center of mass and does not change direction.

When a torque Γ acts on a rigid body that undergoes an angular displacement $d\theta$, the work dW done by the torque is given by

$$dW = \Gamma\, d\theta. \tag{9–32}$$

When the body rotates with angular velocity ω, the power P (rate of doing work) is given by

$$P = \Gamma\omega. \tag{9–34}$$

The angular momentum L with respect to a point O of a particle of mass m moving with speed v is given by

$$L = mvr, \tag{9–38}$$

where r is the perpendicular distance from the line of motion to point O. When a rigid body having moment of inertia I rotates with angular velocity ω about a stationary axis, its angular momentum with respect to that axis is given by

$$L = I\omega. \tag{9–39}$$

A generalized definition of angular momentum of a particle with respect to a point O, as a vector quantity $\boldsymbol{L}$, is

$$\boldsymbol{L} = \boldsymbol{r} \times m\boldsymbol{v}, \tag{9–45}$$

where $\boldsymbol{r}$ is the position vector of the particle with respect to O. The angular impulse J_θ of a torque acting on a body during a time interval from t_1 to t_2 is given by

$$J_\theta = \Gamma(t_2 - t_1) \tag{9–41}$$

if the torque is constant, and in general by

$$J_\theta = \int_{t_1}^{t_2} \Gamma \, dt \qquad (9\text{--}42)$$

for any torque, constant or not.

The relation between angular impulse and angular momentum is

$$J_\theta = I\omega_2 - I\omega_1 = L_2 - L_1. \qquad (9\text{--}43)$$

In terms of vector torque Γ and angular momentum L, the basic dynamic relation for rotational motion can be restated as

$$\Sigma\Gamma = \frac{dL}{dt}. \qquad (9\text{--}46)$$

If a system consists of bodies that interact with each other but not with anything else, or if the total torque associated with the external forces is zero, the total angular momentum of the system is constant, or conserved.

QUESTIONS

9–1 What is the difference between tangential and radial acceleration, for a point on a rotating body?

9–2 A flywheel rotates with constant angular velocity. Does a point on its rim have a tangential acceleration? A radial acceleration? Are these accelerations constant? In magnitude? In direction?

9–3 A flywheel rotates with constant angular acceleration. Does a point on its rim have a tangential acceleration? A radial acceleration? Are these accelerations constant? In magnitude? In direction?

9–4 Can a single force applied to a body change both its translational and rotational motion?

9–5 How might you determine experimentally the moment of inertia of an irregularly shaped body?

9–6 Can you think of a body that has the same moment of inertia for all possible axes? For all axes passing through a certain point? What point?

9–7 A cylindrical body has mass M and radius R. Is it ever possible for its moment of inertia to be greater than MR^2?

9–8 In order to maximize the moment of inertia of a flywheel while minimizing its weight, what shape should it have?

9–9 In tightening cylinder-head bolts in an automobile engine, the critical quantity is the *torque* applied to the bolts. Why is this more important than the actual *force* applied to the wrench handle?

9–10 The flywheel of an automobile engine is included to increase the moment of inertia of the engine crankshaft. Why is this desirable?

9–11 A solid ball and a hollow ball with the same mass and radius roll down a slope. Which one reaches the bottom first?

9–12 A solid ball, a solid cylinder, and a hollow cylinder roll down a slope. Which reaches the bottom first? Last? Does it matter whether the radii are the same?

9–13 When an electrical motor is turned on, it takes longer to come up to final speed if there is a grinding wheel attached to the shaft. Why?

9–14 Experienced cooks can tell whether an egg is raw or hard-boiled by rolling it down a slope (taking care to catch it at the bottom). How is this possible?

9–15 Consider the idea of an automobile powered by energy stored in a rotating flywheel, which can be "recharged" by using an electric motor. What advantages and disadvantages would such a scheme have, compared to more conventional drive mechanisms? Could as much energy be stored as in a tank of gasoline? What factors would limit the maximum energy storage?

9–16 An electric grinder coasts for a minute or more after the power is turned off, while an electric drill coasts for only a few seconds. Why is there a difference?

9–17 Part of the kinetic energy of a moving automobile is in rotational motion of its wheels. When the brakes are applied hard on an icy street, the wheels "lock" and the car starts to slide. What becomes of the rotational kinetic energy?

EXERCISES

Section 9–1 Angular Velocity

9–1

a) What angle in radians is subtended by an arc 3 m in length, on the circumference of a circle whose radius is 2 m?

b) What angle in radians is subtended by an arc of length 78.54 cm on the circumference of a circle of diameter 100 cm? What is this angle in degrees?

c) The angle between two radii of a circle of radius 2.00 m is 0.60 rad. What length of arc is intercepted on the circumference of the circle by the two radii?

9–2 Compute the angular velocity in rad·s^{-1}, of the crankshaft of an automobile engine that is rotating at 4800 rev·min^{-1}.

9–3 A merry-go-round is being pushed by a child. The angle the merry-go-round has turned through varies with time according to $\theta(t) = (2 \text{ rad·s}^{-1})t + (0.05 \text{ rad·s}^{-3})t^3$.

a) Calculate the angular velocity of the merry-go-round as a function of time.

b) What is the initial value of the angular velocity?

c) Calculate the instantaneous value of the angular velocity ω at $t = 5$ s and the average angular velocity ω_{av} for the time interval $t = 0$ to $t = 5$ s.

Section 9–2 Angular Acceleration

9–4 A rigid object rotates with angular velocity that is given by $\omega(t) = 4 \text{ rad·s}^{-1} - (8 \text{ rad·s}^{-3})t^2$.

a) Calculate the angular acceleration as a function of time.

b) Calculate the instantaneous angular acceleration α at $t = 2$ s, and the average angular acceleration α_{av} for the time interval $t = 0$ to $t = 2$ s.

9–5 The angle θ through which a bicycle wheel turns is given by $\theta(t) = a + bt^2 + ct^3$, where a, b, and c are constants such that for t in seconds θ will be in radians. Calculate the angular acceleration of the wheel as a function of time.

Section 9–3 Rotation with Constant Angular Acceleration

9–6 A circular saw blade 0.6 m in diameter starts from rest and accelerates with constant angular acceleration to an angular velocity of 100 rad·s^{-1} in 20 s. Find the angular acceleration and the angle through which the blade has turned.

9–7 An electric motor is turned off, and its angular velocity decreases uniformly from 1000 rev·min^{-1} to 400 rev·min^{-1} in 5 s.

a) Find the angular acceleration and the number of revolutions made by the motor in the 5-s interval.

b) How many more seconds are required for the motor to come to rest?

9–8 A flywheel requires 3 s to rotate through 234 rad. Its angular velocity at the end of this time is 108 rad·s^{-1}. Find

a) the angular velocity at the beginning of the 3-s interval;

b) the constant angular acceleration.

9–9 A flywheel whose angular acceleration is constant and equal to 2 rad·s^{-2} rotates through an angle of 100 rad in 5 s. How long had it been in motion at the beginning of the 5-s interval if it started from rest?

9–10 A bicycle wheel of radius 0.33 m turns with angular acceleration $\alpha(t) = 1.2 \text{ rad·s}^{-2} - (0.4 \text{ rad·s}^{-3})t$. It is at rest at $t = 0$.

a) Calculate the angular velocity and angular displacement as functions of time.

b) Calculate the maximum positive angular velocity and maximum positive angular displacement of the wheel.

Section 9–4 Relation between Angular and Linear Velocity and Acceleration

9–11

a) A cylinder 0.15 m in diameter rotates in a lathe at 750 rev·min^{-1}. What is the tangential velocity of the surface of the cylinder?

b) The proper tangential velocity for machining cast iron is about 0.60 m·s^{-1}. At how many rev·min^{-1} should a piece of stock 0.05 m in diameter be rotated in a lathe?

9–12 Find the required angular velocity of an ultracentrifuge, in rpm, for the radial acceleration of a point 1 cm from the axis to equal 300,000 g (i.e., 300,000 times the acceleration of gravity).

9–13 A wheel rotates with a constant angular velocity of 10 rad·s^{-1}.

a) Compute the radial acceleration of a point 0.5 m from the axis, from the relation $a_\perp = \omega^2 r$.

b) Find the tangential velocity of the point, and compute its radial acceleration from the relation $a_\perp = v^2/r$.

9–14 An electric fan blade 3.0 ft in diameter is rotating about a fixed axis with an initial angular velocity of 2 rev·s^{-1}. The angular acceleration is 3 rev·s^{-2}.

a) Compute the angular velocity after 1 s.

b) Through how many revolutions has the blade turned in this time interval?

c) What is the tangential velocity of a point on the tip of the blade at $t = 1$ s?

d) What is the resultant acceleration of a point on the tip of the blade at $t = 1$ s?

9–15 A flywheel of radius 0.30 m starts from rest and accelerates with constant angular acceleration of 0.50 rad·s^{-2}. Compute the tangential acceleration, the radial acceleration, and the resultant acceleration of a point on its rim

a) at the start;

b) after it has turned through 120°;

c) after it has turned through 240°.

Section 9–5 Kinetic Energy of Rotation

9–16 Small blocks, each of mass m, are clamped at the ends and at the center of a light rod of length L. Compute the moment of inertia of the system about an axis perpendicular to the rod and passing through a point one-quarter of the length from one end. Neglect the moment of inertia of the light rod.

9–17 Four small spheres, each of mass 3 kg, are arranged in a square 0.5 m on a side and are connected by light rods. Find the moment of inertia of the system about an axis

a) through the center of the square, perpendicular to its plane;

b) bisecting two opposite sides of the square.

9–18 Find the moment of inertia of a rod 4 cm in diameter and 2 m long, of mass 8 kg, for the following axes. Use the formulas of Fig. 9–7.

a) About an axis perpendicular to the rod and passing through its center.

b) About a longitudinal axis passing through the center of the rod.

9–19 A wagon wheel is constructed as in Fig. 9–26. The radius of the wheel is 0.3 m and the rim has mass 1.0 kg. Each of the four spokes, which lie along diameters, has a mass of 0.4 kg. What is the moment of inertia of the wheel about an axis through its center and perpendicular to the plane of the wheel? (Use the formulas given in Fig. 9–7.)

FIGURE 9–26

9–20 A grinding wheel 0.2 m in diameter, of mass 3 kg, is rotating at 3600 rev·min^{-1} about an axis through its center.

a) What is its kinetic energy?

b) How far would it have to drop in free fall to acquire the same kinetic energy?

9–21 The flywheel of a gasoline engine is required to give up 300 J of kinetic energy while its angular velocity decreases from 600 rev·min^{-1} to 540 rev·min^{-1}. What moment of inertia is required?

9–22 A light, flexible rope is wrapped several times around a solid cylinder of weight 40 lb and radius 0.75 ft, which rotates without friction about a fixed horizontal axis. The free end of the rope is pulled with a constant force P for a distance of 15 ft. What must P be for the final speed of the end of the rope to be 12.0 ft·s^{-1}?

Section 9–6 Moment-of-Inertia Calculations

Section 9–7 Parallel-Axis Theorem

9–23 A thin, rectangular sheet of steel is 0.3 m by 0.4 m and has mass 24 kg. Find the moment of inertia about an axis

a) through the center, parallel to the long sides;

b) through the center, parallel to the short sides;

c) through the center, perpendicular to the plane.

9–24 The four objects shown in Fig. 9–27 have equal masses m. Object A is a solid cylinder of radius R. Object B is a hollow, thin cylinder of radius R. Object C is a solid square whose length of side = $2R$. Object D is the same size as C, but hollow (i.e., made up of four thin sticks). The objects have axes of rotation perpendicular to the page and through the center of gravity of each object.

a) Which object has the smallest moment of inertia?

b) Which object has the largest moment of inertia?

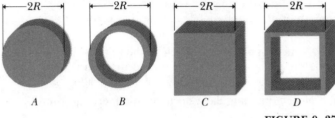

FIGURE 9–27

9–25 Use the parallel-axis theorem to calculate the moment of inertia of a square sheet of metal of side length a and mass M for an axis perpendicular to the sheet and passing through one corner.

9–26

a) Use the integration methods of Section 9–6 to calculate the moment of inertia of a uniform, slender rod of mass M and length l for an axis a distance $l/4$ from one end and perpendicular to the rod.

b) Calculate the same moment of inertia as in (a), but by using Fig. 9–7 and the parallel-axis theorem.

Section 9–8 Torque

9–27 Calculate the torque (magnitude and direction) about point O due to the force F in each of the situations sketched in Fig. 9–28. In each case the object to which the force is applied has length 4 ft, and the force F = 20 lb.

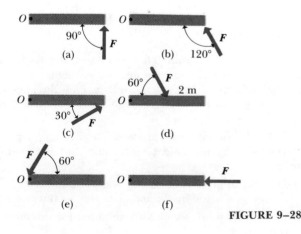

FIGURE 9–28

9–28 Calculate the resultant torque about point O for the two forces applied as in Fig. 9–29.

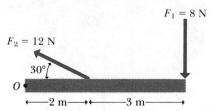

$F_1 = 8$ N

$F_2 = 12$ N

$30°$

O

←—2 m—→←——3 m——→

FIGURE 9–29

9–29 A force F is given by $F = (2 \text{ N})i - (3 \text{ N})j$, and the vector r from the axis to point of application of the force is $r = (-0.3 \text{ m})i + (0.5 \text{ m})j$. Calculate the vector torque produced by this force.

Section 9–9 Torque and Angular Acceleration

9–30 A cord is wrapped around the rim of a flywheel 0.5 m in radius, and a steady pull of 50 N is exerted on the cord. The wheel is mounted in frictionless bearings on a horizontal shaft through its center. The moment of inertia of the wheel is 4 kg·m². Compute the angular acceleration of the wheel.

9–31 A grindstone 1.0 m in diameter, of mass 50 kg, is rotating at 900 rev·min⁻¹. A tool is pressed against the rim with a normal force of 200 N, and the grindstone comes to rest in 10 s. Find the coefficient of friction between the tool and the grindstone. Neglect friction in the bearings.

9–32 A thin rod of length l and mass M is pivoted about a vertical axis at one end. A constant force F is applied to the other end, causing the rod to rotate. The force is maintained perpendicular to the rod. Calculate the angular acceleration α of the rod.

9–33 A bucket of water of mass 20 kg is suspended by a rope wrapped around a windlass in the form of a solid cylinder 0.2 m in diameter, also of mass 20 kg. The cylinder is pivoted on a frictionless axle through its center. The bucket is released from rest at the top of a well and falls 20 m to the water. Neglect the weight of the rope.

a) What is the tension in the rope while the bucket is falling?

b) With what velocity does the bucket strike the water?

c) What was the time of fall?

d) While the bucket is falling, what is the force exerted on the cylinder by the axle?

Section 9–10 Work and Power in Rotational Motion

9–34 What is the power output in horsepower of an electric motor turning at 3600 rpm and developing a torque of 2.5 ft·lb?

9–35 A grindstone in the form of a solid cylinder has a radius of 0.5 m and a mass of 50 kg.

a) What torque will bring it from rest to an angular velocity of 300 rev·min⁻¹ in 10 s?

b) Use Eq. (9–33) to calculate the work done by the torque.

c) What is the grindstone's kinetic energy when it is rotating at 300 rev·min⁻¹? Compare your answer to the result in (b).

9–36 The flywheel of a motor has a mass of 300 kg and a moment of inertia of 675 kg·m². The motor develops a constant torque of 2000 N·m, and the flywheel starts from rest.

a) What is the angular acceleration of the flywheel?

b) What will be its angular velocity after making four revolutions?

c) How much work is done by the motor during the first four revolutions?

9–37

a) Compute the torque developed by an airplane engine whose output is 1.5×10^6 W at an angular velocity of 2400 rev·min⁻¹.

b) If a drum of negligible mass, 0.5 m in diameter, were attached to the motor shaft, and the power output of the motor were used to raise a weight hanging from a rope wrapped around the drum, how large a weight could be lifted?

c) With what velocity would it rise?

Section 9–11 Rotation about a Moving Axis

9–38 A solid cylinder of mass 4 kg rolls, without slipping, down a 30° slope. Find the acceleration, the frictional force, and the minimum coefficient of friction needed to prevent slipping.

9–39 A string is wrapped several times around the rim of a small hoop. The hoop has radius 0.08 m and mass 1.2 kg. If the free end of the string is held in place and the hoop is released from rest, calculate

a) the tension in the string while the hoop is descending;

b) the time it takes the hoop to descend 0.5 m;

c) the angular velocity of the rotating hoop after it has descended 0.5 m.

Section 9–12 Angular Momentum and Angular Impulse

9–40 Calculate the angular momentum of a uniform sphere of radius 0.20 m and mass 4 kg if it is rotating about an axis along a diameter at 6 rad·s⁻¹.

9–41 What is the angular momentum of the hour hand on a clock, about an axis through the center of the clock face, if the clock hand has a length of 25.0 cm and a mass of 20.0 g? Take the hour hand to be a slender rod rotating about one end.

9–42 A solid wood door 1.0 m wide and 2.0 m high is hinged along one side and has a total mass of 50 kg. Initially open and at rest, the door is struck at its center with a hammer. During the blow an average force of 2000 N acts for 0.01 s. Find the angular velocity of the door after the impact.

9–43 A rock of mass 2 kg is thrown with velocity $v = 12$ m·s^{-1}. When it is at point P in Fig. 9–30, what is its angular momentum relative to point O? Assume that the rock travels in a straight line.

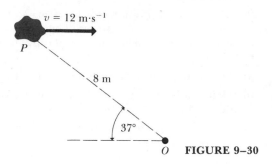

FIGURE 9–30

9–44 A man of mass 70 kg is standing on the rim of a large disk that is rotating at 0.5 rev·s^{-1} about an axis through its center. The disk has mass 120 kg and radius 4 m. Calculate the total angular momentum of the man-plus-disk system.

Section 9–13 Conservation of Angular Momentum

9–45 On an old-fashioned rotating piano stool, a man sits holding a pair of dumbbells at a distance of 0.6 m from the axis of rotation of the stool. He is given an angular velocity of 5 rad·s^{-1}, after which he pulls the dumbbells in until they are only 0.2 m distant from the axis. The man's moment of inertia about the axis of rotation is 5 kg·m^2 and may be considered constant. Each dumbbell has a mass of 5 kg and may be considered a point mass. Neglect friction.

a) What is the initial angular momentum of the system?

b) What is the angular velocity of the system after the dumbbells are pulled in toward the axis?

c) Compute the kinetic energy of the system before and after the dumbbells are pulled in. Account for the difference, if any.

9–46 The outstretched arms of a figure skater preparing for a spin can be considered a slender rod pivoting about an axis through its center. When her arms are brought in and wrapped around her body to execute the spin, they can be considered a thin-walled hollow cylinder. If her original angular velocity is 1 rev·s^{-1}, what is her final angular velocity? Her arms have a combined mass of 8 kg. When outstretched, they span 1.8 m; when wrapped, they form a cylinder of radius 25 cm. The moment of inertia of the remainder of her body is constant and equal to 3 kg·m^2.

9–47 A puck on a frictionless air-hockey table has a mass of 0.05 kg and is attached to a cord passing through a hole in the table surface, as in Fig. 9–31. The puck is originally revolving at a distance of 0.2 m from the hole, with an angular velocity of 3 rad·s^{-1}. The cord is then pulled from below, shortening the radius of the circle in which the puck revolves to 0.1 m. The puck may be considered a point mass.

a) What is the new angular velocity?

b) Find the change in kinetic energy of the puck.

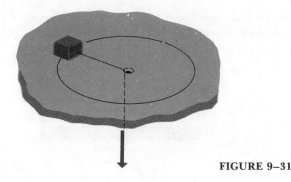

FIGURE 9–31

c) How much work was done by the person who pulled the cord?

9–48 A turntable rotates about a fixed vertical axis, making one revolution in 10 s. The moment of inertia of the turntable about this axis is 1200 kg·m^2. A man of mass 80 kg, initially standing at the center of the turntable, runs out along a radius. What is the angular velocity of the turntable when the man is 2 m from the center?

9–49 In models of stellar evolution, one scenario is for stars to undergo sudden gravitational collapse in which the radius of the star decreases drastically while most of the star's mass is retained. Consider the hypothetical collapse of our sun. Data are given in Appendix F. Also, the present period for rotation on its axis for the sun is about 25 days. If the sun undergoes gravitational collapse to a new radius of 32 km (about 20 mi) while keeping its present mass, calculate its new revolution rate in rev·s^{-1}. (This illustrates one possible model for the formation of pulsars.)

9–50 A large wooden turntable of radius 2.0 m and total mass 120 kg is rotating about a vertical axis through its center, with an angular velocity of 3.0 rad·s^{-1}. From a very small height a sandbag of mass 100 kg is dropped vertically onto the turntable, at a point near the outer edge.

a) Find the angular velocity of the turntable after the sandbag is dropped.

b) Compute the kinetic energy of the system before and after the sandbag is dropped.

Why are these kinetic energies not equal?

9–51 The door in Exercise 9–42 is struck at its center by a handful of sticky mud of mass 0.5 kg, traveling 10 m·s^{-1} just before impact. Find the final angular velocity of the door. Is the moment of inertia of the mud significant?

Section 9–14 Vector Representation of Angular Quantities

9–52 The mass of the rotor of a toy gyroscope is 0.150 kg, and its moment of inertia about its axis is 1.5×10^{-4} kg·m^2. The mass of the frame is 0.030 kg. The gyroscope is supported on a single pivot, as in Fig. 9–32, with its center of gravity a horizontal distance of 4 cm from the pivot. The gyroscope is precessing in a horizontal plane at the rate of one revolution in 6 s.

a) Find the upward force exerted by the pivot.

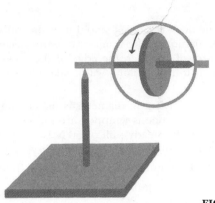

FIGURE 9-32

b) Find the angular velocity with which the rotor is spinning about its axis, expressed in rev·min⁻¹.

c) Copy the diagram, and show by vectors the angular momentum of the rotor and the torque acting on it.

9–53 The stabilizing gyroscope of a ship is a solid disk of mass 50,000 kg; its radius is 2 m, and it rotates about a vertical axis with an angular velocity of 900 rev·min⁻¹.

a) How long a time is required to bring it up to speed, starting from rest, with a constant power input of 7.46×10^4 W?

b) Find the torque needed to cause the axis to precess in a vertical fore-and-aft plane at the rate of $1° \cdot \text{s}^{-1}$.

PROBLEMS

9–54

a) Prove that when a body starts from rest and rotates about a fixed axis with constant angular acceleration, the radial acceleration of a point in the body is directly proportional to its angular displacement.

b) Through what angle has the body turned at the instant when the resultant acceleration of a point makes an angle of 60° with the radial direction?

9–55 The moment of inertia of a sphere of uniform density is $0.400MR^2$ (Fig. 9–7). Recent satellite observations show the earth's moment of inertia to be $0.3308MR^2$. The earth has a mantle and a core. The average density of the core is approximately $11.0 \times 10^3 \text{kg} \cdot \text{m}^{-3}$; the mantle's is $5.0 \times 10^3 \text{kg} \cdot \text{m}^{-3}$. The mantle occurs from the earth's surface to a depth of 3600 km, and the remainder is core. For this two-part model of the earth, what is the earth's moment of inertia? Express your answer in terms of M and R as above.

9–56 A solid uniform disk of mass m and radius R is pivoted about a horizontal axis through its center, and a small object of mass m is attached to the rim of the disk. If the disk is released from rest with the small object at the end of a horizontal radius, find the angular velocity when the small object is at the bottom.

9–57 The flywheel of a punch press has a moment of inertia of 25 kg·m² and runs at 300 rev·min⁻¹. The flywheel supplies all the energy needed in a quick punching operation.

a) Find the speed in rev·min⁻¹ to which the flywheel will be reduced by a sudden punching operation requiring 4000 J of work.

b) What must be the constant power supply to the flywheel (in watts) to bring it back to its initial speed in a time of 5 s?

9–58 A magazine article described a passenger bus in Zurich, Switzerland, that derived its motive power from the energy stored in a large flywheel. The wheel was brought up to speed periodically, when the bus stopped at a station, by an electric motor, which could then be attached to the electric power lines. The flywheel was a solid cylinder of mass 1000 kg and diameter 1.8 m; its top speed was 3000 rev·min⁻¹.

a) At this speed, what was the kinetic energy of the flywheel?

b) If the average power required to operate the bus was 1.86×10^4 W, how long could it operate between stops?

9–59 A meter stick of mass 0.4 kg is pivoted about one end so it can rotate without friction about a horizontal axis. The meter stick is held in a horizontal position and released. As it swings through the vertical, calculate

a) the angular velocity of the stick;

b) the linear velocity of the end of the stick opposite the axis.

9–60 Consider the system sketched in Fig. 9–33. The pulley has radius 0.2 m and moment of inertia. 0.32 kg·m². The rope does not slip on the pulley rim. Use energy methods to calculate the velocity of the 4-kg block just before it strikes the floor.

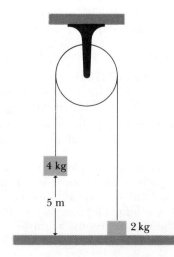

4 kg

5 m

2 kg

FIGURE 9–33

9–61 Consider the system sketched in Fig. 9–34. The pulley has radius R and moment of inertia I. The rope does not slip over the pulley. The coefficient of friction between block A and the table top is μ_k. The system is released from rest, and block B descends. Use energy methods to calculate the velocity of block B as a function of the distance d that it has descended.

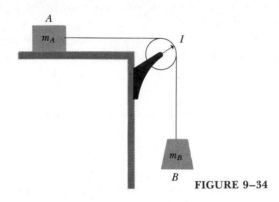

FIGURE 9–34

9–62 A uniform, thin rod is bent into a square of side length a. If the total mass is M, find the moment of inertia about an axis through the center, perpendicular to the plane of the square. (*Hint:* Use the parallel-axis theorem.)

9–63 A constant torque of 20 N·m is exerted on a pivoted wheel for 10 s, during which time the angular velocity of the wheel increases from zero to 100 rev·min^{-1}. The external torque is then removed, and the wheel is brought to rest by friction in its bearings in 100 s. Compute

a) the moment of inertia of the wheel;

b) the friction torque;

c) the total number of revolutions made by the wheel.

9–64 A 60-kg grindstone is 1 m in diameter and has a moment of inertia of 3.75 kg·m^2. A tool is pressed down on the rim with a normal force of 50 N. The coefficient of sliding friction between the tool and the stone is 0.6, and there is a constant friction torque of 5 N·m between the axle of the stone and its bearings

a) How much force must be applied normally at the end of a crank handle 0.5 m long to bring the stone from rest to 120 rev·min^{-1} in 9 s?

b) After attaining a speed of 120 rev·min^{-1}, what must the normal force at the end of the handle become to maintain a constant speed of 120 rev·min^{-1}?

c) How long will it take the grindstone to come from 120 rev·min^{-1} to rest if it is acted on by the axle friction alone?

9–65 Dirk the Dragonslayer is exploring a castle. He is spotted by a dragon who chases him down a hallway. Dirk runs into a room and attempts to swing the heavy door shut before the dragon gets him. The door is initially perpendicular to the wall, so it must be turned through 90°

to close. The door is 3 m tall and 1 m wide and weighs 600 N. The friction at the hinges can be neglected. If Dirk applies a force of 180 N at the edge of the door and perpendicular to it, how long will it take him to close the door?

9–66 A flywheel 1.0 m in diameter is pivoted on a horizontal axis. A rope is wrapped around the outside of the flywheel, and a steady pull of 50 N is exerted on the rope. Ten meters of rope are unwound in 4 s.

a) What was the angular acceleration of the flywheel?

b) What is its final angular velocity?

c) What is its final kinetic energy?

d) What is its moment of inertia?

9–67 A 5-kg block rests on a frictionless horizontal surface. A cord attached to the block passes over a pulley, whose diameter is 0.2 m, to a hanging block also of mass 5 kg. The system is released from rest, and the blocks are observed to move 4 m in 2 s.

a) What was the tension in each part of the cord?

b) What was the moment of inertia of the pulley?

9–68 Figure 9–35 represents an Atwood's machine. Find the linear accelerations of blocks A and B, the angular acceleration of the wheel C, and the tension in each side of the cord

a) if the surface of the wheel is frictionless;

b) if there is no slipping between the cord and the surface of the wheel.

In each case let the masses of the blocks A and B be 4 kg and 2 kg, respectively; the moment of inertia of the wheel about its axis, 0.2 kg·m^2; and the radius of the wheel, 0.1 m.

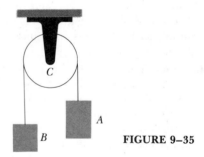

FIGURE 9–35

9–69 A block of mass $m = 5$ kg slides down a surface inclined 37° to the horizontal, as shown in Fig. 9–36. The coefficient of sliding friction is 0.25. A string attached to the

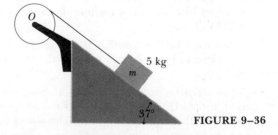

FIGURE 9–36

block is wrapped around a flywheel on a fixed axis at O. The flywheel has a mass $M = 20$ kg, an outer radius $R = 0.2$ m, and a moment of inertia with respect to the axis of 0.2 kg·m².

a) What is the acceleration of the block down the plane?

b) What is the tension in the string?

9–70 A yo-yo is made from two uniform disks, each of mass m and radius R, connected by a light axle of radius b. A string is wound several times around the axle and then held stationary while the yo-yo is released from rest, dropping as the string unwinds. Find the acceleration of the yo-yo and the tension in the string.

9–71 A lawn roller in the form of a hollow cylinder of mass M is pulled horizontally with a constant force F applied by a handle attached to the axle. If it rolls without slipping, find the acceleration and the frictional force.

9–72 A uniform rod of mass 0.03 kg and length 0.2 m rotates in a horizontal plane about a fixed axis through its center. Two small objects, each of mass 0.02 kg, are mounted so that they can slide along the rod. They are initially held by catches at positions 0.05 m on each side of the center of the rod, and the system is rotating at 15 rev·min⁻¹. Without otherwise changing the system, the catches are released and the masses slide outward along the rod and fly off at the ends.

a) What is the angular velocity of the system at the instant when the small masses reach the ends of the rod?

b) What is the angular velocity of the rod after the small masses leave it?

9–73 A small block of mass 4 kg is attached to a cord passing through a hole in a frictionless horizontal surface. The block is originally revolving in a circle of radius 0.5 m about the hole, with a tangential velocity of 4 m·s⁻¹. The cord is then pulled slowly from below, shortening the radius of the circle in which the block revolves. The breaking strength of the cord is 600 N. What will be the radius of the circle when the cord breaks?

9–74 Figure 9–37 shows part of a "fly-ball" governor, a speed-controlling device used in old-fashioned steam engines. Each of the steel balls A and B has a mass of 0.50 kg and is rotating about the vertical axis with an angular velocity of 4 rad·s⁻¹ at a distance of 0.15 m from

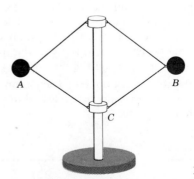

FIGURE 9–37

the axis. Collar C is now forced down until the balls are at a distance of 0.05 m from the axis. How much work must be done to move the collar down?

9–75 Disks A and B are mounted on a shaft SS and may be connected or disconnected by a clutch C, as in Fig. 9–38. The moment of inertia of disk A is one-half that of disk B. With the clutch disconnected, A is brought up to an angular velocity ω_0. The accelerating torque is then removed from A, and A is coupled to disk B by the clutch. (Bearing friction may be neglected.) It is found that 2000 J of heat are developed in the clutch when the connection is made. What was the original kinetic energy of disk A?

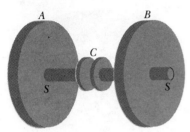

FIGURE 9–38

9–76 A man of mass 60 kg runs around the edge of a horizontal turntable that is mounted on a frictionless vertical axis through its center. The velocity of the man, relative to the earth, is 1 m·s⁻¹. The turntable is rotating in the opposite direction with an angular velocity of 0.2 rad·s⁻¹, relative to the earth. The radius of the turntable is 2 m, and its moment of inertia about the axis of rotation is 400 kg·m². Find the final angular velocity of the system if the man comes to rest, relative to the turntable.

9–77 The moment of inertia of the front wheel of a bicycle is 0.4 kg·m², its radius is 0.5 m, and the forward speed of the bicycle is 5 m·s⁻¹. With what angular velocity must the front wheel be turned about a vertical axis to counteract the capsizing torque due to a mass of 60 kg, 0.02 m horizontally to the right or left of the line of contact of wheels and ground? (Bicycle riders: Compare your own experience and see if your answer seems reasonable.)

9–78 A demonstration gyroscope wheel is constructed by removing the tire from a bicycle wheel 0.70 m in diameter, wrapping lead wire around the rim, and taping it in place. The shaft projects 0.2 m at each side of the wheel, and a man holds the ends of the shaft in his hands. The mass of the system is 5 kg; its entire mass may be assumed to be located at its rim. The shaft is horizontal, and the wheel is spinning about the shaft at 5 rev·s⁻¹. Find the magnitude and direction of the force each hand exerts on the shaft

a) when the shaft is at rest;

b) when the shaft is rotating in a horizontal plane about its center at 0.04 rev·s⁻¹;

c) when the shaft is rotating in a horizontal plane about its center at 0.20 rev·s⁻¹.

d) At what rate must the shaft rotate in order to be supported at one end only?

CHALLENGE PROBLEMS

9–79 A cylinder of radius R and mass M has density that increases linearly with distance r from the cylinder axis, $\rho = \alpha r$. Calculate the moment of inertia of the cylinder about its axis of symmetry, in terms of M and R. Compare the result to that for a cylinder of uniform density. Is the proper qualitative relation obtained?

9–80 Calculate the moment of inertia of a solid cone of uniform density about an axis through the center of the cone, as shown in Fig. 9–39. The cone has mass M and altitude h. The radius of its circular base is R.

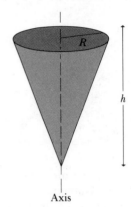

Axis   **FIGURE 9–39**

9–81 A block of mass m is revolving with linear velocity v_1 in a circle of radius r_1 on a frictionless horizontal surface, as shown in Fig. 9–31. The string is slowly pulled from below until the radius of the circle in which the object is revolving is reduced to r_2.

a) Calculate the tension T in the string as a function of r, the distance of the block from the hole. Your answer will be in terms of the initial velocity v_1 and the radius r_1.

b) Use $W = \int_{r_1}^{r_2} \boldsymbol{T}(r) \cdot d\boldsymbol{r}$ to calculate the work done by $\boldsymbol{T}$ when r changes from r_1 to r_2.

c) Compare the results of (b) to the change in the kinetic energy of the block.

9–82 In a binary star system, one star has mass m_A and velocity v_A with respect to an observer, and is located at a position r_A. The other star has mass m_B, velocity v_B, and position r_B. Let R and V be the center of mass position and velocity vectors. Define *relative* positions and velocities by $\boldsymbol{R}_A = \boldsymbol{r}_A - \boldsymbol{R}$ and $\boldsymbol{V}_A = \boldsymbol{v}_A - \boldsymbol{V}$, and similarly for star B.

a) Show that $\boldsymbol{L} = M\boldsymbol{R} \times \boldsymbol{V} + (m_A\boldsymbol{R}_A \times \boldsymbol{V}_A + m_B\boldsymbol{R}_B \times \boldsymbol{V}_B)$. Hence $\boldsymbol{L}$ consists of two parts: $M\boldsymbol{R} \times \boldsymbol{V}$, the angular momentum *of* the center of mass, and $(m_A\boldsymbol{R}_A \times \boldsymbol{V}_A + m_B\boldsymbol{R}_B \times \boldsymbol{V}_B)$, the angular momentum *about* the center of mass.

b) If the only forces acting on the stars are their mutual gravitational attractions, show that $d\boldsymbol{L}/dt = \boldsymbol{0}$. (This result is in fact valid for any number of objects, including planets, asteroids, natural and artificial satellites, comets, and even dust.)

9–83 When an object is rolling without slipping, the rolling friction force is much less than the friction force when the object is sliding; a silver dollar will roll on its edge much farther than it will slide on its flat side. (See Section 5–2.) When an object is rolling without slipping on a horizontal surface, one can therefore take the friction force to be zero so that a and α are approximately zero and v and ω are approximately constant. Rolling without slipping means $v = r\omega$ and $a = r\alpha$. If an object is set into motion on a surface *without* these equalities, sliding (kinetic) friction will act on the object as it slips until rolling without slipping is established.

A solid cylinder of mass M and radius R, rotating with angular velocity ω_0 about an axis through its center, is set on a horizontal surface for which the kinetic friction coefficient is μ_k.

a) Draw a free-body force diagram for the cylinder on the surface. Think carefully about the direction of the kinetic friction force on the cylinder. Calculate the accelerations a of the center of mass (cm) and α of rotation about the cm.

b) The cylinder is initially slipping completely, as initially $\omega = \omega_0$ but $v = 0$. Rolling without slipping sets in when $v = R\omega$. Calculate the *distance* the cylinder rolls before slipping stops.

c) Calculate the work done by the friction force on the cylinder as it moves from where it was set down to where it begins to roll without slipping.

9–84 A uniform sphere is released from rest at the top of an inclined plane of slope θ, as in Example 9–17. The top of the incline is a height h above the horizontal surface on which the incline rests. The coefficient of static friction is *less* than that required to maintain rolling without slipping. The coefficient of sliding friction is $\mu_k = \frac{1}{7}\tan\theta$, and this coefficient determines the friction force.

a) Calculate the linear acceleration a of the center of mass (cm) and the angular acceleration α for rotation about the cm of the sphere.

b) How long does it take the sphere to reach the bottom of the incline, and what are the translational velocity v of the cm and the angular velocity ω of rotation about the cm when it gets there?

c) If the coefficient of kinetic friction between the horizontal surface and the sphere is the same ($\frac{1}{7}\tan\theta$) as between the sphere and the incline, how far will the sphere roll along the horizontal surface before rolling without slipping sets in? What is the (approximately constant) final translational velocity of the sphere as it rolls without slipping along the horizontal surface?

d) Calculate the total work done by friction on the sphere from when the sphere was first released to when it starts to roll without slipping.

10

EQUILIBRIUM OF A RIGID BODY

IN CHAPTER 5 WE DISCUSSED EQUILIBRIUM OF BODIES THAT COULD BE represented as *particles*. Such bodies are in equilibrium whenever the sum of the forces acting on them is zero. Now that we have learned the basic principles of rotational dynamics, we can return to the subject of equilibrium to study the more general problem of equilibrium of a rigid body. We will find that there is a second condition for equilibrium: The sum of the torques about any axis must also be zero for the body not to have any tendency to rotate. Often one of the forces acting on a rigid body is its weight; we show how to compute the torque due to the weight of a body, using the concept of center of mass from Section 8–7 and the related concept of center of gravity, which is introduced in this chapter.

10–1 THE SECOND CONDITION FOR EQUILIBRIUM

We learned in Section 5–3 that for a particle acted on by several forces to be in equilibrium, the vector sum of the forces acting on the particle must be zero, $\Sigma F = 0.$ This is often called the **first condition for equilibrium.** In components,

$$\Sigma F_x = 0, \qquad \Sigma F_y = 0. \tag{10–1}$$

For a rigid body there is a **second condition for equilibrium:**

The sum of the torques of all forces acting on the body, with respect to some specified axis, must be zero.

This condition is needed so the body has no tendency to *rotate*. The body does not actually have to be pivoted about the axis chosen. If a rigid body is to be in equilibrium, it must not have any tendency to rotate about *any* axis; thus the sum of torques must be zero, *no matter what axis is chosen.*

A body in equilibrium has no tendency to rotate about any axis.

$$\Sigma \Gamma = 0 \quad about\ any\ axis. \tag{10–2}$$

Although the choice of axis is arbitrary, we must use the *same* axis to calculate all the torques. An important element of problem-solving strategy is to pick the axis or axes so as to simplify the calculations as much as possible.

The second condition for equilibrium is based on the dynamics of rotational motion in exactly the same way that the first condition is based on Newton's first law. For a rigid body at rest to remain at rest, it must have no angular acceleration about any axis, so the sum of torques with respect to every axis must be zero. Note that the term *equilibrium* is used here in a more restricted sense than for particles. A particle in uniform motion (constant velocity) is in equilibrium, but for rigid bodies we use the word *equilibrium* to mean a body actually *at rest* in an inertial frame of reference. Uniform rotational motion with $\Sigma\Gamma = 0$ is *not* considered to be an equilibrium state because the individual particles of the body have accelerations.

10–2 CENTER OF GRAVITY

Center of gravity: how to find the torque due to the weight of a body

In many equilibrium problems, one of the forces acting on the body is its *weight*. We need to be able to calculate the torque of this force with respect to any axis. This is not a completely trivial problem, because the weight does not act at a single point but is distributed over the entire body. However, we can always calculate the torque due to the body's weight by assuming that the entire force of gravity (weight) is concentrated at the **center of mass** of the body, which in this context is also called the **center of gravity.** We used this result without proof in rigid-body dynamics problems in Section 9–9, and now we return for a detailed proof.

Throughout this discussion the body is assumed to be in a *uniform* gravitational field, so the acceleration due to gravity g has the same magnitude and direction at every point in the body. Every particle in the body experiences a gravitational force, and the total weight of the body is the vector sum (resultant) of a large number of parallel forces.

Let us consider the gravitational torque on a flat body of arbitrary shape in the xy-plane, as shown in Fig. 10–1a. We imagine the body as subdivided into a large number of small particles of weights w_1, w_2, and so on, with coordinates (x_1, y_1), (x_2, y_2), and so on. The total weight W of the body is

$$W = w_1 + w_2 + \cdots = \Sigma w.$$

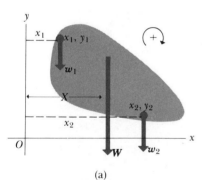

(a)

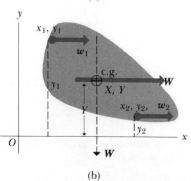

(b)

10–1 The body's weight W is the resultant of a large number of parallel forces. The line of action of W always passes through the center of gravity.

Each particle's weight also contributes to the total torque acting on the body. Computing torques about point O, we see that the torque associated with particle 1 is w_1x_1, that for particle 2 it is w_2x_2, and so on, and that the total torque is

$$w_1x_1 + w_2x_2 + \cdots = \Sigma wx.$$

For convenience, we are taking *clockwise* torques to be positive, the opposite of the convention we usually used in Chapter 9.

We now ask: Along what line must the total weight W act if the total torque is to equal the expression given above? Suppose W acts along a line at a distance X to the right of the origin; then its torque is WX. Hence it must be true that $w_1x_1 + w_2x_2 + \cdots = WX$, or

$$X = \frac{w_1x_1 + w_2x_2 + \cdots}{W} = \frac{\Sigma wx}{W} = \frac{\Sigma wx}{\Sigma w} \qquad (10\text{–}3)$$

That is, in computing the torque due to the weight of a body, we must consider the weight to act at a point whose x-coordinate X is given by Eq. (10–3).

Now we rotate the object and the reference axes 90° clockwise (or equivalently, we rotate the gravitational forces 90° counterclockwise), as in Fig. 10–1b. The total weight W is unaltered, but now it must act along a line a distance Y above the origin, such that $WY = \Sigma wy$, or

$$Y = \frac{w_1 y_1 + w_2 y_2 + \cdots}{w_1 + w_2 + \cdots} = \frac{\Sigma wy}{\Sigma w} = \frac{\Sigma wy}{W}. \qquad (10\text{--}4)$$

The point of intersection of the lines of action of W in the two parts of Fig. 10–1 has coordinates (X, Y) and is called the *center of gravity* of the body. By considering some arbitrary orientation of the object, we can show that the line of action of W *always* passes through the center of gravity, and thus that the torque of W can always be obtained correctly by taking W to act at the center of gravity.

> The center of gravity is identical to the center of mass in a uniform gravitational field. Otherwise center of gravity is meaningless.

If we divide numerator and denominator in Eqs. (10–3) and (10–4) by g and use the fact that $w_1 = m_1 g$, and so on, the right sides of these equations become identical to the equations defining the center of mass in Section 8–7, Eqs. (8–27). Thus *the center of gravity of any body is identical to its center of mass.* Note, however, that the definition of center of gravity makes sense only in a uniform gravitational field, where g has the same value everywhere; otherwise the w's would depend on position. The center of *mass*, conversely, is defined independently of any gravitational effect.

The role of the center of gravity in calculating gravitational torques can be established in a more elegant, though perhaps more acrobatic, way by using the more general vector definition of torque given by Eq. (9–24), $\boldsymbol{\Gamma} = \boldsymbol{r} \times \boldsymbol{F}$. We let the vector $\boldsymbol{g}$ represent the acceleration due to gravity (magnitude and direction). Then the weight $\boldsymbol{w}_1$ of a particle in the body having mass m_1 is given by $\boldsymbol{w}_1 = m_1\boldsymbol{g}$. If $\boldsymbol{r}_1$ is the position vector of this particle with respect to an arbitrary origin O, the torque $\boldsymbol{\Gamma}_1$ of this force with respect to O is

> An elegant vector derivation of the relation of gravitational torque to center of gravity

$$\boldsymbol{\Gamma}_1 = \boldsymbol{r}_1 \times m_1\boldsymbol{g},$$

and the total torque due to the gravitational forces on all the particles is

$$\boldsymbol{\Gamma} = \boldsymbol{r}_1 \times m_1\boldsymbol{g} + \boldsymbol{r}_2 \times m_2\boldsymbol{g} + \cdots$$
$$= (m_1\boldsymbol{r}_1 + m_2\boldsymbol{r}_2 + \cdots) \times \boldsymbol{g}. \qquad (10\text{--}5)$$

When we multiply and divide this by the total mass

$$M = m_1 + m_2 + \cdots,$$

we obtain

$$\boldsymbol{\Gamma} = \frac{m_1\boldsymbol{r}_1 + m_2\boldsymbol{r}_2 + \cdots}{M} \times M\boldsymbol{g}. \qquad (10\text{--}6)$$

The fraction is just the position vector $\boldsymbol{R}$ of the center of mass, with components X and Y as given by Eq. (8–27), and $M\boldsymbol{g}$ is the total weight $\boldsymbol{W} = m\boldsymbol{g}$ of the body. Thus

$$\boldsymbol{\Gamma} = \boldsymbol{R} \times M\boldsymbol{g} = \boldsymbol{R} \times \boldsymbol{W}, \qquad (10\text{--}7)$$

and the total torque is the same as though the total weight were acting at the position $\boldsymbol{R}$ of the center of mass (center of gravity).

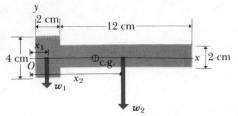

10-2 Machine part for Example 10-1.

Determining the center of gravity of a composite machine part

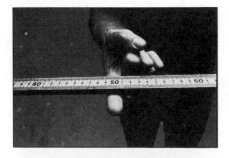

10-3 Locating the center of gravity of a body by balancing. The meter stick balances at its midpoint, but the hammer has a heavy head at one end. (Nancy Rodgers, Exploratorium.)

If the center of gravity of each of a number of bodies has been determined, the coordinates of the center of gravity of the combination can be computed from Eqs. (10–3) and (10–4), letting w_1, w_2, and so on, be the weights of the bodies and (x_1, y_1), (x_2, y_2), and so on, be the coordinates of the center of gravity of each.

Symmetry considerations are often useful in finding the position of the center of gravity. The center of gravity of a homogeneous sphere, cube, circular disk, or rectangular plate is at its geometric center. The center of gravity of a right circular cylinder or cone is on the axis of symmetry, and so on.

EXAMPLE 10–1 Locate the center of gravity of the machine part in Fig. 10–2. The part consists of a disk 4 cm in diameter and 2 cm long, attached to a rod 2 cm in diameter and 12 cm long. Both pieces have the same uniform density.

SOLUTION By symmetry, the center of gravity lies along the axis and the center of gravity of each part is midway between its ends. The volume of the disk is 8π cm^3 and that of the rod is 12π cm^3. Since the weights of the two parts are proportional to their volumes,

$$\frac{w(\text{disk})}{w(\text{rod})} = \frac{w_1}{w_2} = \frac{8\pi}{12\pi} = \frac{2}{3}.$$

Take the origin O at the left face of the disk, on the axis. Then

$$x_1 = 1 \text{ cm}, \qquad x_2 = 8 \text{ cm},$$

and

$$X = \frac{w_1(1 \text{ cm}) + \frac{3}{2}w_1(8 \text{ cm})}{w_1 + \frac{3}{2}w_1} = 5.2 \text{ cm}.$$

The center of gravity is on the axis, 5.2 cm to the right of O.

The center of gravity (center of mass) has several other important properties. First, when a body is suspended in a gravitational field from a single point, the center of gravity always hangs directly below the point of suspension; otherwise the body could not be in rotational equilibrium. This fact can be used for an experimental determination of the location of the center of gravity of an irregular body. Figure 10–3 shows two examples of locating the center of gravity of a body by finding its balance point.

Second, a force applied to a body at its center of gravity does not tend to cause the body to rotate, although a force applied at any other point tends to cause both rotational and translational motion. Third, as discussed in Section 8–7, the total momentum of a body equals the product of its total mass and its center-of-mass velocity. The acceleration of the center of mass is the same as that of a point mass equal to the total mass of the body, acted on by the resultant external force.

10-3 EXAMPLES

As we have seen, the principles in rigid-body equilibrium are few and simple; the vector sum of the forces on the body must be zero, and the sum of torques about any point must be zero. The hard part is *applying* these principles to

A view of the Golden Gate Bridge (in San Francisco Bay) during its construction. Every element of the bridge had to be in equilibrium during each step in the construction. (Golden Gate Bridge, Highway and Transportation District.)

specific problems. As always, careful and systematic problem-solving methodology always pays off! The following strategy is very similar to that outlined in Section 5–3 for equilibrium of a particle.

PROBLEM-SOLVING STRATEGY *Equilibrium of a rigid body*

1. Make a sketch of the physical situation, including dimensions.

2. Choose some appropriate body as the body in equilibrium, and draw a free-body diagram showing the forces acting *on* this body and no others. *Do not* include forces exerted *by* this body on other bodies.

3. Draw coordinate axes and specify a positive sense of rotation for torques. Represent forces in terms of their components with respect to the axes you have chosen.

4. Write equations expressing the equilibrium conditions. Remember that $\Sigma F_x = 0$, $\Sigma F_y = 0$, and $\Sigma \Gamma = 0$ are always separate equations; *never* add x- and y-components in a single equation.

5. In choosing an axis to compute torques, note that if the line of action of a force goes through a particular axis, the torque of the force with respect to that axis is zero. Thus unknown forces or components can sometimes be eliminated from the torque equation by a clever choice of axis location. Also remember that when a force is represented in terms of its components, the torque of that force can be obtained by computing separately the torques of the components, each with its appropriate moment arm and sign, and adding the results. This is often easier than determining the moment arm of the original force.

6. Depending on the number of unknowns, you may need to compute torques with respect to two or more axes to obtain as many equations as unknowns. Often there are several equally good sets of force-and-torque equations in a particular problem. When you have as many equations as unknowns, the equations can be solved simultaneously.

EXAMPLE 10–2 A hanging flower basket B having weight w_2 is hung out over the edge of a balcony railing on a horizontal beam that rests on the balcony railing and is counterbalanced at its other end by a body of weight w_1, as shown in Fig. 10–4a. Find the weight w_1 needed to balance the basket, and the total upward force P exerted on the beam at point O. (This illustrates the principle of *cantilever* construction, by which decks, stairways, and the like can be built over empty space, supported entirely from one side.)

Decorating a balcony: an example of torque calculations and rigid-body equilibrium

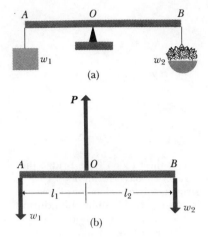

(a)

(b)

10–4 A flower basket cantilevered out over a balcony railing.

SOLUTION Figure 10–4b is a free-body diagram for the beam. The first condition for equilibrium gives

$$\Sigma F_y = P - w_1 - w_2 = 0.$$

If we take torques about an axis through O, perpendicular to the diagram, the second condition for equilibrium gives

$$\Sigma \Gamma_0 = w_1 l_1 - w_2 l_2 = 0.$$

Let $l_1 = 1.2$ m, $l_2 = 1.6$ m, and $w_2 = 15$ N. Then, from the equations above,

$$w_1 = 20 \text{ N}, \qquad P = 35 \text{ N}.$$

To illustrate the point that the total torque about *any* axis is zero, let us compute torques about an axis through point A:

$$\Sigma \Gamma_A = P l_1 - w_2 (l_1 + l_2)$$
$$= (35 \text{ N})(1.2 \text{ m}) - (15 \text{ N})(2.8 \text{ m}) = 0$$

The axis about which torques are computed need not lie on the rod. To verify this, calculate the total torque about a point 0.4 m to the left of A and 0.4 m above it.

A ladder leaning against a wall: using equilibrium conditions to find forces on a body in equilibrium

EXAMPLE 10–3 The ladder in Fig. 10–5a is 10 m long and has a weight of 400 N that we can consider as concentrated at its center. It leans in equilibrium against a frictionless vertical wall and makes an angle of 53.1° with the horizontal (conveniently forming a 3–4–5 right triangle). Find the magnitudes and directions of the forces F_1 and F_2.

SOLUTION Figure 10–5b shows a free-body diagram. If the wall is frictionless, F_1 is horizontal. The direction of F_2 is unknown; except in special cases, its direction does *not* lie along the ladder. Instead of considering its magnitude and direction as unknowns, it is simpler to resolve the force F_2 into (unknown) x- and y-components and solve for these. The magnitude and direction of F_2 may then be computed. The first condition of equilibrium therefore provides the equations

$$\left.\begin{array}{l} \Sigma F_x = F_{2x} - F_1 = 0, \\[2mm] \Sigma F_y = F_{2y} - 400 \text{ N} = 0. \end{array}\right\} \quad \text{(First condition)}$$

10–5 (a) A ladder in equilibrium leaning against a frictionless wall. (b) Free-body diagram for the ladder. The force at the base is represented in terms of its x- and y-components.

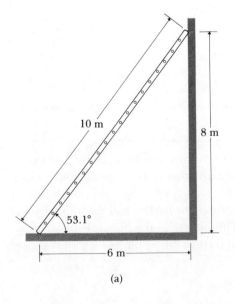

(a)

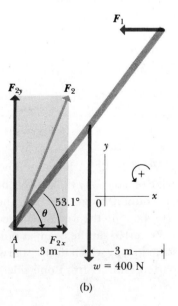

(b)

In writing the second condition, we may compute torques about an axis through any point. The resulting equation is simplest if we choose a point through which the lines of action of two or more forces pass, since these forces then do not appear in the equation. Let us therefore take torques about an axis through point A.

$$\Sigma\Gamma_A = F_1(8 \text{ m}) - (400 \text{ N})(3 \text{ m}) = 0. \qquad \text{(Second condition)}$$

From the second equation, $F_{2y} = 400$ N, and from the third,

$$F_1 = \frac{1200 \text{ N·m}}{8 \text{ m}} = 150 \text{ N.}$$

Then from the first equation,

$$F_{2x} = 150 \text{ N.}$$

Hence

$$F_2 = \sqrt{(400 \text{ N})^2 + (150 \text{ N})^2} = 427 \text{ N,}$$

$$\theta = \arctan\frac{400 \text{ N}}{150 \text{ N}} = 69.4°.$$

EXAMPLE 10–4 Figure 10–6a shows a human arm lifting a dumbbell, and Fig. 10–6b is a free-body diagram for the forearm, showing the forces involved. The forearm is in equilibrium under the action of the weight w of the dumbbell, the tension T in the tendon connected to the biceps muscle, and the forces exerted on the forearm by the upper arm, at the elbow joint. For clarity, the tendon force has been displaced away from the elbow farther than its actual position. The weight w and the angle θ are given; we want to find the tendon tension and the two components of force at the elbow (three unknown scalar quantities in all).

The human arm: forces in joints and tendons under equilibrium conditions

SOLUTION First we represent the tendon force in terms of its components T_x and T_y, using the given angle θ and the unknown magnitude T:

$$T_x = T \cos\theta, \qquad T_y = T \sin\theta.$$

We also represent the force at the elbow in terms of its components E_x and E_y. Next we note that if we take torques about the elbow joint, the resulting torque equation does not contain E_x, E_y, or T_x, because the lines of action of all these forces pass through this point. The torque equation is then simply

$$lw - dT_y = 0.$$

From this we immediately find

$$T_y = \frac{lw}{d} \qquad \text{and} \qquad T = \frac{lw}{d \sin\theta}.$$

To find E_x and E_y, we could now use the first conditions for equilibrium, $\Sigma F_x = 0$ and $\Sigma F_y = 0$. Instead, for added practice in using torques, we take torques about the point A where the tendon is attached:

$$(l - d)w + dE_y = 0 \qquad \text{and} \qquad E_y = -\frac{(l - d)w}{d}.$$

The negative sign of this result shows that our initial guess for the direction of E_y was wrong; it is actually vertically *downward*.

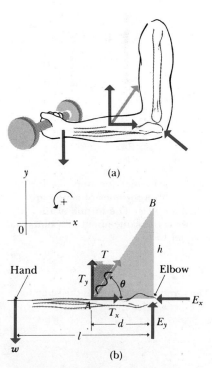

10–6 Force diagrams for human forearm.

Finally, we take torques about point B in the figure:

$$lw - hE_x = 0 \quad \text{and} \quad E_x = \frac{lw}{h}.$$

We note that each stage of this calculation is simplified by choosing the point for calculating torques so as to eliminate one or more of the unknown quantities. In the last step, the force T has no torque about point B; thus when the torques of T_x and T_y are computed separately, they must add to zero. We invite you to verify this statement in detail.

As a specific example, suppose $w = 50$ N, $d = 0.10$ m, $l = 0.50$ m, and $\theta = 80°$. Then $\tan \theta = h/d$, and we find

$$h = d \tan \theta = (0.10 \text{ m})(5.67) = 0.567 \text{ m}.$$

From the previous general results, we find

$$T = \frac{(0.5 \text{ m})(50 \text{ N})}{(0.1 \text{ m})(0.985)} = 254 \text{ N}.$$

$$E_y = -\frac{(0.5 \text{ m} - 0.1 \text{ m})(50 \text{ N})}{0.1 \text{ m}} = -200 \text{ N},$$

$$E_x = \frac{(0.5 \text{ m})(50 \text{ N})}{0.567 \text{ m}} = 44.1 \text{ N}.$$

The magnitude of the force at the elbow is

$$E = \sqrt{E_x{}^2 + E_y{}^2} = 205 \text{ N}.$$

As mentioned above, we have not explicitly used the first condition for equilibrium, that the vector sum of the forces be zero. As a check, you should verify that this condition is satisfied by the results.

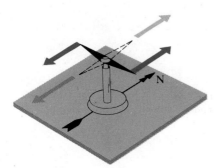

10–7 Forces on the poles of a compass needle.

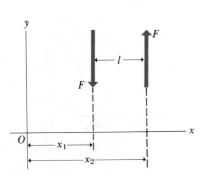

y

F

F

l

O

x_1

x_2

x

10–8 Two equal and opposite forces having different lines of action are called a *couple*. The torque of the couple is the same about all points and is equal to Fl.

The torque of a couple is the same for every axis.

10–4 COUPLES

Sometimes the forces on a body include two forces of equal magnitude and opposite direction, with lines of action that are parallel but do not coincide. Such a pair of forces is called a **couple**. A common example is a compass needle in the earth's magnetic field, as shown in Fig. 10–7. The north and south poles of the needle are acted on by equal forces, one toward the north and the other toward the south. Except when the needle points in the N–S direction, the two forces do not have the same line of action.

Figure 10–8 shows a couple consisting of two forces, each of magnitude F, separated by a perpendicular distance l. The resultant R of the forces is

$$R = F - F = 0.$$

The fact that the resultant force is zero means that a couple has no effect in producing translation (as a whole) of the body on which it acts. The only effect of a couple is to produce rotation.

The resultant torque of the couple in Fig. 10–8 about an arbitrary point O is

$$\begin{aligned} \Sigma\Gamma_0 &= x_2 F - x_1 F \\ &= (x_1 + l)F - x_1 F \\ &= lF. \end{aligned} \qquad (10\text{–}8)$$

The distances x_1 and x_2 do not appear in the result, and we conclude that *the torque of a couple is the same about all points in the plane of the forces; it is equal to the product of the magnitude of either force and the perpendicular distance between their lines of action.*

A plumber tightening a pipe fitting uses the principle of the couple. He applies a force to the end of the wrench handle and holds the pipe close to the fitting with the free hand, exerting an opposite force. Thus there is no net force sideways on the pipe, which would put a strain on the opposite end, but only a torque tending to screw the two parts together.

How plumbers use couples to avoid breaking pipes

SUMMARY

For a rigid body to be in equilibrium, two conditions must be satisfied: The vector sum of forces must be zero, and the sum of torques about any axis must be zero. The force condition is usually expressed most conveniently in terms of components; the equations expressing the equilibrium conditions are then

$$\Sigma F_x = 0, \qquad \Sigma F_y = 0, \qquad (10\text{--}1)$$

and

$$\Sigma \Gamma_A = 0, \qquad (10\text{--}2)$$

where A represents an arbitrarily chosen axis.

The torque due to the weight of a body can be obtained by assuming the entire weight to be concentrated at the center of mass, also called in this context the center of gravity. The coordinates of the center of gravity are given by

$$X = \frac{w_1 x_1 + w_2 x_2 + \cdots}{W} = \frac{\Sigma wx}{W} = \frac{\Sigma wx}{\Sigma w} \qquad (10\text{--}3)$$

and

$$Y = \frac{w_1 y_1 + w_2 y_2 + \cdots}{w_1 + w_2 + \cdots} = \frac{\Sigma wy}{\Sigma w} = \frac{\Sigma wy}{W}. \qquad (10\text{--}4)$$

The torque of a force can be computed by finding the torque of each component, using its appropriate moment arm and sign, and adding these.

A couple consists of two forces of equal magnitude F and opposite direction, acting on a body. If the lines of action of the forces are separated by a distance l, the total torque due to the couple is given (in magnitude) by

$$\Gamma = lF \qquad (10\text{--}8)$$

and is the same for all axes.

KEY TERMS

first condition for equilibrium
second condition for equilibrium
center of mass
center of gravity
couple

QUESTIONS

10–1 Mechanics sometimes extend the length of a wrench handle by slipping a section of pipe over the handle. Why is this a dangerous procedure?

10–2 Manuals for car-engine repair always specify the *torque* (moment) to be applied when tightening the cylinder-head bolts. Why is torque specified, rather than the *force* to be applied to the wrench handle?

10–3 Car tires are sometimes "balanced" on a machine that pivots the tire and wheel about the center, by laying weights around the wheel rim until it does not tip from the horizontal plane. Discuss this procedure in terms of the center of gravity.

10–4 Is it possible for a solid body to have no matter at its center of gravity? Consider, for example, a disk with a hole in the center, an empty bottle, and a hollow cylinder.

10–5 A man climbs a tall, old stepladder that has a tendency to sway. He feels much more unstable when standing near the top than when near the bottom. Why?

10–6 What determines whether a body in equilibrium is stable or unstable? As an example, discuss a cone resting on its base on a horizontal floor; balanced on its point; and lying on its side.

10–7 When a heavy load is placed in the back end of a truck, behind the rear wheels, the effect is to *raise* the front end of the truck. Why?

10–8 When a tall, heavy object, such as a refrigerator, is pushed across a rough floor, what determines whether it slides or tips?

10–9 Why is a tapered water glass with a narrow base easier to tip over than one with straight sides? Does it matter whether the glass is full or empty?

10–10 People sometimes prop open a door by wedging some object in the space between the hinged side of the door and the frame. Explain why this often results in the hinge screws being ripped out.

10–11 Discuss the action of a claw hammer in pulling out nails, in terms of torques.

10–12 In trying to lift an object that is much too heavy to be lifted directly, one often speaks of the importance of "leverage." Discuss this term in relation to moments, and explain how it is related to the concept of leverage in trading stock options and other securities.

10–13 Why is it easier to hold a 10-kg body in your hand at your side than to hold it with your arm extended horizontally?

10–14 In pioneer days when a Conestoga wagon was stuck in the mud, a man would grasp a wheel spoke and try to turn the wheel, rather than simply pushing the wagon. Why?

10–15 Does the center of gravity of a solid body always lie within the body? If not, give a counterexample.

EXERCISES

Section 10–2 Center of Gravity

10–1 A ball of mass 1 kg is attached by a light rod 0.4 m in length to a second ball of mass 3 kg. Where is the center of gravity of this system?

10–2 In Exercise 10–1, suppose the rod is uniform and has mass 2 kg. Where is the system's center of gravity?

10–3 Three small objects of equal mass are located in the xy-plane, at points having coordinates (0.1 m, 0), (0, 0.1 m), (0.1 m, 0.1 m). Determine the coordinates of the center of gravity.

Section 10–3 Examples

10–4 Two men are carrying a uniform ladder that is 20 ft long and weighs 200 lb. If one man can lift a maximum of 80 lb and lifts at one end, at what point should the other man lift?

10–5 A heavy electric motor is to be carried by two men by placing it on a light board 2.0 m long. To lift the board with the motor on it, one man must lift at one end with a force of 600 N and the other at the opposite end with a force of 400 N. What is the weight of the motor, and where is it located?

10–6 In Exercise 10–5, suppose the board is not light but weighs 200 N. The two men each exert the same force as before. What is the weight of the motor in this case, and where is it located?

10–7 A uniform plank 15 m long, weighing 400 N, rests symmetrically on two supports 8 m apart, as shown in Fig. 10–9. A boy weighing 640 N starts at point A and walks toward the right.

a) Construct in the same diagram two graphs showing the upward forces F_A and F_B exerted on the plank at points A and B, as functions of the coordinate x of the boy. Let 1 cm = 100 N vertically, and 1 cm = 1.0 m horizontally.

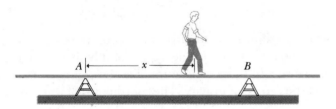

FIGURE 10–9

b) Find from your diagram how far beyond point B the boy can walk before the plank tips.

c) How far from the right end of the plank should support B be placed so that the boy can walk just to the end of the plank without causing it to tip?

10–8 The strut in Fig. 10–10 weighs 200 N, and its center of gravity is at its center. Find

a) the tension in the cable;

b) the horizontal and vertical components of the force exerted on the strut at the wall.

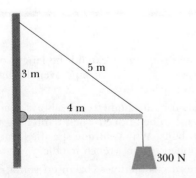

FIGURE 10–10

10–9 Find the tension T in each cable, and the magnitude and direction of the force exerted on the strut by the pivot, in each of the arrangements in Fig. 10–11. Let the weight of the suspended object in each case be w. Neglect the weight of the strut.

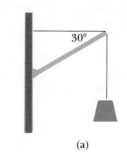

(a)

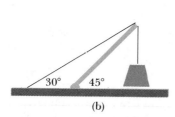

(b)   **FIGURE 10–11**

10–10 The boom in Fig. 10–12 is uniform and weighs 2500 N.

a) Find the tension in the guy wire, and the horizontal and vertical components of the force exerted on the boom at its lower end.

b) Does the line of action of the force exerted on the boom at its lower end lie along the boom?

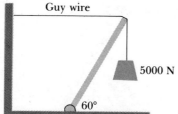

FIGURE 10–12

10–11 A door 1.0 m wide and 2.5 m high weighs 200 N and is supported by two hinges, one 0.5 m from the top and the other 0.5 m from the bottom. Each hinge supports half the total weight of the door. Assuming the door's center of gravity is at its center, find the components of force exerted on the door by each hinge.

Section 10–4 Couples

10–12 Two equal parallel forces of magnitude $F_1 = F_2 = 8$ N are applied to a rod as shown in Fig. 10–13.

a) What should be the distance l between the forces if they are to provide a net torque of 12 N·m about the left-hand end of the rod?

b) Is this torque clockwise or counterclockwise?

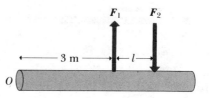

FIGURE 10–13

10–13 Two forces of magnitude $F_1 = F_2 = 3$ N are applied to a rod as shown in Fig. 10–14.

a) Calculate the net torque about point O due to these two forces by calculating the torque due to each separate force.

b) Calculate the net torque about point P due to these two forces by calculating the torque due to each separate force.

c) Compare your results to Eq. (10–8).

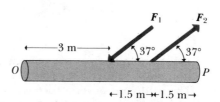

FIGURE 10–14

PROBLEMS

10–14 A certain automobile has a wheelbase (distance between front and rear axles) of 3.0 m. If 60% of the weight rests on the front wheels, how far behind the front wheels is the center of gravity?

10–15 A horizontal boom 4 m long is hinged to a vertical wall at one end, and a 500-N object hangs from its other end. The boom is supported by a guy wire from its outer end to a point on the wall directly above the boom. The weight of the boom can be neglected.

a) If the tension in this wire is not to exceed 1000 N, what is the minimum height above the boom at which it may be fastened to the wall?

b) By how many newtons would the tension be increased if the wire were fastened 0.5 m below this point, the boom remaining horizontal?

10–16 A station wagon weighing 1.96×10^4 N has a wheelbase of 3.0 m. Ordinarily 10,780 N rests on the front wheels and 8820 N on the rear wheels. A box weighing 1960 N is now placed on the tailgate, 1.0 m behind the rear axle. How much total weight now rests on the front wheels? On the rear wheels?

10–17 End A of the bar AB in Fig. 10–15 rests on a frictionless horizontal surface, while end B is hinged. A horizontal force P of 60 N is exerted on end A. Neglect the weight of the bar. What are the horizontal and vertical components of the force exerted by the bar on the hinge at B?

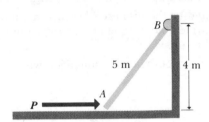

FIGURE 10–15

10–18 A single additional force is to be applied to the bar in Fig. 10–16 to maintain it in equilibrium in the position shown. The weight of the bar can be neglected.

a) What are the horizontal and vertical components of the required force?

b) What is the angle the force must make with the bar?

c) What is the magnitude of the required force?

d) Where should the force be applied?

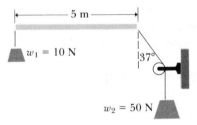

FIGURE 10–16

10–19 A circular disk 0.5 m in diameter, pivoted about a horizontal axis through its center, has a cord wrapped around its rim. The cord passes over a frictionless pulley P and is attached to an object of weight 240 N. A uniform rod 2 m long is fastened to the disk, with one end at the center of the disk. The apparatus is in equilibrium, with the rod horizontal, as shown in Fig. 10–17.

a) What is the weight of the rod?

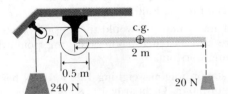

FIGURE 10–17

b) What is the new equilibrium direction of the rod when a second object weighing 20 N is suspended from the other end of the rod, as shown by the broken line? That is, what is then the angle the rod makes with the horizontal?

10–20 The objects in Fig. 10–18 are constructed of uniform wire bent into the shapes shown. Find the position of the center of gravity of each.

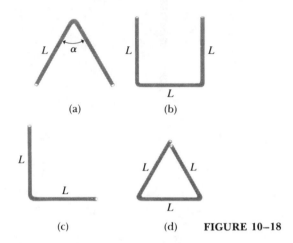

(a)   (b)

(c)   (d)   **FIGURE 10–18**

10–21 A playground equipment company is designing a new swing with the dimensions shown in Fig. 10–19. For safety the company wants to allow for someone as heavy as 1200 N (about 270 lb) sitting on each of the two seats. Adding an additional safety factor of 30%, they consider the downward force at the top of each end of the swing frame to be 1560 N. The two 3.0-m side pieces each weigh 120 N and are joined at the top by a frictionless hinge.

a) What vertical force is exerted by the ground on each side piece?

b) What is the tension in the horizontal rod, whose weight is negligible?

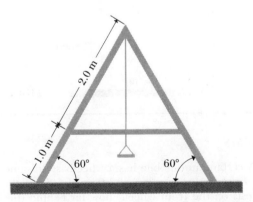

FIGURE 10–19

10–22 A uniform ladder 10 m long rests against a frictionless vertical wall with its lower end 6 m from the wall. The ladder weighs 400 N. The coefficient of static friction between the foot of the ladder and the ground is 0.40. A man weighing 800 N climbs slowly up the ladder.

a) What is the maximum frictional force that the ground can exert on the ladder at its lower end?

b) What is the actual frictional force when the man has climbed 3 m along the ladder?

c) How far along the ladder can the man climb before the ladder starts to slip?

10–23 A gate 4 m long and 2 m high weighs 400 N. Its center of gravity is at its center, and it is hinged at A and B. To relieve the strain on the top hinge, a wire CD is connected as shown in Fig. 10–20. The tension in CD is increased until the horizontal force at hinge A is zero.

a) What is the tension in the wire CD?

b) What is the magnitude of the horizontal component of the force at hinge B?

c) What is the combined vertical force exerted by hinges A and B?

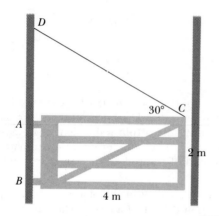

FIGURE 10–20

10–24 A roller whose diameter is 1.0 m weighs 360 N. What is the minimum horizontal force necessary to pull the roller over a brick 0.1 m high when the force is applied

a) at the center?

b) at the top of the roller?

10–25 A uniform rectangular block, 0.5 m high and 0.25 m wide, rests on a plank AB, as in Fig. 10–21. The coefficient of static friction between block and plank is 0.40.

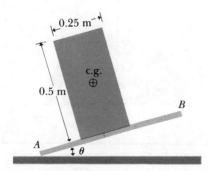

FIGURE 10–21

a) In a diagram drawn to scale, show the line of action of the resultant normal force exerted on the block by the plank when the angle $\theta = 15°$.

b) If end B of the plank is slowly raised, will the block start to slide down the plank before it tips over? Find the angle θ at which it starts to slide, or at which it tips.

c) What would be the answer to part (b) if the coefficient of static friction were 0.60?

10–26 A rectangular block 0.25 m wide and 0.5 m high is dragged to the right along a level surface at constant speed by a horizontal force P, as shown in Fig. 10–22. The coefficient of sliding friction is 0.40; the block weighs 25 N; and its center of gravity is at its center.

a) Find the magnitude of the force P.

b) Find the position of the line of action of the normal force n exerted on the block by the surface, if the height $h = 0.125$ m.

c) Find the value of h at which the block just starts to tip.

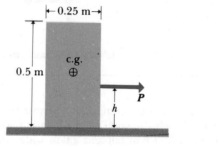

FIGURE 10–22

10–27 A garage door is mounted on an overhead rail, as in Fig. 10–23. The wheels at A and B have rusted so that they do not roll, but rather slide along the track. The coefficient of sliding friction is 0.5. The distance between the wheels is 2 m, and each is 0.5 m in from the vertical sides of the door. The door is symmetrical and weighs 800 N. It is pushed to the left at constant velocity by a horizontal force P.

a) If the distance h is 1.5 m, what is the vertical component of the force exerted on each wheel by the track?

b) Find the maximum value h can have without causing one wheel to leave the track.

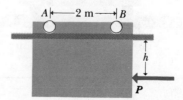

FIGURE 10–23

CHALLENGE PROBLEMS

10–28 One end of a post weighing 500 N rests on a rough horizontal surface with $\mu_s = 0.3$. The upper end is held by a rope fastened to the surface and making an angle of 37° with the post, as in Fig. 10–24. A horizontal force F is exerted on the post as shown.

a) If the force F is applied at the midpoint of the post, what is the largest value it can have without causing the post to slip?

b) How large can the force be, without causing the post to slip, if its point of application is $\frac{7}{10}$ of the way from the ground to the top of the post?

c) Show that if the point of application of the force is too high, the post cannot be made to slip, no matter how great the force. Find the critical height for the point of application.

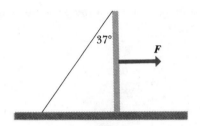

FIGURE 10–24

10–29 A bookcase weighing 1300 N rests on a horizontal surface for which the coefficient of static friction is $\mu_s = 0.3$. The bookcase is 1.8 m tall and 2.0 m wide, and its center of gravity is at its geometrical center. The bookcase rests on four short legs that are each 0.1 m in from the edge of the bookcase. A man pulls on a rope attached to an upper corner of the bookcase with a force P that makes an angle θ with the bookcase, as shown in Fig. 10–25.

a) If $\theta = 90°$, so P is horizontal, show that as P is increased from zero the bookcase will start to slide before it tips, and calculate the magnitude of P that will start the bookcase sliding.

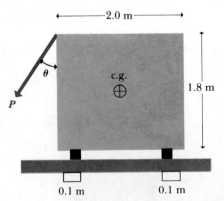

FIGURE 10–25

b) If $\theta = 0°$, so P is vertical, show that the bookcase will tip over rather than slide, and calculate the magnitude of P that will cause the bookcase to start to tip.

c) Calculate as a function of θ the magnitude of P that will cause the bookcase to start to slide and the magnitude that will cause it to start to tip. What is the smallest value θ can have for the bookcase to start to slide before it starts to tip?

10–30 In Section 10–2 the calculation of center of gravity was done by considering objects composed of a *finite* number of point masses or that by symmetry could be represented by a finite number of point masses. For a solid object whose mass distribution does not allow for a simple determination of the center of gravity by symmetry, the sums of Eqs. (10–3) and (10–4) must be generalized to integrals:

$$X = \frac{1}{W} \int xg\, dm$$

and

$$Y = \frac{1}{W} \int yg\, dm,$$

where x and y are the coordinates of the small piece of the object that has mass dm. The integration is over the whole of the object. Consider a thin rod of length L, mass M, and cross-sectional area A. Let the origin of coordinates be at the left-hand end of the rod and the positive x-axis lie along the rod.

a) If the density $\rho = m/V$ of the object is *constant,* perform the integration described above to show that the x-coordinate of the center of gravity of the rod is at its geometrical center.

b) If the density of the object varies linearly with x, $\rho = \alpha x$ where α is a constant, calculate the x-coordinate of the rod's center of gravity.

10–31 Use the methods of Challenge Problem 10–30 to calculate the x- and y-coordinates of the center of gravity of a semicircular metal plate of uniform density ρ and thickness t. Let the radius of the plate be a. The mass of the plate is thus $M = \frac{1}{2}\rho\pi a^2 t$. Use the coordinate system indicated in Fig. 10–26.

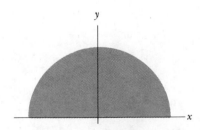

FIGURE 10–26

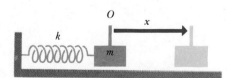

11

PERIODIC MOTION

IN THIS CHAPTER WE CONSIDER A CLASS OF PROBLEMS IN WHICH A BODY undergoes motion that is repetitive or cyclic. In these problems the body has a stable equilibrium position. When it is displaced from equilibrium and released, it moves back toward equilibrium, overshoots, and goes into a back-and-forth motion on both sides of the equilibrium position. A familiar example is the pendulum, a mass suspended from a string. In the equilibrium position it hangs straight down; when we displace and release it, it does not simply return to the straight-down position but instead swings back and forth in a regular, repetitive manner.

Such a motion is called a **periodic motion,** or an **oscillation.** Many other familiar examples come to mind: the balance wheel of a watch, the vibrating strings in a musical instrument, the action of a door buzzer. The molecules in a solid material vibrate around their equilibrium positions. In many types of *wave motion* in a material, the motions of particles of the material are periodic. Thus a study of periodic motion gives us an important foundation for further study in many different areas of physics.

11–1 BASIC CONCEPTS

One of the simplest systems that can undergo periodic motion is a mass attached to a spring, as shown in Fig. 11–1. For example, the body could be a glider on a linear air track, as described in Section 5–2. The body is attached to one end of a spring, and the other end of the spring is held stationary.

Let x be the displacement of the body from its equilibrium position. When $x = 0$, the spring is neither stretched nor compressed. When the body is displaced to the right, x is positive and the spring stretches. The x-component of force F_x that the spring exerts on the body is toward the left (the negative x-direction), toward the equilibrium position, and F_x is negative. When the body is displaced to the left, x is negative and the spring is compressed. The force on the body is toward the right (the positive x-direction), again toward

Restoring force: a force that always pulls a body toward its equilibrium position

11–1 Model for periodic motion. If the spring obeys Hooke's law, the motion is simple harmonic motion.

263

equilibrium, and F_x is positive. Thus the sign of F_x is always opposite to the sign of x itself. We call such a force a *restoring force*.

In Section 7–3 we noted that for some springs the force is *directly proportional* to the deformation, at least for small deformations. This proportionality is called Hooke's law. Thus in Fig. 11–1, if the spring obeys Hooke's law, we may represent the relationship of F_x to x as

$$F_x = -kx, \tag{11-1}$$

where k is the **force constant** for the spring. This relation is valid for both positive and negative x; in both cases F_x and x have opposite signs.

Suppose we displace the body to the right a distance A and release it, with no initial velocity. The spring exerts a force toward the equilibrium position, and the body accelerates in this direction. The acceleration is not constant, because the force decreases as the body approaches the equilibrium position.

When the body reaches $x = 0$, the force and acceleration have decreased to zero, but the velocity that the body has acquired causes it to "overshoot" the equilibrium position and continue to move to the left. The force then reverses direction, and the body's speed starts to decrease. The body comes to rest at some point to the left of O and starts back toward the equilibrium position.

In the next section we will show that the motion is confined to a range $x = \pm A$ on both sides of the equilibrium position, and that each complete back-and-forth trip takes the same amount of time. If there were no loss of mechanical energy due to friction, the motion would continue forever. This specific motion, under the influence of a restoring force proportional to displacement and without any friction, is called **simple harmonic motion,** abbreviated **SHM.**

Here are some terms we will use frequently in this chapter:

A complete vibration, or **cycle,** is one round trip, say from A to $-A$ and back to A, or from O to A to O to $-A$ to O. Note that motion from one side to the other (say A to $-A$) is a half-cycle, not a full cycle.

The **period** of the motion, denoted by τ, is the time required for one cycle.

The **frequency,** f, is the number of cycles per unit time. The frequency is the reciprocal of the period: $f = 1/\tau$. The SI unit of frequency, one cycle per second, is called 1 *hertz* (1 Hz = 1 s^{-1}).

The **amplitude,** A, is the maximum displacement from equilibrium, that is, the maximum value of $|x|$. The total overall range of the motion is therefore $2A$.

Simple harmonic motion is the simplest of all periodic motions to analyze. In more complex examples the force may depend on displacement in a more complicated way; but only when it is *directly proportional* to displacement do we use the expression "simple harmonic motion." However, many more complex periodic motions are *approximately* simple harmonic, if the displacements are small enough. Thus simple harmonic motion is a *model* that serves for an approximate representation of many periodic motions, and as such it is worth analyzing in detail.

11–2 EQUATIONS OF SIMPLE HARMONIC MOTION

In this section we will obtain expressions for position, velocity, and acceleration as functions of time, for a body undergoing simple harmonic motion. Caution: Do not try to use the equations for motion with *constant* acceleration from Chapter 2! In simple harmonic motion, the acceleration is *not* constant.

Figure 11–2 shows the vibrating body of Fig. 11–1 at a distance x from its equilibrium position O. The mass is m; the body moves along a straight line (the x-axis); and the force component F_x along that line is the elastic restoring force: $F_x = -kx$.

From Newton's second law,

$$F_x = -kx = ma_x,$$

or

$$a_x = -\frac{k}{m}x. \qquad (11-2)$$

Thus *at each instant the acceleration is proportional to the negative of the displacement.* When x has its maximum positive value A, the acceleration has its maximum negative value $-kA/m$; at the instants when the body passes the equilibrium position, its acceleration is zero. Of course, its velocity is *not* zero at these times.

We can use conservation of energy to derive expressions for the body's position x and velocity v as functions of time in simple harmonic motion. Because the elastic restoring force is a *conservative* force, we can represent the work done by this force in terms of a potential energy $U = \frac{1}{2}kx^2$, just as in Section 7–6. The kinetic energy is $K = \frac{1}{2}mv^2$, and according to the principle of conservation of energy, the total energy $E = K + U$ is constant. That is,

$$E = \tfrac{1}{2}mv^2 + \tfrac{1}{2}kx^2 = \text{constant}. \qquad (11-3)$$

The total energy E is also directly related to the amplitude A of the motion. When the body reaches its maximum displacement $\pm A$, it stops and turns back toward equilibrium. At this instant $v = 0$, so the body has no kinetic energy. The (constant) total energy is equal to the potential energy at this point, which is $\frac{1}{2}kA^2 = E$. Thus

$$\tfrac{1}{2}mv^2 + \tfrac{1}{2}kx^2 = \tfrac{1}{2}kA^2,$$

or

$$v = \pm\sqrt{\frac{k}{m}}\,\sqrt{A^2 - x^2}. \qquad (11-4)$$

We can use this relation to obtain the velocity (apart from a sign) for any given position.

The significance of Eq. (11–3) is shown by the graph in Fig. 11–3, where energy is plotted vertically and the coordinate x horizontally. The curve represents the potential energy, $U = \frac{1}{2}kx^2$; this curve is a parabola. The horizontal line at height E represents the constant total energy of the body. We see that the body's motion is restricted to values of x lying between the points where the horizontal line intersects the parabola. If x were outside this range, the potential energy would exceed the total energy, which is impossible.

If we draw a vertical line at any value of x within the permitted range, the length of the segment between the x-axis and the parabola represents the

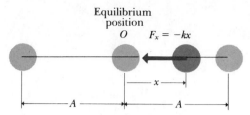

11–2 Coordinate system and free-body diagram for a body in simple harmonic motion.

In simple harmonic motion, the acceleration is proportional to the negative of the displacement.

In simple harmonic motion, energy is conserved; this gives a simple relation between position and speed of the vibrating body.

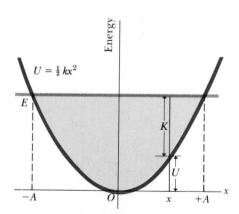

11–3 Relation between total energy E, potential energy U, and kinetic energy K, for a body oscillating with SHM.

The maximum speed and maximum displacement are related to the total energy in a simple way.

potential energy U at that value of x. The length of the segment between the parabola and the horizontal line at height E represents the corresponding kinetic energy K. At the endpoints the energy is all potential, and at the midpoint it is all kinetic. The speed has its maximum value v_{max} at the midpoint:

$$\frac{1}{2}mv_{max}{}^2 = E, \qquad v_{max} = \sqrt{\frac{2E}{m}}. \tag{11-5}$$

The fact that both the maximum potential energy and the maximum kinetic energy are equal to the total energy E (and thus to each other) may be used to relate the maximum velocity to the amplitude:

$$E = \tfrac{1}{2}kA^2 = \tfrac{1}{2}mv_{max}{}^2, \qquad v_{max} = \sqrt{\frac{k}{m}}A. \tag{11-6}$$

The position-velocity relation given by Eq. (11–4) is very useful, but it does not tell us where the body is at any given *time*. To have a complete description of the motion we need to know the position, the velocity, and the acceleration as functions of time. We can obtain these relations either by integrating the energy equation, Eq. (11–4), or by working directly with Newton's second law, Eq. (11–2). Let us take the energy approach first.

In Eq. (11–4) we replace v by dx/dt, temporarily disregarding the $\pm$ sign:

$$\frac{dx}{dt} = \sqrt{\frac{k}{m}}\sqrt{A^2 - x^2}. \tag{11-7}$$

Integrating the energy equation to find position as a function of time

The next step is called *separation of variables*. We write all factors containing x on one side, and all those containing t on the other, so that each side can be integrated:

$$\int \frac{dx}{\sqrt{A^2 - x^2}} = \sqrt{\frac{k}{m}}\int dt.$$

Carrying out the integration, we find

$$\arcsin\frac{x}{A} = \sqrt{\frac{k}{m}}t + C,$$

where C is an integration constant. Solving for x, we obtain

$$x = A\sin\left(\sqrt{\frac{k}{m}}t + C\right). \tag{11-8}$$

The quantity in parentheses plays the role of an angle (measured always in radians). The fact that the sine function is *periodic* shows that x is a periodic function of time, as we expected, and specifically that it is a *sinusoidal* function.

The period and frequency of simple harmonic motion don't depend on amplitude; this is one of the most important properties of SHM.

The *period* of motion, denoted by τ, is the time for one complete cycle, or oscillation. The sine function repeats itself whenever the quantity in parentheses increases by 2π. Thus if we start at time $t = 0$, the time τ at which one cycle has been completed is given by

$$\sqrt{\frac{k}{m}}\,\tau = 2\pi,$$

or

$$\tau = 2\pi\sqrt{\frac{m}{k}}. \tag{11-9}$$

Thus the period of the motion is determined by the mass m and the force constant k. It *does not* depend on the amplitude or on the total energy. For given values of m and k, the time of one complete oscillation is the same whether the amplitude is large or small. A periodic motion in which the period is independent of amplitude is said to be **isochronous.**

The frequency f, which is the number of complete oscillations per unit time, is the reciprocal of the period τ:

$$f = \frac{1}{\tau} = \frac{1}{2\pi}\sqrt{\frac{k}{m}}. \tag{11-10}$$

The basic unit for f is "cycles per second," or simply s^{-1}. In SI units, one cycle per second is given the special name 1 *hertz* (1 Hz), in honor of Heinrich Hertz, one of the pioneers in investigating electromagnetic waves during the latter part of the nineteenth century. Thus

$$1 \text{ hertz} = 1 \text{ Hz} = 1 \text{ cycle·s}^{-1}.$$

We can simplify many of the relationships in simple harmonic motion by using the **angular frequency** ω, defined as $\omega = 2\pi f$ and expressed in radians per second. It follows from either of the two preceding equations that

> Angular frequency is usually more convenient than is ordinary frequency in SHM problems.

$$\omega = \sqrt{\frac{k}{m}}, \tag{11-11}$$

and Eqs. (11–8) and (11–4) can be written more compactly as

$$x = A \sin(\omega t + C), \tag{11-12}$$

$$v = \pm\omega\sqrt{A^2 - x^2}. \tag{11-13}$$

Using ω rather than f saves us from having to write factors of 2π. By convention, the hertz is *not* used as a unit of angular frequency, which is usually given the unit rad·s^{-1}, or simply s^{-1}.

In SI units, m is expressed in kilograms and k in newtons per meter. The frequency f is then in vibrations per second, or hertz, and the period τ is expressed in seconds per vibration.

The general form of Eqs. (11–9) and (11–10) should not be surprising. When m is large, we expect the motion to be slow and ponderous, corresponding to small f and large τ. A large value of k means a very stiff spring, corresponding to large f and small τ. These equations *do not* contain the amplitude A of the motion. Suppose we give a certain spring-mass system an initial displacement, release it, and measure its frequency. Then we stop it, give it a *different* displacement, and release it. We find that the two frequencies are the same! To be sure, the maximum displacement, the maximum speed, and the maximum acceleration are all different in the two cases, but *not* the frequency. Indeed, the most important characteristic of simple harmonic motion is that *the frequency does not depend on the amplitude of the motion.* This is why a tuning fork vibrates with the same frequency, regardless of amplitude. If it were not for simple harmonic motion, it would be impossible to play most musical instruments in tune.

In Eq. (11–8) the sine function can never be greater than $+1$ or less than -1, so x can never be greater than A or less than $-A$; this confirms again that A is the *amplitude* of the motion. The integration constant C determines the position of the body at time $t = 0$. If $C = 0$, then at time $t = 0$, $x = A \sin 0 = 0$, and the body starts at the origin. But if $C = \pi/2$, then at time $t = 0$, $x = A \sin \pi/2 = A$, and the body starts at its maximum positive displacement. In that

> The differential equation for SHM; its solutions are the possible motions.

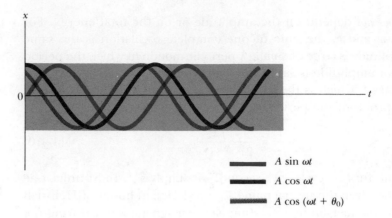

11–4 Graphs of Eqs. (11–16), showing phase relationships.

case we can use the identity $\sin(\theta + \pi/2) = \cos\theta$ to rewrite Eq. (11–12) as

$$x = A \cos \omega t. \tag{11–14}$$

Before exploring further the meaning of Eqs. (11–12) and (11–14), we return to an alternative derivation starting with Newton's second law as expressed by Eq. (11–2). This will give us additional insight into the meaning of these relations. Replacing a_x by d^2x/dt^2 and using $\omega^2 = k/m$, we find

$$\frac{d^2x}{dt^2} = -\omega^2 x. \tag{11–15}$$

This is a *differential equation;* it does not give x as a function of t, but says that x must be a function such that when its second derivative is calculated, the result is a negative constant $(-\omega^2)$ multiplied by the original function. What functions have this property? Sines and cosines immediately come to mind. All the following functions have the required property:

$$x = A \sin \omega t, \tag{11–16a}$$
$$x = A \cos \omega t, \tag{11–16b}$$
$$x = A \cos (\omega t + \theta_0). \tag{11–16c}$$

We invite you to verify that each of Eqs. (11–16) is a *solution* of Eq. (11–15). To do this, calculate the second derivative of each function, substitute it in the left side of Eq. (11–15), substitute the function itself into the right side, and verify that the left and right sides are indeed equal. This in turn shows that each function represents a motion that is consistent with Newton's second law and is thus a physically possible motion.

The three functions in Eqs. (11–16) are plotted as functions of time in Fig. 11–4. We see that the essential difference among them is the position of the body at the instant of time we choose to call $t = 0$.

The function that describes position as a function of time depends on the initial conditions.

Equations (11–16) will form the basis of our further description of simple harmonic motion. For given values of A and θ_0, they differ in the position of the particle at time $t = 0$, that is, in the particular point in the cycle at which $t = 0$. If the body is given an initial displacement A at time $t = 0$ and released with no initial velocity, the motion is described by Eq. (11–16b). If the body is given an initial velocity v_0 at the equilibrium position ($x = 0$) at time $t = 0$, the appropriate equation is Eq. (11–16a). In that case, reference to Eq. (11–6) shows that v_0 and A are related by

$$v_0 = \omega A,$$

since the velocity at $x = 0$ is the maximum velocity.

When the body is given *both* an initial displacement x_0 and an initial velocity v_0 at time $t = 0$, we must use Eq. (11–16c). If x_0 and v_0 are given, we can determine A and θ_0. Angle θ_0 is called a **phase angle,** and its purpose is to describe the point in the cycle at which $t = 0$. To determine A and θ_0, we first note that when $t = 0$ is substituted into Eq. (11–16c), the result must equal x_0; thus

$$x_0 = A \cos \theta_0. \qquad (11\text{–}17)$$

Also, the velocity v at any time t is given by

$$v = \frac{dx}{dt} = -\omega A \sin (\omega t + \theta_0). \qquad (11\text{–}18)$$

Again substituting $t = 0$ and equating the result to v_0, we find

$$v_0 = -\omega A \sin \theta_0. \qquad (11\text{–}19)$$

This and Eq. (11–17) may be solved for A and θ_0. To eliminate A, we divide Eq. (11–19) by Eq. (11–17), obtaining

$$\tan \theta_0 = -\frac{v_0}{\omega x_0}. \qquad (11\text{–}20)$$

To eliminate θ_0, we divide Eq. (11–19) by ω, square it, square Eq. (11–17), and add the two, using the identity $\sin^2 \theta_0 + \cos^2 \theta_0 = 1$, to obtain

$$A^2 = x_0^2 + \frac{v_0^2}{\omega^2}. \qquad (11\text{–}21)$$

The amplitude is not equal to the initial displacement. This is reasonable: If at time $t = 0$ the particle has an initial displacement x_0 in the positive direction and also a positive velocity v_0 in that direction, then it will move *farther* in that direction before returning; hence A must be greater than x_0. The frequency and period relations, Eqs. (11–9) and (11–10), are unchanged.

The phase angle describes what part of the cycle the body is in at the initial time zero.

PROBLEM-SOLVING STRATEGY: Simple harmonic motion

1. It is important to distinguish between quantities that represent basic physical properties of the system and quantities that describe a particular motion that occurs when the system is set in motion in a specific way. The physical properties include the mass m, the force constant k, and the quantities derived from these, including the period τ, the frequency f, and the angular frequency $\omega = 2\pi f$. In some problems m or k, or both, can be determined from other information given about the system. Quantities that describe a particular motion include the amplitude A, the maximum velocity v_{max}, the phase angle θ_0, and any quantity representing the position, the velocity, or the acceleration at a particular time.

2. If the problem involves a relation among position velocity, and acceleration without reference to time, it is usually easiest to use Eqs. (11–2) or (11–4) than to use the general expressions for position as a function of time given by Eqs. (11–16).

3. When detailed information about positions, velocities, and accelerations at various times is required, then Eqs. (11–16) must be used. If the body has an initial velocity v_0 but no initial displacement ($x_0 = 0$), we use Eq. (11–16a), with $v_0 = \omega A$. If $v_0 = 0$—that is, the body has an initial displacement x_0 but no initial velocity—we use Eq. (11–16b) with $A = x_0$. If the initial position x_0 and initial velocity v_0 are both different from zero, we use Eq. (11–16c). The amplitude A and the phase angle θ_0 are related to x_0 and v_0 by Eqs. (11–20) and (11–21).

4. The energy equation, Eq. (11–3), together with the relation $E = \frac{1}{2}kA^2$, sometimes provides a convenient alternative for relations between velocity and position, especially when energy quantities are also required.

EXAMPLE 11–1: A spring is mounted as in Fig. 11–1. By attaching a spring balance to the free end and pulling sideways, we determine that the force is proportional to the displacement and that a force of 4 N causes a displacement of 0.02 m. We attach a 2-kg body to the end, pull it aside a distance of 0.04 m, and release it.

a) Find the force constant of the spring:

$$k = \frac{F}{x} = \frac{4 \text{ N}}{0.02 \text{ m}} = 200 \text{ N·m}^{-1}.$$

b) Find the period and frequency of vibration:

$$\tau = 2\pi\sqrt{\frac{m}{k}} = 2\pi\sqrt{\frac{2 \text{ kg}}{200 \text{ N·m}^{-1}}} = \frac{\pi}{5}\text{s} = 0.628 \text{ s};$$

$$f = \frac{1}{\tau} = \frac{5}{\pi}\text{s}^{-1} = 1.59 \text{ s}^{-1} = 1.59 \text{ Hz};$$

$$\omega = 2\pi f = \sqrt{\frac{k}{m}} = 10 \text{ s}^{-1}.$$

c) Compute the maximum velocity attained by the vibrating body.
     The maximum velocity occurs at the equilibrium position, where $x = 0$. For any x,

$$v = \pm\omega\sqrt{A^2 - x^2},$$

so when $x = 0$,

$$v = v_{max} = \pm\omega A = \pm(10 \text{ s}^{-1})(0.04 \text{ m}) = \pm0.4 \text{ m·s}^{-1}.$$

Also,

$$A = 0.04 \text{ m},$$
$$v_{max} = \pm(10 \text{ s}^{-1})(0.04 \text{ m}) = \pm0.4 \text{ m·s}^{-1}.$$

d) Compute the maximum acceleration.
     From Eq. (11–2) or (11–15),

$$a_x = -\frac{k}{m}x = -\omega^2 x.$$

The maximum acceleration occurs at the ends of the path, where $x = \pm A$. Therefore

$$a_{max} = \mp\omega^2 A = \mp(10 \text{ s}^{-1})^2(0.04 \text{ m}) = \mp4.0 \text{ m·s}^{-2}.$$

e) Compute the velocity and acceleration when the body has moved halfway in to the center from its initial position.
     At this point, from Eq. (11–4),

$$x = \frac{A}{2} = 0.02 \text{ m},$$

$$v = -(10 \text{ s}^{-1})\sqrt{(0.04 \text{ m})^2 - (0.02 \text{ m})^2}$$

$$= -\left(\frac{2\sqrt{3}}{10}\right) \text{ m·s}^{-1} = -0.346 \text{ m·s}^{-1},$$

$$a_x = -\omega^2 x = -(10 \text{ s}^{-1})^2(0.02 \text{ m}) = -2.0 \text{ m·s}^{-2}.$$

f) How much time is required for the body to move halfway in to the center from its initial position?

The position at any time is given by $x = A \cos \omega t$. From this,

$$A/2 = A \cos (10 \text{ s}^{-1})t,$$

$$\cos (10 \text{ s}^{-1})t = \frac{1}{2},$$

$$(10 \text{ s}^{-1})t = \arccos \frac{1}{2} = \frac{\pi}{3},$$

$$t = \frac{\pi}{30} \text{ s}.$$

EXAMPLE 11–2: The system in Example 11–1 is given an initial displacement of 0.05 m and an initial velocity of 2 m·s^{-1}. Find the amplitude, the phase angle, and the total energy of the motion, and write an equation for the position as a function of time.

SOLUTION From Eq. (11–21),

The same system as before, but more complicated initial conditions

$$A = \sqrt{x_0^2 + (v_0/\omega)^2}$$

$$= \sqrt{(0.05 \text{ m})^2 + (2 \text{ m·s}^{-1}/10 \text{ s}^{-1})^2}$$

$$= 0.206 \text{ m}.$$

From Eq. (11–20),

$$\theta_0 = \arctan \frac{-v_0}{\omega x_0}$$

$$= \arctan \frac{-2 \text{ m·s}^{-1}}{(10 \text{ s}^{-1})(0.05 \text{ m})}$$

$$= -76.0°$$

$$= -1.33 \text{ rad}.$$

From Eq. (11–3) and the following discussion,

$$E = \tfrac{1}{2}kA^2 = \tfrac{1}{2}(200 \text{ N·m}^{-1})(0.206 \text{ m})^2$$

$$= 4.24 \text{ J}.$$

Alternatively, from the initial conditions,

$$E = \tfrac{1}{2}mv_0^2 + \tfrac{1}{2}kx_0^2$$

$$= \tfrac{1}{2}(2 \text{ kg})(2 \text{ m·s}^{-1})^2 + \tfrac{1}{2}(200 \text{ N·m}^{-1})(0.05 \text{ m})^2$$

$$= 4.24 \text{ J}.$$

The position is given by Eq. (11–16c):

$$x = (0.206 \text{ m}) \cos [(10 \text{ s}^{-1})t - 1.33 \text{ rad}].$$

Suppose we turn the system of Fig. 11–1 by 90°, so the mass hangs vertically from the spring, in a uniform gravitational field. The motion does not change in any essential way. In Fig. 11–5a a body of mass m hangs in equilibrium from a spring with force constant k. In this position the spring is

A body hanging from a spring. The equilibrium point is shifted, but all the essential characteristics of SHM are the same as before.

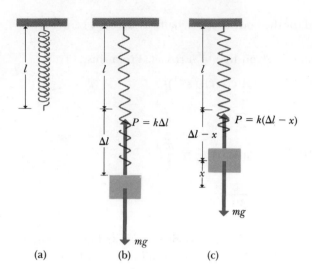

11–5 The restoring force on a body suspended by a spring is proportional to the coordinate measured from the equilibrium position.

stretched an amount Δl just great enough so that the spring's upward vertical force $k\,\Delta l$ on the body balances its weight mg. In that case,

$$k\,\Delta l = mg.$$

When the body is at a distance x *above* its equilibrium position, as in Fig. 11–5c, the extension of the spring is $\Delta l - x$. The upward force it exerts on the body is then $k(\Delta l - x)$, and the resultant force F on the body is

$$F = k(\Delta l - x) - mg = -kx,$$

that is, a net downward force of magnitude kx. Similarly, when the body is *below* the equilibrium position, there is a net upward force proportional to x. If the body is set in vertical motion, it oscillates with SHM, with the same angular frequency as though it were horizontal, $\omega = (k/m)^{1/2}$.

EXAMPLE 11–3: A body of mass 5 kg is suspended by a spring, which stretches 0.1 m when the body is attached. The body is then displaced downward an additional 0.05 m and released. Find the amplitude, the period, and the frequency of the resulting simple harmonic motion.

SOLUTION: Since the initial position is 0.05 m from equilibrium and there is no initial velocity, $A = 0.05$ m. To find the period we first find the force constant of the spring. The spring is stretched 0.1 m by a force of (5 kg) (9.8 m·s^{-2}), so

$$k = \frac{mg}{\Delta l} = \frac{(5 \text{ kg})(9.8 \text{ m·s}^{-2})}{0.1 \text{ m}} = 490 \text{ N·m}^{-1},$$

$$\tau = 2\pi\sqrt{\frac{m}{k}} = 2\pi\sqrt{\frac{5 \text{ kg}}{490 \text{ N·m}^{-1}}} = 0.635 \text{ s},$$

$$f = \frac{1}{\tau} = 1.57 \text{ Hz}.$$

11–3 CIRCLE OF REFERENCE

We can gain additional insight into simple harmonic motion through a geometric representation called the **circle of reference.** This representation makes use of a close relationship between SHM and uniform circular motion, which we studied in Section 3–5. The basic idea is shown in Fig. 11–6. Point Q moves counterclockwise around a circle with a radius A that is equal to the amplitude of the actual simple harmonic motion, with constant angular velocity ω (measured in rad·s^{-1}). Thus ω is the rate of change of the angle θ; $\omega = d\theta/dt$.

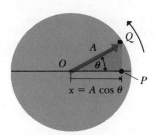

11–6 Coordinate of a body in simple harmonic motion.

The vector from O to Q is the position vector of point Q relative to O. This vector has constant magnitude A and at time t is at an angle θ, measured counterclockwise from the positive x-axis. As Q moves, this vector rotates counterclockwise with constant angular velocity $\omega = d\theta/dt$. The horizontal component of this vector represents the actual motion of the body under study. Such a rotating vector is called a **phasor.** This representation is also useful in many other areas of physics where we encounter quantities that vary sinusoidally with time, including ac-circuit analysis (Chapter 34) and interference phenomena in optics (Chapter 39).

Uniform circular motion of a particle is closely related to SHM.

In Fig. 11–6, point P lies on the horizontal diameter of the circle, directly below Q. We call Q the *reference point,* the circle the *reference circle,* and P the *projection* of Q onto the diameter. The location of P is that of a *shadow* of Q on the x-axis, cast by a light beam parallel to the y-axis. As Q revolves, P moves back and forth along the diameter, staying always directly below (or above) Q. We are about to show that the motion of P is *simple harmonic motion.*

The displacement of P from the origin O at any time t is the distance OP or x. From Fig. 11–6 we see that

$$x = A \cos \theta.$$

If point Q is at the extreme right end of the diameter at time $t = 0$, then $\theta = 0$ when $t = 0$, and the time variation of θ is given by

$$\theta = \omega t.$$

Hence

$$x = A \cos \omega t. \tag{11–22}$$

Now ω, the angular velocity of Q in radians per second, is related to f, the number of complete revolutions of Q per second, by

$$\omega = 2\pi f,$$

since there are 2π radians in one complete revolution. Furthermore, the point P makes one complete back-and-forth vibration for each revolution of Q. Hence f is also the number of vibrations per second, or the *frequency* of vibration of point P. Thus Eq. (11–22) may also be written

$$x = A \cos 2\pi ft. \tag{11–23}$$

We can find the instantaneous velocity of P with the aid of Fig. 11–7. The reference point Q moves with a tangential velocity given by Eq. (9–9):

$$v_\| = \omega A = 2\pi fA.$$

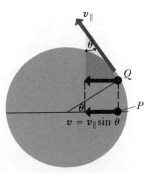

11–7 Velocity in simple harmonic motion.

Since point P is always directly below or above the reference point, the velocity of P at each instant must equal the x-component of the velocity of Q. That is, from Fig. 11–7,

$$v = -v_{\parallel} \sin \theta = -\omega A \sin \theta; \qquad (11\text{–}24)$$
$$v = -\omega A \sin \omega t.$$

The minus sign is needed because the direction of the velocity is toward the left. When Q is below the horizontal diameter, the velocity of P is toward the right; but since $\sin \theta$ is negative at such points, the minus sign is still needed. Equation (11–24) gives the velocity of point P at any time.

We can also find the acceleration of point P by making use again of the fact that since P is always directly below or above Q, its acceleration must equal the x-component of the acceleration of Q. Because point Q moves in a circular path with constant angular velocity ω, it has at each instant an acceleration toward the center given by Eq. (9–11):

$$a_{\perp} = \frac{v_{\parallel}^{2}}{A} = \omega^{2} A.$$

From Fig. 11–8, the x-component of this acceleration is

$$a_x = -a_{\perp} \cos \theta; \qquad (11\text{–}25)$$
$$a_x = -\omega^{2} A \cos \omega t.$$

The minus sign is needed because the acceleration is toward the left. When Q is to the left of the center, the acceleration of P is toward the right; but since $\cos \theta$ is negative at such points, the minus sign is still required. Equation (11–25) gives the acceleration of P at any time.

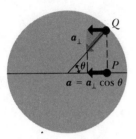

11–8 Acceleration in simple harmonic motion.

Now comes the crucial step in showing that the motion of P is simple harmonic. We combine Eqs. (11–22) and (11–25), obtaining

$$a = -\omega^{2} x. \qquad (11\text{–}26)$$

Because ω is constant, the acceleration a at each instant equals a negative constant times the displacement x at that instant. But this is just the essential feature of simple harmonic motion, as given by Eq. (11–2): Force and acceleration are proportional to displacement from equilibrium. Hence the motion of P is indeed simple harmonic!

In order to make Eqs. (11–2) and (11–26) agree precisely, we must choose an angular velocity ω for the reference point Q such that

$$\omega^{2} = \frac{k}{m}.$$

Thus the angular *velocity* of point Q is identical with the angular *frequency* of motion defined by Eq. (11–11).

Throughout this discussion we have assumed that the initial position of the particle (at time $t = 0$) is its maximum positive displacement A, but this is not an essential restriction. Different initial positions correspond to different initial positions of the reference point Q. For example, if at time $t = 0$ the phasor OQ makes an angle θ_0 with the positive x-axis, then the angle θ at time t is given not by $\theta = \omega t$ as before, but by

$$\theta = \theta_0 + \omega t. \qquad (11\text{–}27)$$

The only change in the discussion is to replace (ωt) in Eqs. (11–22), (11–24), and (11–25) by ($\omega t + \theta_0$). These equations then become

$$x = A \cos (\omega t + \theta_0),$$
$$v = -\omega A \sin (\omega t + \theta_0), \qquad \text{(11–28)}$$
$$a_x = -\omega^2 A \cos (\omega t + \theta_0) = -\omega^2 x.$$

The initial position x_0 and initial velocity v_0 (at time $t = 0$) are then given by

$$x_0 = A \cos \theta_0, \qquad v_0 = -\omega A \sin \theta_0,$$

in agreement with Eqs. (11–17) and (11–19).

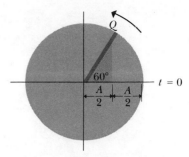

11–9 Phasor diagram for Example 11–4.

EXAMPLE 11–4: In Example 11–1, use the circle of reference to find the time for the displacement from $x = A$ to $x = A/2$, requested in part (f).

SOLUTION While the body moves halfway in, the reference point revolves through an angle of 60° (Fig. 11–9). Since the reference point moves with constant angular velocity and in this example makes one complete revolution during one period ($\tau = \pi/5$ s), the time to rotate through 60° is

$$\left(\frac{1}{6}\right)\left(\frac{\pi}{5} \text{ s}\right) = \frac{\pi}{30} \text{ s} = 0.105 \text{ s}.$$

11–4 ANGULAR SIMPLE HARMONIC MOTION

Simple harmonic motion has a direct *rotational* analog. Suppose a body is pivoted so it can rotate about an axis and is acted on by a torque that is directly proportional to its angular displacement from some equilibrium position. Such a torque might be supplied by a coil spring coaxial with the rotation axis, for example. The resulting rotational motion is directly analogous to linear SHM, and the corresponding equations can be written down immediately from our previous analogies between linear and angular quantities. Caution: In the following discussion, be careful not to confuse the rotational motion of a body in angular SHM with the motion of point Q in the circle-of-reference representation of SHM.

A restoring torque Γ proportional to angular displacement θ is expressed by

$$\Gamma = -k'\theta, \qquad \text{(11–29)}$$

where k' is a proportionality constant called the *torque constant*. The equation of motion, from $\Gamma = I\alpha$, is

$$-k'\theta = I\frac{d^2\theta}{dt^2}, \qquad \text{or} \qquad \frac{d^2\theta}{dt^2} = -\frac{k'}{I}\theta = -\omega^2\theta.$$

This has the same form as Eq. (11–15); the moment of inertia of the pivoted body corresponds to the mass of a body in linear motion. Hence the formula for the period of angular harmonic motion is

$$\tau = 2\pi\sqrt{\frac{I}{k'}}, \qquad \text{(11–30)}$$

When torque is proportional to angular displacement, the motion is angular simple harmonic motion.

where k' is the constant in Eq. (11–29). The angular frequency is now given by

$$\omega = \sqrt{\frac{k'}{I}}. \tag{11–31}$$

We need to be cautious in our choice of symbols; we must not use ω for the angular velocity of the body (a quantity that varies with time) because it has already been used for the angular frequency of the motion (a constant for any given system). A natural symbol for the body's angular velocity is Ω. Then the appropriate formulas are obtained by replacing x everywhere by θ, v by Ω, and a_x by α.

The balance wheel of a watch or mechanical clock is an example of angular harmonic motion. If the hairspring behaves according to Eq. (11–29), the motion is *isochronous:* The period is constant even though the amplitude decreases somewhat as the mainspring unwinds.

11–5 THE SIMPLE PENDULUM

The simple pendulum: a mass on a string. The motion is not precisely simple harmonic.

A **simple pendulum** is an idealized model consisting of a point mass suspended by a weightless, unstretchable string in a uniform gravitational field. When the mass is pulled to one side of its straight-down equilibrium position and released, it oscillates about the equilibrium position. We can now analyze the motion of this system, asking in particular whether it is simple harmonic.

As Fig. 11–10 shows, the path is not a straight line but the arc of a circle of radius L, equal to the length of the string. We use as our coordinate the distance x measured along the arc. If the motion is simple harmonic, the restoring force must be directly proportional to x or, since $x = L\theta$, to θ. Is it?

In Fig. 11–10 the forces on the mass are represented in terms of tangential and radial components; the restoring force F is

$$F = -mg \sin \theta \tag{11–32}$$

The restoring force is therefore proportional *not* to θ but to $\sin \theta$, so the motion is *not* simple harmonic. However, *if the angle θ is small*, $\sin \theta$ is very nearly equal to θ. For example, when $\theta = 0.1$ rad (about 6°), $\sin \theta = 0.0998$, a difference of only 0.2%. With this approximation, Eq. (11–32) becomes

$$F = -mg\theta = -mg\frac{x}{L},$$

or

$$F = -\frac{mg}{L}x. \tag{11–33}$$

The restoring force is then proportional to the coordinate *for small displacements,* and the constant mg/L represents the force constant k. The angular frequency of a simple pendulum when its amplitude is small is therefore

$$\omega = \sqrt{\frac{k}{m}} = \sqrt{\frac{mg/L}{m}} = \sqrt{\frac{g}{L}}. \tag{11–34}$$

The corresponding frequency and period relations are

$$f = \frac{\omega}{2\pi} = \frac{1}{2\pi}\sqrt{\frac{g}{L}}, \tag{11–35}$$

$$\tau = \frac{2\pi}{\omega} = \frac{1}{f} = 2\pi\sqrt{\frac{L}{g}}. \tag{11–36}$$

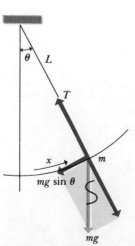

11–10 Forces on the bob of a simple pendulum.

Note that these expressions do not contain the *mass* of the particle; this is because the restoring force, a component of the particle's weight, is proportional to m. Thus the mass appears on both sides of $F = ma$ and may be canceled out. For small oscillations the period of a pendulum for a given value of g is determined entirely by its length.

An elegant argument invented by Galileo four hundred years ago also leads to this conclusion. Make a simple pendulum, said Galileo, measure its period, and then split it down the middle, string and all. Splitting it should not change the motion, so each half must swing with the same period as the original! Thus the period should not depend on the mass.

The form of the dependence on L and g in Eqs. (11–34) through (11–36) is generally what we should expect. It is a familiar fact that long pendulums have longer periods than shorter ones. Increasing g would increase the restoring force, which would cause the frequency to increase and the period to decrease.

We emphasize again that the motion of a pendulum is only *approximately* simple harmonic; when the amplitude is not small, the departures from simple harmonic motion can be substantial. But how small is "small"? The period can be expressed by an infinite series; when the maximum angular displacement is Θ, the period τ is given by

$$\tau = 2\pi\sqrt{\frac{L}{g}}\left(1 + \frac{1^2}{2^2}\sin^2\frac{\Theta}{2} + \frac{1^2 \cdot 3^2}{2^2 \cdot 4^2}\sin^4\frac{\Theta}{2} + \cdots\right). \quad (11\text{–}37)$$

We can compute the period to any desired degree of precision by taking enough terms in the series. When $\Theta = 15°$ (on either side of the central position), the true period differs from that given by the approximate Eq. (11–36) by less than 0.5%.

The usefulness of the pendulum as a timekeeper is based on the fact that the motion is *very nearly* isochronous (period independent of amplitude). So as a pendulum clock runs down and the amplitude of the swings becomes slightly smaller, the clock still keeps very nearly correct time.

The simple pendulum is also a precise and convenient method for measuring the acceleration of gravity g, since L and τ may readily be measured. Such measurements are often used in the field of geophysics. Local deposits of ore or oil affect the local value of g because their density differs from that of their surroundings. Precise measurements of this quantity over an area being prospected often furnish valuable information regarding the nature of underlying deposits.

11–6 THE PHYSICAL PENDULUM

A **physical pendulum** is any *real* pendulum, as contrasted with the idealized model of the *simple* pendulum, where all the mass is concentrated at a point. For small oscillations, analyzing the motion of a real pendulum is almost as easy as for a simple pendulum. Figure 11–11 shows a body of irregular shape pivoted so it can turn without friction about an axis through point O. In the equilibrium position the center of gravity is directly below the pivot; in the position shown in the figure, the body is displaced from equilibrium by an angle θ, which serves as a coordinate for the system. The distance from O to the center of gravity is h, the moment of inertia of the body about the axis through O is I, and the total mass is m. When the body is displaced as shown,

The period and frequency of a simple pendulum are independent of its mass.

Pendulum motion is approximately simple harmonic for small amplitudes, but the period does depend on amplitude.

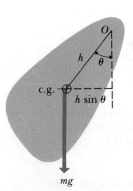

11–11 A physical pendulum.

A physical pendulum is any rigid body suspended from an axis that does not pass through its center of gravity.

the weight mg causes a restoring torque

$$\Gamma = -(mg)(h \sin \theta). \qquad (11-38)$$

When released, the body oscillates about its equilibrium position. As with the simple pendulum, the motion is not simple harmonic, since the torque Γ is proportional not to θ but to $\sin \theta$. However, if θ is small, we can again approximate $\sin \theta$ by θ, and the motion is approximately harmonic. With this approximation,

Motion of a physical pendulum is approximately simple harmonic motion, for small amplitudes.

$$\Gamma \approx -(mgh)\theta,$$

and the effective torque constant is

$$k' = -\frac{\Gamma}{\theta} = mgh.$$

The angular frequency is

$$\omega = \sqrt{\frac{k'}{I}} = \sqrt{\frac{mgh}{I}}, \qquad (11-39)$$

and the period is

$$\tau = 2\pi\sqrt{\frac{I}{k'}} = 2\pi\sqrt{\frac{I}{mgh}}. \qquad (11-40)$$

EXAMPLE 11-5: Let the body in Fig. 11-11 be a meterstick pivoted at one end, and let L be the total length (1 m). Find the period of oscillation.

SOLUTION: We worked out the moment of inertia in Example 9-9 (Section 9-7): $I = \frac{1}{3}mL^2$. The distance h from the center of gravity to the point of suspension is $h = L/2$. From Eq. (11-40), we obtain

$$I = \frac{1}{3}mL^2, \qquad h = \frac{L}{2}, \qquad g = 9.8 \text{ m·s}^{-2},$$

$$\tau = 2\pi\sqrt{\frac{\frac{1}{3}mL^2}{mgL/2}} = 2\pi\sqrt{\frac{2L}{3g}}$$

$$= 2\pi\sqrt{(2/3)(1 \text{ m})/9.8 \text{ m·s}^{-2}} = 1.64 \text{ s}.$$

EXAMPLE 11-6: How can the period of a physical pendulum be used to determine its moment of inertia?

SOLUTION: Equation (11-40) may be solved for the moment of inertia I, giving

$$I = \frac{\tau^2 mgh}{4\pi^2}.$$

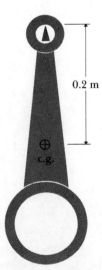

11-12 The moment of inertia of this connecting rod can be determined by measuring its period of oscillation.

Measuring moment of inertia by measuring the period of a physical pendulum

The quantities on the right of the equation can all be measured directly. Hence the moment of inertia of a body of any complex shape may be found by suspending the body as a physical pendulum and measuring its period of vibration. We can find the center of gravity by balancing. Since τ, m, g, and h are known, we can compute I. For example, Fig. 11-12 shows a connecting rod pivoted about a horizontal knife edge. Suppose that when the rod is set into oscillation, it is found to make 100 complete vibrations in 120 s, so $\tau = 120 \text{ s}/100 = 1.2 \text{ s}$. Therefore,

$$I = \frac{(1.2 \text{ s})^2(2 \text{ kg})(9.8 \text{ m·s}^{-2})(0.2 \text{ m})}{4\pi^2} = 0.143 \text{ kg·m}^2.$$

11–7 DAMPED OSCILLATIONS

In the idealized oscillating systems we have discussed so far, there is no friction. Thus the systems are *conservative;* the total mechanical energy is constant, and a system set into motion continues oscillating forever with no decrease in amplitude.

Real-world systems always have some friction, however, and oscillations do die out with time unless some means is provided for replacing the mechanical energy lost to friction. A pendulum clock continues to run because the potential energy stored in the spring is used to replace the mechanical energy lost due to friction in the pendulum and the gears. But when the spring "runs down" and no more energy is available, the pendulum swings decrease in amplitude and stop.

The decrease in amplitude caused by dissipative forces is called **damping,** and the corresponding motion is called **damped oscillation.** The simplest case to analyze in detail is that of a frictional damping force directly proportional to the *velocity* of the oscillating body. This behavior occurs in friction involving viscous fluid flow, such as sliding between oil-lubricated surfaces, shock absorbers, and many other systems of practical importance. We then have an additional force on the body due to friction, $F_x = -bv$, where $v = dx/dt$ is the velocity and b a constant that describes the strength of the damping force. (Why is the negative sign needed?) The total force on the body is then

$$F_x = -kx - bv, \tag{11–41}$$

and the Newton's-second-law formulation of Eq. (11–2) becomes

$$-kx - bv = ma_x,$$

or

$$-kx - b\frac{dx}{dt} = m\frac{d^2x}{dt^2}. \tag{11–42}$$

Finding solutions of this equation is a straightforward problem in differential equations, but we will not go into the details here. If the damping force is relatively small and the body is given an initial displacement A, the motion is described by

$$x = Ae^{-(b/2m)t} \cos \omega't, \tag{11–43}$$

where the frequency of oscillation ω' is given by

$$\omega' = \sqrt{\frac{k}{m} - \frac{b^2}{4m^2}}. \tag{11–44}$$

This motion differs from that of the undamped case in two ways. First, the amplitude is not constant but decreases with time because of the exponential factor $e^{-(b/2m)t}$. The larger the value of b, the more quickly the amplitude decreases. Second, the angular frequency of oscillation is no longer equal to $(k/m)^{1/2}$ but is somewhat smaller. Figure 11–13 shows graphs of Eq. (11–43) for two different values of the constant b.

Note that in Eq. (11–44) the frequency becomes zero when b becomes so large that

$$\frac{k}{m} - \frac{b^2}{4m^2} = 0, \quad \text{or} \quad b = \sqrt{4km}. \tag{11–45}$$

When friction is present, oscillations die out rather than repeating themselves forever.

A damping force proportional to velocity is easy to analyze.

Damping decreases the frequency of a harmonic oscillator.

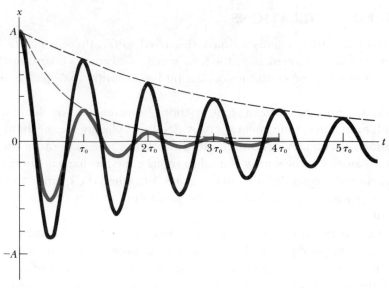

11–13 Graphs of Eq. (11–43), showing damped harmonic motion. The period when there is no damping ($b = 0$) is τ_0. The black curve shows the motion when $b = 0.1\sqrt{km}$, and the color curve is for $b = 0.4\sqrt{km}$. The broken lines show the exponential factor $Ae^{-(b/2m)t}$ for each case. The amplitude decreases more rapidly for the larger value of b. Close inspection of the points where the curves cross the t-axis also shows that the period increases slightly with increasing b. The critical-damping condition is $b = 2\sqrt{km}$.

If the damping force is too large, the system doesn't oscillate at all.

When b exceeds this value, the system no longer oscillates but returns to its equilibrium position without oscillation when it is displaced and released. When Eq. (11–45) is satisfied, the condition is called **critical damping.** The nonoscillating motion that occurs when b is even larger is called **overdamping.** In these cases the solutions of Eq. (11–42) are decreasing exponential functions without any sinusoidal factors. When b is less than the critical value, the situation is called **underdamping.** In all cases, both overdamping and underdamping, the mechanical energy of the system continuously decreases, approaching zero after a sufficiently long time.

Damping can be desirable or undesirable, depending on the situation.

The suspension system of an automobile is a familiar example of damped oscillations. The shock absorbers provide a velocity-dependent damping force so that when the car goes over a bump, it does not continue bouncing forever. For optimal passenger comfort, the system should be critically damped or perhaps slightly underdamped. As the shocks get old and worn, the value of b decreases and the bouncing persists longer. Not only is this nauseating, but it is bad for steering because the front wheels have less positive contact with the ground. Thus damping is an advantage in this system. Conversely, in a system such as a clock or an electrical oscillating system of the type found in radio transmitters, it is usually desirable to minimize damping.

11–8 FORCED OSCILLATIONS

A force that varies periodically with time can make a system vibrate with the same frequency as the force.

There are many practical situations where we would like to maintain oscillations of constant amplitude in a damped oscillating system. A familiar example is a child sitting on a swing. We set the system into motion by pulling the child back from the straight-down equilibrium position and releasing him. If that is all we do, the system oscillates with decreasing amplitude and eventually comes to rest. But by giving the system a little push once each cycle, we can maintain a nearly constant amplitude. More generally, we can maintain a constant-amplitude oscillation in a damped harmonic oscillator by applying an oscillating force, that is, a force that varies with time in a periodic or cyclic way. We call this additional force a **driving force.**

Furthermore, the frequency of variation of the force need not be the same as the natural oscillation frequency of the system. If we apply a periodically varying driving force to the mass of the harmonic oscillator system of Fig. 11–1, the mass undergoes a periodic motion *with the same frequency as that of the driving force.* We call this motion a **forced oscillation,** or a *driven oscillation;* it is different from the motion that occurs when the system is simply set into motion and then left alone to oscillate with a natural frequency determined by m, k, and b.

When the frequency of the driving force is *equal* to the natural frequency of the system, we would expect the amplitude of the resulting oscillation to be larger than when the two are very different, and this expectation is borne out by more detailed analysis and experiment. The easiest case to analyze is that of a *sinusoidally* varying force, say $F_x = F_{max} \sin \omega_d t$, where ω_d is not necessarily equal to the natural frequency ω' of the system given by Eq. (11–44). If we vary the frequency ω_d of the driving force, the amplitude of the resulting forced oscillation varies in an interesting way, as shown in Fig. 11–14. When there is very little damping, the amplitude goes through a sharp peak as the driving frequency passes through the natural oscillation frequency. For increased damping, the peak becomes broader and smaller in height and shifts toward lower frequencies.

The fact that there is an amplitude peak at driving frequencies close to the natural frequency of the system is called **resonance.** Physics is full of examples of resonance; building up the oscillations of a child on a swing is one. A vibrating rattle in a car that occurs only at a certain engine speed is another familiar example. You have probably heard of the dangers of a band marching

The amplitude of a forced oscillation is greatest when the driving frequency is close to the natural frequency of the system.

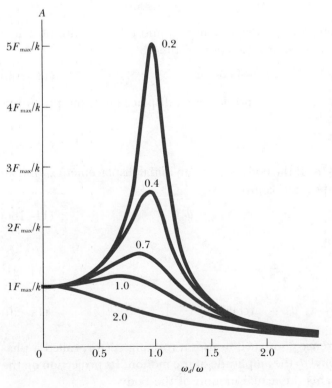

11–14 Graph of the amplitude A of forced oscillation of a damped harmonic oscillator, as a function of the frequency ω_d of the driving force, plotted on the horizontal axis as the ratio of ω_d to the angular frequency $\omega = \sqrt{k/m}$ of an undamped oscillator. Each curve is labeled with the value of the dimensionless quantity $b/\sqrt{km}$, which characterizes the amount of damping. The highest curve has $b = 0.2\sqrt{km}$, the next has $b = 0.4\sqrt{km}$, and so on. As b increases, the peak becomes broader and less sharp and shifts toward lower frequencies. When b is as large as $2\sqrt{km}$, the peak disappears completely.

across a bridge; if the frequency of their steps is close to a natural vibration frequency of the bridge, dangerously large oscillations can build up. A tuned circuit in a radio or television receiver responds strongly to waves having frequencies near its resonant frequency, and this is used to select a particular station and reject the others. We will study resonance in electric circuits in detail in Chapter 34.

SUMMARY

When a body of mass m is acted on by a force $F_x = -kx$ that is directly proportional to the body's displacement from its equilibrium position ($x = 0$), the resulting motion is called simple harmonic motion. The amplitude A is the maximum displacement from equilibrium. The period τ is the time for one complete cycle; it is given by

$$\tau = 2\pi\sqrt{m/k}. \tag{11-9}$$

The frequency f is the number of cycles per unit time and is the reciprocal of the period:

$$f = \frac{1}{\tau} = \frac{1}{2\pi}\sqrt{\frac{k}{m}}. \tag{11-10}$$

The angular frequency ω is given by

$$\omega = 2\pi f = \sqrt{\frac{k}{m}}. \tag{11-11}$$

In simple harmonic motion the acceleration at any instant is proportional to the negative of the displacement.

Conservation of energy leads to the following relation among the position and velocity at any time and the amplitude and total energy:

$$E = \tfrac{1}{2}kA^2 = \tfrac{1}{2}mv^2 + \tfrac{1}{2}kx^2 = \text{constant}. \tag{11-3}$$

If the body is given an initial displacement A and released with no initial velocity, the position is given as a function of time by

$$x = A\cos\omega t. \tag{11-16b}$$

If it is given an initial velocity v_0 and no initial displacement, the position is given as a function of time by

$$x = A\sin\omega t, \tag{11-16a}$$

with A given by $A = v_0/\omega$. If the body is given an initial displacement x_0 and an initial velocity v_0, the position is given by

$$x = A\cos(\omega t + \theta_0), \tag{11-16c}$$

with A and θ_0 given by

$$A^2 = x_0^2 + \frac{v_0^2}{\omega^2}, \tag{11-21}$$

$$\theta_0 = \arctan\left(-\frac{v_0}{\omega x_0}\right). \tag{11-20}$$

The circle-of-reference construction uses a rotating vector, called a phasor, having a length equal to the amplitude of the motion. Its projection on the horizontal axis represents the actual motion of the body.

Angular simple harmonic motion occurs when a pivoted body experiences a torque $\Gamma = -k'\theta$ proportional to its angular displacement θ from an equilibrium position. For a body having moment of inertia I, the angular frequency is given by

$$\omega = \sqrt{\frac{k'}{I}}. \qquad (11\text{–}31)$$

A simple pendulum consists of a point mass m at the end of a string of length L. Its motion is approximately simple harmonic for sufficiently small amplitude; the period, the frequency, and the angular frequency are then given by

$$\tau = 2\pi\sqrt{\frac{L}{g}}, \qquad (11\text{–}36)$$

$$f = \frac{1}{2\pi}\sqrt{\frac{g}{L}}, \qquad (11\text{–}35)$$

$$\omega = \sqrt{\frac{g}{L}}. \qquad (11\text{–}34)$$

These quantities are independent of m.

A physical pendulum is a body suspended from an axis of rotation a distance h from its center of gravity. If the moment of inertia about the axis of rotation is I, the angular frequency and period are given by

$$\omega = \sqrt{\frac{mgh}{I}}, \qquad (11\text{–}39)$$

$$\tau = 2\pi\sqrt{\frac{I}{mgh}}. \qquad (11\text{–}40)$$

When a damping force $F = -bv$ proportional to velocity is added to a simple harmonic oscillator, the motion is described as a damped oscillation:

$$x = Ae^{-(b/2m)t}\cos\omega't, \qquad (11\text{–}43)$$

provided that $b^2 < 4km$. This condition is called underdamping. When $b^2 = 4km$, the system is critically damped and no longer oscillates. When b is still larger, the system is overdamped.

When a sinusoidally varying driving force is added to a damped harmonic oscillator, the resulting motion is called a forced oscillation. Its amplitude reaches a peak at driving frequencies close to the natural oscillation frequency of the system. This behavior is called resonance.

QUESTIONS

11–1 Think of several examples in everyday life of motion that is at least approximately simple harmonic. In what respects does each differ from SHM?

11–2 Does a tuning fork or similar tuning instrument undergo simple harmonic motion? Why is this a crucial question to musicians?

11–3 If a spring is cut in half, what is the force constant of each half? How would the frequency of SHM using a half-spring differ from that using the same mass and the entire spring?

11–4 The analysis of SHM in this chapter neglected the mass of the spring. How does the spring mass change the characteristics of the motion?

11–5 The system shown in Fig. 11–5 is mounted in an elevator, which accelerates upward with constant acceleration. Does the period increase, decrease, or remain the same?

11–6 A highly elastic "superball" bouncing on a hard floor has a motion that is approximately periodic. In what ways is the motion similar to SHM? In what ways is it different?

11–7 How could one determine the force constants of a car's springs by bouncing each end up and down?

11–8 Do the pistons in an automobile engine undergo simple harmonic motion?

11–9 Why is the "springiness" of a diving board adjusted for different dives and different weights of divers? How is the adjustment made?

11–10 In Fig. 11–1, suppose the stationary end of the spring is connected instead to another mass, equal to the original mass and free to slide along the same line. Could such a system undergo SHM? How would the period compare with that of the original system?

11–11 For the mass-spring system of Fig. 11–1, is there any point during the motion at which the mass is in equilibrium?

11–12 In any periodic motion, unavoidable friction always causes the amplitude to decrease with time. Does friction also affect the *period* of the motion? Give a qualitative argument to support your answer.

11–13 If a pendulum clock is taken to a mountain top, does it gain or lose time, assuming it is correct at a lower elevation?

11–14 When the amplitude of a simple pendulum increases, should its period increase or decrease? Give a qualitative argument; do not rely on Eq. (11–37). Is your argument also valid for a physical pendulum?

11–15 A pendulum is mounted in an elevator that accelerates upward with constant acceleration. Does the period increase, decrease, or remain the same?

11–16 At what point in the motion of a simple pendulum is the string tension greatest? Least?

11–17 A child on a swing can increase his amplitude by "pumping up." Where does the extra energy come from? (The answer is not "from the child.")

11–18 Could a standard of time be based on the period of a certain standard pendulum? What advantages and disadvantages would such a standard have, compared to the actual present-day standard discussed in Section 1–2?

EXERCISES

Section 11–1 Basic Concepts

11–1 A vibrating object goes through five complete vibrations in 1 s. Find the angular frequency and the period of the motion.

11–2 In Fig. 11–1 the mass is displaced 0.12 m from its equilibrium position and released with no initial velocity. After 2.0 s its displacement is found to be 0.12 m on the opposite side, and it has passed the equilibrium position once during this interval. Find

a) the amplitude,   b) the period,

c) the frequency,   d) the angular frequency.

Section 11–2 Equations of Simple Harmonic Motion

11–3 A harmonic oscillator has a mass of 4 kg and a spring with force constant 100 N·m^{-1}. Find the period, the frequency, and the angular frequency.

11–4 A harmonic oscillator is made with a block of mass 0.5 kg and a spring of unknown force constant. It is found to have a period of 0.20 s. Find the force constant of the spring.

11–5 A block of unknown mass is attached to a spring of force constant 200 N·m^{-1}, in the arrangement shown in Fig. 11–1. It is found to vibrate with a frequency of 3 Hz. Find the period, the angular frequency, and the mass.

11–6 A tuning fork labeled 512 Hz has the tip of each of its two prongs vibrating through a maximum displacement of 0.80 mm.

a) What is the maximum speed of the tip of a prong?

b) What is the maximum acceleration of the tip of a prong?

c) If the maximum displacement were cut in half, what would then be the answers to (a) and (b)?

11–7 An object is vibrating with simple harmonic motion of amplitude 15 cm and frequency 4 Hz. Compute

a) the maximum values of the acceleration and velocity,

b) the acceleration and velocity when the coordinate is 9 cm,

c) the time required to move from the equilibrium position to a point 12 cm distant from it.

11–8 An object of mass 4 kg is attached to a spring of force constant $k = 100$ N·m^{-1}. The object is given an initial velocity of $v_0 = 12$ m·s^{-1} and an initial displacement of $x_0 = 0$. Find the amplitude, the phase angle, and the total energy of the motion, and write an equation for the position as a function of time.

11–9 Repeat Exercise 11–8, but assume the object is given an initial velocity of $v_0 = -6$ m·s^{-1} and an initial displacement of $x_0 = +0.2$ m.

11–10 An object of mass 0.25 kg is acted on by an elastic restoring force of force constant $k = 25$ N·m^{-1}.

a) Construct the graph of elastic potential energy U as a function of displacement x, over a range of x from -0.3 m to $+0.3$ m. Let 1 cm = 0.1 J vertically, and 1 cm = 0.05 m horizontally.

The object is set into oscillation with an initial potential energy of 0.6 J and an initial kinetic energy of 0.2 J. Answer the following questions by reference to the graph:

b) What is the amplitude of oscillation?

c) What is the potential energy when the displacement is one-half the amplitude?

d) At what displacement are the kinetic and potential energies equal?

e) What is the speed of the object at the midpoint of its path (that is, at $x = 0$)?

Find

f) the period τ, **g)** the frequency f,

h) the angular frequency ω.

i) What is the initial phase angle θ_0 if the initial velocity v_0 is negative?

11–11 A block of mass 2 kg is suspended from a spring of negligible mass and is found to stretch the spring 0.20 m.

a) What is the force constant of the spring?

b) What is the period of oscillation of the block if pulled down and released?

c) What would be the period of a block of mass 4 kg hanging from the same spring?

11–12 The scale of a spring balance reading from zero to 180 N is 9 cm long. A fish suspended from the balance is observed to oscillate vertically at 1.5 Hz. What is the mass of the fish? Neglect the mass of the spring.

11–13 A block of mass 5 kg hangs from a spring and oscillates with a period of 0.5 s. How much will the spring shorten when the block is removed?

11–14

a) A block suspended from a spring vibrates with simple harmonic motion. At an instant when the displacement of the block is equal to one-half the amplitude, what fraction of the total energy of the system is kinetic and what fraction is potential? Assume $U = 0$ at equilibrium.

b) When the block is in equilibrium, the length of the spring is an amount s greater than in the unstretched state. Prove that $\tau = 2\pi\sqrt{s/g}$.

11–15 A body of mass 4 kg is attached to a coil spring and oscillates vertically in simple harmonic motion. The amplitude is 0.5 m, and at the highest point of the motion the spring has its natural unstretched length. Calculate the elastic potential energy of the spring, the kinetic energy of the body, its gravitational potential relative to the lowest point of the motion, and the sum of these three energies, when the body is

a) at its lowest point,

b) at its equilibrium position,

c) at its highest point.

Section 11–3 Circle of Reference

11–16 An object is undergoing simple harmonic motion with period $\tau = 0.4$ s. Use the circle of reference to calculate the time it takes the object to go from $x = 0$ to $x = A/4$.

11–17 An object is undergoing simple harmonic motion with period $(\pi/2)$s and amplitude $A = 0.2$ m. At $t = 0$ the object is at $x = 0$. How far is the object from the equilibrium position when $t = (\pi/10)$ s?

Section 11–4 Angular Simple Harmonic Motion

11–18 The balance wheel of a watch vibrates with an angular amplitude of π rad and a period of 0.5 s.

a) Find its maximum angular velocity.

b) Find its angular velocity when its displacement is one-half its amplitude.

c) Find its angular acceleration when its displacement is 45°.

11–19 A certain alarm clock ticks four times each second, each tick representing half a period. The balance wheel consists of a thin rim of radius 1.5 cm, connected to the balance staff by thin spokes of negligible mass. The total mass of the balance wheel is 0.8 g.

a) What is the moment of inertia of the balance wheel?

b) What is the torque constant of the hairspring?

Section 11–5 The Simple Pendulum

11–20 A simple pendulum 4 m long swings with an amplitude of 0.2 m.

a) Compute the linear velocity v of the pendulum at its lowest point.

b) Compute its linear acceleration a at the end of its path.

11–21 Find the length of a simple pendulum whose period is exactly 1 s at a point where $g = 9.80$ m·s^{-2}.

11–22 A certain simple pendulum has a period on earth of 2.0 s. What is its period on the surface of the moon, where $g = 1.7$ m·s^{-2}?

Section 11–6 The Physical Pendulum

11–23 A thin, uniform rod of length L and mass m is pivoted about a perpendicular axis through the rod at a distance $L/4$ from one end.

a) Find the moment of inertia about this axis.

b) Find the period of oscillation of the rod.

11–24 A monkey wrench is pivoted at one end and allowed to swing as a physical pendulum. The period is 0.9 s, and the pivot is 0.20 m from the center of gravity.

a) What is the ratio of moment of inertia to mass for the wrench, about an axis through the pivot?

b) If the wrench was initially displaced 0.1 rad from its equilibrium position, what is the angular velocity of the wrench as it passes through the equilibrium position?

Section 11–7 Damped Oscillations

11–25 A mass 0.4 kg is moving on the end of a spring of force constant $k = 300$ N·m^{-1} and is acted on by a damping force $F_x = -bv$.

a) If the constant b has the value 5 kg·s^{-1}, what is the frequency of oscillation of the mass?

b) For what value of the constant b will the motion be critically damped?

11–26 A mass 0.2 kg moving on the end of a spring of force constant $k = 250$ N·m^{-1} has an initial displacement of 0.3 m. There is a damping force $F_x = -bv$ acting on the mass. It is observed that the amplitude of the motion has decreased to 0.1 m in 5 s. Calculate the magnitude of the damping constant b.

PROBLEMS

11–27 The motion of the piston of an automobile engine is approximately simple harmonic.

a) If the stroke of an engine (twice the amplitude) is 0.10 m, and the engine runs at 3600 rev·min^{-1}, compute the acceleration of the piston at the end of its stroke.

b) If the piston has a mass of 0.5 kg, what resultant force must be exerted on it at this point?

c) What is the velocity of the piston, in meters per second, at the midpoint of its stroke?

11–28 Four passengers whose combined mass is 300 kg are observed to compress the springs of an automobile by 5 cm when they enter the automobile. If the total load supported by the springs is 900 kg, find the period of vibration of the loaded automobile.

11–29 A small block is executing simple harmonic motion in a horizontal plane with an amplitude of 0.10 m. At a point 0.06 m away from equilibrium, the velocity is 0.24 m·s^{-1}.

a) What is the period?

b) What is the displacement when the velocity is ±0.12 m·s^{-1}?

c) If a small object placed on the oscillating block is just on the verge of slipping at the endpoint of the path, what is the coefficient of static friction?

11–30 An object of mass 0.010 kg moves with simple harmonic motion of amplitude 0.24 m and period 4 s. The coordinate is +0.24 m when $t = 0$. Compute

a) the position of the object when $t = 0.5$ s,

b) the magnitude and direction of the force acting on the object when $t = 0.5$ s,

c) the minimum time required for the object to move from its initial position to the point where $x = -0.12$ m,

d) the velocity of the object when $x = -0.12$ m.

11–31 A rubber raft bobs up and down, executing simple harmonic motion due to the waves on a lake. The amplitude of the motion is 2.0 ft, and the period is 5.0 s. A stable dock is next to the raft and is at a level equal to the highest level of the raft. People wish to step off the raft onto the dock but can do so comfortably only if the level of the raft is within 1.0 ft of the dock level. How much time do the people have to get off comfortably during each period of the simple harmonic motion?

11–32 A force of 30 N stretches a vertical spring 0.15 m.

a) What mass must be suspended from the spring so that the system will oscillate with a period of $(\pi/4)$ s?

b) If the amplitude of the motion is 0.05 m, where is the object and in what direction is it moving $(\pi/12)$ s after it has passed the equilibrium position, moving downward?

c) What force does the spring exert on the object when it is 0.03 m below the equilibrium position, moving upward?

11–33 An object of mass 0.100 kg hangs from a long spiral spring. When pulled down 0.10 m below its equilibrium position and released, it vibrates with a period of 2 s.

a) What is its velocity as it passes through the equilibrium position?

b) What is its acceleration when it is 0.05 m above the equilibrium position?

c) When it is moving upward, how much time is required for it to move from a point 0.05 m below its equilibrium position to a point 0.05 m above it?

d) How much will the spring shorten if the object is removed?

11–34 A very interesting, though impractical, example of simple harmonic motion occurs when considering the motion of a particle dropped down a hole that extends from one side of the earth, through its center, to the other side. Prove that the motion is simple harmonic and find the period. (*Note.* Use the following property of the gravitational force: The force on an object of mass m due to a spherical mass M of radius R is toward the center of the sphere and has magnitude GmM'/r^2, where r is the distance from the center of M out to the location of point mass m and M' is the mass of that part of M inside a sphere of radius r. Hence if $r > R$ then $M' = M$. But if for example, $r = \frac{1}{2}R$ then $M' = \frac{1}{8}M$ because only $\frac{1}{8}$ the volume and hence $\frac{1}{8}$ the mass of M is inside the sphere of radius r. [This property of the gravitational force follows directly from what was shown in Challenge Problem 6–42.])

11–35 A block of mass 0.50 kg sits on top of a block of mass 5.0 kg. The larger block is attached to a horizontal spring that has a force constant of 15 N·m^{-1}. What is the largest amplitude the 5.0 kg mass can have for the smaller mass to remain at rest relative to the larger block? The coefficient of static friction between the two blocks is 0.20. There is no friction between the larger block and the floor.

11–36 The general equation of simple harmonic motion,

$$x = A \cos (\omega t + \theta_0),$$

can be written in the equivalent form

$$x = B \sin \omega t + C \cos \omega t.$$

a) Find the expressions for the amplitudes B and C in terms of the amplitude A and the initial phase angle θ_0.

b) Interpret these expressions in terms of a phasor diagram.

11–37 To measure g in an unorthodox manner, a student places a ball bearing on the concave side of a lens, as shown in Fig. 11–15. She attaches the lens to a simple harmonic oscillator (actually a small stereo speaker) whose amplitude is A and whose frequency f can be varied. She can measure both A and f with a strobe light.

a) If the ball bearing has a mass m, find the normal force exerted by the lens on the ball bearing as a function of time. Your result should be in terms of A, f, m, g, and a phase angle θ_0.

b) The frequency is slowly increased. When it reaches a value f_b, the ball is heard to bounce. What is g in terms of A and f_b?

FIGURE 11–15

11–38 A block of mass m_1 attached to a horizontal spring of force constant k is moving with simple harmonic motion of amplitude A. At the instant it passes through its equilibrium position, a lump of putty of mass m_2 is dropped vertically onto the block from a very small height and sticks to it.

a) Find the new period and amplitude.

b) Was there a loss of mechanical energy? If so, where did it go? Calculate the ratio between the final and the initial mechanical energy.

c) Would the answers be the same if the putty had been dropped on the block when it was at one end of its path?

11–39 A solid disk of radius $R = 12$ cm oscillates as a physical pendulum about an axis perpendicular to the plane of the disk at a distance r from its center. (See Fig. 11–16.)

a) Calculate the period of oscillation (for small amplitudes) for the following values of r: 0, $R/4$, $R/2$, $3R/4$, R, and $3R/2$. (*Hint:* Use the parallel-axis theorem, Section 9–7.)

b) Let τ_0 represent the period when $r = R$, and τ the period at any other value of r. Construct a graph of the dimensionless ratio τ/τ_0 as a function of the dimensionless ratio r/R. (Note that the graph then describes the behavior of *any* solid disk, whatever its radius.)

c) Find the value of r/R that minimizes the period, and calculate the minimum value of the period.

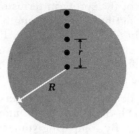

FIGURE 11–16

11–40 Show that $x(t)$ as given in Eq. (11–43) is a solution of Newton's second law for damped oscillations (Eq. 11–42), if ω' is as defined in Eq. (11–44).

11–41 It is desired to construct a pendulum of period 4 s.

a) What is the length of a *simple* pendulum having this period?

b) Suppose the pendulum must be mounted in a case not over 0.50 m high. Can you devise a pendulum, having a period of 4 s, that will satisfy this requirement?

CHALLENGE PROBLEMS

11–42 Two springs with the same unstretched length but different force constants k_1 and k_2 are attached to a block of mass m on a level, frictionless surface. Calculate the effective force constant in each of the three cases (a), (b), and (c) depicted in Fig. 11–17.

d) A body of mass m, suspended from a spring with a force constant k, vibrates with a frequency f_1. When the spring is cut in half and the same body is suspended from one of the halves, the frequency is f_2. What is the ratio f_2/f_1?

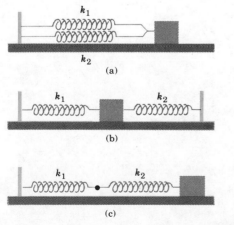

(a)

(b)

(c)

FIGURE 11–17

11–43 Two springs, each of unstretched length 0.2 m, but having different force constants k_1 and k_2, are attached to opposite ends of a block of mass m on a level, frictionless surface. The outer ends of the springs are now attached to two pins P_1 and P_2, 0.10 m from the original positions of the ends of the springs. Let

$$k_1 = 1 \text{ N·m}^{-1}, \qquad k_2 = 3 \text{ N·m}^{-1}, \qquad m = 0.1 \text{ kg}.$$

(See Fig. 11–18.)

a) Find the length of each spring when the block is in its new equilibrium position, after the springs have been attached to the pins.

b) Find the period of vibration of the block if it is slightly displaced from its new equilibrium position and released.

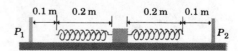

FIGURE 11–18

11–44 All the previous problems in this chapter have assumed that the springs had negligible mass. But of course no spring is completely massless. To find the effect of the spring's mass, consider a spring of mass M, equilibrium length L_0, and spring constant k. When stretched or compressed to a length L, the potential energy is $\frac{1}{2}kx^2$, where $x = L - L_0$.

a) Consider a spring as described above that has one end fixed and the other end moving with speed v. Assume that the speed of points along the length of the spring varies linearly with distance l from the fixed end. Assume also that the mass M of the spring is distributed uniformly along the length of the spring. Calculate the kinetic energy of the spring in terms of M and v. (*Hint:* Divide the spring into pieces of length dl; find the speed of each piece in terms of l, v, and L; find the mass of each piece in terms of dl, M, and L; and integrate from 0 to L. The result is *not* $\frac{1}{2}Mv^2$, since not all of the spring moves with the same speed.)

b) Take the time derivative of the conservation of energy equation, Eq. (11–3), for a mass m moving on the end of a *massless* spring. By comparing your results to Eq. (11–15) that defines ω, show that the angular frequency of oscillation is $\omega = \sqrt{k/m}$.

c) Apply the procedure of part (b) to obtain the angular frequency of oscillation ω of the spring considered in part (a). If the *effective mass M'* of the spring is defined by $\omega = \sqrt{k/M'}$, what is M' in terms of M?

11–45

a) What is the change $\Delta\tau$ in the period of a simple pendulum when the acceleration of gravity g changes by Δg? (*Hint:* The new period $\tau + \Delta\tau$ is obtained by substituting $g + \Delta g$ for g:

$$\tau + \Delta\tau = 2\pi\sqrt{\frac{L}{g + \Delta g}}.$$

To obtain an approximate expression, expand the factor $[g + \Delta g]^{-1/2}$, using the binomial theorem [Appendix B]

and keeping only the first two terms:

$$[g + \Delta g]^{-1/2} = g^{-1/2} - \tfrac{1}{2}g^{-3/2}\,\Delta g + \cdots.$$

The other terms contain higher powers of Δg and are very small if Δg is small.)

b) Find the *fractional* change in period $\Delta\tau/\tau$ in terms of the fractional change $\Delta g/g$.

c) A pendulum clock, which keeps correct time at a point where $g = 9.8000$ m $\cdot$ s^{-2}, is found to lose 10 s each day at a higher elevation. Use the result of part (a) or (b) to find approximately the value of g at this new location.

11–46 An experimenter pivots a thin, uniform rod of length l about a perpendicular axis through the rod and measures its period as a physical pendulum. The rod is then inverted, and by trial and error he finds two other points for which the period is the same as for the first point. One is the point symmetrically opposite the first, of course, but he finds a second such point as well.

a) Show that in this case the period depends only on the distance L between the first point and the second nonsymmetric point and is given by $\tau = 2\pi(L/g)^{1/2}$.

b) For this to happen, what is the *least* distance the first point could have been from the center?

11–47 A meterstick hangs from a horizontal axis at one end and oscillates as a physical pendulum. A body of small dimensions, and of mass equal to that of the meterstick, can be clamped to the stick at a distance d below the axis. Let τ represent the period of the system with the body attached, and τ_0 the period of the meterstick alone.

a) Find the ratio τ/τ_0. Evaluate your expression for d ranging from 0 to 1.0 m in steps of 0.1 m, and sketch a graph of τ/τ_0 versus d.

b) Is there any value of d for which $\tau = \tau_0$? If so, find it and explain why the period is unchanged when d has this value.

PART THREE

MECHANICAL AND THERMAL PROPERTIES OF MATTER

PERSPECTIVE

In the last several chapters we introduced the concepts of momentum and energy; we have seen how these concepts and the associated conservation principles can be used to obtain needed information in problems where the forces and motion of a particle or of a system are too complicated to describe in full detail. We have also applied Newton's laws of motion to the analysis of rotational motion of a *rigid body*, an idealized model to describe a body that does not deform in any way when forces are applied. This analysis involved two new dynamic concepts: *torque* and *angular momentum*. The concept of torque is also useful in the analysis of equilibrium of rigid bodies. Angular momentum and its associated conservation principle extend further our ability to analyze systems that include rotating rigid bodies. Finally, we studied a variety of examples of *periodic motion* in mechanical systems. The importance of periodic motion is not by any means limited to macroscopic mechanical vibrations. We will see later that this discussion lays the foundation for understanding molecular vibrations and molecular spectra, heat capacities of molecules and solids, alternating currents in electric circuits, and many other areas of physics that have great practical as well as fundamental significance.

With this background in the fundamentals of mechanics, we are now ready for an extensive discussion of mechanical and thermal properties of matter. Moving beyond the rigid-body model of a solid body, we consider two kinds of generalizations. The first includes *elastic deformations* of solids and fluids that occur when forces are applied; the second concerns specifically the equilibrium and motion of *fluids*, substances having no definite shape. Included in this discussion are such diverse topics as buoyancy, surface tension, energy relations in fluid flow, and turbulent flow. Then we undertake an extensive discussion of *thermal phenomena* and their associated energy relations. The relation of heat to mechanical energy is vital to the operation of such familiar systems as automobile engines, refrigerators, and electric-power plants, so important practical applications of these principles are always close at hand.

We begin by defining *temperature* and *heat,* and then we study quantitative relationships involving transfer of energy in the form of heat from one body to another. We return to describing the behavior of matter, this time in situations where the temperature, volume, and pressure of a substance may all change.

The heart of this section, though, is the study of the first and second laws of *thermodynamics*. The first law expresses quantitatively the relation of heat to mechanical energy, using the concept of *internal energy* of a substance or a system. The second law is a formulation of fundamental restrictions on processes in which conversion of heat to mechanical energy occurs, including the maximum theoretical efficiency of engines and refrigerators. Another new concept, *entropy*, helps us formulate this second principle concisely and shows its relation to the basic one-way character of natural processes, which tend to proceed always toward states of greater randomness or disorder.

Finally, we look at some aspects of the relationship between the *macroscopic* properties of matter and its *microscopic* structure. We find that many properties of gases, such as the relation of pressure, volume, and temperature, and heat capacities, can be derived from a microscopic model. Although the principles of thermodynamics are developed first without reference to a molecular model, the study of their relation to molecular processes gives us added insight. The relationship between macroscopic properties of a material and its microscopic structure permits understanding of mechanical, thermal, and other properties on the basis of molecular structure. Hence we are sometimes able to design materials having specific desired properties. This ability is a vital part of contemporary technology.

12

ELASTICITY

CHAPTERS 9 AND 10 WERE CONCERNED WITH MOTION AND EQUILIBRIUM OF rigid bodies. The rigid body is an idealized *model* to represent a body that has a definite size and shape and that does not stretch, squeeze, or twist when forces act on it. Of course, real materials always do deform to some extent when forces act on them. In this chapter we consider how the forces and deformations are related. The concepts of stress, strain, and elastic modulus can be used to describe the elastic properties of materials in a way that does not depend on the dimensions of a particular specimen. For sufficiently small forces, the deformation of a material is proportional to the force, but with larger forces the behavior is more complex. For sufficiently large forces, materials can deform irreversibly or break, and we can use the concept of stress to characterize the *strength* of a material. The study of the elastic properties of materials is of tremendous practical importance in the design of buildings, automobiles, and many of the necessities of present-day life.

12–1 TENSILE STRESS AND STRAIN

The simplest elastic behavior to understand is the stretching of a bar, rod, or wire when its ends are pulled. Figure 12–1a shows a bar with uniform cross-sectional area A, subjected to equal and opposite forces F pulling at its ends. We say that the bar is in **tension.** Tensions in ropes and strings were discussed in earlier chapters; the concept is the same here. Imagine a cross section through the bar perpendicular to its length, as shown by the broken line. Every portion of the bar is in equilibrium, so the portion to the right of the section must be pulling on the portion to the left with a force F, and vice versa. If the forces at the ends are applied uniformly over the end surfaces, then the forces at every other cross section are also distributed uniformly over the section, as shown by the short arrows in Fig. 12–1b. We define the **stress** at this section as the ratio of the force F to the cross-sectional area A:

$$\text{Stress} = \frac{F}{A}. \tag{12–1}$$

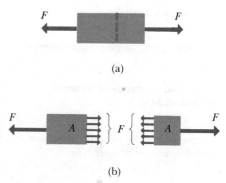

12–1 (a) A bar in tension. (b) The stress at a perpendicular section equals F/A.

When you pull on something, it stretches: tensile stress and strain.

291

This stress is called a **tensile stress** because each portion exerts tension on the other.

The SI unit of stress is the newton per square meter ($N \cdot m^{-2}$). This unit is also given a special name, the *pascal* (abbreviated Pa):

$$\text{One pascal} = 1 \text{ Pa} = 1 \text{ N} \cdot m^{-2}.$$

The units of stress are always force per unit area.

Other units of stress are the dyne per square centimeter ($dyn \cdot cm^{-2}$) and the pound per square foot ($lb \cdot ft^{-2}$). In the British system the hybrid unit, the pound per square inch ($lb \cdot in^{-2}$, or psi), is commonly used. The units of stress are the same as those of *pressure*, which we will encounter frequently in later chapters. The pascal is a fairly small unit; air pressure in automobile tires is typically of the order of 2×10^5 Pa, and steel cables are commonly used with tensile stresses of the order of 10^8 Pa.

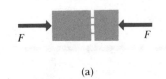

(a)

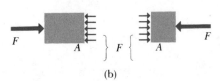

(b)

12–2 A bar in compression.

EXAMPLE 12–1   A human biceps (upper-arm muscle) may exert a force of the order of 600 N on the bones to which it is attached. If the muscle has a cross-sectional area at its center of 50 cm² = 0.005 m², and the tendon attaching its lower end to the bones below the elbow joint has a cross section of 0.5 cm² = 5×10^{-5} m², find the tensile stress in each of these cross sections.

SOLUTION   In each case the stress is the force per unit area. For the muscle,

$$\text{Tensile stress} = \frac{600 \text{ N}}{0.005 \text{ m}^2} = 120{,}000 \text{ N} \cdot m^{-2} = 1.2 \times 10^5 \text{ Pa.}$$

For the tendon,

$$\text{Tensile stress} = \frac{600 \text{ N}}{5 \times 10^{-5} \text{ m}^2} = 1.2 \times 10^7 \text{ Pa.}$$

When the forces acting on the ends of a bar are pushes rather than pulls, as in Fig. 12–2a, we say that the bar is in **compression**. The stress on the cross section shown by a broken line is now a **compressive stress**: Each portion pushes rather than pulls on the other.

The fractional change of length (stretch) of a body subjected to a tensile stress is called the **tensile strain**. Figure 12–3 shows a bar of natural length l_0 that stretches to a length $l = l_0 + \Delta l$ when equal and opposite forces F are applied to its ends. The elongation Δl does not occur only at the ends; every part of the bar stretches in the same proportion as does the bar as a whole. The tensile strain is defined as the ratio of the elongation Δl to the original length l_0:

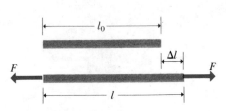

12–3 The longitudinal strain is defined as $\Delta l / l_0$.

Compression is the opposite of tension— a squeeze instead of a stretch.

$$\text{Tensile strain} = \frac{l - l_0}{l_0} = \frac{\Delta l}{l_0}. \tag{12–2}$$

Tensile strain is amount of stretch compared to original length. It has no units.

Strain is a ratio of two lengths, and the two lengths are always measured in the same units. Thus strain is a pure (dimensionless) number with no units. The **compressive strain** of a bar in compression is defined in the same way as tensile strain; it is the ratio of the decrease in length to the original length.

TABLE 12-1 Approximate Elastic Constants

Material	Young's Modulus, Y		Shear Modulus, S		Bulk Modulus, B		Poisson's Ratio, σ
	Pa	lb·in^{-2}	Pa	lb·in^{-2}	Pa	lb·in^{-2}	
Aluminum	0.70×10^{11}	10×10^6	0.30×10^{11}	3.4×10^6	0.70×10^{11}	10×10^6	0.16
Brass	0.91×10^{11}	13×10^6	0.36×10^{11}	5.1×10^6	0.61×10^{11}	8.5×10^6	0.26
Copper	1.1×10^{11}	16×10^6	0.42×10^{11}	6.0×10^6	1.4×10^{11}	20×10^6	0.32
Glass	0.55×10^{11}	7.8×10^6	0.23×10^{11}	3.3×10^6	0.37×10^{11}	5.2×10^6	0.19
Iron	1.9×10^{11}	26×10^6	0.70×10^{11}	10×10^6	1.0×10^{11}	14×10^6	0.27
Lead	0.16×10^{11}	2.3×10^6	0.056×10^{11}	0.8×10^6	0.077×10^{11}	1.1×10^6	0.43
Nickel	2.1×10^{11}	30×10^6	0.77×10^{11}	11×10^6	2.6×10^{11}	34×10^6	0.36
Steel	2.0×10^{11}	29×10^6	0.84×10^{11}	12×10^6	1.6×10^{11}	23×10^6	0.19
Tungsten	3.6×10^{11}	51×10^6	1.5×10^{11}	21×10^6	2.0×10^{11}	29×10^6	0.20

The tensile or compressive *strain* clearly depends on the tensile or compressive *stress:* The harder you pull on something, the more it stretches. Robert Hooke (1635–1703), a contemporary of Newton, discovered that when the forces are not too large, this relation can be represented approximately as a direct proportion; stress is proportional to strain, and the ratio of stress to strain is then constant. This proportionality is called **Hooke's law.**

The quotient of any stress and the corresponding strain is called an **elastic modulus.** For the particular case of tensile or compressive stress and strain, it is called **Young's modulus,** denoted by Y:

An elastic modulus describes a basic property of a material, not just a particular piece of the material.

$$Y = \frac{\text{tensile stress}}{\text{tensile strain}} = \frac{\text{compressive stress}}{\text{compressive strain}},$$

$$Y = \frac{F/A}{\Delta l/l_0} = \frac{l_0}{A}\frac{F}{\Delta l}. \tag{12–3}$$

Since a strain is a pure number, the units of Young's modulus are the same as those of stress, namely, force per unit area. Some typical values are listed in Table 12–1.

When a material stretches under tensile stress, the dimensions *perpendicular* to the direction of stress become *smaller* by an amount proportional to the fractional change in length. When you stretch a wire or a rubber band, it gets a little thinner as well as longer. If w_0 is the original width and Δw is the change in width, then

When a piece of material is stretched, it gets thinner as well as longer.

$$\frac{\Delta w}{w_0} = -\sigma\frac{\Delta l}{l_0}, \tag{12–4}$$

where σ is a dimensionless constant, different for different materials, called **Poisson's ratio.** For many common materials, σ has a value between 0.1 and 0.3. Similarly, a material under compressive stress "bulges" at the sides, and again the fractional change in width is given by Eq. (12–4).

Experiments have shown that for many materials the ratio of compressive strain to compressive stress is the same as the ratio of tensile strain to tensile stress. Hence Young's modulus describes the behavior of a material in both tension and compression.

EXAMPLE 12-2 In a small elevator, a load of 500 kg hanging from a steel cable of length 3 m and cross section 0.20 cm^2 was found to stretch the cable 0.4 cm above its no-load length. What were the stress, the strain, and the value of Young's modulus for the steel in the cable?

SOLUTION We use the definitions of stress, strain, and Young's modulus as given by Eqs. (12-1), (12-2), and (12-3):

$$\text{Stress} = \frac{F}{A} = \frac{(500 \text{ kg})(9.8 \text{ m·s}^{-2})}{2.0 \times 10^{-5} \text{ m}^2} = 2.45 \times 10^8 \text{ Pa};$$

$$\text{Strain} = \frac{\Delta l}{l_0} = \frac{0.004 \text{ m}}{3 \text{ m}} = 0.00133;$$

$$Y = \frac{\text{stress}}{\text{strain}} = \frac{2.45 \times 10^8 \text{ Pa}}{0.00133} = 1.84 \times 10^{11} \text{ Pa}.$$

Young's modulus is related to the force constant of a stretched wire or cable.

Young's modulus characterizes the elastic properties of a material under tension or compression in a way that is independent of the size or shape of the particular specimen. It does not indicate directly how a particular rod, cable, or spring made of the material will distort under given forces. We may solve Eq. (12-3) for F, obtaining

$$F = \frac{YA}{l_0} \Delta l.$$

Now suppose we represent the quantity YA/l_0 by a single letter k, which we call the **force constant;** we also call the elongation x instead of Δl. Then we have

$$F = kx.$$

We have already encountered this relationship in Sections 7-3 and 7-6 in connection with work and potential energy for a spring. It appeared again in our discussion of simple harmonic motion in Chapter 11. The elongation of a body in tension is *directly proportional* to the stretching force, and the shortening of a body in compression is directly proportional to the compressing force. Hooke's law was originally stated in this form; it was reformulated much later, by other physicists, in terms of stress and strain.

When a helical or coil spring is stretched or compressed, the stress and strain in the wire are nearly pure *shear* (to be discussed in Section 12-3), but the elongation or compression is still proportional to the stretching or compressing force, within certain limits of maximum force. That is, the equation $F = kx$ is still valid. Of course, a coil spring can be compressed only if there are spaces between the coils!

The units of the force constant are newtons per meter, dynes per centimeter, or pounds per foot. The reciprocal of the force constant—that is, the ratio of elongation or compression to force—is called the *compliance* of the spring.

12-2 BULK STRESS AND STRAIN

We have discussed tensile and compressive stresses and strains. A different stress–strain situation occurs when a solid or fluid material is subjected to a uniform pressure over its whole surface. The resulting deformation can be described in terms of the volume change of the material. A familiar example is the compression of a gas under pressure. The term *fluid* means a substance that can *flow*, so the term applies to both liquids and gases.

A force transmitted through a cross section of a fluid at rest must always be perpendicular to that section; if we tried to exert a force parallel to a section, the fluid would slip sideways to counteract the effort. In the language to be introduced in Section 12–3, there can be no *shear stress* in a fluid at rest. Similarly, when a solid is immersed in a fluid and both are at rest, the forces that the fluid exerts on the surface of the solid are always perpendicular to the surface at each point.

Figure 12–4 shows a fluid in a cylinder with a piston. We apply a downward force to the piston, and this is transmitted throughout the fluid. The triangle is a side view of a wedge-shaped portion of the fluid. We will show that the force per unit area (pressure) is the same on all surfaces of this wedge of fluid. If we neglect the weight of the fluid, the only forces on this wedge are those exerted on its imaginary surfaces by the surrounding fluid, and each force must be perpendicular to the corresponding surface. Let F_x, F_y, and F represent the forces on the three faces, as shown. Since the fluid is in equilibrium, it follows that

$$F \sin \theta = F_x, \qquad F \cos \theta = F_y.$$

Also,

$$A \sin \theta = A_x, \qquad A \cos \theta = A_y.$$

Dividing the upper equations by the lower, we find

$$\frac{F}{A} = \frac{F_x}{A_x} = \frac{F_y}{A_y}.$$

Hence the force per unit area is the *same* on all these surfaces. It does not depend on their orientation, and it is always a compression. The force per unit area on any of these surfaces is called the **pressure** p in the fluid:

$$p = \frac{F}{A}, \qquad F = pA. \tag{12–5}$$

The fact that pressure applied to the surface of a fluid is transmitted unchanged to all parts of the fluid is called Pascal's law.

Pressure has the same units as stress. Commonly used units include 1 Pa ($= 1 \text{ N·m}^{-2}$), 1 dyn·cm^{-2}, 1 lb·ft^{-2}, and 1 lb·in^{-2}. Also in common use is the *atmosphere*, abbreviated atm. One atmosphere is defined to be the average pressure of the earth's atmosphere at sea level:

One atmosphere = 1 atm = 1.013×10^5 Pa = 14.7 lb·in^{-2}.

Pressure is a scalar quantity, not a vector quantity; no direction can be assigned to it. The force against any area within (or at the boundary surface of) a fluid at rest and under pressure is perpendicular to the area, regardless of the orientation of the area. This is what is meant by the statement that the pressure in a fluid is the same in all directions.

The stress within a solid material is a pressure, if the force per unit area is the same at *all* points of the surface, and if the force at each point is normal to the surface and directed inward. This is *not* the case in Fig. 12–2, where forces

A fluid at rest cannot have shear stress.

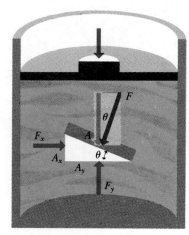

12–4 A fluid under hydrostatic pressure. The force on a surface in any direction is normal to the surface.

Pressure is force per unit area.

If weight can be neglected, the pressure in a fluid at rest is the same everywhere in the fluid.

are applied at the ends of the bar only, but it is automatically the case if a solid is immersed in a fluid under pressure.

Pressure is the *stress* in a volume deformation. The corresponding strain is **volume strain;** it is defined as the fractional change in volume, the ratio of the volume change ΔV to the original volume V_0:

$$\text{Volume strain} = \frac{\Delta V}{V_0}. \qquad (12-6)$$

Like tensile or compressive strain, volume strain is a pure number, without units.

Experimental evidence shows that for sufficiently small pressures, the volume strain is proportional to the stress (pressure). The corresponding elastic modulus (ratio of stress to strain) is called the **bulk modulus,** denoted by B. The general definition of the bulk modulus is the (negative) ratio of an infinitesimal pressure change dp to the volume strain dV/V_0 (fractional change in volume) that it causes:

$$B = -\frac{dp}{dV/V_0} = -V_0 \frac{dp}{dV}. \qquad (12-7)$$

The minus sign is included in the definition of B because an *increase* in pressure always causes a *decrease* in volume. That is, if dp is positive, dV is negative. By including a minus sign in its definition, we make B itself a positive quantity.

If the pressure change is not too great, the ratio dp/dV is constant, the bulk modulus is constant, and we can replace dp and dV in Eq. (12–7) by finite changes Δp and ΔV in pressure and volume. The bulk modulus of a *gas*, however, changes markedly with pressure, and the general definition of B must be used for gases.

The reciprocal of the bulk modulus is called the **compressibility** k. From Eq. (12–7),

$$k = \frac{1}{B} = -\frac{dV/V_0}{dp} = -\frac{1}{V_0}\frac{dV}{dp}. \qquad (12-8)$$

The compressibility of a material thus equals the *fractional decrease in volume,* $-dV/V_0$, *for a small increase dp in pressure.*

Table 12–1 includes values of the bulk modulus. Its units are the same as those of pressure, and the units of compressibility are those of *reciprocal pressure.* The compressibility of water, from Table 12–2, is 46.4×10^{-6} atm^{-1}. This means that the volume decreases by 46.4-millionths of the original volume for each atmosphere increase in pressure.

TABLE 12–2 Compressibilities of Liquids

Liquid	Compressibility, k		
	Pa^{-1}	(lb·in^{-2})$^{-1}$	atm^{-1}
Carbon disulfide	93×10^{-11}	64×10^{-7}	94×10^{-6}
Ethyl alcohol	110×10^{-11}	76×10^{-7}	111×10^{-6}
Glycerine	21×10^{-11}	14×10^{-7}	21×10^{-6}
Mercury	3.7×10^{-11}	2.6×10^{-7}	3.8×10^{-6}
Water	45.8×10^{-11}	31.6×10^{-7}	46.4×10^{-6}

Volume strain is change in volume, compared to original volume.

Bulk modulus might be called incompressibility: The larger the bulk modulus, the less compressible the material is.

EXAMPLE 12–3 The volume of oil contained in a certain hydraulic press is 0.2 m³ = 200 liters. Find the decrease in volume of the oil when subjected to a pressure increase of 2.04×10^7 Pa. The compressibility of the oil is 20×10^{-6} atm⁻¹.

SOLUTION Because we are given the compressibility in atm⁻¹, we first convert the pressure to atmospheres:

$$2.04 \times 10^7 \text{ Pa} = 201 \text{ atm.}$$

From Eq. (12–8),

$$\Delta V = -kV \Delta p = -(20 \times 10^{-6} \text{ atm}^{-1})(0.2 \text{ m}^3)(201 \text{ atm})$$
$$= -8.04 \times 10^{-4} \text{ m}^3 = -0.804 \text{ L.}$$

This represents a substantial compression of the oil under the action of a very large pressure, nearly 3000 pounds per square inch.

Oil in a hydraulic press is often under great pressure, and the compression can be quite appreciable.

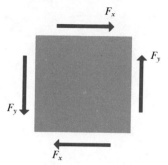

12–5 A body under shear stress.

12–3 SHEAR STRESS AND STRAIN

A third kind of stress is shown in Fig. 12–5. This stress is called **shear stress,** and we define it as the force tangent to a material surface, divided by the area on which the force acts. Shear stress, like the other two types of stress, is a force per unit area.

A shear deformation is shown in Fig. 12–6. The lightly shaded outline *abcd* represents an unstressed block of material, and the lines *a'b'c'd'* represent the block under stress. The centers of the stressed and unstressed block coincide in part (a). In part (b) the deformation is the same as in (a), but the edges *ad* and *a'd'* coincide. Under shear stress the lengths of the faces remain very nearly constant; all dimensions parallel to the diagonal *ac* increase in length, and those parallel to the diagonal *bd* decrease in length. This type of strain is called a **shear strain;** it is defined as the ratio of the displacement *x* of corner *b* to the transverse dimension *h*:

$$\text{Shear strain} = \frac{x}{h} = \tan \phi. \qquad (12\text{–}9)$$

In practice, *x* is nearly always much smaller than *h*, $\tan \phi$ is very nearly equal to ϕ, and the strain is simply the angle ϕ (measured in radians, of course). Like all strains, shear strain is a pure number with no units because it is a ratio of two lengths.

Once again, we find experimentally that if the forces are not too large, the shear strain is proportional to the shear stress. The corresponding elastic modulus (ratio of shear stress to shear strain) is called the **shear modulus,** denoted by *S*:

$$S = \frac{\text{shear stress}}{\text{shear strain}}$$
$$= \frac{F/A}{x/h} = \frac{h}{A}\frac{F}{x} = \frac{F/A}{\phi}, \qquad (12\text{–}10)$$

with *x* and *h* defined as in Fig. 12–6.

Shear stress results when forces are applied tangent to the surfaces of a solid material.

Shear strain is distortion of shapes and angles without volume change.

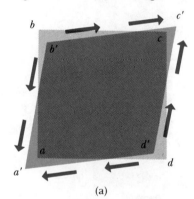

(a)

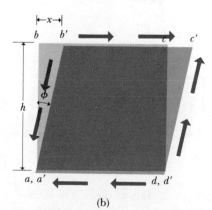

(b)

12–6 Change in shape of a block in shear. The shear strain is defined as *x/h*.

For most materials, the shear modulus is one-third to one-half as large as Young's modulus. The shear modulus is also called the *modulus of rigidity,* or the *torsion modulus.* Representative values of shear modulus are given in Table 12–1. The shear modulus has significance only for *solid* materials. A liquid or gas flows freely under the action of a shear stress, and a fluid at rest cannot sustain such a stress.

EXAMPLE 12–4 Suppose the object in Fig. 12–6 is a brass plate 1.0 m square and 0.5 cm thick. How large a force F must be exerted on each of its edges if the displacement x in Fig. 12–6b is 0.02 cm? The shear modulus of brass is 0.36×10^{11} Pa.

SOLUTION The shear stress on each edge is

$$\text{Shear stress} = \frac{F}{A} = \frac{F}{(1.0 \text{ m})(0.005\text{m})} = (200 \text{ m}^{-2})F.$$

The shear strain is

$$\text{Shear strain} = \frac{x}{h} = \frac{2 \times 10^{-4} \text{ m}}{1.0 \text{ m}} = 2.0 \times 10^{-4}.$$

$$\text{Shear modulus } S = \frac{\text{stress}}{\text{strain}} = 0.36 \times 10^{11} \text{ Pa} = \frac{(200 \text{ m}^{-2})F}{2.0 \times 10^{-4}},$$

$$F = 3.6 \times 10^4 \text{ N}.$$

Although the three elastic moduli and Poisson's ratio have been discussed separately, these four quantities are not completely independent. For materials having no distinction between various directions (i.e., *isotropic* materials), only two of these are really independent. For example, the bulk and shear moduli may be expressed in terms of Young's modulus and Poisson's ratio:

$$B = \frac{Y}{3(1 - 2\sigma)}, \qquad S = \frac{Y}{2(1 + \sigma)}. \qquad (12\text{–}11)$$

For materials having directional properties, such as wood (which has a grain direction) or a single crystal of a material, these relations do not hold, and the elastic behavior is more complex.

Another aspect of the relations among the elastic constants is that different cross sections in a body have different states of stress. As an example, consider the body under shear stress in Fig. 12–7, acted on by the pairs of forces F_x and F_y distributed over its surfaces. The block is in equilibrium, and every portion of it must also be in equilibrium. Thus if we consider the triangular section shown in Fig. 12–7b, the distributed forces over the diagonal face must have a resultant F whose components are equal in magnitude to F_x and F_y. Thus the stress at the diagonal face is a pure *compression,* even though the stresses at the right and bottom faces are shear stresses. Similarly, the diagonal face shown in Fig. 12–7c is in pure *tension.*

The same thing happens with the stretched bar we used in Section 12–1 to introduce tensile stress. A cross section perpendicular to the length of the bar

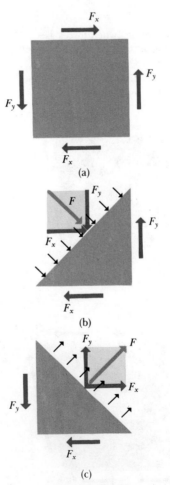

12–7 (a) A body in shear. The stress on one diagonal, part (b), is a pure compression; that on the other, part (c), is a pure tension.

TABLE 12–3   Stresses and Strains

Type of Stress	Stress	Strains	Elastic Modulus	Name of Modulus
Tension or compression	$\dfrac{F_\perp}{A}$	$\dfrac{\Delta l}{l_0}$	$Y = \dfrac{F_\perp/A}{\Delta l/l_0}$	Young's modulus
Hydrostatic pressure	$p\left(= \dfrac{F_\perp}{A}\right)$	$\dfrac{\Delta V}{V_0}$	$B = -\dfrac{p}{\Delta V/V_0}$	Bulk modulus
Shear	$\dfrac{F_\parallel}{A}$	$\tan\phi \approx \phi$	$S = \dfrac{F_\parallel/A}{\phi}$	Shear modulus

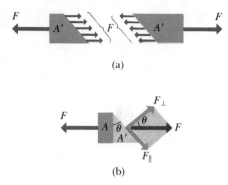

12–8 A bar in tension. The stress at an inclined section (a) can be resolved into (b) a normal stress $F_\perp/A'$, and a tangential or shear stress $F_\parallel/A'$.

has a purely tensile stress, as shown in Fig. 12–1. But if we take a cross section at an angle, as in Fig. 12–8a, the stress at this face can be represented as having both tensile and shear components, as shown in Fig. 12–8b. The force acting at a particular cross section has a definite direction and magnitude and can be represented by means of its components, but these depend also on the orientation of the section. To describe completely the state of stress in a material, we must describe three mutually perpendicular cross-sectional orientations, perhaps using three unit vectors, and then describe the three components of force (per unit area) at each cross section. The resulting set of nine numbers is called the *stress tensor* and is an example of a class of physical quantities called *tensors*.

The various types of stress, strain, and elastic moduli are summarized in Table 12–3.

12–4 ELASTICITY AND PLASTICITY

We are now ready to examine the limitations of Hooke's law. Suppose we plot a graph of stress as a function of the corresponding strain. If Hooke's law is obeyed, stress is directly proportional to strain and the graph is a straight line. Real materials show several types of departures from this idealized behavior.

Figure 12–9 shows a typical stress–strain graph for a metal such as copper or soft iron. The stress in this case is a simple tensile stress, and the strain is shown as the percent elongation. The first portion of the curve, up to a strain of less than 1%, is a straight line, indicating Hooke's-law behavior with stress directly proportional to strain. This straight-line portion ends at point *a*; the stress at this point is called the **proportional limit.**

From *a* to *b*, stress and strain are no longer proportional, but if the load is removed at any point between *O* and *b*, the curve is retraced and the material returns to its original length. In the entire region *Ob* the material is said to be *elastic* or to show *elastic behavior.* Point *b*, the end of this region, is called the **yield point,** and the corresponding stress is called the **elastic limit.** Up to this point the forces exerted by the material are *conservative.* When the load is removed, the material returns to its original shape, and the energy put into the material in causing the deformation is recovered. The deformation is said to be *reversible.*

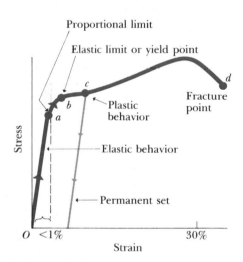

12–9 Typical stress–strain diagram for a ductile metal under tension.

When the stress exceeds the proportional limit, stress and strain are no longer proportional.

A marble cylinder is deformed by compressive stresses applied at its ends, with different confining pressures on the side surface. (a) The original undeformed shape; (b) 20% compressive strain with a confining pressure of 270 atm; (c) 20% strain with a confining pressure of 445 atm. The sample is brittle at the lower confining pressure but becomes ductile at higher confining pressure. (Photo by M. S. Paterson, Australian National University.)

When the stress exceeds the elastic limit, the material does not snap back all the way when the stress is removed.

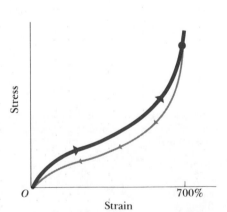

12–10 Typical stress–strain diagram for vulcanized rubber, showing elastic hysteresis.

If the stress is increased further, the strain increases rapidly, but when the load is removed at some point beyond *b*, say *c*, the material does not come back to its original length but traverses the thin line in Fig. 12–9. The length at zero stress is now *greater* than the original length, and the material is said to have a *permanent set*. Further increase of load beyond *c* produces a large increase in strain (even if the stress decreases) until a point *d* is reached at which *fracture* takes place. From *b* to *d*, the material is said to undergo *plastic flow*, or *plastic deformation*. A plastic deformation is *irreversible;* when the stress is removed, the material does not return to its original state. If a large amount of plastic deformation takes place between the elastic limit and the fracture point, the metal is said to be *ductile;* but if fracture occurs soon after the elastic limit is passed, the metal is said to be *brittle*. A soft iron wire that can have considerable permanent stretch without breaking is ductile, while a spring-steel wire that breaks soon after its elastic limit is reached is brittle.

Figure 12–10 shows a stress–strain curve for a typical sample of vulcanized rubber that has been stretched to over seven times its original length. During *no* portion of this curve is the stress proportional to the strain! The substance, however, is elastic, in the sense that when the load is removed, the rubber returns to its original length. On decreasing the load, the stress–strain curve is *not* retraced but follows the thin curve of Fig. 12–10.

The lack of coincidence of the curves for increasing and decreasing stress is known as *elastic hysteresis*. When the stress–strain relation has this behavior, the associated forces are *not* conservative, since the work done by the material in returning to its original shape is *less* than the work required to deform it. It can be shown that the area bounded by the two curves—that is, the area of the *hysteresis loop*—is proportional to the energy dissipated within the elastic material.

Some types of rubber have large elastic hysteresis, and these materials are very useful as vibration absorbers. If a block of such material is placed between a piece of vibrating machinery and the floor, elastic hysteresis takes place during each cycle of vibration. Mechanical energy is converted to a form known as internal energy, causing a rise in temperature of the material. As a result, only a small amount of energy of vibration is transmitted to the floor.

The stress required to cause actual fracture of a material is called the **breaking stress,** or the *ultimate strength.* Two materials, such as two steels, may have very similar elastic constants but vastly different breaking stresses. Table 12–4 gives a few typical values of breaking stress for several materials in tension.

When the stress exceeds the breaking stress, the material cracks or breaks.

TABLE 12–4 Breaking Stresses of Materials

Material	Breaking Stress (Pa or N·m^{-2})
Aluminum	2.2×10^8
Brass	4.7×10^8
Glass	10×10^8
Iron	3.0×10^8
Phosphor bronze	5.6×10^8
Steel	11.0×10^8

SUMMARY

Tensile stress is tensile force per unit area, F/A. Tensile strain is fractional change in length, $\Delta l/l_0$. Young's modulus Y is the ratio of tensile stress to tensile strain:

$$Y = \frac{F/A}{\Delta l/l_0}. \qquad (12\text{–}3)$$

Compressive stress and strain are defined the same way as tensile stress and strain; for many materials, Young's modulus has the same value for both tensile and compressive stresses and strains. Poisson's ratio σ is the negative of the ratio of the fractional change of width $\Delta w/w_0$ of a bar under tension or compression and the fractional change of length $\Delta l/l_0$:

$$\frac{\Delta w}{w_0} = -\sigma \frac{\Delta l}{l_0}. \qquad (12\text{–}4)$$

The force constant k is the ratio of force F to elongation or compression x for a specific specimen of material or a specific spring: $F = kx$. The reciprocal of the force constant is called compliance.

The bulk modulus B is the negative of the ratio of pressure change dp (bulk stress) to fractional volume change dV/V_0:

$$B = -\frac{dp}{dV/V_0}. \qquad (12\text{–}7)$$

Compressibility k is the reciprocal of bulk modulus: $k = 1/B$.

Shear stress is force per unit area F/A for a force applied parallel to a surface. Shear strain is the angle ϕ shown in Fig. 12–6. The shear modulus S is the ratio of shear stress to shear strain:

$$S = \frac{F/A}{\phi}. \qquad (12\text{–}10)$$

The proportional limit is the maximum stress for which stress and strain are proportional, that is, for which Hooke's law is valid. The elastic limit is the stress beyond which irreversible deformation occurs. The breaking stress, or ultimate strength, is the stress at which the material breaks.

KEY TERMS

tension
stress
tensile stress
tensile strain
compression
compressive stress
compressive strain
Hooke's law
elastic modulus
Young's modulus
Poisson's ratio
force constant
pressure
volume strain
bulk modulus
compressibility
shear stress
shear strain
shear modulus
proportional limit
yield point
elastic limit
breaking stress

QUESTIONS

12–1 When a wire is bent back and forth, it becomes hot. Why?

12–2 When a wire is stretched, the cross section decreases somewhat from the undeformed value. How does this affect the definition of tensile stress? Should the original or the decreased value be used?

12–3 Is the work required to stretch a metal rod proportional to the amount of stretch? Explain.

12–4 Why is concrete with steel reinforcing rods embedded in it stronger than plain concrete?

12–5 Is the shear modulus of steel greater or less than that of jelly? By roughly what factor?

12–6 Is the bulk modulus for steel greater or less than that of air? By roughly what factor?

12–7 Looking at the molecular structure of matter, discuss why gases are generally more compressible than liquids and solids.

12–8 Climbing ropes used by mountaineers are usually made of nylon. Would a steel cable of equal strength be just as good? What advantages and disadvantages would it have, compared to nylon?

12–9 How could you measure the force constants of the springs in an automobile?

12–10 In a nylon mountaineering rope, is a lot of mechanical hysteresis desirable or undesirable?

12–11 A spring scale for measuring weight has a spring and a scale that indicates how much the spring stretches under a given weight. Such scales are often illegal in commerce. (Cf. the inscription "no springs, honest weight" found on some grocery-store scales.) Why? (What property of a metal spring could affect its accuracy?)

12–12 Coil springs found in automobile suspension systems are sometimes designed *not* to obey Hooke's law. How can a spring be made so as to achieve this result? Why is it desirable?

12–13 Compare the mechanical properties of a steel cable, made by twisting many thin wires together, with those of a solid steel wire of the same diameter. What advantages does each have?

12–14 Electric power lines are sometimes made by using wires with steel core and copper jacket, or strands of copper and steel twisted together. Why?

12–15 A spring is compressed, clamped in its compressed position, and then dissolved in acid. What becomes of the elastic potential energy?

12–16 When rubber mounting blocks are used to absorb machine vibrations through mechanical hysteresis, as discussed in Section 12–4, what becomes of the energy associated with the vibrations?

EXERCISES

Section 12–1 Tensile Stress and Strain

12–1 A certain metal rod 4 m long and 0.5 cm^2 in cross section is found to stretch 0.2 cm under a tension of 12,000 N. What is Young's modulus for this metal?

12–2 A nylon rope used by mountaineers elongates 1.5 m under the weight of an 80-kg climber.

a) If the rope is 50 m in length and 9 mm in diameter, what is Young's modulus for this material?

b) If Poisson's ratio for nylon is 0.2, find the change in diameter under this stress.

12–3 A steel bar 0.2 cm square and 5 m long is stretched with a force of 400 N at each end. Find the stress, the strain, the total elongation, and the fractional change in thickness of the bar. Use the data for steel given in Table 12–1.

12–4 Two round rods, one of steel, the other of brass, are joined end to end. Each rod is 0.5 m long and 2 cm in diameter. The combination is subjected to tensile forces of 5000 N.

a) What is the strain in each rod?

b) What is the elongation of each rod?

c) What is the change in diameter of each rod?

Use the data for steel given in Table 12–1.

12–5 A 5-kg mass hangs on a vertical steel wire 0.5 m long and 0.004 cm^2 in cross section. Hanging from the bottom of this mass is a similar steel wire that supports a 10-kg mass. Compute the

a) longitudinal strain   b) elongation

of each wire. Use the data for steel given in Table 12–1.

12–6 A copper wire of length 2 m has a diameter of 3 mm. What is the force constant k for this wire?

12–7 A circular steel wire 2 m long is to stretch no more than 0.2 cm when a tensile force of 300 N is applied to each end of the wire. What minimum diameter of wire is required?

12–8 A mass of 100 kg suspended from a wire whose unstretched length l_0 is 4 m is found to stretch the wire by 0.004 m. The cross-sectional area of the wire, which can be assumed constant, is 0.1 cm^2.

a) If the load is pulled down a small additional distance and released, find the frequency at which it will vibrate.

b) Compute Young's modulus for the wire.

Section 12–2 Bulk Stress and Strain

12–9 A specimen of oil having an initial volume of 1000 cm^3 is subjected to a pressure increase of 12×10^5 Pa, and the volume is found to decrease by 0.3 cm^3. What is the bulk modulus of the material? The compressibility?

12–10 In the Challenger Deep of the Marianas Trench, the depth of sea water is 10.9 km, and the pressure is 1.10×10^8 Pa (about 1.09×10^3 atm).

a) If a cubic meter of water is taken from the surface to this depth, what is the change in its volume? (Normal atmospheric pressure is about 1.0×10^5 Pa.)

b) What is the density of sea water at this depth? (At the surface, sea water has a density of 1.03×10^3 kg·m^{-3}.)

Section 12–3 Shear Stress and Strain

12–11 Two strips of metal are riveted together at their ends by four rivets, each of diameter 0.5 cm. What is the maximum tension that can be exerted by the riveted strip if the shearing stress on each rivet is not to exceed 6×10^8 Pa? Assume each rivet to carry one-quarter of the load.

12–12 In Fig. 12–5, suppose the object is a square steel plate, 10 cm on a side and 1 cm thick. Find the magnitude of force required on each of the four sides to cause a shear strain of 0.01.

Section 12–4 Elasticity and Plasticity

12–13 The elastic limit of a steel elevator cable is 2.75×10^8 N·m^{-2}. Find the maximum upward acceleration that can be given a 900-kg elevator when supported by a cable whose cross section is 3 cm^2, if the stress is not to exceed $\frac{1}{4}$ of the elastic limit.

12–14 A steel wire has the following properties:

> Length = 5 m
>
> Cross section = 0.05 cm^2
>
> Young's modulus = 1.8×10^{11} Pa
>
> Shear modulus = 0.6×10^{11} Pa
>
> Proportional limit = 3.6×10^8 Pa
>
> Breaking stress = 7.2×10^8 Pa

The wire is fastened at its upper end and hangs vertically.

a) How great a load can be supported without exceeding the proportional limit?

b) How much will the wire stretch under this load?

c) What is the maximum load that can be supported?

PROBLEMS

12–15 A 15-kg mass, fastened to the end of a steel wire of unstretched length 0.5 m, is whirled in a vertical circle with an angular velocity of 2 rev·s^{-1} at the bottom of the circle. The cross section of the wire is 0.02 cm^2. Calculate the elongation of the wire when the weight is at the lowest point of the path.

12–16 A copper wire 4 m long and 1.0 mm in diameter was given the test below. A load of 20 N was originally hung from the wire to keep it taut. The position of the lower end of the wire was read on a scale:

Added load, N	Scale reading, cm
0	3.02
10	3.07
20	3.12
30	3.17
40	3.22
50	3.27
60	3.32
70	4.27

a) Make a graph of these values, plotting the increase in length horizontally and the added load vertically.

b) Calculate the value of Young's modulus.

c) What was the stress at the proportional limit?

12–17 A rod 1.05 m long, whose weight is negligible, is supported at its ends by wires A and B of equal length, as shown in Fig. 12–11. The cross section of A is 1 mm^2; that

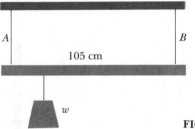

FIGURE 12–11

of B, 2 mm^2. Young's modulus for wire A is 2.4×10^{11} Pa; and for B, 1.6×10^{11} Pa. At what point along the bar should a weight w be suspended in order to produce

a) equal stresses in A and B?

b) equal strains in A and B?

12–18 One area where compressive strength is of everyday importance is in our bones. Young's modulus for bone is about 1.0×10^{10} N·m^{-2}. Bone can take only about a 1.0% change in its length before fracturing.

a) What is the maximum force that can be applied to a bone whose minimum cross-sectional area is 3.0 cm^2, which is approximately the cross-sectional area of a tibia at its narrowest point in the human leg?

b) Estimate from what maximum height a 70-kg person (one weighing about 150 lb) could jump and not fracture the tibia. Take the time between when the person first touches the floor and when he has stopped to be 0.02 s.

12–19 A student has a sling-shot made of a rubber band whose unstretched shape is shown by the shorter band in Fig. 12–12. The student pulls back on the band so it is then in the longer position. To do this the student pulls straight back with a force of 20.0 N. What is Young's modulus for the rubber bands if their cross-sectional area is 10.0 mm²?

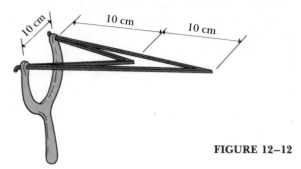

FIGURE 12–12

12–20 An amusement park ride consists of seats attached to cables as shown in Fig. 12–13. Each steel cable has a length of 20.0 m and cross-sectional area of 7.00 cm².

a) What is the amount the cable is stretched when the ride is at rest? Assume the seats plus two people seated in them have a total weight of 4000 N.

b) The ride, when turned on, has a maximum angular velocity of 0.90 rad·s⁻¹. How much will the cable then be stretched?

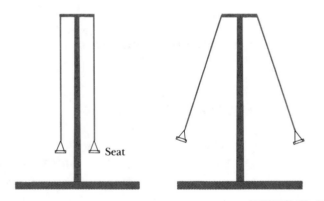

FIGURE 12–13

12–21 A copper rod of length 2 m and cross-sectional area 2.0 cm² is fastened end to end to a steel rod of length L and cross-sectional area 1.0 cm². The compound rod is subjected to equal and opposite pulls of magnitude 3 × 10⁴ N at its ends.

a) Find the length L of the steel rod if the elongations of the two rods are equal.

b) What is the stress in each rod?

c) What is the strain in each rod?

12–22 A moonshiner produces pure ethanol (ethyl alcohol), late at night, and stores it in a stainless steel tank in the form of a cylinder 0.20 m in diameter with a tight-fitting piston at the top. The total volume of the tank is 200 L (0.2 m³). In an attempt to squeeze a little more into the tank, he piles lead bricks on the piston, so the total mass of bricks and piston is 120 kg. What additional volume of ethanol can he squeeze into the tank?

12–23 A bar of cross section A is subjected to equal and opposite tensile forces F at its ends. Consider a plane through the bar making an angle θ with a plane at right angles to the bar (Fig. 12–14).

a) What is the tensile (normal) stress at this plane, in terms of F, A, and θ?

b) What is the shear (tangential) stress at the plane, in terms of F, A, and θ?

c) For what value of θ is the tensile stress a maximum?

d) For what value of θ is the shear stress a maximum?

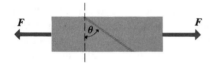

FIGURE 12–14

12–24 In Fig. 12–5, suppose the object is a square aluminum plate 0.2 m on a side. When forces are applied to the four edges, of equal magnitude 1.0 × 10⁶ N each, we want the resulting shear strain to be no greater than 0.01. What minimum thickness of plate is required?

CHALLENGE PROBLEMS

12–25 The compressibility of sodium is to be measured by observing the displacement of the piston in Fig. 12–4 when a force is applied. The sodium is immersed in an oil that fills the cylinder below the piston. Assume that the piston and walls of the cylinder are perfectly rigid and that there is no friction and no oil leak. Compute the compressibility of the sodium in terms of the applied force F, the piston displacement x, the piston area A, the initial volume of oil V_0, the initial volume of sodium v_0, and the compressibility of oil k_0.

12–26 The equation of state (the equation relating pressure, volume, and temperature) for an ideal gas is $pV = nRT$, where n and R are constants.

a) Show that if the gas is compressed while the temperature T is held constant, the bulk modulus is equal to the pressure.

b) When an ideal gas is compressed without the transfer of any heat in or out of it, the pressure and volume are related by pV^γ = constant, where γ is a constant having different values for different gases. Show that in this case the bulk modulus is given by $B = \gamma p$.

12–27 A 5-kg mass is hung from a vertical steel wire 2 m long and 0.004 cm^2 in cross section. The wire is securely fastened to the ceiling. Calculate

a) the amount the wire is stretched by the hanging mass;

b) the external force P needed to pull the mass very slowly downward 0.05 cm from its equilibrium position;

c) the work done by gravity when the mass moves downward 0.05 cm;

d) the work done by the force P;

e) the work done by the force the wire exerts on the mass;

f) the change in the elastic potential energy (the potential energy associated with the tensile stress in the wire) when the mass moves downward 0.05 cm.

13

FLUID MECHANICS

IN THIS CHAPTER WE RESUME OUR STUDY OF THE MECHANICAL PROPERTIES OF real materials. Chapter 12 involved the *elastic* properties of materials, and now we study the mechanical properties of fluids. The behavior of fluids is of great current interest both in pure science and in industry. A *fluid* is any substance that can flow; we use the term for both liquids and gases. In most circumstances a gas is easily compressed, while liquids are nearly incompressible. In this chapter we neglect the small volume changes that occur in a liquid under pressure.

A fluid has no definite shape.

Fluid *statics* is the study of fluids at rest, and fluid *dynamics* is the study of fluids in motion. We begin with equilibrium situations, including the concepts of density, pressure, buoyancy, and surface tension. The conditions for equilibrium are based on Newton's first law. Fluid dynamics is much more complex—indeed, it is one of the most complex branches of mechanics. Fortunately, many important situations can be represented by idealized models that are simple enough to permit detailed analysis. Even so, we will barely scratch the surface of this broad and interesting topic.

A fluid at rest is an equilibrium situation.

13–1 DENSITY

The **density** of a material is defined as its mass per unit volume. A homogeneous material has the same density throughout. The SI unit of density is one kilogram per cubic meter (1 kg·m^{-3}). The cgs unit, one gram per cubic centimeter (1 g·cm^{-3}), is also widely used. We use the Greek letter ρ (rho) for density. If a mass m of material has volume V, the density ρ is given by

$$\rho = \frac{m}{V}, \qquad m = \rho V. \qquad (13\text{–}1)$$

Densities of several common solids and liquids at ordinary temperatures are listed in Table 13–1. The conversion factor

$$1 \text{ g·cm}^{-3} = 1000 \text{ kg·m}^{-3}$$

is useful with this table. The densest material found on earth is the metal

TABLE 13–1 Densities

Material	Density g·cm^{-3}
Aluminum	2.7
Brass	8.6
Copper	8.9
Gold	19.3
Ice	0.92
Iron	7.8
Lead	11.3
Platinum	21.4
Silver	10.5
Steel	7.8
Mercury	13.6
Ethyl alcohol	0.81
Benzene	0.90
Glycerin	1.26
Water	1.00
Sea water	1.03

osmium (22.5 g·cm^{-3}). The density of air is about 0.0012 g·cm^{-3}, but the density of white-dwarf stars is of the order of 10^6 g·cm^{-3}, and neutron stars 10^{15} g·cm^{-3}!

The **specific gravity** of a material is the ratio of its density to that of water; it is a pure (unitless) number. "Specific gravity" is a poor term, since it has nothing to do with gravity; "relative density" would be preferable.

Density measurements are an important analytical technique. For example, we can determine the charge condition of a storage battery by measuring the density of its electrolyte, a sulfuric acid solution. As the battery discharges, the sulfuric acid (H_2SO_4) combines with lead in the battery plates to form insoluble lead sulfate ($PbSO_4$), decreasing the concentration of the solution. The density decreases from about 1.30 g·cm^{-3} for a fully charged battery to 1.15 g·cm^{-3} for a discharged battery. Similarly, permanent-type antifreeze is usually a solution of ethylene glycol (density 1.12 g·cm^{-3}) in water, with small quantities of additives to retard corrosion. The glycol concentration, which determines the freezing point of the solution, can be found from a simple density measurement. Both these measurements are performed routinely in service stations with the aid of a hydrometer, which measures density by observation of the level at which a calibrated body floats in a sample of the solution. The hydrometer is discussed in Section 13–3.

13-2 PRESSURE IN A FLUID

When we introduced the concept of fluid pressure in Section 12–2, we neglected the *weight* of the fluid and assumed that the pressure was the same everywhere in the fluid. This is not really true, of course. Atmospheric pressure is greater at sea level than on high mountains, and the pressure of water in a lake or in the ocean increases with increasing depth below the surface. Thus we need to refine our concept of pressure. We define the **pressure** p at a point in a fluid as the ratio of the normal force dF on a small area dA around that point, to the area:

$$p = \frac{dF}{dA}, \qquad dF = p\, dA. \tag{13-2}$$

If the pressure is the same at all points of a finite plane surface of area A, these equations reduce to Eq. (12–5):

$$p = \frac{F}{A}, \qquad F = pA.$$

We can now derive a general relation between the pressure p at any point in a fluid in a gravitational field and the elevation of the point y. If the fluid is in equilibrium, every volume element is in equilibrium. Consider an element in the form of a thin slab, shown in Fig. 13–1, with thickness dy and face area A. If ρ is the density of the fluid (assumed to be uniform throughout), the mass of the volume element is $\rho A\, dy$ and its weight dw is $\rho g A\, dy$. The force exerted on the element by the surrounding fluid is everywhere normal to its surface. By symmetry, the resultant horizontal force on its vertical sides is zero. The upward force on its lower face is pA, and the downward force on its upper face is $(p + dp)A$. Since it is in equilibrium,

$$\Sigma F_y = 0, \qquad pA - (p + dp)A - \rho g A\, dy = 0,$$

Specific gravity compares the density of a material to the density of water.

Some practical uses for density measurements

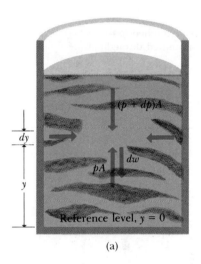

(a)

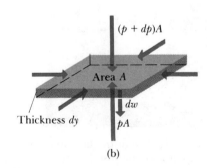

(b)

13-1 Forces on an element of fluid in equilibrium.

Because of the weight of a fluid, pressure varies with depth.

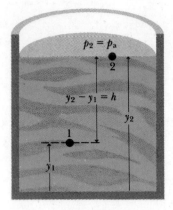

13–2 The pressure at a depth h in a liquid is greater than the surface pressure p_a by $\rho g h$.

Pressure applied to the surface of a fluid is transmitted throughout the fluid.

How a hydraulic jack works: lifting a heavy weight with a small force

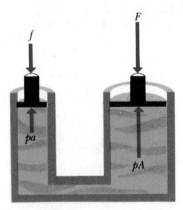

13–3 Principle of the hydraulic jack, an application of Pascal's law.

Gauge pressure: the difference between inside and outside pressures

and therefore

$$\frac{dp}{dy} = -\rho g. \qquad (13\text{–}3)$$

Since ρ and g are both positive quantities, it follows that a positive dy (an increase of elevation) is accompanied by a negative dp (decrease of pressure). If p_1 and p_2 are the pressures at elevations y_1 and y_2 above some reference level, and ρ and g are constant, then

$$p_2 - p_1 = -\rho g(y_2 - y_1) \qquad (13\text{–}4)$$

Let us apply this equation to a liquid in an open container, as shown in Fig. 13–2. Take point 1 at any level and let p represent the pressure at this point. Take point 2 at the surface of the liquid, where the pressure is atmospheric pressure, p_a. Then

$$p_a - p = -\rho g(y_2 - y_1),$$
$$p = p_a + \rho g h. \qquad (13\text{–}5)$$

The pressure depends only on depth; the *shape* of the container does not matter. It also follows from Eq. (13–5) that if the pressure p_a is increased in any way, say by inserting a piston on the top surface and pressing down on it, the pressure p at any depth must increase by exactly the same amount. This fact was recognized in 1653 by the French scientist Blaise Pascal (1623–1662) and is called **Pascal's law:** *Pressure applied to an enclosed fluid is transmitted undiminished to every portion of the fluid and the walls of the containing vessel.*

The hydraulic jack shown schematically in Fig. 13–3 illustrates Pascal's law. A piston of small cross-sectional area a is used to exert a small force f directly on a liquid such as oil. The pressure $p = f/a$ is transmitted through the connecting pipe to a larger piston of area A. Since the pressure is the same in both cylinders,

$$p = \frac{f}{a} = \frac{F}{A} \quad \text{and} \quad F = \frac{A}{a}f.$$

Thus the hydraulic jack is a force-multiplying device with a multiplication factor equal to the ratio of the areas of the two pistons. Barber chairs, dentist chairs, car lifts and jacks, and hydraulic brakes all use this principle.

In deriving Eq. (13–4) we have assumed that the density ρ of the fluid is constant; this is a reasonable assumption for liquids, which are relatively incompressible, but it is *not* realistic for gases. In calculating the variation with altitude of the pressure in the earth's atmosphere, we must include this variation in density. We will carry out this calculation in Section 17–2. However, the density of gases is usually very much less than that of liquids, so over *small* changes of elevation (say, a few meters) the change in pressure in a gas is usually negligibly small.

In many practical situations the significant quantity is the *difference* between the pressure in a container and atmospheric pressure. For example, if the pressure inside a tire is just equal to atmospheric pressure, the tire is flat. When we say the pressure in a car tire is 32 pounds (actually 32 lb·in^{-2}), we mean it is *greater* than atmospheric pressure (14.7 lb·in^{-2}) by this amount. The *total* pressure in the tire is then 46.7 lb·in^{-2}. The excess pressure above atmospheric pressure is usually called **gauge pressure,** and the total pressure is called **absolute pressure.** Engineers use the abbreviations psig and psia for "pounds per square inch gauge" and "pounds per square inch absolute."

Normal atmospheric pressure at sea level is

$$p_a = 1.013 \times 10^5 \text{ Pa} = 14.7 \text{ lb·in}^2 = 1 \text{ atm}.$$

EXAMPLE 13–1 A household hot-water heating system has an expansion tank in the attic, 12 m above the boiler. If the tank is open to the atmosphere, what is the gauge pressure in the boiler? The absolute pressure?

SOLUTION From Eq. (13–5), the absolute pressure is

$$p = p_a + \rho g h$$
$$= (1.01 \times 10^5 \text{ Pa}) + (1000 \text{ kg·m}^{-3})(9.8 \text{ m·s}^{-2})(12 \text{ m})$$
$$= 2.19 \times 10^5 \text{ Pa} = 2.16 \text{ atm} = 31.8 \text{ lb·in}^{-2}$$

The gauge pressure is

$$p - p_a = (2.19 - 1.01) \times 10^5 \text{ Pa} = 1.18 \times 10^5 \text{ Pa}$$
$$= 1.16 \text{ atm} = 17.1 \text{ lb·in}^{-2}.$$

If the furnace has a pressure gauge, it is always calibrated to read gauge rather than absolute pressure.

We can also estimate the variation in atmospheric pressure over this height. The density of air at sea level and a temperature of 20°C is about 1.2 kg·m^{-3}; that is, about 0.12% of the density of water. If this density were constant over the 12-m height, then the atmospheric pressure at furnace level would be greater than that in the attic by

$$\rho g h = (1.2 \text{ kg·m}^{-3})(9.8 \text{ m·s}^{-2})(12 \text{ m}) = 141 \text{ Pa} = 0.0014 \text{ atm},$$

or a very small fraction of the pressure calculated above. Thus the variation in air pressure over this height is negligible.

The simplest pressure gauge is the open-tube manometer, shown in Fig. 13–4a. The U-shaped tube contains a liquid; one end of the tube is connected to the container where the pressure is to be measured, and the other end is open to the atmosphere at pressure p_a. The pressure at the bottom of the left column is $p + \rho g y_1$, and the pressure at the bottom of the right column (the same point) is $p_a + \rho g y_2$, where ρ is the density of the liquid in the manometer. Since these pressures must be equal,

$$p + \rho g y_1 = p_a + \rho g y_2,$$

and

$$p - p_a = \rho g(y_2 - y_1) = \rho g h. \tag{13–6}$$

The pressure p is the *absolute pressure;* the difference $p - p_a$ between absolute and atmospheric pressure is the *gauge pressure.* Thus the gauge pressure is proportional to the difference in height of the liquid columns.

The **mercury barometer** is a long glass tube, closed at one end, that has been filled with mercury and then inverted in a dish of mercury, as shown in Fig. 13–4b. The space above the mercury column contains only mercury

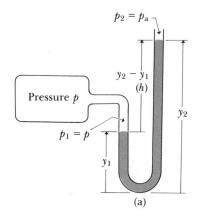

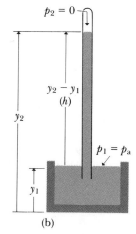

13–4 (a) The open-tube manometer. (b) The barometer.

In a hot-water heating system, how much pressure must the pipes withstand?

A simple pressure gauge using a liquid in a tube

Mercury barometers are used to measure atmospheric pressure for weather forecasting.

vapor; the pressure of the vapor at room temperature is negligibly small. From Eq. (13–5),

$$p_a = \rho g(y_2 - y_1) = \rho g h. \qquad (13-7)$$

Several different units are used to describe pressure.

Thus the mercury barometer reads atmospheric pressure directly from the height of the mercury column. The SI unit of pressure is *one pascal* (1 Pa), equal to one newton per square meter (1 N·m^{-2}). Two related units are the *bar*, defined as 10^5 Pa, and the *millibar*, defined as 10^{-3} bar or 10^2 Pa. Atmospheric pressures are of the order of 1000 millibars; the National Weather Service normally uses this unit. A pressure of 1.013×10^5 Pa = 1.013 bar is called one **atmosphere.** Because mercury manometers and barometers are commonly used laboratory instruments, pressures are also sometimes described in terms of the height of the corresponding mercury column, as so many "inches of mercury" or "millimeters of mercury" (abbreviated mm Hg). The pressure due to a column of mercury one millimeter high has even been given a special name: one *Torr* (after Torricelli, inventor of the mercury barometer). Such units depend on the density of mercury, which varies with temperature, and on the value of g, which varies with location. For these and other reasons, they are gradually passing out of common use in favor of the pascal.

What do blood-pressure readings mean?

One common type of blood-pressure gauge, called a *sphygmomanometer*, includes a manometer similar to that shown in Fig. 13–4a. Blood-pressure readings, such as 130/80, refer to the maximum and minimum gauge pressures, measured in millimeters of mercury or Torrs. Because of height differences, the hydrostatic pressure varies at different points in the body. The standard reference point is the upper arm, level with the heart. Pressure is also affected by the viscous nature of blood flow, by valves throughout the vascular system, and by the body's changing the diameters of blood vessels to regulate pressure.

Measuring atmospheric pressure with a mercury barometer

EXAMPLE 13–2 Compute the atmospheric pressure on a day when the height of mercury in a barometer is 76.0 cm.

SOLUTION The height of the mercury column depends on ρ and g as well as on the atmospheric pressure. As mentioned, ρ varies with temperature, and g varies with latitude and elevation above sea level. If we assume $g = 9.8$ m·s^{-2} and $\rho = 13.6 \times 10^3$ kg·m^{-3},

$$p_a = \rho g h = (13.6 \times 10^3 \text{ kg·m}^{-3})(9.8 \text{ m·s}^{-2})(0.76 \text{ m})$$
$$= 101{,}300 \text{ N·m}^{-2} = 1.013 \times 10^5 \text{ Pa}.$$

In British units,

$$0.76 \text{ m} = 30 \text{ in.} = 2.5 \text{ ft},$$
$$\rho g = 850 \text{ lb·ft}^{-3},$$
$$p_a = 2120 \text{ lb·ft}^{-2} = 14.7 \text{ lb·in}^{-2.}$$

A pressure gauge with a dial and no moving liquid

Another type of pressure gauge, often more convenient than a liquid manometer, is the Bourdon pressure gauge. It consists of a flattened brass tube closed at one end and bent into a circular or spiral shape. The closed end of the tube is connected by a gear and pinion to a pointer that moves over a scale. The open end is connected to the container where the pressure is to be measured. When pressure increases within the flattened tube, it straightens slightly, just as a bent rubber hose straightens when the water is turned on. The resulting motion of the closed end is transmitted to the pointer.

13–3 BUOYANCY

Buoyancy is a familiar phenomenon. A body immersed in water seems to have less weight than when immersed in air, and a body with average density less than that of the fluid in which it is immersed can *float* in that fluid. Examples are the human body in water or a helium-filled balloon in air.

Archimedes' principle states: *When a body is immersed in a fluid, the fluid exerts an upward force on the body equal to the weight of the fluid that is displaced by the body.* To prove this principle, we consider an arbitrary portion of fluid at rest. In Fig. 13–5, the irregular outline is the surface bounding this portion of fluid. The arrows represent the forces exerted by the surrounding fluid on small elements of the boundary surface.

Since the entire fluid is at rest, the x-component of the resultant of these surface forces is zero. The y-component of their resultant, F_y, must be equal in magnitude to the weight mg of the fluid inside the surface, and its line of action must pass through the center of gravity of this fluid.

Now suppose that the fluid inside the surface is removed and is replaced by a solid body having exactly the same shape. The pressure at every point is exactly the same as before. Therefore the force exerted on the body by the surrounding fluid is unaltered and is equal in magnitude to the weight mg of fluid displaced. This upward force is called the buoyant force on the solid body. The line of action of this force again passes through the center of gravity of the displaced fluid.

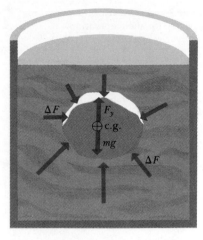

13–5 Archimedes' principle. The buoyant force F_y equals the weight of the displaced fluid.

Buoyancy: Why do bodies float?

Ice is less dense than liquid water (left), so ice cubes float in water. Solid benzene is more dense than liquid benzene, so benzene cubes sink in liquid benzene. (Photograph by Richard Byrnes.)

Balloons and submarines depend on buoyancy for their operation.

A simple density-measuring instrument using buoyancy

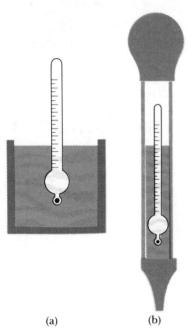

(a)   (b)

13–6 (a) A simple hydrometer. (b) Hydrometer used as a tester for battery acid or antifreeze.

When a balloon floats in air, its weight must be the same as the weight of an amount of air having the same volume as the balloon. That is, the average density of a floating balloon must be the same as that of the surrounding air. Similarly, when a submerged submarine is in equilibrium, its average density must be equal to that of the surrounding water.

A body whose average density is less than that of a liquid can float partially submerged at the free upper surface of the liquid. A familiar example is the hydrometer, shown in Fig. 13–6. The instrument sinks in the fluid until the weight of the fluid it displaces is exactly equal to its own weight. In a fluid of greater density, less fluid is displaced for the same weight, so less of the instrument is immersed; that is, in denser fluids the hydrometer floats *higher*. Similarly, when you swim in sea water (density 1.03 g·cm^{-3}), your body floats higher than in fresh water.

The hydrometer is weighted at its bottom end so the upright position is stable, and a scale in the top stem permits direct density readings. Figure 13–6b shows a hydrometer commonly used to measure density of battery acid or antifreeze. The bottom of the tube is immersed in the liquid; the bulb is then squeezed to expel air and then released (like a giant medicine dropper). The resulting pressure difference causes the liquid to rise into the outer tube, and the hydrometer floats in this sample of the liquid.

EXAMPLE 13–3 A brass block with mass 0.5 kg and density 8.0×10^3 kg·m^{-3} is suspended from a string. What is the tension in the string if the block is in air? If it is completely immersed in water?

SOLUTION If the very small buoyant force of air can be neglected, the tension when the block is in air is just equal to the weight of the block:

$$mg = (0.5 \text{ kg})(9.8 \text{ m·s}^{-2}) = 4.9 \text{ N}.$$

When immersed in water, the block experiences an upward buoyant force equal to the weight of water displaced. To find this, we first find the volume of the block:

$$V = \frac{m}{\rho} = \frac{0.5 \text{ kg}}{8.0 \times 10^3 \text{ kg·m}^{-3}} = 6.25 \times 10^{-5} \text{ m}^3.$$

The weight of this volume of water is

$$\begin{aligned} w = mg &= \rho V g \\ &= (1.0 \times 10^3 \text{ kg·m}^{-3})(6.25 \times 10^{-5} \text{ m}^3)(9.8 \text{ m·s}^{-2}) \\ &= 0.612 \text{ N}. \end{aligned}$$

The tension in the string is the actual weight less the upward buoyant force, or

$$4.9 \text{ N} - 0.612 \text{ N} = 4.29 \text{ N}.$$

A shortcut leading to the same result is as follows: The density of water is less than that of brass by a factor

$$\frac{1.0 \times 10^3 \text{ kg·m}^{-3}}{8.0 \times 10^3 \text{ kg·m}^{-3}} = \frac{1}{8}.$$

Therefore the weight of an amount of water with volume equal to that of the brass will be less than the weight of the brass by the same factor. Thus the weight of the displaced water is $\frac{1}{8}(4.9 \text{ N}) = 0.612 \text{ N}$.

EXAMPLE 13–4 Suppose you place the water container in Example 13–3 on a scale. How does the scale reading change when the brass is immersed in the water?

SOLUTION Taking the water and the brass together as a system, we note that its total weight does not change when the brass is immersed. Thus the sum of the two supporting forces, the string tension and the upward force of the scale on the container, must be the same in both cases. But we found that the string tension decreases by 0.612 N when the brass is immersed, so the scale reading must *increase* by 0.612 N. Qualitatively, we can see that the scale force *must* increase because when the brass is immersed, the water level in the container rises, increasing the pressure of water on the bottom of the container.

13–4 SURFACE TENSION

There are many interesting phenomena associated with the *boundary surface* that exists between a liquid and some other substance. A liquid flowing slowly from the tip of a medicine dropper emerges not as a continuous stream but as a succession of drops. A sewing needle, if placed carefully on a water surface, makes a small depression in the surface and rests there without sinking, even though its density may be as much as ten times that of water. Some insects can walk on the surface of water, their feet making indentations in the surface but not penetrating it. When a clean glass tube of small diameter is dipped into water, the water rises in the tube; but if the tube is dipped in mercury, the mercury is depressed.

> How can insects walk on water? Physics or magic?

In all these phenomena the surface behaves as though it were under *tension*. For any line on the surface, the portions of surface on the two sides of the line exert pulls on it. The situation of Fig. 13–7 shows this effect. A wire ring has a loop of thread attached to it as shown. When the ring and thread are dipped in a soap solution and removed, a thin film of liquid is formed in which the thread "floats" freely, as shown in part (a). If we then puncture the film inside the loop of thread, the thread springs out into a circular shape as in part (b); the surfaces of the liquid pull radially outward on it, as shown by the arrows. These same forces were acting even before the film was punctured,

> The surface of a liquid acts as though it were under tension.

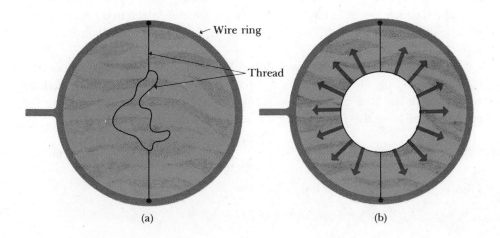

13–7 A wire ring with a flexible loop of thread, dipped in a soap solution, (a) before and (b) after puncturing the surface films inside the loop.

(a)   (b)

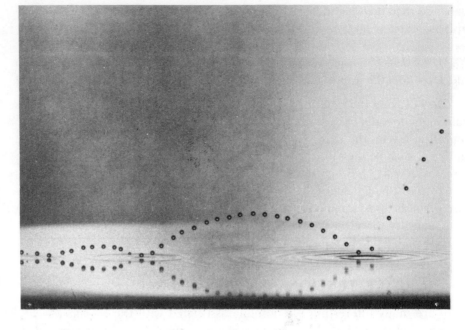

A multiple-flash photograph of a drop of water bouncing off a water surface and creating waves. Surface tension prevents the droplet from coalescing with the water surface, and waves spread out from its point of impact with the surface. (Courtesy of Sir John Mason, Centre for Environmental Technology.)

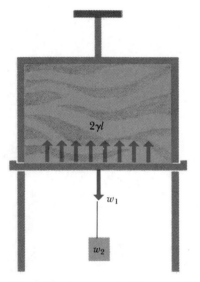

13–8 The horizontal slide wire is in equilibrium under the action of the upward surface force $2\gamma l$ and downward pull $w_1 + w_2$.

Surface tension is surface force per unit length.

but then there was film on *both* sides of the thread, and the net force on a section of thread was zero.

Another simple apparatus for demonstrating surface tension is shown in Fig. 13–8. A piece of wire is bent into a U shape and a second piece of wire is used as a slider. When the apparatus is dipped in a soap solution and removed, the slider (if its weight w_1 is not too great) is quickly pulled up to the top of the U. The slider may be held in equilibrium by adding a second weight w_2. Surprisingly, the same total force $F = w_1 + w_2$ will hold the slider at rest in *any* position, regardless of the area of the liquid film, provided the film remains at constant temperature. That is, the force does not increase as the surface is stretched farther. This is very different from the elastic behavior of a sheet of rubber, for which the force required would be greater as the sheet was stretched. Figure 13–9 shows a similar situation.

Although a soap film like that in Fig. 13–8 is very thin, its thickness is still enormous compared with the size of a molecule. Hence we can consider a soap film as made up chiefly of bulk liquid, bounded by two surface layers a few molecules thick. When the slider in Fig. 13–8 is pulled down and the area of the film is increased, molecules formerly in the main body of the liquid move into the surface layers. That is, these layers are not "stretched" as a rubber sheet would be; more surface is created by molecules moving from the bulk liquid.

Let l be the length of the wire slider. Since the film has two surfaces, the total length along which the surface force acts on the slider is $2l$. The **surface tension** γ in the film is defined as *the ratio of the surface force F to the length d (perpendicular to the force) along which the force acts:*

$$\gamma = \frac{F}{d}. \tag{13–8}$$

Hence, in this case, $d = 2l$ and

$$\gamma = \frac{F}{2l}.$$

TABLE 13–2   Experimental Values of Surface Tension

Liquid in Contact with Air	T, °C	Surface tension, dyn·cm^{-1}
Benzene	20	28.9
Carbon tetrachloride	20	26.8
Ethyl alcohol	20	22.3
Glycerin	20	63.1
Mercury	20	465.0
Olive oil	20	32.0
Soap solution	20	25.0
Water	0	75.6
Water	20	72.8
Water	60	66.2
Water	100	58.9
Oxygen	−193	15.7
Neon	−247	5.15
Helium	−269	0.12

13–9 A paper clip "floating" on water. The clip is supported not by buoyant forces (Section 13–3) but by surface tension of the water. (Nancy Rodgers, Exploratorium.)

Note that although the term *tension* has previously been used in reference to a force, surface tension is a *force per unit length*. The SI unit of surface tension is the newton per meter (1 N·m^{-1}). This unit is not in common use; the usual unit is the cgs unit, the dyne per centimeter (1 dyn·cm^{-1}). The conversion factor is

$$1 \text{ N·m}^{-1} = 1000 \text{ dyn·cm}^{-1}.$$

Some typical values of surface tension are shown in Table 13–2. Surface tension usually decreases as temperature increases; Table 13–2 shows this behavior for water.

A surface under tension contracts until it has the minimum area consistent with any fixed boundaries that may be present and with any pressure differences that may be present on opposite sides of the surface. For example, a drop of liquid under no external forces, or in free fall in vacuum, is always spherical in shape because the sphere has smaller surface area for a given volume than has any other geometric shape. Figure 13–10 is a beautiful example of formation of spherical droplets in a very complex phenomenon, the impact of a drop on a rigid surface.

13–10 A drop of milk splashes on a hard surface. (Dr. Harold Edgerton, M.I.T., Cambridge, Massachusetts.)

Surface tension causes increased pressure inside a bubble or liquid drop.

Surface tension causes a pressure difference between the inside and outside of a soap bubble or a liquid drop. A soap bubble consists of two spherical surface films very close together, with a thin layer of liquid between. Surface tension causes the films to tend to contract; but as the bubble contracts, it compresses the inside air, increasing the interior pressure to a point that prevents further contraction. We can derive a relation between this pressure and the radius of the bubble.

We assume first that there is no external pressure. Then we consider the equilibrium state of one-half the soap bubble, as shown in Fig. 13–11. At the surface where this half joins the upper half, two forces act: the upward force of surface tension and the downward force due to the pressure of air in the upper half. If the radius of the spherical surface is R, the circumference of the circle along which the surface tension acts is $2\pi R$. (Assuming the thickness of the bubble is very small, we neglect the difference between inner and outer radii.) The total force of surface tension for each surface (inner and outer) is $\gamma(2\pi R)$, so the total force of surface tension is $(2\gamma)(2\pi R)$. The force due to air pressure is the pressure p times the area πR^2 of the circle over which it acts. For the resultant of these forces to be zero, we must have

The pressure increase inside a bubble or liquid drop is inversely proportional to its radius.

$$(2\gamma)(2\pi R) = p(\pi R^2),$$

or

$$p = \frac{4\gamma}{R}. \qquad (13\text{–}9)$$

Now the pressure outside the bubble is in general *not* zero. But from the derivation we can see that Eq. (13–9) gives the *difference* between inside and outside pressure. If the outside pressure is p_a, then

$$p - p_a = \frac{4\gamma}{R} \qquad \text{(soap bubble)}. \qquad (13\text{–}10)$$

For a liquid drop, which has only *one* surface film, the difference between pressure of the liquid and that of the outside air is half that for a soap bubble:

$$p - p_a = \frac{2\gamma}{R} \qquad \text{(liquid drop)}. \qquad (13\text{–}11)$$

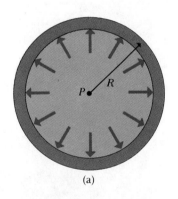

(a)

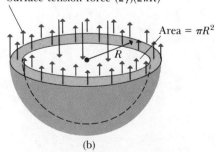

Surface tension force $(2\gamma)(2\pi R)$

Area $= \pi R^2$

(b)

13–11 (a) Cross section of a soap bubble, showing the two surfaces, the thin layer of liquid, and inside air. (b) Equilibrium of half a soap bubble. The surface-tension force exerted by the other half is $(2\gamma)(2\pi R)$, and the force exerted by the air inside the bubble is the pressure p times the area πR^2.

EXAMPLE 13–5 Calculate the excess pressure inside a drop of mercury of diameter 4 mm at 20°C.

SOLUTION From Table 13–2,

$$\gamma = 465 \text{ dyn·cm}^{-1} = 465 \times 10^{-3} \text{ N·m}^{-1}.$$

From Eq. (13–11),

$$p - p_a = \frac{2\gamma}{R} = \frac{(2)(465 \times 10^{-3} \text{ N·m}^{-1})}{0.002 \text{ m}}$$

$$= 465 \text{ N·m}^{-2}$$

$$\cong 0.005 \text{ atm}.$$

Thus far we have talked about surface films at the boundary between a liquid and a gas. Additional surface-tension effects occur when such a film meets a solid surface, such as the wall of a container, as shown in Fig. 13–12.

In general, the liquid-gas surface curves up or down as it approaches the solid surface, and the angle θ at which it meets the surface is called the **contact angle.** When the surface curves up, as with water and glass, θ is less than 90°; when it curves down, as with mercury and glass, θ is greater than 90°. The first case is characteristic of a liquid that tends to "wet" or adhere to the solid surface; the second, of a nonwetting liquid.

An important surface-tension phenomenon is the elevation or depression of the liquid in a narrow tube, as shown in Fig. 13–13. We call this effect **capillarity;** this term stems from the description of such tubes as *capillary,* which means "hairlike." For a liquid such as water, which wets the tube, the contact angle is less than 90° and the liquid rises until it reaches an equilibrium height y. This height is determined by the requirement that the total surface-tension force around the line where the liquid-gas surface meets the solid should just balance the extra weight of the liquid in the tube. The curved liquid surface is called a *meniscus.* Figure 13–13b shows the situation with a nonwetting liquid such as mercury. In this case the contact angle is greater than 90°. Again a meniscus forms, but the surface is depressed, pulled down by the surface-tension forces.

Capillarity is responsible for the rise of ink in blotting paper, the rise of fuel in the wick of a cigarette lighter, and many other everyday phenomena. It is also an essential part of many life processes. A familiar example is the rising of water (actually a dilute aqueous solution) from the roots of a plant to its foliage, due partly to capillarity and partly to osmotic pressure developed in the roots. In the higher animals, including humans, blood is pumped through the arteries and veins, but capillarity is still important in the smallest blood vessels, which indeed are called capillaries.

A related phenomenon is that of *negative pressure.* Ordinarily the stress in a liquid is compressive in nature, but in some circumstances liquids can sustain actual *tensile* stresses. Imagine a cylindrical tube, closed at one end and having a tight-fitting piston at the other. Suppose the tube is completely filled with a liquid that wets both the inner surface of the tube and the piston face; that is, the molecules of liquid adhere to the surfaces. If we pull the piston out, we expect that the liquid will pull away from the piston, leaving an empty space. If the surfaces are very clean and the liquid very pure, however, an actual tensile stress, accompanied by an increase in volume, can be observed. The liquid behaves as though we were stretching it; adhesive forces prevent it from pulling away from the walls of the container.

With water, tensile stresses as large as 300 atm have been observed. In the laboratory this situation is highly unstable; a liquid under tension tends to break up into many small droplets. In tall trees, however, negative pressures are a regular occurrence. Negative pressure is thought to be an important mechanism for transport of water and nutrients from the roots to the leaves in the small xylem tubes (diameter of the order of 0.1 mm) in the growing layers of the tree.

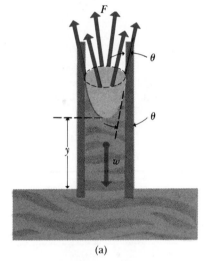

13–12 Surface films exist at the solid–vapor boundary as well as at the liquid–vapor boundary.

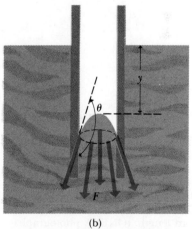

13–13 Surface-tension forces on a liquid in a capillary tube. The liquid (a) rises if $\theta < 90°$ and (b) is depressed if $\theta > 90°$.

13–5 FLUID FLOW

We are now ready to consider *motion* of a fluid. This motion can be extremely complex, as shown by the flow of a river in flood or the swirling flames of a campfire. Despite this, some situations can be represented by relatively simple idealized models. We begin by introducing one of these models, along with

An ideal fluid has constant density and no internal friction.

13–14 A flow tube bounded by flow lines.

Flow lines describe the paths of fluid particles.

A bundle of flow lines forms a flow tube.

several new terms that are useful in our discussion. An **ideal fluid** is a fluid that is *incompressible* and has no internal friction or viscosity. The assumption of incompressibility is usually a good approximation for liquids, and we may also treat a gas as incompressible whenever the pressure differences associated with the flow are not too great. Internal friction in a fluid causes shear stresses when two adjacent layers of fluid move relative to each other, as when the fluid flows inside a tube or around an obstacle. In some cases, these shear forces are negligible compared with forces arising from gravitation and pressure differences.

The path of an individual element of a moving fluid is called a **flow line.** In general, the velocity of a fluid element changes in both magnitude and direction as it moves along its line of flow. If every element passing through a given point follows the same flow line as that of preceding elements, the flow is said to be *steady* or *stationary*. In steady flow, the fluid velocity at each point of space remains constant in time, although the velocity of a particular particle of the fluid may change as it moves from one point to another.

A **streamline** is a curve whose tangent, at any point, is in the direction of the fluid velocity at that point. In steady flow, the streamlines coincide with the flow lines.

If we construct all the streamlines passing through the edge of an imaginary element of area, such as the area A in Fig. 13–14, these lines enclose a tube called a **flow tube.** From the definition of a streamline, no fluid can cross the side walls of a flow tube; in steady flow there can be no mixing of the fluids in different flow tubes.

Figure 13–15 shows fluid flow from left to right around a number of obstacles and in a channel of varying cross section. The photographs were made by using an apparatus in which alternate streams of clear and colored water flow around opaque obstacles between two closely spaced glass plates. Each obstacle is surrounded by a flow tube. The tube splits into two portions at a *stagnation point* (a point where the velocity is zero) on the upstream (left) side of the obstacle. These portions rejoin at a second stagnation point on the downstream (right) side. The velocity at the stagnation points is zero. The

Streamlines in symmetric laminar flow of water past an airfoil. This photograph was made by injecting dye into the water upstream from the airfoil. The left and right edges are stagnation points, where the flow velocity is zero. The Reynolds number is 7000, based on the left-to-right chord length. (ONERA photograph, Werlé 1974.)

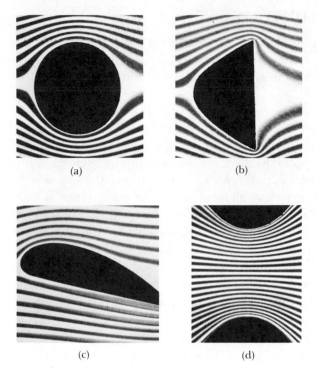

(a)   (b)

(c)   (d)

13–15 (a), (b), (c): Streamline flow around obstacles of various shapes. (d) Flow in a channel of varying cross-sectional area.

cross sections of all flow tubes decrease at a constriction and increase again when the channel widens.

The patterns of Fig. 13–15 are typical of **laminar flow,** in which adjacent layers of fluid slide smoothly past each other. At sufficiently high flow rates, or when boundary surfaces cause abrupt changes in velocity, the flow becomes irregular and much more complex; this is called **turbulent flow.** In turbulent flow *there is no steady-state pattern,* since the flow pattern continuously changes.

We can use the principle of conservation of mass to derive an important relationship in fluid flow called the **continuity equation.** For an incompressible fluid in steady flow, the derivation takes the following form. Figure 13–16 shows a portion of a flow tube, between two fixed cross sections of area A_1 and A_2. Let v_1 and v_2 be the speeds at these sections. No fluid flows across the side wall of the tube because the fluid velocity is tangent to the wall at every point on the wall. The volume of fluid that flows into the tube across A_1 in a time interval Δt is the fluid contained in the short cylindrical element of base A_1 and height $v_1 \Delta t$, that is, $A_1 v_1 \Delta t$. If the density of the fluid is ρ, the *mass* flowing in is $\rho A_1 v_1 \Delta t$. Similarly, the mass that flows out across A_2 in the same time is $\rho A_2 v_2 \Delta t$. The volume between A_1 and A_2 is constant, and since the flow is steady, the mass flowing out equals that flowing in. Hence

$$\rho A_1 v_1 \, \Delta t = \rho A_2 v_2 \, \Delta t,$$

or

$$A_1 v_1 = A_2 v_2. \tag{13–12}$$

That is, the product Av is constant along any given flow tube. It follows that when the cross section of a flow tube decreases, as in the constriction in Fig. 13–15d, the velocity increases. We can observe this by introducing small particles into the fluid and watching their motion.

Sometimes there is a smooth, steady flow pattern; sometimes not.

Conservation of mass in a fluid: What goes in must come out.

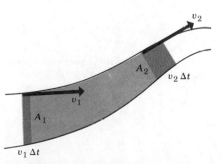

13–16 Flow into and out of a portion of a tube of flow.

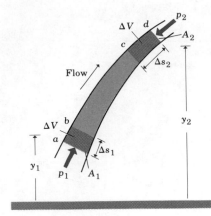

13–17 The net work done on the shaded element equals the increase in its kinetic and potential energy.

Bernoulli's equation: the work–energy theorem for a moving fluid

Work done on a fluid element by pressure differences

Change in kinetic energy for a fluid element

Change in gravitational potential energy for a fluid element

13–6 BERNOULLI'S EQUATION

We can now use the concepts introduced in Section 13–5, along with the energy principle, to derive an important relationship called **Bernoulli's equation.** This equation relates the pressure, flow velocity, and height for flow of an ideal fluid.

First we note that when an incompressible fluid flows along a horizontal flow tube of varying cross section, its velocity must change; Eq. (13–12) shows this. Each element of fluid must have an acceleration. The force that causes this acceleration has to be applied by the fluid surrounding the element. Thus the pressure must be different in different regions; if it were the same everywhere, the net force on every fluid element would be zero. When a fluid element speeds up, it must move from a region of larger pressure to a region of smaller pressure in order to experience a net forward force that makes it speed up. Thus when the cross section of a flow tube varies, the pressure *must* vary along the tube, even when there is no difference of elevation. If the elevation also changes, there is an additional pressure difference. Bernoulli's equation relates the pressure difference between two points in a flow tube to changes in both velocity and elevation.

To derive the Bernoulli equation, we apply the work–energy theorem to the fluid in a section of a flow tube. In Fig. 13–17, we consider the fluid that at some initial time lies between the two cross sections a and c. In a time interval Δt, the fluid initially at a moves to b, a distance $\Delta s_1 = v_1 \Delta t$, where v_1 is the speed at this end of the tube. In the same time interval the fluid initially at c moves to d, a distance $\Delta s_2 = v_2 \Delta t$. The cross-sectional areas at the two ends are A_1 and A_2, as shown. Because of the continuity relation, the volume of fluid ΔV passing any cross section in time Δt is $\Delta V = A_1 \Delta s_1 = A_2 \Delta s_2$.

Let us compute the *work* done on this fluid during Δt. The force on the cross section at a is $p_1 A_1$, and the force at c is $p_2 A_2$, where p_1 and p_2 are the pressures at the two ends. The net work done on the element during this displacement is therefore

$$W = p_1 A_1 \, \Delta s_1 - p_2 A_2 \, \Delta s_2 = (p_1 - p_2) \, \Delta V \qquad (13\text{–}13)$$

The negative sign in the second term arises because the force at c is opposite in direction to the displacement.

We now equate this work to the total change in energy, kinetic and potential, of the element. The kinetic energy of the fluid between sections b and c does not change. At the beginning of Δt, the fluid between a and b has volume $A_1 \Delta s_1$, mass $\rho A_1 \Delta s_1$, and kinetic energy $\frac{1}{2}\rho(A_1 \Delta s_1)v_1^2$. Similarly, at the end of Δt, the fluid between c and d has kinetic energy $\frac{1}{2}\rho(A_2 \Delta s_2)v_2^2$. Thus the net change in kinetic energy is

$$\Delta K = \tfrac{1}{2}\rho \, \Delta V (v_2^2 - v_1^2). \qquad (13\text{–}14)$$

The change in potential energy is obtained similarly. The potential energy of the mass entering at a in time Δt is $\Delta m \, gy_1 = \rho \, \Delta V \, gy_1$, and that of the mass leaving at c is $\Delta m \, gy_2 = \rho \, \Delta V \, gy_2$. The net change in potential energy is

$$\Delta U = \rho \, \Delta V \, g(y_2 - y_1). \qquad (13\text{–}15)$$

Combining Eqs. (13–13), (13–14), and (13–15) in the work–energy theorem $W = \Delta K + \Delta U$, we obtain

$$(p_1 - p_2) \, \Delta V = \tfrac{1}{2}\rho \, \Delta V \, (v_2^2 - v_1^2) + \rho \, \Delta V \, g(y_2 - y_1)$$

or

$$p_1 - p_2 = \tfrac{1}{2}\rho(v_2{}^2 - v_1{}^2) + \rho g(y_2 - y_1). \qquad (13\text{–}16)$$

In this form, Bernoulli's equation represents the equality of the work per unit volume of fluid ($p_1 - p_2$) to the sum of the changes in kinetic and potential energies per unit volume that occur during the flow. Or we may interpret Eq. (13–16) in terms of pressures. The second term on the right is the pressure difference arising from the weight of the fluid and the difference in elevation of the two ends of the fluid element. The first term on the right is the additional pressure difference associated with the change of velocity of the fluid.

Equation (13–16) can also be written

$$p_1 + \rho g y_1 + \tfrac{1}{2}\rho v_1{}^2 = p_2 + \rho g y_2 + \tfrac{1}{2}\rho v_2{}^2, \qquad (13\text{–}17)$$

and since the subscripts 1 and 2 refer to *any* two points along the tube of flow, Bernoulli's equation may also be written

$$p + \rho g y + \tfrac{1}{2}\rho v^2 = \text{constant}. \qquad (13\text{–}18)$$

Note carefully that p is the *absolute* (not gauge) pressure, and that a consistent set of units must be used. In SI units, pressure is expressed in Pa, density in kg·m^{-3}, and velocity in m·s^{-1}.

PROBLEM-SOLVING STRATEGY: *Bernoulli's equation*

Bernoulli's equation is derived from the work–energy relationship, so it isn't surprising that much of the problem-solving strategy suggested in Sections 7–5 and 7–6 is equally applicable here. In particular:

1. Always begin by identifying clearly points 1 and 2, with reference to Bernoulli's equation.

2. Make lists of the known and unknown quantities in Eq. (13–17). The variables are p_1, p_2, v_1, v_2, y_1, and y_2, and the constants are ρ and g. What is given? What do you need to determine?

3. In some problems you will need to use the continuity equation, Eq. (13–12), to get a relation

between the two velocities in terms of cross-sectional areas of pipes or containers. Or perhaps you will know both velocities and need to determine one of the areas.

4. The volume flow rate dV/dt across any area A is given by $dV/dt = Av$, and the corresponding mass flow rate dm/dt is $dm/dt = \rho Av$. These relations sometimes come in handy.

5. As you read the following example and the applications discussed in Section 13–7, watch for applications of this strategy.

EXAMPLE 13–6 Water enters a house through a pipe 2.0 cm in inside diameter, at an absolute pressure of 4×10^5 Pa (about 4 atm). The pipe leading to the second-floor bathroom 5 m above is 1.0 cm in diameter. When the flow velocity at the inlet pipe is 4 m·s^{-1}, find the flow velocity and pressure in the bathroom.

Variation in water pressure in a house: an application of Bernoulli's equation

SOLUTION Let point 1 be at the inlet pipe and point 2 at the bathroom. The velocity v_2 at the bathroom is obtained from the continuity equation:

$$v_2 = \frac{A_1}{A_2}v_1 = \frac{\pi(1.0 \text{ cm})^2}{\pi(0.5 \text{ cm})^2}(4 \text{ m·s}^{-1}) = 16 \text{ m·s}^{-1}.$$

We take $y_1 = 0$ (at the inlet) and $y_2 = 5$ m (at the bathroom). We are given p_1 and v_1; we can find p_2 from Bernoulli's equation:

$$p_2 = p_1 - \tfrac{1}{2}\rho(v_2{}^2 - v_1{}^2) - \rho g(y_2 - y_1)$$
$$= 4 \times 10^5 \text{ Pa} - \tfrac{1}{2}(1.0 \times 10^3 \text{ kg·m}^{-3})(256 \text{ m}^2\text{·s}^{-2} - 16 \text{ m}^2\text{·s}^{-2})$$
$$- (1.0 \times 10^3 \text{ kg·m}^{-3})(9.8 \text{ m·s}^{-2})(5 \text{ m})$$
$$= 2.3 \times 10^5 \text{ Pa} \cong 2.3 \text{ atm.}$$

Note that when the water is turned off, the second term on the right vanishes and the pressure rises to 3.5×10^5 Pa.

13–7 APPLICATIONS OF BERNOULLI'S EQUATION

Here are several applications of Bernoulli's equation to situations of practical interest.

1. *Fluid statics.* Just as equilibrium of a particle is a special case of $\boldsymbol{F} = m\boldsymbol{a}$, the equations of fluid statics are special cases of Bernoulli's equation when the velocity is zero everywhere. Thus when v_1 and v_2 are zero, Eq. (13–16) reduces to

$$p_1 - p_2 = \rho g(y_2 - y_1), \qquad (13\text{–}19)$$

which is the same as Eq. (13–4).

How fast does fluid run out of a container?

2. *Speed of efflux: Torricelli's theorem.* Figure 13–18 shows a tank with cross-sectional area A_1, filled to a depth h with a liquid of density ρ. The space above the top of the liquid contains air at pressure p, and the liquid flows out of an orifice of area A_2. Let us consider the entire volume of moving fluid as a single flow tube, and let v_1 and v_2 be the speeds at points 1 and 2. The quantity v_2 is called the *speed of efflux*. The pressure at point 2 is atmospheric, p_a. Applying Bernoulli's equation to points 1 and 2, and taking the bottom of the tank as our reference level, we find

$$p + \tfrac{1}{2}\rho v_1{}^2 + \rho g h = p_a + \tfrac{1}{2}\rho v_2{}^2,$$

or

$$v_2{}^2 = v_1{}^2 + 2\frac{p - p_a}{\rho} + 2gh. \qquad (13\text{–}20)$$

If the tank is open to the atmosphere, then

$$p = p_a \qquad \text{and} \qquad p - p_a = 0.$$

Also, in Fig. 13–18, if A_2 is much smaller than A_1, then $v_1{}^2$ is very much less than $v_2{}^2$ and can be neglected. Equation (13–20) then becomes

$$v_2 = \sqrt{2gh}. \qquad (13\text{–}21)$$

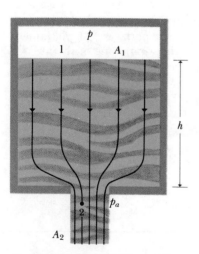

13–18 Flow of a liquid out of an orifice.

That is, *the speed of efflux is the same as the speed a body would acquire in falling freely through a height h.* This is *Torricelli's theorem.* It is not restricted to an opening in the bottom of a vessel, but applies also to a hole in a side wall at a depth h below the surface.

Rocket propulsion: What pushes a rocket forward?

3. *Rocket propulsion.* A flow of fluid out of an opening in a container causes a *reaction force* on the remainder of the system. This is the principle of rocket propulsion. Returning to Fig. 13–18, let us suppose that the vessel contains

only a gas, having density ρ and pressure p. Again we assume that A_2 is much smaller than A_1, so that the $v_1{}^2$ term in Bernoulli's can be neglected. Suppose also that p is so large that the term $2gh$ is negligible compared to the pressure term. The speed of efflux is then given by

$$_2 = \sqrt{2(p - p_a)/\rho}. \qquad (13\text{–}22)$$

We can now derive an expression for the reaction force, or *thrust*, acting on the system. We drop the subscripts, letting A be the orifice area and v the speed of efflux. Then the volume of fluid flowing out in a time interval Δt is $Av\,\Delta t$, the corresponding mass is $\rho Av\,\Delta t$, and the momentum (mass × velocity) is $\rho Av^2\,\Delta t$. The escaping fluid acquires this momentum during the time Δt, so its *rate of change* of momentum is ρAv^2. (This analysis uses the same principles we used in Section 8–8.) From Newton's second law, this equals the force acting on the escaping fluid, and from Newton's third law, an equal and opposite reaction force acts on the remainder of the system. Using Eq. (13–22) for the speed of efflux, we can write the reaction force as

$$F = \rho A v^2 = \rho A \frac{2(p - p_a)}{\rho},$$

or

$$F = 2A(p - p_a). \qquad (13\text{–}23)$$

Thus, while the *speed* of efflux is inversely proportional to the fluid density, the *thrust* is independent of the density and depends only on the area of the orifice and the gauge pressure $p - p_a$.

4. The *Venturi tube*, shown in Fig. 13–19, consists of a constriction or throat inserted in a pipeline, with inlet and outlet tapers to avoid turbulence. Bernoulli's equation, applied to the wide and constricted portions of the pipe, respectively (with $y_1 = y_2$), becomes

$$p_1 + \tfrac{1}{2}\rho v_1{}^2 = p_2 + \tfrac{1}{2}\rho v_2{}^2.$$

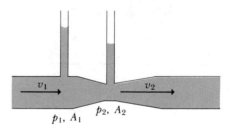

13–19 The Venturi tube.

From the equation of continuity, the speed v_2 is greater than the speed v_1, and hence the pressure p_2 in the throat is *less* than the pressure p_1. Thus a net force to the right accelerates the fluid as it enters the throat, and a net force to the left slows it as it leaves. The pressures p_1 and p_2 can be measured by attaching vertical side tubes, as shown in the diagram. When we know these pressures and the areas A_1 and A_2, we can compute the velocities and the mass rate of flow. When used for this purpose, the device is called a *Venturi meter*.

The effect of reduced pressure at a constriction has many practical applications. In the carburetor of a gasoline engine, gasoline is sprayed from a nozzle into a low-pressure region produced in a Venturi throat. The *aspirator pump* forces water through a Venturi throat. Air is pushed into the low-pressure water rushing through the constricted portion by the higher pressure outside.

5. *Measurement of pressure in a moving fluid.* The pressure p in a fluid flowing in an enclosed channel can be measured with an open-tube manometer, as shown in Fig. 13–20. A *probe* is inserted in the stream. The probe should be small enough so that the flow is not appreciably disturbed and should be shaped so as to avoid turbulence. The difference h in height of the liquid in the arms of the manometer is proportional to the difference between atmos-

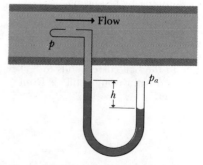

13–20 Pressure gauge for measuring the static pressure p in a fluid flowing in an enclosed channel.

pheric pressure p_a and the fluid pressure p. That is,

$$p_a - p = \rho_m gh,$$
$$p = p_a - \rho_m gh, \qquad (13-24)$$

where ρ_m is the density of the manometer liquid. A variation of this setup is used as an airspeed indicator in airplanes.

6. *Lift on an aircraft wing.* Figure 13–21 shows flow lines around a cross section of an aircraft wing or an airfoil. The orientation of the wing relative to the flow direction causes the flow lines to crowd together above the wing. This corresponds to increased flow velocity in this region, just as in the throat of a Venturi. Hence the region above the wing has increased velocity and reduced pressure, while below the wing the pressure remains nearly atmospheric. Because the upward force on the underside of the wing is greater than the downward force on the top side, there is a net upward force, or *lift*.

We can also understand this phenomenon on the basis of Newton's laws. In Fig. 13–21, the air flowing past the wing has a net *downward* change in the vertical component of its momentum, corresponding to the downward force the wing exerts on it. The reaction force *on* the wing is thus *upward*, in agreement with our discussion above. As the angle of the wing relative to the air flow increases, turbulent flow occurs in a larger and larger region above the wing. When this happens, the pressure drop is no longer as great as that predicted by Bernoulli's equation. The lift on the wing decreases, and in extreme cases the airplane stalls.

7. *The curved flight of a spinning ball.* This effect is somewhat more subtle than the lift on an airplane wing, but the same basic principles are involved. Figure 13–22a represents a stationary ball in a blast of air moving from right to left. The motion of the airstream around and past the ball is the same as though the ball were moving through still air from left to right. Because of the large velocities ordinarily involved, a region of turbulent flow occurs behind the ball, as shown.

When the ball is spinning, as in Fig. 13–22b, the viscosity of air causes layers of air near the ball's surface to be pulled around in the direction of spin. The velocity of air relative to the ball's surface becomes greater at the top of the ball than at the bottom. The region of turbulence becomes asymmetric, with turbulence occurring farther forward on the top side than on the bottom. This asymmetry causes a pressure difference. The average pressure at the top of the ball becomes greater than that at the bottom. The corresponding net downward force deflects the ball as shown. In a baseball curve pitch, the ball spins at an angle, and the actual deflection is sideways. In that case, Fig. 13–22 shows a *top* view of the situation. Tennis balls with spin behave in much the same way.

A similar effect occurs with golf balls, which always have "backspin" from impact with the slanted face of the golf club. The resulting pressure difference between top and bottom of the ball causes a "lift" force that keeps the ball in the air considerably longer than would be possible without spin. A well-hit drive appears from the tee to "float" or even curve *upward* during the initial portion of its flight, and this is a real effect, not an illusion. The dimples on the ball play an essential role: An undimpled ball has much shorter trajectory than a dimpled one given the same initial velocity and spin. Some manufacturers even claim that the *shape* of the dimples is significant; one states that polygonal

How does an airplane stay in the air?

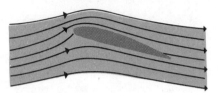

13–21　Flow lines around an airfoil.

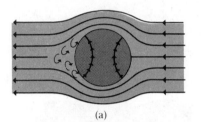

(a)

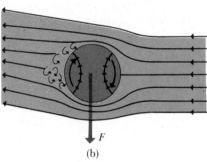

F

(b)

13–22　(a) Air flow past a stationary ball, showing symmetric region of turbulence behind ball. (b) Air flow past a spinning ball, showing asymmetric region of turbulence and deflection of airstream. The net force on the ball is in the direction shown. Motion of the air from right to left relative to the ball corresponds to motion of a ball through still air from left to right.

Spin is important for golf balls as well as baseballs.

dimples are superior to round ones! Figure 13–23 shows the backspin of a golf ball just after being struck by a club.

13–8 VISCOSITY

Another important characteristic of fluids is **viscosity,** which is internal friction of a fluid. Because of viscosity, a force is needed whenever one layer of a fluid slides past another, or when one surface slides past another with a layer of fluid between the surfaces. Viscosity plays a vital role in the flow of fluids in pipes, the flow of blood, the lubrication of engine parts, and many other areas of practical importance. Both liquids and gases have viscosity, but liquids are much more viscous than gases.

The simplest example of flow of a viscous fluid is flow between two parallel plates, as shown in Fig. 13–24. The bottom plate is stationary, while the top plate moves with constant speed v. We find that the fluid in contact with each surface has the same speed as that surface; thus at the top surface the fluid has speed v, while the fluid adjacent to the bottom surface is at rest. The speeds of intermediate layers of fluid increase uniformly from one surface to the other, as shown by the arrows.

Flow of this type is called *laminar*. (A lamina is a thin sheet.) In laminar flow, the layers of liquid slide over one another, and a portion of the liquid, which at some instant has the shape *abcd*, will a moment later take the shape *abc'd'* and will become more and more distorted as the motion continues. That is, the liquid is in a state of continually increasing shear strain.

To maintain the motion, we must apply a constant force to the right on the upper, moving plate, and hence indirectly on the upper liquid surface. This force tends to drag the fluid, and the lower plate as well, to the right. Therefore we must apply an equal force toward the left on the lower plate to hold it stationary. These forces are labeled F in Fig. 13–24. If A is the area of the fluid over which the forces are applied (i.e., the surface area of the plates), the ratio F/A is the shear stress exerted on the fluid.

When we apply a shear stress to a *solid*, the effect is to produce a certain displacement of the solid, such as *dd'*. The shear strain is defined as the ratio of this displacement to the transverse dimension l, and within the elastic limit the shear stress is proportional to the shear strain. With a *fluid*, however, the shear strain increases without limit as long as the stress is applied. Experiments show that the stress depends not on the shear strain but on its *rate of change*. The strain in Fig. 13–24 (at the instant when the volume of fluid has the shape *abc'd'*) is *dd'*/*ad*, or *dd'*/*l*. Since l is constant, the *rate of change* of strain equals $1/l$ times the rate of change of *dd'*. But the rate of change of *dd'* is simply the *speed* of point *d'*; that is, the speed v of the moving wall. Hence

$$\text{Rate of change of shear strain} = \frac{v}{l}. \qquad (13\text{–}25)$$

The rate of change of shear strain is also referred to simply as the *strain rate*.

The *coefficient of viscosity* of the fluid, or simply its *viscosity* η, is defined as the ratio of the shear stress, F/A, to the rate of change of shear strain:

$$\eta = \frac{\text{shear stress}}{\text{rate of change of shear strain}} = \frac{F/A}{v/l},$$

13–23 Stroboscopic photograph of a golf ball being struck by a club. The picture was taken at 1000 flashes per second; the ball rotates once in four pictures, corresponding to an angular velocity of 250 revolutions per second. (Dr. Harold Edgerton, M.I.T., Cambridge, Massachusetts.)

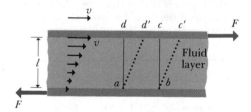

13–24 Laminar flow of a viscous fluid.

Forces are needed to make layers of a viscous fluid slide past each other.

Shear stress in a fluid depends on the rate of change of shear strain.

As a drop of milk falls freely through air, the forces of gravity, surface tension, and air resistance combine to deform and flatten the drop. Note that the drop does not have the classical teardrop shape often drawn by artists. (Dr. Harold Edgerton, M.I.T., Cambridge, Massachusetts.)

Viscosities of liquids increase with decreasing temperature: one reason it's harder to start your car in the winter

Flow of a viscous fluid in a pipe: The fluid sticks to the walls.

TABLE 13–3   Typical Values of Viscosity

Temperature, C°	Viscosity of Castor Oil, Poise	Viscosity of Water, Centipoise	Viscosity of Air, Micropoise
0	53.00	1.792	171
20	9.86	1.005	181
40	2.31	0.656	190
60	0.80	0.469	200
80	0.30	0.357	209
100	0.17	0.284	218

or

$$F = \eta A \frac{v}{l}. \tag{13–26}$$

For a liquid that flows readily, such as water or kerosene, the shear stress needed for a given strain rate is relatively small. For a liquid such as molasses or glycerine, a greater shear stress is necessary for the same strain rate, and the viscosity is correspondingly greater. Viscosities of all fluids are markedly dependent on temperature, increasing for gases and decreasing for liquids as the temperature increases—hence the expression "as slow as molasses in January." An important goal in the design of oils for engine lubrication is to *reduce* the temperature variation of viscosity as much as possible.

From Eq. (13–26), the unit of viscosity is that of force times distance, divided by area times velocity. In SI units it is

$$1 \ \text{N·m·m}^{-2}(\text{m·s}^{-1})^{-1} = 1 \ \text{N·s·m}^{-2}.$$

The corresponding cgs unit, $1 \ \text{dyn·s·cm}^{-2}$, is the only viscosity unit in common use and is called 1 *poise* in honor of the French scientist Poiseuille. Thus

$$1 \ \text{poise} = 1 \ \text{dyn·s·cm}^{-2} = 10^{-1} \ \text{N·s·m}^{-2}.$$

Small viscosities are expressed in *centipoises* ($1 \ \text{cp} = 10^{-2}$ poise) or *micropoises* ($1 \ \mu\text{p} = 10^{-6}$ poise). A few typical values are given in Table 13–3.

Not all fluids behave according to the direct proportionality of force and velocity predicted by Eq. (13–26). An interesting exception is blood, for which velocity increases more rapidly than force. This behavior results from the fact that blood is not a homogeneous fluid but rather a suspension of solid particles in a liquid. As the strain rate increases, these particles deform and become preferentially oriented to facilitate flow. The fluids that lubricate human joints show similar behavior. Some paints cling to the brush but flow when brushed on. A fluid for which Eq. (13–26) does hold is called a **Newtonian fluid;** it is a useful though approximate model for the behavior of many pure substances. Fluids that are suspensions or dispersions are often non-Newtonian in their viscous behavior.

When a viscous fluid flows in a stationary tube or pipe, the flow velocity is different at different points of a cross section. The outermost layer of fluid clings to the walls of the tube, and its velocity is zero. The walls exert a backward drag on this layer, which in turn drags backward on the next layer beyond it, and so on. If the velocity is not too great, the flow is *laminar*, with a velocity that is greatest at the center and decreases to zero at the walls. The flow is like that of a number of telescoping tubes sliding relative to one another, the central tube advancing most rapidly and the outer tube remaining at rest.

We can now derive an equation for the variation of velocity with radius for a cylindrical pipe of inner radius R. We start by considering the flow of a cylindrical element of fluid coaxial with the pipe, of radius r and length L, as shown in Fig. 13–25a. The force on the left end is $p_1\pi r^2$, and that on the right end $p_2\pi r^2$, as shown. The net force is thus

$$F = (p_1 - p_2)\pi r^2.$$

Since the element does not accelerate, this force must just balance the viscous retarding force at the surface of this element. The force is given by Eq. (13–26), but since the velocity does not vary uniformly with distance from the center, we must replace v/l in this expression with $-dv/dr$, where dv is the small change of velocity when we go from distance r to $r + dr$ from the axis. The area over which the viscous force acts is $A = 2\pi rL$. Thus the viscous force is

$$F = \eta 2\pi rL\frac{dv}{dr}.$$

Equating this to the net force due to pressure on the ends and rearranging, we find that

$$\frac{dv}{dr} = -\frac{(p_1 - p_2)r}{2\eta L}.$$

This relation shows that the velocity changes more and more rapidly as we go from the center ($r = 0$) to the pipe wall ($r = R$). The negative sign must be introduced because v decreases as r increases. Integrating, we find

$$-\int_v^0 dv = \frac{p_1 - p_2}{2\eta L}\int_r^R r\, dr,$$

and

$$v = \frac{p_1 - p_2}{4\eta L}(R^2 - r^2). \tag{13–27}$$

The velocity decreases from a maximum value $(p_1 - p_2)R^2/4\eta L$ at the center to zero at the wall. Thus the maximum velocity is proportional to the pressure change per unit length $(p_1 - p_2)/L$, called the *pressure gradient*. The curve in Fig. 13–25b is a graph of Eq. (13–27) with the v-axis horizontal and the r-axis vertical.

We can use Eq. (13–27) to find the *total* rate of fluid flow through the pipe. Note that the velocity at each point is proportional to the pressure gradient $(p_1 - p_2)/L$, so the total flow rate must also be proportional to this quantity. Now consider the thin-walled fluid element in Fig. 13–25c. The volume of fluid dV crossing the ends of this element in a time dt is $v\, dA\, dt$, where v is the velocity at the radius r and dA is the shaded area, equal to $2\pi r\, dr$. Taking the expression for v from Eq. (13–27), we get

$$dV = \frac{p_1 - p_2}{4\eta L}(R^2 - r^2)2\pi r\, dr\, dt.$$

The volume flowing across the entire cross section is obtained by integrating

Flow velocity in a cylindrical pipe depends on the distance from the axis.

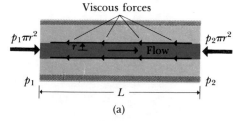

(a)

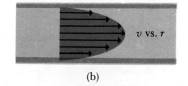

(b)

(c)

13–25 (a) Forces on a cylindrical element of a viscous fluid. (b) Velocity distribution for viscous flow.

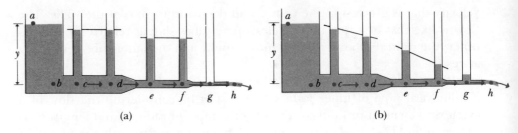

13–26 Pressures along a horizontal tube in which is flowing (a) an ideal fluid, (b) a viscous fluid.

over all elements between $r = 0$ and $r = R$:

$$dV = \frac{\pi(p_1 - p_2)}{2\eta L} \int_0^R (R^2 - r^2)r \, dr \, dt$$

$$= \frac{\pi}{8} \frac{R^4}{\eta} \frac{p_1 - p_2}{L} dt.$$

Volume flow rate of a viscous fluid in a pipe is proportional to the fourth power of the radius!

The total volume of flow per unit time, dV/dt, is given by

$$\frac{dV}{dt} = \frac{\pi}{8} \frac{R^4}{\eta} \frac{p_1 - p_2}{L}. \tag{13–28}$$

This relation was first derived by Poiseuille and is called **Poiseuille's law.** It shows that the volume rate of flow is inversely proportional to viscosity, as might be expected. The volume flow rate is also proportional to the pressure gradient along the pipe, and it varies as the *fourth power* of the radius. For example, if the radius is halved, the flow rate is reduced by a factor of 16. This relation is familiar to physicians in connection with the selection of needles for hypodermic syringes. Needle size is much more important than thumb pressure in determining the flow rate from the needle; doubling the needle diameter has the same effect as increasing the thumb force sixteenfold. Similarly, blood flow in arteries and veins can be controlled over a wide range by relatively small changes in diameter, an important temperature-control mechanism in warm-blooded animals.

Some examples of the differences in behavior between viscous and nonviscous liquids

The difference between the flow of an ideal nonviscous fluid and one having viscosity is shown in Fig. 13–26, where fluid is flowing along a horizontal tube of varying cross section. The height of the fluid in each small vertical tube is proportional to the gauge pressure at the corresponding point in the horizontal tube.

In part (a), the fluid has no viscosity. The pressure at b is very nearly the static pressure $\rho g y$, since the velocity is small in the large tank. The pressure at c is less than at b because the fluid must accelerate between these points. The pressures at c and d are equal, however, since the velocity and elevation at these points are the same. Further pressure drops occur between d and e, and between f and g. The pressure at g is atmospheric, and the gauge pressure at this point is zero.

Figure 13–26b shows the effect of viscosity. Again, the pressure at b is nearly the static pressure $\rho g y$. The pressure drops from b to c, due now in part to viscous effects, and there is a further drop from c to d. The pressure gradient in this part of the tube is represented by the slope of the dotted line. The drop from d to e results in part from acceleration and in part from viscosity. The pressure gradient between e and f is greater than between c and d because of the smaller radius in this portion. Finally, the pressure at g is somewhat above atmospheric, since there is now a pressure gradient between this point and the end of the tube.

One more useful relation in viscous fluid flow is the expression for the force F exerted on a sphere of radius r moving with speed v through a fluid with viscosity η. This relation is called **Stokes's law** and is as follows:

$$F = 6\pi\eta rv. \tag{13-29}$$

We have encountered this kind of velocity-proportional force before, in Example 5–17 (Section 5–4).

A sphere falling in a viscous fluid reaches a *terminal velocity* v_T at which the viscous retarding force plus the buoyant force equals the weight of the sphere. Let ρ be the density of the sphere and ρ' the density of the fluid. The weight of the sphere is then $\frac{4}{3}\pi r^3 \rho g$, and the buoyant force is $\frac{4}{3}\pi r^3 \rho' g$; when the terminal velocity is reached, the total force is zero, and

$$\tfrac{4}{3}\pi r^3 \rho' g + 6\pi\eta\, rv_T = \tfrac{4}{3}\pi r^3 \rho g,$$

or

$$v_T = \frac{2}{9}\frac{r^2 g}{\eta}(\rho - \rho'). \tag{13-30}$$

When the terminal velocity of a sphere of known radius and density is measured, the viscosity of the fluid in which it is falling can be found from Eq. (13–30). Conversely, if the viscosity is known, the radius of the sphere can be determined by measuring the terminal velocity. This method was used by Millikan to determine the radius of very small electrically charged oil drops by observing their free fall in air. He used these drops to measure the charge of an individual electron, as we will discuss further in Section 26–6.

The Stokes's-law force is what keeps clouds in the air. The terminal velocity for a water droplet of radius 10^{-5} m is of the order of 1 mm·s^{-1}. Similarly, raindrops have terminal velocities of a few meters per second; otherwise they would smash everything in their path.

13–9 TURBULENCE

When the speed of a fluid flowing in a pipe exceeds a certain critical value, which depends on the properties of the fluid and the diameter of the pipe, the nature of the flow becomes extremely complicated. Within a thin layer adjacent to the pipe walls, called the *boundary layer*, the flow is still laminar. Beyond the boundary layer, the motion is highly irregular. Random local circular currents called *vortices* develop within the fluid, with a large increase in the resistance to flow. Flow of this sort is called *turbulent*.

Experiments show that a combination of four factors determines whether the flow of a fluid through a pipe is laminar or turbulent. This combination is known as the **Reynolds number**, N_R, and is defined as

$$N_R = \frac{\rho v D}{\eta}, \tag{13-31}$$

where ρ is the density of the fluid, v is its average forward velocity, η its viscosity, and D the diameter of the pipe. (The average velocity of the fluid is defined as the uniform velocity over the entire cross section of the pipe that would result in the same volume rate of flow.) The Reynolds number, $\rho v D/\eta$, is a *dimensionless* quantity and has the same numerical value in any consistent system of units. For example, for water at 20°C flowing in a pipe of diameter

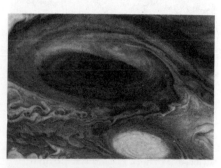

Photograph of the Great Red Spot of Jupiter, sent back by the Voyager spacecraft. This feature, first observed in the seventeenth century, is a large-scale atmospheric disturbance similar to a cyclone. (Courtesy of NASA.)

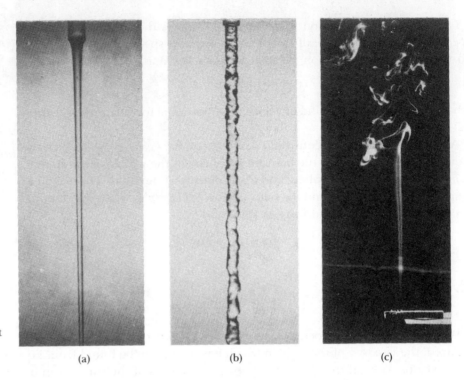

(a)   (b)   (c)

13–27 (a) Laminar flow. (b) Turbulent flow. (c) First laminar, then turbulent flow.

1 cm with an average velocity of $10 \text{ cm} \cdot \text{s}^{-1}$, the Reynolds number is

$$N_{\text{R}} = \frac{\rho v D}{\eta} = \frac{(1 \text{ g} \cdot \text{cm}^{-3})(10 \text{ cm} \cdot \text{s}^{-1})(1 \text{ cm})}{0.01 \text{ dyn} \cdot \text{s} \cdot \text{cm}^{-2}} = 1000.$$

The Reynolds number: predicting when turbulence will occur

Had the four quantities been expressed originally in SI units, the same value of 1000 would have been obtained.

A variety of experiments have shown that for fluid flow in a cylindrical pipe, the flow is laminar when the Reynolds number is less than about 2000. When it is above about 3000 the flow is turbulent. In the transition region between 2000 and 3000 the flow is unstable and may change from one type to the other. Thus for water at 20°C flowing in a pipe 1 cm in diameter, the flow is laminar when

$$\frac{\rho v D}{\eta} \leq 2000,$$

or when

$$v \leq \frac{(2000)(0.01 \text{ dyn} \cdot \text{s} \cdot \text{cm}^{-2})}{(1 \text{ g} \cdot \text{cm}^{-3})(1 \text{ cm})} = 20 \text{ cm} \cdot \text{s}^{-1}.$$

13–28 An example of the onset of turbulence produced by a cylindrical obstacle. Upstream flow is laminar, but downstream flow shows the formation of eddies or vortices. The alternating series of vortices is called a Karman vortex street. The pattern is made visible by injecting smoke filaments into the air upstream from the obstacle. (Courtesy Department of Aeronautics, Imperial College of Science and Technology.)

Above about $30 \text{ cm} \cdot \text{s}^{-1}$ the flow is turbulent. If air at the same temperature were flowing at $30 \text{ cm} \cdot \text{s}^{-1}$ in the same pipe, the Reynolds number would be

$$N_{\text{R}} = \frac{(0.0013 \text{ g} \cdot \text{cm}^{-3})(30 \text{ cm} \cdot \text{s}^{-1})(1 \text{ cm})}{181 \times 10^{-6} \text{ dyn} \cdot \text{s} \cdot \text{cm}^{-2}} = 215.$$

Since this is much less than 3000, the flow would be laminar; it would not become turbulent unless the velocity were at least $420 \text{ cm} \cdot \text{s}^{-1}$.

The distinction between laminar and turbulent flow is shown in the photographs of Fig. 13–27. In (a) and (b) the fluid is water; in (c), air and smoke particles. Figure 13–28 shows the onset of turbulence in the flow of a liquid past a cylindrical obstacle. The study of turbulent flow, and of chaotic behavior in general, is an area of great current interest in theoretical physics.

SUMMARY

Density is mass per unit volume. If a mass m of material has volume V, its density ρ is given by

$$\rho = m/V. \tag{13-1}$$

Specific gravity is the ratio of the density of a material to the density of water. It is a dimensionless number.

Pressure is force per unit area. Pascal's law: Pressure applied to the surface of an enclosed fluid is transmitted undiminished to every portion of the fluid. Absolute pressure is the total pressure in a fluid; gauge pressure is the difference between absolute pressure and atmospheric pressure.

The SI unit of pressure is the pascal (Pa). $1 \text{ Pa} = 1 \text{ N·m}^{-2}$. Other units of pressure are:

1 bar = 10^5 Pa

1 millibar = 10^{-3} bar = 10^2 Pa.

1 Torr = 1 millimeter of mercury (mmHg)

1 atmosphere = 1.013×10^5 Pa = 760 Torr = 1.013 bar

Archimedes' principle: When a body is immersed in a fluid, the fluid exerts an upward force on the body equal to the weight of the fluid that is displaced by the body.

A boundary surface between a liquid and a gas behaves like a surface under tension; the force per unit length across a line on such a surface is called the surface tension, denoted by γ. Surface tension is responsible for pressure differences between the interior and exterior of drops and bubbles, and for the phenomenon of capillarity.

An ideal fluid is an incompressible fluid with no viscosity. A flow line is the path followed by an individual fluid particle; a streamline is a curve tangent at each point in the fluid to the velocity vector at that point. A flow tube is a tube bounded at its sides by flow lines. A stagnation point is a point where the flow velocity is zero. In laminar flow, layers of fluid slide smoothly past each other, while in turbulent flow there is great disorder and a constantly changing flow pattern.

Conservation of mass in an incompressible fluid is expressed by the equation of continuity: For two cross sections A_1 and A_2 in a flow tube, the velocities v_1 and v_2 are related by

$$A_1 v_1 = A_2 v_2. \tag{13-12}$$

Bernoulli's equation relates the pressure, flow speed, and elevation at different points in an ideal fluid; if subscripts 1 and 2 refer to any two points in the fluid,

$$p_1 + \rho g y_1 + \tfrac{1}{2}\rho v_1^{\,2} = p_2 + \rho g y_2 + \tfrac{1}{2}\rho v_2^{\,2}. \tag{13-17}$$

The viscosity of a fluid characterizes its resistance to shear strain. A fluid in which the viscous force is proportional to strain rate is called a Newtonian fluid. When such a fluid flows in a cylindrical pipe of inner radius R, the flow speed v at a distance r from the axis is given by

$$v = \frac{p_1 - p_2}{4\eta L}(R^2 - r^2), \tag{13-27}$$

KEY TERMS

density
specific gravity
pressure
Pascal's law
gauge pressure
absolute pressure
mercury barometer
atmosphere
buoyancy
Archimedes' principle
surface tension
contact angle
capillarity
ideal fluid
flow line
streamline
flow tube
laminar flow
turbulent flow
continuity equation
Bernoulli's equation
viscosity
Newtonian fluid
Poiseuille's law
Stokes's law
Reynolds number

where L is the length of pipe, p_1 and p_2 are the pressures at the two ends, and η is the viscosity. The total volume rate in this situation is given by Poiseuille's equation:

$$\frac{dV}{dt} = \frac{\pi}{8}\frac{R^4}{\eta}\frac{p_1 - p_2}{L}. \tag{13-28}$$

A sphere of radius r moving with speed v through a fluid having viscosity η experiences a resisting force given by Stokes's law:

$$F = 6\pi\eta rv. \tag{13-29}$$

The Reynolds number for fluid flow in a pipe is defined as

$$N_R = \rho v D/\eta. \tag{13-31}$$

The transition from laminar to turbulent flow occurs at a value of N_R in the range of 2000 to 3000.

QUESTIONS

13-1 A rubber hose is attached to a funnel, and the free end is bent around to point upward. When water is poured into the funnel, it rises in the hose to the same level as in the funnel, despite the fact that the funnel has a lot more water in it than the hose. Why?

13-2 Is the pressure inside a siphon tube greater or less than atmospheric pressure? Does this suggest a limitation on the maximum height over which a siphon can be used?

13-3 How can one instrument function as both a barometer and an altimeter?

13-4 In describing the size of a large ship, one uses such expressions as "it displaces 20,000 tons." What does this mean? Can the weight of the ship be obtained from this information?

13-5 A popular, though expensive, sport is hot-air ballooning, in which a large nylon balloon is filled with air heated by a gas burner at the bottom. Why must the air be heated? How does the balloonist control ascent and descent?

13-6 A rigid lighter-than-air balloon filled with helium cannot continue to rise indefinitely. Why not? What determines the maximum height it can attain?

13-7 Does the buoyant force on an object immersed in water change if the object and the container of water are placed in an elevator that accelerates upward?

13-8 If a person can barely float in sea water, will he or she float more easily in fresh water, or be unable to float at all? Why?

13-9 The purity of gold can be tested by weighing it in air and in water. How? Why is it not feasible to make a fake gold brick by gold-plating some cheaper material?

13-10 A possible new technology made feasible by the "zero-gravity" environment of an orbiting space station is the making of metal foam. Why is it easier to make metal foam under zero-g conditions than on earth?

13-11 Why can't a skin diver obtain an air supply at any desired depth by breathing through a "snorkel," a tube connected to a face mask and having its upper end above the water surface?

13-12 Suppose the door of a room makes an airtight but frictionless fit in its frame. Do you think you could open the door if the air pressure on one side were standard atmospheric pressure and that on the other side differed from standard by 1%?

13-13 Is the continuity relation, Eq. (13-12), valid for compressible fluids? If not, is there a similar relation that *is* valid?

13-14 If the velocity at each point in space in steady-state fluid flow is constant, how can a fluid particle accelerate?

13-15 Does the "lift" of an airplane wing depend on altitude?

13-16 How does a baseball pitcher give the ball the spin that makes it curve? Can he make it curve in either direction? Does it matter whether he is right-handed or left-handed? What is a spitball? Why is it illegal?

13-17 In a store-window vacuum cleaner display, a table-tennis ball is suspended in midair in a jet of air blown from the outlet hose of a tank-type vacuum cleaner. The ball bounces around a little but always returns to the center of the jet, even if it is tilted. How does this behavior illustrate the Bernoulli relation?

13-18 Why do jet airplanes usually fly at altitudes of about 30,000 ft, though it takes a lot of fuel to climb that high?

13-19 What causes the sharp hammering sound sometimes heard from a water pipe when an open faucet is suddenly turned off?

13–20 When a smooth-flowing stream of water comes out of a faucet, it narrows as it falls, and if it falls far enough it eventually breaks up into drops. Why does it narrow? Why does it break up?

13–21 A tornado consists of a rapidly whirling air vortex. Why is the pressure always much lower in the center than at the outside? How does this condition account for the destructive power of a tornado?

13–22 When paddling a canoe, one can attain a certain critical speed with relatively little effort, and then a much greater effort is required to make the canoe go even a little faster. Why?

13–23 Why does air escaping from the open end of a small pipe make a lot more noise than air escaping at the same volume rate from a large pipe?

EXERCISES

Section 13–1 Density

13–1 A rectangular block of unidentified material has dimensions $5 \times 15 \times 30$ cm and a mass of 2.0 kg.

a) What is the density?

b) Will the material float in water?

13–2 A lead cube has a total mass of 50 kg. What is the length of a side?

Section 13–2 Pressure in a Fluid

13–3 A diving bell is to be designed to withstand the pressure of sea water at a depth of 600 m.

a) What is the gauge pressure at this depth?

b) What is the total force on a circular glass window 15 cm in diameter, if the pressure inside the dividing bell equals the pressure at the surface of the water?

13–4 What gauge pressure must a pump produce in order to pump water from the bottom of Grand Canyon (730-m elevation) to Indian Gardens (1370 m)? Express your results in pascals and in atmospheres.

13–5 A barrel contains a 0.15-m layer of oil floating on water that is 0.25 m deep. The density of oil is 600 kg·m^{-3}.

a) What is the gauge pressure at the oil–water interface?

b) What is the gauge pressure at the bottom of the barrel?

13–6 The piston of a hydraulic automobile lift is 0.30 m in diameter. What gauge pressure, in pascals, is required to lift a car of mass 1500 kg? Also express this pressure in atmospheres.

13–7 The liquid in the open-tube manometer in Fig. 13–4a is mercury, and $y_1 = 3$ cm, $y_2 = 8$ cm. Atmospheric pressure is 970 millibars.

a) What is the absolute pressure at the bottom of the U-tube?

b) What is the absolute pressure in the open tube, at a depth of 5 cm below the free surface?

c) What is the absolute pressure of the gas in the tank?

d) What is the gauge pressure of the gas, in pascals?

e) What is the gauge pressure of the gas, in "cm of mercury"?

f) What is the gauge pressure in "cm of water"?

Section 13–3 Buoyancy

13–8 A solid brass statue has a weight in air of 125 N.

a) What is its volume?

b) What is its apparent weight when immersed in water?

13–9 A certain object weighs 12 N in air and 10 N when immersed in water. Find its total volume and its density.

13–10 When an iceberg floats in sea water, what fraction of its volume is submerged? (The ice, being of glacial origin, is fresh water.)

13–11 A slab of ice floats on a freshwater lake. What minimum volume must the slab have in order for an 80-kg man to be able to stand on it without getting his feet wet?

13–12 A large hollow plastic sphere is held below the surface of a freshwater lake by a cable anchored to the bottom of the lake. The sphere has a volume of 0.3 m^3, and the tension in the cable is observed to be 600 N.

a) Calculate the buoyant force exerted by the water on the sphere.

b) What is the mass of the sphere?

c) The cable breaks, and the sphere rises to the surface of the lake. When it comes to rest again, what fraction of its volume will be submerged?

13–13 A cubical block of wood 10 cm on a side floats at the interface between oil and water as in Fig. 13–29, with its lower surface 2 cm below the interface. The density of the oil is 600 kg·m^{-3}.

a) What is the gauge pressure at the upper face of the block?

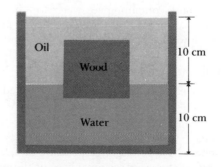

FIGURE 13–29

b) What is the gauge pressure at the lower face of the block?

c) What is the mass of the block?

Section 13–4 Surface Tension

13–14 Find the excess pressure

a) inside a water drop of radius 2 mm;

b) inside one of radius 0.01 mm, typical of water droplets in fog.

Assume $T = 20°C$.

13–15 Find the gauge pressure, in pascals, in a soap bubble 5 cm in diameter. The surface tension is 25×10^{-3} N·m^{-1}.

Section 13–5 Fluid Flow

13–16 Water is flowing in a pipe of varying cross section. At point 1 the cross-sectional area of the pipe is 0.08 m^2 and the fluid velocity is 4 m·s^{-1}.

a) What is the fluid velocity at points in the pipe where the cross-sectional area is

  i) 0.06 m^2?   ii) 0.112 m^2?

b) Calculate the volume of water discharged from the open end of the pipe in 1 s.

13–17 Water is flowing in a circular pipe of varying cross section.

a) At one place in the pipe the radius of the pipe is 0.2 m. What must the water velocity be at this point if the volume flow rate in the pipe is 0.8 m^3·s^{-1}?

b) At a second point in the pipe the water velocity is 3.8 m·s^{-1}. What is the radius of the pipe at this point?

Section 13–7 Applications of Bernoulli's Equation

13–18 A circular hole 2 cm in diameter is cut in the side of a large standpipe, 10 m below the water level in the standpipe. The top of the standpipe is open to the air. Find

a) the velocity of efflux;

b) the volume discharged per unit time.

Neglect the contraction of the streamlines after they emerge from the hole.

13–19 A sealed tank containing sea water to a height of 2 m also contains air above the water at a gauge pressure of 40 atm. Water flows out from a hole at the bottom. The cross-sectional area of the hole is 10 cm^2.

a) Calculate the efflux velocity of the water.

b) Calculate the reaction force on the tank exerted by the water in the emergent stream.

13–20 At a certain point in a horizontal pipeline, the velocity is 2 m·s^{-1} and the gauge pressure is 1.0×10^4 Pa above atmospheric. Find the gauge pressure at a second point in the line if the cross-sectional area at the second point is one-half that at the first. The liquid in the pipe is water.

13–21 Water flowing in a horizontal pipe discharges at the rate of 0.004 m^3·s^{-1}. At a point in the pipe where the cross section is 0.001 m^2, the absolute pressure is 1.2×10^5 Pa. What must be the cross section of a constriction in the pipe such that the pressure there is reduced to 1.0×10^5 Pa?

13–22 At a certain point in a pipeline the velocity is 4 m·s^{-1} and the gauge pressure is 3.0×10^5 Pa. Find the gauge pressure at a second point in the line 20 m lower than the first, if the cross section at the second point is one-half that at the first. The liquid in the pipe is water.

13–23 What gauge pressure is required in the city mains for a stream from a fire hose connected to the mains to reach a vertical height of 20 m?

13–24 An airplane of mass 6000 kg has a wing area of 60 m^2. If the pressure on the lower wing surface is 0.60×10^5 Pa during level flight at an elevation of 4000 m, what is the pressure on the upper wing surface?

13–25 Assume that air is streaming horizontally past a small airplane's wings such that the velocity is 40 m·s^{-1} over the top surface and 30 m·s^{-1} past the bottom surface. If the plane has a mass of 300 kg and a wing area of 5 m^2, what is the net force (including the effects of gravity) on the airplane? The density of air is 1.3 kg·m^{-3}.

Section 13–8 Viscosity

13–26 Water at 20°C flows through a pipe of radius 1.0 cm. If the flow velocity at the center is 0.10 m·s^{-1}, find the pressure drop along a 2-m section of pipe due to viscosity.

13–27 Water at 20°C is flowing in a pipe of radius 20 cm. If the fluid velocity in the center of the pipe is 3 m·s^{-1}, what is the fluid velocity

a) 10 cm from the center of the pipe (i.e., halfway between the center and the walls)?

b) at the walls of the pipe?

13–28 A viscous liquid flows through a tube with laminar flow, as in Fig. 13–25b. Prove that the volume rate of flow is the same as if the velocity were uniform at all points of a cross section and equal to half the velocity at the axis.

13–29 Water at 20°C is flowing in a horizontal pipe that is 15 m long. A pump maintains a gauge pressure of 800 Pa at a large tank at one end of the pipe. The other end of the pipe is open to the air.

a) If the pipe has a diameter of 8 cm, what is the volume flow rate?

b) What gauge pressure must the pump provide to achieve the same volume flow rate for a pipe of diameter 4 cm?

c) For the pipe in part (a), what does the volume flow rate become if the water is at a temperature of 60°C?

13–30 A copper sphere of mass 0.20 g is observed to fall with a terminal velocity of 0.08 m·s^{-1} in an unknown liquid. If the density of copper is 8900 kg·m^{-3} and that of the liquid is 2800 kg·m^{-3}, what is the viscosity of the liquid?

13–31 What velocity must a gold sphere of radius 4 mm have in water at 20°C for the viscous drag force to be one-fourth the weight of the sphere?

Section 13–9 Turbulence

13–32 Water at 20°C flows with a speed of 0.35 m·s⁻¹ through a pipe of diameter 3 mm.

a) What is the Reynolds number?

b) What is the nature of the flow?

13–33 Water at a temperature of 80°C is flowing at a speed of 2 m·s⁻¹ through a tube of diameter 4 mm.

a) What is the Reynolds number?

b) What is the nature of the flow?

PROBLEMS

13–34 A swimming pool measures $25 \times 8 \times 3$ m deep. Compute the force exerted by the water against

a) the bottom, and

b) either end, which has width 8 m. (*Hint:* Calculate the force on a thin horizontal strip at a depth h, and integrate this over the end of the pool.)

Do not include the force due to air pressure.

13–35. The upper edge of a gate in a dam runs along the water surface. The gate is 2 m high and 3 m wide and is hinged along a horizontal line through its center. Calculate the torque about the hinge arising from the force due to the water. (*Hint:* Use a procedure similar to that used in Problem 13–34: Calculate the torque of a thin horizontal strip at a depth h and integrate this over the gate.)

13–36 A U-shaped tube open to the air at both ends contains some mercury. A quantity of water is carefully poured into the left arm of the tube until the vertical height of the water column is 15 cm, as shown in Fig. 13–30.

a) What is the gauge pressure at the water–mercury interface?

b) Calculate the vertical distance h from the top of the mercury in the right-hand arm of the tube to the top of the water in the left-hand arm.

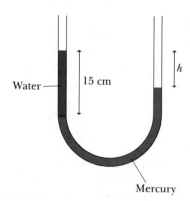

FIGURE 13–30

13–37 The deepest point known in any of the earth's oceans is in the Marianas Trench and is 10.92 km deep.

a) Assuming water to be incompressible, what is the pressure at this depth? Use the density of sea water.

b) The actual pressure is 1.17×10^8 Pa. Your calculated value of the pressure will be less because the density actually varies with depth. Take your value for the pressure and, using the compressibility of water, find its

density at the bottom of the Marianas Trench. What is the percent change in the density of the water?

13–38 According to advertising claims, a certain small car will float in water.

a) If the mass of the car is 1000 kg and its interior volume 4.0 m³, what fraction of the car is immersed when it floats? The buoyancy of steel and other materials may be neglected.

b) As water gradually leaks in and displaces the air in the car, what fraction of the interior volume is filled with water when the car sinks?

13–39 The densities of air, helium, and hydrogen (at standard conditions) are, respectively, 1.29 kg·m⁻³, 0.178 kg·m⁻³, and 0.0899 kg·m⁻³.

a) What is the volume in cubic meters displaced by a hydrogen-filled dirigible that has a total "lift" of 10,000 kg? (The "lift" is the total mass of the dirigible that can be supported by the buoyant force, in addition to the mass of the gas with which it is filled.)

b) What would be the "lift" if helium were used instead of hydrogen? In view of your answer, why is helium the gas actually used? (*Hint:* Remember the *Hindenberg.*)

13–40 A hydrometer consists of a spherical bulb and a cylindrical stem of cross section 0.4 cm². The total volume of bulb and stem is 13.2 cm³. When immersed in water, the hydrometer floats with 8 cm of the stem above the water surface. In alcohol, 1 cm of the stem is above the surface. Find the density of alcohol. (*Note.* This illustrates the accuracy of such a hydrometer. Relatively small density differences give rise to relatively large differences in hydrometer readings.)

13–41 A barge is 20 m wide and has a longitudinal cross section as shown in Fig. 13–31. (The dimension not shown in the figure is 20 m.) If the barge is made out of 7.5-cm-thick steel plate on each of its four sides and its bottom, what mass of coal can the barge carry without sinking? Is there enough room in the barge to hold this amount of coal? (The density of coal is about 1.5 g·cm⁻³.)

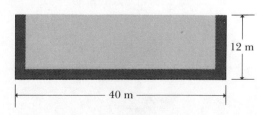

FIGURE 13–31

13–42 Consider an object of height h, mass M, and uniform cross section A floating upright in a liquid of density ρ.

a) Calculate the vertical distance from the surface of the liquid to the bottom of the floating object, at equilibrium.

b) A downward force F is applied to the top of the object. At the new equilibrium position, how much farther below the surface of the liquid is the bottom of the object than it was in part (a)? (Assume that F is small enough for some of the object to remain above the surface of the liquid.)

c) Your result in (b) shows that if F is suddenly removed, the object will oscillate up and down in simple harmonic motion. Calculate the period of this motion, in terms of the density ρ of the liquid and the mass M and cross section A of the object.

d) A 1500-kg cylindrical can buoy floats vertically in salt water (specific gravity = 1.03). The diameter of the buoy is 1.0 m. Calculate the additional distance the buoy will sink when a 100-kg man stands on top.

e) Calculate the period of the resulting vertical simple harmonic motion when the man dives off.

f) Calculate the period of oscillation of an ice cube 3 cm on a side floating in water of density 1×10^3 kg·m^{-3}, if it is pushed down and released.

13–43 A piece of wood is 0.5 m long, 0.2 m wide, and 0.02 m thick. Its specific gravity is 0.6. What volume of lead must be fastened underneath to sink the wood in calm water so that its top is just even with the water level?

13–44 When a life preserver of volume 0.03 m^3 is immersed in sea water (specific gravity 1.03), it will just support an 80-kg man (specific gravity 1.2) with $\frac{2}{10}$ of his volume above water. What is the density of the material composing the life preserver?

13–45 A block of balsa wood placed in one scale pan of an equal-arm balance is found to be exactly balanced by a 0.100-kg brass mass in the other scale pan. Find the true mass of the balsa wood, if its specific gravity is 0.15. Neglect the buoyancy of the brass in air. (The density of air is 1.29 kg·m^{-3}.)

13–46 Block A in Fig. 13–32 hangs by a cord from spring balance D and is submerged in a liquid C contained in beaker B. The mass of the beaker is 1 kg; the mass of the

liquid is 1.5 kg. Balance D reads 2.5 kg and balance E reads 7.5 kg. The volume of block A is 0.003 m^3.

a) What is the mass per unit volume of the liquid?

b) What will each balance read if block A is pulled up out of the liquid?

13–47 A piece of gold-aluminum alloy weighs 45 N. When suspended from a spring balance and submerged in water, the balance reads 36 N. What is the weight of gold in the alloy if the specific gravity of gold is 19.3 and the specific gravity of aluminum is 2.5?

13–48 A cubical block of wood 0.10 m on a side and of density 500 kg·m^{-3} floats in a jar of water. Oil of density 800 kg·m^{-3} is poured on the water until the top of the oil layer is 0.04 m below the top of the block.

a) How deep is the oil layer?

b) What is the gauge pressure at the lower face of the block?

13–49 A cubical block of wood 0.30 m on a side is weighted so that its center of gravity is at the point shown in Fig. 13–33a, and it floats in water with one-half its volume submerged. Compute the restoring torque when the block is "heeled" at an angle of 45° as in Fig. 13–33b.

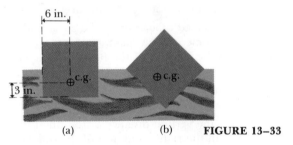

(a)   (b)   **FIGURE 13–33**

13–50 A glass tube of inside diameter 1 mm is dipped vertically into a container of mercury, with its lower end 1 cm below the mercury surface. What must be the gauge pressure of air in the tube to blow a hemispherical bubble at its lower end?

13–51 A cylindrical vessel, open at the top, is 0.20 m high and 0.10 m in diameter. A circular hole whose cross-sectional area is 1 cm^2 is cut in the center of the bottom of the vessel. Water flows into the vessel from a tube above it at the rate of 1.40×10^{-4} m^3·s^{-1}. How high will the water in the vessel rise?

13–52 Water stands at a depth H in a large open tank whose side walls are vertical (Fig. 13–34). A hole is made in one of the walls at a depth h below the water surface.

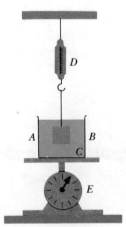

FIGURE 13–32

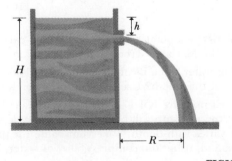

FIGURE 13–34

a) At what distance R from the foot of the wall does the emerging stream of water strike the floor?

b) Let $H = 12$ m and $h = 3$ m. At what height below the water surface could a second hole be cut so that the stream emerging from it would have the same range as for the first hole?

13–53 Water flows through a horizontal pipe of cross-sectional area 10 cm². At one section the cross-sectional area is reduced to 5 cm². The pressure difference between the two sections is 300 Pa. How many cubic meters of water will flow out of the pipe in 1 min?

13–54 Water flows steadily from a reservoir, as in Fig. 13–35. The elevation of point 1 is 10 m; of points 2 and 3, 1 m. The cross section at point 2 is 0.04 m²; at point 3, 0.02 m². The area of the reservoir is very large compared with the cross sections of the pipe. Assuming Bernoulli's equation applies, compute

a) the gauge pressure at point 2;

b) the discharge rate in cubic meters per second.

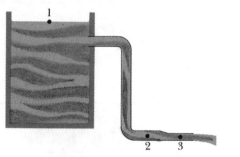

FIGURE 13–35

13–55 Modern airplane design calls for a "lift" of about 1000 N per square meter of wing area. Assume that air flows past the wing of an aircraft with streamline flow. If the velocity of flow past the lower wing surface is 100 m·s⁻¹, what is the required velocity over the upper surface to give a "lift" of 1000 N·m⁻²? The density of air is 1.29 kg·m⁻³.

13–56 The section of pipe shown in Fig. 13–36 has a cross section of 40 cm² at the wider portions and 10 cm² at the constriction. Water is flowing in the pipe, and the discharge from the pipe is 3.0×10^{-3} m³·s⁻¹.

a) Find the velocities at the wide and the narrow portions.

b) Find the pressure difference between these portions.

c) Find the difference in height between the mercury columns in the U-shaped tube.

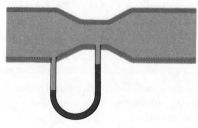

FIGURE 13–36

13–57 Two very large open tanks, A and F (Fig. 13–37), both contain the same liquid. A horizontal pipe BCD, having a constriction at C, leads out of the bottom of tank A, and a vertical pipe E opens into the constriction at C and dips into the liquid in tank F. Assume streamline flow and no viscosity. If the cross section at C is one-half that at D, and if D is a distance h_1 below the level of the liquid in A, to what height h_2 will liquid rise in pipe E? Express your answer in terms of h_1. Neglect changes in atmospheric pressure with elevation.

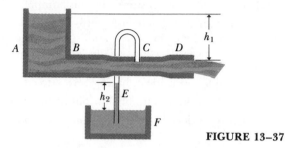

FIGURE 13–37

13–58 Water stands at a depth of 1 m in an enclosed tank whose side walls are vertical. The space above the water surface contains air at a gauge pressure of 8×10^5 Pa. The tank rests on a platform 2 m above the floor. A hole of cross-sectional area 1 cm² is made in one of the side walls just above the bottom of the tank.

a) Where does the stream of water from the hole strike the floor?

b) What is the vertical force exerted on the floor by the stream?

c) What is the horizontal force exerted on the tank?

Assume that the water level and pressure in the tank remain constant, and neglect any effect of viscosity.

13–59 Figure 13–27a shows a liquid flowing from a vertical pipe. Notice that the vertical stream of the liquid has a very definite shape as it flows from the pipe.

a) We will now get the equation for this shape. Assume that the liquid is in free fall once it leaves the pipe and then find an equation for the velocity of the liquid as a function of the distance it has fallen. Combining this with the equation of continuity, find an expression for the radius of the stream of liquid.

b) If water flows out of a vertical pipe at 1.0 m·s⁻¹, how far below the outlet will the radius be one half the original radius of the stream?

13–60 The densities of steel and of glycerine are 8500 kg·m⁻³ and 1320 kg·m⁻³, respectively. The viscosity of glycerine is 8.3 poise.

a) With what velocity is a steel ball 1 mm in radius falling in a tank of glycerine at an instant when its acceleration is one half that of a freely falling body?

b) What is the terminal velocity of the ball?

13–61

a) With what terminal velocity will an air bubble 1 mm in diameter rise in a liquid of viscosity 1.5 poise and density 900 kg·m⁻³?

b) What is the terminal velocity of the same bubble in water at 20°C?

13-62 Oil having a viscosity of 3.0 poise and a density of 900 kg·m^{-3} is to be pumped from one large open tank to another through 1 km of smooth steel pipe 0.15 m in diameter. The line discharges into the air at a point 30 m above the level of the oil in the supply tank.

a) What gauge pressure, in Pa and in atmospheres, must the pump exert in order to maintain a flow of 0.05 m^3·s^{-1}?

b) What is the power consumed by the pump?

13-63 The tank at the left in Fig. 13–26a has a very large cross section and is open to the atmosphere. The depth $y = 0.4$ m. The cross sections of the horizontal tubes leading out of the tank are, respectively, 1 cm^2, 0.5 cm^2, and 0.2 cm^2. The liquid is ideal, having zero viscosity.

a) What is the volume rate of flow out of the tank?

b) What is the velocity in each portion of the horizontal tube?

c) What are heights of the liquid in the vertical side tubes?

Suppose that the liquid in Fig. 13–26b has a viscosity of 0.5 poise and a density of 800 kg·m^{-3}, and that the depth of liquid in the large tank is such that the volume rate of flow is the same as in part (a) above. The distance between the side tubes at c and d, and between those at e and f, is 0.20 m. The cross sections of the horizontal tubes are the same in both diagrams.

d) What is the difference in level between the tops of the liquid columns in tubes c and d?

e) In tubes e and f.

f) What is the flow velocity on the axis of each part of the horizontal tube?

g) Is it reasonable to suppose that the flow of this second liquid is laminar?

h) Would the flow be laminar if the liquid were water at 20°C, with the same volume flow rate?

CHALLENGE PROBLEMS

13-64 The following is quoted from a letter. How would you reply?

> It is the practice of carpenters hereabouts, when laying out and leveling up the foundations of relatively long buildings, to use a garden hose filled with water, into the ends of the hose being thrust glass tubes 10 to 12 inches long.
>
> The theory is that water, seeking a common level, will be of the same height in both the tubes and thus effect a level. Now the question rises as to what happens if a bubble of air is left in the hose. Our greybeards contend the air will not affect the reading from one end to the other. Others say that it will cause important inaccuracies.

Can you give a relatively simple answer to this question, together with an explanation? Figure 13–38 gives a rough sketch of the situation that caused the dispute.

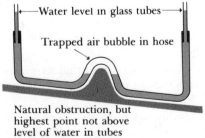

—Water level in glass tubes—

Trapped air bubble in hose

Natural obstruction, but highest point not above level of water in tubes

FIGURE 13–38

13-65 Suppose a piece of styrofoam, $\rho = 1.0 \times 10^2$ kg·m^{-3}, is held completely submerged in water, as shown in Fig. 13–39.

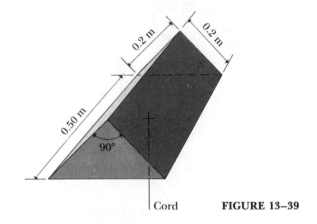

0.2 m 0.2 m

0.50 m

90°

Cord   **FIGURE 13–39**

a) What is the tension in the cord? Find this by using Archimedes' principle.

b) Use $p = p_a + \rho g h$ to calculate directly the force on the two sloped sides and the bottom of the styrofoam; then show that the sum of these forces is the buoyant force.

13-66 In designing a large sea aquarium, the engineer needs to know the force on the circular windows that are planned as observation windows. Each window's center is 4.0 m below the water's surface, and each window has a diameter of 1.0 m. What is the force exerted on each window? This is a marine aquarium, so the density of sea water should be used. (Refer to Table 13–1.)

13-67 A U-shaped tube of length l (see Fig. 13–40) contains a liquid. What is the difference in height between the liquid columns in the vertical arms

a) if the tube has an acceleration a toward the right?

b) if the tube is mounted on a horizontal turntable rotating with an angular velocity ω, with one of the vertical arms on the axis of rotation?

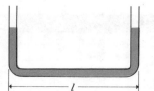

FIGURE 13–40

c) Explain why the difference in height does not depend on the density of the liquid or on the cross-sectional area of the tube. Would it be the same if the vertical tubes did not have equal cross sections? Would it be the same if the horizontal portion were tapered from one end to the other?

13–68 A rock of mass $m = 5$ kg is suspended from the roof of an elevator by a light cord. The rock is totally immersed in a bucket of water that sits on the floor of the elevator.

a) When the elevator is at rest, the tension in the cord is observed to be 36 N. Calculate the volume of the rock.

b) Derive an expression for the tension in the cord when the elevator is accelerating *upward* with an acceleration a. Calculate the tension when $a = 6$ m·s^{-2} upward.

c) Derive an expression for the tension in the cord when the elevator is accelerating *downward* with an acceleration a. Calculate the tension when $a = 6$ m·s^{-2} downward.

d) What is the tension when the elevator is in free fall, with a downward acceleration equal to g?

13–69

a) Derive the expression for the height of capillary rise in the space between two parallel plates dipping in a liquid of density ρ. Assume a contact angle θ. Let the separation between the plates be x.

b) Two glass plates, parallel to each other and separated by 0.5 mm, are dipped in water at 20°C. To what height will the water rise between them? Assume zero angle of contact.

13–70 A large tank of diameter D and open to the air contains water to a height H. A small hole of diameter d ($d \ll D$) is made at the base of the tank. Neglecting any effects of viscosity, calculate the time it takes for the tank to drain completely.

13–71 A siphon, as depicted in Fig. 13–41, is a convenient device for removing liquids from containers. To establish the flow the tube must initially be filled with fluid. Let the

fluid have density ρ. Assume that the cross-sectional area of the tube is the same at all points along the tube.

a) If the lower end of the siphon is at a distance h below the surface of the liquid in the container, what is the velocity of the fluid as it flows out the lower end of the siphon? (Assume that the container has a very large diameter, and neglect any effects of viscosity.)

b) A curious feature of a siphon is that the fluid initially flows "uphill." What is the greatest height H that the high point of the tube can have for the siphon flow still to occur?

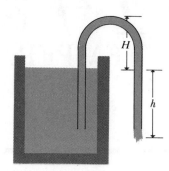

FIGURE 13–41

13–72 For an object falling in air, Stokes's law is obeyed only for small velocities. The air drag on an object falling through the earth's atmosphere is more accurately given by $\frac{1}{2}CA\rho v^2$, where A is the area of the object perpendicular to the motion, ρ is the density of the air, and C is the drag coefficient. (Refer to Section 7–10.) C is determined experimentally; for a spherical object having the velocities of this problem, it is approximately 0.5. (*Note.* For a graph of the values of C over a wide range of values of velocity, see J. E. A. John and W. Haberman, *Introduction to Fluid Mechanics* [Englewood Cliffs, N.J.: Prentice-Hall], p. 198.

A ball bearing of radius 0.5 cm and mass 5.0 g is dropped from the top of the Sears Tower in Chicago, which is 443 m tall.

a) Calculate the velocity the ball bearing would have at the ground if air resistance were negligible.

b) Taking air resistance into account, what is the terminal velocity of the ball bearing?

c) What is its actual velocity just before reaching the ground? Take the density of air to be 1.29 kg·m^{-3}. Note that $\int \tanh x \, dx = \ln(\cosh x)$. (Refer to Challenge Problem 5–66.)

14
TEMPERATURE AND EXPANSION

WITH THIS CHAPTER WE BEGIN A STUDY OF A CLASS OF PHENOMENA ASSO-
ciated with *heat*. For this study we need to introduce a new fundamental physical quantity called *temperature*, in addition to the three fundamental quantities
(mass, length, and time) used in our study of mechanical systems in preceding
chapters. We discuss the concept of thermal equilibrium and various tempera-
ture-measuring instruments—that is, *thermometers*—along with the various
temperature scales in common use. Finally, we consider the expansion and
contraction of materials caused by temperature changes.

This chapter serves as an introduction to the next several chapters, where
we study the broad area known as *thermodynamics*. Such diverse topics as expansion joints in bridges, the efficiency of an automobile engine, solar energy,
and the design of better refrigerators are all directly related to this subject
area.

14–1 TEMPERATURE AND THERMAL EQUILIBRIUM

Temperature: using numbers to describe hot and cold

The concept of **temperature** originates with qualitative ideas of "hot" and
"cold" based on our sense of touch. A body that feels hot to the touch is said to
have a higher temperature than the same body when it feels cold. More generally, many properties of matter depend on temperature. The length of a metal
rod increases when the rod becomes hotter. The pressure of gas in a container
increases when the gas becomes hotter; a steam boiler may explode if it becomes too hot. The electrical properties of materials change with temperature; most metals are poorer electrical conductors when hot than when cold.
When a material is extremely hot, it glows "red-hot" or "white-hot," with a
color that depends on its temperature.

All of this behavior can be understood in greater depth on the basis of the
molecular structure of materials and the dependence of molecular configura-
tion and motion on temperature. In Chapter 20 we will consider the relationship between temperature and molecular motion in more detail. It is

important to understand, however, that temperature and heat are inherently *macroscopic* concepts, and that they can and should be defined independently of any detailed consideration of molecular motion.

To define temperature *quantitatively,* we must devise a scheme to assign numbers to various degrees of hotness or coldness. To do this, we can make use of some measurable property of a system that varies with its hotness or coldness. A simple example is a liquid such as mercury or ethanol in a bulb attached to a thin tube, as in Fig. 14–1a. When the system becomes hotter, the liquid rises in the tube and the value of L increases. Another simple system is a quantity of gas in a constant-volume container, as shown in Fig. 14–1b. The pressure p, measured by the gauge, increases or decreases as the gas becomes hotter or colder. A third example is the electrical resistance R of a conducting wire, which also varies when the wire becomes hotter or colder.

In each of these examples a significant quantity that changes with the state of hotness or coldness of the system, such as the length L, the pressure p, or the resistance R, can be used to assign a number to a system to describe how hot it is. Thus each device can be used as a **thermometer.**

In each of the above examples, the quantity measured indicates the temperature *of the thermometer.* How can such a thermometer be used to measure the temperature of any other body? The answer to this question leads us to the concept of *thermal equilibrium* and an important related principle. Suppose we want to measure the temperature of a cup of hot coffee, using a mercury thermometer initially at room temperature. We immerse the thermometer in the coffee; the thermometer becomes hotter, and the coffee cools off a little. After we observe that no further change is taking place as a result of this interaction, we read the thermometer. This *equilibrium* condition, in which no further change in the parts of the system occurs as a result of their interaction with each other, is called a state of **thermal equilibrium.** But how do we know that at thermal equilibrium the temperature of the thermometer is the same as that of the coffee? To establish this, we need some additional experiments.

First we consider the interaction of two systems. Let system A be the tube-and-liquid system of Fig. 14–1a and system B the container of gas and pressure gauge of Fig. 14–1b. When the two systems are brought into contact, in general the values of L and p change as the systems move toward thermal equilibrium. Initially one system is hotter than the other, and each system changes the state of the other.

However, if the systems are separated by an insulating material, or **insulator,** such as wood, plastic, foam, or fiberglass, they influence each other more slowly. An ideal insulator is defined as a material that permits *no* interactions at all between the two systems. Such an ideal insulator *prevents* the attainment of thermal equilibrium if the two systems are not initially in thermal equilibrium. This, incidentally, is exactly what the insulating material in a camping cooler does. We do not want the cold food inside to warm up and attain thermal equilibrium with the hot summer air outside; the insulation prevents this, although not forever.

Now consider three systems, A, B, and C, as shown in Fig. 14–2. We do not need to describe in detail what they are, but suppose that initially they are *not* in thermal equilibrium. Perhaps one is very hot, one very cold, and the third lukewarm. We surround the three systems with an insulating box so they cannot interact with anything except each other. We also separate systems A and B by an insulating wall, the gray slab in Fig. 14–2a, but we let system C interact

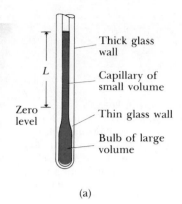

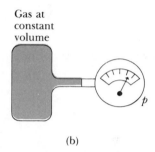

Thick glass wall

Capillary of small volume

Thin glass wall

Bulb of large volume

L

Zero level

(a)

Gas at constant volume

p

(b)

14–1 (a) A system whose state is specified by the value of the length L. (b) A system whose state is given by the value of the pressure p.

Two bodies in thermal equilibrium have the same temperature: a fundamental truth.

An insulator is a material that prevents or slows down thermal interaction.

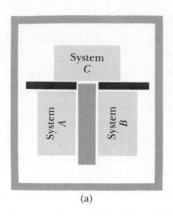 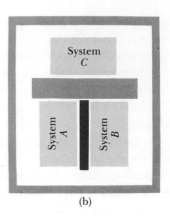

(a)   (b)

14–2   The zeroth law of
thermodynamics. (a) If *A* and *B* are
each in thermal equilibrium with *C*,
then (b) *A* and *B* are in thermal
equilibrium with each other. Thick
gray layers represent insulating walls;
thinner colored layers represent
conducting walls.

**A conductor is a material that permits
thermal interaction.**

with both *A* and *B*, as indicated symbolically by a thin colored slab, a thermal
conductor. (A conducting material is one that permits thermal interactions of
the sort we are discussing between the systems it separates.) We wait until no
further change is occurring as a result of this interaction, that is, until thermal
equilibrium is attained. We then remove *C* from *A* and *B* by inserting an
insulating wall, as in Fig. 14–2b, and we also remove the insulating wall between *A* and *B*, replacing it by a conducting wall. What happens?

Experiment shows that *nothing* happens; the states of *A* and *B* do not
change! Thus the experiment shows that if *C* is initially in thermal equilibrium
with both *A* and *B*, then when *A* and *B* are placed in contact they are also in
thermal equilibrium with each other. This outcome may have seemed obvious,
but we must verify our intuitive expectations experimentally.

**The zeroth law: a strange name for an
essential principle**

We may state these experimental facts concisely as follows: *Two systems in
thermal equilibrium with a third system are in thermal equilibrium with each other.* This
principle is called the **zeroth law of thermodynamics.** This odd-sounding
name refers to the fact that the basic principles of thermodynamics are called
the first, second, and third laws of thermodynamics. The principle just discussed is fundamental to all of these, although it was not clearly recognized
until later. Thus it deserves to be called the zeroth law.

The property of any system that determines whether or not two systems
are in thermal equilibrium is their *temperature. Two systems in thermal equilibrium
must have the same temperature.* As we have seen, the temperature of a system
may be represented by a number. Establishing a temperature scale, such as the
Fahrenheit or Celsius scale, will be discussed in Section 14–3; this is simply a
specific scheme for assigning numbers to temperatures. Once this is done, the
condition for thermal equilibrium of two systems is that they have the same
temperature. When the temperatures of two systems are different, they *cannot*
be in thermal equilibrium.

14–2 THERMOMETERS

One form of thermometer commonly used in everyday life is the tube-and-
liquid type shown in Fig. 14–1a. The liquid is usually mercury or colored
ethanol. If we label the liquid level at the freezing temperature of water "zero"
and the level at the boiling temperature "100," and if we divide the distance

between these two points into 100 equal intervals, we obtain the familiar Celsius scale, discussed in detail in the next section.

Another type of thermometer in common use contains a metallic strip consisting of two dissimilar metals welded together, as shown in Fig. 14–3a. Because one metal expands more than the other with temperature increases, the strip bends with temperature changes. In practical thermometers this strip is usually made in the form of a spiral, with the outer end anchored to the thermometer case and the inner end attached to a pointer, as shown in Fig. 14–3b. The pointer rotates in response to temperature changes, and a circular scale is provided.

Various other systems serve as thermometers in situations having special requirements, such as great precision, speed of attaining equilibrium with the body being measured, or extreme temperature ranges.

In a *resistance thermometer* we measure the changing electrical resistance of a coil of fine wire. Resistance can be measured with great precision, and so the resistance thermometer is one of the most precise instruments for the measurement of temperature. At extremely low temperatures, a small carbon cylinder or a small piece of germanium crystal is used instead of a coil of wire.

To measure very high temperatures, an *optical pyrometer* can be used. It measures the brightness of light emitted by a red-hot or white-hot substance. As shown in Fig. 14–4, it consists of a telescope T with a small electric light bulb L mounted in the tube. The bulb is connected to a variable power supply P, and the brightness of the bulb is varied until it matches that of the substance being observed. From previous calibration at known temperatures, the power-supply adjustment can be marked to read temperature directly. No part of the instrument touches the hot substance, so the optical pyrometer can be used at temperatures that would destroy most other thermometers.

Among all temperature-dependent properties, or **thermometric properties,** the pressure of a gas kept at constant volume is outstanding in its sensitivity, accuracy of measurement, and reproducibility. The constant-volume gas

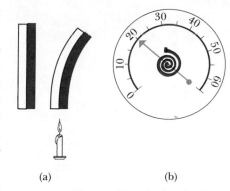

(a)   (b)

14–3   (a) A bimetallic strip bends when heated because one metal expands more than the other. (b) A bimetallic strip, usually in the form of a spiral, may be used as a thermometer.

Some thermometers depend on electrical resistance.

Thermometers that measure very high temperatures by using the emitted light from a hot body

The gas thermometer: a cumbersome, high-precision instrument with fundamental significance

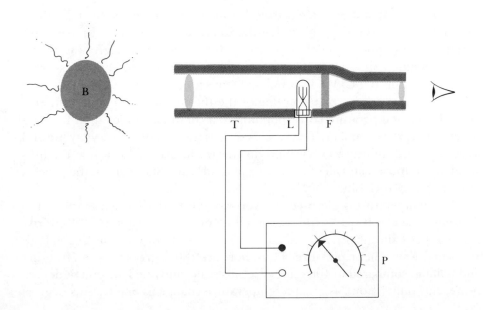

14–4   An optical pyrometer. The telescope T is pointed toward the hot body B, and the power supply P is adjusted until the brightness of the filament in the lamp L is the same as that of the hot body. The filter F reduces the overall brightness level. The power supply can be calibrated to read temperature directly.

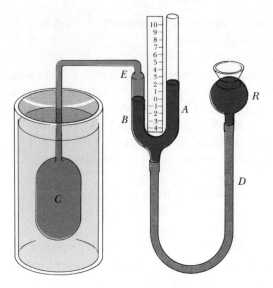

14–5 The constant-volume gas thermometer.

thermometer is shown schematically in Fig. 14–5. The gas, usually helium, is contained in bulb C, and its pressure is measured by the open-tube mercury manometer. As the temperature of the gas increases, the gas expands, forcing the mercury down in tube B and up in tube A. Tubes A and B are connected through a rubber tube D to a mercury reservoir R. By raising R, we can bring the mercury level in B back to a reference mark E. The gas is thus kept at constant volume. The pressure is determined by the difference in mercury levels in tubes A and B.

Gas thermometers are used mainly in bureaus of standards and in research laboratories. They are large, bulky, and slow in coming to thermal equilibrium.

14–3 THE CELSIUS AND FAHRENHEIT SCALES

Two familiar temperature scales and how they are related

The **Celsius temperature scale** (formerly called the *centigrade* scale in English-speaking countries) is defined so that the freezing temperature of pure water at normal atmospheric pressure is zero or 0°C (read "zero degrees Celsius") and the boiling temperature of pure water at normal atmospheric pressure is one hundred or 100°C. We interpolate between or extrapolate beyond these reference temperatures by using one of the thermometers described in Section 14–2. A temperature corresponding to a condition colder than freezing water is a negative number. The Celsius scale is used, both in everyday life and in science and industry, everywhere in the world except for a few English-speaking countries, and all of these except the United States are in the process of converting to Celsius.

The temperature scale used in everyday life in the United States is the **Fahrenheit scale.** It places the freezing temperature of water at 32°F (thirty-two degrees Fahrenheit) and the boiling temperature of water at 212°F, both at normal atmospheric pressure. Thus there are 180 degrees between freezing and boiling, compared to 100 on the Celsius scale, and one Fahrenheit degree represents only $\frac{100}{180}$ or $\frac{5}{9}$ as great a temperature change as one Celsius degree.

To convert temperatures from Celsius to Fahrenheit, we note that a Celsius temperature T_C is the number of Celsius degrees above freezing; to obtain the number of Fahrenheit degrees above freezing we must multiply this by $\frac{9}{5}$. But freezing on the Fahrenheit scale is at 32°F, so to obtain the actual Fahrenheit temperature we must first multiply the Celsius value by $\frac{9}{5}$ and then add 32°. Symbolically,

$$T_F = \tfrac{9}{5}T_C + 32°. \tag{14–1}$$

To convert Fahrenheit to Celsius, we solve this equation for T_C, obtaining

$$T_C = \tfrac{5}{9}(T_F - 32°). \tag{14–2}$$

That is, we first subtract 32° to obtain the number of Fahrenheit degrees above freezing, and then multiply by $\frac{5}{9}$ to obtain the number of Celsius degrees above freezing, that is, the Celsius temperature.

We do not recommend memorizing Eqs. (14–1) and (14–2) to convert between Celsius and Fahrenheit temperatures. Instead, try to understand the reasoning that led to them well enough so you can derive them on the spot when you need them.

We often need to distinguish between an actual temperature on a certain scale and a temperature *interval*, representing a *difference* in the temperature of two bodies or a *change* of temperature of a body. It is customary to represent a temperature *interval* of 10° as 10 C° (ten Celsius degrees) and an actual temperature of 20° as 20°C (twenty degrees Celsius). Thus a beaker of water heated from 20°C to 30°C undergoes a temperature change of 10 C°. This notation will be followed throughout this book.

A fundamental problem in defining temperature scales is that when two thermometers, such as a liquid-in-tube system and a resistance thermometer, are calibrated so that they agree at 0°C and 100°C, they *do not* necessarily agree at intermediate temperatures. Different thermometric properties such as length and resistance do not necessarily change with temperature in precisely the same way. Thus a temperature scale defined in this way always depends somewhat on the specific properties of the material used in the thermometer.

The closest approach to consistency is obtained by using a constant-volume gas thermometer with the lowest practical pressures. Low-pressure gas thermometers using various gases are found to agree very closely, and such thermometers are used for high-precision standards. But we are still at the mercy of properties of specific materials. We might well wish for a temperature scale that is *completely* independent of material properties. It is, in fact, possible to define such a scale; we will return to this fundamental problem in Chapter 19, after developing the necessary thermodynamic principles.

The bimetallic strip shown in Fig. 14–3a is also used in most thermostats for controlling heating systems. One end of the bimetallic strip is attached to an electrical contact set, which makes or breaks contact as the bimetallic strip bends or straightens in response to temperature changes. This electrical contact turns the furnace on or off so as to maintain a constant temperature. This is a rudimentary example of a *feedback control system;* a physical quantity is measured, and if it does not have the desired value, a system is activated to correct the situation.

A notation for distinguishing between temperatures and temperature intervals

Thermometers aren't always consistent with each other.

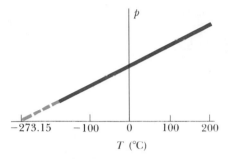

14–6 Graph of pressure versus temperature for a constant-volume gas thermometer. A straight-line extrapolation predicts that the pressure would become zero at −273.15°C if that temperature could be reached and the proportionality of pressure to temperature held exactly.

14–4 THE KELVIN AND RANKINE SCALES

Calibration of a constant-volume gas thermometer might well make use of a graph similar to Fig. 14–6, showing the relation between gas pressure and temperature. Such a graph suggests that there is a hypothetical temperature at which the gas pressure would become zero. Surprisingly, this temperature turns out to be the same for a wide variety of gases: namely, −273.15°C. It is not possible actually to observe this zero-pressure condition, for several reasons. One is that most gases liquefy and solidify at very low temperatures, and the proportionality of pressure to temperature no longer holds.

The discovery that the extrapolated zero-pressure temperature is the same for many gases raises the possibility that there may be a more fundamental way to define the zero point on a temperature scale than the freezing point of water. This is in fact the basis of the **Kelvin temperature scale,** named after Lord Kelvin (1824–1907). The degrees are the same size as on the Celsius scale, but the zero is shifted so that 0 K = −273.15°C and 273.15 K = 0°C; that is,

$$T_K = T_C + 273.15°. \tag{14–3}$$

Thus ordinary room temperature, 20°C, becomes 20 + 273.15, or about 293 K.

In SI nomenclature, "degree" is not used with the Kelvin scale; the temperature mentioned above is read "293 kelvins," not "degrees Kelvin." Also, Kelvin is capitalized when it refers to the temperature scale, but the *unit* of temperature is the *kelvin*. This is not capitalized but is abbreviated K. (The phrase "degrees Kelvin," although officially obsolete, is still fairly common.)

The Kelvin scale has been defined with reference to the Celsius scale, which has two fixed points, the normal freezing and boiling temperatures of water. However, it may also be defined with reference to a gas thermometer (at very low pressure) by means of a single reference temperature. We define the ratio of any two temperatures T_1 and T_2 on the Kelvin scale as the ratio of the corresponding gas-thermometer pressures p_1 and p_2:

$$\frac{T_2}{T_1} = \frac{p_2}{p_1}. \tag{14–4}$$

14–7 A cell used to establish the triple-point temperature of water. The cylindrical pyrex container is nearly filled with water of the highest possible purity and then permanently sealed. Partial freezing of the sealed water leads to the triple-point solid-liquid-vapor equilibrium condition and establishes the triple-point temperature of 0.0100°C or 273.1600 K. Calibrations with a single triple-point cell are precise to within 0.0001 K, and triple-point cells agree with each other within 0.0002 K. (Courtesy National Bureau of Standards.)

To complete the definition, we need only specify the Kelvin temperature of a single specific state. For reasons of precision and reproducibility, the state chosen is not a freezing or boiling point but the *triple point* of water. This is the unique condition under which solid water, liquid, and vapor can all coexist together. This occurs at a temperature of 0.01°C and a pressure of 610 Pa (about 0.006 atm). The triple-point temperature of water is assigned the value 273.16 K. A cell used to establish the triple-point temperature is shown in Fig. 14–7. Thus, with reference to Eq. (14–4), if p_3 is the pressure in a gas thermometer at the triple point and p is the pressure at some other temperature T, then T is given on the Kelvin scale by

$$T = T_3 \frac{p}{p_3}, \tag{14–5}$$

with T_3 *defined* to be 273.16 K.

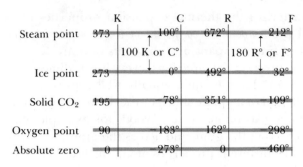

	K	C	R	F
Steam point	373	100°	672°	212°
		100 K or C°		180 R° or F°
Ice point	273	0°	492°	32°
Solid CO_2	195	−78°	351°	−109°
Oxygen point	90	−183°	162°	−298°
Absolute zero	0	−273°	0	−460°

14–8 Relations among Kelvin, Celsius, Rankine, and Fahrenheit temperature scales. Temperatures have been rounded off to the nearest degree.

EXAMPLE 14–1 Suppose a constant-volume gas thermometer has a pressure of 1.50×10^4 Pa at the triple point of water and a pressure of 1.95×10^4 Pa at some unknown temperature T. What is T?

SOLUTION From Eq. (14–5),

$$T = (273 \text{ K})\frac{1.95 \times 10^4 \text{ Pa}}{1.50 \times 10^4 \text{ Pa}} = 355 \text{ K}$$

$$= 82°C.$$

The **Rankine temperature scale,** named after William Rankine (1820–1872), has the same relation to the Fahrenheit scale as Kelvin has to Celsius. The zero point on the Kelvin scale is −273.15°C, which on the Fahrenheit scale is $(\frac{9}{5})(-273.15°) + 32° = -459.7°F$. Thus we define the Rankine scale so that $0°R = -459.7°F$ and $0°F = 459.7°R$. The size of a degree is the same on the Rankine and Fahrenheit scales, and

$$T_R = T_F + 459.7°.$$

The relationships of the four temperature scales discussed above are shown graphically in Fig. 14–8.

The Kelvin and Rankine scales are called **absolute temperature scales,** and their zero point is called **absolute zero.** To define completely what is meant by absolute zero, we need to use thermodynamic principles that are developed in the next several chapters. We will return to the concept of absolute zero in Chapter 19.

What is absolute zero?

14–5 THERMAL EXPANSION

Most solid materials expand when heated. Suppose a rod of material has a length L_0 at some initial temperature T_0, and that, when the temperature increases by an amount ΔT, the length increases by ΔL. Experiment shows that if ΔT is not too large, ΔL is *directly proportional* to ΔT. But ΔL is also proportional to L_0; if two rods of the same material have the same temperature change, but one rod is initially twice as long as the other, then the *change* in its length is also twice as great. Introducing a proportionality constant α (which is different for different materials), we may summarize this relation as follows:

Thermal expansion: When things get hotter, they get bigger.

$$\Delta L = \alpha L_0 \, \Delta T. \qquad (14-6)$$

The horizontal black lines on the roadway of the George Washington suspension bridge in New York City are expansion joints; these are needed to accommodate changes in the lengths of the sections that result from thermal expansion. (Courtesy of New York State Department of Commerce.)

The constant α, which characterizes the thermal expansion properties of a particular material, is called the *temperature coefficient of linear expansion* or, more briefly, the **coefficient of linear expansion.** The units of α are K^{-1} or $(C°)^{-1}$. For materials having no preferential directions, every linear dimension changes according to Eq. (14–6). Thus L could equally well represent the thickness of the rod, the side length of a square sheet, or the diameter of a hole in the material. There are some exceptional cases. Wood, for example, has different expansion properties along the grain and across the grain; and single crystals of material can have different properties along different crystal-axis directions. We exclude these exceptional cases from the present discussion.

We emphasize that the direct proportionality expressed by Eq. (14–6) is not exact, but that it is *approximately* correct for sufficiently small temperature changes. For any temperature, we can define a coefficient of thermal expansion by the equation

$$\alpha = \frac{1}{L} \frac{dL}{dT}. \qquad (14-7)$$

It is found that α for a given material varies somewhat with the initial temperature T_0 and the size of the temperature interval. Because Eq. (14–6) is at best an approximation, we will ignore this variation. Average values of α for several materials are listed in Table 14–1. Within the precision of these values, we do not need to worry about whether T_0 is 0°C or 20°C or some other temperature.

PROBLEM-SOLVING STRATEGY: *Linear expansion*

1. Identify which of the quantities in Eq. (14–6) are known and which are unknown. Often you will be given two temperatures and will have to compute ΔT. Or you may be given an initial temperature T_0 and may have to find a final temperature corresponding to a given length change. In this case, find ΔT first; then the final temperature is $T_0 + \Delta T$.

2. Unit consistency is crucial, as always. L_0 and ΔL must have the same units, and if you use a value of α in $(C°)^{-1}$, then ΔT must be in Celsius degrees $(C°)$.

3. Remember that sizes of holes in a material expand with temperature just the same way as any other linear dimension.

4. Points 1 and 2 are applicable also to volume expansion, which we will discuss following this first example.

Correcting a measuring tape for temperature variations

EXAMPLE 14–2 A carpenter uses a steel measuring tape 5 m long calibrated at a temperature of 20°C. What is its length on a hot summer day when the temperature is 35°C?

SOLUTION From Eq. (14–6),

$$\Delta L = [5 \text{ m}][1.2 \times 10^{-5}(C°)^{-1}][35°C - 20°C]$$
$$= 0.9 \times 10^{-3} \text{ m} = 0.9 \text{ mm}.$$

Thus the length at 35°C is 5.0009 m.

Increasing temperature usually causes increases in *volume*, for both solid and liquid materials. Experiments show that if the temperature change ΔT is not too great, the increase in volume ΔV is approximately *proportional* to the temperature change. The volume change is also proportional to the initial volume V_0, just as with linear expansion. The relationship can be expressed as follows:

$$\Delta V = \beta V_0 \, \Delta T. \qquad (14\text{–}8)$$

The constant β, which characterizes the volume expansion properties of a particular material, is called the *temperature coefficient of volume expansion,* or the **coefficient of volume expansion.** The units of β are K^{-1} or $(C°)^{-1}$. Like the coefficient of linear expansion, β varies somewhat with temperature, and Eq. (14–8) should be regarded as an approximate relation, valid for sufficiently small temperature changes. For many substances, β decreases as the temperature is lowered, approaching zero as the Kelvin temperature approaches zero. Also, metals with high melting temperatures usually have small volume-expansion coefficients.

Some values of β in the neighborhood of room temperature are listed in Table 14–2. Note that the values for liquids are much larger than those for solids.

Fluids usually expand when they get hotter.

TABLE 14–1 Coefficient of Linear Expansion

Material	α, $(C°)^{-1}$
Aluminum	2.4×10^{-5}
Brass	2.0×10^{-5}
Copper	1.7×10^{-5}
Glass	$0.4\text{–}0.9 \times 10^{-5}$
Steel	1.2×10^{-5}
Invar	0.09×10^{-5}
Quartz (fused)	0.04×10^{-5}

TABLE 14–2 Coefficient of Volume Expansion

Solids	β, $(C°)^{-1}$	Liquids	β, $(C°)^{-1}$
Aluminum	7.2×10^{-5}	Ethanol	75×10^{-5}
Brass	6.0×10^{-5}	Carbon disulfide	115×10^{-5}
Copper	5.1×10^{-5}	Glycerin	49×10^{-5}
Glass	$1.2\text{–}2.7 \times 10^{-5}$	Mercury	18×10^{-5}
Steel	3.6×10^{-5}		
Invar	0.27×10^{-5}		
Quartz (fused)	0.12×10^{-5}		

If there is a hole in a solid body, the volume of the hole increases when the body expands, just as if the hole were a solid of the same material as the body. This remains true even if the hole is so large that the surrounding body is only a thin shell. Thus the volume enclosed by a thin-walled bottle or thermometer bulb increases just as would a solid body of glass of the same size.

Holes in solid bodies expand too.

EXAMPLE 14–3 A glass flask of volume 200 cm³ is just filled with mercury at 20°C. How much mercury will overflow when the temperature of the system is raised to 100°C? The coefficient of volume expansion of the glass is $1.2 \times 10^{-5}(C°)^{-1}$.

SOLUTION The increase in the volume of the flask is

$$\Delta V = (1.2 \times 10^{-5}(C°)^{-1})(200 \text{ cm}^3)(100° - 20°) = 0.192 \text{ cm}^3.$$

The increase in the volume of the mercury is

$$\Delta V = (18 \times 10^{-5}(C°)^{-1})(200 \text{ cm}^3)(100° - 20°) = 2.88 \text{ cm}^3.$$

The volume of mercury overflowing is therefore

$$2.88 \text{ cm}^3 - 0.19 \text{ cm}^3 = 2.69 \text{ cm}^3.$$

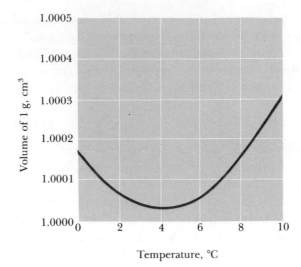

14-9 Volume of 1 gram of water, in the temperature range from 0°C to 10°C. If Eq. (14-8) were obeyed, the curve would be a straight line.

Water sometimes shrinks when it gets warmer. Why is this important?

Water, in the temperature range from 0°C to 4°C, *decreases* in volume with increasing temperature, contrary to the behavior of most substances. That is, between 0°C and 4°C the coefficient of expansion of water is *negative*. Above 4°C, water expands when heated. Since the volume of a given mass of water is smaller at 4°C than at any other temperature, the density of water is maximum at 4°C. Water also expands upon freezing, unlike most materials.

This anomalous behavior of water has an important effect on plant and animal life in lakes. When a lake cools, the cooled water at the surface flows to the bottom because of its greater density. But when the temperature reaches 4°C, this flow ceases and the water near the surface remains colder (and less dense) than that at the bottom. As the surface freezes, the ice floats because it is less dense than water. The water at the bottom remains at 4°C until nearly the entire lake is frozen. If water behaved like most substances, contracting continuously on cooling and freezing, lakes would freeze from the bottom up. Circulation due to density differences would continuously carry warmer water to the surface for efficient cooling, and lakes would freeze solid much more easily. This would destroy all plant and animal life that can withstand cold water but not freezing.

The anomalous expansion of water in the temperature range 0°C to 10°C is shown in Fig. 14-9. Table 14-3 covers a wider range of temperatures.

The volume-expansion coefficient for a solid material is related to the linear-expansion coefficient. To obtain the relation, consider a solid body in the form of a rectangular block with dimensions L_1, L_2, and L_3. Then the volume is

$$V_0 = L_1 L_2 L_3.$$

According to Eq. (14-6), when the temperature increases by ΔT, each linear dimension changes, and the new volume is

$$V_0 + \Delta V = [L_1(1 + \alpha\,\Delta T)][L_2(1 + \alpha\,\Delta T)][L_3(1 + \alpha\,\Delta T)]$$
$$= L_1 L_2 L_3 (1 + \alpha\,\Delta T)^3$$
$$= V_0 (1 + \alpha\,\Delta T)^3$$
$$= V_0[1 + 3\alpha\,\Delta T + 3\alpha^2(\Delta T)^2 + \alpha^3(\Delta T)^3].$$

TABLE 14-3   Volume and Density of Water

T, °C	Volume of 1 g, cm³	Density, g·cm⁻³
0	1.0002	0.9998
4	1.0000	1.0000
10	1.0003	0.9997
20	1.0018	0.9982
50	1.0121	0.9881
75	1.0258	0.9749
100	1.0434	0.9584

If ΔT is small, the terms containing $(\Delta T)^2$ and $(\Delta T)^3$ are very small and may be neglected. Dropping these and subtracting V_0 from both sides, we find

$$\Delta V = (3\alpha)V_0\,\Delta T.$$

Linear- and volume-expansion coefficients are related in a very simple way.

Comparing this with Eq. (14–8), we obtain

$$\beta = 3\alpha. \qquad (14\text{–}9)$$

We invite you to check the validity of this relation for some of the materials listed in Tables 14–1 and 14–2.

14–6 THERMAL STRESSES

If the ends of a rod are rigidly held so as to prevent expansion or contraction and the temperature of the rod is changed, tensile or compressive stresses, called **thermal stresses,** are set up in the rod. These stresses may become large enough to stress the rod beyond its elastic limit or even beyond its breaking strength. Hence in the design of any structure that is subject to changes in temperature, provision must be made for expansion. Long steam pipes have expansion joints or U-shaped sections of pipe. In bridges, one end may be rigidly fastened to its abutment while the other rests on rollers.

Thermal stress: What happens when a material wants to expand and you don't let it?

We can compute the thermal stress set up in a rod that is not free to expand or contract. We simply compute the amount the rod would contract if not held, and then compute the stress needed to stretch it back to its original length. Suppose that a rod of length L_0 and cross-sectional area A has its ends rigidly fastened while the temperature is reduced by an amount ΔT, causing a tensile stress. The fractional change in length if the rod were free to contract would be

$$\frac{\Delta L}{L_0} = \alpha\,\Delta T. \qquad (14\text{–}10)$$

Both ΔL and ΔT are *not* free to contract. Since the rod is *not* free to contract, the tension must increase enough to produce an equal and opposite fractional change in length. But from the definition of Young's modulus, Eq. (12–3),

$$Y = \frac{F/A}{\Delta L/L_0}, \qquad \frac{\Delta L}{L_0} = \frac{F}{AY}. \qquad (14\text{–}11)$$

The tensile force F is determined by the requirement that the *total* fractional change in length, thermal expansion plus elastic strain, must be zero:

$$\alpha\,\Delta T + \frac{F}{AY} = 0,$$

$$F = -AY\alpha\,\Delta T. \qquad (14\text{–}12)$$

Since ΔT represents a decrease in temperature, it is negative, so F is positive. The tensile *stress* F/A in the rod is

$$\frac{F}{A} = -Y\alpha\,\Delta T. \qquad (14\text{–}13)$$

If, instead, ΔT represents an *increase* in temperature, then F and F/A become negative, corresponding to *compressive* force and stress, respectively.

A photograph of the surface of Mars, showing flaking of rock as a result of thermal stresses created by alternating extremes of hot and cold. (Courtesy of NASA.)

Thermal stress: how to break a thick-walled glass container

Thermal stresses can also be induced by nonuniform temperature. Even if a solid body at uniform temperature has no internal stresses, stress may be induced by nonuniform expansion due to temperature differences. Have you ever broken a thick glass container such as a vase by pouring very hot water into it? Thermal stress caused by temperature differences exceeded the breaking stress of the material, and it cracked. Heat-resistant glasses such as Pyrex have exceptionally low expansion coefficients and usually also have high strength to permit thin-wall construction to minimize temperature difference.

Similar phenomena occur with volume expansion. If you fill a bottle completely with water, tightly cap it and then heat it, it will break because the thermal expansion coefficient for water is greater than that for glass. If a material is enclosed in a very rigid container so that its volume cannot change, then a rise in temperature ΔT is accompanied by an increase in pressure Δp. An analysis similar to that leading to Eq. (14–13) shows that the pressure increase is given by

$$\Delta p = B\beta \, \Delta T, \qquad (14\text{–}14)$$

where B is the bulk modulus for the material.

KEY TERMS

temperature

thermometer

thermal equilibrium

insulator

conductor

zeroth law of thermodynamics

thermometric properties

Celsius scale

Fahrenheit scale

Kelvin scale

Rankine scale

absolute temperature scales

absolute zero

coefficient of linear expansion

coefficient of volume expansion

thermal stresses

SUMMARY

A thermometer measures temperature. Any thermometric property can be used to make a thermometer. Two bodies in thermal equilibrium must have the same temperature. A conducting material between two bodies permits interaction leading to thermal equilibrium; an insulating material prevents or impedes this interaction.

The Celsius temperature scale is based on the freezing (0°C) and boiling (100°C) temperatures of water, with one hundred degrees (100 C°) between the two. The Fahrenheit scale is also based on the freezing (32°F) and boiling (212°F) temperatures, but with 180 F° between the two. Celsius and Fahrenheit temperatures are related by

$$T_F = \tfrac{9}{5} T_C + 32°, \qquad (14\text{–}1)$$

$$T_C = \tfrac{5}{9}(T_F - 32°). \qquad (14\text{–}2)$$

The Kelvin scale has its zero at the extrapolated zero-pressure temperature for a gas thermometer, which is $-273.15°C$. Thus $0\,K = -273.15°C$, and

$$T_K = T_C + 273.15. \qquad (14\text{--}3)$$

The Rankine scale is related similarly to the Fahrenheit scale; $0°\,R = -459.7°F$, and

$$T_R = T_F + 459.7°. $$

In the gas-thermometer scale, the ratio of two temperatures is defined to be equal to the ratio of the two corresponding gas-thermometer pressures:

$$\frac{T_2}{T_1} = \frac{p_2}{p_1}. \qquad (14\text{--}4)$$

The definition of this scale is completed by defining the triple-point temperature of water $(0.01°C)$ to be $273.16\,K$.

The change ΔL of any linear dimension L_0 of a solid body with a temperature change ΔT is given approximately by

$$\Delta L = \alpha L_0\,\Delta T, \qquad (14\text{--}6)$$

where α is the coefficient of linear expansion.

The change ΔV in the volume V_0 of any solid or liquid material with a temperature change ΔT is given approximately by

$$\Delta V = \beta V_0\,\Delta T, \qquad (14\text{--}8)$$

where β is the coefficient of volume expansion. For solid materials,

$$\beta = 3\alpha. \qquad (14\text{--}9)$$

When a material is cooled and held so it cannot contract, the tensile stress F/A is given by

$$F/A = -Y\alpha\,\Delta T, \qquad (14\text{--}13)$$

where Y is Young's modulus, α is the coefficient of linear expansion, and ΔT is the temperature change. This expression also gives the compressive stress accompanying a rise in temperature if expansion is prevented.

The pressure Δp resulting from thermal stress when a solid or liquid material is heated by an amount ΔT and prevented from expanding is given by

$$\Delta p = B\beta\,\Delta T, \qquad (14\text{--}14)$$

where B is the bulk modulus and β the volume-expansion coefficient for the material.

QUESTIONS

14–1 Does it make sense to say that one body is twice as hot as another?

14–2 A student claimed that thermometers are useless because a thermometer always registers *its own* temperature. How would you respond?

14–3 What other properties of matter, in addition to those mentioned in the text, might be used as thermometric properties? How could they be used to make a thermometer?

14–4 A thermometer is laid out in direct sunlight. Does it measure the temperature of the air, or of the sun, or what?

14–5 Thermometers sometimes contain red or blue liquid, which is often ethanol. What advantages and disadvantages does this have compared with mercury?

14–6 Could a thermometer similar to that shown in Fig. 14–1a be made by using water as the liquid? What difficulties would this thermometer present?

14–7 What is the temperature of vacuum?

14–8 Is there any particular reason for constructing a temperature scale with higher numbers corresponding to hotter bodies, rather than the reverse?

14–9 If a brass pin is a little too large to fit in a hole in a steel block, should you heat the pin and cool the block, or the reverse?

14–10 When a block with a hole in it is heated, why doesn't the material around the hole expand into the hole and make it smaller?

14–11 Many automobile engines have cast-iron cylinders and aluminum pistons. What kinds of problems could occur if the engine gets too hot?

14–12 When a hot-water faucet is turned on, the flow often decreases gradually before it settles down. This is an annoyance in the shower. Why does it happen?

14–13 Two bodies made of the same material have the same external dimensions and appearance, but one is solid and the other is hollow. When they are heated, is the overall volume expansion the same or different?

14–14 A thermostat for controlling household heating systems often contains a bimetallic element consisting of two strips of different metals, welded together face to face. When the temperature changes, this composite strip bends one way or the other. Why?

14–15 Why is it sometimes possible to loosen caps on screw-top bottles by dipping the cap briefly in hot water?

14–16 The rate at which a pendulum clock runs depends on the length of the pendulum. Would a pendulum clock gain time in hot weather and lose in cold, or the reverse? Could one design a pendulum, perhaps using two different metals, that would *not* change length with temperature?

14–17 When a rod is cooled but prevented from contracting, as in Section 14–6, thermal tension develops. Under these circumstances, does the *thickness* of the rod change? If so, how would the change be calculated?

EXERCISES

Section 14–3 The Celsius and Fahrenheit Scales

14–1

a) If you feel sick in France and are told you have a temperature of 40°C, should you be concerned?

b) What is normal body temperature on the Celsius scale?

c) At what temperature do the Fahrenheit and Celsius scales coincide?

14–2 When the United States finally converts officially to metric units, the Celsius temperature scale will replace the Fahrenheit scale for everyday use. As a familiarization exercise, find the Celsius temperatures corresponding to

a) a cool room (68°F);

b) a hot summer day (95°F);

c) a cold winter day (5°F).

Section 14–4 The Kelvin and Rankine Scales

14–3 The normal boiling point of liquid oxygen is −182.97°C. What is this temperature on the Kelvin and Rankine scales?

14–4 The ratio of the pressure of a gas at the melting point of lead and at the triple point of water, when the gas is kept at constant volume, is found to be 2.19816. What is the Kelvin temperature of the melting point of lead?

14–5 In a rather primitive experiment with a constant-volume gas thermometer, the pressure at the triple point of water was found to be 4.0×10^4 Pa, and the pressure at the normal boiling point 5.4×10^4 Pa. According to these data, what is the temperature of absolute zero on the Celsius scale?

Section 14–5 Thermal Expansion

14–6 A building with a steel framework is 200 m tall when the temperature is 0°C. How much taller is the building on a hot summer day when the temperature is 30°C?

14–7 A steel bridge is built in the summer when the temperature is 25°C. At the time of construction its length is 80.00 m. What is the length of the bridge on a cold winter day when the temperature is −10°C?

14–8 The pendulum shaft of a clock is made of aluminum. What is the fractional change in length of the shaft when it is cooled from 25°C to 10°C?

14–9 A metal rod is observed to be 80.000 cm long at 20°C and 80.024 cm long at 40°C. From these data calculate the average coefficient of linear expansion of the rod for this temperature range.

14–10 To ensure a tight fit, the aluminum rivets used in airplane construction are made slightly larger than the rivet holes and cooled by "dry ice" (solid CO_2) before being driven. If the diameter of a hole is 0.6000 cm, what should be the diameter of a rivet at 20°C if its diameter is to equal that of the hole when the rivet is cooled to −78°C, the temperature of dry ice? Assume the expansion coefficient remains constant at the value given in Table 14–1.

14–11 A machinist bores a hole 2.500 cm in diameter in a brass plate at a temperature of 20°C. What is the diameter of the hole when the temperature of the plate is increased to 200°C? Assume the expansion coefficient remains constant.

14–12 An area is measured on the surface of a solid body. If the area is A_0 at some initial temperature and then changes by ΔA when the temperature changes by ΔT, show

that
$$\Delta A = (2\alpha)A_0 \, \Delta T.$$

14–13 A brass cylinder is initially at 20°C. At what temperature will its volume be 0.2% larger?

14–14 Use the data of Table 14–3 to find the average coefficient of volume expansion of water in the temperature range

a) 0°C to 4°C;

b) 75°C to 100°C.

14–15 An underground tank with a capacity of 500 L (0.5 m³) is filled with ethanol that has an initial temperature of 25°C. After the ethanol has cooled off to the temperature of the tank and ground, which is 10°C, how much air space will there be above the ethanol in the tank?

Section 14–6 Thermal Stresses

14–16 The cross section of a steel rod is 10 cm². What force is needed to prevent it from contracting while cooling from 520°C to 20°C?

14–17

a) A wire that is 3 m long at 20°C is found to increase in length by 1.5 cm when heated to 520°C. Compute its average coefficient of linear expansion.

b) Find the stress in the wire if it is stretched taut at 520°C and cooled to 20°C without being allowed to contract. Young's modulus for the wire is 2×10^{11} Pa.

14–18 Steel railroad rails 18 m long are laid on a winter day when the temperature is 3°C.

a) How much space must be left between adjacent rails if they are to just touch on a summer day when the temperature is 39°C?

b) If the rails were originally laid in contact, what would be the stress in them on a summer day when the temperature is 39°C?

14–19 What hydrostatic pressure is necessary to prevent a copper block from expanding when its temperature is increased from 20°C to 39°C?

14–20

a) A block of steel at an initial pressure of 1 atm and a temperature of 20°C is kept at constant volume. If the temperature is raised to 32°C, what is the final pressure?

b) If the block is maintained at constant volume by rigid walls that can withstand a maximum pressure of 1200 atm (1.22×10^8 Pa), what is the highest temperature to which the system may be raised? Assume B and β remain practically constant at the values 1.6×10^{11} Pa and 3.6×10^{-5}(C°)$^{-1}$, respectively.

PROBLEMS

14–21 An aluminum cube 0.10 m on a side is heated from 10°C to 30°C.

a) What is the change in its volume?

b) In its density?

14–22 A surveyor's 30-m steel tape is correct at a temperature of 20°C. The distance between two points, as measured by this tape on a day when the temperature is 35°C, is 25.970 m. What is the true distance between the points?

14–23 A steel ring of 3.000 in. inside diameter at 20°C is to be heated and slipped over a brass shaft measuring 3.002 in. in diameter at 20°C.

a) To what temperature should the ring be heated?

b) If the ring and shaft together are cooled by some means such as liquid air, at what temperature will the ring just slip off the shaft?

14–24 Suppose that a steel hoop could be constructed around the earth's equator, just fitting it at a temperature of 20°C. What would be the thickness of the space between the hoop and the earth if the temperature of the hoop were increased by 1 C°?

14–25 A metal rod 30.0 cm long expands by 0.075 cm when its temperature is raised from 0°C to 100°C. A rod of a different metal and of the same length expands by 0.045 cm for the same rise in temperature. A third rod, also 30.0 cm long, is made up of pieces of each of the above metals placed end to end, and expands 0.065 cm between 0°C and 100°C. Find the length of each portion of the composite bar.

14–26 At a temperature of 20°C, the volume of a certain glass flask, up to a reference mark on the stem of the flask, is exactly 100 cm³. The flask is filled to this point with a liquid whose coefficient of volume expansion is 40×10^{-5}(C°)$^{-1}$, with both flask and liquid at 20°C. The coefficient of volume expansion of the glass is 2×10^{-5}(C°)$^{-1}$. The cross section of the stem is 6 mm² and can be considered constant. How far will the liquid rise or fall in the stem when the temperature is raised to 40°C?

14–27 A glass flask whose volume is exactly 1000 cm³ at 0°C is completely filled with mercury at this temperature. When flask and mercury are heated to 100°C, 15.2 cm³ of mercury overflow. If the coefficient of volume expansion of mercury is 18×10^{-5} per Celsius degree, compute the coefficient of volume expansion of the glass.

14–28 A steel rod of length 0.40 m and a copper rod of length 0.36 m, both of the same diameter, are placed end to end between rigid supports, with no initial stress in the rods. The temperature of the rods is now raised by 50 C°. What is the stress in each rod?

14–29 A heavy brass bar has projections at its ends, as in Fig. 14–10. Two fine steel wires fastened between the projections are just taut (zero tension) when the whole system is at 0°C. What is the tensile stress in the steel wires

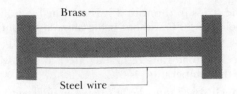
Brass

Steel wire

FIGURE 14–10

when the temperature of the system is raised to 300°C? Make any simplifying assumptions you think are justified, but state what they are.

14–30 Prove that if a body under hydrostatic pressure is raised in temperature but not allowed to expand, the increase in pressure is

$$\Delta p = B\beta \, \Delta T,$$

where the bulk modulus B and the average coefficient of volume expansion β are both assumed positive and constant.

14–31 Table 14–3 lists the density of water, and the volume of 1 g, at atmospheric pressure. A steel bomb is

filled with water at 10°C and atmospheric pressure, and the system is heated to 75°C. What is then the pressure in the bomb? Assume the bomb to be sufficiently rigid so that its volume is not affected by the increased pressure.

14–32 A liquid is enclosed in a metal cylinder provided with a piston of the same metal. The system is originally at atmospheric pressure and at a temperature of 80°C. The piston is forced down until the pressure on the liquid is increased by 100 atm (1.013×10^7 Pa), and is then clamped in this position. Find the new temperature at which the pressure of the liquid is again 1 atm (1.013×10^5 Pa). Assume that the cylinder is sufficiently strong so that its volume is not altered by changes in pressure but only by changes in temperature.

Compressibility of liquid:
$k = 50 \times 10^{-6}$ atm^{-1}.

Coefficient of volume expansion of liquid:
$\beta = 53 \times 10^{-5}$(C°)$^{-1}$.

Coefficient of volume expansion of metal:
$\beta = 3.0 \times 10^{-5}$(C°)$^{-1}$.

CHALLENGE PROBLEMS

14–33 A clock whose pendulum makes one vibration in 2 s is correct at 25°C. The pendulum shaft is of steel and its mass may be neglected compared with that of the bob.

a) What is the fractional change in length of the shaft when it is cooled to 15°C?

b) How many seconds per day will the clock gain or lose at 15°C? (The calculation technique outlined for Challenge Problem 11–45 may be used.)

c) How closely must the temperature be controlled if the clock is not to gain or lose more than 1 s a day? Does the answer depend on the period of the pendulum?

14–34

a) The pressure p, volume V, number of moles n, and Kelvin temperature T of an ideal gas are related by the equation $pV = nRT$. Prove that the coefficient of volume expansion is equal to the reciprocal of the Kelvin temperature, if the expansion occurs at constant pressure.

b) Compare the coefficients of volume expansion of copper and air at a temperature of 20°C. Assume that air may be treated as an ideal gas and that the pressure remains constant.

14–35

a) For any material, density ρ, mass m, and volume V are related by $\rho = m/V$. Prove that

$$\beta = -\frac{1}{\rho}\frac{d\rho}{dT}.$$

b) The density of rock salt in units of g·cm^{-3} between −193°C and −13°C is given by the empirical formula

$$\rho = 2.1680(1 - 11.2 \times 10^{-5}T - 0.5 \times 10^{-7}T^2),$$

with T measured on the Celsius scale. Calculate β at −100°C.

14–36 In going from Eq. (14–7) to (14–6), we make two assumptions: that ΔL is small compared to L_0 and that $\alpha(T)$ can be taken to be constant over the temperature range in question. If ΔT is sufficiently large, or if a high degree of accuracy is required, both these approximations can break down.

a) Eq. (14–7) gives $dL/L = \alpha(T) \, dT$. Assuming $\alpha(T)$ to be constant, integrate this expression from an initial temperature T_0, where the length is L_0, to a final temperature T, where the length is L, and thereby obtain an equation from which L can be calculated.

b) Consider a copper rod 80.000 cm long at 0°C. Calculate its length at 800°C both from the equation you derived in (a) and also from Eq. (14–6). In each case assume that α remains constant at the value given in Table 14–1. What percentage error does Eq. (14–6) make in the calculation of ΔL?

c) If $\alpha(T)$ is given by $\alpha(T) = A + BT + CT^2$, where A, B, and C are constants, derive an expression analogous to Eq. (14–6) for ΔL, assuming ΔL is small.

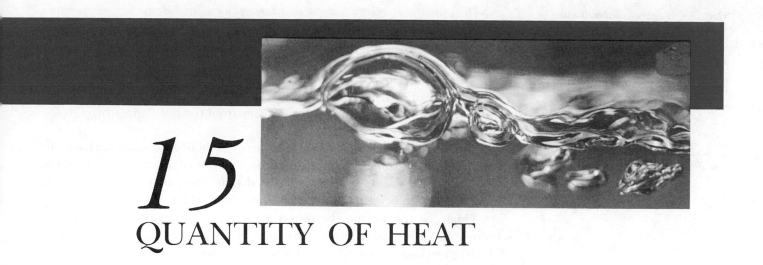

15 QUANTITY OF HEAT

IN THE PRECEDING CHAPTER WE DISCUSSED THE CONCEPT OF TEMPERATURE in connection with thermal equilibrium; two bodies in thermal equilibrium must have the same temperature. When two bodies that are *not* initially in thermal equilibrium are placed in contact, their temperatures change until they reach thermal equilibrium. The study of the interaction that takes place during the approach to thermal equilibrium leads us to the concept of *heat*, the subject of this chapter. We define what we mean by quantity of heat, and we introduce units to measure quantity of heat. Then we apply this concept to a study of the quantities of heat involved in temperature changes and changes of phase of materials.

15–1 HEAT TRANSFER

Suppose we have two systems, *A* and *B*, with *A* initially at higher temperature than *B*. We place them in contact; when they have reached thermal equilibrium, *A*'s temperature has decreased and *B*'s has increased. It is natural to speculate that during this process *A* loses something and that this "something" flows into *B*. Indeed, the interaction that takes place while the temperatures are changing is called a **heat transfer** or a *heat flow* from *A* to *B*.

The process of heat transfer was formerly thought to be a flow of an invisible, weightless fluid called *caloric*. But during the eighteenth and nineteenth centuries the relationship of heat to mechanical energy and work gradually emerged. Count Rumford (1753–1814) studied the heat evolved during the drilling of cannon barrels, especially with dull drills. Sir James Joule (1818–1889) discovered that water can be heated by vigorous stirring with a paddle wheel and that there is a correlation between the temperature rise and the work needed to turn the wheel. These and many other similar experimental studies established that heat flow is really *energy transfer* and that there is an equivalence between heat and work. The relation of heat to work is at the core of the branch of physics called *thermodynamics*, the subject of the next several chapters. The concept of caloric did not include this equivalence, and it has been discarded.

The decline and fall of the caloric theory

Heat flow is energy transfer caused by temperature differences.

In many systems, energy can be transferred by both heat and work.

How you describe energy transfer depends on what system you're talking about: the importance of defining your system carefully.

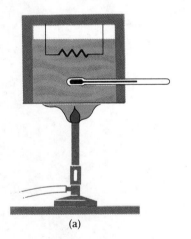

(a)

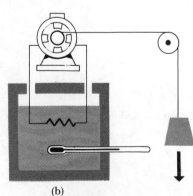

(b)

15–1 The same temperature change of the same system may be accomplished by either (a) a heat flow or (b) the performance of work.

Heat transfer is energy transfer that takes place solely because of a temperature difference. For example, water is converted to steam in a steam boiler by contact with a metal pipe or container that is kept at a high temperature by a hot flame from burning coal or gas. Water in the form of steam has greater ability to do *work* (by pushing against a turbine blade or the piston in a steam engine) than in the form of cold liquid water. Therefore the water must have received *energy* in a process involving a heat transfer from the hot flame (through the container) to the cooler water.

Of course, energy transfer can also occur without heat transfer. In an air compressor, a moving piston pushes against a mass of air, doing work on the air that it compresses. In this compressed state, the gas is capable of doing more work than before and hence has acquired energy.

Finally, if the air and the piston in the air compressor are at different temperatures, heat flow between the piston and the air can occur. This process involves two kinds of energy transfer simultaneously: transfer of *heat* and performance of *work*. Such processes are of great importance in many practical devices, such as internal-combustion engines. We will study them in detail in Chapters 18 and 19.

In many cases where a particular change of state of a quantity of material involves energy exchange, the change can be brought about in various ways. In Fig. 15–1a, the temperature of a quantity of water is raised by a gas flame. In Fig. 15–1b, the same change of state is accomplished by a falling weight that turns a generator and powers an electric heating coil in the water. We can regard the change as caused either by heat transfer or by the performance of mechanical work, depending on whether or not we consider the generator and weight as part of our system. Thus in describing various energy-transfer processes, we must always be careful to specify what is and is not included in the system under discussion.

This example shows again that there must be an equivalence between heat and mechanical work, since the same change of state of a system can be produced by *either* heat flow *or* work. A detailed study of this equivalence leads to the *first law of thermodynamics,* which we will study in Chapter 18.

15–2 QUANTITY OF HEAT

We will use the term *heat* only in reference to *transfer* or *flow*. A heat transfer is an energy transfer brought about solely by a temperature difference. The amount of energy transferred may be described as a quantity of heat, but we must be careful *not* to take the view that a certain body *contains* a certain quantity of heat. Such a statement has no meaning. This is a somewhat subtle point, and a thorough discussion of it leads to the concept of *internal energy,* to be introduced in Chapter 18.

We can define a unit of quantity of heat with reference to temperature change of any particular material. The *calorie* (abbreviated cal) was originally defined in the eighteenth century as the amount of heat required to raise the temperature of one gram of water by one Celsius degree (one kelvin). It was found later that this is ambiguous because more heat is required to raise the temperature from, say 90° to 91° than from 20° to 21°. In one of several refinements of the definition, the calorie was defined to be the amount of heat required to heat one gram of water from 14.5°C to 15.5°C. This leads to the

15–2 Sugar cubes obtained in a restaurant in Germany. A rough translation is "This package has 22 Calories (i.e., kilocalories), equal to 92 kilojoules" and "Get into the swing with sugar." By law, foods marketed in Germany must show the energy content in joules; the equivalent in calories is optional.

"15-degree calorie." Several alternative and slightly different definitions are unfortunately also in current use, so the ambiguity persists even today.

A corresponding unit defined in terms of Fahrenheit degrees and British units is the *British thermal unit*, or Btu. By definition, 1 Btu is the quantity of heat required to raise the temperature of one pound of water from 63°F to 64°F. A third unit in common use, especially in measuring food energy, is the *kilocalorie* (kcal), equal to 1000 cal. The relations among these three units are

$$1 \text{ Btu} = 252 \text{ cal} = 0.252 \text{ kcal.}$$

The food-value calorie is really 1000 calories.

The unit of quantity of heat is fundamentally a unit of energy; thus there must be a relation between the above units and the familiar mechanical-energy units such as the joule. It has been found experimentally that

$$1 \text{ cal} = 4.186 \text{ joules} = 4.186 \text{ J,}$$
$$1 \text{ kcal} = 4186 \text{ J,}$$
$$1 \text{ Btu} = 778 \text{ ft·lb} = 252 \text{ cal} = 1055 \text{ J.}$$

Figure 15–2 shows an example of this equivalence.

The International Committee on Weights and Measures no longer recognizes the calorie as a fundamental unit but recommends instead that the joule be used for quantity of heat as well as for all other forms of energy. Although the calorie is a convenient unit for heat-transfer problems involving water, it is awkward when both heat and other forms of energy are involved. Eventually the joule will probably become the universal unit of energy; we use it in most of the examples and problems of this and the following chapters.

Using the joule for all forms of energy, including heat

15–3 HEAT CAPACITY

We use the symbol Q for quantity of heat. When the quantity is associated with a temperature change ΔT, we usually call it ΔQ, and for an infinitesimal temperature change dT, we use dQ. The quantity of heat ΔQ required to increase the temperature of a mass m of a certain material by an amount ΔT is found to be approximately proportional to ΔT. It is also proportional to m; twice as much heat is needed to heat two cups of water to make tea than to heat only one cup to the same temperature. The quantity of heat needed also depends

How to calculate the heat needed for a certain temperature change

on the nature of the material; to raise the temperature of one gram of water 1 C° requires over five times as much heat as the same temperature increase for the same mass of aluminum.

Specific heat capacity: Different materials have different relationships of heat to temperature change.

Thus the relationship among all these quantities can be expressed as

$$\Delta Q = mc \, \Delta T, \qquad (15-1)$$

where c is a constant, different for different materials, called the **specific heat capacity** for the material. Strictly speaking, it is not precisely constant but depends somewhat on temperature. A more precise definition, obtained by rearranging Eq. (15-1) and replacing ΔQ by dQ and ΔT by dT, is

$$c = \frac{1}{m} \frac{dQ}{dT}. \qquad (15-2)$$

With this definition, c for a given material depends somewhat on the temperature at which the derivative dQ/dT is evaluated; in solving problems we will usually ignore this variation. The specific heat capacity of water is approximately

$$4.19 \, \text{J} \cdot \text{g}^{-1} \cdot (\text{C}°)^{-1}, \qquad 4190 \, \text{J} \cdot \text{kg}^{-1} \cdot (\text{C}°)^{-1},$$
$$1 \, \text{cal} \cdot \text{g}^{-1} \cdot (\text{C}°)^{-1}, \qquad \text{or} \qquad 1 \, \text{Btu} \cdot \text{lb}^{-1} \cdot (\text{F}°)^{-1}.$$

When you have a fever, how much extra heat does your body produce to warm itself up?

EXAMPLE 15–1 During a bout with the flu, an 80-kg man ran a fever of 2 C° above normal, that is, a body temperature of 39°C or 102.2°F. To raise his temperature by that amount, his body had to produce extra heat by means of chemical reactions in his cells. Assuming the human body is mostly water, how much heat was needed? Express the result in joules and calories.

SOLUTION From Eq. (15–1),

$$\Delta Q = (80 \, \text{kg})(4190 \, \text{J} \cdot \text{kg}^{-1} \cdot (\text{C}°)^{-1})(2 \, \text{C}°)$$
$$= 6.70 \times 10^5 \, \text{J},$$

or

$$\Delta Q = (80{,}000 \, \text{g})(1 \, \text{cal} \cdot \text{g}^{-1} \cdot (\text{C}°)^{-1})(2 \, \text{C}°)$$
$$= 1.60 \times 10^5 \, \text{cal}$$
$$= 160 \, \text{kcal} \, (160 \, \text{food-value calories}).$$

Describing a quantity of material by the number of moles instead of the number of kilograms

It is often convenient to describe a quantity of substance by use of the number of *moles* rather than the *mass* of material. One mole (1 mol) of any substance is a quantity of matter such that its mass in grams is numerically equal to the *molecular mass M*. (This quantity is often called *molecular weight,* but *molecular mass* is preferable because the quantity depends on the mass of a molecule, not its weight.) To calculate the number of moles n, we divide the mass m in grams by the molecular mass M; thus $n = m/M$. Replacing the mass m in Eq. (15–1) by the product nM, we find

$$Mc = \frac{\Delta Q}{n \, \Delta T}.$$

The product Mc is called the **molar heat capacity** and is represented by the symbol C. Hence, by definition

$$C = Mc = \frac{\Delta Q}{n \, \Delta T}, \qquad (15-3)$$

$$\Delta Q = nC \, \Delta T. \qquad (15-4)$$

The molecular mass of water is $M = 18 \text{ g·mol}^{-1}$; its molar heat capacity is approximately

$$C = 75.3 \text{ J·mol}^{-1} \cdot (\text{C}°)^{-1} \quad \text{or} \quad 18 \text{ cal·mol}^{-1} \cdot (\text{C}°)^{-1}.$$

The quantity defined by Eq. (15–2) is sometimes called simply *specific heat*, and the molar heat capacity defined by Eq. (15–3) is often called the *molar specific heat*. However, we will use the terms *specific heat capacity* and *molar heat capacity* in this book. Representative values of specific and molar heat capacity are given in Table 15–1.

From Eq. (15–1), the total quantity of heat Q that must be supplied to a body of mass m to change its temperature from T_1 to T_2 is

$$Q = mc \, (T_2 - T_1), \qquad (15-5)$$

assuming c may be considered constant through this temperature interval. If T_2 is less than T_1, Q is negative, indicating transfer of heat *out of* the body rather than *into* it.

When the specific heat capacity of a material varies appreciably with temperature, the value used in Eq. (15–5) should be the *mean* specific heat capacity over the interval $(T_2 - T_1)$. For small temperature changes, this variation may often be ignored.

The specific or molar heat capacity of a substance is not the only physical property involving quantity of heat. Heat conductivity, heat of fusion, heat of vaporization, and heat of combustion are a few other examples of *thermal properties* of matter. We will discuss some of these in Section 15–5. The field of physics and physical chemistry concerned with the measurement of thermal properties is called **calorimetry.**

TABLE 15–1 **Mean Specific and Molar Heat Capacities (Constant Pressure, Temperature Range 0°C to 100°C)**

| Metal | Specific (c) | | M, | Molar (C) |
	$J \cdot kg^{-1} \cdot (C°)^{-1}$	$cal \cdot g^{-1} \cdot (C°)^{-1}$	$g \cdot mol^{-1}$	$J \cdot mol^{-1} \cdot (C°)^{-1}$
Aluminum	910	0.217	27.0	24.6
Beryllium	1970	0.471	9.01	17.7
Copper	390	0.093	63.5	24.8
Ethanol	2428	0.58	46.0	112.0
Ethylene glycol	2386	0.57	62.0	148.0
Ice ($-25°C$ to $0°C$)	2000	0.48	18.0	36.5
Iron	470	0.112	55.9	26.3
Lead	130	0.031	207.0	26.9
Marble ($CaCO_3$)	879	0.21	100.0	87.9
Mercury	138	0.033	201.0	27.7
Salt	879	0.21	58.5	51.4
Silver	234	0.056	108.0	25.3
Water	4190	1.00	18.0	75.4

Do you know the difference between temperature and heat?

Finally, we remark that it is absolutely essential to distinguish carefully between the two physical quantities *temperature* and *heat*. Temperature depends on the physical state of a material and is a quantitative description of its hotness or coldness; heat is energy in transit from one body to another. We can change the temperature of a body by adding heat to it or taking heat away, but the temperature can also be changed in other ways that do not involve heat transfer.

15–4 MEASUREMENT OF HEAT CAPACITY

An experimental setup to measure heat capacities.

To measure a heat capacity we need to add a measured quantity of heat to a measured quantity of a material and observe the resulting temperature change. For maximum precision, the measurements are often made electrically. In a typical laboratory procedure, we supply heat input by passing a current through a heater wire wound around the specimen. By measuring the voltage, current, and time interval, we can determine the total energy (heat) input Q. We measure the temperature change ΔT by using a resistance thermometer or thermocouple embedded in the specimen, and we measure the mass m by weighing. We can then use Eq. (15–1) to determine the specific heat capacity c. This sounds simple, but great experimental skill is needed to avoid or compensate for unwanted heat transfer between the sample and its surroundings.

Figure 15–3 shows the results of specific heat capacity measurements for water. The quantity of heat needed to raise the temperature of 1 g of water from 14.5°C to 15.5°C is 4.186 J, and this is defined to be one 15° calorie. Two other calories are (unfortunately) frequently used. The *international table calorie* (IT cal) is defined to be precisely

$$1 \text{ IT cal} = \frac{1 \text{ W·hr}}{860} = \frac{3600 \text{ J}}{860} = 4.186 \text{ J}.$$

The terrible ambiguity of the calorie, and how to avoid it

It is almost identical with the 15° calorie. The *thermochemical calorie,* however, is defined as 4.1840 J. From Fig. 15-3 we see that this corresponds to about a 17° calorie. This variety of definitions of the calorie is an additional argument in favor of eliminating the calorie completely and using the joule as the fundamental unit of quantity of heat as well as of all other forms of energy. There are also several different definitions of the Btu, differing by as much as 0.5%.

15–3 Specific heat capacity of water as a function of temperature.

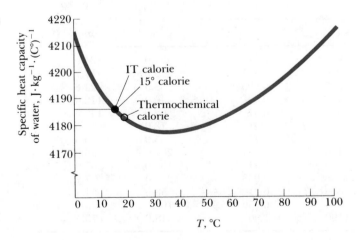

The heat capacity of a material also depends on the conditions imposed during the heat transfer. The two conditions of greatest practical usefulness are for the system to be kept at *constant pressure* or *constant volume;* the corresponding specific heat capacities are denoted as c_p and c_v, respectively, and the molar heat capacities as C_p and C_v, respectively. Laboratory measurements are most often made under constant-pressure conditions. The two heat capacities differ because if the system can expand during the heating, additional energy exchange occurs through the performance of *work* by the system on its surroundings. If the volume is held constant, the system does no work. For gases the difference between c_p and c_v is substantial. We will study heat capacities of gases in detail in Section 18–7.

When measuring heat capacities, do you keep the volume or the pressure constant?

A few values of heat capacities of metals and some familiar compounds are listed in Table 15–1 (Section 15–3). All the specific heat capacities are less than that of water, and they generally decrease with increasing molecular mass. The last column is of particular interest; it shows that the molar heat capacities for all metals except those with very small molecular mass are about the same, about $25 \, \text{J} \cdot \text{mol}^{-1} \cdot (\text{C}°)^{-1}$. This correlation is called the **law of Dulong and Petit,** after its discoverers. Although only an approximate rule, it contains the germ of a very important idea.

Molar heat capacities of some materials obey a surprisingly simple rule.

The number of molecules in one mole is the same for all substances. Thus about the same amount of heat is required *per molecule* to raise the temperature of each of these metals by a given amount, even though the *mass* of a molecule of lead (for example) is nearly ten times as great as that of a molecule of aluminum. The heat required for a given temperature increase depends only on *how many* molecules the sample contains, and not on the mass of an individual molecule. Peter Debye (1884–1966) showed that the behavior of the molar heat capacity of *nonmetals* over the entire temperature range can be accounted for by the vibrational motion of the molecules in the crystal lattice. We will study the molecular basis of heat capacities in greater detail in Chapter 20.

A molecular viewpoint on heat capacities

15–5 PHASE CHANGES

The term **phase,** as we use it here, refers to a specific state of matter, such as a solid, liquid, or gaseous state. For example, the chemical compound H_2O exists in the *solid phase* as ice, in the *liquid phase* as water, and in the *gaseous phase* as steam. All substances that do not decompose at high temperatures can exist in any of these phases under proper conditions of temperature and pressure. A transition from one phase to another is called a **phase change** or phase transition. For any given pressure, a phase change takes place at a definite temperature. Usually a phase change is accompanied by absorption or liberation of heat and a change of volume and density.

Phase changes: melting, boiling, and all that

A familiar example of a phase change is the melting of ice. When heat is added to ice at 0°C and normal atmospheric pressure, the temperature of the ice *does not* increase; instead some of it melts to form liquid water. If the heat is added slowly, so that thermal equilibrium is maintained between the ice and liquid water, then the temperature remains at 0°C as the heat is added, until all the ice is melted. Thus the effect of adding heat to this system is not to raise its temperature but to change its *phase* from solid to liquid.

Phase changes require heat exchange.

Experiments show that to change 1 kg of ice at 0°C to 1 kg of liquid water at 0°C requires the addition of 3.34×10^5 J of heat. This quantity of heat is called the **heat of fusion** of water. The term *latent* heat of fusion is sometimes used; this is somewhat redundant and will not be used here, but we will use the symbol L, with a subscript, for quantities of heat associated with phase transitions. Thus the heat of fusion L_F of water is

$$L_F = 3.34 \times 10^5 \text{ J·kg}^{-1}.$$

In other units the value is

$$L_F = 79.7 \text{ cal·g}^{-1}$$
$$= 143 \text{ Btu·lb}^{-1}.$$

More generally, to melt a mass m of material requires the addition of a quantity of heat Q given by

$$Q = mL_F. \tag{15–6}$$

The heat of fusion is, of course, different for different materials, and it also varies somewhat with pressure.

Phase equilibrium: when two phases can exist together

The phase change described above is *reversible*. The conversion of liquid water to ice at 0°C requires the *removal* of heat; the amount of heat is again given by Eq. (15–6), but in this case it is considered negative because it is removed rather than added. We also note that at a given pressure it is possible for liquid water and ice to coexist only at one very specific temperature, which we call the *melting temperature*. This coexistence of two phases is called **phase equilibrium.**

This entire discussion can be repeated for boiling, a phase transition between liquid and gaseous phases. The corresponding heat of the phase transition is called the **heat of vaporization L_V**; both L_V and the boiling temperature of a material depend on pressure. Water boils at a lower temperature in Denver than in Pittsburgh because the average atmospheric pressure in Denver is less due to its higher elevation. The heat of vaporization is somewhat greater at this lower temperature. At normal atmospheric pressure, the heat of vaporization of water is

$$L_V = 2.26 \times 10^6 \text{ J·kg}^{-1}$$
$$= 539 \text{ cal·g}^{-1}$$
$$= 970 \text{ Btu·lb}^{-1}.$$

It is interesting to note that over five times as much heat is required to boil a quantity of water at 100°C as to raise its temperature from 0° to 100°C.

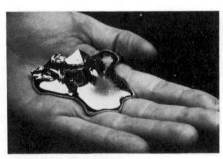

15–4 The metal gallium is one of the few *elements* that melt in the vicinity of room temperature; its melting temperature is 29.8°C, its heat of fusion is 8.04×10^4 J·kg^{-1}, or 19.2 cal·g^{-1}. A crystal of gallium is shown melting in a person's hand. (Photo by Chip Clark.)

Sublimation: a direct solid-to-vapor transition

Table 15–2 lists heats of fusion and vaporization for several materials. Also given are "normal" melting and boiling temperatures, that is, the melting and boiling temperatures at normal atmospheric pressure. Very few *elements* have melting temperatures in the vicinity of ordinary room temperatures; one of the few is the metal gallium, as shown in Fig. 15–4.

Under some conditions of temperature and pressure, a substance can change directly from the solid to the gaseous phase without passing through the liquid phase. The transfer from solid to vapor is called **sublimation,** and the solid is said to *sublime.* "Dry ice" (solid carbon dioxide) sublimes at atmospheric pressure. Liquid carbon dioxide cannot exist at a pressure lower than about 5×10^5 Pa (about 5 atm). Heat is absorbed in the process of sublimation

TABLE 15–2 Heats of Fusion and Vaporization

Substance	Normal Melting Point		Heat of Fusion, L_F, $J \cdot kg^{-1}$	Normal Boiling Point		Heat of Vaporization, L_V, $J \cdot kg^{-1}$
	K	°C		K	°C	
Helium	3.5	−269.65	5.23×10^3	4.216	−268.93	20.9×10^3
Hydrogen	13.84	−259.31	58.6×10^3	20.26	−252.89	452×10^3
Nitrogen	63.18	−209.97	25.5×10^3	77.34	−195.81	201×10^3
Oxygen	54.36	−218.79	13.8×10^3	90.18	−182.97	213×10^3
Ethyl alcohol	159	−114	104.2×10^3	351	78	854×10^3
Mercury	234	−39	11.8×10^3	630	357	272×10^3
Water	273.15	0.00	334×10^3	373.15	100.00	2256×10^3
Sulfur	392	119	38.1×10^3	717.75	444.60	326×10^3
Lead	600.5	327.3	24.5×10^3	2023	1750	871×10^3
Antimony	903.65	630.50	165×10^3	1713	1440	561×10^3
Silver	1233.95	960.80	88.3×10^3	2466	2193	2336×10^3
Gold	1336.15	1063.00	64.5×10^3	2933	2660	1578×10^3
Copper	1356	1083	134×10^3	1460	1187	5069×10^3

and liberated in the reverse process. The quantity of heat absorbed per unit mass is called the **heat of sublimation.** In freeze-drying of food, the food is first frozen, and then water is removed by sublimation. The heat of sublimation is often supplied by infrared or microwave radiation.

Under some conditions a material can be cooled below the normal phase-change temperature without a phase change occurring. The resulting state is unstable and is described as *supercooled*. Very pure water can be cooled several degrees below the normal freezing point under ideal conditions; when a small ice crystal is dropped in or the water is agitated, it crystallizes very quickly. Similarly, supercooled water vapor condenses quickly into fog droplets when a disturbance, such as dust particles or ionizing radiation, is introduced. This phenomenon is used in the *cloud chamber,* where charged particles such as protons and electrons induce condensation of supercooled vapor along their path, thus making the path visible. The same principle is involved in "seeding" clouds, which often contain supercooled water vapor, to cause condensation and rain.

Similarly, the temperature of a liquid can sometimes be raised above its normal boiling temperature, and the liquid is then said to be *superheated*. Again, any small disturbance—such as agitation, a dust particle, or the passage of a charged particle through the material—causes local boiling with bubble formation. This phenomenon is used in the bubble chamber, to be discussed in Chapter 44, to show the tracks of high-energy particles.

As an illustration of phase changes, suppose we take crushed ice from a freezer at −25°C, place it in a container with a thermometer, and surround it with a heating coil that supplies heat at a constant rate. We insulate the system from its surroundings so no other heat enters it, and we observe the rise in temperature with time. The result is shown in Fig. 15–5, a graph of temperature as a function of time. We find that the temperature of the ice increases steadily, as shown by the portion of the graph from *a* to *b*, until it has risen to 0°C. In this temperature range the specific heat capacity of ice is approximately

$$2.0 \times 10^3 \text{ J} \cdot kg^{-1} \cdot (C°)^{-1} \quad \text{or} \quad 0.48 \text{ cal} \cdot g^{-1} \cdot (C°)^{-1}.$$

Superheating and supercooling: postponing the inevitable

An experiment showing energy relations in phase transitions

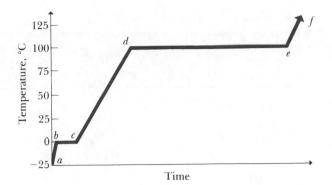

15–5 The temperature remains constant during each change of phase, provided the pressure remains constant.

As soon as this temperature is reached, the ice begins to *melt*, a *change of phase*, from the solid phase to the liquid phase. The thermometer, however, shows *no increase in temperature*, and although heat is being supplied at the same rate as before, the temperature remains at 0°C until all the ice is melted (point *c*).

As soon as the last of the ice has melted, the temperature begins to rise again at a uniform rate (from *c* to *d*), although this rate is *slower* than for ice. When a temperature of 100°C is reached (point *d*), bubbles of steam (gaseous water or water vapor) start to escape from the liquid surface; that is, the water begins to *boil*. The temperature remains constant at 100°C until all the water has boiled away. Another change of phase has taken place, from the liquid phase to the gaseous phase.

If all the water vapor had been trapped and not allowed to diffuse away (a very large container would be needed), the heating process could be continued as from *e* to *f*. The gas would now be called "superheated steam."

An essential point in this discussion is that when heat is added slowly (to maintain thermal equilibrium) to a substance that can exist in different phases, *either* the temperature rises *or* some of the substance undergoes a phase change, but *never* both at the same time. Once the temperature for a phase change (e.g., the melting or boiling temperature) has been reached, no further temperature change occurs until *all* the substance has undergone the phase change.

Although we have used water as an example, the same type of curve as in Fig. 15–5 is obtained for many other substances. Some, of course, decompose before reaching a melting or boiling point, and others, such as glass or tar, do not change phase at a definite temperature but become gradually softer as their temperature is raised. Crystalline substances, such as ice, or a metal melt at a definite temperature. Glass and tar are amorphous solids with no definite crystal structure, and they behave like liquids of very high viscosity.

When heat is removed from a gas, its temperature falls; at the same temperature at which it boiled, it returns to the liquid phase, or *condenses*. In so doing it gives up to its surroundings the same quantity of heat that was required to vaporize it. The heat so given up, per unit mass, is equal to the heat of vaporization. Similarly, a liquid returns to the solid phase, or freezes, when cooled to the temperature at which it melted, and it gives up heat equal to the heat of fusion.

Steam-heating systems use a boiling-condensing process to transfer heat from the furnace to the radiators. Each kilogram of water that is turned to steam in the furnace absorbs 2.26×10^6 J (the heat of vaporization of water)

Amorphous solids: When is a solid not really solid?

Steam heating systems: how to get the heat from the furnace to the radiators

from the furnace, and gives up this same amount when it condenses in the radiators. (This figure is correct if the steam pressure is 1 atm. It is slightly smaller at higher pressures.) Thus a steam-heating system does not need to circulate as much water as a hot-water heating system. If water leaves a hot-water furnace at 60°C and returns at 40°C, dropping 20°C, about 27 kg of water must circulate to carry the same amount of heat as is transferred in the form of heat of vaporization by 1 kg of steam.

The temperature-control mechanisms of many warm-blooded animals operate on a similar principle. As sweat (chiefly water) evaporates from the surface of the body, it removes heat from the body as heat of vaporization. At normal human body temperature (37°C) the heat of vaporization of water is 2.41×10^6 J·kg^{-1}, so each gram of sweat that evaporates carries away 2410 J of heat with it. Evaporative cooling makes it possible for humans to maintain normal body temperature in hot, dry desert climates where the air temperature may reach 55°C (about 130°F). Evaporation keeps the skin temperature as much as 20 C° cooler than the surrounding air. Adequate water intake is essential under these conditions; a normal person may perspire several liters per day, and unless this lost water is replaced regularly, dehydration, heat stroke, and death result. Old-time desert rats (such as the author) state that in the desert any canteen that holds less than a gallon is to be regarded as only a toy!

Temperature control in warm-blooded animals: how to stay cool in the desert

The same principle is applied in evaporative cooling systems. Hot, dry air is pushed through wet filters. As some of the water evaporates, it takes its heat of vaporization from the air, which thus becomes cooler by as much as 20 C° in hot, dry climates. Evaporative cooling is also used to condense and recirculate "used" steam in coal-fired or nuclear-powered electric power plants.

Finally, we mention that definite quantities of heat are also involved in chemical reactions. A familiar example is combustion; complete combustion of one gram of gasoline produces about 46,000 J or about 11,000 cal, and the **heat of combustion** of gasoline is

Heat of combustion: an example of heat associated with a chemical reaction

$$46,000 \text{ J·g}^{-1} \qquad \text{or} \qquad 46 \times 10^6 \text{ J·kg}^{-1}.$$

Energy values of foods are defined similarly; the unit of food energy, although called a calorie, is really a kilocalorie, equal to 1000 cal or 4186 J. When we say a gram of peanut butter "contains" 12 calories, we mean that, when it reacts with oxygen, with the help of enzymes, to convert the carbon and hydrogen completely to CO_2 and H_2O, the total energy liberated as heat is 12 kcal, 12,000 cal, or 50,200 J. Not all this energy is directly useful for mechanical work; we will take up the matter of *efficiency* of utilization of energy in Chapter 19.

Heat equivalents of food energy, and commonly used units

15–6 EXAMPLES

The basic principle in heat calculations is very simple: When heat flow occurs between two bodies, the amount of heat lost by one body must equal the amount gained by the other. Heat is energy in transit, so this is really just conservation of energy. We take each quantity of heat *added to* a body as *positive*, and each quantity *leaving* a body as *negative*. Then in general when several bodies interact, the *algebraic sum* of the quantities of heat transferred to each body must be zero. This is the basic principle of calorimetry, in some respects the simplest of all physical theories.

Calorimetry problems involve a simple conservation principle.

PROBLEM-SOLVING STRATEGY: *Calorimetry problems*

1. To avoid confusion about algebraic signs when calculating quantities of heat, use Eqs. (15–1) and (15–6) consistently for each body, noting that each Q (or ΔQ) is positive when heat enters a body and negative when it leaves. Then the algebraic sum of all the Q's must be zero.

2. Often you will need to find an unknown temperature. Represent it by an algebraic symbol such as T. Then if a body has an initial temperature of 20°C and an unknown final temperature T, the temperature

change for the body is $\Delta T = T_{final} - T_{initial} = T - 20°C$ (*not* 20°C − T!). And so on.

3. In problems where a phase change takes place, as when ice melts, you may not know in advance whether *all* the material undergoes a phase change or only part of it. You can always assume one or the other, and if the resulting calculation gives an absurd result (such as a final temperature higher or lower than *any* of the initial temperatures), you know the initial assumption was wrong. Back up and try again!

How much does your coffee cool when you pour it into a cold cup?

EXAMPLE 15–2 An ethnic restaurant down by the docks serves coffee in copper mugs. A waiter fills a cup of mass 0.1 kg, initially at 20°C, with 0.2 kg of coffee initially at 70°C. What is the final temperature after the coffee and the cup attain thermal equilibrium? (Assume coffee has the same specific heat capacity as water.)

SOLUTION Let the final temperature be T. The (negative) heat added to the coffee is

$$Q_{coffee} = mc_{water}\,\Delta T$$
$$= (0.2\ kg)(4190\ J \cdot kg^{-1} \cdot (C°)^{-1})(T - 70°C).$$

Similarly, the heat gained by the copper cup is

$$Q_{copper} = mc_{copper}\,\Delta T$$
$$= (0.1\ kg)(390\ J \cdot kg^{-1} \cdot (C°)^{-1})(T - 20°C).$$

We equate the sum of these two quantities of heat to zero, obtaining an algebraic equation for T: $Q_{coffee} + Q_{copper} = 0$, or

$$(0.2\ kg)(4190\ J \cdot kg^{-1} \cdot (C°)^{-1})(T - 70°C)$$
$$+ (0.1\ kg)(390\ J \cdot kg^{-1} \cdot (C°)^{-1})(T - 20°C) = 0.$$

Solution of this equation gives $T = 67.8°C$. The final temperature is much closer to the initial temperature of the coffee than that of the cup because of the much larger specific heat capacity of water. We can also find the quantities of heat by substituting this value for T back into the original equations. We leave it for you to show that $Q_{coffee} = -1860\ J$, and $Q_{copper} = +1860\ J$; Q_{coffee} is negative, as expected.

Cooling a soft drink: How much ice do you need?

EXAMPLE 15–3 A physics student wants to cool 0.25 kg of Omni-Cola (mostly water), which is initially at 20°C, by adding ice initially at −20°C. How much ice should be added so the final temperature will be 0°C with all the ice melted, if the heat capacity of the container can be neglected?

SOLUTION The (negative) heat added to the water is

$$Q = mc_{water}\,\Delta T$$
$$= (0.25\ kg)(4190\ J \cdot kg^{-1} \cdot (C°)^{-1})(0°C - 20°C)$$
$$= -20,900\ J.$$

The specific heat capacity of ice is approximately $2000 \text{ J·kg}^{-1}\text{·(C°)}^{-1}$. Let the mass of ice be m; then the heat needed to heat it from $-20°C$ to $0°C$ is

$$Q = mc_{\text{ice}}\,\Delta T$$
$$= m(2000 \text{ J·kg}^{-1}\text{·(C°)}^{-1})(0°C - (-20°C))$$
$$= m(40{,}000 \text{ J·kg}^{-1}).$$

The additional heat needed to melt the ice is the heat of fusion times the mass:

$$Q = mL_F = m(334{,}000 \text{ J·kg}^{-1}).$$

The sum of these three quantities must equal zero:

$$-20{,}900 \text{ J} + m(40{,}000 \text{ J·kg}^{-1}) + m(334{,}000 \text{ J·kg}^{-1}) = 0,$$

from which $m = 0.056 \text{ kg} = 56 \text{ g}$, or roughly two medium-sized ice cubes.

EXAMPLE 15–4   A certain gasoline camping lantern emits as much light as a 25-watt electric light bulb. Assuming that the efficiency of converting heat into light is the same for the lantern and the bulb (which is not actually correct), how much gasoline does the lantern burn in 10 hours?

Energy relations in a gasoline lantern

SOLUTION   The rate of energy conversion in the lantern is 25 W, so in 10 hours (36,000 s) the total energy needed is

$$(25 \text{ J·s}^{-1})(36{,}000 \text{ s}) = 0.9 \times 10^6 \text{ J}.$$

As mentioned in Section 15–6, combustion of one gram of gasoline produces 46,000 J, so the mass of gasoline required is

$$\frac{0.9 \times 10^6 \text{ J}}{46{,}000 \text{ J·g}^{-1}} = 19.6 \text{ g}.$$

Actual lanterns are much less efficient than this, and typically require on the order of 300 to 400 g of gasoline (roughly one pint) to operate for ten hours.

SUMMARY

Heat is energy in transit; the term is properly used only in reference to transfer of energy from one body to another as a result of a temperature difference. The quantity of heat ΔQ required to raise the temperature of a mass m of material by a small amount ΔT is

$$\Delta Q = mc\,\Delta T, \tag{15–1}$$

where c is the specific heat capacity of the material. If c can be considered constant over a finite temperature range T_1 to T_2, then the amount of heat Q needed to cause this temperature change is given by

$$Q = mc(T_2 - T_1) = mc\,\Delta T. \tag{15–5}$$

When the quantity of material is represented in terms of the number of moles n, the corresponding relation is

$$\Delta Q = nC\,\Delta T, \tag{15–4}$$

where C is the molar heat capacity; $C = Mc$, where M is the molecular mass. The number of moles n and the mass m of material are related by $m = nM$.

KEY TERMS
heat transfer
specific heat capacity
molar heat capacity
calorimetry
law of Dulong and Petit
phase
phase change
heat of fusion
phase equilibrium
heat of vaporization
sublimation
heat of sublimation
heat of combustion

The molar heat capacities of most metals at sufficiently high temperatures approach the value 25 $J \cdot mol^{-1} \cdot (C°)^{-1}$; this is the law of Dulong and Petit.

A phase change that occurs at constant temperature usually requires the addition or removal of heat. To change a mass m of solid material to liquid at the same temperature requires the addition of a quantity of heat Q given by

$$Q = mL_F, \qquad (15-6)$$

where L_F is called the heat of fusion. When the material changes back to the solid state, an equal quantity of heat must be removed. The corresponding quantities associated with boiling and sublimation are called the heat of vaporization and heat of sublimation, respectively.

When heat is added to a body, the corresponding Q is positive; when it is removed, Q is negative. The basic principle of calorimetry, stemming from conservation of energy, is that in a system whose parts interact by heat exchange the algebraic sum of the Q's for all parts of the system must be zero.

QUESTIONS

15-1 Suppose you have a thermos bottle half full of cold coffee. Can you warm it up to drinking temperature by shaking it? Is this possible *in principle*? Is it feasible in practice? Are you adding heat to the coffee?

15-2 When the oil in an automatic transmission is churned up by the turbine blades, it becomes hot, and usually an oil-cooling system is required. Is the engine adding heat to the oil?

15-3 A student asserted that a suitable unit for specific heat capacity was 1 $m^2 \cdot s^{-2} \cdot (C°)^{-1}$. Is this correct?

15-4 The specific heat capacity of water has the same numerical value when expressed in $cal \cdot g^{-1} \cdot (C°)^{-1}$ as when expressed in $Btu \cdot lb^{-1} \cdot (F°)^{-1}$. Is this a coincidence? Does the same relation hold for heat capacities of other materials?

15-5 A student claimed that when two bodies not initially in thermal equilibrium are placed in contact, the temperature rise of the cooler body must always equal the temperature drop of the warmer. Do you agree? Is there a principle of conservation of temperature, or something like that?

15-6 In choosing a fluid to circulate inside a gasoline engine to cool it (such as water or antifreeze), should you choose a material with a large or a small specific heat capacity? Why? What other considerations are important?

15-7 Any heat of a phase transition has a numerical value that is $\frac{5}{9}$ as great when expressed in $cal \cdot g^{-1}$ as when expressed in $Btu \cdot lb^{-1}$. Why is the conversion so simple?

15-8 Why do you think the heat of vaporization for water is so much larger than the heat of fusion?

15-9 Some household air conditioners used in dry climates cool air by blowing it through a water-soaked filter, evaporating some of the water. How does this work? Would such a system work well in a high-humidity climate?

15-10 Why does food cook faster in a pressure cooker than in boiling water?

15-11 How does the human body maintain a temperature of 37.0°C (98.6°F) in the desert where the temperature is 50°C (122°F)?

15-12 Desert travelers sometimes keep water in a canvas bag. Some water seeps through the bag and evaporates. How does this cool the water inside?

15-13 When water is placed in ice-cube trays in a freezer, why does the water not freeze all at once when the temperature has reached 0°C? In fact, it freezes first in a layer adjacent to the sides of the tray. Why?

15-14 When an automobile engine overheats and the radiator water begins to boil, the car can still be driven some distance before catastrophic engine damage occurs. Why? What determines the onset of really disastrous overheating?

15-15 Why do automobile manufacturers recommend that antifreeze (typically a 50% solution of ethylene glycol in water) be kept in the engine in summer as well as in winter?

15-16 When you step out of the shower you feel cold, but as soon as you are dry you feel warmer, even though the room temperature is the same. Why?

15-17 Suppose the heat of fusion of ice were only 10 $J \cdot g^{-1}$ instead of 334 $J \cdot g^{-1}$. Would this change the way we make iced tea? Martinis? Lemonade?

15-18 Why is the climate of regions adjacent to large bodies of water usually more moderate than that of regions far from large bodies of water?

EXERCISES

Section 15–2 Quantity of Heat

15–1 A taxi driver drives an automobile of mass 1500 kg at a speed of 5 m·s⁻¹. How many joules of heat are generated in the brake mechanism when the automobile is brought to rest?

15–2 A crate of mass 50 kg slides down a ramp inclined at 53° below the horizontal. The ramp is 6 m long. If the crate was at rest at the top of the incline and has a velocity of 6 m·s⁻¹ at the bottom, how much heat is generated by friction? Express your answer in joules, calories, and Btu.

Section 15–3 Heat Capacity

15–3

a) How much heat is required to raise the temperature of 0.20 kg of water from 20°C to 30°C?

b) If this amount of heat is added to an equal mass of mercury initially at 20°C, what is the final temperature?

c) If this amount of heat is added to an equal volume of mercury initially at 20°C, what is the final temperature?

15–4 A student uses a 100-watt electric immersion heater to heat 0.2 kg of water from 20°C to 100°C to make tea.

a) How much heat must be added to the water?

b) How much time is required?

15–5 A copper tea kettle of mass 2.0 kg, containing 4.0 kg of water, is placed on a stove. How much heat must be added to raise the temperature from 20°C to 80°C?

15–6 A copper cup of mass 0.20 kg contains 0.40 kg of water. The water is heated by a friction device that dissipates mechanical energy, and it is observed that the temperature of the system rises at the rate of 3 C°·min⁻¹. Neglect heat losses to the surroundings. What power in watts is being dissipated in the water?

15–7 A technician measures the specific heat capacity of an unidentified liquid by immersing an electrical resistor in it. Electrical energy is dissipated for 100 s at a constant rate of 50 W. The mass of the liquid is 0.530 kg, and its temperature increases from 17.64°C to 20.77°C. Find the mean specific heat capacity of the liquid in this temperature range.

Section 15–5 Phase Changes

15–8 An ice-cube tray contains 0.8 kg of water at 20°C. How much heat must be removed to cool the water to 0°C and freeze it?

15–9 How much heat is required to convert 1 g of ice at −10°C to steam at 100°C? Express you answer in joules, calories, and Btu.

15–10 An open vessel contains 0.50 kg of ice at −20°C. The mass of the container can be neglected. Heat is supplied to the vessel at the constant rate of 420 J·min⁻¹ for 500 min.

a) After how many minutes does the ice *start* to melt?

b) After how many minutes does the temperature start to rise above 0°C?

c) Plot a curve showing the elapsed time as abscissa and the temperature as ordinate.

15–11 An automobile engine whose output is 3.0×10^4W (about 40 hp) uses 4.5 gal of gasoline per hour. The heat of combustion is 12×10^7 J per gallon. What is the efficiency of the engine? That is, what fraction of the heat of combustion is converted to mechanical work?

15–12 The nominal food-energy value of butter is about 6.0 kcal·g⁻¹. If all this energy could be converted completely to mechanical energy, how much butter would be required to power an 80-kg mountaineer on his journey from Lupine Meadows (elevation 2070 m) to the summit of Grand Teton (4196 m)?

15–13 What must be the initial velocity of a lead bullet at a temperature of 25°C so that the heat developed when it is brought to rest will be just sufficient to melt it?

15–14 Evaporation of sweat is an important mechanism for temperature control in warm-blooded animals. What mass of water must evaporate from the surface of an 80-kg human body to cool it 1C°? The specific heat capacity of the human body is approximately 1.0 cal·g⁻¹·(C°)⁻¹, and the heat of vaporization of water at body temperature (37°C) is 577 cal·g⁻¹.

15–15 The capacity of air conditioners is sometimes expressed in "tons," the number of tons of ice that can be frozen from water at 0°C in 24 hr by the unit. Express the capacity of a one-ton air conditioner in watts and in Btu·hr⁻¹.

Section 15–6 Examples

15–16 In a physics lab experiment, a student immersed 100 copper pennies (mass 3.0 g each) in boiling water. After they reached thermal equilibrium, she fished them out and dropped them into 0.2 kg of water at 20°C. What was the final temperature?

15–17 An aluminum can of mass 0.500 kg contains 0.118 kg of water at a temperature of 20°C. A 0.200-kg block of iron at 75°C is dropped into the can. Find the final temperature, assuming no heat loss to the surroundings.

15–18 A 0.050-kg sample of unknown material, at a temperature of 100°C, is dropped into a calorimeter containing 0.200 kg of water initially at 20°C. The calorimeter is of copper, and its mass is 0.100 kg. The final temperature of the calorimeter is 22°C. Compute the specific heat capacity of the sample.

15–19 A 2-kg iron block is taken from a furnace where its temperature was 650°C and placed on a large block of ice at 0°C. Assuming that all the heat given up by the iron is used to melt the ice, how much ice is melted?

15–20 A copper calorimeter of mass 0.100 kg contains 0.150 kg of water and 0.008 kg of ice in thermal equilibrium at atmospheric pressure. If 0.500 kg of lead at a temperature of 200°C are dropped into the calorimeter, what is the final temperature, assuming no heat is lost to the surroundings?

15–21 A beaker of very small mass contains 0.500 kg of water at a temperature of 80°C. How many grams of ice at a temperature of −20°C must be dropped into the water so that the final temperature of the system will be 50°C?

15–22 A thirsty farmer cools a 1-L bottle of soft drink (mostly water) by pouring the contents into a large copper mug of mass 0.278 kg and adding 0.050 kg of ice initially at

−16°C. If soft drink and mug are initially at 20°C, what is the final temperature of the system, assuming no heat losses?

15–23 A vessel whose walls are thermally insulated contains 2.10 kg of water and 0.200 kg of ice, all at a temperature of 0°C. The outlet of a tube leading from a boiler in which water is boiling at atmospheric pressure is inserted into the water. How many grams of steam must condense to raise the temperature of the system to 20°C? Neglect the heat capacity of the container.

PROBLEMS

15–24 An artificial satellite, constructed of aluminum, encircles the earth at a speed of 9000 m·s⁻¹.

a) Find the ratio of its kinetic energy to the energy required to raise its temperature by 600 C°. (The melting point of aluminum is 660°C.) Assume a constant specific heat capacity of 910 J·kg⁻¹·(C°)⁻¹.

b) Discuss the bearing of your answer on the problem of the reentry of a satellite into the earth's atmosphere.

15–25 A piece of ice at 0°C falls from rest into a lake whose temperature is 0°C, and one-half of 1% of the ice melts. Compute the minimum height from which the ice falls.

15–26 The windlass is a rotating drum or cylinder over which a rope or cord slides in order to provide a great amplification of the rope's tension while keeping both ends free. See Fig. 15–6. Since the added tension in the rope is due to friction, the windlass generates heat.

a) If the difference in tension between the two parts of the rope is 100 N, and the windlass has a diameter of 10 cm and turns once in 1 s, find the rate at which heat is being generated. Why does the number of turns not matter?

b) If the windlass is made of iron and has a mass of 5 kg, at what rate does its temperature rise? Assume that the temperature in the windlass is uniform.

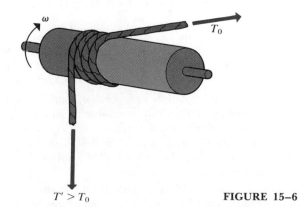

$T' > T_0$ **FIGURE 15–6**

15–27 At very low temperatures, the molar heat capacity of rock salt varies with temperature according to "Debye's T^3 law"; thus

$$C = k\frac{T^3}{\Theta^3},$$

where $k = 1940$ J·mol⁻¹·K⁻¹ and $\Theta = 281$ K.

a) How much heat is required to raise the temperature of 2 mol of rock salt from 10 K to 50 K? (*Hint:* Use Eq. (15–2) for dQ and integrate.)

b) What is the mean molar heat capacity in this range?

c) What is the true molar heat capacity at 50 K?

15–28 The molar heat capacity at constant pressure of a certain substance varies with temperature according to the empirical equation

$$C_p = 27.2 \text{ J·mol}^{-1}\cdot\text{K}^{-1} + (4 \times 10^{-3} \text{ J·mol}^{-1}\cdot\text{K}^{-2})T,$$

where T is in kelvins. How much heat is necessary to change the temperature of 10 mol of this substance from 27°C to 527°C? (*Hint:* Use Eq. (15–2) for dQ and integrate.)

15–29

a) A home owner in a cold climate has a coal-burning furnace that burns 9000 kg (about 10 tons) of coal during a winter. The heat of combustion of coal is 2.50×10^7 J·kg⁻¹. If stack losses are 15% (the amount of heat energy lost up the chimney), how many joules were actually used to heat the house?

b) The home owner proposes to install a solar heating system, heating large tanks of water by solar radiation during the summer and using the stored energy for heating during the winter. Find the required dimensions of the storage tank, assuming it to be a cube, to store a quantity of energy equal to that computed in part (a). Assume that the water is raised to 49°C (120°F) in the summer and cooled to 27°C (80°F) in the winter.

15–30 In a household hot-water heating system, water is delivered to the radiators at 60°C (140°F) and leaves at 38°C (100°F). The system is to be replaced by a steam system in which steam at atmospheric pressure condenses in the radiators, the condensed steam leaving the radiators at 82°C (180°F). How many kilograms of steam will supply the same heat as was supplied by 1 kg of hot water in the first installation?

15–31 A "solar house" has storage facilities for 4.2×10^9 J (about 4 million Btu). Compare the space requirements for this storage on the assumption

a) that the heat is stored in water heated from a minimum temperature of 27°C (80°F) to a maximum of 49°C (120°F);

b) that the heat is stored in Glauber salt ($Na_2SO_4 \cdot 10\ H_2O$) heated in the same temperature range.

Properties of Glauber Salt	
Specific heat capacity	
Solid	$1930\ J \cdot kg^{-1} \cdot (C°)^{-1}$
Liquid	$2850\ J \cdot kg^{-1} \cdot (C°)^{-1}$
Specific gravity	1.6
Melting point	32°C
Heat of fusion	$2.42 \times 10^5\ J \cdot kg^{-1}$

15–32 A calorimeter contains 0.100 kg of water at 0°C. A 1-kg copper cylinder and a 1-kg lead cylinder, both at 100°C, are placed in the calorimeter. Find the final temperature if there is no loss of heat to the surroundings. Neglect the heat capacity of the calorimeter itself.

15–33 An ice cube whose mass is 0.050 kg is taken from a refrigerator where its temperature was −10°C and dropped into a glass of water at 0°C. If no heat is gained or lost from outside, how much water will freeze onto the cube?

15–34 A tube leads from a flask in which water is boiling under atmospheric pressure to a calorimeter. The mass of the calorimeter is 0.150 kg, its specific heat capacity is $420 \cdot J \cdot kg^{-1} \cdot (C°)^{-1}$, and it contains originally 0.340 kg of water at 15°C. Steam is allowed to condense in the calorimeter until its temperature increases to 71°C, after which the total mass of calorimeter and contents is found to be 0.525 kg. Compute the heat of condensation of steam from these data.

15–35 A copper calorimeter can of mass 0.322 kg contains 0.050 kg of ice. The system is initially at 0°C. If 0.012 kg of steam at 100°C and 1 atm pressure are admitted into the calorimeter, what will be the final temperature of the calorimeter and its contents?

CHALLENGE PROBLEMS

15–36 An electric heater is to provide a continuous supply of hot water. One trial design is shown in Fig. 15–7. Water is flowing at the rate of 0.300 kg·min^{-1}, the inlet thermometer registers 15°C, the voltmeter reads 120 V, and the ammeter reads 10 A (corresponding to a power input of [120 V][10 A] = 1200 W).

a) When a steady state is finally reached, what will be the reading of the outlet thermometer?

b) Why is it unnecessary to take into account the heat capacity mc of the apparatus itself?

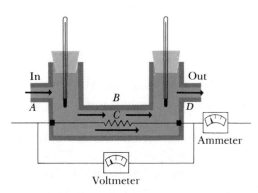

FIGURE 15–7

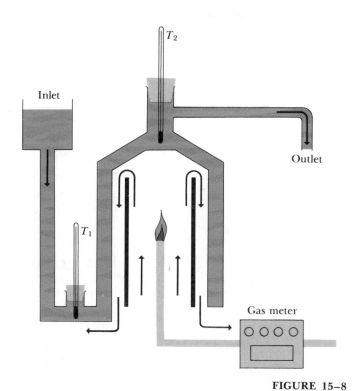

FIGURE 15–8

15–37 An engineer has designed a continuous-flow calorimeter to measure the heat of combustion of a gaseous fuel. A sketch of his design is shown in Fig. 15–8. Water is supplied at the rate of 5.67 kg·min^{-1} and natural gas at 5.67×10^{-4} m^3·min^{-1}. In the steady state, the inlet and outlet thermometers register 15.6°C and 24.4°C, respectively. What is the heat of combustion of natural gas in J·m^{-3}? Why should the gas flow be made as small as possible?

15–38 An aluminum rod of cross-sectional area 0.04 cm^2 and length 80.00 cm at a temperature of 140°C is laid alongside a copper rod of cross-sectional area 0.02 cm^2 and length 79.94 cm at temperature T. The two rods are laid alongside each other so they are in thermal contact. No heat is lost to the surroundings, and after they have come to thermal equilibrium they are observed to be of the same length. Calculate the original temperature T of the copper rod and the final temperature of the rods after they come to equilibrium.

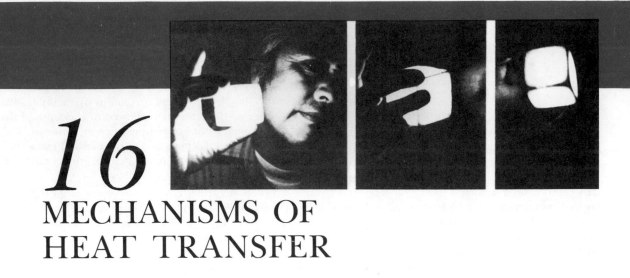

16 MECHANISMS OF HEAT TRANSFER

WE DISCUSSED CONDUCTION OF HEAT QUALITATIVELY IN CHAPTER 15 IN connection with the transfer of heat between two bodies at different temperatures. We did not need to be concerned there with the *rate* of transfer of heat from one body to another. In many situations, however, we need to know how quickly or slowly heat is transferred, and how the rate of heat transfer depends on the properties of the system. In this chapter we study the three mechanisms of heat transfer: conduction, convection, and radiation. Conduction occurs within a body and between two bodies in actual contact with each other. Convection involves motion of mass from one region of space to another. Radiation is heat transfer by electromagnetic radiation, with no need for matter to be present in the space between bodies.

16–1 CONDUCTION

Conduction: energy transfer through a material

If we place one end of a metal rod in a flame and hold the other end, the end we are holding gets hotter and hotter, even though it is not in direct contact with the flame. Heat reaches the cooler end by **conduction** through the material of the rod. Microscopically, the molecules at the hot end increase the energy of their vibrations as the temperature increases. They then interact with their more slowly moving neighbors farther from the flame; they share some of their energy with these neighbors, who in turn pass it on to those still farther from the flame. Thus energy associated with thermal motion is passed along from one molecule to the next, from the hotter end to the cooler, while each individual molecule remains at its original position.

Metals are good conductors.

Most metals are good conductors of electricity and also good conductors of heat. The ability of a metal to conduct an electric current is due to the fact that some electrons in the material have become detached from their parent atoms. These "free" electrons also provide an effective mechanism for heat transfer from the hotter to the cooler portions of the metal. The best electrical conductors (silver, copper, aluminum, and gold) are also the best thermal conductors.

Conduction of heat takes place in a body only when different parts of the body are at different temperatures, and the direction of heat flow is always from points of higher temperature to points of lower temperature. Figure 16–1 shows a rod of material with cross-sectional area A and length L. Let the left end of the rod be kept at a temperature T_2, and the right end at a lower temperature T_1. The direction of the flow of heat is then from left to right through the rod. We assume that the sides of the rod are covered by an insulating material, so no heat transfer occurs at the sides. This is, of course, an idealized model; all real materials, even the best thermal insulators, conduct heat to some extent.

Experiments show that the *rate* of flow of heat through the rod is proportional to the area A, proportional to the temperature difference $(T_2 - T_1)$, and inversely proportional to the length L. We can express these proportions as an equation by introducing a constant k whose numerical value depends on the material of the rod. The quantity k is called the **thermal conductivity** of the material:

$$\frac{dQ}{dt} = H = \frac{kA(T_2 - T_1)}{L}, \qquad (16-1)$$

where H is the quantity of heat flowing through the rod per unit time, also called the **heat current.** The quantity $(T_2 - T_1)/L$ represents the temperature difference per unit length and is called the **temperature gradient**.

Equation (16–1) may also be used to compute the rate of heat flow through a slab or through *any* homogeneous body having a uniform cross section perpendicular to the direction of flow, provided that the flow has attained steady-state conditions and the ends are kept at constant temperatures.

The units of heat flow or heat current are energy per unit time; thus the SI unit of heat current is the joule per second, or watt. Other units such as the calorie per second or Btu per second are also sometimes used. The unit of k is obtained by solving Eq. (16–1) for k:

$$k = \frac{HL}{A(T_2 - T_1)}. \qquad (16-2)$$

This shows that in SI units, the unit of k is

$$\frac{(1\ \text{J·s}^{-1})(1\ \text{m})}{(1\ \text{m})^2(1\ \text{C}°)} = 1\ \text{J·(s·m·C}°)^{-1}.$$

Values of thermal conductivity are sometimes tabulated by using cgs units, with the calorie as the energy unit. The unit of k then is $1\ \text{cal·(s·cm·C}°)^{-1}$. The conversion is

$$1\ \text{cal·(s·cm·C}°)^{-1} = 419\ \text{J·(s·m·C}°)^{-1}.$$

Some numerical values of k, at temperatures near room temperature, are given in Table 16–1. The properties of materials used commercially as heat insulators are sometimes expressed in a mixed system in which the unit of heat current is $1\ \text{Btu·hr}^{-1}$, the unit of area is $1\ \text{ft}^2$, and the unit of temperature gradient is one Fahrenheit degree per inch ($1\ \text{F}°·\text{in.}^{-1}$)! What better argument could there be for adoption of the metric system?

Heat conduction in a material requires temperature differences between different regions.

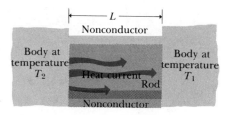

16–1 Steady-state heat flow in a uniform rod.

Thermal conductivity describes how well a material conducts heat.

TABLE 16–1   Thermal Conductivities (k)

	$\text{J}\cdot\text{s}^{-1}\cdot\text{m}^{-1}\cdot(\text{C}°)^{-1}$	$\text{cal}\cdot\text{s}^{-1}\cdot\text{cm}^{-1}\cdot(\text{C}°)^{-1}$
Metals		
Aluminum	205.0	0.49
Brass	109.0	0.26
Copper	385.0	0.92
Lead	34.7	0.083
Mercury	8.3	0.020
Silver	406.0	0.97
Steel	50.2	0.12
Various solids		
(Representative values)		
Insulating brick	0.15	0.00035
Red brick	0.6	0.0015
Concrete	0.8	0.002
Cork	0.04	0.0001
Felt	0.04	0.0001
Fiberglass	0.04	0.0001
Glass	0.8	0.002
Ice	1.6	0.004
Rock wool	0.04	0.0001
Styrofoam	0.01	0.00002
Wood	0.12–0.04	0.0003–0.0001
Gases		
Air	0.024	0.000057
Argon	0.016	0.000039
Helium	0.14	0.00033
Hydrogen	0.14	0.00033
Oxygen	0.023	0.000056

Equation (16–1) shows that the larger the thermal conductivity k, the larger the heat current, if other factors are equal. A good heat conductor has a large value of k, and a good insulator has a small k. A perfect insulator would have $k = 0$. This is an unattainable idealization, but Table 16–1 shows that the metals as a group have much greater thermal conductivities than the non-metals, and that thermal conductivities of gases are extremely small. Figure 16–2 shows a "space-age" ceramic material with very unusual thermal properties.

Thermal resistance is a convenient way to describe the insulating properties of building insulation materials.

For thermal insulation in buildings, engineers use the concept of **thermal resistance,** denoted by R. The thermal resistance R of a slab of material with thickness L is defined to be

$$R = \frac{L}{k}. \tag{16–3}$$

Using this concept, we may rewrite Eq. (16–1) as

$$H = \frac{A(T_2 - T_1)}{R}. \tag{16–4}$$

The heat current H is *inversely* proportional to R; hence the term thermal *resistance*. In the units used for commercial insulating materials, H is expressed in $\text{Btu}\cdot\text{hr}^{-1}$, A in ft^2, and $(T_2 - T_1)$ in $\text{F}°$. The units of R are then $\text{ft}^2\cdot\text{F}°\cdot\text{hr}\cdot\text{Btu}^{-1}$. Values of R are quoted without units; thus a six-inch-thick layer of fiberglass has an R value of 19, a two-inch slab of polyurethane foam a value of 12, and so on. Doubling the thickness doubles the R-value, of course.

16–2 This protective tile, developed for use on the space shuttle *Columbia*, has extraordinary thermal properties. The extremely small thermal conductivity and small heat capacity of the material make it possible to hold the tile by its edges, even though it is hot enough to emit the light for these photographs. (Courtesy Lockheed Corp.)

Common practice in new construction in severe northern climates is to specify *R* values of around 30 for exterior walls and ceilings. When the insulating material is in layers, such as a plastered wall, fiberglass insulation, and wood exterior siding, the *R* values are additive.

PROBLEM-SOLVING STRATEGY: *Heat conduction*

1. Identify the direction of heat flow in the problem; in Eq. (16–1), *L* is always measured along this direction, and *A* is always an area perpendicular to this direction. Sometimes a box or other container of irregular shape but uniform wall thickness can be approximated as a flat slab of the same thickness and total wall area.

2. In some problems the heat flows through two different materials in succession. The temperature at the interface between the two materials is then intermediate between T_1 and T_2; represent it by a symbol such as T_3. The temperature differences for the two materials are then $(T_2 - T_3)$ and $(T_3 - T_1)$. In steady-state heat flow, heat cannot accumulate within either material, so the total rate of heat flow *H* is equal to the heat flow in each material (the same in both materials). This is like an electric circuit with the elements connected in series.

3. If there are two *parallel* heat-flow paths, so that some heat flows through each, then the total *H* is the sum of the quantities H_1 and H_2 for the separate paths. In this case the temperature difference is the same for the two materials, but *L*, *A*, and *k* may be different for the two paths.

4. As always, it is essential to use a consistent set of units. If you use a value of *k* expressed in $J \cdot s^{-1} \cdot m^{-1} \cdot (C°)^{-1}$, don't use distances in cm, heat in calories, or *T* in °F!

EXAMPLE 16–1 A styrofoam box used to keep drinks cold at a picnic has total wall area (including the lid) of 0.8 m² and wall thickness 2.0 cm. It is filled with ice and cans of orange soda at 0°C. What is the rate of heat flow into the box if the outside temperature is 30°C? How much ice melts in one day?

How fast does the ice melt in a styrofoam cooler?

SOLUTION We assume that the total heat flow is approximately the same as it would be through a flat slab of area 0.8 m² and thickness 2.0 cm = 0.02 m. We find k from Table 16–1. The rate of heat flow, from Eq. (16–1), is

$$H = kA\frac{T_2 - T_1}{L} = (0.01 \text{ J·m}^{-1}\text{·s}^{-1}\text{·(C°)}^{-1})(0.8 \text{ m}^2)\frac{30 \text{ C°}}{0.02 \text{ m}} = 12 \text{ J·s}^{-1}.$$

There are 86,400 s in one day, so the total heat flow in one day is

$$(12 \text{ J·s}^{-1})(86,400 \text{ s}) = 1.04 \times 10^6 \text{ J}.$$

The heat of fusion of ice is 334 J·g^{-1}, so the quantity of ice melted by this quantity of heat is

$$m = \frac{Q}{L_{\text{F}}} = \frac{1.04 \times 10^6 \text{ J}}{334 \text{ J·g}^{-1}} = 3110 \text{ g}, \quad \text{or about 3.1 kg.}$$

Heat flow in two bars in series

EXAMPLE 16–2 A steel bar 10 cm long is welded end to end to a copper bar 20 cm long. Each bar has a square cross section, 2 cm on a side. The free end of the steel bar is in contact with steam at 100°C, and the free end of the copper bar is in contact with ice at 0°C. Find the temperature at the junction of the two bars and the total rate of heat flow.

SOLUTION The key to the solution is the fact that the rates of heat flow in the two bars must be equal; otherwise some sections would have more heat flowing in than out, or the reverse, and steady-state conditions could not exist. Let T be the unknown junction temperature; we use Eq. (16–1) for each bar and equate the two expressions:

$$\frac{k_s A(100°C - T)}{L_s} = \frac{k_c A(T - 0°C)}{L_c}.$$

The areas A are equal and may be divided out. Substituting numerical values, we find

$$\frac{(50.2 \text{ J·s}^{-1}\text{·m}^{-1}\text{·(C°)}^{-1})(100°C - T)}{0.1 \text{ m}} = \frac{(385 \text{ J·s}^{-1}\text{·m}^{-1}\text{·(C°)}^{-1})(T - 0°C)}{0.2 \text{ m}}.$$

Rearranging and solving for T, we find

$$T = 20.7°C.$$

Even though the steel bar is shorter, the temperature drop across it is much greater than across the copper bar because steel is a much poorer conductor.
    We now obtain the total heat current by substituting this value for T back into either of the above expressions:

$$H = \frac{(50.2 \text{ J·s}^{-1}\text{·m}^{-1}\text{·(C°)}^{-1})(0.02 \text{ m})^2(100°C - 20.7°C)}{0.1 \text{ m}} = 15.9 \text{ J·s}^{-1} = 15.9 \text{ W},$$

or

$$H = \frac{(385 \text{ J·s}^{-1}\text{·m}^{-1}\text{·(C°)}^{-1})(0.02 \text{ m})^2(20.7°C - 0°C)}{0.2 \text{ m}} = 15.9 \text{ J·s}^{-1} = 15.9 \text{ W}.$$

Heat flow in two bars in parallel

EXAMPLE 16–3 In Example 16–2, suppose the two bars are separated; one end of each bar is placed in contact with steam at 100°C, and the other end of each bar contacts ice at 0°C. What is the *total* rate of heat flow in the two bars?

SOLUTION In this case, the bars are in parallel rather than in series. The total heat flow is the sum of the flows in the two bars, and for each bar $T_2 - T_1 = 100°C - 0°C = 100$ C°. Thus

$$H = \frac{(50.2 \text{ J·s}^{-1}\text{·m}^{-1}\text{·(C°)}^{-1})(0.02 \text{ m})^2(100 \text{ C°})}{0.1 \text{ m}}$$

$$+ \frac{(385 \text{ J·s}^{-1}\text{·m}^{-1}\text{·(C°)}^{-1})(0.02 \text{ m})^2(100 \text{ C°})}{0.2 \text{ m}}$$

$$= 20.1 \text{ J·s}^{-1} + 77.0 \text{ J·s}^{-1} = 97.1 \text{ J·s}^{-1} = 97.1 \text{ W}.$$

The heat flow in the copper bar is much greater than in the steel bar, even though the copper bar is longer, because its thermal conductivity is much larger. The total heat flow is much larger than in Example 16–2 because the full 100 C° temperature difference appears across each bar.

16–2 CONVECTION

In heat conduction, energy is transferred by molecular vibrations and electron motion, but there is no overall bulk motion of the material. In contrast, **convection** is transfer of heat by actual motion of material from one region of space to another. The hot-air furnace, the hot-water heating system, and the flow of blood in the body are examples. If the material is forced to move by a blower or pump, the process is called *forced convection;* if the material flows owing to differences in density caused by thermal expansion, the process is called *natural* or *free convection.*

A simple example of natural convection is shown in Fig. 16–3. In (a) the water is at the same temperature throughout, so the levels are the same in the two sides of the U-tube. In (b) we heat the right side; the water expands, becoming less dense, and a taller column is needed to balance the pressure of the left-hand side. When we open the valve, water flows from the top of the warm column into the cooler column. If we supply heat continuously to the hot side and remove it from the cold side, this circulation continues indefinitely. The net result is a continuous heat transfer from the hot to the cold side. In hot-water heating systems, the cold side corresponds to the radiators and the hot side to the furnace.

Convective heat transfer is a complex process, and there is no simple equation to describe it, as there is for conduction. When a surface at one temperature is in contact with a fluid at a different temperature, the heat transfer depends on the shape and orientation of the surface, the properties of the fluid, and the nature of the fluid flow. An approximate rule of thumb often used for practical calculations is to represent the heat current H by means of the equation

$$H = hA \, \Delta T, \tag{16–5}$$

where A is the surface area, ΔT is the temperature difference between the surface and the main body of the fluid, and h is a quantity called the **convection coefficient.** Unlike the situation with conduction, h is not constant but depends on ΔT. In practice, values of h are determined by experiment and published as tables, which a physicist or engineer may then use for calculations

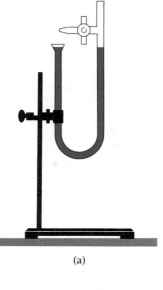

(a)

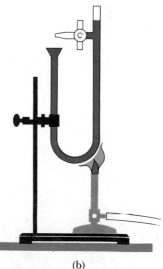

(b)

16–3 Free convection is caused by density differences due to thermal expansion. In (b), the level in the right side is higher than in the left side because the liquid on the right is warmer and therefore less dense.

TABLE 16–2 Coefficients of Natural Convection in Air at Atmospheric Pressure

Equipment	Convection Coefficient h, $\text{J}\cdot\text{s}^{-1}\cdot\text{m}^{-2}\cdot(\text{C}°)^{-1}$
Horizontal plate, facing upward	$2.49\,(\Delta T)^{1/4}$
Horizontal plate, facing downward	$1.31\,(\Delta T)^{1/4}$
Vertical plate	$1.77\,(\Delta T)^{1/4}$
Horizontal or vertical pipe $\left(\dfrac{\text{diameter}}{D}\right)$	$1.32\left(\dfrac{\Delta T}{D}\right)^{1/4}$

in specific problems. A few values are given in Table 16–2. In the last line, D is measured in meters.

A common situation is natural convection from a wall or a pipe that is at a constant temperature and is surrounded by air that is cooler by an amount ΔT. The convection coefficients applicable in this situation are given in Table 16–2. For a horizontal plate, if the air is *hotter* than the plate, the first two values of h should be interchanged.

Calculating heat transfer through a glass window: conduction and convection

EXAMPLE 16–5 The air in a room is at a temperature of 25°C and the outside air is at −15°C. How much heat is transferred per unit area of a glass windowpane of thermal conductivity $0.80\ \text{J}\cdot\text{m}^{-1}\cdot\text{s}^{-1}\cdot(\text{C}°)^{-1}$ and thickness 2 mm?

SOLUTION To assume that the inner surface of the glass is at 25°C and the outer surface is at −15°C is erroneous, as you can verify by touching the inner surface of a glass windowpane on a cold day. We must expect a much *smaller* temperature difference across the windowpane, so that in the steady state the rates of heat transfer (1) by convection in the room, (2) by conduction through the glass, and (3) by convection in the outside air are all equal.

As a first approximation in the solution of this problem, let us assume that the window is at a uniform temperature T. If $T = 5°C$, then the temperature difference between the inside air and the glass is the same as that between the glass and the outside air, or 20 C°. Hence the convection coefficient in both cases is

$$h = 1.77(20)^{1/4}\,\text{J}\cdot\text{s}^{-1}\cdot\text{m}^{-2}\cdot(\text{C}°)^{-1}$$
$$= 3.74\ \text{J}\cdot\text{s}^{-1}\cdot\text{m}^{-2}\cdot(\text{C}°)^{-1},$$

and, from Eq. (16–5), the heat transferred per unit area is

$$\frac{H}{A} = (3.74\ \text{J}\cdot\text{s}^{-1}\cdot\text{m}^{-2}\cdot(\text{C}°)^{-1})(20\ \text{C}°)$$
$$= 74.8\ \text{J}\cdot\text{s}^{-1}\cdot\text{m}^{-2}.$$

The glass, however, is *not* at uniform temperature; there must be a temperature difference ΔT across the glass sufficient to provide heat conduction at the rate of $74.9\ \text{J}\cdot\text{s}^{-1}\cdot\text{m}^{-2}$. Using the conduction equation, Eq. (16–1), we obtain

$$\Delta T = \frac{LH}{kA} = \frac{(0.002\ \text{m})(74.8\ \text{J}\cdot\text{s}^{-1}\cdot\text{m}^{-2})}{0.80\ \text{J}\cdot\text{m}^{-1}\cdot\text{s}^{-1}\cdot(\text{C}°)^{-1}} = 0.19\ \text{C}°.$$

Thus the inner surface is at about 5.09°C and the outer surface at 4.91°C.

16–3 RADIATION

Heat transfer by **radiation** is associated with energy carried by electromagnetic radiation, such as visible light, infrared, and ultraviolet radiation. Every body at a finite temperature (above absolute zero) emits electromagnetic radiation, with associated energy called **radiant energy.** This radiation contains a mixture of different wavelengths. At ordinary temperatures, say 20°C, nearly all the energy is carried by infrared waves having wavelengths longer than those of visible light. As the temperature rises, the wavelengths shift to shorter values. At 800°C a body emits enough visible radiation to be self-luminous and appears "red-hot," although even at this temperature most of the energy is carried by infrared waves. At 3000°C, the temperature of an incandescent lamp filament, the radiation contains enough visible light so the body appears "white-hot."

> Radiation: energy transfer by electromagnetic waves

The total rate of radiation of energy from a surface increases very rapidly with temperature, in proportion to the *fourth power* of the absolute (Kelvin) temperature. For example, a copper block at a temperature of 100°C (373 K) radiates about 0.03 J·s^{-1} or 0.03 W from each square centimeter of its surface. At a temperature of 500°C (773 K), it radiates about 0.54 W from each square centimeter, and at 1000°C (1273 K), it radiates 4 W per square centimeter. This rate is 130 times as great as the rate at a temperature of 100°C.

> Emission of radiation by a hot body increases very rapidly with temperature.

Radiation is the mechanism by which the sun's energy reaches the earth. Near the earth, the rate of energy transfer due to solar radiation is about 1400 W for each square meter of surface area. Not all of this energy reaches the earth's surface; even on a clear day, about one fourth of this energy is absorbed by the earth's atmosphere.

Experiments show that the rate of radiation of energy from a surface is proportional not only to the fourth power of the absolute temperature but also to the surface area A. In addition, it depends on the nature of the surface; this dependence is described by a quantity e called **emissivity,** which is a dimensionless number between 0 and 1. Thus the heat current H due to radiation from a surface area A with emissivity e at absolute temperature T can be expressed as

> Some surfaces are better emitters of radiation than others at the same temperature.

$$H = Ae\sigma T^4, \tag{16–6}$$

where σ is a fundamental physical constant called the **Stefan-Boltzmann constant.** This relation, called the **Stefan-Boltzmann law,** was deduced by Josef Stefan (1835–1893) on the basis of experimental measurements made by John Tyndall (1820–1893) and was later derived from theoretical considerations by Ludwig Boltzmann (1844–1906).

In Eq. (16–6), H has units of power (energy per unit time). Thus in SI units σ has the units W·m^{-2}·K^{-4}. The numerical value is found experimentally to be

$$\sigma = 5.6699 \times 10^{-8} \text{ W·m}^{-2}\text{·K}^{-4}.$$

Emissivity is usually larger for dark, rough surfaces than for light, smooth ones. The emissivity of a smooth copper surface is about 0.3.

EXAMPLE 16–6 A thin square steel plate, 10 cm on a side, is heated in a blacksmith's forge to a temperature of 800°C. If the emissivity is unity, what is the total rate of radiation of energy?

SOLUTION The total surface area, including both sides, is $2(0.01 \text{ m})^2 = 0.02 \text{ m}^2$. The temperature used in Eq. (16–6) must be *absolute* temperature; $800°C = 1073$ K. Then Eq. (16–6) gives

$$H = (0.02 \text{ m}^2)(1)(5.67 \times 10^{-8} \text{ W·m}^{-2}\text{·K}^{-4})(1073 \text{ K})^4 = 1503 \text{ W}.$$

If the plate were heated by an electric heater instead, an electric power input of 1503 W would be required to maintain it at constant temperature.

Why doesn't a radiating body radiate away all its energy?

If the surfaces of all bodies continuously emit radiant energy according to Eq. (16–6), then why do they not eventually radiate away *all* their energy and cool down to a temperature of absolute zero? The answer is that they *would* do so if energy were not supplied to them in some way. In the case of an electric-heater element or the filament of an electric lamp, energy is supplied electrically to make up for the energy radiated. As soon as this energy supply is cut off, these bodies do, in fact, cool down very quickly to the temperature of their surroundings.

But why do they not continue to radiate and cool still more? The reason is that the surroundings are *also* radiating. Some of this radiated energy is intercepted and absorbed. The rate at which a body *radiates* energy is determined by the temperature of the *body,* but the rate at which it *absorbs* energy by radiation depends on the temperature of its *surroundings.* When a body is hotter than its surroundings, the rate of emission is greater than the rate of absorption; there is a net loss of energy, and the body cools down, unless its temperature is maintained by some other means. When a body is colder than its surroundings, the rate of absorption is greater than the rate of emission, and the body's temperature rises. At thermal equilibrium, the two rates must be equal.

Emission and absorption of radiation: a two-way process

In thermal equilibrium, the rate of emission equals the rate of absorption.

Thus when a body at temperature T is surrounded by walls also at temperature T, maintenance of thermal equilibrium requires that the body's rate of *absorption* of radiant energy from the walls be equal to the rate of radiation from its surface, which is $H = Ae\sigma T^4$. Hence, for such a body at a temperature T_1, surrounded by walls at a temperature T_2, the *net* rate of loss (or gain) of energy per unit area by radiation is

$$H_{\text{net}} = Ae\sigma T_1{}^4 - Ae\sigma T_2{}^4 = Ae\sigma(T_1{}^4 - T_2{}^4). \qquad (16-7)$$

EXAMPLE 16–7 Assuming the total surface of the human body is 1.2 m^2 and the surface temperature is $30°C = 303$ K, find the total rate of radiation of energy from the body. If the surroundings are at a temperature of $20°C$, what is the net rate of heat loss from the body by radiation?

SOLUTION For infrared radiation, the emissivity of the body is very close to unity, irrespective of skin pigmentation. The rate of energy transfer per unit area is given by Eq. (16–6). Taking $e = 1$, we find

$$\begin{aligned} H &= Ae\sigma T^4 \\ &= (1.2 \text{ m}^2)(1)(5.67 \times 10^{-8} \text{ W·m}^{-2}\text{·K}^{-4})(303 \text{ K})^4 \\ &= 574 \text{ W}. \end{aligned}$$

This loss is partially balanced by *absorption* of radiation, which depends on the temperature of the surroundings. The *net* rate of radiative energy transfer is given by Eq. (16–7):

$$H = Ae\sigma(T_1{}^4 - T_2{}^4)$$
$$= (1.2 \text{ m}^2)(1)(5.67 \times 10^{-8} \text{ W·m}^{-2}\text{·K}^{-4})[(303 \text{ K})^4 - (293 \text{ K})^4]$$
$$= 72 \text{ W}.$$

From the above discussion, we can see that a body that is a good absorber is also a good emitter. An ideal radiator, with an emissivity of unity, is also an ideal absorber, absorbing *all* the radiation that strikes it. Such an ideal surface is called an ideal black body, or simply a **blackbody.** Conversely, an ideal *reflector*, which absorbs *no* radiation at all, is also a very ineffective radiator. This is the reason for the silver coatings on vacuum ("thermos") bottles. A vacuum bottle is constructed with double glass walls; the air is pumped out of the spaces between the walls, so that heat transfer by conduction and convection is practically eliminated. The silver coating on the walls reflects most of the radiation from the contents back into the container, and the wall itself is a very poor emitter. The Dewar flask, used to store liquefied gases, is exactly the same in principle.

The blackbody: an ideal absorber and radiator

Heat transfer by radiation is more important than one might have guessed. A premature baby in an incubator can be dangerously cooled by radiation, if the walls of the incubator happen to be cold, even when the air in the incubator is warm. Some incubators regulate the air temperature by measuring the baby's skin temperature.

16–4 SOLAR ENERGY AND RESOURCE CONSERVATION

The principles of heat transfer discussed in this chapter have many very practical applications in a civilization such as ours with growing energy consumption and dwindling energy resources. A substantial fraction of all energy consumption in the United States is for heating and cooling of homes and other buildings, to maintain comfortable temperature and humidity inside when the outside temperature is much hotter or much colder.

For space heating, the objective is to prevent as much heat flow as possible from inside to outside; heating units replace the inevitable loss of heat. Walls insulated with material of low thermal conductivity, storm windows, and multiple-layer glass windows all help to reduce heat loss. It has been estimated that if all buildings used such materials, the total energy needed for space heating would be reduced by at least one-third.

Heat transfer considerations in heating and cooling buildings

Air conditioning in summer poses the reverse problem; heat flows from outside to inside, and energy must be expended in refrigerating units to remove it. Again, appropriate insulation can decrease this energy cost considerably.

Direct conversion of solar energy is a promising development in energy technology. In a typical household heating system, large black plates facing the sun are backed with pipes through which water circulates. The black surface absorbs most of the sun's radiation; the heat is transferred by conduction to the water, and then by forced convection to radiators inside the house. Heat loss by convection of air near the solar collecting panels is reduced by covering

Solar energy in household heating systems

them with glass, with a thin air space. An insulated heat reservoir provides for storage of collected energy for use at night and on cloudy days. Many larger-scale solar-energy conversion systems are also currently under study; a few examples are discussed in Section 19–10.

Solar energy has many environmental advantages.

From an environmental standpoint, solar energy has multiple advantages over the use of fossil fuels (burning of coal or oil) or nuclear power. Fossil fuels are being used up; solar power continues indefinitely. Obtaining fossil fuels may involve strip mining, with its associated destruction of landscape and elimination of other useful land functions, such as farming or timber. Offshore oil drilling is a source of ocean water pollution. Air pollution from combustion products is a familiar problem, as is acid rain, directly attributable in many cases to coal smoke. The long-range effects on our climate of excess atmospheric carbon dioxide produced by combustion are unknown, but some scientists believe they may be serious or even catastrophic. With nuclear power there are radiation hazards and the problem of disposal of radioactive waste material. Solar power avoids all these problems.

Intelligent consideration of the effects of solar radiation in the design of buildings can also reduce energy consumption in heating and cooling. An example is the use of movable shades on the outsides of buildings, permitting the sun to enter windows in winter but keeping it out in summer. Architects have unfortunately tended in the past to ignore such solutions, but awareness of the need to design buildings with energy costs in mind is gradually growing.

SUMMARY

KEY TERMS

conduction

thermal conductivity

heat current

temperature gradient

thermal resistance

convection

convection coefficient

radiation

radiant energy

emissivity

Stefan-Boltzmann constant

Stefan-Boltzmann law

blackbody

The three mechanisms of heat transfer are conduction, convection, and radiation. Conduction is transfer of energy of molecular motion within materials without bulk motion of the materials. Convection involves mass motion from one region to another, and radiation is energy transfer through electromagnetic waves.

The heat current H for conduction depends on the area A through which the heat flows, the length L of the heat path, the temperature difference $(T_2 - T_1)$, and the thermal conductivity k of the material, according to

$$H = \frac{kA(T_2 - T_1)}{L}. \tag{16–1}$$

Good thermal conductors have large values of k, and poor conductors have small k's. The temperature change per unit length along the heat-flow path is called the temperature gradient.

Heat transfer by convection is a complex process; an approximate relation used for practical calculations of heat current H is

$$H = hA \, \Delta T, \tag{16–5}$$

where A is the surface area, ΔT is the temperature difference between the surface and the main body of fluid, and h is the convection coefficient, an empirical number that depends on the geometry and on ΔT.

The heat current H due to radiation is given by

$$H = Ae\sigma T^4, \tag{16–6}$$

where A is the surface area, e the emissivity of the surface (a pure number between 0 and 1), T the absolute temperature, and σ a fundamental constant called the Stefan-Boltzmann constant. When a body at temperature T_1 is surrounded by material at temperature T_2, the net heat current from the body to its surroundings is

$$H_{net} = Ae\sigma(T_1{}^4 - T_2{}^4). \qquad (16-7)$$

QUESTIONS

16-1 Why does a marble floor feel cooler to the feet than a carpet at the same temperature?

16-2 A beaker of boiling water can be picked up with bare fingers without burning if it is grasped only at the thin turned-out rim at the top. Why?

16-3 Old-time kitchen lore suggests that things cook better (evenly and without burning) in heavy cast-iron pots. What desirable characteristics would such pots have?

16-4 If you have wet hands and pick up a piece of metal that is below freezing, you may stick to it. This doesn't happen with wood. Why not?

16-5 A cold block of metal feels colder than a block of wood at the same temperature, but a hot block of metal feels hotter than a block of wood at the same temperature. Is there any temperature at which they feel equally hot or cold?

16-6 A person pours a cup of hot coffee, intending to drink it five minutes later. To keep it as hot as possible, should he put cream in it now or wait until just before he drinks it?

16-7 In late afternoon when the sun is about to set, swarms of insects such as mosquitoes are sometimes seen in vertical "plumes" above a tree or above the hood of a car with a hot engine. Why?

16-8 Mountaineers caught in a storm sometimes survive by digging a cave in snow or ice. How in the world can you keep warm in an ice cave?

16-9 Old-time pioneers sometimes kept themselves warm on cold winter nights by heating bricks in the fire, wrapping them in thick layers of cloth, and taking them to bed. Discuss the roles of conduction, convection, and radiation in this technique.

16-10 It is well known that a potato bakes faster if a large nail is stuck through it. Why? Is aluminum better than steel? There is also a gadget on the market to hasten roasting of meat, consisting of a hollow metal tube containing a wick and some water; this is claimed to be much better than a solid metal rod. How does this work?

16-11 The temperature in outer space, far from any solid body, is believed to be about 3 K. If you leave a spaceship and go for a space walk, do you get cold very quickly?

16-12 Aluminum foil used for food cooking and storage sometimes has one shiny surface and one dull surface. When food is wrapped for baking, should the shiny side be in or out? Which side should be out when it is to be frozen?

16-13 Some cooks claim that the bottom crust of a pie gets browner if the pie pan is glass then if it is metal. Why should this be?

16-14 Glider pilots in the midwest know that thermal updrafts are likely to occur above freshly plowed fields. Why?

16-15 We're lucky the earth isn't in thermal equilibrium with the sun. But why isn't it?

16-16 On a chilly fall morning, the grass is covered with frost but the concrete sidewalk isn't. Why?

16-17 Why can a microwave oven cook massive objects such as potatoes and beef roasts much faster than a conventional oven? What disadvantages are associated with the reasons for this greater speed?

16-18 Some folks claim that ice cubes freeze faster if the trays are filled with hot water because hot water cools off faster than cold water. What do you think?

EXERCISES

Section 16-1 Conduction

16-1 A slab of a thermal insulator is 100 cm^2 in cross section and 2 cm thick. Its thermal conductivity is 2.4×10^{-4} cal·s^{-1}·cm^{-1}·(C°)$^{-1}$. If the temperature difference between opposite faces is 100°C, how much heat flows through the slab in one day?

16-2 The oven in a kitchen range has a total wall area of 1.5 m^2 and is insulated with a layer of fiberglass 5 cm thick.

The inside is to be kept at a temperature of 200°C, and the outside is at room temperature, 20°C.

a) What is the heat current through the insulation, assuming it may be treated as a flat slab of area 1.5 m^2?

b) What electric-power input to the heating element is required to maintain this temperature?

16-3 A carpenter builds an outer house wall with a layer of wood 3 cm thick on the outside, and a layer of styrofoam insulation 3 cm thick at the inside surface of the wall. If the wood has $k = 9.5 \times 10^{-5}\,\text{cal·s}^{-1}\text{·cm}^{-1}\text{·(C°)}^{-1}$, the interior temperature is 20°C, and the exterior temperature is −10°C,

a) what is the temperature at the plane where the wood meets the styrofoam?

b) what is the rate of heat flow, per square meter, through this wall?

16–4 One end of an insulated metal rod is maintained at 100°C and the other one is placed in an ice-water mixture. The bar has length 50 cm and cross-sectional area 0.8 cm². It is observed that the heat transported by the rod melts 2 g of ice in 5 min. Calculate the thermal conductivity k of the metal. Express your answer in $\text{J·s}^{-1}\text{·m}^{-1}\text{·(C°)}^{-1}$.

16–5 The ceiling of a room has an area of 200 ft². The ceiling contains insulation such that its R-value is 30. If the room is maintained at 72°F and the attic has a temperature of 105°F, how many Btu of heat flow through the ceiling into the room in 6 hr? Also express your answer in joules.

16–6 A long rod, insulated to prevent heat loss, has one end immersed in boiling water (at atmospheric pressure) and the other end in an ice-water mixture. The rod consists of 1.00 m of copper (one end in steam) and a length L_2 of steel (one end in ice). Both rods are of cross-sectional area 5 cm². The temperature of the copper-steel junction is 60°C, after a steady state has been set up.

a) How much heat per second flows from the steam bath to the ice-water mixture?

b) How long is L_2?

16–7 Suppose that the rod in Fig. 16–1 is of copper, of length 10 cm, and of cross-sectional area 1 cm². Let $T_2 = 100°C$ and $T_1 = 0°C$.

a) What is the final steady-state temperature gradient along the rod?

b) What is the heat current in the rod, in the final steady state?

c) What is the final steady-state temperature at a point in the rod 2 cm from its left end?

16–8 A boiler with a steel bottom 1.5 cm thick rests on a hot stove. The area of the bottom of the boiler is 0.15 m². The water inside the boiler is at 100°C, and 0.75 kg are evaporated every 5 min. Find the temperature of the lower surface of the boiler, which is in contact with the stove.

Section 16–2 Convection

16–9

a) What would be the difference in height between the columns in the U-tube in Fig. 16–3 if the liquid is water and the left arm is 1 m high at 4°C while the other is at 75°C? (Recall that the density of water at various temperatures is given in Table 14–3.)

b) What is the difference between the pressures at the foot of two columns of water, each 10 m high, if the temperature of one is 4°C and that of the other is 75°C?

16–10 A flat plate is maintained at constant temperature of 100°C, and the air on both sides is at atmospheric pressure and at 20°C. How much heat is lost by natural convection from 1 m² of the plate (both sides) in 1 hr if

a) the plate is vertical?

b) the plate is horizontal?

16–11 A vertical steam pipe of outside diameter 7.5 cm and height 4 m has its outer surface at the constant temperature of 95°C. The surrounding air is at atmospheric pressure and at 20°C. How much heat is delivered to the air by natural convection in 1 hr?

Section 16–3 Radiation

16–12 What is the rate of energy radiation per unit area of a blackbody at a temperature of

a) 300 K?   b) 3000 K?

16–13 In Example 16–7, what is the net rate of heat loss by radiation if the temperature of the surroundings is 0°C?

16–14 The emissivity of tungsten is approximately 0.35. A tungsten sphere 1 cm in radius is suspended within a large evacuated enclosure whose walls are at 300 K. What power input is required to maintain the sphere at a temperature of 3000 K if heat conduction along the supports is neglected?

16–15 The operating temperature of a tungsten filament in an incandescent lamp is 2450 K, and its emissivity is 0.30. Find the surface area of the filament of a 25-W lamp.

Section 16–4 Solar Energy and Resource Conservation

16–16 A well-insulated house of moderate size in a temperate climate may require a maximum heat input rate of the order of 20 kW. If this heat is to be supplied by a solar collector with an average (night and day) energy input of 300 W·m⁻² and a collection efficiency of 60%, what area of solar collector is required?

16–17 In Exercise 16–16, suppose a heat reservoir is required that can store enough energy for a week of cloudy days, by means of a large tank of water that is heated to 80°C by solar energy and cooled to 40°C as it circulates through the house. What volume of water is required?

16–18 A solar water heater for domestic hot-water supply uses solar collecting panels with collection efficiency of 50% in a location where the average solar-energy input is 200 W·m⁻². If the water comes into the house at 15°C and is to be heated to 65°C, what volume of water can be heated per hour if the collector area is 20 m²?

16–19 The average rate of energy consumption of all forms in the United States is of the order of 3×10^{11} W. Suppose this energy were to be supplied by solar collectors in the desert. Assume an average solar-energy input of 500 W·m⁻² for 12 hr each day, and a collection efficiency of 50%. What total collector area would be required? Express your result in square kilometers and square miles.

PROBLEMS

16–20 One experimental method of measuring the thermal conductivity of an insulating material is to construct a box of the material and measure the power input to an electric heater, inside the box, that maintains the interior at a measured temperature above the outside surface. Suppose that in such an apparatus a power input of 120 W is required to keep the interior surface of the box 65 C° (about 120 F°) above the temperature of the outer surface. The total area of the box is 2.32 m², and the wall thickness is 3.8 cm. Find the thermal conductivity of the material, in SI units.

16–21 A carpenter builds a solid wood door with dimensions 2.0 m × 0.8 m × 4 cm. Its thermal conductivity is

$$k = 0.04 \text{ J·s}^{-1}\text{·m}^{-1}\text{·(C°)}^{-1}.$$

The inside air temperature is 20°C, and the outside air temperature is −10°C.

a) What is the rate of heat flow through the door, assuming the surface temperatures are those of the surrounding air?

b) By what factor is the heat flow increased if a window 0.5 m square is inserted, assuming the glass is 0.5 cm thick and the surface temperatures are again those of the surrounding air?

16–22

a) A wood ceiling with thermal resistance R_1 is covered with a layer of insulation with thermal resistance R_2. Prove that the effective thermal resistance of the combination is given by $R = R_1 + R_2$.

b) Two metal bars of the same length L and cross-sectional area A are laid side by side, in good thermal contact, as shown in Fig. 16–4. If the bars have thermal conductivities k_1 and k_2, respectively, calculate the total heat current carried by the bars and from this deduce the effective thermal conductivity of the two combined bars. Does your result make sense in the special case where $k_1 = k_2$?

FIGURE 16–4

16–23 An engineer designs a camping icebox having a wall area of 2 m² and a thickness of 5 cm, using insulating material having a thermal conductivity of 0.05 J·s⁻¹·m⁻¹·(C°)⁻¹. The outside temperature is 20°C, and the inside of the box is to be maintained at 5°C by ice. The melted ice drains from the box at a temperature of 5°C. If ice costs 25 cents per kilogram, what will it cost to run the icebox for 1 hr?

16–24 A layer of ice 10 cm thick has formed on the surface of a lake, where the temperature is −10°C. The water under the slab is at 0°C. At what rate does the thickness of

the slab increase if the heat of fusion of water freezing on the underside of the slab is conducted through the slab? Express your answer in units of cm·hr⁻¹.

16–25 Three rods of equal length and cross section are placed end to end, one copper, one steel, one aluminum, in that order. The free end of the copper rod is in contact with ice water at 0°C, and the free end of the aluminum rod is in contact with steam at 100°C. If no heat transfer occurs at the sides, find the temperature at each junction. (*Note:* The concept of thermal resistance R can be used here to reduce the algebra. See Problem 16–22.)

16–26 Rods of copper, brass, and steel are welded together to form a Y-shaped figure. The cross-sectional area of each rod is 2 cm². The free end of the copper rod is maintained at 100°C, and the free ends of the brass and steel rods at 0°C. Assume there is no heat loss from the surfaces of the rods. The lengths of the rods are: copper, 46 cm; brass, 13 cm; steel, 12 cm.

a) What is the temperature of the junction point?

b) What is the heat current in each of the three rods?

16–27 A compound bar 2 m long is constructed of a solid steel core 1 cm in diameter surrounded by a copper casing whose outside diameter is 2 cm. The outer surface of the bar is thermally insulated, and one end is maintained at 100°C, the other at 0°C.

a) Find the total heat current in the bar.

b) What fraction of this total is carried by each material?

16–28 A rod is initially at a uniform temperature of 0°C throughout. One end is kept at 0°C, and the other is brought into contact with a steam bath at 100°C. The surface of the rod is insulated so that heat can flow only lengthwise along the rod. The cross-sectional area of the rod is 2 cm², its length is 100 cm, its thermal conductivity is 335 J·s⁻¹·m⁻¹·(C°)⁻¹, its density is $1.0 \times 10^4 \text{kg·m}^{-3}$, and its specific heat capacity is 419 J·kg⁻¹·(C°)⁻¹. Consider a short cylindrical element of the rod 1 cm in length.

a) If the temperature gradient at the cooler end of this element is 1000C°·m⁻¹, how many joules of heat energy flow across this end per second?

b) If the average temperature of the element is increasing at the rate of 2 C°·s⁻¹, what is the temperature gradient at the other end of the element?

16–29 An electrician installs an electric transformer in a cylindrical tank 0.60 m in diameter and 1 m high, with a flat top and bottom. If the tank transfers heat to the air only by natural convection, and electrical losses are to be dissipated at the rate of 1 kW, how many degrees will the tank surface rise above room temperature?

16–30 The rate at which radiant energy reaches the earth from the sun is about 1.4 kW·m⁻². The distance from earth to sun is about 1.5×10^{11} m, and the radius of the sun is about 7.0×10^8 m.

a) What is the rate of radiation of energy, per unit area, from the sun's surface?

b) If the sun radiates as an ideal blackbody, what is the temperature of its surface?

16–31 A physicist uses a cylindrical metal can 0.10 m high and 0.05 m in diameter to store liquid helium at 4 K, where its heat of vaporization is 2.0×10^4 J·kg^{-1}. Completely surrounding the helium can are walls maintained at the temperature of liquid nitrogen, 80 K, the intervening space being evacuated. How much helium is lost per hour? Assume the emissivity of the helium can to be 0.2.

CHALLENGE PROBLEMS

16–32 A solid cylindrical copper rod 0.10 m long has one end maintained at a temperature of 20.00 K. The other end is blackened and exposed to thermal radiation from a body at 300 K, no energy being lost or gained elsewhere. When equilibrium is reached, what is the temperature of the blackened end? (*Hint:* Since copper is a very good conductor of heat at low temperature, $k = 1670$ J·s^{-1}·m^{-1}·(C°)$^{-1}$, the temperature of the blackened end is slightly greater than 20 K.)

16–33

a) A spherical shell has inner and outer radii a and b, respectively, and the temperatures at the inner and outer surfaces are T_2 and T_1. The thermal conductivity of the material of which the shell is made is k. Derive an equation for the total heat current through the shell.

b) For the shell in part (a), derive an equation for the temperature variation within the shell. That is, calculate T as a function of r, the distance from the center of the shell.

c) A hollow cylinder has length L, inner radius a, and outer radius b, and the temperatures at the inner and outer surfaces are T_2 and T_1. (The cylinder could represent an insulated hot-water pipe, for example.) The thermal conductivity of the material of which the cylinder is made is k. Derive an equation for the total heat current through the walls of the cylinder.

d) For the cylinder of part (c), derive an equation for the temperature variation inside the cylinder walls.

e) For the spherical shell of part (a) and the hollow cylinder of part (c), show that the equation for the total heat current in each case reduces to Eq. (16–1) for linear heat flow when the shell or cylinder is very thin.

16–34 A steam pipe of radius 2 cm, carrying steam at 120°C, is surrounded by a cylindrical jacket of cork with inner and outer radii 2 cm and 4 cm, and this in turn is surrounded by a cylindrical jacket of styrofoam having inner and outer radii 4 cm and 6 cm. The outer surface of the styrofoam is in contact with air at 20°C.

a) What is the temperature at a radius of 4 cm, where the two insulating layers meet?

b) What is the total rate of transfer of heat out of a 2 m length of pipe?

(*Hint:* Use the expression derived in part (c) of Problem 16–33.)

16–35 Suppose that both ends of the rod in Fig. 16–1 are kept at a temperature of 0° C, and that the initial temperature distribution along the rod is given by $T = 100°C \sin \pi x / L$, where T is in °C. Let the rod be copper, of length $L = 0.10$ m and of cross section 1 cm^2.

a) Show the initial temperature distribution in a diagram.

b) What is the final temperature distribution after a very long time has elapsed?

c) Sketch curves that you think would represent the temperature distribution at intermediate times.

d) What is the initial temperature gradient at the ends of the rod?

e) What is the initial heat current from the ends of the rod into the bodies making contact with its ends?

f) What is the initial heat current at the center of the rod? Explain. What is the heat current at this point at any later time?

g) What is the value of $k/\rho c$ for copper, and in what unit is it expressed? Here k is the thermal conductivity, ρ is the density, and c is the specific heat capacity. (This quantity is called the *thermal diffusivity*.)

h) What is the initial time rate of change of temperature at the center of the rod?

i) How long a time would be required for the rod to reach its final temperature, if the temperature continued to decrease at this rate? (This time can be described as the *relaxation time* of the rod.)

j) From the graphs in part (c), would you expect the rate of change of temperature at the midpoint to remain constant, to increase, or to decrease?

k) What is the initial rate of change of temperature at a point in the rod 2.5 cm from its left end?

17
THERMAL PROPERTIES OF MATTER

IN PREVIOUS CHAPTERS WE HAVE DISCUSSED SEVERAL PROPERTIES OF materials, including elasticity, thermal expansion, specific heat capacity, and thermal conductivity. All these properties describe how a material behaves when some aspect of its environment, such as temperature or pressure, changes. In describing the quantity and condition of the material itself, we have used several physical quantities, including total mass m, number of moles n, pressure p, volume V, and temperature T. We now want to consider a more general formulation of the relationships among these quantities, using the concept of the *equation of state* of a material. We also look in greater detail at the conditions that determine the *phase* of a material (e.g., solid, liquid, or gas) and the conditions under which *phase transitions* occur. This discussion provides some of the language needed for our study of *thermodynamics*, the relation of heat to other forms of energy, in the following chapters.

17–1 EQUATIONS OF STATE

Physical quantities such as pressure, volume, temperature, and amount of substance describe the conditions in which a particular specimen of material exists. They determine the *state* of the material, and they are called *state variables* or **state coordinates.** For some physical phenomena we need additional state variables, such as magnetization or electric polarization, but those named above are sufficient for our present discussion.

State variables are interrelated; ordinarily we cannot change one without also causing a change in one or more of the others. For example, when we heat a gas such as steam in a closed container of constant volume, the pressure *must* increase as the temperature increases. Indeed, if the temperature becomes high enough, the pressure may cause the container, such as a boiler, to explode. The volume V of a substance is determined by its pressure p, temperature T, and amount of substance, m or n.

In some cases the relationship among p, V, and T is simple enough that it can be expressed in the form of an equation, which we call the **equation of**

State coordinates: describing the conditions in which a material exists

The equation of state is an expression of the interrelation of the state coordinates.

state. We can construct a simple though approximate equation of state for a solid material. Recall that the temperature coefficient of volume expansion β is the fractional volume change per unit temperature change, and that the isothermal compressibility k is the fractional volume change per unit pressure change. Thus if a specimen of material has volume V_0 when the pressure is p_0 and the temperature T_0, the volume V at pressure p and temperature T is given approximately by

$$V = V_0[1 + \beta(T - T_0) - k(p - p_0)],$$

provided the temperature and pressure changes are not too great.

Another simple equation of state is that of an *ideal gas*, to be discussed in Section 17–2. In many cases, however, the relationship among p, V, and T is so complex that it can be described only by use of graphs or tables of numerical values. Even then, the relation among the variables still exists, and we call it an equation of state even when it is too complex to be expressed in actual equation form.

For the present we consider only **equilibrium states,** that is, states in which the system is in mechanical and thermal equilibrium. For such states the temperature and pressure are uniform throughout the system. We note that when a system changes from one state to another, nonequilibrium states must occur during the transition. For example, when a material expands or is compressed, there must be mass in motion; this change requires acceleration and nonuniform pressure. After the system has returned to mechanical equilibrium, the pressure is again the same throughout the system. Similarly, when heat conduction occurs within a system, different regions must have different temperatures. Only when thermal equilibrium is once again attained is the temperature again the same throughout.

> An equation of state has meaning only for systems that are in mechanical and thermal equilibrium.

17–2 IDEAL GASES

> The quantity of material in a system can be described by either the mass or the number of moles.

Gases at low pressures have particularly simple equations of state. We can study the behavior of a gas by means of a cylinder with a movable piston, equipped with a pressure gauge and a thermometer, as shown in Fig. 17–1.

17–1 A hypothetical setup for studying the behavior of gases. The pressure, volume, temperature, and number of moles of gas can be controlled and measured.

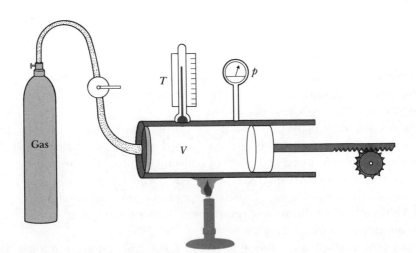

We can vary the pressure, volume, and temperature, and pump any desired mass of any gas into the cylinder, in order to explore the relationships between pressure, volume, temperature, and quantity of substance. It is often convenient to measure the amount of gas in terms of the number of *moles n*, rather than the mass *m*. Since the molecular mass *M* is the mass per mole, the total mass *m* is given by

$$m = nM. \qquad (17\text{–}1)$$

That is, the total mass *m* of material is the mass of one mole *M* multiplied by the number of moles *n*.

From measurements of the pressure, volume, temperature, and number of moles of various gases, several conclusions emerge. First, the volume *V* is proportional to the number of moles *n*. If we double the number of moles, keeping pressure and temperature constant, the volume doubles. Second, the volume varies *inversely* with pressure *p*; if we double the pressure, holding the temperature *T* and quantity of material *n* constant, the gas is compressed to one-half its initial volume. Thus

$$pV = \text{constant} \qquad (T \text{ constant}), \qquad (17\text{–}2)$$

or, if p_1 and V_1 are the pressure and volume in the initial state and p_2 and V_2 those of the final state, then

$$p_1 V_1 = p_2 V_2 \qquad (T \text{ constant}). \qquad (17\text{–}3)$$

This relation is called Boyle's law, after Robert Boyle (1627–1691), a contemporary of Newton.

Third, the pressure is proportional to the absolute temperature. If we double the absolute temperature, keeping the volume and quantity of material constant, the pressure doubles. Thus

$$p = (\text{constant})T \qquad (V \text{ constant}), \qquad (17\text{–}4)$$

or, if p_1 and T_1 are the pressure and absolute temperature in the initial state and p_2 and T_2 those of the final state, then

$$\frac{p_1}{T_1} = \frac{p_2}{T_2}. \qquad (17\text{–}5)$$

This is called Charles's law, after Jacques Charles (1746–1823).

These three relationships can be combined neatly into a single equation of state:

$$pV = nRT. \qquad (17\text{–}6)$$

The constant of proportionality *R* might be expected to have different values for different gases, but instead it turns out to have the same value for *all* gases, at least at sufficiently high temperature and low pressure. This quantity is called the **ideal gas constant.** The numerical value of *R* depends, of course, on the units in which *p*, *V*, *n*, and *T* are expressed. In SI units, where the unit of *p* is the Pa or N·m^{-2} and the unit of *V* is m^3, the numerical value of *R* is found to be

$$R = 8.314 \ (\text{N·m}^{-2})\text{·m}^3\text{·mol}^{-1}\text{·K}^{-1} = 8.314 \ \text{J·mol}^{-1}\text{·K}^{-1}.$$

A launch balloon used to lift a three-ton optical telescope. This balloon is used to lift the main balloon to an altitude of about 3000 m, where it inflates. The inflated main balloon system is 200 m tall, holds 150,000 m^3 of hydrogen gas, and lifts the telescope to a final altitude of 24,000 m. The behavior of the hydrogen gas in the balloon is very close to that of an ideal gas. (Courtesy of NASA.)

Charles's law: At constant volume, pressure is directly proportional to absolute temperature.

The ideal-gas equation: a simple equation of state for an idealized system

In cgs units, where the unit of p is dyn·cm^{-2} and the unit of V is cm^3,

$$R = 8.314 \times 10^7 \ (\text{dyn·cm}^{-2}) \cdot \text{cm}^3 \cdot \text{mol}^{-1} \cdot \text{K}^{-1}$$
$$= 8.314 \times 10^7 \ \text{erg·mol}^{-1} \cdot \text{K}^{-1}.$$

Note that the units of pressure times volume are the same as units of energy, so in *all* systems of units, R has units of energy per mole, per unit of absolute temperature. In terms of calories,

$$R = 1.99 \ \text{cal·mol}^{-1} \cdot \text{K}^{-1}.$$

In chemical calculations, volumes are commonly expressed in liters (L), pressures in atmospheres, and temperatures in kelvins. In this system,

$$R = 0.08207 \ \text{L·atm·mol}^{-1} \cdot \text{K}^{-1}.$$

We now define an **ideal gas** as one for which Eq. (17–6) holds precisely for *all* pressures and temperatures. As the term suggests, an ideal gas is an idealized model, which represents the behavior of gases very well in some circumstances, less well in others. Generally, gas behavior approximates the ideal-gas model most closely at very low pressures, when the gas molecules are far apart. However, the deviations are not very great at moderate pressures (e.g., 1 atm) and at temperatures not too near those at which the gas liquefies.

For a *fixed mass* (or fixed number of moles) of an ideal gas, the product nR is constant, and hence pV/T is constant also. Thus if the subscripts 1 and 2 refer to two states of the same mass of a gas, but at different pressures, volumes, and temperatures,

$$\frac{p_1 V_1}{T_1} = \frac{p_2 V_2}{T_2} \tag{17–7}$$
$$= \text{constant}.$$

If the temperatures T_1 and T_2 are the same, then

$$p_1 V_1 = p_2 V_2 \tag{17–8}$$
$$= \text{constant},$$

which is again Boyle's law.

Now you may recall that in Chapter 14 our most satisfactory definition of temperature was stated in terms of the constant-volume gas thermometer, for which pressure was proportional to absolute temperature. Hence the proportionality of pressure to absolute temperature, as expressed by Charles's law, Eqs. (17–4) and (17–5), is not at all surprising; indeed, it is inevitable!

You may have the feeling that you have been swindled into accepting as natural law something that is actually only a matter of definition. To this charge we plead guilty, but with mitigating circumstances. Toward the end of Chapter 19 we will arrive at a definition of temperature that really *is* independent of the properties of any particular material (including ideal gases). The temperature scale based on the gas thermometer will then emerge as a very close approximation of this truly absolute scale. Until then, consider Charles's law and the ideal-gas equation of state as based on this genuinely material-independent temperature scale, even though we have not actually defined it yet!

PROBLEM-SOLVING STRATEGY: *Ideal gases*

1. In some problems you will be concerned with only one state of the system; some of the quantities in Eq. (17–6) will be known, some unknown. Make a list of what you know and what you have to find: $p = 1.0 \times 10^6$ Pa, $V = 4$ m^3, $T = ?$, $n = 2$ mol. Or something like that.

2. In other problems there will be two different states of the same quantity of gas. Decide which is state 1, which state 2, and make a list of the quantities for each: p_1, p_2, V_1, V_2, T_1, T_2, and so on. You may be able to use Eq. (17–7) if all but one of the quantities in that equation are known; otherwise you may first have to use Eq. (17–6), for example, if p_1, V_1, and n are given.

3. As always, be sure to use a consistent set of units. First decide which value of the gas constant R you are going to use, and then convert the units of the other quantities accordingly. You may have to convert

atmospheres to pascals or liters to cubic meters. Sometimes the problem statement will make one system of units clearly more convenient than others. Decide on your system, and stick to it.

4. Don't forget that T must always be an *absolute* temperature. If you are given temperatures in °C, be sure to add 273 to convert to Kelvin.

5. You may sometimes have to convert between mass m and number of moles n. The relationship is $m = Mn$, where M is the molecular mass. Here's a tricky point: If you replace n in Eq. (17–6) by (m/M), you *must* use the same mass units for m and M. So if M is in grams per mole (the usual units for molecular mass), then m must also be in grams. If you want to use m in kg, then you must convert M to kg·mol^{-1}. For example, the molecular mass of oxygen is 32 g·mol^{-1} or 32×10^{-3} kg·mol^{-1}. Be careful!

EXAMPLE 17–1 The condition called **standard temperature and pressure (STP)** for a gas is defined to be a temperature of 0°C = 273 K and a pressure of 1 atm = 1.013×10^5 Pa. Find the volume of one mole of any gas at STP.

Standard temperature and pressure: a convenient reference condition

SOLUTION From Eq. (17–6), using R in J·mol^{-1}·K^{-1},

$$V = \frac{nRT}{p} = \frac{(1 \text{ mol})(8.314 \text{ J·mol}^{-1}\cdot\text{K}^{-1})(273 \text{ K})}{1.01 \times 10^5 \text{ Pa}}$$
$$= 0.0224 \text{ m}^3$$
$$= 22,400 \text{ cm}^3$$
$$= 22.4 \text{ L.}$$

Alternatively, using R in L·atm·mol^{-1}·K^{-1},

$$V = \frac{nRT}{p} = \frac{(1 \text{ mol})(0.0821 \text{ L·atm·mol}^{-1}\cdot\text{K}^{-1})(273 \text{ K})}{1.00 \text{ atm}}$$
$$= 22.4 \text{ L.}$$

EXAMPLE 17–2 A tank attached to an air compressor contains 20.0 L of air at a temperature of 30°C and gauge pressure of 4.00×10^5 Pa (about 60 lb·in^{-2}). What is the mass of air, and what volume would it occupy at normal atmospheric pressure and 0°C? Air is a mixture of gases, consisting of about 78% nitrogen and 21% oxygen, with small percentages of other gases. The *average* molecular mass is $M = 28.8$ g·mol^{-1}.

SOLUTION Gauge pressure is the amount in excess of atmospheric pressure (1.01×10^5 Pa), so the total pressure is

$$p = 4.00 \times 10^5 \text{ Pa} + 1.01 \times 10^5 \text{ Pa} = 5.01 \times 10^5 \text{ Pa.}$$

The temperatures must be expressed in kelvins: 30°C = 303 K. We express the volume in cubic meters: 20 L = 20×10^{-3} m³. To find the mass of gas, we first use Eq. (17–6) to find the number of moles, and then use Eq. (17–1):

$$n = \frac{pV}{RT} = \frac{(5.01 \times 10^5 \text{ Pa})(20 \times 10^{-3} \text{ m}^3)}{(8.314 \text{ J·mol}^{-1}\text{·K}^{-1})(303 \text{ K})} = 3.98 \text{ mol};$$

$$m = nM = (3.98 \text{ mol})(28.8 \text{ g·mol}^{-1}) = 115 \text{ g} = 0.115 \text{ kg}.$$

The volume at 1 atm and 0°C (273 K) is

$$V = \frac{nRT}{p} = \frac{(3.98 \text{ mol})(8.314 \text{ J·mol}^{-1}\text{·K}^{-1})(273 \text{ K})}{1.01 \times 10^5 \text{ Pa}}$$

$$= 89.4 \times 10^{-3} \text{ m}^3 = 89.4 \text{ L}.$$

How does atmospheric pressure vary with elevation?

EXAMPLE 17–3 Find the variation of atmospheric pressure with elevation in the earth's atmosphere, assuming the temperature to be uniform throughout (which is not actually the case).

SOLUTION We begin with the hydrostatic relation of Eq. (13–3), $dp/dy = -\rho g$. The density ρ is not constant but varies with pressure. Replacing n in the ideal-gas equation by m/M, we obtain

$$pV = \frac{m}{M}RT \qquad \text{and} \qquad \rho = \frac{m}{V} = \frac{pM}{RT}.$$

We substitute this expression for ρ into Eq. (13–3), separate variables, and integrate, letting p_1 be the pressure at elevation y_1 and p_2 that at y_2:

$$\frac{dp}{dy} = -\frac{pMg}{RT}, \qquad \int_{p_1}^{p_2} \frac{dp}{p} = -\frac{Mg}{RT}\int_{y_1}^{y_2} dy, \qquad \ln\frac{p_2}{p_1} = -\frac{Mg}{RT}(y_2 - y_1).$$

If the pressure at $y = 0$ is p_0, then the pressure p at any y is given by

$$p = p_0 e^{-Mgy/RT} \tag{17–9}$$

At the summit of Mount Everest, where $y = 8882$ m,

$$\frac{Mgy}{RT} = \frac{(28.8 \times 10^{-3} \text{kg·mol}^{-1})(9.8 \text{ m·s}^{-2})(8882 \text{ m})}{(8.314 \text{ J·mol}^{-1}\text{·K}^{-1})(273 \text{ K})} = 1.10;$$

$$p = (1.01 \times 10^5 \text{ Pa})e^{-1.10} = 0.336 \times 10^5 \text{ Pa} = 0.333 \text{ atm}.$$

We have used a constant temperature of 0°C = 273 K, which is not completely realistic, but this example shows why mountaineers have so much difficulty breathing on Mount Everest; air pressure is only one-third its sea-level value. For similar reasons, jet airplanes, which typically fly at altitudes of 8000 to 10,000 m, must have pressurized cabins for passenger comfort and health.

17–3 pV-DIAGRAMS

Representing the behavior of a material by use of pressure–volume graphs

For a given quantity of a material, the equation of state is a relation among the three state coordinates p, V, and T. We could in principle represent this relationship graphically as a *surface* in a three-dimensional space with coordinates p, V, and T. This representation sometimes helps in grasping the overall behavior of the material, but usually ordinary two-dimensional graphs are more convenient. One of the most useful of these is a set of graphs of pressure as a

function of volume for each of several constant temperatures. Such a diagram is called a **pV-diagram.** Each curve, representing a particular constant temperature, is called an **isotherm,** or a *pV-isotherm.*

A *pV*-diagram for an ideal gas is shown in Fig. 17–2. Each curve is an isotherm, showing the relation of pressure to volume at a certain constant temperature. We can think of this as a graphical representation of Boyle's law, Eq. (17–2), for each of several temperatures, and it is also a graphical representation of the ideal-gas equation of state. From these curves we can read off the volume V corresponding to any given pressure p and temperature T.

Figure 17–3 shows a *pV*-diagram for a material that *does not* obey the ideal-gas equation. At a high enough temperature, say T_4, the behavior is approximately that of an ideal gas, but at lower temperatures the behavior is rather different. At temperatures below a critical temperature T_c, the isotherms develop flat portions. In these portions, as the material is compressed at constant temperature, the pressure does not increase at all. As the volume decreases and the end of the flat portion of an isotherm is reached, the pressure increases very rapidly with further decrease of volume.

Visual inspection of the material shows that in the flat portion of the isotherm the material is *condensing* from the vapor to the liquid phase; this flat portion represents a condition of phase equilibrium. As compression continues, more and more material goes from vapor to liquid, but the pressure does not change. During this process, heat must be removed in order to keep the temperature constant; this is the heat of vaporization, discussed in Section 15–6.

Thus when a real (nonideal) gas is compressed at a constant temperature T_2, it remains in the gaseous phase until a in the diagram is reached. At point a it begins to liquefy, and with further decrease in volume more and more material liquefies, with both pressure and temperature remaining constant. At point b all the material is in the liquid state; any further compression is accompanied by a rapid rise of pressure, because in general liquids are much less compressible than gases. At a lower constant temperature T_1 similar behavior occurs, but the onset of condensation occurs at lower pressure and greater volume than at the constant temperature T_2.

The shaded area in Fig. 17–3 is the region of liquid-vapor phase equilibrium for this material. At temperatures greater than T_c, no phase transition occurs as the material is compressed. The temperature T_c is called the *critical temperature* for this material. We will explore its significance in greater detail in the next section.

The next two chapters provide many more opportunities to use *pV*-diagrams in connection with energy calculations. As we will see, the area under a *pV*-curve, whether or not it is an isotherm, represents the *work* done by or on the system during a volume change. This work, in turn, is directly related to heat transfer and changes in the *internal energy* of the system, which will be defined in Chapter 18.

17–4 PHASE DIAGRAMS, TRIPLE POINT, CRITICAL POINT

We have discussed *pV*-diagrams and their usefulness in describing the behavior of materials, including the equation of state. Another useful diagram, especially in reference to phase transitions, is a graph with axes p and T, called a **phase diagram.**

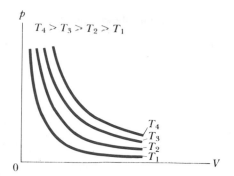

17–2 A *pV*-diagram for an ideal gas. Each curve represents the relation between pressure and volume at the indicated constant temperature. This is a graphical representation of Boyle's law; for each curve the product pV is constant; the value of that constant is different from different temperatures, increasing with increasing temperature. These curves are called *pV*-isotherms for the substance.

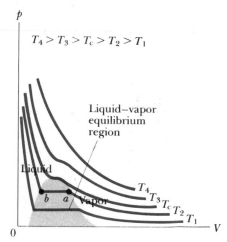

17–3 A *pV*-diagram for a nonideal gas, showing *pV*-isotherms for temperatures above and below the critical temperature T_c. The liquid–vapor equilibrium region is shown as a shaded area. At still lower temperature the material might undergo phase transitions from liquid to solid or from gas to solid; these are not shown on this diagram.

Each phase of a material can exist only in certain regions of temperature and pressure.

The phase diagram shows what phase exists for any given temperature and pressure.

In the discussion of phases of matter in Section 15–5 we found that each phase is stable only in certain ranges of temperature and pressure. A transition from one phase to another takes place ordinarily under conditions of **phase equilibrium** between the two phases, and for a given pressure this occurs at only one specific temperature. A phase diagram shows what phase occurs for each possible combination of temperature and pressure. A typical example is shown in Fig. 17–4. At each point on the diagram, only a single phase can exist, except for points on the lines, where two phases can coexist in phase equilibrium. Thus any point on a line represents a set of conditions under which phase changes can occur.

Curves on the phase diagram show conditions of phase equilibrium.

If we increase the temperature of a substance at constant pressure, it passes through a sequence of states represented by points on the broken horizontal line labeled (a). The melting and boiling temperatures at this pressure are the temperatures at which the broken line crosses the fusion and vaporization curves, respectively. Similarly, if we compress a material by increasing the pressure while we hold the temperature constant, as represented by the vertical broken line labeled (b), the material passes from vapor to liquid and then to solid at the points where the broken line crosses the vaporization curve and fusion curve, respectively.

When the pressure is low enough, a material may be transformed by constant-pressure heating from solid directly to vapor, as represented by line (c). This process, called *sublimation*, occurs where the line crosses the sublimation curve. At atmospheric pressure, solid carbon dioxide undergoes sublimation; there is no liquid phase at this pressure. Similarly, when a substance with initial temperature lower than the temperature at the point labeled "triple point" is compressed by increasing its pressure, it goes directly from the vapor to the solid phase, as shown by line (d).

Figure 17–4 is characteristic of materials that expand on melting. A few materials *contract* on melting; the most familiar examples are water and the metal antimony. For such a substance the fusion curve always slopes upward to the left, rather than to the right. The melting temperature decreases when the pressure increases, so these solid materials can be melted by an increase of pressure. This is part of the reason ice cubes stick together.

At the triple point, gas, liquid, and solid phases can all exist together.

The intersection point of the equilibrium curves in Fig. 17–4 is called the **triple point.** For any substance there is only one pressure and temperature at which *all three* phases can coexist, namely, the triple-point pressure and tem-

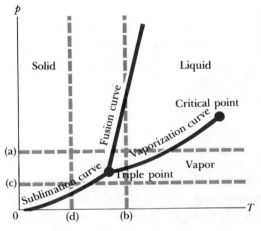

17–4 A pT phase diagram, showing regions of temperature and pressure at which the various phases exist and where phase changes occur.

TABLE 17–1 Triple-Point Data

Substance	Temperature, K	Pressure, Pa
Hydrogen	13.84	0.0704×10^5
Deuterium	18.63	0.171×10^5
Neon	24.57	0.432×10^5
Nitrogen	63.18	0.125×10^5
Oxygen	54.36	0.00152×10^5
Ammonia	195.40	0.0607×10^5
Carbon dioxide	216.55	5.17×10^5
Sulfur dioxide	197.68	0.00167×10^5
Water	273.16	0.00610×10^5

perature. Triple-point data for a few substances are given in Table 17–1. In Section 14–4 we saw the role of the triple-point temperature of water in defining a temperature scale.

Figure 17–3 shows that the liquid and gas (or vapor) phases can exist together only when the temperature and pressure are less than those at the point lying at the top of the tongue-shaped area labeled "liquid–vapor equilibrium region." This point also corresponds to the endpoint at the top of the vaporization curve in Fig. 17–4. It is called the **critical point,** and the corresponding values of p and T are called the critical pressure and temperature. A gas at a temperature *above* the critical temperature does not separate into two phases when compressed isothermally. Its properties change gradually and continuously from those we ordinarily associate with a gas (low density, large compressibility) to those of a liquid (high density, small compressibility), *without a phase transition.*

If this description stretches credibility, here is another point of view. When we look at liquid–vapor phase transitions at successively higher points on the vaporization curve, we find that as we approach the critical point, the changes in density, bulk modulus, index of refraction, viscosity, and all other physical properties become smaller and smaller. At the critical point they all become zero, and at this point the distinction between liquid and vapor disappears. The heat of vaporization also grows smaller and smaller as we approach the critical point, and it too becomes zero at the critical point.

Table 17–2 lists critical constants for a few substances. The very low critical temperatures of hydrogen and helium make it evident why these gases defied attempts to liquefy them for many years.

The information presented by the pT phase diagram is also contained in the pV-diagram. In Fig. 17–3, for example, we can read off for each temperature the pressure at which liquid and vapor are in equilibrium. This set of pairs of values of p and T can then be used to plot the vaporization curve of Fig. 17–4, and we can also obtain the critical-point temperature and pressure.

Many substances can exist in more than one solid phase. A familiar example is carbon, which exists as noncrystalline lampblack and crystalline graphite and diamond. Transitions from one solid phase to another occur at definite temperatures and pressures, just as with phase changes from liquid to solid, and so on. Water is one such substance; at least eight types of ice, differing in their crystal structure and physical properties, have been observed at very high pressures.

Above the critical point, the distinction between liquid and vapor disappears.

How can a liquid turn into a vapor without a phase transition?

Some materials have more than one solid phase.

TABLE 17–2 Critical Constants

Substance	Critical Temperature, K	Critical Pressure, Pa	Critical Volume, m³·mol⁻¹	Critical Density, kg·m⁻³
Helium (4)	5.3	2.29×10^5	57.8×10^{-6}	69.3
Helium (3)	3.34	1.17×10^5	72.6×10^{-6}	41.3
Hydrogen (normal)	33.3	13.0×10^5	65.0×10^{-6}	31.0
Deuterium (normal)	38.4	16.6×10^5	60.3×10^{-6}	66.3
Nitrogen	126.2	33.9×10^5	90.1×10^{-6}	311
Oxygen	154.8	50.8×10^5	78×10^{-6}	410
Ammonia	405.5	112.8×10^5	72.5×10^{-6}	235
Freon 12	384.7	40.1×10^5	218×10^{-6}	555
Carbon dioxide	304.2	73.9×10^5	94.0×10^{-6}	468
Sulfur dioxide	430.7	78.8×10^5	122×10^{-6}	524
Water	647.4	221.2×10^5	56×10^{-6}	320
Carbon disulfide	552.0	79.0×10^5	170×10^{-6}	440

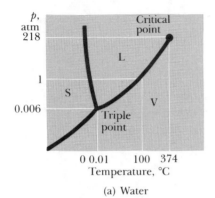

(a) Water

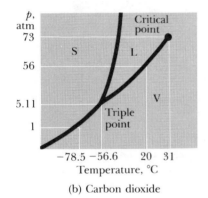

(b) Carbon dioxide

17–5 Pressure–temperature diagrams (not to a uniform scale).

A *pVT*-surface gives a bird's-eye view of the behavior of a material.

Examples of processes represented on a *pVT*-surface

As examples of phase diagrams, consider the *pT*-diagrams for water and carbon dioxide shown in Fig. 17–5. In (a), a horizontal line at a pressure of 1 atm intersects the freezing-point curve at 0°C and the boiling-point curve at 100°C. The boiling point increases with increasing pressure up to the critical temperature of 374°C. Solid, liquid, and vapor can remain in equilibrium only at the triple point, where the vapor pressure is 0.006 atm and the temperature is 0.01°C.

For carbon dioxide, the triple-point temperature is -56.6°C, and the corresponding pressure is 5.11 atm. Hence at atmospheric pressure CO_2 can exist only as a solid or vapor. When heat is supplied to solid CO_2, open to the atmosphere, it changes directly to vapor without passing through the liquid phase, whence the name "dry ice." Liquid CO_2 can exist only at a pressure greater than 5.11 atm. The steel tanks in which CO_2 is commonly stored contain liquid and vapor (both saturated). The pressure in these tanks is the vapor pressure of CO_2 at the temperature of the tank. If this temperature is 20°C, the vapor pressure is 56 atm or about 830 lb·in⁻².

We remarked at the beginning of Section 17–3 that the equation of state of any material can be represented graphically as a surface in a three-dimensional space with coordinates *p*, *V*, and *T*. Such a surface is seldom useful in representing detailed quantitative information, but it can enhance our general understanding of the behavior of materials at various temperatures and pressure. Figure 17–6 shows a typical *pVT*-surface. The light lines represent *pV*-isotherms, and we see that their projections on the *pV*-plane yield a diagram similar to Fig. 17–3. The *pV*-isotherms represent contour lines on the *pVT*-surface, just as contour lines on a topographic map represent the elevation (the third dimension) at each point. The projections of the edges of the surface onto the *pT*-plane yield the *pT*-phase diagram of Fig. 17–5b.

Three lines on the surface show specific processes. Line *abcdef* shows constant-volume heating, with melting at *bc* and vaporization at *de*. Note the volume changes that occur as *T* increases along this line. Line *ghjklm* is an isothermal compression; *hj* shows liquefaction and *kl* solidification. Between these, sections *gh* and *jk* represent constant-temperature compression with

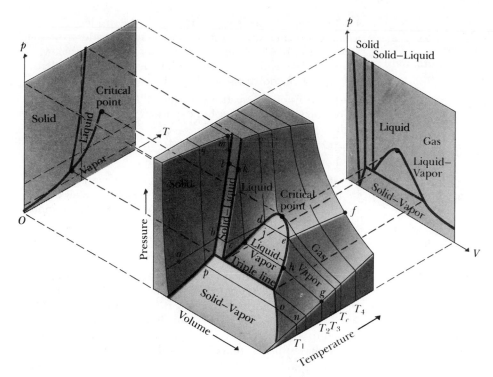

17–6 pVT-surface for a substance that expands on melting. Projections of the surface on the pT- and pV-planes are also shown.

increase in pressure; the pressure increases are much greater in the liquid region *jk* and the solid region *lm* than in the vapor region *gh*. Finally, *nopq* represents isothermal solidification directly from the vapor phase. These three lines on the pVT-surface are worth careful study.

SUMMARY

The pressure, volume, and temperature of a given quantity of a substance are called state coordinates. They cannot all vary independently but are related by an equation of state. The equation of state pertains only to equilibrium states, where p and T are uniform throughout the system.

The ideal gas equation of state is

$$pV = nRT, \qquad (17-6)$$

where T is always absolute temperature, n is the number of moles, and R is a fundamental constant called the ideal gas constant.

A pV-diagram is a set of graphs of pressure as a function of volume for several constant temperatures. The curves are called isotherms. A pT phase diagram shows what phase is stable for any given values of p and T and the conditions for which phase equilibrium can occur. A pVT-surface is a geometric representation of the equation of state of a substance.

KEY TERMS

state coordinates

equation of state

equilibrium states

ideal gas constant

ideal gas

standard temperature and pressure (STP)

pV-diagram

isotherm

phase diagram

phase equilibrium

triple point

critical point

QUESTIONS

17–1 If the ideal-gas model were perfectly valid at all temperatures, what would be the volume of a gas when the temperature is absolute zero?

17–2 In the ideal-gas equation of state, could Celsius temperature be used instead of Kelvin if an appropriate numerical value of the gas constant R is used?

17–3 Comment on this statement:

> Equal masses of two different gases, placed in containers of equal volume at equal temperature, must exert equal pressures.

17–4 When a car is driven some distance, the air pressure in the tires increases. Why? Should you let some air out to reduce the pressure?

17–5 Section 17–1 states that pressure, volume, and temperature cannot change individually without one affecting the others. Yet when a liquid evaporates, its volume changes, even though its pressure and temperature are constant. Is this inconsistent?

17–6 At low pressure or high temperature, gas behavior deviates from the ideal-gas model. One reason is the finite size of gas molecules. From these deviations, how could we estimate what fraction of the total volume in a gas is occupied by the molecules themselves?

17–7 Radiator caps in automobile engines will release coolant when the pressure reaches a certain value, typically 15 psi or so. Why is it desirable to have the coolant under pressure? Why not just seal the system completely?

17–8 People speak of being angry enough to make their blood boil. Could your blood boil if you went to a high enough altitude? Is the pressure of your blood the same as atmospheric pressure? Does this matter?

17–9 Can it be too cold for good ice skating? What about skiing? Does a thin layer of ice under the skis melt?

17–10 Why does the air cool off more on a clear night than on a cloudy night? Desert areas often have exceptionally large day–night temperature variations. Why?

17–11 On a chilly evening you can "see your breath." Can you really? What are you really seeing? Does this phenomenon depend on the temperature of the air, the humidity, or both?

17–12 Why do frozen water pipes burst? Would a mercury thermometer break if the temperature went below the freezing temperature of mercury?

17–13 Why does moisture condense on the insides of windows on cold days? If it is cold enough outside, frost will form, but only when the outside temperature is far below freezing. Why is this so? Why are the bottoms frostier than the tops? How do storm windows change the situation?

17–14 "It's not the heat; it's the humidity!" Why are you more uncomfortable on a hot day when the humidity is high than when it is low?

17–15 Why does a shower room get steamy or foggy when the hot water runs for a long time?

17–16 The critical volumes of several substances are listed in Table 17–2, and with a few exceptions they are all in the range of 50 to 100×10^{-6} $m^3 \cdot mol^{-1}$. Why is there not more variation?

17–17 Why do flames from a fire always go up rather than down?

17–18 Unwrapped food placed in a freezer experiences dehydration, known as "freezer burn." Why? The same process, carried out intentionally, is called "freeze-drying." For freeze-drying, the food is usually frozen first and then placed in a partial vacuum. Why? What advantages might freeze-drying have, compared to ordinary drying?

17–19 An old or burned-out incandescent light bulb usually has a dark gray area on part of the inside of the bulb. What is this? Why is it not a uniform deposit all over the inside of the bulb?

17–20 Why do glasses containing cold liquids sweat in warm weather? Glasses containing alcoholic drinks sometimes form frost on the outside. How can this happen?

17–21 Why does a person perspire more when working hard than when resting?

17–22 Hiking down a steep trail ought to be much easier than hiking up, yet people perspire a lot when descending a steep trail. Why?

17–23 In a hot desert climate a person can survive a lot longer without food than without water. Why?

EXERCISES

Section 17–2 Ideal Gases

17–1 A cylindrical tank has a tight-fitting piston that allows the volume of the tank to be changed. The tank originally contains 0.12 m^3 of air at a pressure of 1 atm. The piston is slowly pushed in until the volume of the gas is decreased to 0.04 m^3. If the temperature remains constant, what is the final value of the pressure?

17–2 A 2-L tank contains air at 1 atm and 20°C. The tank is sealed and heated until the pressure is 3 atm.

a) What is the temperature now, in degrees Celsius? Assume that the volume of the tank is constant.

b) If the temperature is kept at the value found in (a) and the gas is permitted to expand, what is the volume when the pressure again becomes 1 atm?

17–3 A 20-L tank contains 0.20 kg of helium at 27°C.

a) What is the number of moles of helium?

b) What is the pressure in pascals? In atmospheres?

17–4 A room with dimensions 3 m × 4 m × 5 m is filled with pure oxygen at 20°C and 1 atm.

a) What is the number of moles of oxygen?

b) What is the mass of oxygen, in kilograms?

17–5 A tank contains 0.5 m^3 of nitrogen at 27°C and 1.5×10^5 Pa (absolute pressure). What is the pressure if the volume is increased to 5.0 m^3 and the temperature is increased to 327°C?

17–6 A welder using a tank of volume 0.10 m^3 fills it with oxygen at a gauge pressure of 4.0×10^5 Pa and temperature of 47°C. Later it is found that, because of a leak, the gauge pressure has dropped to 3.0×10^5 Pa and the temperature has decreased to 27°C. Find

a) the initial mass of oxygen;

b) the mass that has leaked out.

17–7 At the beginning of the compression stroke, a cylinder of a diesel engine contains 800 cm³ of air at atmospheric pressure and a temperature of 27°C. At the end of the stroke, the air has been compressed to a volume of 50 cm³ and the gauge pressure has increased to 40×10^5 Pa. Compute the temperature.

17–8 A diver observes a bubble of air rising from the bottom of a lake, where the pressure is 3 atm, to the surface, where the pressure is 1 atm. The temperature at the bottom is 7°C, and the temperature at the surface is 27°C. What is the ratio of the volume of the bubble as it reaches the surface to its volume at the bottom?

17–9

a) Derive from the equation of state of an ideal gas an equation for the density of an ideal gas in terms of pressure, temperature, and appropriate constants.

b) What is the density of air, in kg·m⁻³, at 1 atm and 20°C? The average molecular mass of air is 28.8 g·mol⁻¹.

17–10 What is the atmospheric pressure at an altitude of 5000 m, if the air temperature is a uniform 0°C? This is roughly the maximum altitude usually attained by aircraft with nonpressurized cabins.

Section 17–4 Phase Diagrams, Triple Point, Critical Point

17–11 A physicist places a piece of ice at 0°C alongside a beaker of water at 0°C in a glass vessel, from which all the air has been removed. If the ice, water, and vessel are all maintained at a temperature of 0°C by a suitable thermostat, describe the final equilibrium state inside the vessel.

17–12 Calculate the volume of 1 mol of liquid water at a temperature of 20°C (use Table 14–3), and compare this to the volume occupied by 1 mol of water at the critical point.

17–13 Solid nitrogen is slowly heated from a very low temperature.

a) What minimum external pressure p_1 must be applied to the solid if a melting-phase transition is to be observed? Describe the sequence of phase transitions that occur if the applied pressure p is such that $p < p_1$.

b) Above a certain maximum pressure p_2 no liquid-to-vapor (boiling) transition is observed. What is this pressure? Describe the sequence of phase transitions that occur if $p_1 < p < p_2$.

PROBLEMS

17–14 Partial vacuums where the residual gas pressure is about 10^{-10} atm are not difficult to obtain with modern technology. Calculate the mass of nitrogen (N₂) present in a volume of 2000 cm³ at this pressure. Assume that the temperature of the gas is 27°C.

17–15 An automobile tire has a volume of about 0.015 m³ on a cold day when the temperature of the air in the tire is 0°C. Under these conditions the gauge pressure is measured to be 2 atm (about 30 lb·in⁻²). After the car is driven on the highway for 30 min, the temperature of the air in the tires has risen to 47°C and the volume to 0.016 m³. What then is the gauge pressure?

17–16 A glassblower makes a barometer from a tube 0.90 m long and of cross section 1.5 cm². Mercury stands in this tube to a height of 0.75 m. The room temperature is 27°C. A small amount of nitrogen is introduced into the evacuated space above the mercury, and the column drops to a height of 0.70 m. How many grams of nitrogen were introduced?

17–17 A bicyclist uses a tire pump whose cylinder is initially full of air at an absolute pressure of 1.01×10^5 Pa. The length of stroke of the pump (the length of the cylinder) is 45.7 cm. At what part of the stroke (i.e., what is the length of air column) does air begin to enter a tire in which the gauge pressure is 2.76×10^5 Pa? Assume that the temperature remains constant during the compression.

17–18 The submarine *Squalus* sank at a point where the depth of water was 73 m. The temperature at the surface is 27°C, and at the bottom it is 7°C. The density of sea water is 1030 kg·m⁻³.

a) If a diving bell in the form of a circular cylinder 2.4 m high, open at the bottom and closed at the top, is lowered to this depth, to what height will water rise within it when it reaches the bottom?

b) At what gauge pressure must compressed air be supplied to the bell while on the bottom to expel all the water from it?

17–19 A flask of volume 2 L (2×10^{-3} m³), provided with a stopcock, contains oxygen at 300 K and atmospheric pressure. The system is heated to a temperature of 400 K, with the stopcock open to the atmosphere. The stopcock is then closed and the flask cooled to its original temperature.

a) What is the final pressure of the oxygen in the flask?

b) How many grams of oxygen remain in the flask?

17–20 A hot-air balloon makes use of the fact that hot air at atmospheric pressure is less dense than cooler air at the same pressure; the buoyant force is calculated as discussed in Chapter 13. If the volume of the balloon is 500 m³ and the surrounding air is at 0°C, what must be the temperature of the air in the balloon in order for it to lift a total load of 200 kg (in addition to the mass of the hot air)? The density of air at 0°C and atmospheric pressure is 1.29 kg·m⁻³.

17–21 A balloon whose volume is 500 m³ is to be filled with hydrogen at atmospheric pressure.

a) If the hydrogen is stored in cylinders of volume 2.5 m³ at an absolute pressure of 15×10^5 Pa, how many cylinders are required? Assume that the temperature of the hydrogen remains constant.

b) What is the total weight (in addition to the weight of the gas) that can be supported by the balloon, in air at standard conditions?

c) What weight could be supported if the balloon were filled with helium instead of hydrogen?

17–22 A vertical cylindrical tank 1 m high has its top end closed by a tightly fitting frictionless piston of negligible

weight. The air inside the cylinder is at an absolute pressure of 1 atm. The piston is depressed by pouring mercury on it slowly. How far will the piston descend before mercury spills over the top of the cylinder? The temperature of the air is maintained constant.

CHALLENGE PROBLEMS

17–23 A large tank of water has a hose connected to it, as shown in Fig. 17–7. The tank is sealed at the top and has compressed air between the water surface and the top. When the water height h_2 has the value 3.0 m, the absolute pressure p_1 is 2.00×10^5 Pa. Assume that the air above the water expands isothermally (i.e., at constant T). Take the atmospheric pressure to be 1.00×10^5 Pa.

a) What is the velocity of flow out of the hose when $h_2 = 3.0$ m?

b) As water flows out of the tank, h_2 decreases. Calculate the velocity of flow for $h_2 = 2.5$ and 2.0 m.

c) At what value of h_2 will the flow stop?

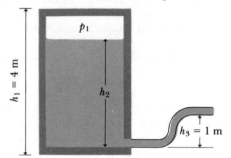

FIGURE 17–7

17–24 In the lower part of the atmosphere (the troposphere) the temperature is not uniform but decreases with increasing elevation.

a) Show that if the temperature variation is approximated by the linear relation

$$T = T_0 - \alpha y,$$

where T_0 is the temperature at the earth's surface and T is the temperature at height y, the pressure is given by

$$\ln\left(\frac{p}{p_0}\right) = \frac{Mg}{R\alpha} \ln\left(\frac{T_0 - \alpha y}{T_0}\right),$$

where p_0 is the pressure at the earth's surface and M is the molecular mass for air. The coefficient α is called the "lapse rate of temperature." Although it varies with atmospheric conditions, an average value is about 0.6 C°/100 m.

b) Show that the above result reduces to Eq. (17–9) in the limit that $\alpha \to 0$.

c) With $\alpha = 0.6$ C°/100 m, calculate p for $y = 8882$ m and compare your answer to the result of Example 17–3. Take $T_0 = 273$ K and $p_0 = 1$ atm, as in the example.

17–25 In pV-diagrams, the slope dp/dV along an isotherm is never positive. (You should be able to explain why.) As

shown in Section 17–3, and indicated in Figs. 17–3 and 17–6, regions where $dp/dV = 0$ represent equilibrium between two phases; volume can change with no change in pressure. We can use this fact to determine critical constants if we have an equation of state. With $p = p(V, T, n)$, if $T > T_c$, then $p(V)$ has no maximum along an isotherm, whereas if $T < T_c$, then $p(V)$ has a maximum. This leads to the following condition for determining the critical point:

$$\frac{dp}{dV} = 0 \quad \text{and} \quad \frac{d^2p}{dV^2} = 0 \quad \text{at the critical point.}$$

An equation of state that has isotherms resembling those in Fig. 17–3 is the van der Waals equation:

$$\left(p + \frac{an^2}{V^2}\right)(V - nb) = nRT,$$

where a and b are constants.

a) Solve the van der Waals equation for p. That is, find $p(V, T, n)$. Find dp/dV and d^2p/dV^2. Set these equal to zero, and thereby obtain two equations for V, T, and n.

b) Simultaneous solution of the two equations obtained in (a) gives the critical values T_c and $(V/n)_c$. Find these constants in terms of a and b. (*Hint:* Divide one equation by the other to eliminate T.)

c) Substitute these values into the equation of state to find p_c, the critical pressure.

d) Find the ratio

$$\frac{RT_c}{p_c(V/n)_c}.$$

This should not contain either a or b.

e) Compute the ratio in part (d) for the gases H_2, N_2, O_2, and H_2O by using the critical-point data of Table 17–2. Compare the results to the prediction you made based on the van der Waals equation.

f) The experimentally determined values of a and b for some gases are given below. Use your results of parts (b) and (c) to predict T_c, p_c, and $(V/n)_c$ for these gases. Compare these values to those in Table 17–2.

Gas	$a(m^6 \cdot Pa \cdot mol^{-2})$	$b(m^3 \cdot mol^{-1})$
H_2	0.0247	2.66×10^{-5}
N_2	0.141	3.91×10^{-5}
O_2	0.138	3.18×10^{-5}
H_2O	0.553	3.05×10^{-5}

18 THE FIRST LAW OF THERMODYNAMICS

THERMODYNAMICS IS THE STUDY OF ENERGY RELATIONSHIPS INVOLVING heat, mechanical work, and other aspects of energy and energy transfer. The first law of thermodynamics extends the principle of conservation of energy to include not only the forms of mechanical energy encountered in Chapter 7 but also energy associated with heat. Conservation of energy plays a vital role in every area of physical science, so the first law is widely applicable. Energy relationships are of central importance in the study of energy-conversion devices such as engines, batteries, and refrigerators, and also in the functioning of living organisms. To state these relationships precisely, we introduce the concept of a *thermodynamic system* and discuss *heat* and *work* as two means of transferring energy into or out of such a system. Emerging from this discussion and from a generalized view of the principle of conservation of energy is the concept of *internal energy* of a system.

18–1 ENERGY, HEAT, AND WORK

We have studied the transfer of energy through mechanical work (Chapter 7) and through heat transfer (Chapters 15 and 16); now we are ready to combine and generalize these principles. Most of our analysis will refer to a *system*, which will usually be a specified quantity of material, perhaps in the form of a device or an organism. A **thermodynamic system** is a system that can interact with its surroundings in at least two ways, one of which must be heat transfer. A familiar example of a thermodynamic system is a quantity of a gas confined in a cylinder with a piston. Energy can be added to the system by conduction of heat, and the system can also do *work*, since the gas exerts a force on the piston and can move it through a displacement.

We have already discussed the concept of a system. In Chapters 4, 5, and 9 we made extensive use of free-body diagrams to help identify the forces acting on a particular mechanical system so as to include only these forces in our analysis. Similarly, in Chapter 8 we studied the principle of conservation of momentum for an isolated mechanical system. In both examples, it is essential

Energy exchange in a thermodynamic system by means of heat and work

The concept of a system occurs throughout physics; a thermodynamic system is a specific example.

to define clearly at the outset precisely what is and is not included in the system. The same is true of thermodynamic systems; it is essential in solving problems to identify the system under discussion and to describe unambiguously the energy transfers into and out of that system.

The concept of a system can be broadened to include a variable quantity of matter; in that case the transport of *mass* into or out of a system provides an additional mechanism for energy exchange with the surroundings. This is a complication we will not consider here; all the systems discussed in this chapter have a constant amount of matter.

As already mentioned, thermodynamics has its roots in practical problems. For example, a steam engine or steam turbine uses the heat of combustion of coal or other fuel to perform mechanical work such as driving an electric generator, pulling a train, or performing some other useful function. The gasoline engine in an automobile and the jet engines in an airplane have similar functions. Muscle tissue in living organisms metabolizes chemical energy in food and performs mechanical work on the surroundings of the organism.

In all these situations, we will describe the energy relations in terms of the quantity of heat Q added *to* the system and the work W done *by* the system. Both Q and W may be positive or negative. A positive value of Q represents heat flow *into* the system, with a corresponding input of energy to it; negative Q represents heat flow *out of* the system. A positive value of W represents work done *by* the system against its surroundings, such as an expanding gas, and hence corresponds to energy leaving the system. Negative W, such as compression of a gas in which work is done *on it* by its surroundings, represents energy entering the system. These conventions will be clarified with many examples in this chapter and the next.

Heat can be understood on the basis of *microscopic* mechanical energy, that is, the kinetic and potential energies of individual molecules in a material; it is also possible to develop the principles of thermodynamics from a microscopic viewpoint. In this chapter we deliberately avoid that development in order to emphasize that the central principles and concepts of thermodynamics can be treated in a wholly *macroscopic* way, without reference to microscopic models. Indeed, part of the great power and generality of thermodynamics springs from the fact that it *does not* depend on details of the structure of matter. In Chapter 20, however, we will return to microscopic considerations and will examine their relation to the principles of thermodynamics.

18–2 WORK DONE DURING VOLUME CHANGES

A gas in a cylinder with a movable piston is a simple example of a thermodynamic system. In the next several sections this system will be used as an example to explore several kinds of processes involving energy considerations. First we consider the *work* done by the system during a volume change. When a gas expands, it pushes out on its boundary surfaces as they move outward; hence an expanding gas always does positive work. The same is true of any solid or fluid material confined under pressure. Figure 18–1 shows a solid or fluid in a cylinder with a movable piston. Suppose that the cylinder has a cross-sectional area A and that the pressure exerted by the system at the piston face is p. The force F exerted by the system is therefore $F = pA$. If the piston moves out a

Energy conversions involving heat and work have many practical applications.

Sign rules are an essential part of energy bookkeeping.

The relation of heat to molecular mechanical energy: not essential in thermodynamics, but useful later for additional insight

When a system under pressure expands, it does work against its surroundings.

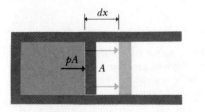

18–1 Force exerted *by* a system during a small expansion.

The catastrophic eruption of Mt. St. Helens on May 18, 1980, a dramatic illustration of the enormous pressure developed in a confined hot gas. As the gas expands, it does enough work on the overlying rock and soil to lift roughly a cubic mile of material several thousand feet into the air. (Courtesy of USGS.)

small distance dx, the work dW done by this force is equal to

$$dW = pA \, dx.$$

But

$$A \, dx = dV,$$

where dV is the change of volume of the system. Therefore

$$dW = p \, dV, \tag{18–1}$$

and in a finite change of volume from V_1 to V_2,

$$W = \int_{V_1}^{V_2} p \, dV. \tag{18–2}$$

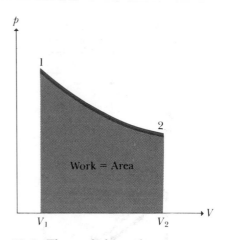

18–2 The work done when a substance changes its volume and pressure is the area under the curve on a pV-diagram.

In general, the pressure of the system may vary during the volume change, and the integral can be evaluated only when the pressure is known as a function of volume. It is customary to represent this relationship graphically by plotting p as a function of V, as in Fig. 18–2. Then Eq. (18–2) may be interpreted graphically as the *area* under the curve between the limits V_1 and V_2.

According to the convention stated above, work is *positive* when a system *expands*. Thus, if a system expands from 1 to 2 in Fig. 18–2, the area is regarded as positive. A *compression* from 2 to 1 gives rise to a *negative* area; when a system is compressed, its volume decreases and it does *negative* work on its surroundings.

If the pressure remains constant while the volume changes, then the work is

An expanding system does positive work; a contracting system does negative work.

$$W = p(V_2 - V_1) \quad \text{(constant pressure only).} \tag{18–3}$$

EXAMPLE 18–1 An ideal gas at constant temperature T undergoes an *isothermal expansion* in which its volume changes from V_1 to V_2. How much work does the gas do?

SOLUTION From Eq. (18–2),

$$W = \int_{V_1}^{V_2} p \, dV.$$

For an ideal gas,

$$p = \frac{nRT}{V}.$$

Since n, R, and T are constant,

$$W = nRT \int_{V_1}^{V_2} \frac{dV}{V} = nRT \ln \frac{V_2}{V_1}. \tag{18–4}$$

In an expansion, $V_2 > V_1$ and W is positive. At constant T,

$$p_1 V_1 = p_2 V_2, \qquad \text{or} \qquad \frac{V_2}{V_1} = \frac{p_1}{p_2},$$

so the isothermal work may be expressed also in the form

$$W = nRT \ln \frac{p_1}{p_2}. \tag{18–5}$$

When a system goes from an initial state to a final state, there are many different possible sequences of intermediate states.

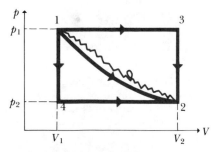

18–3 Work depends on the path.

On the pV-diagram in Fig. 18–3, an initial state 1 (characterized by pressure p_1 and volume V_1) and a final state 2 (characterized by pressure p_2 and volume V_2) are represented by the two points 1 and 2. To progress from the initial state to the final state, the system must go through a series of intermediate states, and infinitely many such series are possible. For example, the pressure might be kept constant at p_1 while the system expands to volume V_2 (point 3 on the diagram), and then the pressure might be reduced to p_2 with the volume remaining constant at V_2 (to point 2 on the diagram). The work during this process is the area under the line $1 \to 3$; no work is done during the constant-volume process $3 \to 2$. Or the system might traverse the path $1 \to 4 \to 2$, in which case the work is the area under the line $4 \to 2$. The wiggly line and the smooth curve from 1 to 2 are two other possibilities, and the work is different for each one.

We conclude that *the work depends not only on the initial and final states but also on the intermediate states, that is, on the path.* For this reason, we cannot speak of the amount of work contained in a system in a given state; such a concept has no meaning. In this respect, the character of work is very different from that of the state coordinates p, V, and T.

18–3 HEAT TRANSFER DURING VOLUME CHANGES

As explained in Chapter 15, heat is energy transferred into or out of a system as a result of a temperature difference between the system and its surroundings. The performance of work, discussed in the preceding section, and the transfer of heat are means of energy transfer, that is, processes by which the energy of a system may increase or decrease.

We have seen in Section 18–2 that in a change of state of a thermodynamic system, the work done by the system depends not only on the initial and final states, but also on the *path* taken, that is, the series of intermediate states. This is also true of the *heat* added to the system in such a process. The following example illustrates this point.

In Fig. 18–4a, a quantity of gas is contained in a cylinder with a piston, with initial volume of two liters; it is kept at a constant temperature of 300 K by an electric stove. We then let the gas expand slowly; heat flows from the stove to the gas, keeping the temperature of the gas at 300 K. Suppose that the gas expands in this slow, controlled, isothermal manner until its volume becomes, say, seven liters. A finite amount of heat is absorbed by the gas during this process.

Figure 18–4b shows a container surrounded by insulating walls and divided into two compartments (the lower one having volume two liters, the upper one five liters) by a thin, breakable partition. In the lower compartment, the same gas has the same initial volume and temperature as in (a). Now we break the partition; the gas in (b) undergoes a rapid, uncontrolled expansion, with no heat passing through the insulating walls. The final volume is seven liters, as in (a). The rapid, uncontrolled expansion of a gas into a vacuum is called a **free expansion;** we will discuss it further in Section 18–6. Experiments have shown that in a free expansion of an ideal gas, no temperature change takes place; therefore the final state of the gas is the same as the final state in (a). But the states traversed by the gas (the pressures and volumes) while proceeding from state 1 to state 2 are entirely different, so that (a) and (b) represent *two different paths* connecting the *same states* 1 and 2. For path (b), *no* heat is transferred and no work is done. Like work, *heat depends not only on the initial and final states but also on the path.*

Thus it would *not* be correct to say that a body *contains* a certain amount of heat. Suppose we assigned an arbitrary value to "the heat in a body" in some standard reference state. The "heat in the body" in some other state would then equal the "heat" in the reference state plus the heat added when the body is carried to the second state. But the heat added depends on the path by which the body goes from one state to the other. There are infinitely many paths that might be followed, so there are also infinitely many values that might equally well be assigned to the "heat in the body" in the second state. Because there is no consistent way to define "heat in a body," this is not a useful concept. It *does* make sense, however, to speak of the amount of *internal energy* in a body, and this important concept is our next topic.

Heat depends on the path.

There is no such thing as the amount of heat in a body.

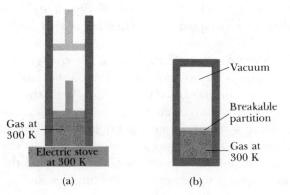

18–4 (a) Slow, controlled isothermal expansion of a gas from an initial state 1 to a final state 2. (b) Rapid, uncontrolled expansion of the same gas starting at the same state 1 and ending at the same state 2.

18–4 INTERNAL ENERGY AND THE FIRST LAW

Internal energy is associated with
molecular kinetic and potential energies.

The concept of internal energy of a thermodynamic system can be viewed in various ways, some simple, some subtle. To begin with the simple, we note that matter consists of atoms and molecules; these, in turn, are made up of particles having kinetic and potential energies. We tentatively define the **internal energy** of a system as the sum of all the kinetic and potential energies of its constituent particles. We use the symbol U for internal energy. During a change of state of the system, the internal energy may change from an initial value U_1 to a final value U_2; in that case we denote the change in internal energy in the usual way as $\Delta U = U_2 - U_1$.

The internal energy of a system changes when heat is added and when the system does work.

We also know that heat transfer is really energy transfer. Thus when a quantity of heat Q is added to a system, the result should be to increase the internal energy by an amount equal to Q; that is, $\Delta U = Q$. Also, when a system does work W by expanding against its surroundings, this represents energy leaving the system. When W is positive, ΔU is negative, and conversely, so $\Delta U = -W$. Finally, if both heat transfer and work occur, the total change in internal energy is

$$U_2 - U_1 = \Delta U = Q - W. \qquad (18\text{–}6)$$

This equation may be rearranged to the form

$$Q = \Delta U + W, \qquad (18\text{–}7)$$

which can be read as follows: When heat Q is added to a system, some of this added energy remains within the system, increasing its internal energy by an amount ΔU; the remainder leaves the system again as the system does work W against its surroundings. Of course, as explained above, W and Q are algebraic quantities that may be either positive or negative, and correspondingly we expect ΔU to be positive for some processes and negative for others.

The first law of thermodynamics: energy conservation in a thermodynamic system

Equation (18–6) or (18–7) is the **first law of thermodynamics.** In the context of the above discussion, the law simply represents conservation of energy, generalized to include energy transfer through heat as well as mechanical work. It seems eminently reasonable that conservation of energy, which we encountered in Chapter 7 in a purely mechanical context, should have this more general validity; nevertheless, it is worth a somewhat closer look.

First of all, our definition of internal energy in terms of microscopic kinetic and potential energies is not entirely convincing. Actually calculating this total energy for any real system would be hopelessly complicated. Thus our definition is not an *operational* definition because it does not describe how to obtain the defined quantity from physical quantities that can be measured directly. We need to backtrack a little in defining internal energy, in order to show clearly the empirical and experimental basis of the first law of thermodynamics.

An operational definition of internal energy

Starting over, we *define* the internal energy change ΔU in any change of a system by means of Eq. (18–6). This *is* an operational definition, because Q and W can be measured. It does not define U itself, only ΔU. This is not a serious shortcoming; we can define the internal energy of a system to have a specified value in a certain standard state, and then use Eq. (18–6) to define the internal energy in any other state. This procedure is analogous to the definition of potential energy in Chapter 6, where we arbitrarily defined the potential energy of a mechanical system to be zero at a certain position.

This new definition trades one difficulty for another. If we define ΔU via Eq. (18–6), then we must ask the following question: If the system goes from some initial state to some final state by two different paths, is ΔU necessarily the same for the two paths? We have already seen that in such a case Q and W are in general both path-dependent. If ΔU, which is just $Q - W$, is also path-dependent, then ΔU is ambiguous, and the concept of internal energy in a system is subject to the same criticism as the erroneous concept of quantity of heat in a system, as discussed at the end of Section 18–3.

At this point the only recourse is to experiment. We study the properties of a variety of materials; in particular, we measure Q and W for various changes of state and various paths, in order to learn whether ΔU is or is not path-dependent. The results of many such investigations are clear and unambiguous: Within experimental error, ΔU is found to be path-independent. The change in internal energy in any thermodynamic process is found to depend only on the initial and final states and *not* on the path leading from one to the other.

Experiment, then, is the ultimate justification for believing that in any state, a thermodynamic system has a unique internal energy that depends only on the state the system is in. An equivalent statement is that the internal energy U of a system is a function of the state coordinates p, V, and T (or actually of any two of these, since the three variables are related by the equation of state).

Internal energy depends only on the state of the system: the real significance of the first law.

Let us now return to the first law of thermodynamics. It is perfectly correct to say that the first law, given by Eq. (18–6) or (18–7), is a statement of conservation of energy for thermodynamic processes. But we must also recognize that part of the content of the first law is the experimental verification that internal energy is a property of the state of a system; in changes of state, the change in internal energy is path-independent.

All this discussion may seem a little abstract to a reader who is satisfied to regard internal energy as microscopic mechanical energy. There is nothing wrong with that view. But in the interest of precise *operational* definitions, internal energy, like heat, can and must be defined in a way that is independent of the detailed microscopic structure of the material.

If a system is carried through a process that eventually returns it to its initial state (a *cyclic* process), then

In any cyclic process, the internal energy change is zero.

$$U_2 = U_1 \quad \text{and} \quad Q = W.$$

Thus, although net work W may be done by the system in the process, energy has not been created, since an equal amount of energy must have flowed into the system as heat Q.

An *isolated* system is one that does no external work and into which there is no heat flow. Then, for any process taking place in such a system,

$$W = Q = 0$$

and

$$U_2 - U_1 = 0 \quad \text{and} \quad \Delta U = 0.$$

That is, *the internal energy of an isolated system remains constant.* This is the most general statement of the principle of conservation of energy. The internal energy of an *isolated* system cannot be changed by any process (mechanical, electrical, chemical, nuclear, or biological) taking place *within* the system. The energy of a system can be changed only by a flow of heat across its boundary,

The internal energy of an isolated system cannot change.

or by the performance of work. If either of these takes place, the system is no longer isolated. The increase in energy of the system is then equal to the energy flowing in as heat, minus the energy flowing out as work.

PROBLEM-SOLVING STRATEGY: First law of thermodynamics

1. The internal energy change ΔU in any thermodynamic process or series of processes is independent of the path, whether or not the substance is an ideal gas. This is of utmost importance in the problems in this chapter and the next. Sometimes you will be given enough information about one path between given initial and final states to calculate ΔU for that path. Then you can use the fact that ΔU is the same for every other path between the same two states to relate the various energy quantities for other paths.

2. As usual, consistent units are essential. If p is in Pa and V in m³, then W is in joules. Otherwise you may want to convert the pressure and volume units into Pa and m³. If a heat capacity is given in terms of calories,

usually the simplest procedure is to convert it to joules. Be especially careful with moles. When using $n = m/M$ to convert between mass and number of moles, remember that if m is in kilograms, M must be in kilograms per mole. The usual units for M are *grams* per mole, so be careful.

3. When a process consists of several distinct steps, it is often helpful to make a chart showing Q, W, and ΔU for each step. Put these quantities for each step on a different line, and arrange them so the Q's, W's, and ΔU's form columns. Then you can apply the first law to each line; in addition, you can add each column and apply the first law to the sums. Do you see why?

An example of the usefulness of internal energy and the first law

EXAMPLE 18–2 A thermodynamic process is shown in the pV-diagram of Fig. 18–5. In process ab, 600 J of heat are added to the system, and in process bd 200 J of heat are added. Find (a) the internal energy change in process ab; (b) the internal energy change in process abd; and (c) the total heat added in process acd.

SOLUTION

a) No volume change occurs in process ab, so $W = 0$ and $\Delta U = Q = 600$ J.

b) Process bd occurs at constant pressure, so the work done by the system during this expansion is

$$W = p(V_2 - V_1) = (8 \times 10^4 \text{ Pa})(5 \times 10^{-3} \text{ m}^3 - 2 \times 10^{-3} \text{ m}^3)$$
$$= 240 \text{ J}.$$

Thus the total work for abd is $W = 240$ J, and the total heat is $Q = 800$ J. Equation (18–6) gives

$$\Delta U = Q - W = 800 \text{ J} - 240 \text{ J} = 560 \text{ J}.$$

c) Because ΔU is independent of path, the internal energy change is the same for path acd as for abd, that is, 560 J. The total work for the path acd is

$$W = p(V_2 - V_1)$$
$$= (3 \times 10^4 \text{ Pa})(5 \times 10^{-3} \text{ m}^3 - 2 \times 10^{-3} \text{ m}^3)$$
$$= 90 \text{ J}.$$

Equation (18–7) then gives

$$Q = \Delta U + W = 560 \text{ J} + 90 \text{ J} = 650 \text{ J}.$$

We see that although ΔU is the same for abd and acd, W and Q are quite different for the two processes.

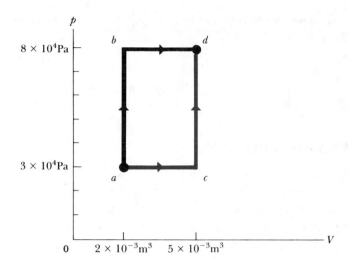

18–5 pV-diagram for Example 18–2, showing the various thermodynamic processes.

EXAMPLE 18–3   One gram of water (1 cm^3) becomes 1671 cm^3 of steam when boiled at a constant pressure of 1 atm. The heat of vaporization at this pressure is 2256 J·g^{-1}. Compute (a) the work done by the water when it vaporizes and (b) its increase in internal energy.

The first law applied to boiling: Where does the heat of vaporization go?

SOLUTION

a)   For a constant-pressure process we may use Eq. (18–3) to compute the work done by the water:

$$W = p(V_2 - V_1)$$
$$= (1.013 \times 10^5 \text{ Pa})(1671 \times 10^{-6} \text{ m}^3 - 1 \times 10^{-6} \text{ m}^3)$$
$$= 169 \text{ J}.$$

b)   The heat added is the heat of vaporization:

$$Q = mL_V$$
$$= (1 \text{ g})(2256 \text{ J·g}^{-1}) = 2256 \text{ J}.$$

From the first law of thermodynamics, Eq. (18–6), the change in internal energy is

$$\Delta U = Q - W = 2256 \text{ J} - 169 \text{ J} = 2087 \text{ J}.$$

That is, when one gram of water vaporizes, 2256 J of heat are added. Of this added energy, 2087 J remain in the system as an increase in internal energy, and the remaining 169 J leave the system again as it does work against the surroundings while expanding from liquid to vapor.

In the examples above we have applied the first law of thermodynamics to processes in which the initial and final states differ in pressure, volume, and temperature by a finite amount. Later we will often want to consider an *infinitesimal* change of state in which a small amount of heat dQ is transferred into the system, the system does a small amount of work dW, and its internal energy changes by an amount dU. For such a process, we state the first law as

The first law for an infinitesimal process

$$dU = dQ - dW. \tag{18–8}$$

Also, in the systems we have been considering, the work dW is given by $dW = p\, dV$, and we can write the first law as

$$dU = dQ - p\, dV. \tag{18-9}$$

Equations (18–8) and (18–9) are often called the *differential form* of the first law of thermodynamics; we will use them several times later in this chapter.

18–5 THERMODYNAMIC PROCESSES

Here are four specific kinds of thermodynamic processes that occur often enough in practical problems to be worth special mention. Each one permits a particular simplification in applying the first law of thermodynamics, at least in some cases. Their characteristics can be summarized briefly as "no heat transfer," "constant volume," "constant pressure," and "constant temperature," respectively.

In an adiabatic process, no heat is transferred.

1. **Adiabatic process**   An **adiabatic process** is one in which there is no heat transfer either into or out of a system; in other words, $Q = 0$. We can prevent heat flow either by surrounding the system with thermally insulating material or by carrying out the process so quickly that there is not enough time for appreciable heat flow. From the first law, we find for every adiabatic process

$$U_2 - U_1 = \Delta U = -W \qquad \text{(adiabatic process).} \tag{18-10}$$

   If a system expands under adiabatic conditions, W is positive, ΔU is negative, and the internal energy decreases; when a system is compressed adiabatically, W is negative and U increases. An increase of internal energy is usually, although not always, accompanied by a rise in temperature.

   The compression stroke in an internal-combustion engine is an example of a process that is approximately adiabatic. The temperature rises as the air-fuel mixture in the cylinder is compressed. Similarly, the expansion of the burned fuel during the power stroke is an approximately adiabatic expansion with a drop in temperature.

In an isochoric process, no work is done.

2. **Isochoric process**   **Isochoric** means simply *constant-volume*. When the volume of a thermodynamic system is constant, it does no work on its surroundings. Thus $W = 0$, and

$$U_2 - U_1 = \Delta U = Q \qquad \text{(isochoric process).} \tag{18-11}$$

   In an isochoric process, all heat added remains in the system as an increase in internal energy. Heating a gas in a closed constant-volume container is an example of an isochoric process.

In an isobaric process, it is easy to calculate the work.

3. **Isobaric process**   **Isobaric** means *constant-pressure*. In general, *none* of the three quantities ΔU, Q, and W in the first law is zero, but calculating W is easy, as we have already seen in Section 18–2. For an isobaric process the integral in Eq. (18–2) becomes simply

$$W = p(V_2 - V_1) \qquad \text{(isobaric process).} \tag{18-12}$$

Example 18–3 (Section 18–4) is an example of an isobaric process.

4. **Isothermal process** **Isothermal** means *constant-temperature.* For a process to be isothermal, the system must remain in thermal equilibrium; this means that the pressure and volume changes must occur slowly enough to maintain thermal equilibrium. In general, *none* of the quantities ΔU, Q, and W is zero. Example 18–1 (Section 18–2) is an example of an isothermal process.

There are some special cases, considered later in this chapter, in which the internal energy of a system depends *only* in its temperature, not on its pressure or volume. The most familiar system having this property is an ideal gas. For such a system, if the temperature is constant, the internal energy is also constant; in this case, $\Delta U = 0$ and $Q = W$. That is, any energy entering the system as heat Q must leave it again as work W done by the system. This does *not* hold for systems other than ideal gases, because in general the internal energy of a system depends on pressure as well as temperature, and it may vary even when T is constant.

Isothermal means constant-temperature.

18–6 INTERNAL ENERGY OF AN IDEAL GAS

We mentioned in Section 18–5 that for an ideal gas the internal energy U depends only on temperature, not on pressure or volume. To explore this special property in more detail, imagine a thermally insulated container with rigid walls, divided into two compartments by a partition, as in Fig. 18–4b. Suppose there is a gas in one compartment and that the other contains only vacuum. When the partition is removed or broken, the gas expands to fill both parts of the container. This process is called a *free expansion;* the gas does no work on its surroundings because the total volume of the container does not change. No heat is transferred because of the thermal insulation. From the first law, since both Q and W are zero, it follows that *the internal energy remains unchanged during a free expansion.* This is true of any substance, whether it is an ideal gas or not.

In a free expansion the internal energy is constant.

Whether or not the *temperature* of a gas changes during a free expansion is an important question, for it provides information concerning the dependence of internal energy on other quantities. If the temperature should change while the internal energy stays the same, we would have to conclude that the internal energy depends on both the temperature and the volume, or both the temperature and the pressure, but certainly not on the temperature alone. If, on the other hand, T remains unchanged during a free expansion in which we know U remains unchanged, then the only admissible conclusion is that *U is a function of T only,* not of p or V.

Experiments have shown that *in a free expansion an ideal gas undergoes no temperature change.* That is, when the behavior of a gas is described well by the ideal-gas equation of state, it is found that in a free expansion its temperature does not change. Thus *the internal energy of an ideal gas depends only on its temperature, not on its pressure or volume.* We may regard this property as part of the ideal-gas model, in addition to the ideal-gas equation of state. This property will be applied several times in the following sections.

The internal energy of an ideal gas does not depend on its pressure or volume.

For real gases that do not precisely obey the ideal-gas equation, experiments have shown that there is some temperature change during free expansions. For such gases, the internal energy does depend to some extent on the pressure as well as the temperature.

18–7 HEAT CAPACITIES OF AN IDEAL GAS

A material has many different heat capacities. Which one do you mean?

The temperature of a substance may be changed under a variety of conditions. The volume may be kept constant, or the pressure may be kept constant, or both may be allowed to vary in some definite way. In general, the amount of heat per mole needed to cause a unit rise in temperature is different for each type of process. In other words, a substance has *many different* molar heat capacities, depending on the conditions under which it is heated or cooled.

The basis of this variation is the first law of thermodynamics. In a constant-volume temperature change, no work is done, and the change in internal energy equals the heat added. In a constant-pressure temperature change, however, the volume *must* increase—otherwise the pressure could not remain constant—and as the material expands it does work. Thus for a given temperature change, the heat input for a constant-pressure process must be *greater* than for a constant-volume process because, in the former, additional energy must be supplied to account for the work done in the expansion. If the volume were to *decrease* during heating, the heat input would be *less* than in the constant-volume case.

The discussion of heat capacities in Chapter 15 ignored this distinction. Indeed, for solid and liquid materials the difference is not usually of great practical importance. The coefficients of volume expansion of such materials are quite small, and ordinarily the work done against the surroundings is so small compared to the total heat added that it may be ignored. The distinction *is* important in theoretical studies of heat capacities.

Heat capacities for constant-volume and constant-pressure processes

For gases the situation is quite different, and the dependence of heat capacity on the details of the process is significant. For air, for example, the heat capacity under constant-pressure conditions is 40 percent greater than under constant-volume conditions. Fortunately, for ideal gases we can derive a simple relation between two heat capacities of special importance.

From a practical standpoint, the two conditions of greatest interest are raising the temperature of a gas at *constant volume*, as in a container whose volume may be considered constant, and at *constant pressure,* where the gas expands just enough so the pressure remains constant. The two corresponding molar heat capacities are called the **molar heat capacity at constant volume,** denoted as C_v, and the **molar heat capacity at constant pressure,** denoted as C_p. The discussion above shows that ordinarily C_p should be larger than C_v. We will now derive the relation between these two quantities for an ideal gas.

First consider the constant-volume process. We place n moles of an ideal gas at temperature T in a rigid container and bring it in contact with a body at a slightly higher temperature $T + dT$. A quantity of heat dQ flows into the gas, and by definition of the molar heat capacity at constant volume, C_v,

$$dQ = nC_v \, dT \qquad \text{(constant volume)}. \qquad (18\text{--}13)$$

In a constant-pressure process the system expands and does work.

The pressure of the gas increases during this process, but no *work* is done because the volume is constant. Hence, $W = 0$; from the first law in differential form, Eq. (18–8),

$$dU = dQ - dW = dQ,$$

we also have

$$dU = nC_v \, dT. \qquad (18\text{--}14)$$

Now consider a constant-pressure process. We enclose the gas in a cylinder with a piston that moves so as to maintain constant pressure. Again the system is brought in contact with a body at a temperature $T + dT$. As heat flows into the gas, it expands at constant pressure and does work. By definition of the molar heat capacity at constant pressure, C_p, the heat dQ flowing into the gas is

$$dQ = nC_p \, dT \qquad \text{(constant pressure).} \qquad (18\text{--}15)$$

The work dW done by the gas is

$$dW = p \, dV.$$

This may also be expressed in terms of the temperature change dT by use of the ideal-gas equation of state $pV = nRT$. Since p is constant, we have

$$p \, dV = nR \, dT,$$

so

$$dW = nR \, dT. \qquad (18\text{--}16)$$

Then, from the differential form of the first law of thermodynamics,

$$dQ = dU + dW,$$

or, using Eqs. (18–15) and (18–16),

$$nC_p \, dT = dU + nR \, dT. \qquad (18\text{--}17)$$

Now, here comes the crux of the calculation. The internal energy change dU is again given by Eq. (18–14), that is, $dU = nC_v \, dT$, *even though now the volume is not constant!* Why is this so? Recall the discussion of Section 18–6: One of the special properties of an ideal gas is that its internal energy depends *only* on temperature. Thus the *change* in internal energy in any process must be determined only by the temperature change. That is, if Eq. (18–14) is valid for an ideal gas for one particular kind of process, it must be valid for an ideal gas for *every* kind of process.

> Internal energy change for an ideal gas depends only on temperature change, even when the volume is not constant.

To complete our derivation, we replace dU in Eq. (18–17) by $nC_v \, dT$ and divide each term by the common factor $n \, dT$ to obtain

$$nC_p \, dT = nC_v \, dT + nR \, dT,$$

and

$$C_p = C_v + R. \qquad (18\text{--}18)$$

As predicted, the molar heat capacity of an ideal gas at constant pressure is *greater* than the molar heat capacity at constant volume; the difference is the gas constant R. Of course, R must be expressed in the same units as C_p and C_v, such as $J \cdot mol^{-1} \cdot K^{-1}$.

Although Eq. (18–18) was derived for an ideal gas, it is very nearly true for many real gases at moderate pressures, where their behavior approaches that of an ideal gas. Measured values of C_p and C_v are given in Table 18–1 for some real gases at low pressures; the difference in most cases is approximately $8.31 \, J \cdot mol^{-1} \cdot K^{-1}$.

TABLE 18–1   Molar Heat Capacities of Gases at Low Pressure

Type of Gas	Gas	C_p, $J \cdot mol^{-1} \cdot K^{-1}$	C_v, $J \cdot mol^{-1} \cdot K^{-1}$	$C_p - C_v$	$\gamma = \dfrac{C_p}{C_v}$
Monatomic	He	20.78	12.47	8.31	1.67
	A	20.78	12.47	8.31	1.67
Diatomic	H_2	28.74	20.42	8.32	1.41
	N_2	29.07	20.76	8.31	1.40
	O_2	29.41	21.10	8.31	1.40
	CO	29.16	20.85	8.31	1.40
Polyatomic	CO_2	36.94	28.46	8.48	1.30
	SO_2	40.37	31.39	8.98	1.29
	H_2S	34.60	25.95	8.65	1.33

The last column of Table 18–1 lists the values of the ratio C_p/C_v, denoted by the Greek letter γ (gamma):

$$\gamma = \frac{C_p}{C_v}. \tag{18–19}$$

The ratio of two heat capacities: a quantity that will come in handy later

The quantity γ plays an important role in adiabatic processes for an ideal gas, which we will study in the next section. We see that γ is 1.67 for monatomic gases and about 1.40 for diatomic gases. There is no simple regularity for polyatomic gases. Table 18–1 also shows that the molar heat capacity of a gas seems to be related to its molecular structure; all the *monatomic* gases have approximately equal values of C_v, and all the *diatomic* gases have about equal values. This is not an accident; in Chapter 20 we will undertake a more detailed theoretical study of heat capacities of gases.

Solids and liquids also expand with increasing temperature, if free to do so, and hence perform work. The coefficients of volume expansion of solids and liquids are, however, so much smaller than those of gases that the work is small. For solids and liquids, C_p is usually greater than C_v, but the difference is small and is not expressible as simply as for a gas. Because of the large stresses set up when solids or liquids are heated and *not* allowed to expand, most heating processes involving them take place at constant pressure. Hence C_p (rather than C_v) is the quantity usually measured for a solid or liquid.

The relation of heat capacities as seen on a pV-diagram

Finally, we emphasize once more that for an ideal gas, the internal energy change in *any* process is given by Eq. (18–14).

$$dU = nC_v \, dT,$$

whether the volume is constant or not. This relation holds for other substances *only* when the volume is constant.

This discussion may be represented graphically by using isotherms on a pV-diagram. Figure 18–6 shows two isotherms for an ideal gas, one representing possible states of the system at temperature T, the other at $T + dT$. Since the internal energy of an ideal gas depends only on the temperature, the internal energy is constant if the temperature is constant; the isotherms are also curves of *constant internal energy.* The internal energy therefore has a constant value U at every point on the isotherm at temperature T, and a constant value $U + dU$ at every point on the isotherm at $T + dT$. It follows that the *change* in internal energy, dU, is the same in all processes in which the gas

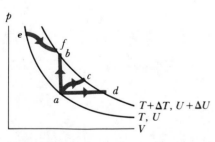

18–6 The change in internal energy of an ideal gas is the same in all processes between two given temperatures.

is taken from *any* point on one isotherm to *any* point on the other. The dU is the same for all the processes *ab*, *ac*, *ad*, and *ef* in Fig. 18–6. In particular, *ab* represents constant-volume heating from temperature T to $T + dT$, and *ad* represents constant-pressure heating through the same temperature interval. No work is done in process *ab*; the work in process *ad* is the area under the curve.

18–8 ADIABATIC PROCESSES FOR AN IDEAL GAS

An adiabatic process, defined in Section 18–5, is a process in which no heat transfer takes place between a system and its surroundings. A completely adiabatic process is an idealization, but a process may be approximately adiabatic if the system is well insulated or if the process takes place so rapidly that there is no time for appreciable heat flow to occur. The compression stroke in a gasoline or diesel engine is thus approximately adiabatic.

An **adiabatic process for an ideal gas** is shown on the pV-diagram of Fig. 18–7. As the gas expands from volume V_a to V_b, its temperature drops because of the decrease in internal energy. If the point representing the initial state lies on an isotherm at temperature $T + \Delta T$, then the point for the final state is on a different isotherm at a lower temperature T. Thus an adiabatic curve or *adiabat* at any point is always steeper than the isotherm passing through the same point. For an adiabatic *compression* from V_b to V_a, the situation is reversed and the temperature rises.

The air in the output pipe of an air compressor, as used in gasoline stations and paint-spraying equipment, is always warmer than that entering the compressor, a result of the approximately adiabatic compression. When air is compressed in the cylinders of a diesel engine during the compression stroke, it becomes so hot that fuel injected into the cylinders during the power stroke ignites spontaneously.

It is useful to have quantitative relations between volume and temperature, or volume and pressure, for an adiabatic process. Consider first an infinitesimal change of state in which the temperature changes by dT and the volume by dV. Equation (18–14) gives the internal energy change for *any* process for an ideal gas, adiabatic or not, so we have $dU = nC_v\,dT$. Also, the work done by the gas during the process is given by $dW = p\,dV$. Then, since $dU = -dW$ for an adiabatic process, we have

$$nC_v\,dT = -p\,dV. \qquad (18\text{–}20)$$

To obtain a relation containing only the variables T and V, we may eliminate p by using the ideal-gas equation in the form $p = nRT/V$. Substituting this in Eq. (18–20) and rearranging, we obtain

$$nC_v\,dT = -\frac{nRT}{V}\,dV,$$

$$\frac{dT}{T} + \frac{R}{C_v}\frac{dV}{V} = 0,$$

or, since $R/C_v = (C_p - C_v)/C_v = C_p/C_v - 1 = \gamma - 1$,

$$\frac{dT}{T} + (\gamma - 1)\frac{dV}{V} = 0. \qquad (18\text{–}21)$$

In an adiabatic process, there is no heat transfer into or out of the system.

Adiabatic compression causes a rise in temperature. How much?

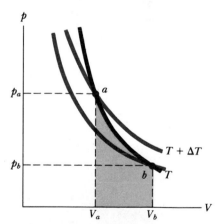

18–7 pV-diagram of an adiabatic process for an ideal gas. As the gas expands from V_a to V_b, its temperature drops from $T + \Delta T$ to T, corresponding to the decrease in internal energy due to the work done by the gas (indicated by shaded area). For an ideal gas, when an isotherm and an adiabat pass through the same point, the adiabat is always steeper.

This gives the relation between dT and dV for an infinitesimal adiabatic process. For a finite change, we integrate both sides of Eq. (18–21), obtaining

$$\ln T + (\gamma - 1) \ln V = \text{constant},$$
$$\ln T + \ln V^{\gamma-1} = \text{constant},$$
$$\ln [TV^{\gamma-1}] = \text{constant},$$

and finally

$$TV^{\gamma-1} = \text{constant}. \qquad (18\text{–}22)$$

Thus for an initial state (T_1, V_1) and a final state (T_2, V_2),

$$T_1 V_1^{\gamma-1} = T_2 V_2^{\gamma-1}. \qquad (18\text{–}23)$$

Temperature–volume relation for an adiabatic process

In applications of Eqs. (18–22) and (18–23) it is important to remember that the T's must be *absolute* temperatures (Kelvin or Rankine).

We can also convert Eq. (18–22) into a relation between pressure and volume by eliminating T, using the ideal-gas equation in the form $T = pV/nR$. We obtain

$$\left(\frac{pV}{nR}\right) V^{\gamma-1} = \text{constant},$$

or, since n and R are constants,

$$pV^{\gamma} = \text{constant}. \qquad (18\text{–}24)$$

Thus for an initial state (p_1, V_1) and a final state (p_2, V_2),

$$p_1 V_1^{\gamma} = p_2 V_2^{\gamma}. \qquad (18\text{–}25)$$

Why don't diesel engines need spark plugs? What ignites the fuel?

EXAMPLE 18–4   The compression ratio of a certain diesel engine is 15. This means that air in the cylinders is compressed to $\frac{1}{15}$ of its initial volume. If the initial pressure is 1.0×10^5 Pa and the initial temperature is 27°C (= 300 K), find the final pressure and temperature after compression. Air is mostly a mixture of oxygen and nitrogen, and $\gamma = 1.40$.

SOLUTION   From Eq. (18–23),

$$T_2 = T_1\left(\frac{V_1}{V_2}\right)^{\gamma-1} = (300 \text{ K})(15)^{0.40} = 886 \text{ K} = 613°C.$$

From Eq. (18–25),

$$p_2 = p_1\left(\frac{V_1}{V_2}\right)^{\gamma} = (1.0 \times 10^5 \text{ Pa})(15)^{1.40}$$
$$= 44.3 \times 10^5 \text{ Pa} \cong 44 \text{ atm}.$$

If the compression had been isothermal, the final pressure would have been 15 atm, but since the temperature also increases, the final pressure is much greater. The high temperature attained during compression causes the fuel to ignite spontaneously without the need for spark plugs, when it is injected into the cylinders at the end of the compression stroke.

How much work does a gas do during an adiabatic expansion?

Because $W = -dU$ for an adiabatic process, it is easy to calculate the work done by an ideal gas during an adiabatic process. If the initial and final temperatures are known, we have simply

$$W = nC_v(T_1 - T_2). \tag{18–26}$$

Or, if the pressure or volumes are known, we may use $pV = n\,RT$ to obtain

$$W = \frac{C_v}{R}(p_1V_1 - p_2V_2) = \frac{1}{\gamma - 1}(p_1V_1 - p_2V_2). \tag{18–27}$$

Note that if the process is an expansion, the temperature drops, T_1 is greater than T_2, p_1V_1 is greater than p_2V_2, and the work is *positive*, as we should expect. If it is a compression, the work is negative.

EXAMPLE 18–5 In Example 18–4, how much work does the gas do during the compression if the initial volume of the cylinder is $1.0\ \text{L} = 1.0 \times 10^{-3}\ \text{m}^3$?

SOLUTION We may determine the number of moles and then use Eq. (18–26), or we may use Eq. (18–27). In the first case we have

Work in a diesel-engine cylinder

$$n = \frac{p_1V_1}{RT_1} = \frac{(1.0 \times 10^5\ \text{Pa})(1.0 \times 10^{-3}\ \text{m}^3)}{(8.314\ \text{J}\cdot\text{mol}^{-1}\cdot\text{K}^{-1})(300\ \text{K})} = 0.040\ \text{mol}.$$

C_v for air is $20.8\ \text{J}\cdot\text{mol}^{-1}\cdot\text{K}^{-1}$; so

$$\begin{aligned}
W &= nC_v(T_1 - T_2)\\
&= (0.040\ \text{mol})(20.8\ \text{J}\cdot\text{mol}\cdot\text{K}^{-1})(300\ \text{K} - 886\ \text{K})\\
&= -488\ \text{J}.
\end{aligned}$$

In the second case,

$$\begin{aligned}
W &= \frac{1}{1.40 - 1}\Big[(1.0 \times 10^5\ \text{Pa})(1.0 \times 10^{-3}\ \text{m}^3)\\
&\qquad\qquad - (44.3 \times 10^5\ \text{Pa})\Big(\frac{1.0}{15} \times 10^{-3}\ \text{m}^3\Big)\Big]\\
&= -488\ \text{J}.
\end{aligned}$$

The work is negative because the gas is compressed.

In the analysis above we have used the ideal-gas equation of state, which holds only when the state of the gas changes slowly enough so that at each step the pressure and temperature are *uniform* throughout the gas. Thus the validity of our results is limited to situations where the process is rapid enough to prevent appreciable heat exchange with the surroundings, yet slow enough so that the system does not depart very much from thermal and mechanical equilibrium.

SUMMARY

A thermodynamic system is a system that can exchange energy with its surroundings by heat transfer, through mechanical work, and in some cases by other mechanisms. When a system at pressure p expands from volume V_1 to V_2, it does an amount of work W given by

$$W = \int_{V_1}^{V_2} p\,dV. \tag{18–2}$$

KEY TERMS

thermodynamic system
free expansion
internal energy
first law of thermodynamics
adiabatic process
isochoric process

If the pressure p is constant during the expansion, the work W is given by

$$W = p(V_2 - V_1). \tag{18-3}$$

In a thermodynamic process leading from an initial state to a final state, the heat added to the system and the work done by the system depend not only on the initial and final states but also on the path, that is, the series of intermediate states through which the system passes during the process.

In a thermodynamic process where heat Q is added to a system while it does work W, the internal energy U changes by an amount

$$\Delta U = U_2 - U_1 = Q - W. \tag{18-6}$$

In an infinitesimal process,

$$dU = dQ - dW = dQ - p \, dV. \tag{18-8}$$

The internal energy of any thermodynamic system depends only on its state, not on the processes by which it reached that state. The change in internal energy in any process depends only on the initial and final states, not on the path (the intermediate states through which the system passes). The internal energy of an isolated system is constant.

Four kinds of thermodynamic processes that occur are:

Adiabatic process: A process where there is no heat transfer in or out of a system. For any adiabatic process, $Q = 0$.

Isochoric process: A process that takes place at constant volume. For any isochoric process, $W = 0$.

Isobaric process: A process that takes place at constant pressure. For any isobaric process, $W = p(V_2 - V_1)$.

Isothermal process: A process that takes place at constant temperature.

The internal energy of an ideal gas depends only on its temperature, not on its pressure or volume. For other substances, the internal energy in general depends on both pressure and temperature.

The molar heat capacities of an ideal gas at constant volume (C_v) and at constant pressure (C_p) are related by

$$C_p = C_v + R, \tag{18-18}$$

where R is the ideal gas constant. The ratio of C_p to C_v is denoted by γ:

$$\gamma = \frac{C_p}{C_v}. \tag{18-19}$$

For an adiabatic process for an ideal gas, the quantities $TV^{\gamma-1}$ and pV^{γ} are constant. For an initial state (p_1, V_1, T_1) and a final state (p_2, V_2, T_2), the corresponding relations are

$$T_1 V_1^{\gamma-1} = T_2 V_2^{\gamma-1} \tag{18-23}$$

and

$$p_1 V_1^{\gamma} = p_2 V_2^{\gamma}. \tag{18-25}$$

QUESTIONS

18–1 It is not correct to say that a body contains a certain amount of heat, yet a body can transfer heat to another body. How can a body give away something it does not have in the first place?

18–2 Discuss the application of the first law of thermodynamics to a mountaineer who eats food, gets warm and sweats a lot during the climb, and does a lot of mechanical work in raising himself to the summit. What about the descent? One also gets warm during the descent. Is the source of this energy the same as during the ascent?

18–3 How can you cool a room (i.e., take heat out of it) by adding energy to it in the form of electric energy supplied to an air conditioner?

18–4 If you are told the initial and final states of a system and the associated change in internal energy, can you determine whether the internal energy change was due to work or to heat transfer?

18–5 Household refrigerators always have arrays or coils of tubing on the outside, usually at the back or bottom. When the refrigerator is running, the tubing becomes quite hot. Where does the heat come from?

18–6 There are a few materials that contract when heated, such as water between 0°C and 4°C. Would you expect C_p for such a material to be greater or less than C_v?

18–7 When ice melts (decreasing its volume), is the internal energy change greater or less than the heat added?

18–8 When one drives a car in cool, foggy weather, ice sometimes forms in the throat of the carburetor, even though the outside air temperature is above freezing. Why?

18–9 On a warm summer day a large cylinder of compressed gas (propane or butane) was used to supply several large gas burners at a cookout. After a while, frost formed on the outside of the tank. Why?

18–10 Air escaping from an air hose at a gas station always feels cold. Why?

18–11 The prevailing winds blow across the central valley of California and up the western slopes of the Sierra Nevada. They cool as they reach the slopes, and the precipitation there is much greater than in the valley. But what makes them cool?

18–12 Applying the same considerations as in Question 18–11, can you explain why Death Valley, on the opposite side of the Sierra from the central valley, is so hot and dry?

18–13 In the situation of Question 18–11, during certain seasons the wind blows in the opposite direction. Although the mountains are cool, the wind in the valley (called the "Santa Ana," after the notorious Mexican general) is always very hot. What heats it? A similar phenomenon in the Alps is called the "Foehn"; by local legend it is blamed for irrational behavior in man and beast, and has even been used as a defense in murder trials.

18–14 When a gas expands adiabatically, it does work on its surroundings. But if there is no heat input to the gas, where does the energy come from?

18–15 In a constant-volume process, $dU = nC_v\,dT$, but in a constant-pressure process it is *not* true that $dU = nC_p\,dT$. Why not?

18–16 Since C_v is defined with specific reference to a constant-volume process, how can it be correct that for an ideal gas $dU = nC_v\,dT$ even when the volume is not constant?

EXERCISES

Section 18–2 Work Done during Volume Changes

18–1 Two moles of oxygen are in a container with rigid walls. The gas is heated until the pressure doubles. Neglect the thermal expansion of the container. Calculate the work done by the gas.

18–2 A gas under a constant pressure of 3×10^5 Pa and an initial volume of 0.05 m³ is cooled until its volume becomes 0.04 m³. Calculate the work done by the gas.

18–3 Three moles of an ideal gas are heated at constant pressure from $T = 27°C$ to 127°C. Calculate the work done by the gas.

18–4 One mole of an ideal gas has an initial temperature of 27°C. While the temperature is kept constant, the volume is decreased until the pressure doubles. Calculate the work done by the gas.

Section 18–4 Internal Energy and the First Law

18–5 A student performs a combustion experiment by burning a mixture of fuel and oxygen in a constant-volume "bomb" surrounded by a water bath. During the experiment the temperature of the water is observed to rise. Regarding the mixture of fuel and oxygen as the system,

a) has heat been transferred?

b) has work been done?

c) what is the sign of ΔU?

18–6 A liquid is irregularly stirred in a well-insulated container and thereby undergoes a rise in temperature. Regarding the liquid as the system,

a) has heat been transferred?

b) has work been done?

c) what is the sign of ΔU?

18–7 A gas in a cylinder expands from a volume of 0.4 m³ to 0.6 m³. Heat is added just rapidly enough to keep the pressure constant at 2.0×10^5 Pa during the expansion. The total heat added is 1.2×10^5 J.

a) Find the work done by the gas.

b) Find the change in internal energy of the gas.

c) Does it matter whether or not the gas is ideal?

18–8 A gas in a cylinder is cooled and compressed at a constant pressure of 2.0×10^5 Pa, from 1.2 m³ to 0.8 m³. A quantity of heat of magnitude 2.4×10^5 J is removed from the gas.

a) Find the work done by the gas.

b) Find the change in internal energy of the gas.

c) Does it matter whether or not the gas is ideal?

18–9 When a system is taken from state a to state b, in Fig. 18–8, along the path acb, 80 J of heat flow into the system, and 30 J of work are done.

a) How much heat flows into the system along path adb if the work is 10 J?

b) When the system is returned from b to a along the curved path, the magnitude of the work is 20 J. Does the system absorb or liberate heat, and how much?

c) If $U_a = 0$ and $U_d = 40$ J, find the heat absorbed in the processes ad and db.

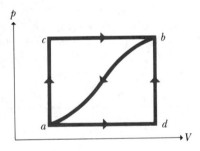

FIGURE 18–8

18–10 When water is boiled under a pressure of 2 atm, the heat of vaporization is 2.20×10^6 J·kg⁻¹ and the boiling point is 120°C. At this pressure, 1 kg of water has a volume of 10^{-3} m³, and 1 kg of steam a volume of 0.824 m³.

a) Compute the work done when 1 kg of steam is formed at this temperature.

b) Compute the increase in internal energy.

Section 18–6 Internal Energy of an Ideal Gas

Section 18–7 Heat Capacities of an Ideal Gas

18–11 Consider the isothermal compression of 0.10 mol of an ideal gas at $T = 0$°C. The initial pressure is 1 atm and the final volume is ⅕ the initial volume.

a) Determine the work required.

b) What is the change in internal energy?

c) Does the gas exchange heat with its surroundings? If so, how much and in what direction?

18–12 A cylinder contains 1 mol of oxygen gas at a temperature of 27°C. The cylinder is provided with a frictionless piston, which maintains a constant pressure of 1 atm on the gas. The gas is heated until its temperature increases to 127°C.

a) Draw a diagram representing the process in the pV-plane.

b) How much work is done by the gas in this process?

c) On what is this work done?

d) What is the change in internal energy of the gas?

e) How much heat was supplied to the gas?

f) How much work would have been done if the pressure had been 0.5 atm?

18–13 A certain ideal gas has $\gamma = 1.33$. Determine the molar heat capacities at constant volume and at constant pressure.

Section 18–8 Adiabatic Processes for an Ideal Gas

18–14 An ideal gas initially at 10 atm and 300 K is permitted to expand adiabatically until its volume doubles. Find the final pressure and temperature if the gas is

a) monatomic;

b) diatomic.

18–15 A gasoline engine takes in air at 20°C and 1 atm, and compresses it adiabatically to $\frac{1}{10}$ the original volume. Find the final temperature and pressure.

18–16 An engineer is designing an engine that runs on compressed air. Air enters the engine at a pressure of 2×10^6 Pa and leaves at a pressure of 2×10^5 Pa. What must be the temperature of the compressed air so that there is no possibility of frost forming in the exhaust ports of the engine? Assume the expansion to be adiabatic. (*Note.* Frost frequently forms in the exhaust ports of an air-driven engine. This happens when the moist air is cooled below 0°C by the expansion that takes place in the engine.)

18–17 During an adiabatic expansion, the temperature of 0.1 mol of oxygen drops from 30°C to 10°C.

a) How much work does the gas do?

b) How much heat is added?

PROBLEMS

18–18 Nitrogen gas in an expandable container is raised from 0°C to 50°C, with the pressure held constant at 4×10^5 Pa. The total heat added is 3.0×10^4 J.

a) Find the number of moles of gas.

b) Find the change in internal energy of the gas.

c) Find the work done by the gas.

d) How much heat would be needed to cause the same temperature change if the volume were constant?

18–19 In a certain process, 2.11×10^5 J of heat are supplied to a system and at the same time the system expands against a constant external pressure of 6.89×10^5 Pa. The internal energy of the system is the same at the beginning and end of the process. Find the increase in volume of the system.

18–20 A chemical engineer studying the properties of glycerin uses a steel cylinder of cross-sectional area 0.01 m² that contains 1×10^{-2} m³ of glycerin. The cylinder is equipped with a tightly fitting piston that supports a load of 3×10^4 N. The temperature of the system is increased from 20°C to 70°C. The coefficient of volume expansion of glycerin is given in Table 14–2. Neglect the expansion of the steel cylinder. Find

a) the increase in volume of the glycerin;

b) the mechanical work of the 3×10^4 N force;

c) the amount of heat added to the glycerin [c_p for glycerin is 243×10^3 J·kg^{-1}·(C°)$^{-1}$];

d) the change in internal energy of the glycerin.

18–21 A pump compressing air from atmospheric pressure into a very large tank at 4.13×10^5 Pa gauge pressure has a cylinder 0.254 m long.

a) At what position in the stroke will air begin to enter the tank? Assume the compression to be adiabatic. (You are being asked to calculate the distance the piston has moved in the cylinder.)

b) If the air is taken into the pump at 27°C, what is the temperature of the compressed air?

18–22 Initially at a temperature of 60°C, 0.283 m³ of air expands at a constant gauge pressure of 1.38×10^5 Pa to a volume of 1.42 m³, and then expands further adiabatically to a final volume of 2.27 m³ and a final gauge pressure of 2.07×10^4 Pa. Sketch the process in the pV-plane and compute the total work done by the air. (Take atmospheric pressure to be 1.01×10^5 Pa.)

18–23 An ideal gas expands slowly to twice its original volume, doing 500 J of work in the process. Find the heat added and the change in internal energy if the process is

a) isothermal;   b) adiabatic.

18–24 In a cylinder, 2.0 mol of an ideal monatomic gas initially at 1.0×10^6 Pa and 300 K expands until its volume doubles. Compute the work done if the expansion is

a) isothermal;   b) adiabatic;   c) isobaric.

d) Show each process on a pV-diagram.

e) In which case is the magnitude of the heat transfer greatest? Least?

f) In which case is the magnitude of the internal-energy change greatest? Least?

18–25 Two moles of helium are initially at a temperature of 27°C and occupy a volume of 0.020 m³. The helium first expands at constant pressure until the volume has doubled, and then adiabatically until the temperature returns to its initial value.

a) Draw a diagram of the process in the pV-plane.

b) What is the total heat supplied in the process?

c) What is the total change in the internal energy of the helium?

d) What is the total work done by the helium?

e) What is the final volume?

18–26 A cylinder with a piston contains 0.5 mol of oxygen at 2×10^5 Pa and 300 K. The gas first expands at constant pressure to twice its original volume; it is then compressed isothermally back to its original volume, and finally it is cooled at constant volume to its original pressure.

a) Show the series of processes on a pV-diagram.

b) Compute the temperature during the isothermal compression.

c) Compute the maximum pressure.

18–27 Use the conditions and processes of Problem 18–26 to compute

a) the work done, the heat added, and the internal-energy change during the initial expansion;

b) the work done, the heat added, and the internal-energy change during the final cooling;

c) the internal-energy change during the isothermal compression.

CHALLENGE PROBLEM

18–28 The van der Waals equation of state, an approximate representation of the behavior of gases at high pressure, is

$$\left(p + \frac{an^2}{V^2}\right)(V - nb) = nRT,$$

where a and b are constants having different values for different gases. (In the special case of $a = b = 0$, this is the ideal-gas equation.) Calculate the work done by a van der Waals gas in an isothermal expansion from V_1 to V_2.

19

THE SECOND LAW OF THERMODYNAMICS

THUS FAR OUR DISCUSSION OF THERMODYNAMICS HAS CENTERED ON THE first law, which expresses conservation of energy in thermodynamic processes. But there is a whole family of questions that the first law cannot answer, having to do with the *directions* of thermodynamic processes. A study of the inherently one-way nature of such processes as the flow of heat from hotter to colder regions and the irreversible conversion of work into heat by friction leads to the *second law of thermodynamics*. This law places fundamental limitations on the efficiency with which an engine can convert heat into useful mechanical work; it also places limitations on the minimum energy input for a refrigerator. Hence the second law is directly relevant for many extremely important practical problems. We can also use the second law to define a temperature scale that is independent of the properties of any specific material. Finally, we restate the second law in terms of the concept of *entropy*, a quantitative measure of the degree of disorder or randomness of a system.

19–1 DIRECTIONS OF THERMODYNAMIC PROCESSES

Some energy-conversion processes go more easily in one direction than in the other.

Many thermodynamic processes proceed naturally in one direction but not the opposite. For example, heat always flows from a hot body to a cooler body, never the reverse. If heat were to flow from cooler to hotter regions, the first law would still be obeyed, but such flow does not occur in nature. Or suppose all the air in a box were to rush to one side, leaving vacuum in the other side, the reverse of the free expansion described in Section 18–3. This does not occur in nature either, although the first law does not forbid it. It is easy to convert mechanical energy completely into heat; we do this every time we use a car's brakes to stop it. It is not easy to convert heat into mechanical energy, and no one has ever built a machine that converts heat *completely* into mechanical energy. The world is full of would-be inventors who would like to cool the air a little, extracting heat from it, and convert that heat to mechanical energy to propel an airplane. No one has ever succeeded with such a scheme.

What these examples have in common is a preferred *direction*. In each case a process proceeds spontaneously in one direction but not in the other. This fact suggests that there must be some physical law that determines what the preferred direction is for a given process. As we will see in the following sections, that law is the *second law of thermodynamics*.

Closely related to the concept of directionality is another concept, **reversibility.** We call a process *reversible* if it involves a system that is always very close to being in thermodynamic equilibrium within itself and with its surroundings. When this is the case, any change of state that takes place can be reversed (i.e., made to go the other way) by making only an infinitesimal change in the conditions of the system. For example, heat flow between two bodies whose temperatures differ only infinitesimally can be reversed by making only a very small change in one temperature or the other. A gas expanding slowly and adiabatically can be compressed slowly by an infinitesimal increase in pressure.

> A reversible process can be made to change direction by an infinitesimal change in conditions.

Reversible processes are thus **equilibrium processes.** By contrast, the processes cited above—heat flow with finite temperature difference, free expansion of a gas, and conversion of work to heat by friction—are all **irreversible processes;** no small change in conditions could make any of them go the other way. They are also all *nonequilibrium* processes.

> An irreversible process is a nonequilibrium process.

You may notice an apparent contradiction in this discussion. If a system is *really* in thermodynamic equilibrium, how can any change of state at all take place? How can heat flow into or out of a system if it is at uniform temperature throughout? How can it start to move in order to expand and do work against its surroundings if it is really in mechanical equilibrium?

The answer to all these questions is that an equilibrium process is an idealization; strictly speaking, such a process can never be precisely attained in the real world. But by carrying out a process *slowly* enough, we can make the temperature gradients and the pressure differences in the substance as small as we like, and thus come as close as we like to the goal of keeping the system in equilibrium states. The term *quasi-equilibrium process* is sometimes used for such a process. We will not use that term; instead, we suggest you keep this discussion in mind when you read the words *equilibrium process*.

Finally, we will find that there is a relation between the direction of a process and the *disorder* or *randomness* of the resulting state. For example, imagine a tedious sorting job, such as alphabetizing a thousand book titles written on file cards. Throw the alphabetized stack of cards into the air. Do they come down in alphabetical order? No; the tendency is for them to come down in a random or disordered state. In the free-expansion example, the air is more disordered after it has expanded into the entire box than when it was confined to one side, because the molecules are scattered over more space.

> Things always tend to get more mixed up.

Similarly, macroscopic kinetic energy is energy associated with organized, coordinated motions of many molecules. Energy associated with heat is random, disordered molecular motion, as we will see in Chapter 20. Hence conversion of mechanical energy into heat involves an increase of randomness or disorder.

In the following section we will explore the implications of the second law of thermodynamics by considering two specific classes of devices, namely, *heat engines,* which are partly successful in converting heat into work, and *refrigerators,* which are partly successful in transporting heat from cooler to hotter bodies.

19–2 HEAT ENGINES

The essence of a technological society is its ability to use sources of energy other than muscle power. Sometimes, as with water power, mechanical energy is directly available. Most of our energy, however, comes from the burning of fossil fuels (coal, oil, and gas) and from nuclear reactions; both of these supply energy that is transferred as *heat*. Some heat can be used directly for heating buildings, for cooking, and for chemical and metallurgical processing; but to operate a machine or propel a vehicle, we need *mechanical* energy.

How do we turn heat into mechanical work?

Thus a problem of the utmost practical importance is how to take heat from a source and convert as much of it as possible into mechanical energy or work. This is exactly what happens in gasoline engines in automobiles, jet engines in airplanes, steam turbines in electric power plants, and many other systems. Closely related processes occur in the animal kingdom, where food energy is "burned" (i.e., carbohydrates combine with oxygen to yield water, carbon dioxide, and energy) and partly converted to mechanical energy as the animal's muscles do work on their surroundings.

Any device for transforming heat into work or mechanical energy is called a **heat engine.** Typically, a certain quantity of matter in the engine undergoes addition and subtraction of heat, expansion and compression, and sometimes change of phase. We call this matter the **working substance** of the engine. In internal-combustion engines the working substance is a mixture of air and burnt fuel; in a steam turbine it is water.

In a cyclic process, a substance goes through the same series of states over and over.

For the sake of simplicity, we will discuss an engine in which the working substance undergoes a **cyclic process,** that is, a sequence of processes that eventually leaves the substance in the same state in which it started. In a steam turbine the water actually is recycled and used over and over. Internal-combustion engines do not use the same air over and over, but we can still analyze them in terms of cyclic processes that approximate the actual operation.

All the heat engines mentioned absorb heat from a source at a relatively high temperature, perform some mechanical work and discard some heat at a lower temperature. As far as the engine is concerned, the discarded heat is wasted. In internal-combustion engines the waste heat is in the hot exhaust gases and the cooling system; in a steam turbine it is heat that must be taken from the used steam to condense and recycle it.

The total internal-energy change in any cyclic process is zero.

When a system is carried through a cyclic process, its initial and final internal energies are equal. From the first law, for any number of complete cycles,

$$U_2 - U_1 = 0 = Q - W,$$
$$Q = W.$$

That is, the *net* heat flowing into the engine in a cyclic process equals the net work done by the engine.

Heat reservoirs: constant-temperature environments where a system can gain or lose heat

In the analysis of heat engines, it is useful to think of two bodies with which the working substance of the engine can interact. One of these, called the *hot reservoir*, can give the working substance large amounts of heat without appreciably changing its own temperature. The other body, called the *cold reservoir*, can absorb large amounts of discarded heat from the engine, also without appreciable change in its temperature. Thus in a steam-turbine system the flames and hot gases in the boiler are the hot reservoir, and the cold water and air used to condense and cool the used steam are the cold reservoir.

Heat transferred from the hot and cold reservoirs will be denoted by Q_H and Q_C, respectively. Each Q is considered positive when heat is transferred *from* a reservoir *to* the working substance, negative when the reverse. Thus in a heat engine, Q_H is positive but Q_C is negative, representing heat *leaving* the working substance. In the following discussion this sign convention may sometimes seem awkward; the alternatives are not any better, however, and it is best to keep the sign convention we have already learned for quantity of heat added to a body.

We can represent the energy transformations in a heat engine by the *flow diagram* of Fig. 19–1. The engine itself is represented by the circle. The heat Q_H supplied to the engine by the hot reservoir is proportional to the cross section of the incoming "pipeline" at the top of the diagram. The cross section of the outgoing pipeline at the bottom is proportional to the magnitude $|Q_C|$ of the heat rejected in the exhaust. The branch line to the right represents the portion of the heat supplied that the engine converts to mechanical work, W.

As mentioned above, we are talking about engines that repeat the same cycle over and over. In this case Q_H and Q_C represent the quantities of heat absorbed and rejected by the engine *during one cycle*. The *net* heat absorbed per cycle is

$$Q = Q_H + Q_C = Q_H - |Q_C|, \qquad (19\text{–}1)$$

where Q_C is a negative number. The useful output of the engine is the net work W done by the working substance; from the first law,

$$W = Q = Q_H + Q_C. \qquad (19\text{–}2)$$

Ideally we would like to convert *all* the heat Q_H into work; in that case we would have $Q_H = W$ and $Q_C = 0$. Experience shows that this is impossible; some heat is always wasted, and Q_C is never zero. We define the **thermal efficiency** of an engine, denoted by e, as the quotient

$$e = \frac{W}{Q_H}. \qquad (19\text{–}3)$$

Thus e represents the fraction of Q_H that *is* converted to work.

To put it another way, e is what you get, divided by what you pay for. This value is always less than unity, an all-too-familiar experience! In terms of the flow diagram of Fig. 19–1, the most efficient engine is the one for which the branch pipeline representing the work output is as *large* as possible and the exhaust pipeline representing the heat thrown away is as small as possible.

Using Eq. (19–2) and keeping in mind the signs of Q_H and Q_C, we can write the following equivalent expressions:

$$\begin{aligned}
e = \frac{W}{Q_H} &= \frac{Q_H + Q_C}{Q_H} \\
&= 1 + \frac{Q_C}{Q_H} \\
&= 1 - \left| \frac{Q_C}{Q_H} \right|. \qquad (19\text{–}4)
\end{aligned}$$

Note that e is a quotient of two energy quantities and thus is a pure number, without units. Of course, W and Q_H must always be expressed in the same units.

Be careful about algebraic signs of heat quantities.

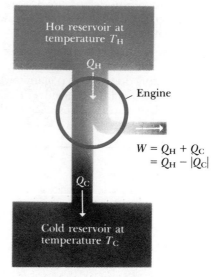

$$W = Q_H + Q_C$$
$$= Q_H - |Q_C|$$

19–1 Schematic flow diagram of a heat engine.

Thermal efficiency: What fraction of the heat you pay for can you convert to useful work?

PROBLEM-SOLVING STRATEGY: *Heat engines*

We recommend you reread the strategy in Section 18–4; these suggestions are equally useful throughout the present chapter. The following points may need additional emphasis.

1. Be very careful with the sign conventions for W and the various Q's; W is positive when the system expands and does work, negative when it is compressed. Each Q is positive if it represents heat entering the working substance of the engine or other system, negative when heat leaves the system. When in doubt, use the first law if possible, to check consistency.

2. Some problems deal with power rather than energy quantities. Keep in mind that power is work per unit time ($P = W/t$,) and that rate of heat transfer (heat current) H is heat transfer per unit time ($H = Q/t$). Sometimes it helps to ask: "What is W or Q in one second (or one hour)?"

Thermal efficiency of a gasoline engine: How much heat from burning gasoline is wasted?

EXAMPLE 19–1 A gasoline engine takes in 2500 J of heat and delivers 500 J of mechanical work per cycle. The heat is obtained by burning gasoline having a heat of combustion of $L_C = 5.0 \times 10^4$ J·g^{-1}.

a) What is the thermal efficiency of this engine?

b) How much heat is discarded in each cycle?

c) How much gasoline is burned in each cycle?

d) If the engine goes through 100 cycles per second, what is its power output in watts? In horsepower?

e) How much gasoline is burned per second? per hour?

SOLUTION We have $Q_H = 2500$ J and $W = 500$ J.

a) From Eq. (19–3) the thermal efficiency is

$$e = \frac{W}{Q_H} = \frac{500 \text{ J}}{2500 \text{ J}} = 0.20 = 20\%.$$

b) From Eq. (19–2)

$$W = Q_H + Q_C,$$
$$500 \text{ J} = 2500 \text{ J} + Q_C, \qquad Q_C = -2000 \text{ J}.$$

That is, 2000 J of heat leave the engine during each cycle.

c) If m is the mass of gasoline burned, then

$$Q_H = mL_C,$$
$$2500 \text{ J} = m(5.0 \times 10^4 \text{ J·g}^{-1}),$$
$$m = 0.05 \text{ g}.$$

d) The rate of doing work P is the work per cycle multiplied by the number of cycles per second:

$$P = (500 \text{ J})(100 \text{ s}^{-1}) = 50,000 \text{ W} = 50 \text{ kW}$$
$$= (50,000 \text{ W})\left(\frac{1 \text{ hp}}{746 \text{ W}}\right)$$
$$= 67 \text{ hp}.$$

e) The mass of gasoline burned per second is the mass per cycle multiplied by the number of cycles per second:

$$(0.05 \text{ g})(100 \text{ s}^{-1}) = 5 \text{ g·s}^{-1}.$$

The mass burned per hour is

$$(5 \text{ g·s}^{-1})\left(\frac{3600 \text{ s}}{1 \text{ hr}}\right) = 18,000 \text{ g·hr}^{-1} = 18 \text{ kg·hr}^{-1}.$$

The density of gasoline is about 0.70 g·cm^{-3}, so this is about 25,700 cm^3, 25.7 L, or 6.8 gal of gasoline per hour.

19–3 INTERNAL-COMBUSTION ENGINES

For our first example of a heat engine and a calculation of thermal efficiency, let us consider the common gasoline engine, as found in automobiles and many other types of machinery. The sequence of processes in the cycle is shown in Fig. 19–2. First a mixture of air and gasoline vapor flows into a cylinder through an open intake valve while the piston descends, increasing the volume of the cylinder from a minimum of V (when the piston is all the way up) to a maximum of rV (when it is all the way down). The ratio r is called the **compression ratio,** and for present-day automobile engines it is typically about 8. At the end of this *intake stroke* the intake valve closes, and the mixture is compressed to volume V during the *compression stroke*. The mixture is then ignited by the spark plug, and the heated gas expands back to volume rV, pushing on the piston and doing work; this is the *power stroke*. Finally, the exhaust valve opens and the combustion products are pushed out (during the *exhaust stroke*) to prepare the cylinder for the next intake stroke.

Figure 19–3 is a pV-diagram showing an idealized model of the corresponding thermodynamic processes. This model is called the **Otto cycle.** At point *a* the gasoline-air mixture has entered the cylinder. The mixture is com-

How does a gasoline engine work? The intake, compression, power, and exhaust strokes.

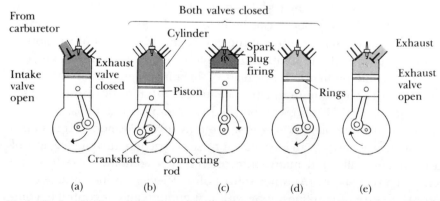

19–2 Cycle of a four-stroke internal-combustion engine. (a) Intake stroke: piston moves down, causing a partial vacuum in cylinder; gasoline and air are mixed in carburetor and flow through open intake valve into cylinder. (b) Compression stroke: intake valve closes and mixture is compressed as piston moves up. (c) Ignition: spark plug ignites mixture. (d) Power stroke: hot burned mixture pushes piston down, doing work. (e) Exhaust stroke: exhaust valve opens and piston moves up, pushing burned mixture out of cylinder. Engine is now ready for next intake stroke, and the cycle repeats.

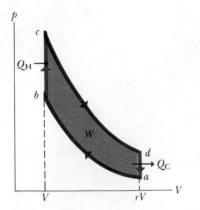

19–3 pV-diagram for the Otto cycle, an idealized model of the thermodynamic processes in a gasoline engine.

pressed adiabatically along line ab and is then ignited. Heat Q_H is added to the system by the burning gasoline (line bc), and the power stroke is the adiabatic expansion cd. The gas is cooled to the temperature of the outside air (da); during this process heat Q_C is exchanged. In practice, of course, this same air does not enter the engine again but, since an equivalent amount does enter, we may consider the process to be cyclic.

It is not hard to calculate the efficiency of this idealized cycle. Processes bc and da are constant-volume, so the heats Q_H and Q_C are related simply to the temperatures:

$$Q_H = nC_v(T_c - T_b),$$
$$Q_C = nC_v(T_a - T_d).$$

The thermal efficiency is given by Eq. (19–4); inserting the above expressions and canceling out the common factor nC_v, we obtain

$$e = \frac{T_c - T_b + T_a - T_d}{T_c - T_b}. \tag{19–5}$$

This equation may be further simplified by using the temperature–volume relation for adiabatic processes, Eq. (18–23). For the two adiabatic processes ab and cd, we find

$$T_a(rV)^{\gamma-1} = T_b V^{\gamma-1},$$
$$T_d(rV)^{\gamma-1} = T_c V^{\gamma-1}.$$

The thermal efficiency of an idealized gasoline engine depends on its compression ratio.

We divide out the common factor $V^{\gamma-1}$ and substitute the resulting expressions for T_b and T_c back into Eq. (19–5). The result is

$$e = \frac{T_d r^{\gamma-1} - T_a r^{\gamma-1} + T_a - T_d}{T_d r^{\gamma-1} - T_a r^{\gamma-1}} = \frac{(T_d - T_a)(r^{\gamma-1} - 1)}{(T_d - T_a)r^{\gamma-1}}.$$

Finally, dividing out the common factor $(T_d - T_a)$ yields the simple result

$$e = 1 - \frac{1}{r^{\gamma-1}}. \tag{19–6}$$

The thermal efficiency given by Eq. (19–6) is always less than unity, even for this idealized model. Using $r = 8$ and $\gamma = 1.4$ (the value for air), we find $e = 0.56$, or 56%. The efficiency can be increased by increasing r. However, this also increases the temperature at the end of the adiabatic compression of the air-fuel mixture. If the temperature is too high, the mixture explodes spontaneously and prematurely, instead of burning evenly after the spark plug ignites it. The maximum practical compression ratio for ordinary gasoline is about 10. Higher ratios can be used with more exotic fuels.

The cycle just described is of course a highly idealized model. It assumes that the mixture behaves as an ideal gas; it neglects friction, turbulence, loss of heat to cylinder walls, and many other effects that combine to reduce the efficiency of a real engine. Another source of inefficiency is incomplete combustion. A mixture of gasoline vapor with just enough air for complete combustion of the hydrocarbons to H_2O and CO_2 does not ignite readily. Reliable ignition requires a mixture "richer" in gasoline; the resulting incomplete combustion leads to CO and unburned hydrocarbons in the exhaust. The heat obtained from the gasoline is then less than the total heat of combustion; the difference is wasted, and the exhaust products contribute to air pollution. One attack on this problem is the stratified-charge engine, in which the concentration of gasoline vapor near the spark plug is greater than in the remainder of the combustion chamber. Efficiencies of real gasoline engines are typically around 20%.

The operation of the diesel engine is similar to that of the gasoline engine; the principal difference is that there is no fuel in the cylinder during compression. At the beginning of the power stroke, fuel is injected into the cylinder just rapidly enough to keep the pressure approximately constant during the first part of the power stroke. The fuel ignites spontaneously because of the high temperature developed during the adiabatic compression; no spark plugs are needed.

The idealized **Diesel cycle** is shown in Fig. 19–4. Starting at point a, air is compressed adiabatically to point b, heated at constant pressure to point c, expanded adiabatically to point d, and cooled at constant volume to point a.

Since there is no fuel in the cylinder of a Diesel engine during the compression stroke, preignition cannot occur, and the compression ratio r may be much higher than for a gasoline engine. Values of 15 to 20 are typical; with these values and $\gamma = 1.4$, the efficiency of the idealized Diesel cycle is about 0.65 to 0.70. As with the Otto cycle, the efficiency of any actual engine is substantially less than this. Although usually more efficient, Diesel engines are heavier (per unit power output) and often harder to start than gasoline engines. They need no carburetor or ignition system, but the fuel-injection system requires expensive high-precision machining.

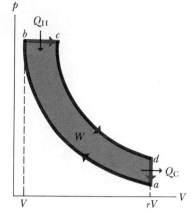

19–4 A pV-diagram of the Diesel cycle.

Why do diesel engines have higher compression ratios than gasoline engines?

19–4 REFRIGERATORS

We can think of a **refrigerator** as a heat engine operating in reverse. A heat engine takes heat from a hot place and gives off heat to a colder place. A refrigerator does the opposite; it takes heat from a cold place (the inside of the refrigerator) and gives it off to a warmer place (usually the air in the room where the refrigerator is located). A heat engine has a net *output* of mechanical work; the refrigerator requires a net *input* of mechanical work. Thus, with the symbols used in Section 19–1, Q_C is positive for a refrigerator, but both W and Q_H are negative.

A flow diagram for a refrigerator is shown in Fig. 19–5. From the first law for a cyclic process,

$$Q_H + Q_C - W = 0, \quad \text{or} \quad -Q_H = Q_C - W,$$

or, since both Q_H and W are negative,

$$|Q_H| = Q_C + |W|. \tag{19–7}$$

Thus, as the diagram shows, the heat Q_H leaving the working substance of the engine and given to the hot reservoir is always *greater* than the heat Q_C taken from the cold reservoir.

From an economic point of view, the best refrigeration cycle is one that removes the greatest amount of heat Q_C from the refrigerator for the least expenditure of mechanical work, W. We therefore define the **performance coefficient** (rather than the efficiency) of a refrigerator as the ratio $K = -Q_C/W$. Since $W = Q_H + Q_C$,

$$\text{Performance coefficient} = K = -\frac{Q_C}{W} = -\frac{Q_C}{Q_H + Q_C}. \tag{19–8}$$

A refrigerator transfers heat from a cool place to a warmer place.

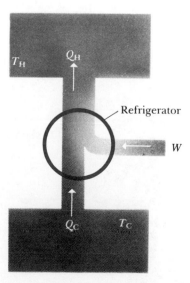

19–5 Schematic flow diagram of a refrigerator.

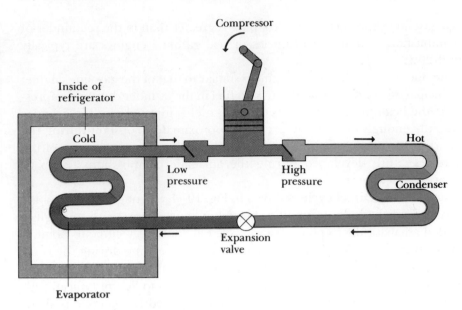

19–6 Principle of the mechanical refrigerator cycle.

The nuts and bolts of refrigerator operation

As always, we assume that Q_H, Q_C, and W are all measured in the same energy units; K is then a dimensionless number.

The principles of the common refrigeration cycle are illustrated schematically in Fig. 19–6. The fluid "circuit" contains a refrigerant fluid (the working substance), typically CCl_2F_2 or another member of the "Freon" family. The left side is at low temperature and low pressure, the right side at high temperature and high pressure; ordinarily both sides contain liquid and vapor in phase equilibrium. The compressor takes in fluid, compresses it adiabatically, and delivers it to the condenser coil at high pressure. The fluid temperature is then higher than that of the air surrounding the condenser, so the refrigerant gives off heat (Q_H) and partially condenses to liquid. When the expansion valve opens, fluid expands adiabatically into the evaporator. As it does so, it cools considerably, enough so that the fluid in the evaporator coil is colder than its surroundings. It absorbs heat (Q_C) from its surroundings, cooling them and partially vaporizing. The fluid then enters the compressor to begin another cycle. The compressor, usually driven by an electric motor, requires energy input and does work $|W|$ on the working substance during each cycle.

What's the difference between a refrigerator and an air conditioner?

An air conditioner operates on exactly the same principle. In this case the refrigerator box is a room or an entire building. The evaporator coils are inside, the condenser is outside, and fans are used to circulate air through these. In large installations the condenser coils are often cooled by water.

For air conditioners the quantity of greatest practical importance is the *rate* of heat removal (i.e., the heat current H) from the region being cooled. If heat Q_C is removed in time t, then $H = Q_C/t$. Similarly, the work W is usually expressed in terms of the *power* input $P = W/t$ to the compressor. The performance coefficient can then be expressed as

$$K = \frac{Q_C}{|W|} = \frac{Ht}{Pt} = \frac{H}{P}.$$

Typical room air conditioners used in homes have heat removal rates H of 5000 to 10,000 Btu·hr^{-1}, or about 1500 to 3000 W, and require electric power input of about 600 to 1500 W. Typical performance coefficients are roughly 2 to 3, with somewhat larger values for larger-capacity units.

Unfortunately, K is usually expressed commercially in mixed units, with H in Btu per hour and P in watts. In these units H/P is called the **energy efficiency rating** (EER); for room air conditioners it typically has a numerical value of 7 to 10. The units, customarily omitted, are Btu·hr^{-1}·W^{-1}.

A variation on this theme is the **heat pump,** used to heat buildings by cooling the outside air. It functions like a refrigerator turned inside out: The evaporator coils are outside, where they take heat from cold air, and the condenser coils are inside, where they give off heat to the warmer air. With proper design the heat Q_H delivered to the inside per cycle can be considerably greater than the work W required to get it there.

> The heat pump: how to heat your house by cooling the outside. Are you getting something for nothing?

If *no* work were needed to operate a refrigerator, the performance coefficient (heat removed divided by work done) would be infinite. Performance coefficients of actual refrigerators are typically in the range from 2 to 6. Experience shows that some work is *always* needed to transfer heat from a colder to a hotter body. Heat flows spontaneously from a hotter to a colder body, and to reverse this flow requires the addition of work from the outside. It is impossible to make a refrigerator that transports heat from a cold body to a hotter body without the addition of work. Such a mythical device is called a *workless refrigerator.*

> A refrigerator can't function without some mechanical work input.

19–5 THE SECOND LAW OF THERMODYNAMICS

We have discussed the fact that no one has ever been able to build a heat engine that converts heat completely to work, that is, an engine with 100% thermal efficiency. Experimental evidence suggests strongly that it is *impossible* to build such an engine. This impossibility forms the basis of one form of the **second law of thermodynamics,** as follows:

> It is impossible for any system to undergo a process in which it absorbs heat from a reservoir at a single temperature and converts it completely into mechanical work, while ending in the same state in which it began.

In other words, it is impossible *in principle* for any heat engine to have a thermal efficiency of 100%.

> No engine can be 100% efficient. It's the (second) law!

The basis of the second law of thermodynamics lies in the difference between the nature of internal energy and that of macroscopic mechanical energy. In a moving body the molecules have random motion, but superimposed on this is a coordinated motion of every molecule in the direction of the velocity of the body. The kinetic energy associated with this coordinated macroscopic motion is what we call the kinetic energy of the moving body. The kinetic and potential energies associated with the *random* motion constitute the internal energy.

When a moving body makes an inelastic collision or comes to rest as a result of friction, the organized part of the motion is converted to random motion. Since we cannot control the motions of individual molecules, we cannot convert this random motion completely back to organized motion. We can, however, convert *part* of it, and this is what a heat engine does.

If the second law were *not* true, it would be possible to power an automobile or to run a power plant by extracting heat from the surrounding air. Neither of these impossibilities violates the *first* law of thermodynamics. The second law, therefore, is not a deduction from the first but stands by itself as a separate law of nature, referring to an aspect of nature different from that

> The second law of thermodynamics limits energy-conversion processes.

19–7 Energy-flow diagrams for equivalent forms of the second law. (a) A workless refrigerator (left), if it existed, could be used in combination with an ordinary heat engine (right) to form a composite device that functions as an engine with 100% efficiency, converting heat Q_H completely to work. (b) An engine with 100% efficiency (left), if it existed, could be used in combination with an ordinary refrigerator (right) to form a workless refrigerator, transferring heat Q_C from the cold reservoir to the hot with no net input of work. Thus if either of these is impossible, the other must be also.

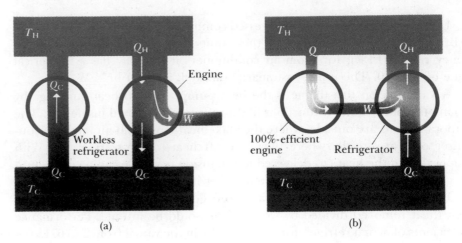

The "engine" and "refrigerator" statements of the second law are logically equivalent.

considered in the first law. The first law denies the possibility of creating or destroying energy; the second law limits the ways energy may be used and converted.

The analysis of refrigerators forms the basis for an alternative statement of the second law of thermodynamics. Heat flows spontaneously from hotter to colder bodies, never the reverse. A refrigerator does take heat from a colder to a hotter body, but its operation depends on input of mechanical energy or work. Generalizing this observation, we state:

> It is impossible for any process to have as its sole result the transfer of heat from a cooler to a hotter body.

This statement may not seem to be very closely related to the previous statement about heat engines. In fact, though, the two statements are completely equivalent. For example, if we could build a workless refrigerator, violating the second or "refrigerator" statement of the second law, we could use it in conjunction with a heat engine, pumping the heat rejected by the engine back to the hot reservoir to be reused. The composite machine (Fig. 19–7a) would violate the first or "engine" statement of the second law.

Alternatively, if we could make an engine with 100% thermal efficiency, in violation of the first statement, we could run it by using heat from the hot reservoir, and use the work output to drive a refrigerator that pumps heat from the cold reservoir to the hot (Fig. 19–7b). This would violate the "refrigerator" statement. Thus any device that violates one form of the second law can also be used to violate the other form. We conclude that if violations of the first form are impossible, so are violations of the second.

We have mentioned that heat flow across a finite temperature gradient and conversion of mechanical energy to heat, as in friction or turbulent fluid flow, are irreversible processes. Other examples can be cited. Gases always seep through an opening spontaneously from a region of high pressure to a region of low pressure; gases and liquids left by themselves always tend to mix, not to unmix. The second law of thermodynamics is an expression of the inherent one-way aspect of these and many other irreversible processes that take place naturally in only one direction. An irreversible process is always a *nonequilibrium* process. Irreversible heat flow accompanies departures from thermal equilibrium, free expansion of a gas involves states that are not in mechanical equilibrium, and so on. In each case the process tends to move the system toward an equilibrium state.

19–6 THE CARNOT CYCLE

According to the second law, no heat engine can have 100% efficiency. But what is the theoretical *maximum* possible efficiency of an engine, given two heat reservoirs at temperatures T_H and T_C? This question was answered in 1824 by the French engineer Sadi Carnot, who developed a hypothetical, idealized heat engine that has the maximum possible efficiency consistent with the second law. The cycle of this engine is called the **Carnot cycle.**

To understand the rationale of the Carnot cycle, we return to the matter of reversibility, discussed at the end of Section 19–5. Conversion of work to heat is an irreversible process; the purpose of a heat engine is a *partial* reversal of this process, the conversion of heat to work with as great efficiency as possible. For maximum heat-engine efficiency, therefore, we must *avoid* all irreversible processes.

> The Carnot cycle: the best possible thermal efficiency for a heat engine

Heat flow through a finite temperature drop is an irreversible process. Therefore, during heat transfer in the Carnot cycle there must be *no* finite temperature difference. When the engine takes heat from the hot reservoir at T_H, the engine itself must also be at T_H; otherwise irreversible heat flow would occur. Similarly, when heat is rejected to the cold reservoir at T_C, the engine itself must be at T_C. That is, every process that involves heat transfer must be *isothermal* at either T_H or T_C. Conversely, there must be *no* heat transfer between the engine and either reservoir in any process where the temperature of the engine changes; such heat transfer could not be reversible. In short, *every process* in our idealized cycle must be either *isothermal* or *adiabatic*. In addition, not only thermal but mechanical equilibrium must be maintained at all times, so that each process is completely reversible.

> Complete reversibility in the Carnot cycle requires that every process be either isothermal or adiabatic.

The Carnot cycle consists of two isothermal and two adiabatic processes. A Carnot cycle using an ideal gas as the working substance is shown on a *pV*-diagram in Fig. 19–8. It comprises the following steps:

1. The gas expands isothermally at temperature T_H, absorbing heat Q_H (*ab*).
2. It expands adiabatically until its temperature drops to T_C (*bc*).
3. It is compressed isothermally at T_C, rejecting heat Q_C (*cd*).
4. It is compressed adiabatically back to its initial state at temperature T_H (*da*).

When the working substance in a Carnot engine is an ideal gas, it is easy to calculate the thermal efficiency *e*. To carry out this calculation we will first find the ratio Q_C/Q_H of the quantities of heat transferred in the two isothermal processes, and then use Eq. (19–4) to find *e*. For an ideal gas the internal energy depends only on temperature and is thus constant in any isothermal process. For example, Q_H is equal to the work W_{ab} done by the gas during its isothermal expansion at temperature T_H. This work in turn is given by Eq. (18–4), and we find

$$Q_H = nRT_H \ln \frac{V_b}{V_a}. \tag{19–9}$$

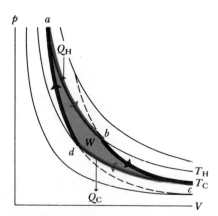

19–8 The Carnot cycle for an ideal gas. Color lines are isothermals; black lines are adiabatics.

Similarly,

$$Q_C = nRT_C \ln \frac{V_d}{V_c} = -nRT_C \ln \frac{V_c}{V_d}. \tag{19-10}$$

This quantity is negative because V_d is less than V_c. The ratio of the two quantities of heat is thus

$$\frac{Q_C}{Q_H} = -\frac{T_C}{T_H} \frac{\ln (V_c/V_d)}{\ln (V_b/V_a)}. \tag{19-11}$$

This equation can be simplified further by use of the temperature–volume relation for an adiabatic process, Eq. (18–23). We find for the two adiabatic processes:

$$T_H V_b{}^{\gamma-1} = T_C V_c{}^{\gamma-1}$$

and

$$T_H V_a{}^{\gamma-1} = T_C V_d{}^{\gamma-1}.$$

Dividing the first of these equations by the second, we find

$$\frac{V_b{}^{\gamma-1}}{V_a{}^{\gamma-1}} = \frac{V_c{}^{\gamma-1}}{V_d{}^{\gamma-1}} \quad \text{and} \quad \frac{V_b}{V_a} = \frac{V_c}{V_d}.$$

The thermal efficiency of a Carnot engine depends only on the temperatures of the two heat reservoirs.

Thus the two logarithms in Eq. (19–11) are equal, and that equation reduces to

$$\frac{Q_C}{Q_H} = -\frac{T_C}{T_H}. \tag{19-12}$$

Then from Eq. (19–3) the efficiency of the Carnot engine is

$$e = 1 - \frac{T_C}{T_H} = \frac{T_H - T_C}{T_H}. \tag{19-13}$$

This surprisingly simple result says that the efficiency of a Carnot engine depends only on the temperatures of the two heat reservoirs. The efficiency is large when the temperature *difference* is large, and it is very small when the temperatures are nearly equal. The efficiency can never be exactly unity, unless $T_C = 0$, and as we will see, this too is impossible.

EXAMPLE 19–2   A Carnot engine takes 2000 J of heat from a reservoir at 500 K, does some work, and discards some heat to a reservoir at 350 K. How much work does it do, how much heat is discarded, and what is the efficiency?

SOLUTION   From Eq. (19–12),

$$Q_C = -Q_H \frac{T_C}{T_H} = -(2000 \text{ J}) \frac{350 \text{ K}}{500 \text{ K}} = -1400 \text{ J}.$$

Then from the first law,

$$W = Q_H + Q_C = 2000 \text{ J} + (-1400 \text{ J})$$
$$= 600 \text{ J}.$$

From Eq. (19–13) the efficiency is

$$e = 1 - \frac{350 \text{ K}}{500 \text{ K}} = 0.30 = 30\%.$$

Alternatively, from the basic definition of efficiency,

$$e = \frac{W}{Q_H} = \frac{600 \text{ J}}{2000 \text{ J}} = 0.30 = 30\%.$$

EXAMPLE 19–3 Suppose 0.2 mol of an ideal diatomic gas ($\gamma = 1.40$) undergoes a Carnot cycle with temperatures $T_H = 400$ K and $T_C = 300$ K. The initial pressure is $p_a = 10.0 \times 10^5$ Pa, and during the isothermal expansion at temperature T_H the volume doubles.

A detailed example of a specific Carnot cycle using an ideal gas

a) Find the pressure and volume at each of points a, b, c, and d in Fig. 19–8.

b) Find Q, W, and ΔU for each step in the cycle and for the entire cycle.

c) Determine the efficiency directly from the results of (b) and compare with the result from Eq. (19–13).

SOLUTION

a) First,

$$V_a = \frac{nRT_H}{p_a} = \frac{(0.2 \text{ mol})(8.314 \text{ J·mol}^{-1}\text{·K}^{-1})(400 \text{ K})}{10.0 \times 10^5 \text{ Pa}}$$

$$= 6.65 \times 10^{-4} \text{ m}^3,$$

$$V_b = 2V_a = 2(6.65 \times 10^{-4} \text{ m}^3) = 13.3 \times 10^{-4} \text{ m}^3.$$

For the isothermal expansion $a \rightarrow b$, $p_a V_a = p_b V_b$, so

$$p_b = \frac{p_a V_a}{V_b} = 5.0 \times 10^5 \text{ Pa}.$$

For the adiabatic expansion $b \rightarrow c$, $T_H V_b^{\gamma-1} = T_C V_c^{\gamma-1}$, so

$$V_c = V_b \left(\frac{T_H}{T_C}\right)^{1/(\gamma-1)} = (13.3 \times 10^{-4} \text{ m}^3)\left(\frac{4}{3}\right)^{2.5} = 27.3 \times 10^{-4} \text{ m}^3.$$

So

$$p_c = \frac{nRT_C}{V_c} = \frac{(0.2 \text{ mol})(8.314 \text{ J·mol}^{-1}\text{·K}^{-1})(300 \text{ K})}{27.3 \times 10^{-4} \text{ m}^3}$$

$$= 1.83 \times 10^5 \text{ Pa}.$$

For the adiabatic compression $d \rightarrow a$, $T_C V_d^{\gamma-1} = T_H V_a^{\gamma-1}$, and

$$V_d = V_a \left(\frac{T_H}{T_C}\right)^{1/\gamma-1} = (6.65 \times 10^{-4} \text{ m}^3)\left(\frac{4}{3}\right)^{2.5} = 13.65 \times 10^{-4} \text{ m}^3;$$

$$p_d = \frac{nRT_C}{V_d} = \frac{(0.2 \text{ mol})(8.314 \text{ J·mol}^{-1}\text{·K}^{-1})(300 \text{ K})}{13.65 \times 10^{-4} \text{ m}^3}$$

$$= 3.65 \times 10^5 \text{ Pa}.$$

b) For the isothermal expansion $a \rightarrow b$, $\Delta U = 0$ and, from Eq. (18–4),

$$W = Q_H = nRT_H \ln \frac{V_b}{V_a}$$

$$= (0.2 \text{ mol})(8.314 \text{ J·mol}^{-1}\text{·K}^{-1})(400 \text{ K}) \ln 2 = 461 \text{ J}.$$

For the adiabatic expansion $b \rightarrow c$, $Q = 0$ and, from Eq. (18–26),

$$W = -\Delta U = nC_v(T_H - T_C)$$

$$= (0.2 \text{ mol})(20.78 \text{ J·mol}^{-1}\text{·K}^{-1})(400 \text{ K} - 300 \text{ K}) = 416 \text{ J}.$$

For the isothermal compression $c \rightarrow d$, $\Delta U = 0$, and

$$W = Q_C = nRT_C \ln \frac{V_d}{V_c}$$

$$= (0.2 \text{ mol})(8.314 \text{ J·mol}^{-1}\text{·K}^{-1})(300 \text{ K}) \ln \frac{13.65 \times 10^{-4} \text{ m}^3}{27.3 \times 10^{-4} \text{ m}^3}$$

$$= -346 \text{ J}.$$

For the adiabatic compression $d \rightarrow a$, $Q = 0$; and

$$W = -\Delta U = nC_v(T_C - T_H)$$

$$= (0.2 \text{ mol})(20.78 \text{ J·mol}^{-1}\text{·K}^{-1})(300 \text{ K} - 400 \text{ K})$$

$$= -416 \text{ J}$$

Alternatively Eq. (18–27) may be used for either adiabatic process. For $d \rightarrow a$,

$$W = \frac{1}{\gamma - 1}(p_d V_d - p_a V_a)$$

$$= (2.5)[(3.65 \times 10^5 \text{ Pa})(13.65 \times 10^{-4} \text{ m}^3)$$

$$- (10.0 \times 10^5 \text{ Pa})(6.65 \times 10^{-4} \text{ m}^3)]$$

$$= -416 \text{ J}.$$

The results may be tabulated as follows:

Process	Q	W	ΔU
$a \rightarrow b$	461 J	461 J	0
$b \rightarrow c$	0	416 J	−416 J
$c \rightarrow d$	−346 J	−346 J	0
$d \rightarrow a$	0	−416 J	416 J
Total	115 J	115 J	0

Note that for the entire cycle, $Q = W$ and $\Delta U = 0$. Also note that the quantities of work in the two adiabatic processes are negatives of each other; it is easy to prove from the analysis leading to Eq. (19–13) that this must always be the case.

c)   From the table, the total work is 115 J and $Q_H = 461$ J. Thus

$$e = \frac{W}{Q_H} = \frac{115 \text{ J}}{461 \text{ J}} = 0.250.$$

From Eq. (19–13)

$$e = \frac{T_H - T_C}{T_H} = \frac{400 \text{ K} - 300 \text{ K}}{400 \text{ K}} = 0.250.$$

A Carnot engine run backward is a Carnot refrigerator.

Because each step in the Carnot cycle is reversible, the *entire cycle* may be reversed, thereby converting the engine into a refrigerator. The performance coefficient of the Carnot refrigerator is obtained by combining Eqs. (19–8) and (19–12), first rewriting Eq. (19–8) as

$$K = -\left(\frac{Q_H + Q_C}{Q_C}\right)^{-1}.$$

We invite you to fill in the details; the result is

$$K = \frac{T_C}{T_H - T_C}. \tag{19–14}$$

When the temperature difference is small, K is much larger than unity; in this case a lot of heat can be "pumped" from the lower to the higher temperature with only a little expenditure of work. But the greater the temperature difference, the smaller is K and the more work is required to transfer a given quantity of heat.

EXAMPLE 19–4 If the cycle described in Example 19–3 is run backward as a refrigerator, the performance coefficient is given by Eq. (19–8):

$$K = -\frac{Q_C}{W} = -\frac{(-346 \text{ J})}{115 \text{ J}} = 3.00.$$

Because the cycle is a Carnot cycle, we may also use Eq. (19–14):

$$K = \frac{T_C}{T_H - T_C} = \frac{300 \text{ K}}{400 \text{ K} - 300 \text{ K}} = 3.00.$$

For a Carnot cycle, e and K depend only on the temperatures, and it is not necessary to calculate Q and W. For cycles containing irreversible processes, however, more detailed calculations are necessary.

It is easy to prove that *no engine can be more efficient than a Carnot engine operating between the same two temperatures.* The key to the proof is the observation that because each step in the Carnot cycle is reversible, the *entire cycle* may be reversed. Run backward, the engine becomes a refrigerator. If any engine is more efficient than a Carnot engine, its work output may be used to drive a Carnot refrigerator and pump the rejected heat back to the reservoir, thus violating the engine statement of the second law. Hence the above statement is still another equivalent statement of the second law of thermodynamics. It also follows directly that *all Carnot engines operating between the same two temperatures have the same efficiency, irrespective of the nature of the working substance.* Thus, although Eqs. (19–13) and (19–14) were derived for a Carnot engine using an ideal gas as its working substance, they are in fact valid for *any* Carnot engine, no matter what its working substance.

No engine can be more efficient than a Carnot engine, for given temperatures.

Equation (19–13) points the way to the conditions that a real engine, such as a steam turbine, must fulfill to approach as closely as possible its maximum attainable efficiency. These conditions are that the intake temperature T_H must be made as high as possible and the exhaust temperature T_C as low as possible.

For given temperatures, all Carnot engines have the same efficiency.

The exhaust temperature cannot be lower than the lowest temperature available for cooling the exhaust. This is usually the temperature of the air, or perhaps of river water if available at the plant. The only recourse then is to raise the boiler temperature T_H. Because the vapor pressure of all liquids increases rapidly with increasing temperature, a limit is set by the mechanical strength of the boiler. At 500°C the vapor pressure of water is about 240 ×

10^5 Pa, 235 atm, or 3450 lb·in^{-2}, which is about the maximum practical pressure in large present-day steam boilers.

The unavoidable exhaust heat loss in electric-power plants creates a serious environmental problem. When a lake or river is used for cooling, the temperature of the body of water may be raised several degrees. Such a temperature change has a severely disruptive effect on the overall ecological balance, inasmuch as relatively small temperature changes can have significant effects on metabolic rates in plants and animals. Since **thermal pollution,** as this effect is called, is an inevitable consequence of the second law of thermodynamics, careful planning is essential to minimize the ecological impact of new power plants.

The waste heat from power plants can have a major impact on the ecology.

19–7 ENTROPY

The second law of thermodynamics, as stated above, is rather different in form from other familiar physical laws; it is not a quantitative relationship but rather a statement of *impossibility*. We can also express the second law in quantitative form, using the concept of **entropy,** the subject of this section.

In this chapter we have seen several examples of processes that proceed naturally in the direction of increasing disorder, although we have not yet given a quantitative definition of disorder. Irreversible heat flow increases disorder because initially the molecules are sorted into hotter and cooler regions; this sorting is lost when the system comes to thermal equilibrium. Adding heat to a body increases its disorder because it increases the randomness of molecular motion. Free expansion of a gas increases its disorder because the molecules have greater randomness of position after the expansion than before.

Entropy: how to measure mixed-up-ness

Entropy provides a quantitative measure of disorder. Consider first the entropy change of a substance undergoing a reversible addition of heat. We use the symbol S for the entropy of the system, and ΔS for the change in entropy during any process. Adding heat increases molecular motion and thus disorder, but the effect is greater if the substance is cold (with little molecular motion) at the beginning than if it is already hot (with a lot of molecular motion). A suitable definition of entropy change in such a process is

$$\Delta S = \frac{Q}{T} \quad \text{(reversible isothermal process)}, \quad (19\text{–}15)$$

where Q is the heat added and T is the *absolute* temperature. This definition holds only for reversible isothermal equilibrium processes.

We can generalize the definition of entropy change to include *any* reversible process leading from one state to another, whether it is isothermal or not. We represent the process as a series of infinitesimal reversible steps. During a typical step an infinitesimal quantity of heat dQ is added to the system at absolute temperature T. Then we sum (integrate) the quotients dQ/T for the entire process; that is,

$$\Delta S = \int_1^2 \frac{dQ}{T} \quad \text{(reversible process)}. \quad (19\text{–}16)$$

The limits 1 and 2 refer to the initial and final states.

Ink mixing with water. There is greater disorder and more entropy after mixing than before; the water and ink mix spontaneously but will never unmix spontaneously. (Photo by Chip Clark.)

The entropy of a system depends only on its state.

Because entropy is a measure of the disorder of a system in any specific state, we expect it to depend only on the current state of the system, not on its past history. It can be proved, by use of the second law of thermodynamics, that when a system proceeds from an initial state with entropy S_1 to a final state with entropy S_2, the change in entropy $\Delta S = S_2 - S_1$ defined by Eq. (19–16) does not depend on the path leading from the initial to the final state but is the same for *all possible* processes leading from state 1 to state 2. Thus the entropy of a system must also have a definite value for any given state of the system. We recall that *internal energy*, introduced in Chapter 18, also has this property, although entropy and internal energy are very different quantities. The unit of entropy is $1\ \text{J}\cdot\text{K}^{-1}$, $1\ \text{cal}\cdot\text{K}^{-1}$, $1\ \text{Btu}\cdot(\text{R}°)^{-1}$, and so on.

The fact that entropy is a function only of the state of a system enables us to compute entropy changes in nonequilibrium processes, where Eq. (19–16) is not applicable. We can simply invent a path connecting the given initial and final states that *does* consist entirely of reversible equilibrium processes and compute the total entropy change for that path. It is not the actual path, but the entropy change must be the same as for the actual path.

Entropy changes are independent of the path.

As with internal energy, this discussion does not define entropy itself but only the change in entropy in any given process. To complete the definition we may arbitrarily assign a value to the entropy of a system in a specified reference state and then calculate the entropy of any other state with reference to this value.

EXAMPLE 19–5   One kilogram of ice at 0°C is melted and converted to water at 0°C. Compute its change in entropy.

SOLUTION   The temperature is constant at 273 K, and

$$\Delta S = S_2 - S_1 = \frac{Q}{T}.$$

What is the entropy change of melting ice?

But Q is simply the total heat of fusion that must be supplied to melt the ice, or $334 \times 10^3\ \text{J}$. Hence

$$S_2 - S_1 = \frac{334 \times 10^3\ \text{J}}{273\ \text{K}} = 1223\ \text{J}\cdot\text{K}^{-1},$$

and the increase in entropy of the system is $1223\ \text{J}\cdot\text{K}^{-1}$. In any *isothermal* reversible process, the entropy change equals the heat added divided by the absolute temperature.

EXAMPLE 19–6   One kilogram of water at 0°C is heated to 100°C. Compute its change in entropy.

SOLUTION   The temperature is not constant; to carry out the integral in Eq. (19–16) we replace dQ by $mc\,dT$, obtaining

$$\Delta S = S_2 - S_1 = \int_{T_1}^{T_2} mc\,\frac{dT}{T} = mc\ln\frac{T_2}{T_1}$$

$$= (1000\ \text{g})(4.19\ \text{J}\cdot\text{g}^{-1}\cdot\text{K}^{-1})\ln\frac{373\ \text{K}}{273\ \text{K}}$$

$$= 1308\ \text{J}\cdot\text{K}^{-1}.$$

EXAMPLE 19–7   A gas expands adiabatically and reversibly. What is its change in entropy?

SOLUTION In an adiabatic process, no heat enters or leaves the system. Hence $Q = 0$ and there is *no* change in entropy. Every *reversible* adiabatic process is a constant-entropy process.

EXAMPLE 19–8 A thermally insulated box is divided by a partition into two compartments, each having volume V. Initially one compartment contains n moles of an ideal gas at temperature T, and the other is evacuated. We then break the partition, and the gas expands to fill both compartments. What is its entropy change?

When a gas undergoes a free expansion, its entropy increases.

SOLUTION For this process $Q = 0$, $W = 0$, $\Delta U = 0$, and therefore (since it is an ideal gas) $\Delta T = 0$. We might think that the entropy change is zero because there is no heat exchange. But Eq. (19–16) is valid only for *reversible* processes; this free expansion is *not* reversible, and there *is* an entropy change. To calculate ΔS we can use the fact that the entropy change depends only on the initial and final states. We can devise a *reversible* process having the same endpoints, use Eq. (19–16) to calculate its entropy change, and thus determine the entropy change in the original process. The appropriate reversible process in this case is an isothermal expansion from V to $2V$ at temperature T. The gas does work during this expansion, so heat must be supplied in order to keep the internal energy constant. The total heat equals the total work, which is given by Eq. (18–4):

$$W = Q = nRT \ln 2.$$

Thus the entropy change is

$$\Delta S = \frac{Q}{T} = nR \ln 2,$$

which is also the entropy change for the free expansion. For one mole,

$$\Delta S = (1 \text{ mol})(8.314 \text{ J·mol}^{-1}\text{·K}^{-1})(0.693) = 5.76 \text{ J·K}^{-1}.$$

EXAMPLE 19–9 For the Carnot engine in Example 19–2 (Section 19–6), find the total entropy change in the engine during one cycle.

Entropy changes in a Carnot engine

SOLUTION During the isothermal expansion at 500 K the engine takes in 2000 J, and its entropy change is

$$\Delta S = \frac{Q}{T} = \frac{2000 \text{ J}}{500 \text{ K}} = 4.0 \text{ J·K}^{-1}.$$

During the isothermal compression at 350 K the engine gives off 1400 J of heat, and its entropy change is

$$\Delta S = \frac{-1400 \text{ J}}{350 \text{ K}} = -4.0 \text{ J·K}^{-1}.$$

Thus the total entropy change is $4.0 \text{ J·K}^{-1} - 4.0 \text{ J·K}^{-1} = 0$. This is to be expected, of course, since the final state is the same as the initial state. The total entropy change of the two heat reservoirs is also zero. This cycle contains no irreversible processes, and the total entropy change is zero.

There is no such thing as conservation of entropy.

Unlike energy, entropy is *not* a conserved quantity. In fact, the reverse is true; the entropy of an isolated system *can* change but, as we will see, it can never decrease. An entropy increase occurs in every natural process, if all systems taking part in the process are included. In an idealized, completely reversible process involving only equilibrium states, no entropy change occurs, but all natural (i.e., irreversible) processes take place with an increase in entropy.

EXAMPLE 19–10 Suppose 1 kg of water at 100°C is placed in thermal contact with 1 kg of water at 0°C. What is the total change in entropy?

SOLUTION This process involves irreversible heat flow. We assume the specific heat capacity of water is constant in this temperature range. Then the first 4186 J of heat transferred cool the hot water to 99°C and warm the cold water from 0°C to 1°C. The net change of entropy is approximately

Mixing hot and cold water increases the entropy.

$$\Delta S = -\frac{4190 \text{ J}}{373 \text{ K}} + \frac{4190 \text{ J}}{273 \text{ K}} = 4.1 \text{ J} \cdot \text{K}^{-1}.$$

Further increases in entropy occur as the system approaches thermal equilibrium at 50°C. The *total* increase in entropy can be calculated by the same method as in Example 19–6. The entropy change of the hot water is

$$\Delta S = (1 \text{ kg})(4190 \text{ J} \cdot \text{kg}^{-1} \cdot \text{K}^{-1}) \int_{373 \text{ K}}^{323 \text{ K}} \frac{dT}{T}$$

$$= (4190 \text{ J} \cdot \text{K}^{-1}) \ln \frac{323 \text{ K}}{373 \text{ K}} = -602 \text{ J} \cdot \text{K}^{-1};$$

the entropy change of the cold water is

$$\Delta S = (4190 \text{ J} \cdot \text{K}^{-1}) \ln \frac{323 \text{ K}}{273 \text{ K}} = +704 \text{ J} \cdot \text{K}^{-1};$$

and the *total* entropy change of the system is

$$\Delta S = +704 \text{ J} \cdot \text{K}^{-1} - 602 \text{ J} \cdot \text{K}^{-1} = +102 \text{ J} \cdot \text{K}^{-1}.$$

Thus an irreversible heat flow in an isolated system is accompanied by an increase in entropy. The same end state could have been achieved by simply mixing the two quantities of water. This too is an irreversible process, and because entropy depends only on the state of the system, the total entropy change would be the same.

These examples of the mixing of substances at different temperatures, or the flow of heat from a higher to a lower temperature, are characteristic of *all* natural (i.e., irreversible) processes. When all the entropy changes in the process are included, the increases in entropy are always greater than the decreases. In the special case of a *reversible* process, the increases and decreases are equal. Hence we can state the general principle that *when all systems taking part in a process are included, the entropy either remains constant or increases.* In other words, *no process is possible in which the total entropy decreases,* when all systems taking part in the process are included. This is an alternative statement of the second law of thermodynamics, in terms of entropy. This statement is equivalent to the "engine" and "refrigerator" statements discussed earlier.

The increase of entropy that accompanies every natural (irreversible) process measures the increase of disorder or randomness in the universe in

Entropy always increases in an irreversible process.

that process. Consider again the example of mixing hot and cold water. We *might* have used the hot and cold water as the high- and low-temperature reservoirs of a heat engine. In the course of removing heat from the hot water and giving heat to the cold water, we could have obtained some mechanical work. But once the hot and cold water have been mixed and have come to a uniform temperature, this opportunity of converting heat to mechanical work is lost, and it is lost irretrievably. The lukewarm water will never *unmix* itself and separate into hotter and colder portions. No decrease in *energy* occurs when the hot and cold water are mixed; what has been lost in the mixing process is not *energy,* but *opportunity*—the opportunity to convert part of the heat from the hot water into mechanical work. Hence when entropy increases, energy becomes less *available,* and the universe becomes more random or "run down."

19–8 THE KELVIN TEMPERATURE SCALE

The Kelvin temperature scale does not depend on the properties of any particular material; it is truly absolute.

The Carnot cycle can be used to define a temperature scale that is completely independent of the properties of any particular material. As we have seen, the efficiency of a Carnot engine operating between reservoirs at two given temperatures is independent of the nature of the working substance and is a function only of the temperatures. If we consider a number of Carnot engines using different working substances and absorbing and rejecting heat to the same two reservoirs, the thermal efficiency is the same for all:

$$e = \frac{Q_H + Q_C}{Q_H} = 1 + \frac{Q_C}{Q_H}$$
$$= \text{constant.}$$

Hence the ratio Q_H/Q_C is the same for all Carnot engines operating between two given temperatures T_H and T_C. Kelvin proposed that the ratio of the reservoir temperatures be *defined* as equal to this constant ratio of the magnitudes of the quantities of heat absorbed and rejected or, since Q_C is a negative quantity, as equal to the negative of the ratio Q_H/Q_C. Thus

$$\frac{T_H}{T_C} = \frac{|Q_H|}{|Q_C|} = -\frac{Q_H}{Q_C}. \tag{19–17}$$

Equation (19–17) appears identical to Eq. (19–12), but there is a subtle and crucial difference. The temperatures in Eq. (19–12) are based on an ideal-gas thermometer, as defined in Sections 14–2 and 14–4, while Eq. (19–17) defines a temperature scale, based on the Carnot cycle and the second law of thermodynamics, that is independent of the behavior of any particular substance. Thus the **Kelvin temperature scale** is truly *absolute.* To complete the definition of the Kelvin scale we proceed, as in Chapter 14, to assign the arbitrary value of 273.16 K to the temperature of the triple point of water. When a substance is taken around a Carnot cycle, the ratio of the heats absorbed and rejected, $|Q_H|/|Q_C|$, is equal to the ratio of the temperatures of the reservoirs *as expressed on the gas scale,* defined in Chapter 14. Since, in both scales, the triple point of water is chosen to be 273.16 K, it follows that *the Kelvin and the ideal gas scales are identical.*

Absolute zero is as cold as you can get.

The zero point on the Kelvin scale is called **absolute zero.** There are theoretical reasons for believing that absolute zero cannot be attained experimentally, although temperatures as low as 10^{-6} K have been achieved. The more closely we approach absolute zero, the more difficult it is to get closer. Absolute zero can also be interpreted on a molecular level, although this must be done with some care. Because of quantum effects, it is *not* correct to say that at $T = 0$ all molecular motion ceases. Rather, at absolute zero the system has its *minimum* possible total energy (kinetic plus potential), although this minimum amount is in general not zero.

19–9 ENERGY CONVERSION

The laws of thermodynamics place very general limitations on conversion of energy from one form to another. In this day of increasing energy demand and diminishing resources, these matters are of the utmost practical importance. We conclude this chapter with a brief discussion of a few energy-conversion systems, present and proposed.

Over half the electric power generated in the United States is obtained from coal-fired steam-turbine generating plants. Modern boilers can transfer about 80% to 90% of the heat of combustion of coal into steam. The theoretical thermal efficiency of the turbine, given by Eq. (19–13), is usually limited to about 0.55, and the actual efficiency is typically 90% of the theoretical value, or about 0.50. The efficiency of large electrical generators in converting mechanical power to electrical is very large, typically 99%. Thus the overall thermal efficiency of such a plant is roughly (0.85)(0.50)(0.99), or about 40%.

In 1970 a generator with a capacity of about one gigawatt (=1000 MW = 10^9 W) went into operation at the Tennessee Valley Authority's Paradise power plant. The steam is heated to 540°C (1003°F) at a pressure of 248 atm (2.51×10^7 Pa or 3650 lb·in^{-2}). The plant consumes 10,500 tons of coal per day, and the overall thermal efficiency is 39.3%.

Nuclear-power plants have the same theoretical efficiency limit as coal-fired plants. Because at present is is not practical to run nuclear reactors at as

Efficiencies of modern electric-power plants

Interior of the Paradise power plant, showing turbine housings and steam ducts. For scale, note the men in the right foreground.

Exterior of the Paradise power plant. The cooling towers on the left cool and condense the exhaust steam from the turbines so it can be recycled into the boilers. Waste heat is given off to the air flowing through the towers. (Photographs courtesy of Tennessee Valley Authority.)

(a)

(b)

19–9 (a) A solar-energy installation in the Mojave Desert, near Barstow, California. An array of mirrors concentrates the sun's energy on a boiler (central tower) to generate steam for turbines. The mirrors move continuously, controlled by a computer, to track the sun's motion. (b) An array of photovoltaic cells for direct conversion of sun power to electric power. Such an array supplies the electrical needs of the Headquarters and Visitors Center of Natural Bridges National Monument, Utah. (Photos by Dan McCoy/Rainbow.)

high temperatures and pressures as coal boilers, the theoretical thermal efficiency is usually lower. The overall thermal efficiency of a nuclear plant is typically 30%.

In both coal-fired and nuclear plants, the energy not converted to electrical energy is wasted and must be disposed of. A common practice is to locate such a plant near a lake or river and use the water for disposal of excess heat. This can raise the water temperature several degrees, often with serious ecological consequences.

Solar energy is an inviting possibility. The power in the sun's radiation (before it passes through the earth's atmosphere) is about 1.4 kW per square meter. A maximum of about 1.0 kW·m^{-2} reaches the surface of the earth on a clear day, and the time average, over a 24-hour period, is about 0.2 kW·m^{-2}. This radiation can be collected and focused with mirrors and used to generate steam for a heat engine, as in Fig. 19–9a. A different scheme is to use large banks of photocells for direct conversion of solar energy to electricity. Such a process is not a heat engine in the usual sense and is not limited by the Carnot efficiency. There are other fundamental limitations on photocell efficiency, but 50% seems attainable in multilayer semiconductor photocells. The energy of wind, which is fundamentally solar in origin, can be gathered and converted by "forests" of windmills, as shown in Fig. 19–10.

An indirect scheme for collection and conversion of solar energy would use the temperature gradient in the ocean. In the Caribbean, for example, the water temperature near the surface is about 25°C, while at a depth of a few hundred meters it may be 10°C. Although the second law of thermodynamics forbids taking heat from the ocean and converting it completely into work, nothing forbids running a heat engine between these two temperatures. The thermodynamic efficiency would be very low, but with such a vast reservoir of energy available, this would not be a serious problem.

This is only a small sample of present-day activity in energy-conversion research. Many other processes are under discussion or development, and the principles of thermodynamics outlined in this chapter are of central importance in all of them.

19–10 An array of windmills to collect and convert wind energy. The propellerlike blades turn electric generators, converting the kinetic energy of moving air into electrical energy. (Photo by Kurt Rogers, *San Francisco Examiner*.)

SUMMARY

A cyclic process is one in which the initial and final states are the same; the total internal-energy change is always zero.

A heat engine takes heat Q_H from a source, converts part of it into work W, and discards the remainder Q_C at a lower temperature. The thermal efficiency e of a heat engine is defined as

$$e = \frac{W}{Q_H} = \frac{Q_H + Q_C}{Q_H} = 1 + \frac{Q_C}{Q_H} = 1 - \left| \frac{Q_C}{Q_H} \right|. \qquad (19\text{--}4)$$

A gasoline engine operating on the Otto cycle with compression ratio r has a theoretical maximum thermal efficiency e given by

$$e = 1 - \frac{1}{r^{\gamma - 1}}. \qquad (19\text{--}6)$$

A refrigerator takes heat Q_C from a cold place, has a work input W, and discards heat Q_H at a warmer place. The performance coefficient K is defined as

$$\text{Performance coefficient} = K = -\frac{Q_C}{W} = -\frac{Q_C}{Q_H + Q_C}. \qquad (19\text{--}8)$$

A reversible process is one whose direction can be reversed by an infinitesimal change in the conditions of the process. A reversible process is also an equilibrium process.

The second law of thermodynamics describes the directionality of natural thermodynamic processes. It can be stated in several equivalent forms; the two simplest are (1) the impossibility of a cyclic process in which heat is converted completely to work; (2) the impossibility of a cyclic process in which heat is transferred from a cold place to a hotter place with no input of mechanical work.

The Carnot cycle is a theoretical heat-engine cycle operating between two heat reservoirs at temperatures T_H and T_C and using only reversible processes. Its thermal efficiency is given by

$$e = 1 - \frac{T_C}{T_H} = \frac{T_H - T_C}{T_H}. \qquad (19\text{--}13)$$

No engine operating between the same two temperatures can be more efficient than a Carnot engine, and all Carnot engines operating between the same two temperatures have the same efficiency.

A Carnot engine run backward makes a Carnot refrigerator. Its performance coefficient is given by

$$K = \frac{T_C}{T_H - T_C}. \qquad (19\text{--}14)$$

No refrigerator operating between the same two temperatures can have a larger performance coefficient than a Carnot refrigerator, and all Carnot refrigerators operating between the same two temperatures have the same performance coefficient.

Entropy is a quantitative measure of the disorder of a system. The entropy change in any reversible thermodynamic process is defined as

$$\Delta S = \int_1^2 \frac{dQ}{T} \qquad \text{(reversible process)}, \qquad (19\text{--}16)$$

where T must always be the absolute temperature. Entropy depends only on

KEY TERMS

reversibility
equilibrium processes
irreversible processes
heat engine
working substance
cyclic process
thermal efficiency
compression ratio
Otto cycle
Diesel cycle
refrigerator
performance coefficient
energy efficiency rating
heat pump
second law of thermodynamics
Carnot cycle
thermal pollution
entropy
Kelvin temperature scale
absolute zero

the state of the system, and the change in entropy between given initial and final states is the same for all processes leading from one to the other. This fact can be used to find the entropy change in an irreversible process, where Eq. (19–16) is not applicable.

The second law of thermodynamics can be restated in terms of entropy. The entropy of an isolated system may increase but can never decrease. When a system interacts with its surroundings, the total entropy change of system and surroundings can never decrease. If the interaction involves only reversible processes, the total entropy is constant; if there is any irreversible process, the total entropy increases.

The Kelvin temperature scale is based on the efficiency of the Carnot cycle and is independent of the properties of any specific material. The zero point on the Kelvin scale is called absolute zero.

QUESTIONS

19–1 Suppose you want to increase the efficiency of a heat engine. Would it be better to increase T_H or to decrease T_C by an equal amount?

19–2 If an energy-conversion process involves two steps, each with its own efficiency, such as heat to work and work to electric energy, is the efficiency of the composite process equal to the product of the two efficiencies, or the sum, or the difference, or what?

19–3 In some climates it is practical to heat a house by using a heat pump, which acts as an air conditioner in reverse, cooling the outside air and heating the inside air. Can the heat delivered to the house ever exceed the electric-energy input to the pump?

19–4 What irreversible processes occur in a gasoline engine?

19–5 A housewife tries to cool her kitchen on a hot day by leaving the refrigerator door open. What happens? Would the result be different if an old-fashioned ice box were used?

19–6 Is it a violation of the second law to convert mechanical energy completely into heat? To convert heat completely into work?

19–7 A growing plant creates a highly complex and organized structure out of simple materials, such as air, water, and trace minerals. Does this violate the second law of thermodynamics? What is the plant's ultimate source of energy?

19–8 An electric motor has its shaft coupled to that of a generator. The motor drives the generator, and the current from the generator is used to run the motor. The excess current is used to power a television set. What is wrong with this scheme?

19–9 Think of some reversible and some irreversible processes in purely mechanical systems, such as blocks sliding on planes, springs, pulleys, and strings.

19–10 Why must a room air conditioner be placed in a window? Why can't it just be set on the floor and plugged in?

19–11 Discuss the following examples of increasing disorder or randomness: mixing of hot and cold water; free expansion of a gas; irreversible heat flow; and development of heat through mechanical friction. Are entropy increases involved in all these?

19–12 When the sun shines on a glass-roofed greenhouse, the temperature becomes higher inside than outside. Does this phenomenon violate the second law?

19–13 When a wet cloth is hung up in a hot wind in the desert, it is cooled by evaporation to a temperature that may be 20 C° or so below that of the air. Discuss this process in light of the second law.

19–14 Are the earth and sun in thermal equilibrium? Are there entropy changes associated with the transmission of energy from sun to earth? Does radiation differ from other modes of heat transfer with respect to entropy changes?

19–15 Discuss the entropy changes involved in the preparation and consumption of a hot-fudge sundae.

EXERCISES

Section 19–2 Heat Engines

19–1 A large diesel engine takes in 8000 J of heat and delivers 3000 J of work per cycle. The heat is obtained by burning diesel fuel with a heat of combustion of $5.0 \times 10^4 \, \text{J} \cdot \text{g}^{-1}$.

a) What is the thermal efficiency?

b) How much heat is discarded in each cycle?

c) What mass of fuel is burned in each cycle?

d) If the engine goes through 50 cycles per second, what is its power output in watts? In horsepower?

19–2 A gasoline engine has a power output of 20 kW (about 27 hp). Its thermal efficiency is 20%.

a) How much heat must be supplied to the engine per second?

b) How much heat is discarded by the engine per second?

19–3 A nuclear-power plant has a mechanical-power output (used to drive an electric generator) of 200 MW. Its rate of heat input from the nuclear reactor is 800 MW.

a) What is the thermal efficiency of the system?

b) At what rate is heat discarded by the system?

19–4 A coal-fired steam-turbine power plant has a mechanical-power output of 500 MW and a thermal efficiency of 40%.

a) At what rate must heat be supplied by burning coal?

b) If the heat of combustion of coal is 2.5×10^4 J·g^{-1}, what mass of coal is burned per second? Per day?

c) At what rate is heat discarded by the system?

d) If the discarded heat is given to water in a river, and its temperature rises by 5 C°, what volume of water is needed per second?

e) In part (d), if the river is 100 m wide and 5 m deep, what must be the minimum flow velocity of the water?

19–5 A heat engine takes 0.1 mol of an ideal gas around the cycle shown in the pV-diagram of Fig. 19–11. Process 1–2 is at constant volume, process 2–3 is adiabatic, and process 3–1 is at a constant pressure of 1 atm. The value of γ for this gas is $\frac{5}{3}$.

a) Find the pressure and volume at points 1, 2, and 3.

b) Find the net work done by the gas in the cycle.

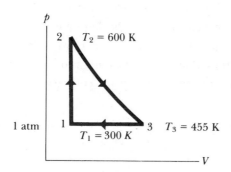

FIGURE 19–11

Section 19–3 Internal-Combustion Engines

19–6 For a gas with $\gamma = 1.4$, what compression ratio r must an Otto cycle have to achieve an ideal efficiency of 70%?

19–7 For an Otto cycle with $\gamma = 1.4$ and $r = 8$, the temperature of the gasoline-air mixture when it enters the cylinder is 22°C (point a of Fig. 19–3). What is the temperature at the end of the compression stroke (point b)?

Section 19–4 Refrigerators

19–8 A window air-conditioner unit absorbs 4000 J of heat per minute from the room being cooled and in the same time period deposits 12,000 J of heat to the outside air.

a) What is the power consumption of the unit, in watts?

b) What is the performance coefficient of the unit?

19–9 A freezer has a performance coefficient $K = 5$. The freezer is to convert 2 kg of water at $T = 20$°C into 2 kg of ice at $T = -10$°C in one hour.

a) What amount of heat must be removed from the water at 20°C to convert it into ice at −10°C?

b) How much electrical energy will be consumed by the freezer?

c) How much waste heat will be rejected to the room in which the freezer sits?

Section 19–6 the Carnot Cycle

19–10 Show that the efficiency e of a Carnot engine and the performance coefficient K of a Carnot refrigerator are related by $K = (1 - e)/e$.

19–11 A Carnot engine whose high-temperature reservoir is at 400 K takes in 420 J of heat at this temperature in each cycle and gives up 335 J to the low-temperature reservoir.

a) What is the temperature of the low-temperature reservoir?

b) What is the thermal efficiency of the cycle?

19–12 A Carnot engine is operated between two heat reservoirs at temperatures of 400 K and 300 K.

a) If the engine receives 5000 J of heat energy from the reservoir at 400 K in each cycle, how many joules per cycle does it reject to the reservoir at 300 K?

b) If the engine is operated in reverse, as a refrigerator, and receives 5000 J from the reservoir at 300 K, how many joules does it deliver to the reservoir at 400 K?

c) How many joules of mechanical work are required to operate the refrigerator in part (b)?

19–13 An ice-making machine operates in a Carnot cycle; it takes heat from water at 0°C and rejects heat to a room at 27°C. Suppose that 50 kg of water at 0°C are converted into ice at 0°C.

a) How much heat is rejected to the room?

b) How much energy must be supplied to the refrigerator?

Section 19–7 Entropy

19–14 A sophomore with nothing better to do adds heat to 0.5 kg of ice at 0°C until it is all melted.

a) What is the change in entropy of the water?

b) If the source of heat is a very massive body at a temperature of 20°C, what is the change in entropy of this body?

c) What is the total change in entropy of the water and the heat source?

19–15 Calculate the entropy change that occurs when 1 kg of water at 20°C is mixed with 2 kg of water at 80°C.

19–16 A block of aluminum of mass 1 kg, initially at 100°C, is dropped into 1 kg of water initially at 0°C.

a) What is the final temperature?

b) What is the total change in entropy of the system?

19–17 Two moles of an ideal gas undergo a reversible isothermal expansion from 0.02 m³ to 0.04 m³ at a temperature of 300 K. What is the change in entropy of the gas?

Section 19–9 Energy Conversion

19–18 An engine is to be built to extract power from the temperature gradient of the ocean. If the surface and deep-water temperatures are 25°C and 10°C, respectively, what is the maximum theoretical efficiency of such an engine?

19–19 A solar-power plant is to be built with a power output capacity of 1000 MW. What land area must the solar energy collectors occupy if they are

a) photocells with 60% efficiency?

b) mirrors that generate steam for a turbine-generator unit with overall efficiency of 30%?

Take the average power in the sun's radiation to be 200 W·m⁻² at the earth's surface. Express your answers in square kilometers and square miles.

PROBLEMS

19–20 A cylinder contains oxygen at a pressure of 2 atm. The volume is 3 L and the temperature is 300 K. The oxygen is carried through the following processes:

1. Heated at constant pressure to 500 K.
2. Cooled at constant volume to 250 K.
3. Cooled at constant pressure to 150 K.
4. Heated at constant volume to 300 K.

a) Show these four processes in a pV-diagram, giving the numerical values of p and V at the end of each process.

b) Calculate the net work done by the oxygen.

c) What is the efficiency of this device as a heat engine?

19–21 What is the thermal efficiency of an engine that operates by taking n moles of an ideal gas through the following cycle? Let $C_v = 12$ J·mol⁻¹·K⁻¹.

1. Start with n moles at P_0, V_0, T_0.
2. Change to $2P_0$, V_0, at constant volume.
3. Change to $2P_0$, $2V_0$, at constant pressure.
4. Change to P_0, $2V_0$, at constant volume.
5. Change to P_0, V_0, at constant pressure.

19–22 A Carnot engine operates between two heat reservoirs at temperatures T_H and T_C. An inventor proposes to increase the efficiency by running one engine between T_H and an intermediate temperature T', and a second engine between T' and T_C, using the heat expelled by the first engine. Compute the efficiency of this composite system and compare it to that of the original engine.

19–23 An 0.08 kg cube of ice at an initial temperature of -15°C is placed in 0.50 kg of water at $T = 60$°C in an insulated container of negligible mass. Calculate the entropy change of the system.

19–24 A physics student performing a heat-conduction experiment immerses one end of a copper rod in boiling water at 100°C, the other end in an ice-water mixture at 0°C. The sides of the rod are insulated. During a certain time interval, 0.5 kg of ice melts. Find

a) the entropy change of the boiling water;

b) the entropy change of the ice-water mixture;

c) the entropy change of the copper rod;

d) the total entropy change of the entire system.

CHALLENGE PROBLEMS

19–25 Consider a Diesel cycle that starts (point a in Fig. 19–4) with 2.0 L of air at a temperature of 300 K and a pressure of 1.0×10^5 Pa. If the temperature at point c is $T_c = 1200$ K, derive an expression for the efficiency of the cycle in terms of the compression ratio r. What is the efficiency when $r = 20$?

19–26

a) Draw a graph of a Carnot cycle, plotting Kelvin temperature vertically and entropy horizontally (a temperature–entropy or TS diagram).

b) Show that the area under any curve in a temperature–entropy diagram represents the heat absorbed by the system.

c) Derive from your diagram the expression for the thermal efficiency of a Carnot cycle.

20

MOLECULAR PROPERTIES OF MATTER

WE HAVE STUDIED SEVERAL PROPERTIES OF MATTER IN BULK, INCLUDING elasticity, density, surface tension, equations of state, heat capacities, phase changes, internal energy, entropy, and others. We have discussed qualitatively the relationships of these properties to molecular structure, but have deliberately avoided any detailed discussion of these relationships. There is a good reason for this; all of the above macroscopic (bulk) properties can be used in practical calculations without any detailed understanding of their microscopic (molecular) basis. Even more important, quantities such as heat, internal energy, and entropy *must* be defined in a way that does not depend on the details of any microscopic picture. Indeed, much of the power and usefulness of thermodynamics lies in its generality and its lack of dependence on microscopic models.

Understanding bulk properties of matter on the basis of molecular structure

Nevertheless, we can gain a lot of additional insight into the behavior of matter by looking at the relation of bulk behavior to microscopic structure. We will study a few examples of molecular models that enable us actually to *predict* some properties of matter. We begin with a general discussion of the molecular structure of matter. Then we develop the kinetic-molecular model of a gas, which helps us understand the equation of state and heat capacities of gases on the basis of a molecular model. Finally, we look briefly at the heat capacities of solids and their relation to molecular structure.

20–1 MOLECULAR STRUCTURE OF MATTER

An abundance of physical and chemical evidence has shown conclusively that matter in all phases is made up of particles called **molecules.** For any specific chemical compound, the molecules are all identical. The smallest molecules are of the order of 10^{-10} m in size; the largest are at least 10,000 times this large. In some materials the molecules are dissociated into electrically charged substructures called *ions*.

In liquids and solids molecules are held together by intermolecular forces that are *electrical* in nature, arising from interactions of the electrically charged fundamental particles that make up the molecules. *Gravitational* forces be-

Forces between atoms and molecules are basically electrical in nature.

451

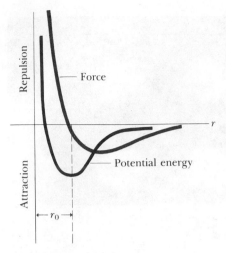

20-1 The force between two molecules (color curve) changes from an attraction when the separation is large to a repulsion when the separation is small. The potential energy (black curve) is minimum at r_0, where the force is zero

Forces between molecules are attractive at some distances and repulsive at others.

Molecules are always in motion, and the amount of motion increases with temperature.

20-2 Crystal of necrosis virus protein. The actual size of the entire crystal is about two-thousandths of a millimeter (0.002 mm). (Courtesy of Ralph W. G. Wyckoff. Reprinted with the permission of Educational Services, Inc., from *Physics*, D. C. Heath and Co., Boston, 1960.)

tween molecules are so weak compared with electrical forces that they are completely negligible.

The interaction of two point electric charges is described by a force (repulsive for like charges, attractive for unlike charges) whose magnitude is proportional to $1/r^2$, where r is the distance between the points. This relationship is called *Coulomb's law;* we will study it in detail in Chapter 24. Forces between *molecules,* however, do not follow a simple inverse-square law. Molecules are not point charges but complex structures containing both positive and negative charges, and their interactions are correspondingly complex. When molecules are far apart, as in a gas, the intermolecular force is very small and usually attractive. As a gas is compressed and its molecules are brought closer together, the force increases. In liquids, relatively large pressures are required to compress the substance appreciably. We conclude that at separations between molecules that are only slightly *less* than their normal spacing, the force becomes *repulsive* and relatively large.

Thus the intermolecular force must vary with the distance r between molecules somewhat as shown in Fig. 20–1. At large distances the force is small and attractive. As the molecules come closer together, the force of attraction becomes larger, passes through a maximum, and then decreases to zero at an equilibrium separation r_0. When the distance is less than r_0, the force becomes repulsive and increases quite rapidly. Figure 20–1 also shows the potential energy as a function of r. This function has a *minimum* at r_0, where the force is zero. Such a potential-energy function is often called a **potential well.** The force F and potential energy U are related by $F = -dU/dr$.

In view of these attractive intermolecular forces, why do all molecules not eventually coalesce into matter in the liquid or solid phase? The answer is that molecules are always in *motion*. There is kinetic energy associated with this motion, and it usually increases with temperature. At very low temperatures the average kinetic energy of a molecule may be much *less* than the maximum magnitude of potential energy, which is the "depth" of the potential well in Fig. 20–1. The molecules then condense into the liquid or solid phase with average intermolecular spacing of about r_0. But at higher temperatures the average kinetic energy becomes larger than the depth of the potential well; molecules can then escape the intermolecular force and become free to move independently, as in the gaseous phase of matter.

In *solids,* molecules execute vibratory motion about more-or-less fixed centers. The potential well is usually approximately parabolic in shape near its minimum, and the motion is then approximately simple harmonic. The amplitudes of the vibratory motions are relatively small, and the fixed centers form a space lattice, corresponding to the repeated spatial patterns and symmetry of crystals. A photograph of the individual, very large molecules of necrosis virus protein is shown in Fig. 20–2. It was taken with an electron microscope at a magnification of about 80,000. The molecules look like neatly stacked oranges. Each molecule is about 1.2×10^{-8} m in diameter.

In a *liquid* the intermolecular distances are usually only slightly greater than in the solid phase of the same substance. The molecules have vibratory motions of greater energy about centers that are free to move, but they remain at approximately the same distances from each other. Liquids show a certain regularity of structure only in the immediate neighborhood of a few molecules. This is called **short-range order,** in contrast to the **long-range order** of a solid crystal.

The molecules of a gas have greater average kinetic energy than those of liquids and solids. The molecules are usually widely separated and have only very small attractive forces. A molecule of a gas therefore moves in a straight line until it collides either with another molecule or with a wall of the container. In molecular terms, an *ideal gas* is a gas whose molecules exert *no* forces of attraction on each other. The mathematical analysis of a collection of such idealized molecules in random linear motion between collisions is called the *kinetic-molecular theory of gases*. The analysis of an ideal gas given in this chapter is a simple example of kinetic theory.

Most common substances exist in the solid phase at low temperatures. When the temperature is raised beyond a definite value, the liquid phase results, and when the temperature of the liquid is raised further, the substance exists in the gaseous phase. That is, from a large-scale, or *macroscopic*, point of view, the transition from solid to liquid to gas is in the direction of increasing temperature. From a *molecular* point of view, this transition is in the direction of increasing molecular kinetic energy. Thus temperature and molecular kinetic energy are closely related.

20–2 AVOGADRO'S NUMBER

In 1811 the chemist John Dalton suggested that the molecules of a given chemical substance are all alike. This hypothesis has since been confirmed by an overwhelming variety of evidence; it explains why, in chemical reactions, elements and compounds combine with each other in definite proportions by mass. It also leads directly to the concepts of *molecular mass* and *number of moles*. Specifically, one **mole** of any pure chemical element or compound contains a definite number of molecules, the same number for all elements and compounds. The official SI definition of the mole is as follows:

> The mole is the amount of substance that contains as many elementary entities as there are atoms in 0.012 kilogram of carbon 12.

In our discussion the "elementary entities" referred to are atoms or molecules.

The number of atoms or molecules in a mole is called **Avogadro's number,** denoted by N_A. (Avogadro was a contemporary of Dalton.) The most precise measurements of N_A have been obtained by using x-rays to measure the distance between layers of molecules in a crystal. The numerical value of N_A is now known with an uncertainty of less than six parts per million; to four significant figures, it is

$$N_A = 6.022 \times 10^{23} \text{ molecules·mol}^{-1}.$$

The **molecular mass** M of a compound is the mass of one mole; the mass m of a single molecule is thus given by

$$M = N_A m. \qquad (20-1)$$

When the molecule consists of a single atom, the term *atomic mass* is often used.

Once Avogadro's number has been determined, it can be used to compute the mass of a molecule. For example, the mass of 1 mol of atomic hydrogen (i.e., the atomic mass) is 1.008 g. The mass of 1 mol of diatomic hydrogen molecules is 2.016 g. Since, by definition, the number of atoms or molecules in

High-resolution electron micrographs showing the surface of a small crystal of gold. Most of the atoms are bound in a crystal lattice, but a few columns of gold atoms (shown by an arrow) are "hopping" with respect to the crystal lattice. (Courtesy of J.-O. Bovin, R. Wallenberg (Univ. of Lund, Sweden) and D. J. Smith (Arizona State Univ.).)

A mole is a definite number of atoms or molecules.

Avogadro's number is the number of atoms or molecules in a mole.

Finding the mass of a single molecule from the molecular mass and Avogadro's number

20–3 How much is one mole? The photograph shows one mole each of sucrose (ordinary sugar, rear pile), iodine (metallic-looking chips), water, mercury, iron (cube), acetylsalicylic acid (aspirin tablets), and sodium chloride (ordinary salt, front pile). (Photo by Chip Clark.)

The physical properties of materials are determined by their molecular structure.

1 mol is Avogadro's number, it follows that the mass of a single atom of hydrogen is

$$m_{\mathrm{H}} = \frac{1.008 \text{ g·mol}^{-1}}{6.022 \times 10^{23} \text{ molecules·mol}^{-1}} = 1.674 \times 10^{-24} \text{ g·molecule}^{-1}.$$

For an oxygen molecule of molecular mass 32 g·mol^{-1},

$$m_{\mathrm{O}_2} = \frac{32 \text{ g·mol}^{-1}}{6.022 \times 10^{23} \text{ molecules·mol}^{-1}} = 53.14 \times 10^{-24} \text{ g·molecule}^{-1}.$$

Figure 20–3 shows one mole of each of several familiar materials.

20–3 MOLECULAR BASIS OF PROPERTIES OF MATTER

The usual goal of any molecular theory of matter is to understand the *macroscopic* properties of matter in terms of the properties, behavior, and interactions of the molecules of which it is composed. Such theories are of tremendous practical importance; once this understanding has been gained, it becomes possible actually to *design* materials with specific desired properties. Thus this kind of molecular analysis has led to the development of high-strength steels, glasses with special optical properties for use in optical instruments, semiconductor materials for solid-state electronic devices, and countless other materials essential to contemporary technology.

In the following sections we will consider a few simple examples of molecular theories of matter. For example, we can represent a monatomic gas with a model consisting of a large number of particles described completely by their

Installing heat-shield tiles on the Space Shuttle. The material must have exceptional strength and heat resistance to protect the space vehicle during re-entry into the earth's atmosphere. (Courtesy of Rockwell International.)

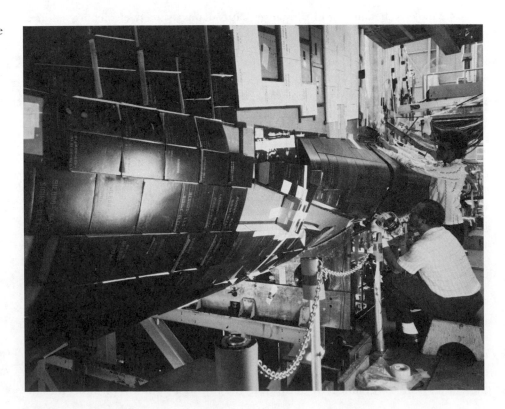

mass and velocities. We can then derive from this model the ideal-gas equation of state of a gas at low pressure and show that C_v for the gas should equal 12.5 J·mol^{-1}·K^{-1} (see Table 18–1). If we add the hypothesis that the "particles" are not simply points but have a finite size, then the general features of the viscosity, thermal conductivity, and coefficient of diffusion of a gas can be understood, as well as the fact that polyatomic gases have larger values of C_v than monatomic gases. By assuming that there are interaction forces between the particles, we can refine the equation of state to bring it into better agreement with the behavior of a real gas, and we can begin to understand the phenomena of liquefaction and solidification at low temperatures.

The electrical and magnetic properties of matter, and the emission and absorption of light by matter, call for a molecular model in which the molecules themselves are aggregates of subatomic particles. Some of these particles are electrically charged, and the forces between molecules originate in these electric charges. This development of molecular theory will take us into the area of atomic and molecular structure in Chapter 43. In the next section we return to the starting point and see what properties of a gas at low pressure can be explained by the simplest possible molecular model, an aggregate of particles having mass and velocity.

20–4 KINETIC-MOLECULAR THEORY OF AN IDEAL GAS

We are now ready to develop in detail the relation between the kinetic-molecular model of an ideal gas and the ideal-gas equation of state, discussed in Section 17–2. Consider a container of volume V, containing N identical molecules each of mass m. The molecules are in constant motion; each molecule collides from time to time with a wall of the container. During such collisions, the molecules exert forces on the walls, and this is the origin of the macroscopic *pressure* the gas exerts on the container walls. We assume that each collision of a molecule with a wall of the container is perfectly elastic, as shown in Fig. 20–4. In each collision the component of velocity parallel to the wall is unchanged and the component perpendicular to the wall is reversed.

Our program will be to determine for a given wall area A the number of collisions per unit time, the associated momentum change, and the force needed for the momentum change. Then we can obtain an expression for the pressure, which is force per unit area. Let v_x be the *magnitude* of the x-component of velocity of a molecule. At first we will assume that all molecules have the same v_x. This assumption, though unrealistic, helps clarify the basic ideas; and we will show soon that it is not really necessary.

In each collision the change in the x-component of momentum is $2mv_x$. To find the number of collisions with a given wall area A during a time interval Δt, we note that, in order to experience a collision during Δt, a molecule must be within a distance $v_x \Delta t$ from the wall at the beginning of Δt, as shown in Fig. 20–5, and must be headed toward the wall. Thus to collide with A during Δt, a molecule must, at the beginning of Δt, be within a cylinder of base area A and length $v_x \Delta t$. The volume of such a cylinder is $Av_x \Delta t$.

Assuming the number of molecules per unit volume (N/V) is uniform, the *number* of molecules in this cylinder is $(N/V)(Av_x \Delta t)$. But, on average, half of these molecules are moving *away from* the wall. Thus the number of collisions

The pressure of a gas results from collisions of molecules with the walls of the container.

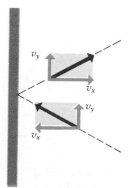

20–4 Elastic collision of a molecule with container wall. The component v_y parallel to the wall does not change; the component v_x perpendicular to the wall reverses direction. The speed v does not change.

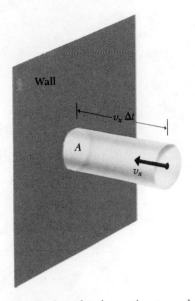

20–5 A molecule moving toward the wall with speed v_x collides with the area A during the time interval Δt only if it is within a distance $v_x \Delta t$ of the wall at the beginning of the interval. All such molecules are contained within a volume $A v_x \Delta t$.

with A during Δt is

$$\frac{1}{2} \left(\frac{N}{V} \right) (A v_x \Delta t). \qquad (20\text{–}2)$$

The total momentum change ΔP_x due to all these collisions is $2 m v_x$ times the *number* of collisions. (We are using capital P for momentum, and we will use small p for pressure; be careful!)

$$\Delta P_x = \frac{1}{2} \left(\frac{N}{V} \right) (A v_x \, \Delta t)(2 m v_x) = \frac{N A m v_x^2 \, \Delta t}{V}, \qquad (20\text{–}3)$$

and the *rate* of change of momentum is

$$\frac{\Delta P_x}{\Delta t} = \frac{N A m v_x^2}{V}. \qquad (20\text{–}4)$$

According to Newton's second law, this rate of change of momentum equals the average force exerted by the wall area A on the molecules. From Newton's *third* law, this is the negative of the force exerted *on* the wall *by* the molecules. Finally, pressure p is force per unit area, and we obtain

$$p = \frac{F}{A} = \frac{N m v_x^2}{V}. \qquad (20\text{–}5)$$

Now in fact v_x is *not* the same for all molecules. But we could have sorted the molecules into groups having the same v_x within each group and added up the resulting contributions to the pressure. The net effect is simply to replace v_x^2 in Eq. (20–5) by the *average* value of v_x^2, which we denote by $(v_x^2)_{av}$. Furthermore, $(v_x^2)_{av}$ is related simply to the *speeds* of the molecules. The speed v (magnitude of velocity) of any molecule is related to the velocity components v_x, v_y, and v_z by

$$v^2 = v_x^2 + v_y^2 + v_z^2.$$

We can average this relation over all molecules:

$$(v^2)_{av} = (v_x^2)_{av} + (v_y^2)_{av} + (v_z^2)_{av}.$$

But since the x-, y-, and z-directions are all equivalent,

$$(v_x^2)_{av} = (v_y^2)_{av} = (v_z^2)_{av}.$$

Hence

$$(v_x^2)_{av} = \frac{1}{3} (v^2)_{av},$$

and Eq. (20–5) becomes

$$pV = \frac{1}{3} N m (v^2)_{av} = \frac{2}{3} N \left[\frac{1}{2} m (v^2)_{av} \right]. \qquad (20\text{–}6)$$

The pressure of a gas is proportional to the average kinetic energy of its molecules.

But $\frac{1}{2} m (v^2)_{av}$ is the average kinetic energy of a single molecule, and the product of this energy and the total number of molecules N equals the total random kinetic energy, or internal energy U. Hence the product pV equals two-thirds of the internal energy:

$$pV = \frac{2}{3} U. \qquad (20\text{–}7)$$

By experiment, at low pressures, the equation of state of a gas is

$$pV = nRT.$$

The theoretical and experimental laws will therefore be in complete agreement if we set

$$U = \frac{3}{2}nRT. \qquad (20\text{–}8)$$

The average kinetic energy of a single molecule is then

$$\frac{U}{N} = \frac{1}{2}m(v^2)_{av} = \frac{3nRT}{2N}.$$

But the number of moles, n, equals the total number of molecules, N, divided by Avogadro's number N_A, the number of molecules per mole:

$$n = \frac{N}{N_A}, \qquad \frac{n}{N} = \frac{1}{N_A}.$$

Hence

$$\frac{1}{2}m(v^2)_{av} = \frac{3}{2}\frac{R}{N_A}T. \qquad (20\text{–}9)$$

The ratio R/N_A occurs frequently in molecular theory. It is called the **Boltzmann constant**, k:

$$k = \frac{R}{N_A} = \frac{8.31 \text{ J·mol}^{-1}\text{·K}^{-1}}{6.02 \times 10^{23} \text{ molecules·mol}^{-1}}$$
$$= 1.38 \times 10^{-23} \text{ J·molecule}^{-1}\text{·K}^{-1}.$$

Since R and N_A are fundamental physical constants, the same is true of k. Then

$$\frac{1}{2}m(v^2)_{av} = \frac{3}{2}kT. \qquad (20\text{–}10)$$

Thus the average kinetic energy *per molecule* depends only on the temperature, not on the pressure, volume, or molecular species. An equivalent statement can also be obtained from Eq. (20–9) by using the relation $M = N_A m$:

$$N_A\left[\frac{1}{2}m(v^2)_{av}\right] = \frac{1}{2}M(v^2)_{av} = \frac{3}{2}RT. \qquad (20\text{–}11)$$

That is, the kinetic energy of a *mole* of molecules depends only on T.

From Eqs. (20–10) and (20–11) we can obtain expressions for the square root of $(v^2)_{av}$, called the *root-mean-square speed* v_{rms}:

$$v_{rms} = \sqrt{(v^2)_{av}} = \sqrt{\frac{3kT}{m}} = \sqrt{\frac{3RT}{M}}. \qquad (20\text{–}12)$$

Finally, it is sometimes convenient to rewrite the ideal-gas equation on a molecular basis. Since $N = N_A n$ and $R = N_A k$, an alternative form of the ideal-gas equation is

$$pV = NkT. \qquad (20\text{–}13)$$

Thus k may be regarded as a gas constant on a "per molecule" basis instead of the usual "per mole" basis for R.

The total kinetic energy of gas molecules is proportional to the temperature.

The Boltzmann constant is a gas constant on a "per molecule" basis.

The rms speed of gas molecules depends on the molecular mass and on temperature.

PROBLEM-SOLVING STRATEGY: *Kinetic-molecular theory*

1. As always, using a consistent set of units is essential. We list below several places where caution is needed.

2. The most common units for molecular mass M are grams per mole; the molecular mass of oxygen is 32 g·mol^{-1}, for example. These units are often omitted in tables. In equations with SI units, such as Eq. (20–12), M *must* be converted to kilograms per mole by dividing by 10^3. Thus in SI units $M = 32 \times 10^{-3}$ kg·mol^{-1} for oxygen.

3. Are you working on a "per molecule" basis or a "per mole" basis? Remember that m is the mass of a molecule, and M is the mass of a mole. Similarly, N is the number of molecules, and n is the number of moles; k is the gas constant per molecule, and R is the gas constant per mole. Although N, the number of molecules, is in one sense a dimensionless number, you can do a complete unit check if you think of N as having the unit "molecules"; m has units "mass per molecule," and k has units "joules per molecule per kelvin."

4. Remember that T is always *absolute* temperature. In the problems in this chapter, it is always in kelvins.

EXAMPLE 20–1   What is the average kinetic energy of a molecule of a gas at a temperature of 300 K?

SOLUTION   From Eq. (20–10),

$$\frac{1}{2}m(v^2)_{av} = \frac{3}{2}kT = \left(\frac{3}{2}\right)(1.38 \times 10^{-23}\,\text{J·K}^{-1})(300\,\text{K})$$
$$= 6.21 \times 10^{-21}\,\text{J}.$$

EXAMPLE 20–2   What is the total random kinetic energy of the molecules in one mole of a gas at a temperature of 300 K?

SOLUTION   From Eq. (20–8),

$$U = \frac{3}{2}nRT = \frac{3}{2}(1\,\text{mol})(8.314\,\text{J·mol}^{-1}\text{·K}^{-1})(300\,\text{K})$$
$$= 3741\,\text{J} = 894\,\text{cal}.$$

EXAMPLE 20–3   What is the root-mean-square speed of a hydrogen molecule at 300 K?

Finding the rms speed of hydrogen molecules

SOLUTION   The mass of a hydrogen molecule (see Section 20–2) is
$$m_{\text{H}_2} = (2)(1.674 \times 10^{-27}\,\text{kg}) = 3.348 \times 10^{-27}\,\text{kg}.$$

Hence, from Eq. (20–12),

$$v_{\text{rms}} = \sqrt{\frac{3kT}{m}} = \sqrt{\frac{3(1.38 \times 10^{-23}\,\text{J·K}^{-1})(300\,\text{K})}{3.348 \times 10^{-27}\,\text{kg}}}$$
$$= 1927\,\text{m·s}^{-1}.$$

Alternatively,

$$v_{\text{rms}} = \sqrt{\frac{3RT}{M}} = \sqrt{\frac{3(8.314\,\text{J·mol}^{-1}\text{·K}^{-1})(300\,\text{K})}{2(1.008 \times 10^{-3}\,\text{kg·mol}^{-1})}}$$
$$= 1927\,\text{m·s}^{-1}.$$

Note that in Eq. (20–12), when we use the value of R in SI units, M must be expressed in *kilograms* per mole, not grams per mole. In this example, $M = 2.016 \times 10^{-3}$ kg·mol^{-1}, not 2.016 g·mol^{-1}.

EXAMPLE 20–4 Five gas molecules chosen at random are found to have speeds of 500, 600, 700, 800, and 900 m·s^{-1}. Find the rms speed. Is it the same as the *average* speed?

An exercise in calculating rms speed.

SOLUTION The average value of v^2 for the five molecules is

$$(v^2)_{av} = \frac{(500 \text{ m·s}^{-1})^2 + (600 \text{ m·s}^{-1})^2 + (700 \text{ m·s}^{-1})^2 + (800 \text{ m·s}^{-1})^2 + (900 \text{ m·s}^{-1})^2}{5}$$

$$= 510,000 \text{ m}^2\text{·s}^{-2},$$

and v_{rms} is the square root of this value:

$$v_{rms} = 714 \text{ m·s}^{-1}.$$

The *average* speed v_{av} is given by

$$v_{av} = \frac{500 \text{ m·s}^{-1} + 600 \text{ m·s}^{-1} + 700 \text{ m·s}^{-1} + 800 \text{ m·s}^{-1} + 900 \text{ m·s}^{-1}}{5}$$

$$= 700 \text{ m·s}^{-1}.$$

Clearly, v_{rms} and v_{av} are not, in general, the same.

EXAMPLE 20–5 Find the number of molecules in one cubic meter of air at atmospheric pressure and 0°C.

SOLUTION From Eq. (20–13),

$$N = \frac{pV}{kT} = \frac{(1.013 \times 10^5 \text{ Pa})(1 \text{ m}^3)}{(1.38 \times 10^{-23} \text{ J·K}^{-1})(273 \text{ K})} = 2.69 \times 10^{25}.$$

When a gas expands against a moving piston, it does work. This work is accompanied by a decrease in the random kinetic energy of the gas molecules. Conversely, when work is done on a gas during compression, the random kinetic energy of its molecules increases. But if the collisions with the walls are perfectly elastic, as we have assumed, how can a molecule gain or lose energy in a collision with a piston? To understand this phenomenon, we must consider the collision of a molecule with a *moving* wall.

When a molecule collides with a *stationary* wall, it exerts a momentary force on the wall but does no work, because the wall does not move. But if the wall is in motion, work *is* done during the collision. Thus if the wall in Fig. 20-4 is moving to the left, work is done on it by the molecules that strike it, and their speeds (and kinetic energies) after colliding are smaller than they were before the collision. The collision is still completely elastic because the work done on the moving piston is just equal to the decrease in the kinetic energy of the molecules. Similarly, if the piston is moving toward the right, the kinetic energy of a colliding molecule *increases*, by an amount equal to the work done on the molecule.

When a molecule collides with a moving wall, it does work on the wall.

The assumption that individual molecules undergo elastic collisions with the container wall is not strictly correct. More detailed investigation has shown that in most cases, molecules actually adhere to the wall for a short time, and then leave again with speeds characteristic of the temperature *of the wall*. The gas and the wall are ordinarily in thermal equilibrium, however, and the validity of our conclusions is not altered by this discovery.

20–5 MOLAR HEAT CAPACITY OF A GAS

The analysis in Section 20–4 can be extended to include a discussion of the molar heat capacities of an ideal gas. We begin with the molar heat capacity at constant volume, C_v. When we add heat to a gas under constant-volume conditions, the gas does no work. According to the first law, then, *all* the added energy goes to increase the internal energy. That is, $\Delta U = Q - W$, but $W = 0$; so $\Delta U = Q$. From the *molecular* viewpoint, the internal energy is the sum of the kinetic and potential energies of the molecules. If we know how this total internal energy depends on temperature, then from its rate of change with temperature we can make a theoretical prediction of the molar heat capacity of the system.

The simplest system is a monatomic ideal gas. We represent each molecule (a single atom) as a point particle. There is no potential energy of interaction of the molecules, and the only energy is the translational kinetic energy of random motion of the molecules. For n molecules of gas at absolute temperature T, this energy is given by Eq. (20-8):

$$\text{Kinetic energy} = U = \frac{3}{2}nRT.$$

If the temperature increases by ΔT, the kinetic energy increases by

$$\Delta U = \frac{3}{2}nR\,\Delta T.$$

In a process in which the temperature increases by ΔT at constant volume, the energy flowing into a system is, by definition of C_v,

$$\Delta Q = nC_v\,\Delta T = \Delta U.$$

Hence

$$C_v = \frac{3}{2}R. \tag{20–14}$$

In SI units,

$$C_v = \frac{3}{2}(8.314\ \text{J}\cdot\text{mol}^{-1}\cdot\text{K}^{-1}) = 12.47\ \text{J}\cdot\text{mol}^{-1}\cdot\text{K}^{-1}.$$

The experimental values of C_v listed in Table 18–1 for monatomic gases are, in fact, almost exactly equal to $\frac{3}{2}R$. This agreement is a striking confirmation of the basic correctness of the kinetic-molecular model of a gas.

Since $C_p = C_v + R$, it follows that the theoretical ratio of molar heat capacities for a monatomic ideal gas is

$$\frac{C_p}{C_v} = \gamma = \frac{\frac{3}{2}R + R}{\frac{3}{2}R} = \frac{5}{3} = 1.67.$$

This is also in good agreement with the experimental values in Table 18–1.

In this analysis we have treated each molecule as a point particle. For a gas whose molecules have two or more atoms each—that is, *polyatomic* gases—the problem is more complicated. For example, a diatomic molecule can be thought of in terms of *two* point masses, with an interaction force of the kind

Predicting the heat capacity of a gas by using the kinetic-molecular model

Gases with more than one atom in a molecule have additional energy associated with rotational and vibrational motion

shown in Fig. 20–1. We can picture such a molecule as a kind of elastic dumb-bell. It can have additional kinetic energy associated with *rotation* about an axis through its center of mass, and the atoms may have a back-and-forth *vibrating* motion along the line joining them, with associated additional kinetic and potential energies.

The *temperature* of a gas is determined by the average random *translational* kinetic energy $\frac{1}{2}mv^2$ of its molecules. When heat flows into a *monatomic* gas, at constant volume, all of this energy goes into an increase in random *transla-tional* molecular kinetic energy, as shown by the agreement between the meas-ured values of C_v and the values computed from the increase in translational kinetic energy. But when heat flows into a *diatomic* or *polyatomic* gas, part of the energy goes into increasing rotational and vibrational motion. Hence for a given temperature change, a greater total amount of energy is required in order to increase the rotational and vibrational as well as translational ener-gies. Thus polyatomic gases have larger molar heat capacities than monatomic gases. The experimental values in Table 18–1 show this effect.

To progress further in our theoretical understanding of heat capacities, we need some principle that will tell us *how much* energy is associated with each additional kind of motion of a complex molecule, compared to the transla-tional kinetic energy we have already discussed. The necessary principle goes by the fancy name of **principle of equipartition of energy.** In the early days of kinetic-molecular theory, this principle was treated as an axiom. Today it can be derived from sophisticated statistical-mechanics considerations; but that derivation is beyond our scope, and we too will treat it as an axiom.

> Equipartition of energy: Each kind of motion has its energy quota.

According to the principle of equipartition of energy, each velocity com-ponent (either linear or angular) has, on the average, an associated kinetic energy per molecule of $\frac{1}{2}kT$. The number of velocity components needed to describe the motion of a molecule completely is called the number of **degrees of freedom.** For a monatomic gas, the number is three. For a diatomic mole-cule, there are two possible axes of rotation, perpendicular to each other and to the molecule's axis. (Rotation about the molecule's own axis is not counted because, in ordinary collisions, there is no way for this rotational motion to change.) Thus if we assign five degrees of freedom to a diatomic molecule, the average total kinetic energy per molecule is $5kT/2$ instead of $3kT/2$. The total internal energy of n moles is $U = 5nRT/2$, and the molar heat capacity (at constant volume) is

> Degrees of freedom: How many kinds of motion can a molecule have?

$$C_v = \frac{5}{2}R. \qquad (20\text{–}15)$$

In SI units,

$$C_v = \frac{5}{2}(8.314 \text{ J·mol}^{-1}\cdot\text{K}^{-1}) = 20.78 \text{ J·mol}^{-1}\cdot\text{K}^{-1}.$$

Reference to Table 18–1 shows that this value is in approximate agreement with measured values for diatomic gases. The corresponding value of γ is

$$\gamma = \frac{C_p}{C_v} = \frac{\frac{5}{2}R + R}{\frac{5}{2}R} = \frac{7}{5} = 1.40,$$

which is again in reasonable agreement with experimental values.

Additional contributions to the heat capacities of gases can arise from *vibrational* motion. Molecules are never perfectly rigid; the molecular bonds can stretch and bend, permitting internal vibrations to occur. There are addi-tional degrees of freedom and energies associated with vibrational motion. For

> Vibrational motion of molecules: Does it contribute to heat capacities, or not, and why?

most diatomic gases, however, vibrational motion does *not* contribute appreciably to heat capacity, for reasons that involve concepts of quantum mechanics. Briefly, the energy of a vibrational motion can change only in finite steps. If the energy change of the first step is much larger than the energy possessed by most molecules, then nearly all the molecules will remain in the minimum-energy state of motion. In this case changing the temperature does not change their average vibrational energy appreciably, and the vibrational degrees of freedom are said to be "frozen out." In more complex molecules, the gaps between permitted energy levels are sometimes much smaller, and then vibration *does* contribute to heat capacity. In Table 18–1 the larger values of C_v for some polyatomic molecules show the contributions of vibrational energy. In addition, a molecule with three or more atoms not in a straight line has three, not two, rotational degrees of freedom.

Energy-level spacing determines whether vibrational motion contributes to heat capacities.

20–6 DISTRIBUTION OF MOLECULAR SPEEDS

Measuring the speeds of molecules in a gas. What is the distribution of speeds?

As mentioned in Section 20–4, the molecules in a gas do not all have the same speed. Direct measurements of the distribution of molecular speeds can be made; one experimental scheme is shown in Fig. 20–6. A substance is vaporized in a hot oven; molecules of the vapor escape through an aperture in the oven wall and into a vacuum chamber. A series of slits blocks all molecules except those in a narrow beam; the beam is aimed at a pair of rotating disks. A molecule passing through the slit in the first disk arrives at the second disk just as *its* slit is lined up with the beam only if the molecule has a certain speed. The setup thus functions as a speed selector that allows only molecules with a certain narrow range of speeds to pass. This speed can be varied by changing the disk speed, and we can measure how many molecules have each of various speeds.

The distribution function: describing how many molecules have how much speed

The results of such measurements can be represented graphically as shown in Fig. 20–7. We define a function $f(v)$ called a *distribution function*, such that if N molecules are observed, the number dN having speeds in any range between v and $v + dv$ is given by

$$dN = Nf(v)\,dv. \tag{20–16}$$

The figure shows distribution functions for several different temperatures; at each temperature the height of the curve for any value of v is proportional to the number of molecules with speeds near v. The peak of the curve represents the *most probable speed* for the corresponding temperature. As the temperature increases, the peak shifts to higher and higher speeds, corresponding to the increase in average molecular kinetic energy with temperature. Figure 20–7 also shows that the area under a curve between any two values of v represents

20–6 Apparatus for producing a molecular beam and observing the distribution of molecular speeds in the beam.

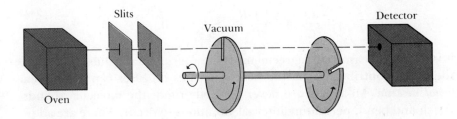

Oven   Slits   Vacuum   Detector

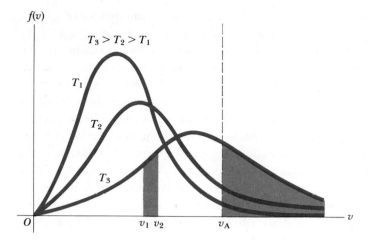

20–7 Maxwell-Boltzmann distribution curves for various temperatures. As the temperature increases, the curve becomes flatter, and its maximum shifts to higher temperature. At temperature T_3, the number of molecules having speeds in the range v_1 to v_2 and the number having speeds greater than v_A are shown by the shaded areas under the T_3 curve.

the number of molecules having speeds in that range. The area corresponding to the range v_1 to v_2 and the area for all speeds greater than v_A are shown for the distribution for temperature T_3.

The function $f(v)$ describing the actual distribution of molecular speeds is called the **Maxwell–Boltzmann distribution.** It can be derived from statistical-mechanics considerations, but that derivation is beyond our scope. Here is the result:

The Maxwell–Boltzmann distribution describes both speed distribution and energy distribution in a gas.

$$f(v) = 4\pi \left(\frac{m}{2\pi kT}\right)^{3/2} v^2 e^{-mv^2/2kT}. \qquad (20\text{--}17)$$

We can also express this function in terms of the translational kinetic energy of a molecule, which we denote by ϵ (the Greek letter epsilon). That is, $\epsilon = \frac{1}{2}mv^2$. We invite you to verify that when this value is substituted into Eq. (20–17), the result is

$$f(v) = \frac{8\pi}{m} \left(\frac{m}{2\pi kT}\right)^{3/2} \epsilon e^{-\epsilon/kT}. \qquad (20\text{--}18)$$

In this form the Maxwell–Boltzmann distribution function shows that the exponent is $-\epsilon/kT$, and that the shape of the curve is determined by the relative magnitude of ϵ and kT at any point. In particular, we invite you to prove that the *peak* of each curve occurs when $\epsilon = kT$, corresponding to a speed v given by

$$v = \sqrt{\frac{2kT}{m}}. \qquad (20\text{--}19)$$

This is therefore the *most probable* speed. Note that it differs from the rms speed given by Eq. (20–12) by a simple numerical factor $(\frac{2}{3})^{1/2}$.

Every molecule must have *some* speed, so the integral of $f(v)$ over all v must have the value unity. Also, the integral of the quantity $v^2 f(v)$ over all v must equal the average value of v^2; its square root is the rms speed, Eq. (20–12). To verify these two statements we have to evaluate some fairly complicated integrals, and the easiest procedure is to look them up in a table of definite integrals. We leave these calculations for problems.

The distribution of molecular speeds in liquids is similar, although not identical, to that for gases. We can understand the vapor pressure of a liquid and the phenomenon of boiling on this basis. Suppose a molecule must have a

Gas–liquid phase equilibrium: when the gas-to-liquid and liquid-to-gas molecular traffic just balance

speed at least as great as v_A in Fig. 20–7 to escape from the surface of a liquid into the adjacent vapor. The number of such molecules, represented by the area under each curve to the right of v_A, increases rapidly with temperature. Thus the rate at which molecules can escape is strongly temperature-dependent. This process is balanced by another one in which molecules in the vapor phase collide inelastically with the surface and are trapped back into the liquid phase. The number of molecules suffering this fate, per unit time, is proportional to the pressure in the vapor phase. Phase equilibrium between liquid and vapor occurs when these two competing processes proceed at exactly the same rate. Hence if we have an accurate collection of curves such as Fig. 20–7 for a variety of temperatures, we can make a theoretical prediction of the vapor pressure of a substance as a function of temperature.

Rates of chemical reactions are often strongly temperature-dependent, and the Maxwell–Boltzmann distribution contains the reason for this dependence. When two reacting molecules collide, the reaction can occur only when the molecules are close enough for the electric-charge distributions of their electrons to interact strongly. This requires a minimum energy, called the *activation energy*, and thus a certain minimum speed. We will call this speed v_A, although it is not necessarily equal to the v_A in the vapor-pressure discussion above. We have noted that the number of molecules whose speeds exceed some value v_A increases rapidly with temperature. Thus we expect the rate of any reaction that depends on an activation energy to increase rapidly with temperature. Similarly, many plant-growth processes have strongly temperature-dependent rates.

20–7 CRYSTALS

Many materials can exist in a variety of solid forms; a familiar example is the element *carbon*. The black soot deposited on a kettle by a smoky campfire is nearly pure carbon. The "lead" in a pencil is not lead at all, but chiefly a

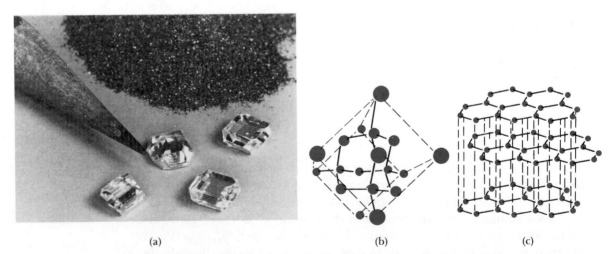

(a)　　　　　　　　(b)　　　　　　　　(c)

20–8　(a) Two different crystalline forms of carbon: graphite and diamond. (Courtesy of General Electric.) (b) Crystal structure of diamond; each atom is located at the center of a regular tetrahedron, with four equidistant nearest-neighbor atoms at the four corners. The diamonds shown are synthetic, made by subjecting graphite to extreme temperature and pressure; they are much more costly than natural diamonds of comparable size (about one carat). (c) Crystal structure of graphite, showing two-dimensional hexagonal arrays with strong bonds between atoms in each plane and much weaker bonds between planes.

20–9 Close packing of the sodium ions (black spheres) and chlorine ions (white spheres) in a sodium chloride crystal. (Courtesy of Alan Holden, reprinted with permission of Educational Services, Inc., from *Crystals and Crystal growing,* Doubleday and Co., New York, 1960.)

different form of carbon called graphite. Diamond is a third form of solid carbon.

The remarkably dissimilar mechanical, thermal, electrical, and optical properties of these three forms of carbon can be understood in terms of the different arrangement of the carbon atoms in the solid structures. In soot the atoms have no regular arrangement and are said to form an **amorphous solid.** A **crystalline solid,** as we discussed in Section 20–1, is characterized by an orderly geometric arrangement of atoms or molecules forming a recurrent pattern, called a crystal lattice, that extends over many molecules. Graphite and diamond are both crystals, but the orderly arrangement of atoms is different in the two crystal lattices, as shown in Fig. 20–8. Crystallography, the study of the spatial arrangements of atoms (or molecules or ions) in the various types of crystal lattices, is an important and interesting branch of solid-state physics. We will be concerned here with only a few basic ideas that are needed to understand some of the mechanical and thermal properties of single crystals or solids composed of an aggregate of crystals, that is, *polycrystalline solids.*

The particles constituting a crystal lattice are *closely packed,* as seen in Fig. 20–2 and as suggested by the model in Fig. 20–9, where the sodium and chloride ions of a sodium chloride crystal are represented as spheres touching one another. Actually, each ion consists of a nucleus surrounded by electrons in motion, so that the boundary of an ion must be thought of as the average positions of the outermost electrons. It follows that the outer electrons of one particle in a crystal lattice come close to and even at times interpenetrate the outer electrons of a neighboring particle. In drawings we often exaggerate the distances between neighboring particles, as in the *face-centered cubic* lattice structure shown in Fig. 20–10.

If the chemical composition and density of a very small volume element of a substance are measured at many different places and are found to be the same at all points, the substance is said to be *homogeneous.* If the physical properties determining the transport of heat, electricity, light, and the like, are *the same in all directions,* the substance is said to be *isotropic.* Cubic crystals are

In a crystalline solid the atoms or molecules form a regular, recurrent pattern.

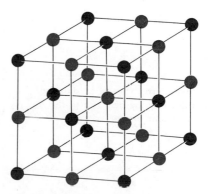

20–10 Symbolic representation of a sodium chloride crystal, with exaggerated distances between ions. Sodium ions are shown as black spheres, chlorine ions as colored spheres.

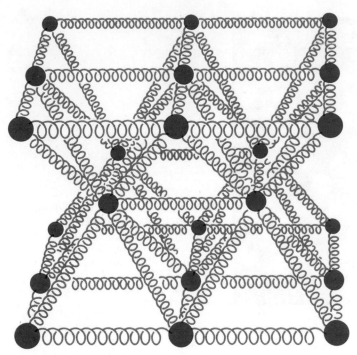

20-11 The forces between neighboring particles in a crystal may be visualized by imagining every particle to be connected to its neighbors by springs. In the case of a cubic crystal, all springs are assumed to have the same spring constant. Anisotropy is associated with differing spring constants in different directions.

both homogeneous and isotropic, whereas all other crystals, although homogeneous, are not isotropic—they are *anisotropic*. Anisotropic crystals have different properties for different directions in the crystal. Such a crystal may have a larger Young's modulus for stretching in one direction than for a perpendicular direction, it may conduct electricity better in some directions than in others; and so on.

The anisotropy of noncubic crystals can be understood in terms of the forces between neighboring particles in a lattice. A convenient way to visualize these forces, as shown in Fig. 20–11, is to use a model that represents a crystal lattice as a large number of spheres, each connected to its neighbors by springs. We imagine the force constants of parallel springs to be equal, but springs pointing in different directions may have different force constants. If we imagine each of the spheres in Fig. 20–11 as vibrating about its equilibrium position, we obtain a rough picture of the dynamic character of a crystal lattice.

The elastic properties of crystals can be partially understood from a diagram such as Fig. 20–11. Young's modulus is simply a measure of the stiffness of the springs in the direction in which the crystal is pulled or pushed. The shear modulus depends on the springs that are stretched and those that are compressed when the crystal is twisted. One of the first tests of the correctness of a lattice structure representing a particular material is to compare a calculated value of an elastic modulus with a measured value.

In Chapter 43 we return to a more detailed study of the electrical and optical properties of crystals. Meanwhile, in the next section we take a brief look at heat capacities of crystals.

The elastic properties of crystals are determined by the behavior of the intermolecular forces.

20–8 HEAT CAPACITY OF CRYSTALS

In Section 20–5 we studied the heat capacities of ideal gases on the basis of a kinetic-molecular model. We can carry out a similar analysis for a crystalline solid. Consider a crystal consisting of N identical atoms. Each atom is bound to an equilibrium position by forces that may be pictured in terms of springs, in the same spirit as Fig. 20–11. Thus each atom can vibrate about its equilibrium position; each atom has three degrees of freedom, corresponding to the three components of velocity needed to describe its vibrational motion. According to the equipartition principle, introduced in Section 20–5, each atom should have an average kinetic energy of $\frac{1}{2}kT$ for each of its three degrees of freedom (corresponding to the three components of velocity). In addition, each atom has on the average some *potential* energy associated with the elastic deformation. Now, for a simple harmonic oscillator (discussed in Chapter 11), it is not hard to show that the average kinetic energy of an atom is *equal* to its average potential energy. In our crystal lattice model, each atom is essentially a three-dimensional harmonic oscillator; and it can be shown that the equality of average kinetic and potential energies also holds here, provided the "spring" forces are proportional to the displacement from the equilibrium position.

Thus we expect each atom to have an average kinetic energy $\frac{3}{2}kT$ and an average potential energy $\frac{3}{2}kT$, or an average total energy $3kT$. The total energy of the whole crystal, which may be expressed in either molecular or molar terms, is

$$U = 3NkT = 3nRT. \tag{20–20}$$

From this relation we conclude that the molar heat capacity of a crystal of N identical atoms (n moles) should be

$$C = 3R. \tag{20–21}$$

In SI units,

$$c = (3)(8{,}314\ \text{J·mol}^{-1}\text{·K}^{-1}) = 24.9\ \text{J·mol}^{-1}\text{·K}^{-1}.$$

But this is just the **law of Dulong and Petit,** which we discussed in Section 15–4. There we regarded it as an *empirical* rule, but now we have *derived* it from kinetic theory. The agreement is only approximate, to be sure, but considering the very simple nature of the model, it is quite significant.

We have ignored the distinction between C_v and C_p. As we remarked in Section 15–4, the difference is usually insignificant for solid materials, unlike gases, where it is always significant.

At low temperatures, the heat capacities of most solids *decrease* with decreasing temperature. As with vibrational energy of molecules, this phenomenon requires quantum-mechanical concepts for its understanding. The vibrating atoms in the crystal lattice can give or lose energy only in certain finite increments. At very low temperatures, the quantity kT is much *smaller* than the smallest energy increment the vibrating atoms can accommodate. As a result, at low T most of the atoms remain in their lowest energy states because the next higher energy level is out of reach. Thus at low temperatures the average vibrational energy per vibrator is *less* than kT, and correspondingly the heat capacity per molecule is *less* than k. But in the other limit, where kT is *large* compared to the minimum energy increment, the classical equipar-

---

The heat capacity of a crystal is determined by the kinetic and potential energies associated with molecular vibrations.

The law of Dulong and Petit can be derived from a molecular picture of a solid.

Temperature variation of heat capacities of solids: an interesting test of quantum theory

tition theorem holds, and the total heat capacity is $3k$ per molecule, or $3R$ per mole, as the Dulong and Petit relation predicts. This behavior is illustrated by Fig. 15–5. Quantitative understanding of the temperature variation of heat capacities was one of the triumphs of quantum mechanics during its initial development, early in the twentieth century.

SUMMARY

Intermolecular forces are small and usually attractive when the molecules are far apart, as in a gas. At shorter distances the force becomes stronger, then weaker again, then zero, and finally becomes repulsive at sufficiently short distances. Crystalline solids have long-range order, a regular periodic structure extending over many molecules. Liquids and amorphous solids have only short-range order.

Avogadro's number N_A is the number of molecules in a mole. Molecular mass M is the mass of a mole of substance. These are related to the mass m of a single molecule by

$$M = N_A m. \qquad (20\text{–}1)$$

The kinetic-molecular model of an ideal gas can be used to derive the ideal-gas equation of state ($pV = nRT$) and molar heat capacities of some gases. Several other useful relations also follow from this analysis. The total kinetic energy of translational motion of n moles of gas at temperature T is

$$U = \frac{3}{2} nRT. \qquad (20\text{–}8)$$

The Boltzmann constant k may be thought of as the gas constant on a "per molecule" basis; $k = R/N_A$. The ideal-gas equation of state may be written in the alternative form

$$pV = NkT, \qquad (20\text{–}13)$$

and Eq. (20–8) can also be expressed as

$$U = \frac{3}{2} NkT, \qquad (20\text{–}10)$$

where N is the number of molecules, related to the number of moles n by $N = N_A n$.

The rms speed of molecules in an ideal gas is given by

$$v_{\text{rms}} = \sqrt{(v^2)_{\text{av}}} = \sqrt{\frac{3kT}{m}} = \sqrt{\frac{3RT}{M}}. \qquad (20\text{–}12)$$

The molar heat capacity at constant volume for an ideal monatomic gas is given by

$$C_v = \frac{3}{2} R. \qquad (20\text{–}14)$$

Diatomic and polyatomic gases have additional energy associated with rotational and sometimes vibrational motion. The principle of equipartition of energy states that on average there is a kinetic energy $\frac{1}{2}kT$ per molecule for each degree of freedom (component of velocity, either linear or angular).

The Maxwell–Boltzmann distribution describes the speed distribution of molecules in a gas. The height of the curve for any value of v is determined by the ratio of the corresponding kinetic energy of translational motion of a molecule, ϵ, to the quantity kT. This distribution function can be used to determine the fraction of all molecules that have speed greater than a certain value; this information can be used in the study of evaporation and chemical reactions, where a certain minimum molecular kinetic energy is required.

The equipartition principle can be used to calculate the molar heat capacity of a crystal; each molecule has, on average, $\frac{3}{2}kT$ of kinetic energy and an additional $\frac{3}{2}kT$ of potential energy. The resulting molar heat capacity is $C = 3R$, or about $25 \text{ J·mol}^{-1}\text{·K}^{-1}$. Thus the law of Dulong and Petit can be derived from the kinetic theory of gases.

QUESTIONS

20–1 Which has more atoms, a kilogram of hydrogen or a kilogram of lead? Which has more mass?

20–2 Chlorine is a mixture of two isotopes, one having molecular mass 35 g·mol^{-1}, the other 37 g·mol^{-1}. Which molecules move faster, on average?

20–3 The proportion of various gases in the earth's atmosphere changes somewhat with altitude. Would you expect the proportion of oxygen at high altitude to be greater or less than at sea level?

20–4 *Comment on this statement:* When two gases are mixed, if they are to be in thermal equilibrium they must have the same average molecular speed.

20–5 In deriving the ideal-gas equation from the kinetic-molecular model, we ignored potential energy due to the earth's gravity. Is this omission justified?

20–6 In deriving the ideal-gas equation, we assumed the number of molecules to be very large, so that we could compute the average force due to many collisions; however, the ideal-gas equation holds accurately only at low pressures, where the molecules are few and far between. Is this inconsistent?

20–7 Some elements of solid crystalline form have molar heat capacities *larger* than $3R$. What effects could account for this?

20–8 Considering molecular speeds, is v_{rms} equal to the *most probable* speed, as indicated by the peak on one of the curves in Fig. 20–7?

20–9 A gas storage tank has a small leak. The pressure in the tank drops more quickly if the gas is hydrogen or helium than if it is oxygen. Why?

20–10 Consider two specimens of gas at the same temperature, both having the same total mass but different molecular masses. Which has the greater internal energy? Does your answer depend on the molecular structure of the gases?

20–11 A process called *gaseous diffusion* is sometimes used to separate isotopes of uranium, i.e., atoms of the element having different masses, such as ^{235}U and ^{238}U. Can you speculate on how this might work?

20–12 When food is preserved by freeze-drying, it is first quick-frozen, then placed in a vacuum chamber and irradiated with infrared radiation. What is the purpose of the vacuum? The radiation?

EXERCISES

Molecular data

N_A = Avogadro's number
  $= 6.02 \times 10^{23}$ molecules·mol^{-1}.
Mass of a hydrogen atom $= 1.67 \times 10^{-27}$ kg.
Mass of a nitrogen molecule $= 28 \times 1.66 \times 10^{-27}$ kg.
Mass of an oxygen molecule $= 32 \times 1.66 \times 10^{-27}$ kg.

Section 20–2 Avogadro's Number

20–1 How many moles are there in a glass of water (0.2 kg)? How many molecules?

20–2 Consider an ideal gas at 0°C and 1 atm pressure. Imagine each molecule to be, on average, at the center of a small cube.

a) What is the length of an edge of this cube?

b) How does this distance compare with the diameter of a molecule?

20–3 Consider 1 mol of liquid water.

a) What volume is occupied by this amount of water?

b) Imagine each molecule to be, on average, at the center of a small cube. What is the length of an edge of this cube?

c) How does this distance compare with the diameter of a molecule?

20–4 What is the length of the side of a cube, in a gas at standard conditions, that contains a number of molecules equal to the population of the United States (about 200 million)?

Section 20–4 Kinetic-Molecular Theory of an Ideal Gas

20–5 At what temperature is the rms speed of oxygen molecules equal to the rms speed of hydrogen molecules at 0°C?

20–6 A flask contains a mixture of mercury vapor, neon, and helium. Compare

a) the average kinetic energies of the three types of atoms;

b) the root-mean-square speeds.

The molecular masses are: helium, 4 g·mol^{-1}; neon, 20 g·mol^{-1}; mercury 201 g·mol^{-1}

20–7 Isotopes of uranium are sometimes separated by gaseous diffusion, using the fact that the rms speeds of the molecules in vapor are slightly different, and hence the vapors diffuse at slightly different rates. If the atomic masses for ^{235}U and ^{238}U are $0.235 \text{ kg·mol}^{-1}$ and $0.238 \text{ kg·mol}^{-1}$, respectively, what is the ratio of the rms speed of the ^{235}U atoms in the vapor to that of the ^{238}U atoms, assuming the temperature is uniform?

20–8 Smoke particles in the air typically have masses of the order of 10^{-16} kg. The Brownian motion of these particles resulting from collisions with air molecules can be observed with a microscope.

a) Find the root-mean-square speed of Brownian motion for such a particle in air at 300 K.

b) Would the speed be different if the particle were in hydrogen gas at the same temperature? Explain.

20–9

a) What is the average translational kinetic energy of a molecule of oxygen at a temperature of 300 K?

b) What is the average value of the square of its speed?

c) What is the root-mean-square speed?

d) What is the momentum of an oxygen molecule traveling at this speed?

e) Suppose a molecule traveling at this speed bounces back and forth between opposite sides of a cubical vessel 0.10 m on a side. What is the average force it exerts on the walls of the container? (Assume that the molecule's velocity is perpendicular to the two sides that it strikes.)

f) What is the average force per unit area?

g) How many molecules traveling at this speed are necessary to produce an average pressure of 1 atm?

h) Compute the number of oxygen molecules actually contained in a vessel of this size, at 300 K and atmospheric pressure.

i) Your answer for (h) should be three times as large as the answer for (g). Where does this discrepancy arise?

Section 20–5 Molar Heat Capacity of a Gas

20–10 Compute the specific heat capacity at constant volume of hydrogen gas, and compare with the specific heat capacity of water.

20–11 Calculate the molar heat capacity at constant volume of water vapor, assuming the triatomic molecule has three translational and three rotational degrees of freedom and that vibrational motion does not contribute. The actual specific heat capacity of water vapor at low pressures is about $2000 \text{ J·kg}^{-1}\text{·K}^{-1}$. Compare this with your calculation, and comment on the actual role of vibrational motion.

Section 20–6 Distribution of Molecular Speeds

20–12 Derive Eq. (20–18) from Eq. (20–17).

20–13 Prove that $f(v)$ as given by Eq. (20–18) is maximum for $\epsilon = kT$.

Section 20–7 Crystals

20–14 The density of crystalline sodium chloride is 2160 kg·m^{-3}, and its crystal structure is shown in Fig. 20–10. Find the spacing of adjacent atoms in the crystal lattice. The atomic mass of sodium is 23.0 g·mol^{-1} and that of chlorine is 35.5 g·mol^{-1}.

PROBLEMS

20–15 The lowest pressures readily attainable in the laboratory are of the order of 10^{-13} atm. At this pressure and ordinary temperature (say $T = 300$ K), how many molecules are present in a volume of 1 cm³?

20–16 Experiment shows that the size of an oxygen molecule is of the order 2×10^{-10} m. Make a rough estimate of the pressure at which the finite volume of the molecules should cause noticeable deviations from ideal-gas behavior at ordinary temperatures ($T = 300$ K).

20–17 The speed of propagation of a sound wave in air at 27°C is about 350 m·s^{-1}. Compare this with

a) v_{rms} for nitrogen molecules;

b) the root-mean-square value of v_x at this temperature.

(If sound propagation were an isothermal process, which it ordinarily is not, the speed of sound would be equal to $(v_x)_{\text{rms}}$. See Section 21–6.)

20–18

a) Compute the increase in gravitational potential energy of an oxygen molecule for an increase in elevation of 1 m near the earth's surface.

b) At what temperature is this equal to the average kinetic energy of oxygen molecules?

20–19

a) For what mass of molecule or particle is v_{rms} equal to 1 m·s^{-1} at a temperature of 300 K?

b) If the particle is an ice crystal, how many molecules does it contain?

c) Estimate the size of the particle. Would it be visible to the naked eye? Through a microscope?

20–20 The surface of the sun has a temperature of about 6000 K and consists largely of hydrogen atoms. (In fact, most of the atoms will be ionized, so the hydrogen "atom" is actually a proton.)

a) Find the rms speed and average kinetic energy of a hydrogen atom at this temperature.

b) Show that the escape velocity for a particle to leave the gravitational influence of the sun is given by $(2GM/R)^{1/2}$, where M is the sun's mass, R its radius, and G the gravitational constant (Example 7–10).

c) Can appreciable quantities of hydrogen escape from the sun's gravitational field?

20–21

a) Show that a projectile of mass m can "escape" from the earth's gravitational field if it is launched vertically upward with a kinetic energy greater than mgR, where g is the acceleration due to gravity at the earth's surface, and R is the earth's radius. (See the preceding problem.)

b) At what temperature would an average oxygen molecule have this much energy? An average hydrogen molecule?

20–22 For each of the triatomic gases in Table 18–1, compute the value of C_v on the assumption that there is no vibrational energy. Compare with the measured values in the table, and compute the fraction of the total heat capacity due to vibration for each of the three gases. (*Note.* CO_2 is linear; SO_2 and H_2S are not.)

CHALLENGE PROBLEMS

20–23 A commonly used potential-energy function for the interaction of two molecules (Fig. 20–1) is the Lennard–Jones 6–12 potential:

$$U(r) = U_0\left[\left(\frac{\sigma}{r}\right)^{12} - \left(\frac{\sigma}{r}\right)^6\right].$$

a) What is the force $F(r)$ that corresponds to this potential? Sketch $U(r)$ and $F(r)$.

b) In terms of σ and U_0, what are the values of r_1 defined by $U(r_1) = 0$ and of r_2 defined by $F(r_2) = 0$? Also, what is the ratio r_1/r_2? Show the location of r_1 and r_2 on your sketch of $U(r)$.

c) If the molecules are located a distance r_2 apart, as calculated in (b), how much work must be done to pull them apart such that $r \to \infty$?

20–24

a) Show that

$$\int_0^\infty f(v)\,dv = 1,$$

where $f(v)$ is the Maxwell–Boltzmann distribution of Eq. (20–17).

b) Calculate

$$\int_0^\infty v^2 f(v)\,dv$$

and compare this result to $(v^2)_{av}$ as given by Eq. (20–12). (*Hint:* You may use the tabulated integral

$$\int_0^\infty x^{2n} e^{-ax}\,dx = \frac{1\cdot3\cdot5\cdots(2n-1)}{2^{n+1}a^n}\sqrt{\frac{\pi}{a}},$$

where n is an integer and a is a positive constant.)

PART FOUR

WAVES

PERSPECTIVE

In the past nine chapters we studied various aspects of the behavior of matter. We began with mechanical properties of deformable materials, including *elastic* deformations and the concepts of *stress* and *strain,* and with several aspects of the equilibrium and flow of *fluids.* We then expanded our study of properties of matter to include *thermal phenomena* and *thermal properties* of matter, including the concepts of *temperature* and *heat,* thermal expansion, and mechanisms for heat transfer. We generalized our description of the mechanical and thermal behavior of materials with the concept of the *equation of state* of a substance.

But the heart of this section of our book is the subject of *thermodynamics.* We studied two very general and very powerful principles. The first law of thermodynamics extends the principle of conservation of energy to include internal energy of matter, and the second law states in very general terms a principle governing the direction in which a thermodynamic process tends to proceed. In formulating these principles we took a strongly empirical and inductive approach, relying heavily on experimental observations of a wide variety of physical phenomena. Both these laws are new generalizations; neither can be derived from principles we have encountered previously. The fact that the principles of thermodynamics can be applied to such a wide variety of physical systems makes them among the most powerful and useful of all the laws of physics. Finally, we have seen that for the simple case of an ideal gas, some of the mechanical and thermal properties of a material can be understood on the basis of its microscopic or molecular structure. With respect to the foundations of thermodynamics, consideration of microscopic structure of a material is not a substitute for the careful observation of and induction from macroscopic behavior, but it does give us additional insight into the nature of such concepts as pressure, heat, internal energy, and entropy.

Next we turn to a discussion of *wave phenomena.* A wave is a disturbance from an equilibrium state of a system that can travel or propagate from one region of space to another. Waves and wave phenomena occur in all areas of physics, including acoustics (a branch of mechanics), electromagnetism, optics, and quantum mechanics. To introduce wave concepts in as simple and familiar a context as possible, we consider various wave phenomena in *mechanical* systems, including waves traveling on stretched strings or ropes and compression waves in solids and fluids. We learn the appropriate mathematical language for describing waves, and we study the special features of *periodic* waves, in which each particle of a material undergoes a periodic motion during wave propagation. Then we examine several applications of these concepts to vibrating systems and acoustical phenomena. Although we concentrate here on mechanical waves, the broad variety of wave phenomena found in all branches of physics makes the wave concept one of the important unifying threads that run through the entire fabric of physics, giving it strength and cohesiveness.

21

MECHANICAL WAVES

A **WAVE** IS ANY DISTURBANCE FROM AN EQUILIBRIUM CONDITION THAT travels or *propagates* with time from one region of space to another. Examples of wave phenomena are everywhere around us; sound, light, ocean waves, radio and television transmission, and earthquakes are all wave phenomena. One of the most familiar kinds of wave is shown in Fig. 21–1. Wave phenomena occur in all branches of physical and biological science, and the concept of waves is one of the most important unifying threads running through the entire fabric of the natural sciences.

In this chapter and the next two we consider **mechanical waves,** which always travel within some material substance called the **medium** for the wave. Some waves are **periodic;** in these the particles of the medium undergo periodic motions during wave propagation. If the periodic motions are *sinusoidal,* the result is a **sinusoidal wave,** a type of periodic wave of special importance. We can also develop relations between the speed of propagation of a wave and the mechanical properties of the medium. The concepts of this chapter form an important part of the foundation for the study of other kinds of wave phenomena, including electromagnetic waves, later in this book.

21–1 A small ball dropped vertically into water produces a wave pattern that moves radially outward from the source of the wave. The wave crests and troughs are concentric circles. (Photo by Fundamental Photographs, New York.)

21–1 MECHANICAL WAVE PHENOMENA

In this chapter we are concerned with *mechanical* waves. Each type of mechanical wave is associated with some material or substance called the *medium* for that type. As the wave travels through the medium, the particles that make up the medium undergo displacements of various kinds, depending on the nature of the wave.

Here are a few examples of mechanical waves. In Fig. 21–2a the medium is a long spring or "Slinky," or even just a wire or cord under tension. If we give the left end a small sideways shake or wriggle, the wriggle travels down the length of the spring, with successive sections of spring going through the same sideways motion that was given at the end, but at successively later times. Because the displacements of the medium are perpendicular (transverse) to

In a transverse wave, the motions of the particles of the medium are perpendicular to the direction of wave travel.

475

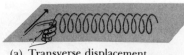

(a) Transverse displacement

(b) Longitudinal displacement

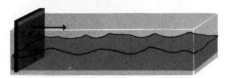

(c) Longitudinal and transverse displacement

21–2   Propagation of disturbances.

Many mechanical waves travel with a definite speed, called the wave speed.

the direction of travel of the wave along the medium, this is called a **transverse wave.**

In Fig. 21–2b the medium is a liquid or a gas in a tube with a rigid wall at the right end and a movable piston at the left end. If the piston is given a back-and-forth motion, again a displacement travels down the length of the medium. This time the motions of the particles of the medium are back and forth along the same direction as the travel of the wave, so this type of wave is called a **longitudinal wave.**

In Fig. 21–2c the medium is water in a trough such as an irrigation ditch or canal. When the flat board at the left end is moved back and forth, a disturbance travels down the length of the trough; careful observation shows that in this case the displacements of the water have both longitudinal and transverse components.

These examples have several things in common. First, the medium itself does not travel through space; its individual particles undergo back-and-forth motions around their equilibrium positions. What *does* travel is the wave disturbance. Second, to set any of these systems in motion requires energy input in the form of mechanical work, and the wave motion has the effect of transporting this energy from one region of space to another. Third, the wave motion can be used to transmit information, a function that is familiar with sound waves. Thus waves transport energy, but not matter, from one region to another. Finally, observation shows that in each case the disturbance travels with a definite speed through the medium. This speed is called the **wave speed,** and it is determined by the mechanical properties of the medium. We will use the symbol c to denote wave speed.

Not all waves are mechanical in nature. Another broad class is *electromagnetic* waves, including light, radio waves, infrared and ultraviolet radiation, x-rays, and gamma rays. There is *no* medium for electromagnetic waves; they can travel through empty space. Still another class of wave phenomena is the wavelike behavior of fundamental particles in some situations; this behavior forms part of the foundation of quantum mechanics, the basic theory used for the analysis of atomic and molecular structure. We will return to electromagnetic waves in Chapter 35 and to the wave nature of particles in Chapter 42. Meanwhile, we can learn the essential language of waves in the context of mechanical waves.

21–2   PERIODIC WAVES

In a periodic wave, each particle of the medium undergoes periodic motion with a definite frequency and period.

To introduce basic concepts as simply as possible, we concentrate our discussion first on one specific kind of mechanical wave, namely, transverse waves on a stretched string or rope. We tie one end of a long, flexible rope to a stationary object and hold the other end, stretching the rope tight. We then give this end some transverse (sideways) motion. If we give it a single "flip" or "wiggle," the result is a single *wave pulse*, which travels down the length of the string.

A more interesting situation develops when we give the free end of the rope a repetitive or *periodic* motion. (We studied periodic motion in Chapter 11, and we suggest you review that discussion now.) In particular, suppose we move it back and forth with *simple harmonic motion* of amplitude A, frequency f, and period τ, where as usual $f = 1/\tau$.

A *continuous succession* of transverse sinusoidal waves then advances along the string. The shape of a portion of the string near the end, at intervals of $\frac{1}{8}$ period, is shown in Fig. 21–3 for a total time of one period. The waveform advances steadily toward the right, as indicated by the short arrow pointing to one particular wave crest, while any one point on the string (see the black dot) oscillates back and forth about its equilibrium position with simple harmonic motion. Be careful to distinguish between the motion of a *waveform*, which moves with constant speed *along* the string, and the motion of a *particle of the string*, which moves with simple harmonic motion *transverse* to the string.

The distance between two successive maxima (or between any two successive points at the same position in the repeating wave shape) is the **wavelength** of the wave, denoted by λ. Since the waveform, traveling with constant speed c, advances a distance of one wavelength in a time interval of one period, it follows that $c = \lambda/\tau$, or since $f = 1/\tau$,

$$c = \lambda f. \qquad (21\text{–}1)$$

That is, *the speed of propagation equals the product of frequency and wavelength.*

To understand the mechanics of a *longitudinal* wave, consider a long tube filled with a compressible fluid, with a plunger at the left end, as shown in Fig. 21–4. Suppose we move the plunger with simple harmonic motion along a line parallel to the direction of the tube. During a part of each oscillation, a region whose pressure is greater than the equilibrium pressure is formed. Such a region is called a *condensation* and is represented in the figure by a darkly shaded area. Following the production of a condensation, a region forms where the pressure is lower than the equilibrium value. This region is called a *rarefaction* and is represented by a lightly shaded area in the figure. The condensations and rarefactions move to the right with constant speed c, as indicated by successive positions of the small vertical arrow. The speed of

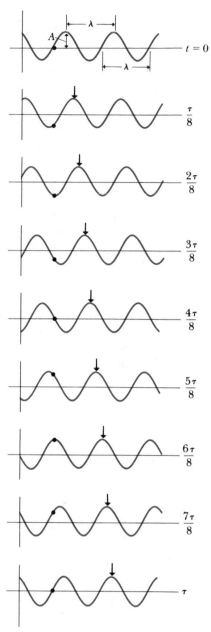

21–3 A sinusoidal transverse wave traveling toward the right. The shape of the string is shown at intervals of one-eighth of a period. The vertical scale is exaggerated.

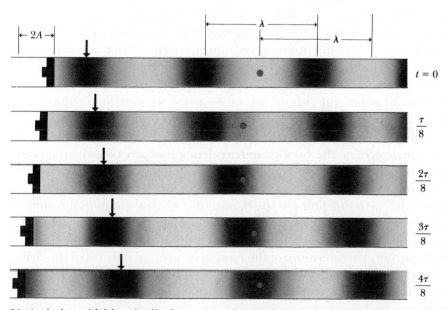

21–4 A sinusoidal longitudinal wave traveling toward the right, shown at intervals of one-eighth of a period.

longitudinal waves in air (i.e., sound) at 20°C is 344 m·s^{-1}, or 1130 ft·s^{-1}. The motion of a single particle of the medium, shown by a colored dot, is simple harmonic, parallel to the direction of propagation.

The wavelength is the distance between two successive condensations or two successive rarefactions. The same fundamental equation, $c = f\lambda$, holds in this example, as in all types of periodic or repetitive waves.

EXAMPLE 21–1   What is the wavelength of a sound wave having a frequency of 262 Hz (the approximate frequency of the note "middle C" on the piano)?

SOLUTION   At 20°C the speed of sound in air is 344 m·s^{-1}, as mentioned above. From Eq. (21–1),

$$\lambda = \frac{c}{f} = \frac{344 \text{ m·s}^{-1}}{262 \text{ s}^{-1}} = 1.31 \text{ m}.$$

The "high C" sung by coloratura sopranos is two octaves above middle C. The corresponding frequency is four times as large, $f = 4(262 \text{ Hz}) = 1048 \text{ Hz}$, and the wavelength is one-fourth as large, $\lambda = (1.31 \text{ m})/4 = 0.328 \text{ m}$.

21–3 MATHEMATICAL DESCRIPTION OF A WAVE

For complete analysis of wave motion we need a mathematical language that provides a detailed description of the motion of the medium during wave propagation. A central element of this language is the concept of **wave function,** which is a function that describes the position of an arbitrary particle in the medium at any time. In this discussion we will concentrate primarily on sinusoidal waves, in which each particle of the medium undergoes simple harmonic motion about its equilibrium position.

As a concrete example, we look first at waves on a stretched string. If we ignore the sag of the string due to gravity, the equilibrium position of the string is along a straight line. We take this to be the x-axis of a coordinate system. Waves on a string are *transverse;* during wave motion a particle with equilibrium position x is displaced some distance y in the direction perpendicular to the x-axis. The value of y depends on which particle we are talking about (that is, on x) and also on the time t when we look at it. Thus y is a *function* of x and t; $y = f(x, t)$. If we know this function for a particular wave motion, we can use it to predict the position of any particle at any time. From this we can determine the velocity and acceleration of any particle, the shape of the string, its slope at any point, and anything else related to the position and motion of the string at any time.

Thus the wave function $y = f(x, t)$, once it is known, contains a complete description of the motion. Let us now consider wave functions for sinusoidal waves. Suppose a wave travels from left to right (the direction of increasing x) along the string. We can compare the motion of any one particle of the string with the motion of a second particle to the right of the first. We find that the second particle has the same motion as the first, but after a time lag that is proportional to the distance between the particles. If one end of a stretched string oscillates with simple harmonic motion, all other points oscillate with

What is the wavelength of "middle C"? Of "high C"?

The wave function is a mathematical description of the displacements of points in the medium at various times.

In a sinusoidal wave, all points in the medium move with the same frequency but with phase differences.

simple harmonic motion of the same amplitude and frequency. The *phase* of the motion, however, is different for different points. This means that the cyclic motions of various points are out of step with each other by various fractions of a cycle. For example, if one point has its maximum positive displacement at the same time another has its maximum negative displacement, the two are a half-cycle out of phase. In Section 11–2 we described phase relationships in terms of a *phase angle* θ_0. A phase angle of π (180°) corresponds to 1/2 cycle, $\pi/2$ to 1/4 cycle, and so on.

Suppose the displacement of a particle at the left end (at $x = 0$), where the motion originates, is given by

$$y = A \sin \omega t = A \sin 2\pi f t. \tag{21–2}$$

The time required for the wave disturbance to travel from $x = 0$ to some point x to the right of the origin is given by x/c, where c is the wave speed. The motion of point x at time t is the same as the motion of point $x = 0$ at the earlier time $(t - x/c)$. Thus the displacement of point x at time t is obtained simply by replacing t in Eq. (21–2) by $(t - x/c)$, and we find

$$y(x, t) = A \sin \omega\left(t - \frac{x}{c}\right)$$

$$= A \sin 2\pi f\left(t - \frac{x}{c}\right). \tag{21–3}$$

The notation $y(x, t)$ is a reminder that the displacement y is a function of both the location x of the point and the time t.

Equation (21–3) can be rewritten in several alternative forms, conveying the same information in different ways. In terms of the period τ and wavelength λ, we find, using Eq. (21–1),

$$y(x, t) = A \sin 2\pi\left(\frac{t}{\tau} - \frac{x}{\lambda}\right). \tag{21–4}$$

Another convenient form is obtained by defining a quantity k, called the **propagation constant** or the **wave number:**

$$k = \frac{2\pi}{\lambda}. \tag{21–5}$$

In terms of k and the angular frequency ω, the wavelength-frequency relation $c = \lambda f$ becomes

$$\omega = ck, \tag{21–6}$$

and we can rewrite Eq. (21–4) as

$$y(x, t) = A \sin (\omega t - kx). \tag{21–7}$$

Which of these various forms we use is a matter of convenience in a specific problem; fundamentally they all say the same thing. We should mention in passing that some authors define the wave number as $1/\lambda$ rather than $2\pi/\lambda$. In that case, our k is still called the propagation constant.

At any given *time* t, Eq. (21–3), (21–4), or (21–7) gives the displacement y of a particle from its equilibrium position, as a function of the *coordinate* x of the particle. If the wave is a transverse wave in a string, the equation repre-

Wave functions for a sinusoidal wave can be written in several alternative forms.

The simplest form of the sinusoidal wave function, in terms of angular frequency and wave number

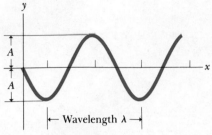

21–5 Waveform at $t = 0$.

Wave functions for a wave traveling in the negative direction

The wave function contains a wealth of information about the wave motion it describes.

sents the *shape* of the string at that instant, as if we have taken a photograph of the string. Thus at time $t = 0$,

$$y = A \sin (-kx) = -A \sin kx = -A \sin 2\pi \frac{x}{\lambda}.$$

This curve is plotted in Fig. 21–5.

At any given *coordinate* x, Eq. (21–3), (21–4), or (21–7) gives the displacement y of the particle at that coordinate, as a function of *time*. That is, it describes the motion of that particle. Thus at the position $x = 0$,

$$y = A \sin \omega t = A \sin 2\pi \frac{t}{\tau}.$$

This curve is plotted in Fig. 21–6.

The above formulas may be used to represent a wave traveling in the *negative* x-direction by making a simple modification. In this case the displacement of point x at time t is the same as the motion of point $x = 0$ at the *later* time $(t + x/c)$. In Eq. (21–2) we must therefore replace t by $(t + x/c)$. Thus, for a wave traveling in the negative x-direction,

$$y = A \sin 2\pi f \left(t + \frac{x}{c}\right) = A \sin 2\pi \left(\frac{t}{\tau} + \frac{x}{\lambda}\right)$$

$$= A \sin (\omega t + kx). \tag{21–8}$$

We must be careful to distinguish between the *speed of propagation* c of the waveform and the *particle speed* v of a particle of the medium in which the wave is traveling. The wave speed c is given by

$$c = \lambda f = \frac{\omega}{k}. \tag{21–9}$$

The particle speed v for any point in a transverse wave (that is, at a fixed value of x) is obtained by taking the derivative of y with respect to t, holding x constant. Such a derivative is called a *partial derivative*, written $\partial y/\partial t$. Thus for a sinusoidal wave given by

$$y = A \sin (\omega t - kx), \tag{21–10}$$

we have

$$v = \frac{\partial y}{\partial t} = \omega A \cos (\omega t - kx). \tag{21–11}$$

The *acceleration* of the particle is the *second* partial derivative:

$$a = \frac{\partial^2 y}{\partial t^2} = -\omega^2 A \sin (\omega t - kx). \tag{21–12}$$

We may also compute partial derivatives with respect to x, holding t constant. The first derivative $\partial y/\partial x$ is the *slope* of the string at any point. The second partial derivative with respect to x is

$$\frac{\partial^2 y}{\partial x^2} = -k^2 A \sin (\omega t - kx). \tag{21–13}$$

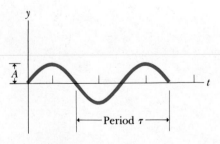

21–6 Vibration at $x = 0$.

It follows from Eqs. (21–12) and (21–13) that

$$\frac{\partial^2 y / \partial t^2}{\partial^2 y / \partial x^2} = \frac{\omega^2}{k^2} = c^2,$$

since $c = \omega / k$. The *partial differential equation*

$$\frac{\partial^2 y}{\partial t^2} = c^2 \frac{\partial^2 y}{\partial x^2} \qquad (21\text{–}14)$$

The wave equation: Every wave function for a physically possible wave must satisfy this equation.

is one of the most important equations in all physics. It is called the **wave equation,** and whenever it occurs, we can conclude immediately that the disturbance described by the function y propagates as a traveling wave along the x-axis with a wave speed c.

Although we have introduced the concept of wave function with reference to *transverse* waves on a string, the concept is equally useful with *longitudinal* waves. The quantity y still measures the displacement of a particle of the medium from its equilibrium position; the difference is that for a longitudinal wave this displacement is *parallel* to the x-axis instead of perpendicular to it. Thus the distance y is measured parallel to the x-axis, and x and y no longer represent distances measured in perpendicular directions, as in the usual xy-coordinate system. If this sounds confusing, don't panic! It will become clear as we study longitudinal waves later in this chapter and in the next.

21–4 SPEED OF A TRANSVERSE WAVE

How is the speed of propagation c of a transverse wave on a string related to the *mechanical* properties of the system? The relevant physical quantities are the *tension* in the string and its *mass per unit length*. Intuition suggests that increasing the tension should increase the speed, while increasing the mass should decrease the speed. We now develop this relationship by two different methods. The first is simple and considers a specific type of waveform; the second is more general but also more formal. Choose whichever you like better!

For our first development, we consider a perfectly flexible string, as shown in Fig. 21–7, having linear mass density (mass per unit length) μ and stretched with a tension S. In Fig. 21–7a the string is at rest. At time $t = 0$, a constant transverse force $\boldsymbol{F}$ is applied at the left end of the string. We might expect that the end would move with constant acceleration; this would certainly occur if the force were applied to a point mass. But here the effect of the force is to set successively more and more mass in motion. The wave travels with constant speed, so the division point between moving and nonmoving portions also travels with definite speed. Hence the total mass in motion is proportional to the time the force has been acting and thus to the impulse of the force. This, in turn, is equal to the total momentum mv of the moving part of the string. The total momentum thus must increase proportionately with time, so the change of momentum must be associated entirely with the increasing amount of mass in motion, not with the increasing velocity of an individual mass element. Force is rate of change of momentum mv, and mv changes because m changes, not v. Hence the end of the string moves upward with constant *velocity* v.

The speed of a wave on a string is determined by the tension and the mass per unit length.

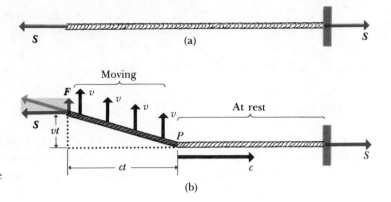

21–7 Propagation of a transverse disturbance in a string.

Figure 21–7b shows the shape of the string after a time t has elapsed. All particles of the string to the left of the point P are moving upward with speed v, and all particles to the right of P are still at rest. The boundary point P between the moving and the stationary portions is traveling to the right with the *speed of propagation* c. The left end of the string has moved up a distance vt, and the boundary point P has advanced a distance ct.

The tension at the left end of the string is the vector sum of the forces $\boldsymbol{S}$ and $\boldsymbol{F}$. Because no motion occurs in a direction along the length of the string, there is no unbalanced horizontal force, so S, the magnitude of the horizontal component, does not change when the string is displaced. As a result of the increased tension, the string clearly stretches somewhat. It can be shown that, for small displacements, the amount of stretch is approximately proportional to the increase in tension, as we would expect from Hooke's law.

We can obtain an expression for the speed of propagation c by applying the impulse–momentum relation to the portion of the string in motion at time t; that is, the darkly shaded portion in Fig. 21–7. We set the transverse *impulse* (transverse force × time) equal to the change of transverse *momentum* of the moving portion (mass × transverse velocity). The impulse of the transverse force F in time t is Ft. By similar triangles,

$$\frac{F}{S} = \frac{vt}{ct}, \qquad F = S\frac{v}{c}.$$

The transverse momentum of the string is equal to the transverse impulse of the force acting at the end.

Hence

$$\text{Transverse impulse} = S\frac{v}{c}t.$$

The mass of the moving portion of the string is the product of the mass per unit length μ and the length ct. (In its displaced position, the section of string is stretched somewhat, making its mass per unit length somewhat *less* than μ and its length somewhat *greater* than ct. But the mass of the section is still μct, the same as in the undisplaced position.) Hence

$$\text{Transverse momentum} = \mu ctv.$$

Note again that the momentum increases with time *not* because mass is moving faster, as was usually the case in Chapter 8, but because *more mass* is brought into motion. Nevertheless, the impulse of the force F is still equal to

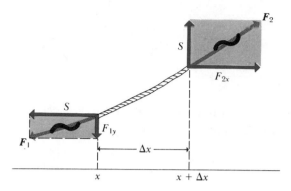

21–8 Free-body diagram for a segment of string whose length in its equilibrium position is Δx. The force at each end of the string is tangent to the string at the point of application; each force is represented in terms of its x- and y-components.

the total change in momentum of the system. Applying this relation, we obtain

$$S\frac{v}{c}t = \mu c t v,$$

and therefore

$$c = \sqrt{\frac{S}{\mu}} \qquad \text{(transverse wave)}. \qquad (21\text{–}15)$$

Hence the speed of propagation of a transverse pulse in a string depends only on the tension (a force) and the mass per unit length. Although this calculation of wave speed considered only a very special kind of pulse, a little thought shows that *any* shape of wave disturbance can be considered as a series of pulses with different rates of transverse displacement. Thus, although derived for a special case, Eq. (21–15) is valid for *any* transverse wave motion on a string, including, in particular, the sinusoidal and other periodic waves discussed in Sections 21–2 and 21–3.

Here is an alternative derivation of Eq. (21–15). We apply Newton's second law, $F = ma$, to a small segment of string whose length in the equilibrium position is Δx, as shown in Fig. 21–8. The mass of the segment is $m = \mu\Delta x$; the forces at the ends are represented in terms of their x- and y-components. The x-components have equal magnitude S and add to zero, since the motion is transverse and there is no component of acceleration in the x-direction. To obtain F_{1y} and F_{2y}, we note that the ratio F_{1y}/S is equal in magnitude to the *slope* of the string at point x, and F_{2y}/S is equal to the slope at point $(x + \Delta x)$. Taking proper account of signs, we find

A derivation of the wave equation for a string, starting with Newton's second law

$$\frac{F_{1y}}{S} = -\left(\frac{\partial y}{\partial x}\right)_x, \qquad \frac{F_{2y}}{S} = \left(\frac{\partial y}{\partial x}\right)_{x+\Delta x}. \qquad (21\text{–}16)$$

The notation reminds us that the derivatives are evaluated at points x and $(x + \Delta x)$, respectively. Thus the net y-component of force is

$$F_{1y} + F_{2y} = S\left[\left(\frac{\partial y}{\partial x}\right)_{x+\Delta x} - \left(\frac{\partial y}{\partial x}\right)_x\right]. \qquad (21\text{–}17)$$

We now equate this to the mass ($\mu\Delta x$) times the y-component of acceleration $\partial^2 y/\partial t^2$ to obtain

$$S\left[\left(\frac{\partial y}{\partial x}\right)_{x+\Delta x} - \left(\frac{\partial y}{\partial x}\right)_x\right] = \mu\Delta x\,\frac{\partial^2 y}{\partial t^2}, \qquad (21\text{–}18)$$

or, dividing by $S\Delta x$,

$$\frac{\left(\frac{\partial y}{\partial x}\right)_{x+\Delta x} - \left(\frac{\partial y}{\partial x}\right)_x}{\Delta x} = \frac{\mu}{S}\frac{\partial^2 y}{\partial t^2}. \tag{21-19}$$

The speed of a wave on a string, derived from the wave equation

We now take the limit as $\Delta x \to 0$. In this limit, the left side of Eq. (21–19) becomes simply the *second* (partial) derivative of y with respect to x:

$$\frac{\partial^2 y}{\partial x^2} = \frac{\mu}{S}\frac{\partial^2 y}{\partial t^2}. \tag{21-20}$$

Now, finally, comes the punch line of our story. Equation (21–20) has exactly the same form as the general differential equation we derived at the end of Section 21–3, Eq. (21–14). Both that equation and Eq. (21–20), just derived, describe the very same wave motion, so they must be identical. Comparing the two equations, we see that for this to be so, we must have

$$c = \sqrt{\frac{S}{\mu}}, \tag{21-21}$$

which is the same expression as Eq. (21–15). We also note that this result confirms our prediction that c should increase when S increases but decrease when μ increases.

PROBLEM-SOLVING STRATEGY: *Mechanical waves*

1. It is helpful to distinguish between *kinematic* problems and *dynamic* problems. In kinematics problems we are concerned only with *describing* motion; the relevant quantities are wave speed, wavelength (or wave number), frequency (or angular frequency), amplitude, and the position, velocity, and acceleration of individual particles. In dynamics problems, concepts such as force and mass enter; the relation of wave speed to the mechanical properties of a system is an example.

2. If λ is given, it is easy to find k, and vice versa. If f is given, it is easy to find τ, and vice versa. If any two of the quantities c, λ, and f are known, it is easy to find the third. In some problems that is all you need. To determine the wave function completely, you need to know A and any two of c, λ, and f. Once this information is known, you can use it in Eq. (21–3),

(21–4), or (21–7) to get the specific wave function for the problem at hand. Once you have that, you can find values of y, v, a, and the slope of the string at any point (values of x) and at any time, by substituting into the wave function.

3. The wave speed c, the tension S, and the mass per unit length μ are related by Eq. (21–15). If you know any two of these, it is easy to find the third. Look to see what you are given in the problem statement. Later we will find similar relations for speeds of longitudinal waves.

4. The vertical component of force at any point in the string can be obtained from Eq. (21–16). This is useful in connection with energy transfer along the string, as we will see later.

Transverse waves on a clothesline: physics in everyday life

EXAMPLE 21–2 A clothesline has a linear mass density of 0.25 kg·m^{-1} and is stretched with a tension of 25 N. One end is given a sinusoidal motion with frequency 5 Hz and amplitude 0.01 m. At time $t = 0$ the end has zero displacement and is moving in the $+y$-direction.

a) Find the wave speed, amplitude, angular frequency, period, wavelength, and wave number.

b) Write a wave function describing the wave.

c) Find the position of the point at $x = 0.25$ m at time $t = 0.1$ s.

d) Find the transverse velocity at the point $x = 0.25$ m at time $t = 0.1$ s.

e) Find the slope of the string at the point $x = 0.25$ m at time $t = 0.1$ s.

SOLUTION

a) From Eq. (21–15), the wave speed is

$$c = \sqrt{\frac{S}{\mu}} = \sqrt{\frac{25 \text{ N}}{0.25 \text{ kg·m}^{-1}}} = 10 \text{ m·s}^{-1}.$$

The amplitude A is just the amplitude of the motion of the endpoint, $A = 0.01$ m. The angular frequency is

$$\omega = 2\pi f = 2\pi(5 \text{ s}^{-1}) = 31.4 \text{ s}^{-1}.$$

The period is $\tau = 1/f = 0.2$ s. The wavelength is obtained from Eq. (21–1):

$$\lambda = \frac{c}{f} = \frac{10 \text{ m·s}^{-1}}{5 \text{ s}^{-1}} = 2 \text{ m}.$$

The wave number is obtained from Eq. (21–5) or (21–6):

$$k = \frac{2\pi}{\lambda} = \frac{2\pi}{2 \text{ m}} = 3.14 \text{ m}^{-1},$$

or

$$k = \frac{\omega}{c} = \frac{31.4 \text{ s}^{-1}}{10 \text{ m·s}^{-1}} = 3.14 \text{ m}^{-1}.$$

b) The wave function, if we use the form of Eq. (21–4), is

$$y = (0.01 \text{ m}) \sin 2\pi\left(\frac{t}{0.2 \text{ s}} - \frac{x}{2 \text{ m}}\right)$$

$$= (0.01 \text{ m}) \sin [(31.4 \text{ s}^{-1})t - (3.14 \text{ m}^{-1})x].$$

This equation can also be obtained from Eq. (21–7), if we use the values of ω and k obtained above.

c) The displacement of the point $x = 0.25$ m at time $t = 0.1$ s is given by substituting these values into either of the above wave equations:

$$y = (0.01 \text{ m}) \sin 2\pi\left(\frac{0.1 \text{ s}}{0.2 \text{ s}} - \frac{0.25 \text{ m}}{2 \text{ m}}\right)$$

$$= (0.01 \text{ m}) \sin 2\pi(0.375) = 0.00707 \text{ m}.$$

d) The transverse velocity at any time is given by

$$v = \frac{\partial y}{\partial t} = \omega A \cos (\omega t - kx).$$

In the present problem,

$$v = (31.4 \text{ s}^{-1})(0.01 \text{ m}) \cos [(31.4 \text{ s}^{-1})(0.1 \text{ s}) - (3.14 \text{ m}^{-1})(0.25 \text{ m})]$$

$$= -0.22 \text{ m·s}^{-1}.$$

e) The slope at any point at any time is given by

$$\frac{\partial y}{\partial x} = -kA \cos (\omega t - kx).$$

In the present problem,

$$\frac{\partial y}{\partial x} = -(3.14 \text{ m}^{-1})(0.01 \text{ m}) \cos \left[(31.4 \text{ s}^{-1})(0.1 \text{ s}) - (3.14 \text{ m}^{-1})(0.25 \text{ m})\right]$$

$$= 0.022.$$

For a wave traveling in the $+x$-direction, the transverse velocity is positive at every point where the slope is negative, and negative where the slope is positive. Can you prove this?

Polarization: an important property of all transverse waves, including light

An important property of transverse waves is **polarization.** When we produce a transverse wave on a string, we have a choice between moving the end up and down or sideways; in either case the wave displacements are perpendicular or *transverse* to the length of the string. If the end moves up and down, the motion of the entire string is confined to a vertical plane; if the end moves sideways, the wave moves in a horizontal plane. In either case the wave is said to be *linearly polarized* because the individual particles move back and forth in straight lines perpendicular to the string.

The motion may also be more complex, containing both vertical and horizontal components. Combining two sinusoidal motions of equal amplitude, with a quarter-cycle phase difference, results in a wave in which each particle moves in a circular path perpendicular to the string. Such a wave is said to be *circularly polarized.*

A polarizing filter sorts out waves with differing states of polarization.

We can make a device to separate the various components of motion. We cut a thin slot in a flat board, thread the string through it, and orient the plane of the board perpendicular to the string. Then any transverse motion parallel to the slot passes through unimpeded, while any motion perpendicular to the slot is blocked. Such a device is called a *polarizing filter.* Analogous optical devices for polarized light form the basis of some kinds of sunglasses and also of polarizing filters used in photography. We will discuss polarization of light in Chapter 36.

21-5 SPEED OF A LONGITUDINAL WAVE

Propagation speeds of longitudinal as well as transverse waves are determined by the mechanical properties of the medium, and we can derive relations for longitudinal waves that are analogous to Eq. (21-15) for transverse waves on a string. Here is an example of such a derivation for longitudinal waves in a fluid in a tube. Our analysis will be quite similar to the derivation of Eq. (21-15), and we invite you to compare the two developments.

Figure 21-9 shows a fluid (either liquid or gas) with density ρ in a tube with cross-sectional area A. In the equilibrium state the fluid is under a uniform pressure p. In Fig. 21-9a the fluid is at rest. At time $t = 0$ we start the piston at the left end moving toward the right with constant speed v. This initiates a wave motion that travels to the right along the length of the tube, in which successive sections of fluid begin to move and become compressed at successively later times.

Figure 21-9b shows the fluid after a time t has elapsed. All portions of fluid to the left of point P are moving with speed v, and all portions to the right of P are still at rest. The boundary between the moving and stationary

21-9 Propagation of a longitudinal disturbance in a fluid confined in a tube.

portions travels to the right with a speed equal to the speed of propagation c. At time t the piston has moved a distance vt, and the boundary has advanced a distance ct. As with a transverse disturbance in a string, we can compute the speed of propagation from the impulse–momentum theorem.

The quantity of fluid set in motion in time t is the amount that originally occupied a volume of length ct and of cross-sectional area A. The mass of this fluid is therefore ρctA, and the longitudinal momentum it has acquired is

$$\text{Longitudinal momentum} = \rho ctAv.$$

We next compute the increase of pressure, Δp, in the moving fluid. The original volume of the moving fluid, Act, has decreased by an amount Avt. From the definition of bulk modulus B (see Chapter 12),

The longitudinal momentum of the medium equals the longitudinal impulse of the force acting at the end.

$$B = \frac{\text{Change in pressure}}{\text{Fractional change in volume}} = \frac{\Delta p}{Avt/Act}.$$

Therefore

$$\Delta p = B\frac{v}{c}.$$

The pressure in the moving fluid is $p + \Delta p$, and the force exerted on it by the piston is $(p + \Delta p)A$. The net force on the moving fluid (see Fig. 21–9b) is ΔpA, and the longitudinal impulse is

$$\text{Longitudinal impulse} = \Delta pAt = B\frac{v}{c}At.$$

Applying the impulse–momentum theorem, we find

$$B\frac{v}{c}At = \rho ctAv;$$

hence

$$c = \sqrt{\frac{B}{\rho}} \qquad \text{(longitudinal wave)}. \qquad (21–22)$$

The speed of propagation of a longitudinal pulse in a fluid therefore depends only on the bulk modulus and the density of the medium. The form of this relation is similar to that of Eq. (21–15); in both cases the numerator is a quantity characterizing the strength of the restoring force, and the denominator is a quantity describing the inertial properties of the medium.

When a longitudinal wave propagates in a *solid* bar, the situation is somewhat different from that of a fluid confined in a tube of constant cross section, because the bar expands slightly sidewise when it is compressed longitudinally. It can be shown, by the same type of reasoning as that just given, that the speed of a longitudinal pulse in the bar is given by

The speed of a longitudinal wave in a solid depends on the density and Young's modulus.

$$c = \sqrt{\frac{Y}{\rho}} \qquad \text{(longitudinal wave)}, \qquad (21–23)$$

where Y is Young's modulus, defined in Chapter 12.

As with the calculation for a transverse wave on a string, Eqs. (21–22) and (21–23) are valid for all wave motions, not just the special case discussed here. In particular, they are valid for sinusoidal and other periodic waves.

The wave equation for longitudinal waves in a fluid: an alternative approach to finding the wave speed

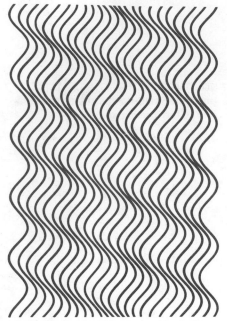

21–10 Diagram for illustrating longitudinal traveling waves.

We can also derive Eq. (21–22) by using a method analogous to the second derivation in Section 21–4 but applying $F = ma$ to a thin slice of fluid and deriving a partial differential equation. In this case the wave function $y(x, t)$ represents the *longitudinal* displacement of a point in the fluid. We omit the details; the result is

$$\frac{\partial^2 y}{\partial x^2} = \frac{\rho}{B} \frac{\partial^2 y}{\partial t^2}. \qquad (21\text{–}24)$$

For this equation to be consistent with Eq. (21–14), c *must* be given by Eq. (21–22). A similar derivation for elastic solids leads to Eq. (21–23).

Longitudinal waves, unlike transverse waves, *do not* exhibit polarization. This concept has no meaning for a longitudinal wave.

Since the *shape* of a fluid does not change when a longitudinal wave passes through it, visualizing the relation between particle motion and wave motion is not as easy as for transverse waves in a string. Figure 21–10 may help in correlating these motions. To use this figure, cut a slit about $\frac{1}{16}$ in. wide and 3 in. long in a 3-in. × 5-in. card (or fasten two cards edge-to-edge with a $\frac{1}{16}$-in. gap). Place the card over the figure with the slit along the top of the diagram, and move the card downward with constant velocity. The portions of the sine curves that are visible through the slit correspond to a row of particles along which a longitudinal, sinusoidal wave is traveling. Each particle undergoes simple harmonic motion about its equilibrium position, with a phase that increases continuously along the slit, while the regions of maximum compression and expansion move from left to right with constant speed. Moving the card upward simulates a wave traveling from right to left.

What does the speed of sound in water have to do with dolphin navigation?

EXAMPLE 21–3 Determine the speed of sound waves in water, and find the wavelength of a wave having a frequency of 262 Hz.

SOLUTION We use Eq. (21–22) to find the wave speed. From Table 12–2, we find that the compressibility of water, which is the reciprocal of the bulk modulus, is $k = 45.8 \times 10^{-11} \, \text{Pa}^{-1}$. Thus $B = (1/45.8) \times 10^{11} \, \text{Pa}$. The density of water is $\rho = 1.00 \times 10^3 \, \text{kg·m}^{-3}$. We obtain

$$c = \sqrt{\frac{B}{\rho}} = \sqrt{\frac{(1/45.8) \times 10^{11} \, \text{Pa}}{1.00 \times 10^3 \, \text{kg·m}^{-3}}} = 1478 \, \text{m·s}^{-1}.$$

This is over four times the speed of sound in air at ordinary temperatures. The wavelength is given by

$$\lambda = \frac{c}{f} = \frac{1478 \, \text{m·s}^{-1}}{262 \, \text{s}^{-1}} = 5.64 \, \text{m}.$$

A wave of this frequency in air has a wavelength of 1.31 m, as obtained in Example 21–1 (Section 21–2).

Dolphins emit high-frequency sound waves (typically 100,000 Hz) and use the echoes for guidance and for hunting. The corresponding wavelength is 1.48 cm. With this high-frequency "sonar" system, they can sense objects of about the size of the wavelength, but not much smaller.

EXAMPLE 21–4 What is the speed of longitudinal sound waves in a steel rod?

SOLUTION We use Eq. (21–23). From Table 12–1, $Y = 2.0 \times 10^{11}$ Pa, and from Table 13–1, $\rho = 7.8 \times 10^3$ kg·m^{-3}. From these data,

$$c = \sqrt{\frac{Y}{\rho}} = \sqrt{\frac{2.0 \times 10^{11} \text{ Pa}}{7.8 \times 10^3 \text{ kg·m}^{-3}}} = 5064 \text{ m·s}^{-1}.$$

21–6 SOUND WAVES IN GASES

In Section 21–5 we derived an expression for the speed of sound in a fluid in a pipe, in terms of its density ρ and bulk modulus B. We learned in Section 18–8 that when a gas is compressed adiabatically, its temperature rises; when it expands adiabatically, its temperature drops. Does this happen when a wave travels through a gas, or does enough heat conduction occur between adjacent layers of gas to maintain a nearly constant temperature throughout? This is a crucial question, because it determines what we use for the bulk modulus B in Eq. (21–22). The bulk modulus is defined in general as in Eq. (12–7): $B = -V \, dp/dV$. If the temperature is constant, then according to Boyle's law the product pV is constant, and we can use this to evaluate dp/dV. But if the process is adiabatic, then Eq. (18–24) states that pV^γ is constant, and we get a different result for B.

The compressions and expansions of a gas during sound wave propagation are adiabatic processes.

Experiments show that for ordinary sound frequencies, say 20 to 20,000 Hz, the thermal conductivity of gases is so small that the propagation of sound is, in fact, very nearly *adiabatic*. Thus we must use the **adiabatic bulk modulus** B_{ad}, derived from the assumption

$$pV^\gamma = \text{constant}. \tag{21–25}$$

We take the derivative of Eq. (21–25) with respect to V:

$$\frac{dp}{dV}V^\gamma + \gamma p V^{\gamma-1} = 0.$$

Dividing by $V^{\gamma-1}$ and rearranging, we find that the adiabatic bulk modulus for an ideal gas is simply

$$-V\frac{dp}{dV} = B_{ad} = \gamma p. \tag{21–26}$$

For an isothermal process, however, $pV = $ constant, and the isothermal bulk modulus is

$$B_{iso} = p. \tag{21–27}$$

In each case the bulk modulus (characterizing the material's resistance to being compressed) is proportional to the pressure, but the adiabatic modulus is *larger* than the isothermal by a factor γ.

Sound speed in a gas can be expressed in terms of the pressure, density, and γ.

Combining Eqs. (21–22) and (21–26), we obtain

$$c = \sqrt{\frac{\gamma p}{\rho}} \quad \text{(ideal gas)}. \tag{21–28}$$

We can obtain an alternative form by using the relation

Sound speed in a gas depends on molecular mass, temperature, and γ.

$$\frac{p}{\rho} = \frac{RT}{M},$$

where R is the gas constant, M is the molecular mass, and T is the absolute temperature. Therefore

$$c = \sqrt{\frac{\gamma RT}{M}} \quad \text{(ideal gas)}. \quad\quad (21\text{--}29)$$

For a given gas, γ, R, and M are constants, so the speed of propagation is proportional to the square root of the absolute temperature. It is interesting to compare this result with Eq. (20–12), which gives the rms speed of molecules in an ideal gas. The two expressions are identical except for the numerical factor of 3 in one and γ in the other.

Computing the speed of sound in air

EXAMPLE 21–5 Compute the speed of longitudinal waves in air at an absolute temperature of 300 K.

SOLUTION The mean molecular mass of air is

$$28.8 \text{ g·mol}^{-1} = 28.8 \times 10^{-3} \text{ kg·mol}^{-1}.$$

Also, $\gamma = 1.40$ for air, and $R = 8.314$ J·mol^{-1}·K^{-1}. At $T = 300$ K we obtain

$$c = \sqrt{\frac{(1.40)(8.314 \text{ J·mol}^{-1}\text{·K}^{-1})(300 \text{ K})}{28.8 \times 10^{-3} \text{ kg·mol}^{-1}}}$$

$$= 348 \text{ m·s}^{-1} = 1142 \text{ ft·s}^{-1} = 779 \text{ mi·hr}^{-1}.$$

This result agrees with the measured speed at this temperature to within 0.3%.

Sound consists of longitudinal waves. The ear is sensitive to a range of sound frequencies from about 20 Hz to about 20,000 Hz. From the relation $c = f\lambda$, the corresponding wavelength range is from about 17 m, corresponding to a 20-Hz note, to about 1.7 cm, corresponding to 20,000 Hz.

Bat and dolphin navigation: Why do they have to use such high frequencies?

Like the dolphin, the bat uses high-frequency sound waves for navigation. A typical frequency is 100,000 Hz; the corresponding wavelength in air is about 3.5 mm, small enough to permit detection of flying insects useful as food.

In this discussion we have ignored the *molecular* nature of a gas and have treated it as a continuous medium. Actually, we know that a gas is composed of molecules in random motion, separated by distances that are large compared with their diameters. The vibrations that constitute a wave in a gas are superposed on the random thermal motion. At atmospheric pressure, a molecule travels an average distance of about 10^{-5} cm between collisions, while the displacement amplitude of a faint sound may be only a few ten-thousandths of this amount. An element of gas in which a sound wave is traveling can be compared to a swarm of bees, where the swarm as a whole can be seen to oscillate slightly while individual insects move about through the swarm, apparently at random.

21–7 ENERGY IN WAVE MOTION

Every wave motion has energy associated with it. To initiate a wave motion, we exert a force on a portion of the wave medium; the point where the force is applied moves, so we do *work* on the system. A wave can transport energy from one region of space to another. For example, transmission of energy by electromagnetic waves is familiar, and the destructive power of ocean surf is a convincing demonstration of energy transported by water waves.

As an example of energy considerations in wave motion, we return to the first example of the chapter, transverse waves on a string. To understand how energy is transferred from one portion of string to another, picture a wave traveling from left to right (the positive *x*-direction) on the string. Considering a particular point on the string, note that according to Eq. (21–16), the portion of string to the left of this point exerts a transverse force F_y on the portion to the right, given by

$$F_y = -S \frac{\partial y}{\partial x}. \qquad (21\text{--}30)$$

> Waves on a string transport energy because of the work done by one section on another by the transverse force.

If the instantaneous vertical velocity of the point is v, then the left side of the string does *work* on the right side. The corresponding *power P* (rate of doing work) is given by

$$P = F_y v = -S \left(\frac{\partial y}{\partial x} \right) \left(\frac{\partial y}{\partial t} \right). \qquad (21\text{--}31)$$

To be more specific, suppose the wave is a sinusoidal wave:

$$y(x, t) = A \sin (\omega t - kx). \qquad (21\text{--}32)$$

Then

$$\frac{\partial y}{\partial x} = -kA \cos (\omega t - kx),$$

$$\frac{\partial y}{\partial t} = \omega A \cos (\omega t - kx),$$

$$P = Sk\omega A^2 \cos^2 (\omega t - kx). \qquad (21\text{--}33)$$

By using the relations $\omega = ck$ and $c^2 = S/\mu$, we can also express this in the alternative form

$$P = \sqrt{\mu S} \omega^2 A^2 \cos^2 (\omega t - kx). \qquad (21\text{--}34)$$

Equation (21–34) gives the *instantaneous* rate of energy transmission along the string; it depends on both x and t. We can obtain the *average* power by using the familiar fact that the *average* value of the $\cos^2$ function, averaged over any whole number of cycles, is $\frac{1}{2}$. Hence the average power is

$$P_{av} = \frac{1}{2} \sqrt{\mu S} \omega^2 A^2. \qquad (21\text{--}35)$$

It is interesting to note that the rate of energy transfer is proportional to the *square* of the amplitude and is also proportional to the square of the frequency.

Analogous relationships can be worked out for longitudinal waves. We will not go into detail; the results are most easily stated in terms of the average power *per unit cross-sectional area* in the wave motion. This power is called the *intensity*, denoted by I. For fluids in a pipe it is given by

> The intensity of a wave in a solid or fluid is the average power transmitted per unit area.

$$I = \frac{1}{2} \sqrt{\rho B} \omega^2 A^2,$$

and for a solid rod by

$$I = \frac{1}{2} \sqrt{\rho Y} \omega^2 A^2.$$

Again the power is proportional to A^2 and to ω^2.

SUMMARY

A wave is any disturbance from an equilibrium condition that propagates from one region to another. A mechanical wave always travels within some material called the medium. In a periodic wave, the motion of each point of the medium is cyclic or periodic; if the motion is sinusoidal, the wave is called a sinusoidal wave. The frequency f of a periodic wave is the number of repetitions per unit time, and the period τ is the time for one cycle. The wavelength λ is the distance over which the spatial pattern repeats. The speed of propagation c is the speed with which the wave disturbance travels. For any periodic wave these quantities are related by

$$c = \lambda f. \tag{21-1}$$

A wave function describes the displacements of individual particles in the medium. It is a function of the undisplaced position x and time t. The wave function for a sinusoidal wave traveling in the $+x$-direction can be written as

$$y(x, t) = A \sin \omega\left(t - \frac{x}{c}\right) = A \sin 2\pi f\left(t - \frac{x}{c}\right), \tag{21-3}$$

or

$$y(x, t) = A \sin 2\pi\left(\frac{t}{\tau} - \frac{x}{\lambda}\right), \tag{21-4}$$

or, using the wave number k, defined as $k = 2\pi/\lambda$, and the angular frequency ω defined as $\omega = 2\pi f$,

$$y(x, t) = A \sin (\omega t - kx). \tag{21-7}$$

In all three forms A is the amplitude, which is the maximum displacement of a particle from its equilibrium position.

The wave function must obey a partial differential equation called the wave equation:

$$\frac{\partial^2 y}{\partial t^2} = c^2 \frac{\partial^2 y}{\partial x^2}. \tag{21-14}$$

The speed of a transverse wave on a string having tension S and mass per unit length μ is given by

$$c = \sqrt{\frac{S}{\mu}} \quad \text{(transverse wave).} \tag{21-15}$$

Transverse waves have the property of polarization; longitudinal waves do not.

The speed of a longitudinal wave in a fluid having bulk modulus B and density ρ is given by

$$c = \sqrt{\frac{B}{\rho}} \quad \text{(longitudinal wave).} \tag{21-22}$$

The speed of a longitudinal wave in a solid rod having Young's modulus Y and density ρ is given by

$$c = \sqrt{\frac{Y}{\rho}} \quad \text{(longitudinal wave).} \tag{21-23}$$

Sound propagation is ordinarily an adiabatic process; the adiabatic bulk modulus for an ideal gas is given by

$$B_{ad} = \gamma p. \qquad (21\text{--}26)$$

The speed of sound in an ideal gas is given by

$$c = \sqrt{\frac{\gamma RT}{M}} \quad \text{(ideal gas).} \qquad (21\text{--}29)$$

Wave motion conveys energy from one region to another. For a transverse wave on a stretched string, a portion of string exerts a transverse force on an adjacent portion while it undergoes a displacement; hence one section does work on another, and energy is transferred along the length of the string.

QUESTIONS

21–1 The term *wave* has a variety of meanings in ordinary language. Think of several, and discuss their relation to the precise physical sense in which the term is used in this chapter.

21–2 What kinds of energy are associated with waves on a stretched string? How could such energy be detected experimentally?

21–3 Is it possible to have a longitudinal wave on a stretched string? A transverse wave on a steel rod?

21–4 The speed of sound waves in air depends on temperature, but the speed of light waves does not. Why?

21–5 For the wave motions discussed in this chapter, does the speed of propagation depend on the amplitude?

21–6 Children make toy telephones by sticking each end of a long string through a hole in the bottom of a paper cup and knotting it so it will not pull out. When the string is pulled taut, sound can be transmitted from one cup to the other. How does this work? Why is the transmitted sound louder than the sound traveling through air for the same distance?

21–7 An echo is sound reflected from a distant object, such as a wall or a cliff. Explain how you can determine how far away the object is by timing the echo.

21–8 Why do you see lightning before you hear the thunder? A familiar rule of thumb is to start counting slowly, once per second, when you see the lightning; when

you hear thunder, divide the number you have reached by 5 to obtain your distance (in miles) from the lightning. Why does this work? Or does it?

21–9 When ocean waves approach a beach, the crests are always nearly parallel to the shore, despite the fact that they must come from various directions. Why?

21–10 The amplitudes of ocean waves increase as they approach the shore, and often the crests bend over and "break." Why?

21–11 When a rock is thrown into a pond and the resulting ripples spread in ever-widening circles, the amplitude decreases with increasing distance from the center. Why?

21–12 In speaker systems designed for high-fidelity music reproduction, the "tweeters" that reproduce the high frequencies are always much smaller than the "woofers" used for low frequencies. Why?

21–13 When sound travels from air into water, does the frequency of the wave change? The wavelength? The speed?

21–14 Which of the quantities describing a sinusoidal wave is most closely related to musical pitch? To loudness?

21–15 Musical notes produced by different instruments (such as a flute and an oboe) may have the same pitch and loudness and yet sound different. What is the difference, in physical terms?

EXERCISES

Section 21–2 Periodic Waves

21–1 The speed of sound in air at 20°C is 344 m·s^{-1}.

a) What is the wavelength of a sound wave of frequency 32 Hz, the lowest pedal note on medium-sized pipe organs?

b) What is the frequency of a wave having a wavelength of 1.22 m (4 ft), corresponding approximately to the note D above middle C on the piano?

21–2 The speed of radio waves in vacuum (equal to the speed of light) is 3.00×10^8 m·s^{-1}. Find the wavelength for

a) an AM radio station with frequency 1000 kHz;

b) an FM radio station with frequency 100 MHz.

21–3 Provided the amplitude is sufficiently great, the human ear can respond to longitudinal waves over a range of frequencies from about 20 Hz to about 20,000 Hz. Compute the wavelengths corresponding to these frequencies

a) for waves in air ($c = 345$ m·s^{-1});

b) for waves in water ($c = 1480$ m·s^{-1}).

21–4 The sound waves from a loudspeaker spread out nearly uniformly in all directions when their wavelengths are large compared with the diameter of the speaker. When the wavelength is small compared with the diameter of the speaker, much of the sound energy is concentrated forward. For a speaker of diameter 25 cm, compute the frequency for which the wavelength of the sound waves, in air for which $c = 345$ m·s^{-1}, is

a) 10 times the diameter of the speaker;

b) equal to the diameter of the speaker;

c) $\frac{1}{10}$ the diameter of the speaker.

Section 21–3 Mathematical Description of a Wave

21–5 Show that Eq. (21–4) may be written

$$y = -A \sin \frac{2\pi}{\lambda}(x - ct).$$

21–6 The equation of a certain traveling transverse wave is

$$y = 2 \sin 2\pi\left(\frac{t}{0.01} - \frac{x}{30}\right),$$

where x and y are in centimeters and t is in seconds. What are the wave's

a) amplitude,

b) wavelength,

c) frequency,

d) speed of propagation?

21–7 A traveling transverse wave on a string is represented by the equation in Exercise 21–5. Let $A = 8$ cm, $\lambda = 16$ cm, and $c = 2$ cm·s^{-1}.

a) At time $t = 0$, compute the transverse displacement y at 2-cm intervals of x (that is, at $x = 0$, $x = 2$ cm, $x = 4$ cm, etc.) from $x = 0$ to $x = 32$ cm. Show the results in a graph. This is the shape of the string at time $t = 0$.

b) Repeat the calculations, for the same values of x, at times $t = 1$ s, $t = 2$ s, $t = 3$ s, and $t = 4$ s. Show on the same graph the shape of the string at these instants. In what direction is the wave traveling?

Section 21–4 Speed of a Transverse Wave

21–8 A steel wire 6 m long has a mass of 0.060 kg and is stretched with a tension of 1000 N. What is the speed of propagation of a transverse wave in the wire?

21–9 One end of a horizontal string is attached to a prong of an electrically driven tuning fork whose frequency of vibration is 240 Hz. The other end passes over a pulley and supports a mass of 5 kg. The linear mass density of the string is 0.02 kg·m^{-1}.

a) What is the speed of a transverse wave in the string?

b) What is the wavelength?

21–10 The equation of a transverse traveling wave on a string is

$$y = 2 \cos [\pi(0.5x - 200t)],$$

where x and y are in cm and t is in seconds.

a) Find the amplitude, wavelength, frequency, period, and velocity of propagation.

b) Sketch the shape of the string at the following values of t: 0; 0.0025 s; and 0.005 s.

c) If the mass per unit length of the string is 0.5 kg·m^{-1}, find the tension.

Section 21–5 Speed of a Longitudinal Wave

Section 21–6 Sound Waves in Gases

21–11 A steel pipe 100 m long is struck at one end. A person at the other end hears two sounds as a result of two longitudinal waves, one traveling in the metal pipe and the other traveling in the air. What is the time interval between the two sounds? Take Young's modulus of steel to be 2×10^{11} Pa, the density of steel to be 7800 kg·m^{-3}, and the speed of sound in air to be 345 m·s^{-1}.

21–12 At a temperature of 27°C, what is the speed of longitudinal waves in

a) argon?

b) hydrogen?

c) Compare your answers to (a) and (b) with the speed in air at the same temperature.

21–13 What is the difference between the speeds of longitudinal waves in air at −3°C and at 57°C?

21–14 Use the definition $B = -V(dp/dV)$ and the relation between p and V for an isothermal process to derive Eq. (21–27).

21–15

a) If the propagation of sound waves in gases were characterized by isothermal rather than adiabatic expansions and compressions, and assuming that the gas behaves as an ideal gas, show that the speed of sound would be given by $(RT/M)^{1/2}$.

b) What would be the speed of sound in air at 27°C in this case?

c) Under what circumstances might the wave propagation be expected to be isothermal?

Section 21–7 Energy in Wave Motion

21–16

a) A string of mass 4 g and length 2 m is stretched with a tension of 30 N. Waves of frequency $f = 60$ Hz and amplitude 8 cm are traveling along the string. Calculate the average power carried by these waves.

b) What happens to the average power if the amplitude of the waves is doubled?

21–17 Show that Eq. (21–35) can also be written as $P_{av} = \frac{1}{2}Sk\omega A^2$, where k is the wave number of the wave.

PROBLEMS

21–18 A student who could not get tickets to a Red Sox–Yankees baseball game is listening to the radio broadcast of the game in her dorm room while doing her physics homework. In the bottom of the fourth inning, however, a thunderstorm approaching from the west makes its presence known in three ways: (1) The student sees a lightning flash (and hears the electromagnetic pulse on her radio receiver); (2) 2.4 s later, she hears the thunder over the radio; (3) 4.0 s after the lightning flash, the thunder rattles her window. By a previous careful measurement, she knows that she is 1.12 km due north of the broadcast booth at the ballpark, and that the speed of sound is 350 m·s^{-1}. Where did the lightning flash occur, relative to the ballpark?

21–19 A transverse sine wave of amplitude 0.10 m and wavelength 2 m travels from left to right along a long horizontal stretched string with a speed of 1 m·s^{-1}. Take the origin at the left end of the undisturbed string. At time $t = 0$, the left end of the string is at the origin and is moving downward.

a) What is the frequency of the wave?

b) What is the angular frequency?

c) What is the propagation constant?

d) What is the equation of the wave?

e) What is the equation of motion of the left end of the string?

f) What is the equation of motion of a particle 1.5 m to the right of the origin?

g) What is the maximum magnitude of transverse velocity of any particle of the string?

h) Find the transverse displacement and the transverse velocity of a particle 1.5 m to the right of the origin, at time $t = 3.25$ s.

i) Make a sketch of the shape of the string, for a length of 4 m, at time $t = 3.25$ s.

21–20 Show that $y(x, t) = A \cos(\omega t + kx)$ satisfies the wave equation, Eq. (21–14).

21–21 One end of a rubber tube 20 m long, of total mass 1 kg, is fastened to a fixed support. A cord attached to the other end passes over a pulley and supports a body of mass 10 kg. The tube is struck a transverse blow at one end. Find the time required for the pulse to reach the other end.

21–22 One end of a stretched rope is given a periodic transverse motion with a frequency of 10 Hz. The rope is 50 m long, has a total mass of 0.5 kg, and is stretched with a tension of 400 N.

a) Find the wave speed and the wavelength.

b) If the tension is doubled, how must the frequency be changed to maintain the same wavelength?

21–23 What must be the stress (F/A) in a stretched wire of a material whose Young's modulus is Y, for the speed of longitudinal waves to equal ten times the speed of transverse waves?

21–24 A metal wire, of density 5×10^3 kg·m^{-3}, with a Young's modulus equal to 2.0×10^{11} Pa, is stretched between rigid supports. At one temperature the speed of a transverse wave is found to be 200 m·s^{-1}. When the temperature is raised 25 C°, the speed decreases to 160 m·s^{-1}. Determine the coefficient of linear expansion.

21–25 What is the ratio of the speed of sound in a diatomic gas to the rms speed of gas molecules at the same temperature?

21–26

a) By how many meters per second, at a temperature of 27°C, does the speed of sound in air increase per Celsius degree rise in temperature?

b) Is the change of speed for a 1 C° change in temperature the same at all temperatures?

CHALLENGE PROBLEMS

21–27 Derive Eq. (21–24) by applying $F = ma$ to a thin slice of fluid in which a longitudinal wave is propagating.

21–28 For longitudinal waves in a solid, prove that the power per unit cross-sectional area in the wave motion (the *intensity*) is given by $\frac{1}{2}\omega^2 A^2 \sqrt{\rho Y}$, where ρ and Y are the density and Young's modulus for the material, and ω and A are the angular frequency and amplitude of the waves.

22

VIBRATING BODIES

Vibrating bodies can have many interesting vibration patterns.

IN CHAPTER 21 WE STUDIED THE PROPAGATION OF MECHANICAL WAVES IN media without ends or boundaries; we were not concerned with what happens when a wave arrives at an end or boundary of the medium in which it propagates. But in many wave phenomena such boundaries do play a significant role. A familiar example is the echo that occurs when a sound wave reflects from a rigid wall. Such reflections lead to overlap, or *superposition,* of two waves—the initial and reflected waves—in the same region of the medium. When there are two or more boundary points or surfaces, repeated reflections can occur. In such cases it turns out that sinusoidal wave motion is possible only for certain special values of the frequency of the wave, determined by the dimensions and mechanical properties of the medium. These special frequencies and their associated wave patterns are called *normal modes.* Many familiar phenomena are associated with normal modes. The pitches of most musical instruments are determined by normal-mode frequencies, and many other mechanical vibrations involve normal-mode motion. This concept will also reappear later in some unexpected places, such as the energy levels of atoms.

22–1 BOUNDARY CONDITIONS FOR A STRING

When a transverse wave reaches the end of a string, it is reflected.

As a simple example of reflections and the role of boundaries, let us return to transverse waves on a stretched string. Consider what happens when a wave pulse or a continuous succession of waves (such as a sinusoidal wave) arrives at the end of the string. If the end is fastened to a rigid support, it must remain at rest. The arriving wave exerts a force on the support; the reaction to this force, exerted *by* the support *on* the string, "kicks back" on the string and sets up a *reflected* pulse or wave traveling in the reverse direction.

At the opposite extreme from a rigidly fixed end is one that is perfectly free to move in the direction transverse to the length of the string. For example, the string might be tied to a light ring that slides on a smooth rod perpendicular to the length of the string. The ring and rod maintain the tension of the string but exert no transverse force. At the free end the arriving pulse or wave train causes the end to "overshoot," and again a reflected wave is set up.

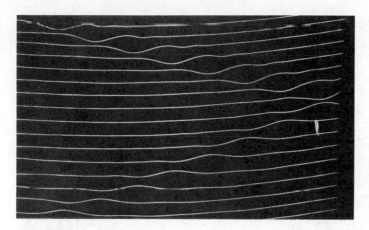

22–1 A pulse starts at the left in the top image, travels to the right, and is reflected from the fixed end of the string at the right.

The conditions imposed on the motion of the end of the string, such as attachment to a rigid support or the complete absence of transverse force, are called **boundary conditions.**

The multiflash photograph of Fig. 22–1 shows the reflection of a pulse at a fixed end of a string. (The camera was tipped vertically while the photographs were taken so that successive images lie one under the other. The "string" is a rubber tube, which sags somewhat.) The pulse is reflected with both its displacement and its direction of propagation reversed. When reflection takes place at a *free* end, the direction of propagation is reversed but the direction of the displacement is unchanged.

A useful way to think of the process of reflection is to imagine that the string extends indefinitely beyond its actual end. Consider the actual pulse to continue on into the imaginary portion as though the support were not there, while at the same time a "virtual" pulse, which has been traveling in the opposite direction in the imaginary portion, moves out into the real string and forms the reflected pulse. The nature of the reflected pulse depends on whether the end is fixed or free. The two cases are shown in Fig. 22–2.

The actual displacement of the string at a point where the actual and virtual pulses cross each other is the *algebraic sum* of the displacements in the

The total displacement of the string is the superposition of the original wave and the reflected wave.

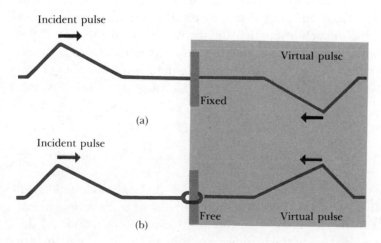

(a)

(b)

22–2 Description of the reflection of a pulse (a) at a fixed end of a string and (b) at a free end, in terms of an imaginary "virtual" pulse.

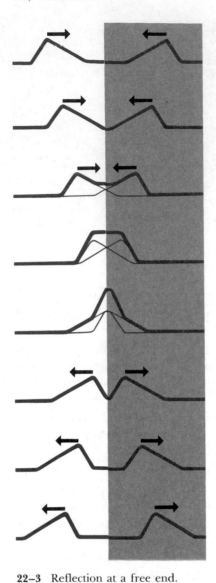

22–3 Reflection at a free end.

22–4 Reflection at a fixed end.

individual pulses. Figures 22–3 and 22–4 show the shape of the region near
the end of the string for both types of reflected pulses, Fig. 22–3 for a free end
and Fig. 22–4 for a fixed end. In the latter case, the incident and reflected
pulses must combine in such a way that the displacement of the end of the
string is always zero. This combining of the displacements of the separate
pulses at each point to obtain the actual displacement is an example of the
principle of superposition, which plays a central role in most of this chapter.

22–2 SUPERPOSITION AND STANDING WAVES

Consider a continuous succession of waves, such as a sinusoidal wave, arriving
at a boundary point on a string. The boundary point may be a fixed end or an
end that it is free to move transversely, as in Fig. 22–2. As we discussed in the
preceding section, a corresponding succession of *reflected* waves originates at
this end and travels in the opposite direction. The resulting motion of the
string is determined by an extremely important principle called the **principle**

*The principle of superposition: What
happens when several waves are present
in a medium at the same time?*

of superposition. This principle states that the actual displacement of any point on the string, at any time, is obtained by adding the displacement that point would have if only the first wave were present, and the displacement it would have with only the second wave.

Mathematically speaking, the principle of superposition states that the wave function describing the resulting motion in the above situation, where a wave combines with a reflected wave, is obtained by adding the two wave functions for the two separate waves. This additive property of wave functions depends, in turn, on the form of the wave equation, Eq. (21–14), which every physically possible wave function must satisfy. Specifically, the wave equation is *linear;* that is, it contains derivatives of the function y only to the first power. As a result, if each of any two functions $y_1(x, t)$ and $y_2(x, t)$ satisfies the wave equation separately, their *sum* $y_1 + y_2$ automatically satisfies the wave equation also and hence is a physically possible motion. In view of this linearity of the wave equation and the corresponding linear-combination property of its solutions, the principle is also called the *principle of linear superposition.*

The principle of superposition is of central importance in all types of wave motion. It applies not only to waves on a string, but also to sound waves, electromagnetic waves (such as light), and all other wave phenomena in which the wave equation is linear. The general term **interference** is used to describe phenomena that result from two or more waves passing through the same region at the same time.

When a sinusoidal wave is reflected by a fixed point on a string, the appearance of the resulting motion gives no evidence of two waves traveling in opposite directions. If the frequency is sufficiently great so that the eye cannot follow the motion, the string appears subdivided into a number of segments, as in the time-exposure photograph of Fig. 22–5a. A multiflash photograph of the same string, in Fig. 22–5b, indicates a few of the instantaneous shapes of the string. At every instant its shape is a sine curve, but in a traveling wave the amplitude remains constant as the wave progresses, while here the waveform remains fixed in position (longitudinally) as the amplitude fluctuates. Certain points, known as the **nodes,** remain always at rest. Midway between these points, at the *loops* or **antinodes,** the fluctuations are maximum. The vibration as a whole is called a **standing wave.**

To understand the formation of a standing wave, consider the separate graphs of the waveform at four instants one-tenth of a period apart, shown in Fig. 22–6. The gray curves represent a wave traveling to the right. The light-color curves represent a wave of the same propagation speed, wavelength, and amplitude traveling to the left. The dark-color curves represent the resultant waveform, obtained by applying the principle of superposition, that is, by adding displacements. At those places marked N at the bottom of Fig. 22–6, the resultant displacements are *always* zero. These are the *nodes.* Midway between the nodes, the vibrations have the *largest* amplitude. These are the *antinodes.* It is evident from the figure that

$$\left\{ \begin{array}{c} \text{Distance between adjacent nodes} \\ \text{or} \\ \text{Distance between adjacent antinodes} \end{array} \right\} = \frac{\lambda}{2}.$$

The wave function for the standing wave of Fig. 22–6 may be obtained by adding the displacements of two waves of equal amplitude, period, and wave-

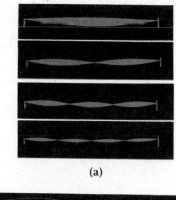

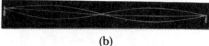

22–5 (a) Standing waves in a stretched string (time exposure). (b) Multiflash photograph of a standing wave, with nodes at the center and at the ends.

A node is a point that never moves during wave motion.

An antinode is a point with maximum amplitude of motion, between two nodes.

A standing wave does not appear to be moving in either the positive or negative direction.

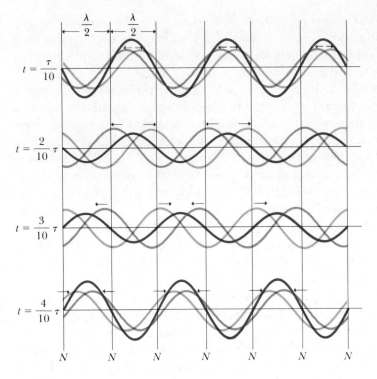

22–6 The formation of a standing wave. A wave traveling to the right (gray lines) combines with a wave traveling to the left (light-color lines) to form a standing wave (dark-color lines). The thin horizontal line in each part shows the equilibrium position of the string.

length, traveling in opposite directions. Thus if

$$y_1 = A \sin (\omega t - kx) \qquad \text{(positive x-direction)},$$
$$y_2 = -A \sin (\omega t + kx) \qquad \text{(negative x-direction)},$$

then

$$y_1 + y_2 = A[\sin (\omega t - kx) - \sin (\omega t + kx)].$$

Introducing the expressions for the sine of the sum and difference of two angles and combining terms, we obtain

A standing wave formed by superposition of two sinusoidal waves moving in opposite directions

$$y_1 + y_2 = -[2A \cos \omega t] \sin kx. \qquad (22\text{–}1)$$

The shape of the string at each instant is, therefore, a sine curve whose amplitude (the expression in brackets) varies with time. The appearance is not that of a traveling wave shape, but of a sinusoidal wave shape in one position with an amplitude that grows larger and smaller with time. Each point in the string still undergoes simple harmonic motion, but all points move *in phase* (or 180° out of phase). This is in contrast to the phase differences between motions of adjacent points that are seen with a wave traveling in one direction.

The nodes are spaced a half-wavelength apart.

The positions of the nodes can be obtained from Eq. (22–1). Wherever $\sin kx = 0$, the displacement is always zero. This occurs when $kx = 0$, π, 2π, 3π, . . . , or

$$x = 0, \pi/k, 2\pi/k, 3\pi/k, \ldots$$
$$= 0, \lambda/2, \lambda, 3\lambda/2, \ldots. \qquad (22\text{–}2)$$

22–3 NORMAL MODES OF A STRING

We have mentioned the reflection or echo of a sound wave from a rigid wall, and the analogous reflection of a transverse wave on a string from a rigidly held end. Now suppose we have two parallel walls. If a sharp sound pulse, such as a hand clap, originates at a point between the walls, the result is a series of regularly spaced echoes caused by repeated back-and-forth reflection between the walls. In room acoustics this phenomenon is called "flutter echo"; it is the bane of acoustical engineers.

The analogous situation with transverse waves on a string is a string with some definite length L, rigidly held at *both* ends. If a sinusoidal wave is produced on such a string, the wave is reflected and re-reflected. Because the string is held at both ends, both ends must be nodes. Since adjacent nodes are one-half wavelength apart, the length of the string may be

$$\frac{\lambda}{2}, \frac{2\lambda}{2}, \frac{3\lambda}{2},$$

or, in general, *any* integer number of half-wavelengths. To put it differently, if we consider a particular string of length L, standing waves may be set up in the string by vibrations of a number of different frequencies—namely, those that give rise to waves of wavelengths

$$\lambda = 2L, \frac{2L}{2}, \frac{2L}{3}, \cdots = \frac{2L}{n} \qquad (n = 1, 2, 3, \ldots). \qquad (22\text{–}3)$$

The wave speed c is the same for all frequencies. From the relation $f = c/\lambda$, the possible frequencies are

$$f = \frac{c}{2L}, \frac{2c}{2L}, \frac{3c}{2L}, \cdots = \frac{nc}{2L} \qquad (n = 1, 2, 3, \ldots). \qquad (22\text{–}4)$$

The lowest frequency, $c/2L$, is called the **fundamental frequency** f_1. All the other frequencies are integer multiples of f_1, such as $2f_1, 3f_1, 4f_1$, and so on. Thus we can rewrite Eq. (22–4) as

$$f_n = n\frac{c}{2L} = nf_1 \qquad (n = 1, 2, 3, \ldots). \qquad (22\text{–}5)$$

This series of frequencies, all integer multiples of the fundamental, is called a **harmonic series.** Musicians sometimes call f_2, f_3, and so on **overtones;** f_2 is the second harmonic or the first overtone, f_3 is the third harmonic or the second overtone, and so on.

These results may also be obtained directly from Eq. (22–1). The boundary conditions require that $y_1 + y_2 = 0$ at the ends of the string, that is, at $x = 0$ and $x = L$. Since the sine of zero is zero, the first condition is satisfied automatically. The second condition requires that $\sin kL = 0$, and this is true only when k has certain special values. The sine of an angle is zero only when the angle is zero or an integer multiple of π (180°). Thus we must have

$$kL = n\pi \qquad (n = 1, 2, 3, \ldots).$$

We do not include the possibility $n = 0$ because that gives $k = 0$, that is, a wave with zero displacement *everywhere* (a possible case, to be sure, but not a very interesting one!).

Replacing k above by $2\pi/\lambda$, we obtain

$$\frac{2\pi L}{\lambda} = n\pi \quad \text{or} \quad \lambda = \frac{2L}{n}, \qquad (22\text{–}6)$$

in agreement with Eq. (22–3).

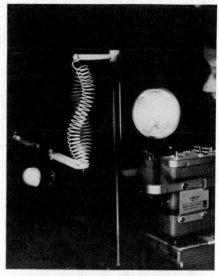

A stroboscopic photograph showing a transverse standing wave on a spring. One end of the spring is held stationary while the other end is vibrated with small amplitude by a motor. The maximum amplitude of the wave is much greater than the amplitude at the driven end, and both ends are approximately node points. (Dr. Harold Edgerton, M.I.T., Cambridge, Massachusetts.)

The frequencies are all integer multiples of the lowest frequency.

What is the difference between harmonics and overtones?

In a normal-mode motion, every particle vibrates sinusoidally with the same frequency. Each mode has its own vibration pattern.

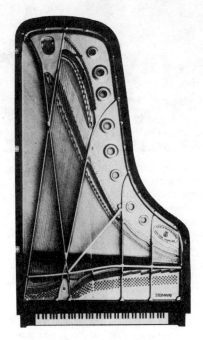

Interior of a concert grand piano, showing the full-length bass strings and the progressively shorter strings in the treble range. (Courtesy of Steinway and Sons.)

Each of the frequencies given by Eq. (22–4) corresponds to a possible **normal mode** of motion, that is, a motion in which each particle of the string moves sinusoidally, all with the same frequency. As this analysis shows, there are an infinite number of normal modes, each with its characteristic frequency. This situation is in striking contrast with the simple harmonic oscillator system, consisting of a single mass and a spring. The harmonic oscillator has only one normal mode and one characteristic frequency, while the vibrating string has an infinite number.

If a string is initially displaced so that its shape is the same as *any one* of the possible harmonics, it will vibrate, when released, at the frequency of that particular harmonic. But when a piano string is struck or a guitar string is plucked, not only the fundamental but many of the overtones are present in the resulting vibration. This motion is therefore a combination or *superposition* of normal modes. Several frequencies and motions are present simultaneously, and the displacement of any point on the string is the sum (or superposition) of displacements associated with the individual modes. Indeed, every possible motion of the string can be represented as some superposition of normal-mode motions.

The fundamental frequency of the vibrating string is $f_1 = c/2L$, where, from Eq. (21–15), $c = \sqrt{S/\mu}$. It follows that

$$f_1 = \frac{1}{2L}\sqrt{\frac{S}{\mu}}. \qquad (22\text{–}7)$$

Stringed instruments provide many examples of the implications of this equation. For example, all such instruments are "tuned" by varying the tension S. An increase of tension increases the frequency or pitch, and vice versa. The inverse dependence of frequency on length L is illustrated by the long strings of the bass section of the piano or the bass viol compared with the shorter strings on the piano treble or the violin. In playing the violin or guitar, the usual means of varying the pitch is to press the strings against the fingerboard with the fingers to change the length of the vibrating portion of the string. One reason for winding the bass strings of a piano with wire is to increase the mass per unit length μ, so as to obtain the desired low frequency without resorting to a string that is inconveniently long.

PROBLEM-SOLVING STRATEGY: *Standing waves*

1. As with the problems in Chapter 21, it is useful to distinguish between the purely kinematic quantities, such as wave speed c, wavelength λ, and frequency f, and the dynamic quantities involving the properties of the medium, including S, μ, and (in the next section) B and ρ. The latter determine the wave speed c. Try to determine, in the problem at hand, whether the properties of the medium are involved or whether the problem is only kinematic in nature.

2. In visualizing nodes and antinodes in standing waves, it is always helpful to draw diagrams. For a string you can draw the shape at one instant and label the nodes N and antinodes A. For longitudinal waves (discussed in the next section), it is not so easy to draw the shape, but you can still label the nodes and antinodes. Remember that the distance between a node and the adjacent antinode is *always* $\lambda/4$, and that the distance between two adjacent nodes or two adjacent antinodes is always $\lambda/2$.

22–4 LONGITUDINAL STANDING WAVES

When longitudinal waves propagate in a fluid in a tube of finite length, the waves are reflected from the ends in the same way that transverse waves on a string are reflected at its ends. The superposition of the waves traveling in opposite directions again forms a standing wave.

When reflection takes place at a *closed* end, the displacement of the particles there must always be zero. This situation is analogous to a fixed end of a string; in both cases there is no displacement at the end, and the end is a *node*. For clarity, in the following discussion we call a closed end of a tube or pipe a *displacement node*. If the end of the tube is open, the nature of the reflection is more complex and depends on whether the tube is wide or narrow compared with the wavelength. If the tube is narrow compared with the wavelength, which is the case in most musical instruments, the open end is a *displacement antinode,* for reasons we will discuss below. (A free end of a stretched string, as discussed in Section 22–1, is also a displacement antinode.) Thus longitudinal waves in a column of fluid are reflected at the closed and open ends of a tube in the same way that transverse waves in a string are reflected at fixed and free ends, respectively.

Reflections of longitudinal waves take place at a closed or open end of a pipe.

We can demonstrate longitudinal standing waves in a column of gas, and also measure the wave speed, by using the apparatus called Kundt's tube, shown in Fig. 22–7. A glass tube a meter or so long is closed at one end and has a flexible diaphragm at the other end that can transmit vibrations. We use a sound source, which might be a small loudspeaker driven by an audio oscillator and amplifier, to vibrate the diaphragm sinusoidally with a variable frequency. A small amount of light powder or cork dust is distributed uniformly along the bottom side of the tube. As we vary the frequency of the sound, we pass through frequencies where the amplitude of the standing waves becomes large enough for the moving gas to sweep the cork dust along the tube at all points where the gas is in motion. The powder therefore collects at the displacement nodes, which can be seen and measured easily.

In a standing wave, the distance between two adjacent nodes is one-half (*not* one!) wavelength. Thus we may measure the wavelength by measuring the distances $\lambda/2$ between adjacent clumps of powder. We read the frequency f from the oscillator dial, and we can calculate the speed c of the waves from the usual relation

Adjacent nodes are a half-wavelength apart; so are adjacent antinodes.

$$c = \lambda f.$$

At a displacement node, the pressure variations above and below the average have their *maximum* value, while at a displacement antinode the pressure does not vary. To understand this, note that two small masses of gas on opposite sides of a displacement *node* vibrate in *opposite phase*. When the masses of

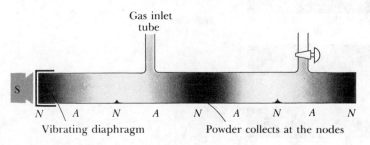

Gas inlet tube

N \ *A* *N* *A* *N* \ *A* *N* *A* *N*

Vibrating diaphragm   Powder collects at the nodes

22–7 Kundt's tube for determining the velocity of sound in a gas. The shading represents the density of the gas molecules at an instant when the pressure at the displacement nodes is a maximum or a minimum.

22–8 Diagram for illustrating longitudinal standing waves.

gas approach each other, the gas between them is compressed and the pressure rises; when they recede from each other, the pressure drops. But two small masses of gas on opposite sides of a displacement *antinode* vibrate *in phase,* and so cause *no* pressure variations at the antinode.

We can describe this relationship in terms of **pressure nodes,** which are the points where the pressure does not vary, and **pressure antinodes,** which are the points where its variation is greatest. *A pressure node is always a displacement antinode, and a pressure antinode is always a displacement node.* An *open* end of a thin tube or pipe is a pressure node because such an end is open to the atmosphere and is thus at constant pressure. But for this reason, an open end is always a displacement *antinode.*

We can visualize longitudinal standing waves with the help of Fig. 22–8, which is analogous to Fig. 21–10 for longitudinal traveling waves. Again use a 3-in. × 5-in. card with a slit $\frac{1}{16}$ in. wide and 3 in. long, or two cards fastened edge-to-edge with a $\frac{1}{16}$-in. gap. Place the card over the diagram with the slit horizontal and move it vertically with constant velocity. The portions of the sine curves that appear in the slit correspond to the oscillations of the particles in a longitudinal standing wave. Each particle moves with simple harmonic motion about its equilibrium positions. The particles at the nodes do not move, and the nodes are regions of maximum compression and expansion. Midway between the nodes are the antinodes, regions of maximum displacement but zero compression and expansion.

22–5 NORMAL MODES OF ORGAN PIPES

The behavior of an open organ pipe is similar to that of a string held at both ends.

Organ pipes provide us with a good example of standing waves and normal modes in vibrating air columns. Air is supplied by a blower, at a gauge pressure typically of the order of 10^3 Pa or 10^{-2} atm, to the left ends of the pipes in Figs. 22–9 and 22–10. A stream of air emerges from the narrow opening formed by the vertical surface and is directed against the right edge of the opening in the top surface of the pipe, called the *mouth* of the pipe. The column of air in the pipe is set into vibration; just as with the stretched string, there is a series of possible normal modes. For a pipe open at both ends, the fundamental frequency f_1 corresponds to a displacement antinode (pressure node) at each end and a displacement node in the middle, as shown at the top of Fig. 22–9. The other two parts of this figure show the second and third harmonics; their vibration patterns have two and three displacement nodes, respectively, and their frequencies are $f_2 = 2f_1$ and $f_3 = 3f_1$, respectively. *In an*

22–9 Modes of vibration of an open organ pipe.

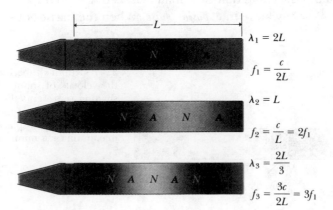

$$\lambda_1 = 2L$$
$$f_1 = \frac{c}{2L}$$

$$\lambda_2 = L$$
$$f_2 = \frac{c}{L} = 2f_1$$

$$\lambda_3 = \frac{2L}{3}$$
$$f_3 = \frac{3c}{2L} = 3f_1$$

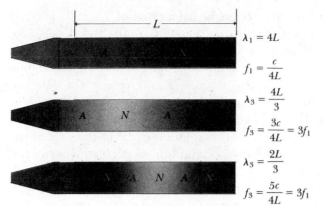

$$\lambda_1 = 4L$$

$$f_1 = \frac{c}{4L}$$

$$\lambda_3 = \frac{4L}{3}$$

$$f_3 = \frac{3c}{4L} = 3f_1$$

$$\lambda_3 = \frac{2L}{3}$$

$$f_3 = \frac{5c}{4L} = 3f_1$$

22–10 Modes of vibration of a stopped organ pipe.

open pipe the fundamental frequency is $c/2L$, and all harmonics in the series are possible.

The properties of a *stopped* pipe (open at one end, closed at the other) are shown in Fig. 22–10. The left (open) end is a displacement antinode (pressure node), but the right (closed) end is a displacement node (pressure antinode). We see that in the lowest-frequency mode the length of the pipe is a quarter-wavelength ($L = \lambda/4$). The fundamental frequency is $f_1 = c/4L$, which is one-half that of an open pipe of the same length. In musical language, the *pitch* of a closed pipe is one octave lower (a factor of two in frequency) than that of an open pipe of the same length. From the other parts of Fig. 22–10 we see that the second, fourth, and all *even* harmonics are missing. *In a stopped pipe the fundamental frequency is $f_1 = c/4L$, and only the odd harmonics in the series ($3f_1$, $5f_1$, . . .) are possible.*

In an organ pipe in actual use, several modes are almost always present at once. This situation is analogous to that of a string that is struck or plucked, producing several modes at the same time. The motion in each case is then a *superposition* of various modes. The extent to which modes higher than the fundamental are present depends on the cross-section of the pipe, the proportion of length to width, the shape of the mouth, and other more subtle factors. The harmonic content of the tone is an important factor in determining the tone quality, or timbre. A very narrow pipe produces a tone rich in higher harmonics, which the ear perceives as thin and "stringy"; a fatter pipe produces principally the fundamental mode, perceived as a softer, more flutelike tone.

Although this discussion has centered on organ pipes, it is also relevant for other wind instruments. The flute and the recorder are directly analogous. The most significant difference is that those instruments have holes along the pipe; opening and closing these with the fingers changes the effective length L of the air column and thus changes the pitch. Any individual organ pipe, by comparison, plays only a single note. The flute and recorder function as *open* pipes, while the clarinet acts as a *closed* pipe (closed at the reed end, open at the bell).

Console and a few of the thousands of pipes in the organ of the Shrine of the Immaculate Conception, Washington, DC. Each pipe produces a single note. The largest pipe is 38 feet long overall and weighs 1050 lb. (Courtesy of M. P. Möller, Inc.)

Several wind instruments are similar to organ pipes in their operation.

22–6 INTERFERENCE OF WAVES

Wave phenomena that occur when two or more waves overlap in the same region of space are grouped under the heading *interference*. As we have seen, standing waves are a simple example of an interference effect. Here are two

What happens when two waves overlap in the same region of space? They can either reinforce or cancel each other.

Source
S

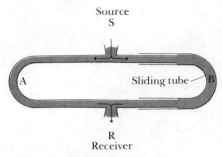

R
Receiver

22–11 Apparatus for demonstrating interference of longitudinal waves.

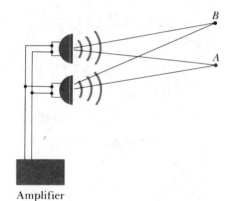

Amplifier

22–12 Two speakers driven by the same amplifier. The waves emitted by the speakers are in phase; they arrive at point A in phase because the two path lengths are the same. They arrive at point B a half-cycle out of phase because the path lengths differ by $\lambda/2$.

Resonance: When an applied force varies with the same frequency as a normal-mode frequency, the amplitude of oscillation builds up.

more examples. Consider the trombonelike apparatus of Fig. 22–11. The source S emits a sinusoidal sound wave that enters the tube and divides into two waves. One wave follows the path SAR, which has constant length, and the other follows path SBR, whose length can be changed by moving the sliding tube. The two waves recombine and emerge at R, the receiver or detector.

Suppose the wavelength of the sound is 1.0 m. If the paths SAR and SBR have the same length, the waves take equal times to reach R and arrive there *in phase*. The resulting displacement has an amplitude equal to the sum of the two individual amplitudes. This addition of amplitudes is called *reinforcement* or *constructive interference.*

Now we slide the tube B out a distance of 0.25 m, making the path SBR 0.5 m longer than SAR. Then the right-hand wave travels a distance $\lambda/2$ farther than the left-hand wave. When the two waves arrive at R, their displacements differ in phase by a half-cycle and thus add to zero. This is called *cancellation* or *destructive interference,* and results in a sharp decrease in sound level at R.

If we now pull the tube B out another 0.25 m, so that the *path difference,* SBR minus SAR, is 1.0 m (1 wavelength), the two vibrations at R again reinforce each other. Thus

$$\left\{ \begin{array}{l} \text{Reinforcement takes place} \\ \text{when the path difference} \end{array} \right\} = 0, \lambda, 2\lambda, \text{ etc.}$$

$$\left\{ \begin{array}{l} \text{Cancellation takes place} \\ \text{when the path difference} \end{array} \right\} = \frac{\lambda}{2}, \frac{3\lambda}{2}, \frac{5\lambda}{2}, \text{ etc.}$$

We can demonstrate a similar phenomenon with two loudspeakers driven by the same amplifier, as shown in Fig. 22–12. Suppose both speakers emit a pure sinusoidal sound wave of constant frequency. When we place a microphone at point A in the figure, equidistant from the speakers, it receives a strong acoustic signal, corresponding to the fact that the two waves emitted from the speakers in phase also arrive at point A in phase and add to each other. But at point B the signal is much *weaker* than when only one speaker is present, corresponding to the fact that the wave from one speaker travels a half-wavelength farther than that from the other. The two waves arrive a half-cycle out of phase and cancel each other out almost completely. An experiment closely analogous to this one, but using light waves, provided the first conclusive evidence of the wave nature of light. This experiment is discussed in detail in Chapter 39.

22–7 RESONANCE

In this chapter we have discussed several examples of mechanical systems having normal modes of oscillation. In each mode, every particle of the system oscillates with simple harmonic motion of the same frequency as the frequency of this mode. The systems we have discussed have an infinite series of normal modes, but the basic concept is closely related to the simple harmonic oscillator, discussed in Chapter 11, which has only a single normal mode.

Now suppose we apply a periodically varying force to a system having normal modes. The system vibrates with a frequency equal to that of the *force.* This motion is called a *forced oscillation.* In general, the amplitude of this motion is relatively small, but if the frequency of the force is close to one of the normal-mode frequencies, the amplitude can become quite large. We discussed forced oscillations of the harmonic oscillator in Section 11–8, and we suggest you review that discussion now.

If the frequency of the force were precisely *equal* to a normal-mode frequency, and if there were no friction or other energy-dissipating mechanism, then the force would continue to add energy to the system, and the amplitude would increase indefinitely. In any real system there is always some dissipation of energy, or damping, as we noted in Section 11–7. Even so, the "response" of the system (that is, the amplitude of the forced oscillation) is greatest when the force frequency is equal to one of the normal-mode frequencies. This behavior is called **resonance.**

Pushing a child's swing is a common example of mechanical resonance, as mentioned in Section 11–8. The swing is a pendulum; it has only a single natural frequency, determined by its length. If we give the swing a series of regularly spaced pushes, with a frequency equal to that of the swing, the motion may be made quite large. If the frequency of the pushes differs from the natural frequency of the swing, or if the pushes occur at irregular intervals, the swing will execute hardly any vibration at all.

Unlike a simple pendulum, which has only one natural frequency, a stretched string (and other systems discussed in this chapter) has an infinite number of natural frequencies. Suppose that one end of a stretched string is held stationary while the other is moved back and forth in a transverse direction. The amplitude at the driven end is fixed by the driving mechanism. Standing waves are set up in the string, whatever the value of the frequency f. If the frequency is *not* equal to one of the natural frequencies of the string, the amplitude at the antinodes is fairly small.

A pendulum has only one normal mode; a stretched string has an infinite number.

However, if the frequency is equal to *any one* of the natural frequencies, the string is in resonance and the amplitude at the antinodes is very much *larger* than that at the driven end. In other words, although the driven end is not a node, it lies much closer to a node than to an antinode when the string is in resonance. In Fig. 22–5a, the right end of the string was held stationary and the left end was forced to oscillate vertically with small amplitude. The photographs show the standing waves of relatively large amplitude that resulted when the frequency of oscillation of the left end was equal to the fundamental frequency or to one of the first three overtones.

A steel bridge or, for that matter, any elastic structure has normal modes and can vibrate with certain natural frequencies. If the regular footsteps of a marching band have a frequency equal to one of the natural frequencies of a bridge that the band is crossing, a vibration of dangerously large amplitude may result. Therefore, when crossing a bridge, a marching group should always "break step."

How to make a bridge collapse by marching a band across it. We don't recommend it.

We can demonstrate the phenomenon of resonance with two identical tuning forks. We place them some distance apart and strike one to set it into vibration. The resulting sound waves set the other fork into vibration, and its vibration can be heard if the first fork is suddenly damped. If we place a small piece of wax or modeling clay on one of the forks, the frequency of that fork will be altered enough to destroy the resonance.

A similar phenomenon can be demonstrated with a piano. Depress the damper pedal (the right-hand pedal) so the dampers are lifted and the strings are free to vibrate, and then sing a steady tone into the piano. When the singing stops, the piano seems to continue to sing the same note. The sound waves from your voice excite vibrations in the strings that have natural fre-

How to sing to a piano and have it sing back to you

quencies close to the frequencies (fundamental and harmonic) present in the note sung initially.

Resonance is a very important concept, not only in mechanical systems but in all areas of physics. Later we will see examples of resonance in electric circuits (Chapter 34) and in atomic systems (Chapter 42).

SUMMARY

KEY TERMS
boundary conditions
principle of superposition
interference
nodes
antinodes
standing wave
fundamental frequency
harmonic series
overtones
normal mode
pressure nodes
pressure antinodes
resonance

A wave that reaches a boundary of the medium in which it propagates is reflected. The total wave displacement at any point where the initial and reflected waves overlap is the sum of the displacements of the individual waves; this statement is the principle of superposition.

When a wave is reflected from a fixed or free end of a stretched string, the incident and reflected waves combine to form a standing wave, which does not appear to travel in either direction. Its pattern contains nodes and antinodes; adjacent nodes are spaced a distance $\lambda/2$ apart, as are adjacent antinodes.

When both ends of a string having length L are held, standing waves can occur only when L is an integer multiple of $\lambda/2$; the corresponding possible frequencies are given by

$$f_n = n\frac{c}{2L} \qquad (n = 1, 2, 3, \ldots). \qquad (22\text{--}5)$$

Each frequency, and its associated vibration pattern, is called a normal mode. The lowest frequency f_1 is called the fundamental frequency. In terms of the mechanical properties S and μ of the string, the fundamental frequency is given by

$$f_1 = \frac{1}{2L}\sqrt{\frac{S}{\mu}}. \qquad (22\text{--}7)$$

Standing waves also occur in wave motion in pipes or tubes. A closed end is a displacement node and a pressure antinode; an open end is a displacement antinode and a pressure node.

For a pipe open at both ends, the normal-mode frequencies are given by

$$f_n = n\frac{c}{2L} \qquad (n = 1, 2, 3, \ldots).$$

For a pipe open at one end and closed at the other, the normal-mode frequencies are

$$f_n = n\frac{c}{4L} \qquad (n = 1, 3, 5, \ldots).$$

When two or more waves overlap in the same region of space, the resulting effects are called interference. The resulting amplitude can be either larger or smaller than the amplitude of each individual wave, depending on whether the waves are in phase or out of phase. If the waves are in phase, the result is called reinforcement or constructive interference; if they are out of phase, it is called cancellation or destructive interference.

When a periodically varying force is applied to a system having normal modes of vibration, the system vibrates with the same frequency as that of the force; this is called a forced oscillation. If the force frequency is equal or close to one of the normal-mode frequencies, the amplitude of the resulting forced oscillation can become very large; this phenomenon is called resonance.

QUESTIONS

22–1 When you inhale helium, your voice becomes high and squeaky. Why? (Don't try this with hydrogen; you might explode like the Hindenburg.) What happens when you inhale carbon dioxide?

22–2 A musical interval of an octave corresponds to a factor of two in frequency. By what factor must the tension in a piano or violin string be increased to raise its pitch one octave?

22–3 The pitch (or frequency) of an organ pipe changes with temperature. Does it increase or decrease with increasing temperature? Why?

22–4 By touching a string lightly at its center while bowing, a violinist can produce a note exactly one octave above the note to which the string is tuned, i.e., a note with exactly twice the frequency. Why is this possible?

22–5 In most modern wind instruments the pitch is changed by changing the length of the vibrating air column with keys or valves. The bugle, however, has no valves or keys, yet it can play many notes. How? Are there restrictions on what notes it can play?

22–6 Kettledrums (tympani) have a pitch determined by the frequencies of the normal modes. How can a kettledrum be tuned?

22–7 Energy can be transferred along a string by wave motion. However, in a standing wave no energy can ever be transferred past a node. Why?

22–8 Can a standing wave be produced on a string by superposing two waves with the same frequency but different amplitudes, traveling in opposite directions?

22–9 In discussing standing longitudinal waves in an organ pipe, we spoke of pressure nodes and antinodes. What physical quantity is analogous to pressure for standing transverse waves on a stretched string?

22–10 A glass of water sits on a kitchen counter just above a dishwasher that is running. The surface of the water has a set of concentric stationary circular ripples. What is happening?

22–11 Consider the peculiar sound of a large bell. Do you think the overtones are harmonic?

22–12 What is the difference between music and noise?

22–13 Is the mass per unit length the same for all strings on a piano? On a guitar? Why?

22–14 When a heavy truck drives up a steep hill, the windows in nearby houses sometimes vibrate. Why?

22–15 A popular children's pastime is lining up a row of dominoes, standing on edge, so that when the first one is pushed over, the whole row falls, one after another. Is this a wave motion? In what respects is it similar to waves on a stretched string? In what respects is it different?

22–16 A series of traffic lights along a street is timed so that a car going at the right speed can hit a succession of green lights and not have to stop. If several cars are on the same street, does this situation have wavelike properties?

22–17 Some opera singers are reputed to be able to break a glass by singing the appropriate note. What physical phenomenon could account for this?

22–18 A pipe organ in a church is tuned when the temperature is 20°C. On a cold winter day when the temperature in the room is 10°C, the organ still sounds "in tune," despite the phenomenon mentioned in Question 22–3. Why?

EXERCISES

Unless otherwise indicated, assume the speed of sound in air to be $c = 345$ m·s^{-1}.

Section 22–2 Superposition and Standing Waves

22–1 Let
$$y_1(x, t) = A_1 \sin (\omega_1 t - k_1 x)$$
and
$$y_2(x, t) = A_2 \sin (\omega_2 t - k_2 x)$$
be two solutions to the wave equation, Eq. (21–14), for the *same c*. Show that $y(x, t) = y_1(x, t) + y_2(x, t)$ is also a solution to the wave equation.

22–2 Give the details of the derivation of Eq. (22–1) from $y_1 + y_2 = A[\sin (\omega t - kx) - \sin (\omega t + kx)]$.

22–3 Prove by direct substitution that
$$y = -[2A \cos \omega t] \sin kx$$
is a solution of the wave equation, for $c = \omega/k$.

Section 22–3 Normal Modes of a String

22–4 A piano tuner stretches a steel piano "string" 0.50 m long, of mass 5 g, with a tension of 400 N.

a) What is the frequency of its fundamental mode of vibration?

b) What is the number of the highest overtone that could be heard by a person capable of hearing frequencies up to 10,000 Hz?

22–5 A physics student observes that a stretched string vibrates with a frequency of 30 Hz in its fundamental mode when the supports are 0.60 m apart. The amplitude at the antinode is 3 cm. The string has a mass of 0.030 kg.

a) What is the speed of propagation of a transverse wave in the string?

b) Compute the tension in the string.

22–6

a) A string is vibrating in its fundamental mode. The waves have velocity c, frequency f, amplitude A, and wavelength λ. Calculate the maximum velocity and acceleration of points located at
   i) $x = \lambda/2$,   ii) $x = \lambda/4$,   iii) $x = \lambda/8$
   from the left-hand end of the string.

b) At each of the points in (a), what is the amplitude of the motion?

c) At each of the points in (a), how much time does it take the string to go from its largest upward displacement to its largest downward displacement?

Section 22–4 Longitudinal Standing Waves

Section 22–5 Normal Modes of Organ Pipes

22–7 The longest pipe in most medium-sized pipe organs is 4.88 m (16 ft) long. What is the frequency of the note corresponding to the fundamental mode if the pipe is

a) open at both ends?

b) open at one end, closed at the other?

22–8 Find the fundamental frequency and the first four overtones of a 15-cm pipe

a) if the pipe is open at both ends;

b) if the pipe is closed at one end.

c) How many overtones can be heard by a person with normal hearing for each of the above cases? (A person with normal hearing can hear frequencies in the range 20 Hz to 20,000 Hz.)

22–9 A certain pipe produces a frequency of 440 Hz in air. If the pipe is filled with helium at the same temperature, what frequency does it produce?

22–10 A standing wave of frequency 1100 Hz in a column of methane (CH_4) at 20°C produces nodes that are 0.20 m apart. What is the ratio of the heat capacity at constant pressure to that at constant volume?

22–11 A student in a physics lab sets up standing waves in a Kundt's tube (Fig. 22–7) by the longitudinal vibration of an iron rod 1 m long, clamped at the center. If the frequency of vibration of the iron rod is 2480 Hz and the powder heaps within the tube are 6.9 cm apart, what is the speed of the waves

a) in the iron rod?   b) in the gas?

PROBLEMS

22–12 A steel wire of length $L = 1.00$ m and density $\rho = 8000$ kg·m^{-3} is stretched tightly between two rigid supports. When the wire vibrates in its fundamental mode, the frequency is $f = 200$ Hz.

a) What is the speed of transverse waves on this wire?

b) What is the longitudinal stress (force per unit area) in the wire?

c) If the maximum acceleration at the midpoint of the wire is 800 m·s^{-2}, what is the amplitude of vibration at the midpoint?

22–13 A cellist tunes the A-string of her instrument to a fundamental frequency of 220 Hz. The vibrating portion of the string is 0.68 m long and has a mass of 1.29 g.

a) With what tension must it be stretched?

b) What percentage increase in tension is needed to increase the frequency from 220 Hz to 233 Hz, corresponding to a rise in pitch from A to A-sharp?

22–14 A solid aluminum sculpture is hung from a steel wire. The fundamental frequency for transverse standing waves on the wire is 300 Hz. The sculpture is then immersed in water so that one-half its volume is submerged. What is the new fundamental frequency?

22–15 A long tube contains air at a pressure of 1 atm and temperature of 77°C. The tube is open at one end and closed at the other by a movable piston. A tuning fork near the open end is vibrating with a frequency of 500 Hz. Resonance is produced when the piston is at distances 18.0, 55.5, and 93.0 cm from the open end.

a) From these measurements, what is the speed of sound in air at 77°C?

b) From the above result, what is the ratio of the specific heat capacities (γ) for air?

22–16 The atomic mass of iodine is 127 g·mol^{-1}. A standing wave in iodine vapor at 400 K produces nodes that are 6.77 cm apart when the frequency is 1000 Hz. Is iodine vapor monatomic or diatomic?

22–17 The frequency of middle C is 262 Hz.

a) If an organ pipe is open at both ends, what length must it have to produce this note at 20°C?

b) At what temperature will the frequency be 6% higher, corresponding to a rise in pitch from C to C-sharp?

22–18 The steel B string of an acoustic guitar is 63.5 cm long and has a diameter of 0.406 mm (16 gauge).

a) Under what tension must the string be placed to give a frequency for transverse waves of 247.5 Hz? Use 7.8 g·cm^{-3} as the density of steel, and assume that the string vibrates in its fundamental mode.

b) If the tension is changed by a small amount ΔS, the frequency changes by Δf. Show that

$$\frac{\Delta f}{f} = \frac{1}{2} \frac{\Delta S}{S}.$$

c) If the string is tuned indoors as in part (a), where the temperature is 25°C, and the guitar is then taken to an outdoor stage where $T = 15$°C, the frequency will change, with unpleasant results. Find Δf if the Young's modulus Y is 2.0×10^{11} Pa and the coefficient of linear expansion α is 1.2×10^{-5}(C°)$^{-1}$. Will the pitch be raised or lowered?

CHALLENGE PROBLEMS

22–19 Two identical loudspeakers are located at points A and B that are 2 m apart. The loudspeakers are driven by the same amplifier and produce sound waves of frequency 440 Hz. Take the speed of sound in air to be 340 m·s^{-1}. A small microphone is moved out from point B along a line perpendicular to the line connecting A and B (line BC in Fig. 22–13). At what distances from B will there be *destructive* interference?

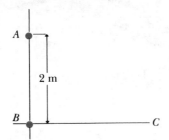

FIGURE 22–13

22–20 A simple example of resonance in a mechanical system is as follows. A mass m is attached to one end of a massless spring of force constant k and unstretched length l_0. The other end of the spring is free to turn about a nail

driven into a frictionless horizontal surface (Fig. 22–14). The mass is then made to revolve in a circle with an angular frequency of revolution ω'. (Refer to Challenge Problem 6–44.)

a) Calculate the length l of the spring as a function of ω'.

b) What happens to the result in (a) when ω' approaches the natural frequency $\omega = \sqrt{k/m}$ of the mass-spring system. (If your result bothers you, remember that precisely massless springs and frictionless surfaces do not exist but can only be approximated. Also, Hooke's law is itself only an approximation to the way real springs behave; the greater the elongation of the spring, the greater the deviation from Hooke's law.)

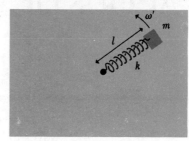

FIGURE 22–14

23

ACOUSTIC PHENOMENA

IN THIS CHAPTER WE ARE CONCERNED PRIMARILY WITH LONGITUDINAL waves in air, commonly called **sound.** The human ear is sensitive to waves in the frequency range from about 20 to 20,000 Hz, although the term *sound* is sometimes used for similar waves with frequencies outside the range of human hearing. Properties of sound waves include frequency, amplitude, and intensity. We will study the relations among displacement, pressure variation, and intensity, and the connections between these quantities and human sound perception. The interference of two sound waves differing in frequency by a few hertz causes phenomena called *beats*. When the source of sound or the observer, or both, are in motion relative to the air, frequency shifts known as the *Doppler effect* are observed. We will examine both of these phenomena.

23–1 SOUND WAVES

The simplest sound waves are sinusoidal waves with definite frequency, amplitude, and wavelength. When such a wave arrives at the ear, the air particles at the eardrum vibrate with definite frequency and amplitude. This vibration may also be described in terms of the variation of *air pressure* at the same point. The pressure fluctuates above and below atmospheric pressure with a sinusoidal variation having the same frequency as the motions of the air particles.

A sinusoidal sound wave in an elastic medium is described by a wave function of the form introduced in Section 21–3, Eq. (21–7):

$$y = A \sin (\omega t - kx), \qquad (23–1)$$

where y is the displacement from equilibrium of a point in the medium, and the amplitude A is, as usual, the *maximum* displacement from equilibrium.

From a practical standpoint it is nearly always easier to measure the *pressure* variations in a sound wave than to measure the displacements, so it is worthwhile to develop a relation between the two. Let p be the instantaneous pressure fluctuation at any point; that is, the amount by which the pressure *differs* from normal atmospheric pressure. If the displacements of two neighboring points x and $x + \Delta x$ are the same, the air between these points is neither compressed nor expanded, there is no volume change, and consequently

$p = 0$. Only when y varies from one point to a neighboring point is there a change of volume and therefore of pressure.

The fractional volume change $\Delta V/V$ in a volume element near point x turns out to be given simply by $\partial y/\partial x$, which is the rate of change of y with x as we go from one point to a neighboring point. To see why this is so, consider an imaginary cylinder of air, as in Fig. 23–1, with cross-sectional area A and axis along the direction of propagation. The color cylinder shows the undisplaced position, and the dashed lines show the displaced position. When no sound disturbance is present, the cylinder's length is Δx and its volume is $V = A\,\Delta x$. When a wave is present, the end of the cylinder initially at x is displaced a distance $y_1 = y(x, t)$, and the end initially at $x + \Delta x$ is displaced a distance $y_2 = y(x + \Delta x, t)$. The change in volume ΔV of this element is

$$A(y_2 - y_1) = A[y(x + \Delta x, t) - y(x, t)],$$

and in the limit as $\Delta x \to 0$, the fractional change in volume $\Delta V/V$ is given by

$$\frac{\Delta V}{V} = \frac{y(x + \Delta x, t) - y(x, t)}{\Delta x} = \frac{\partial y}{\partial x}. \qquad (23\text{–}2)$$

Now from the definition of the bulk modulus B, Eq. (12–7), $p = -B\,\Delta V/V$, and we find

$$p = -B\left(\frac{\partial y}{\partial x}\right). \qquad (23\text{–}3)$$

The negative sign arises because, when $\partial y/\partial x$ is positive, the displacement is greater at $x + \Delta x$ than at x, corresponding to an increase in volume and a *decrease* in pressure. For the sinusoidal wave of Eq. (23–1) we find

$$p = BkA \cos(\omega t - kx). \qquad (23\text{–}4)$$

This expression shows that the quantity BkA represents the maximum pressure variation. This maximum is called the **pressure amplitude** and is denoted by p_{max}. Thus

$$p_{max} = BkA. \qquad (23\text{–}5)$$

The pressure amplitude is directly proportional to the displacement amplitude A, as might be expected, and it also depends on wavelength. Waves of shorter wavelength (larger k) have greater pressure variations for a given amplitude because the maxima and minima are squeezed closer together.

> The pressure variation is proportional to the rate of change of displacement with position.

> The pressure amplitude is proportional to the displacement amplitude.

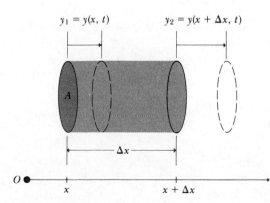

$y_1 = y(x, t)$ $y_2 = y(x + \Delta x, t)$

23–1 A cylindrical volume of gas with cross-sectional area A. The length in the undisplaced position is Δx. During wave propagation along the axis, the left end is displaced to the right a distance y_1, and the right end is displaced a different distance y_2. The resulting change in volume is $A(y_2 - y_1)$.

EXAMPLE 23–1   Measurements of sound waves show that the maximum pressure variations in the loudest sounds that the ear can tolerate without pain are of the order of 30 Pa (above and below atmospheric pressure, which is about 100,000 Pa). Find the corresponding maximum displacement if the frequency is 1000 Hz and $c = 350 \text{ m·s}^{-1}$.

SOLUTION   We have $\omega = (2\pi)(1000 \text{ Hz}) = 6283 \text{ s}^{-1}$ and

$$k = \frac{\omega}{c} = \frac{6283 \text{ s}^{-1}}{350 \text{ m·s}^{-1}} = 18.0 \text{ m}^{-1}.$$

The adiabatic bulk modulus for air is

$$B = \gamma p = (1.4)(1.01 \times 10^5 \text{ Pa}) = 1.42 \times 10^5 \text{ Pa}.$$

The displacement amplitude of even the loudest sound is very small.

From Eq. (23–5) we find

$$A = \frac{p_{max}}{Bk} = \frac{(30 \text{ Pa})}{(1.42 \times 10^5 \text{ Pa})(18.0 \text{ m}^{-1})}$$

$$= 1.18 \times 10^{-5} \text{ m} = 0.0118 \text{ mm}.$$

Thus the displacement amplitude of even the loudest sound is *extremely* small. The maximum pressure variation in the *faintest* audible sound of frequency 1000 Hz is only about 3×10^{-5} Pa. The corresponding displacement amplitude is about 10^{-9} cm. For comparison, the wavelength of yellow light is 6×10^{-5} cm, and the diameter of a molecule is about 10^{-8} cm. The ear is an extremely sensitive organ!

23–2 INTENSITY

Intensity: a concept to describe the energy transported by a sound wave

An essential aspect of wave propagation of all sorts is transfer of energy. A familiar example is the energy supply of the earth, which reaches us from the sun via electromagnetic waves. The **intensity** I of a traveling wave is defined as *the time average rate at which energy is transported by the wave, per unit area,* across a surface perpendicular to the direction of propagation. More briefly, the intensity is the average *power* transported per unit area.

We have seen that the power developed by a force equals the product of force and velocity. Hence the power *per unit area* in a sound wave equals the product of the excess pressure (force per unit area), given by Eq. (23–4), and the *particle* velocity v, obtained by taking the time derivative of Eq. (23–1). We find

$$v = \omega A \cos (\omega t - kx), \tag{23–6}$$

$$pv = \omega B k A^2 \cos^2 (\omega t - kx). \tag{23–7}$$

The intensity is, by definition, the average value of this quantity. The average value of the function $\cos^2 z$ is 1/2, so we find

$$I = \frac{1}{2} \omega B k A^2. \tag{23–8}$$

By using the relations $\omega = ck$ and $c^2 = B/\rho$, we can transform this into the form

$$I = \frac{1}{2} \sqrt{\rho B} \omega^2 A^2,$$

which we cited at the end of Section 21–8.

It is usually more convenient to express I in terms of the pressure amplitude p_{max}. Using Eq. (23–5) and the relation $\omega = ck$, we find

$$I = \frac{\omega p_{max}^2}{2Bk} = \frac{c p_{max}^2}{2B}. \tag{23–9}$$

By using the wave speed relation $c^2 = B/\rho$, we can also write this in the alternative forms

$$I = \frac{p_{max}^2}{2\rho c} = \frac{p_{max}^2}{2\sqrt{\rho B}}. \tag{23–10}$$

The intensity of a sound wave of the largest amplitude tolerable to the human ear (about $p_{max} = 30$ Pa) is

$$I = \frac{(30 \text{ Pa})^2}{2(1.22 \text{ kg·m}^{-3})(346 \text{ m·s}^{-1})}$$
$$= 1.07 \text{ J·s}^{-1}\text{·m}^{-2} = 1.07 \text{ W·m}^{-2}$$
$$= 1.07 \times 10^{-4} \text{ W·cm}^{-2}.$$

The unit 1 W·cm^{-2} is a mixed unit, neither cgs nor mks. We mention it here because it is unfortunately in general use among acousticians.

The pressure amplitude of the *faintest* sound wave that can be heard is about 3×10^{-5} Pa, and the corresponding intensity is about 10^{-12} W·m^{-2} or 10^{-16} W·cm^{-2}.

The *total* power carried across a surface by a sound wave equals the product of the intensity at the surface and the surface area, if the intensity over the surface is uniform. The average total sound power emitted by a person speaking in a conversational tone is about 10^{-5} W, while a loud shout corresponds to about 3×10^{-2} W. If all the residents of New York City were to talk at the same time, the total sound power would be about 100 W, equivalent to the electric-power requirement of a medium-sized light bulb. Yet the power required to fill a large auditorium with loud sound is considerable. Suppose the sound intensity over the surface of a hemisphere 20 m in radius is 1 W·m^{-2}. The area of the surface is about 2500 m^2. Hence the acoustic power output of a speaker at the center of the sphere would have to be

$$(1 \text{ W·m}^{-2})(2500 \text{ m}^2) = 2500 \text{ W},$$

or 2.5 kW. The electric-power input to the speaker would need to be considerably larger, since the efficiency of such devices is not very high.

Because of the extremely large range of intensities over which the ear is sensitive, a *logarithmic* rather than an arithmetic intensity scale is convenient. The **intensity level** β of a sound wave is defined by the equation

$$\beta = 10 \log \frac{I}{I_0}, \tag{23–11}$$

where I_0 is an arbitrary reference intensity, taken as 10^{-12} W·m^{-2}. This value corresponds roughly to the faintest sound that can be heard. Intensity levels are expressed in **decibels,** abbreviated dB. A decibel is 1/10 of a bel, a unit named for Alexander Graham Bell. The bel is inconveniently large for most purposes, and the decibel is the usual unit of sound intensity level.

If the intensity of a sound wave equals I_0 or 10^{-12} W·m^{-2}, its intensity level is 0 dB. The maximum intensity that the ear can tolerate without pain is

It doesn't take very much energy to make a very loud sound.

The decibel: a logarithmic scale that describes a very wide range of sound intensities

TABLE 23–1   Noise Levels Due To Various Sources (Representative Values)

Source or Description of Noise	Noise Level, dB	Intensity, $W \cdot m^{-2}$
Threshold of pain	120	1
Riveter	95	3.2×10^{-3}
Elevated train	90	10^{-3}
Busy street traffic	70	10^{-5}
Ordinary conversation	65	3.2×10^{-6}
Quiet automobile	50	10^{-7}
Quiet radio in home	40	10^{-8}
Average whisper	20	10^{-10}
Rustle of leaves	10	10^{-11}
Threshold of hearing	0	10^{-12}

about 1 $W \cdot m^{-2}$, which corresponds to an intensity level of 120 dB. Table 23–1 gives the intensity levels in decibels of several familiar noises. It is taken from a survey made by the New York City Noise Abatement Commission.

The relation between sound intensity and human hearing: What is the smallest intensity that can be heard? The loudest that can be tolerated?

Within the range of audibility, the sensitivity of the ear varies with frequency. The **threshold of audibility** at any frequency is the minimum intensity of sound at that frequency that can be detected. For a young adult with normal hearing, the threshold of audibility at 1000 Hz is about 0 dB; at 200 and 15,000 Hz it is about 20 dB; and at 50 and 18,000 Hz it is about 50 dB. Thus the ear's sensitivity drops off at the low and high ends of the frequency scale. Frequencies above 20,000 Hz (20 kHz) are not audible to humans at *any* intensity, and such frequencies are referred to as **ultrasonic.**

Is sound still sound if you can't hear it?

PROBLEM-SOLVING STRATEGY

1. Quite a few quantities are involved in characterizing the amplitude and intensity of a sound wave, and it's easy to get lost in the maze of relationships. It helps to put them in categories: The amplitude is described by A or p_{max}; the frequency by f, ω, k, or λ. These quantities are related through the wave speed c, which in turn is determined by the properties of the medium, B and ρ. Take a hard look at the problem at hand; identify which of these quantities are given and which you have to find; then start looking for relationships that take you where you want to go.

2. In using Eq. (23–11) for the sound intensity level, remember that I and I_0 must be in the same units, usually $W \cdot m^{-2}$. If they aren't, convert!

Frequency ratios are the basis of consonance and dissonance in music.

The **pitch** of a musical tone is determined by its frequency, and musical intervals can be defined in terms of frequency ratios. For example, middle C on the piano has a frequency of 262 Hz, and the C an octave higher has a frequency of 524 Hz, a factor of two larger. An octave is always a pair of tones having a frequency ratio of 2:1, and a perfect fifth corresponds to a ratio of 3:2. A major triad consists of three tones with frequency ratios of 4:5:6. These and similar relationships can be used to explore the physical basis of consonance and dissonance of combinations of musical tones.

23–3 BEATS

In Section 22–6 we discussed *interference* effects that occur when two different waves overlap in the same region of space. In that discussion we considered two waves with the same frequency. Now let us examine an interference effect resulting from the overlap of two waves with equal amplitude but slightly different frequency. This occurs, for example, when two tuning forks with slightly different frequencies are sounded together, or when two organ pipes that are supposed to have exactly the same frequency are slightly "out of tune."

Consider a particular point in space where the two waves overlap. The displacements of the individual waves at this point are plotted as functions of time in Fig. 23–2a. If the total length of the time axis represents about one second, the frequencies are 16 Hz and 18 Hz. Applying the principle of superposition, we add the two displacements at each instant of time to find the total displacement at that time, obtaining the graph of Fig. 23–2b. At certain times the two waves are in phase; their maxima coincide and their amplitudes add. But as time goes on, they become more and more out of phase because of their slightly different frequencies. Eventually a maximum of one wave coincides with a maximum in the opposite direction for the other wave. The two waves then cancel each other, and the total amplitude is zero.

Beats: superposing two waves with slightly different frequencies

Thus the amplitude of the composite wave varies from a maximum value to zero and back, as shown in Fig. 23–2b. The appearance is that of a single sinusoidal wave with a varying amplitude. In this example, the amplitude goes through two maxima and two minima in one second, so the frequency of this amplitude variation is 2 Hz. The amplitude variation causes variations of loudness called **beats,** and the frequency with which the amplitude varies is called the **beat frequency.** In this example the beat frequency is the *difference* of the two frequencies. If the beat frequency is a few hertz, it is perceived as a waver or pulsation in the tone.

We can prove that the beat frequency is *always* the difference of the two frequencies f_1 and f_2. Suppose f_1 is larger than f_2; the corresponding periods are τ_1 and τ_2, with $\tau_1 < \tau_2$. If the two waves start out in phase at time $t = 0$, they will again be in phase at a time T such that the first wave has gone through exactly one more cycle than the second. Let n be the number of cycles of the first wave in time T; then the number of cycles of the second wave in the same time is $(n - 1)$, and we have the relations

$$T = n\tau_1 = (n - 1)\tau_2.$$

We solve the second equation for n and substitute the result back into the first

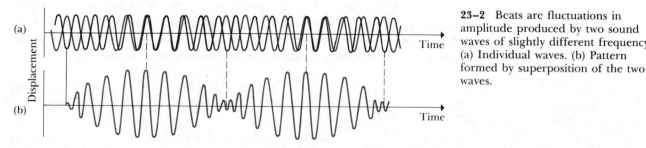

(a)

Displacement

(b)

Time

Time

23–2 Beats are fluctuations in amplitude produced by two sound waves of slightly different frequency. (a) Individual waves. (b) Pattern formed by superposition of the two waves.

equation, obtaining

$$T = \frac{\tau_1 \tau_2}{\tau_2 - \tau_1}.$$

Now T is just the *period* of the beat, and its reciprocal is the beat *frequency*, $f_{\text{beat}} = 1/T$. Thus we find

$$f_{\text{beat}} = \frac{\tau_2 - \tau_1}{\tau_1 \tau_2} = \frac{1}{\tau_1} - \frac{1}{\tau_2},$$

and finally

$$f_{\text{beat}} = f_1 - f_2. \tag{23–12}$$

When a musical instrument sounds wobbly and out of tune, you may be hearing beats.

Beats between two tones can be detected by the ear up to a beat frequency of 6 or 7 Hz. Two piano strings or two organ pipes differing in frequency by 2 or 3 Hz sound wavery and "out of tune," although some organ stops contain two sets of pipes deliberately tuned to beat frequencies of about 1 Hz to 2 Hz for a gently undulating effect. Listening for beats is an important technique in tuning all musical instruments.

At higher-frequency differences, individual beats can no longer be distinguished. The sensation then merges into one of *consonance* or *dissonance*, depending on the frequency ratio of the two tones. In some cases the ear perceives a tone having a pitch corresponding to the difference of frequency of the two tones. Such a tone is called a *difference* tone.

23–4 THE DOPPLER EFFECT

Why does the pitch of a tone sound different when you are moving toward or away from the source?

When a source of sound or a listener, or both, are in motion relative to the air, the pitch of the sound, as heard by the listener, is in general not the same as when source and listener are at rest. This phenomenon is called the **Doppler effect**. A common example is the sudden drop in pitch of the sound from an automobile horn as one meets and passes a car proceeding in the opposite direction.

Let v_L and v_S represent the velocities of a listener and a source, relative to the air. We will consider only the special case in which the velocities lie along the line joining listener and source. Since these velocities may be in the same or opposite directions, and the listener may be either ahead of or behind the source, we need a sign convention. We will take the positive directions of v_L and v_S as that *from* the position of the listener, toward the position of the source. The speed of propagation of sound waves, c, will always be considered positive.

Consider first a listener L moving with velocity v_L toward a stationary source S, as in Fig. 23–3. The source emits a wave with frequency f_S and wavelength $\lambda = c/f_S$. The figure shows several wave crests, separated by equal distances λ. The waves approaching the moving listener have a speed of propagation *relative to the listener* of $(c + v_L)$. Thus the frequency f_L with which the listener encounters wave crests—that is, the frequency heard—is

$$f_L = \frac{c + v_L}{\lambda} = \frac{c + v_L}{c/f_S}, \tag{23–13}$$

or

$$f_L = f_S \left(\frac{c + v_L}{c} \right) = f_S \left(1 + \frac{v_L}{c} \right). \tag{23–14}$$

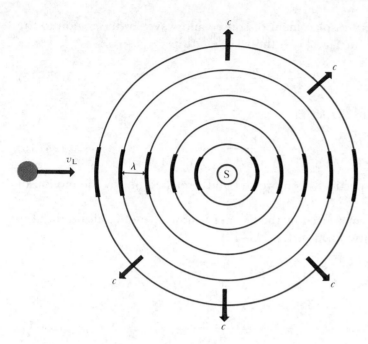

23–3 A listener moving toward a stationary source hears a frequency higher than the source frequency because the relative velocity of listener and wave is greater than c.

Hence an observer moving toward a source hears a larger frequency and higher pitch than a stationary observer. Similarly, a listener moving away from the source ($v_L < 0$) hears a lower pitch.

Now suppose the source is also moving with velocity v_S, as in Fig. 23–4. The wave speed relative to the air is still c; the speed of propagation of a wave is not altered by the motion of the source but is a property of the wave medium alone. But the wavelength is no longer given by c/f_S. The time for emission of one cycle of the wave is the period $\tau = 1/f_S$. During this time the wave travels a distance $c\tau = c/f_S$, and the source moves a distance $v_S\tau = v_S/f_S$. The wavelength is the distance between successive wave crests, and this is deter-

When the source moves relative to the medium, the wavelength changes.

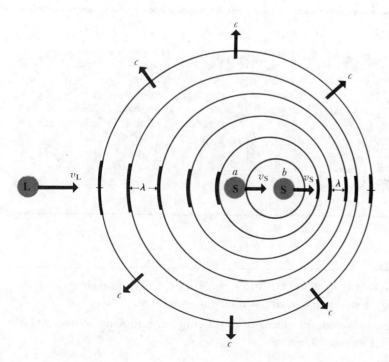

23–4 Wave surfaces emitted by a moving source are crowded together in front of the source and stretched out behind it.

mined by the *relative* displacement of source and wave. In the region to the right of the source in Fig. 23–4, the wavelength is

$$\lambda = \frac{c}{f_S} - \frac{v_S}{f_S} = \frac{c - v_S}{f_S}, \tag{23–15}$$

and in the region to the left it is

$$\lambda = \frac{c + v_S}{f_S}. \tag{23–16}$$

The waves are, respectively, compressed and stretched out by the motion of the source.

The frequency measured by the listener is now given by substituting Eq. (23–16) into the first form of Eq. (23–13):

$$f_L = \left(\frac{c + v_L}{\lambda}\right) = \frac{c + v_L}{(c + v_S)/f_S},$$

or

$$\frac{f_L}{c + v_L} = \frac{f_S}{c + v_S}, \tag{23–17}$$

which expresses the frequency f_L heard by the listener in terms of the frequency f_S of the source.

This general relation includes all possibilities for collinear motion of source and listener relative to the medium. If the listener is at rest in the medium, the corresponding velocity is zero, and of course when both source and listener are at rest or have the same velocity relative to the medium, then $f_L = f_S$. Whenever source or listener is moving in the opposite direction to what we have designated as positive, the corresponding velocity to be used in Eq. (23–17) is negative. The examples illustrate these sign conventions.

Both the source and the listener may be moving with respect to the medium.

PROBLEM-SOLVING STRATEGY: Doppler effect

1. Establish a coordinate system, with the positive direction from the listener toward the source, and make sure you know the signs of all the relevant velocities, using the sign convention described above.

2. Use consistent notation to identify the various quantities: subscript S for source, L for listener.

3. When a wave is reflected from a surface, either stationary or moving, the analysis can be carried out in two stages. First find the frequency with which the wave crests arrive at the surface; this is f_L. Then think of the surface as a new source, emitting waves with this same frequency f_L, which becomes the frequency of the new source. Finally, determine what frequency is heard by a listener detecting this new wave.

Some examples of frequency shifts caused by the Doppler effect

EXAMPLE 23–2 Let $f_S = 300$ Hz and $c = 300$ m·s^{-1}. The wavelength of the waves emitted by a stationary source is then $c/f_S = 1.00$ m.

a) What are the wavelengths ahead of and behind the moving source in Fig. 23–4 if its velocity is 30 m·s^{-1}?

In front of the source,

$$\lambda = \frac{c - v_S}{f_S} = \frac{300 \text{ m·s}^{-1} - 30 \text{ m·s}^{-1}}{300 \text{ Hz}} = 0.90 \text{ m}.$$

Behind the source,

$$\lambda = \frac{c + v_S}{f_S} = \frac{300 \text{ m·s}^{-1} + 30 \text{ m·s}^{-1}}{300 \text{ Hz}} = 1.10 \text{ m}.$$

b) If the listener L in Fig. 23–4 is at rest and the source is moving away from L at 30 m·s^{-1}, what is the frequency as heard by the listener?

Since

$$v_L = 0 \quad \text{and} \quad v_S = 30 \text{ m·s}^{-1},$$

we have

$$f_L = f_S \frac{c}{c + v_S}$$

$$= 300 \text{ Hz} \left(\frac{300 \text{ m·s}^{-1}}{300 \text{ m·s}^{-1} + 30 \text{ m·s}^{-1}} \right) = 273 \text{ Hz}.$$

c) If the source in Fig. 23–4 is at rest and the listener is moving toward the left at 30 m·s^{-1}, what is the frequency as heard by the listener?

The positive direction (from listener to source) is still from left to right, so

$$v_L = -30 \text{ m·s}^{-1}, \qquad v_S = 0,$$

$$f_L = f_S \frac{c + v_L}{c} = 300 \text{ Hz} \left(\frac{300 \text{ m·s}^{-1} - 30 \text{ m·s}^{-1}}{300 \text{ m·s}^{-1}} \right) = 270 \text{ Hz}.$$

Thus, while the frequency f_L heard by the listener is less than the frequency f_S both when the source moves away from the listener and when the listener moves away from the source, the decrease in frequency is not the same for the same speed of recession.

In the preceding equations, the velocities v_L, v_S, and c are all *relative to the air*, or more generally, to the medium in which the waves are traveling. The Doppler effect exists also for electromagnetic waves in empty space, such as light waves or radio waves. In this case there is no "medium" relative to which a velocity can be defined, and we can speak only of the *relative* velocity v *of source and receiver*.

For light waves there is no material medium, but there is still a Doppler effect.

To derive the expression for the Doppler frequency shift for light requires the use of relativistic kinematic relations. These will be derived in Chapter 40, but meanwhile we quote the result without derivation. The wave speed c is the speed of light and is the same for both source and listener. In the frame of reference in which the listener is at rest, the source is moving away from the listener with velocity v. (If, instead, the source is *approaching* the listener, v is negative.) The source frequency is again f_S. The frequency f_L measured by the listener (i.e., the frequency of arrival of the waves at L) is then given by

Working out the Doppler effect for light requires methods of the special theory of relativity.

$$f_L = \left(\sqrt{\frac{c - v}{c + v}} \right) f_S. \qquad (23\text{–}18)$$

When v is positive, the source moves *away* from the listener and f_L is always *less* than f_S; when v is negative, the source moves *toward* the listener and f_L is *greater* than f_S. Thus the qualitative effect is the same as for sound, although the quantitative relationship is different.

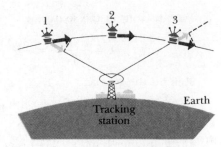

23–5 Change of velocity component along the line of sight of a satellite passing a tracking station.

The Doppler effect provides a convenient means of tracking a satellite that is emitting a radio signal of constant frequency f_S. The frequency f_L of the signal received on the earth decreases as the satellite is passing, since the velocity component *toward* the tracking station decreases from position 1 to position 2 in Fig. 23–5 and then points *away* from the earth from 2 to 3. If the received signal is combined with a constant signal generated in the receiver to give *beats,* then the beat frequency may be such as to produce an audible note whose pitch changes as the satellite passes overhead.

A similar technique is used by law-enforcement officers to measure automobile speeds. An electromagnetic wave is emitted by a source (sometimes called a "radar gun"), typically attached to a police car, at the side of the road. The wave is reflected from a moving car, which thus acts as a moving source; the reflected wave is Doppler-shifted in frequency. Measuring the frequency shift by using beats, as with satellite tracking, permits simple measurement of the speed.

The Doppler effect for *light* is important in astronomy. Analysis of light emitted by elements in distant stars shows shifts in wavelength compared to light from the same elements on earth. These shifts can be interpreted as Doppler shifts due to motion of the stars. The shift is nearly always toward the longer wavelength, or red end of the spectrum, and is therefore called the *red shift*. Such observations have provided practically all the evidence for the "exploding universe" cosmological theories, which represent the universe as having evolved from a great explosion several billion years ago in a relatively small region of space.

> How do policemen use the Doppler effect to discourage drivers from exceeding the speed limit?

> Astronomers use the Doppler effect to observe the motions of distant celestial objects.

SUMMARY

KEY TERMS

sound

pressure amplitude

intensity

intensity level

decibels

threshold of audibility

ultrasonic

pitch

beats

beat frequency

Doppler effect

Sound consists of longitudinal waves in air. A sinusoidal sound wave is characterized by its frequency f, wavelength λ, and amplitude A. The amplitude is also related to the pressure amplitude p_{max} by

$$p_{max} = BkA, \qquad (23–5)$$

where B is the bulk modulus and k is the wave number.

The intensity I of a wave is the time average rate at which energy is transported by the wave, per unit area. The intensity of a sound wave can be expressed in terms of the amplitude A as

$$I = \frac{1}{2}\omega BkA^2, \qquad (23–8)$$

or in terms of the pressure amplitude as

$$I = \frac{cp_{max}^2}{2B} = \frac{p_{max}^2}{2\rho c}. \qquad (23–9)$$

The intensity level β of a sound wave is defined as

$$\beta = 10 \log \frac{I}{I_0}, \qquad (23\text{–}11)$$

where I_0 is an arbitrary intensity defined to be 10^{-12} W·m^{-2}. Intensity levels are expressed in decibels (dB).

The threshold of audibility at any frequency is the minimum intensity of sound that can be heard. The threshold of pain is the intensity at which the perception changes from hearing to feeling or pain.

Beats are heard when two tones with slightly different frequencies are sounded together. The beat frequency is the difference of the two frequencies.

The Doppler effect is the frequency shift that occurs when there is relative motion of a source of sound and a listener. The source and listener frequencies f_S and f_L and their velocities v_S and v_L are related by

$$\frac{f_L}{c + v_L} = \frac{f_S}{c + v_S}. \qquad (23\text{–}17)$$

QUESTIONS

23–1 Ultrasonic cleaners use ultrasonic waves of high intensity in water or a solvent to clean dirt from dishes, engine parts, and so on. How do they work? What advantages and disadvantages does this process have, compared with other cleaning methods?

23–2 Lane dividers on highways sometimes have regularly spaced ridges or ripples. When the tires of a moving car roll along such a divider, a musical note is produced. Why? Could this phenomenon be used to measure the car's speed? How?

23–3 Two tuning forks have identical frequencies, but one is stationary while the other is mounted on a rotating record turntable. What does a listener hear?

23–4 The organist in a cathedral plays a loud chord and then releases it. The sound persists for a few seconds and gradually dies away. Why does it persist? What happens to the energy when it dies away?

23–5 Why do foghorns always have very low pitches?

23–6 Some stereo amplifiers intended for home use have a maximum power output of 600 W or more. What would happen if you had 600 W of actual sound power in a moderate-sized room?

23–7 Why does your voice sound different over the telephone than in person?

23–8 Why does a guitar have a sharper, more metallic sound when played with a hard pick than when plucked with the bare fingertips?

23–9 When a record is being played with the volume control turned down all the way, a faint sound can be heard coming directly from the stylus. It seems to be all treble and no bass. How is the sound produced, and why are the frequencies so unbalanced?

23–10 The tone quality of a violin is different when the bow is near the bridge (the ends of the strings) than when it is nearer the centers of the strings. Why?

23–11 An engineer who likes to express everything in technical terms remarked that the price of gasoline rose by 2 dB in 1979. What do you think he meant?

23–12 When you are shouting at someone a fair distance away, it is easier for him to hear you if the wind is blowing from you to him, than if it is in the opposite direction. What is the physical basis for this difference?

23–13 Two notes an octave apart on a piano have a frequency ratio of 2:1. When one is slightly out of tune, beats are heard. How are they produced?

23–14 A large church has part of the organ in front and part in back. When both divisions are playing at once and a person walks rapidly down the center aisle, the two divisions sound out of tune. Why?

23–15 Can you think of circumstances in which a Doppler effect would be observed for surface waves in water? For elastic waves propagating in a body of water?

23–16 How does a person perceive the direction from which a sound comes? Is our directional perception more acute for some frequency ranges than for others? Why?

23–17 A sound source and a listener are both at rest on the earth, but a strong wind is blowing. Is there a Doppler effect?

EXERCISES

Unless indicated otherwise, assume the speed of sound in air to be $c = 345$ m·s^{-1}.

Section 23–1 Sound Waves

Section 23–2 Intensity

23–1 Consider a sound wave in air that has displacement amplitude 0.01 mm. Calculate the pressure amplitude for frequencies of

a) 500 Hz;

b) 20,000 Hz (not audible).

In each case compare the results to the pain threshold given in Example 23–1.

23–2

a) If the pressure amplitude in a sound wave is tripled, by what factor is the intensity of the wave increased?

b) By what factor must the pressure amplitude of a sound wave be increased in order to increase the intensity by a factor of 16?

23–3

a) Two sound waves of the same frequency, one in air and one in water, are equal in intensity. What is the ratio of the pressure amplitude of the wave in water to that of the wave in air?

b) If the pressure amplitudes of the waves are equal, what is the ratio of their intensities?

23–4 Derive Eq. (23–9) from the preceding equations.

23–5 A sound wave in air has frequency 400 Hz, wave velocity 345 m·s^{-1}, and displacement amplitude 0.005 mm. Calculate the intensity (in W·m^{-2}) and intensity level (in decibels) for this sound wave.

23–6

a) Relative to the arbitrary reference intensity of 10^{-12} W·m^{-2}, what is the intensity level in decibels of a sound wave whose intensity is 10^{-6} W·m^{-2}?

b) What is the intensity level of a sound wave in air whose pressure amplitude is 0.2 Pa?

23–7

a) Show that if β_1 and β_2 are the intensity levels in decibels of sounds of intensities I_1 and I_2, respectively, the difference in intensity levels of the sounds is

$$\beta_2 - \beta_1 = 10 \log \frac{I_2}{I_1}.$$

b) Show that if $(p_{max})_1$ and $(p_{max})_2$ are the pressure amplitudes of two sound waves, the difference in intensity levels of the waves is

$$\beta_2 - \beta_1 = 20 \log \frac{(p_{max})_2}{(p_{max})_1}.$$

c) Show that if the reference level of intensity is $I_0 = 10^{-12}$ W·m^{-2}, the intensity level of a sound of intensity I (in W·m^{-2}) is

$$\beta = 120 + 10 \log I.$$

23–8 The intensity due to a number of independent sound sources is the sum of the individual intensities. How many decibels greater is the intensity level when all five quintuplets cry simultaneously than when a single one cries? How many more crying babies would be required to produce a further increase in the intensity level of an equal number of decibels?

Section 23–3 Beats

23–9 A trumpet player is tuning his instrument by playing an A note simultaneously with the first-chair trumpeter, who has perfect pitch. The first-chair player's note is exactly 440 Hz, and 3.6 beats per second are heard. What is the frequency of the other player's note?

23–10 Two identical piano strings, when stretched with the same tension, have a fundamental frequency of 440 Hz. By what fractional amount must the tension in one string be increased so that four beats per second will occur when both strings vibrate simultaneously?

Section 23–4 The Doppler Effect

23–11 A railroad train is traveling at 30 m·s^{-1} in still air. The frequency of the note emitted by the locomotive whistle is 500 Hz. What is the wavelength of the sound waves

a) in front of the locomotive?

b) behind the locomotive?

What would be the frequency of the sound heard by a stationary listener

c) in front of the locomotive?

d) behind the locomotive?

What frequency would be heard by a passenger on a train moving in the direction opposite to the first at 15 m·s^{-1} and

e) approaching the first?

f) receding from the first?

g) How would each of the preceding answers be altered if a wind speed of 10 m·s^{-1} were blowing in the same direction as that in which the first locomotive was traveling?

23–12 Two whistles, A and B, each have a frequency of 500 Hz. A is stationary and B is moving toward the right (away from A) at a speed of 50 m·s^{-1}. An observer is between the two whistles, moving toward the right with a speed of 25 m·s^{-1}.

a) What is the frequency from A as heard by the observer?

b) What is the frequency from B as heard by the observer?

c) What is the beat frequency heard by the observer?

PROBLEMS

23–13 A certain sound source radiates uniformly in all directions in air. At a distance of 5 m the sound level is 80 db. The frequency is 440 Hz.

a) What is the pressure amplitude at this distance?

b) What is the displacement amplitude?

c) At what distance is the sound level 60 db?

23–14 A window whose area is 1 m² opens on a street where the street noises result in an intensity level, at the window, of 60 db. How much "acoustic power" enters the window via the sound waves?

23–15 A very noisy chain saw operated by a tree surgeon emits a total sound power of 10 W uniformly in all directions. At what distance from the source is the sound level

a) 100 db?

b) 60 db?

23–16 Two loudspeakers, A and B, radiate sound uniformly in all directions. The output of acoustic power from A is 8×10^{-4} W, and from B it is 13.5×10^{-4} W. Both loudspeakers are vibrating in phase at a frequency of 172.5 Hz.

a) Determine the difference in phase of the two signals at a point C along the line joining A and B, 3 m from B and 4 m from A.

b) Determine the intensity at C from speaker A if speaker B is turned off, and the intensity at C from speaker B if speaker A is turned off.

c) With both speakers on, what are the intensity and intensity level at C?

23–17 The frequency ratio of a half-tone interval on the equally tempered scale is 1.059. Find the speed of an automobile passing a listener at rest in still air, if the pitch of the car's horn drops a half-tone between the times when the car is coming directly toward him and when it is moving directly away from him.

23–18

a) Show that Eq. (23–18) can be written

$$f_L = f_S \left(1 - \frac{v}{c}\right)^{1/2} \left(1 + \frac{v}{c}\right)^{-1/2}.$$

b) Use the binomial theorem to show that if $v \ll c$, this

expression is approximately equal to

$$f_L = f_S \left(1 - \frac{v}{c}\right).$$

c) An earth satellite emits a radio signal of frequency 10^8 Hz. An observer on the ground detects beats between the received signal and a local signal also of frequency 10^8 Hz. At a particular moment, the beat frequency is 2400 Hz. What is the component of the satellite's velocity directed toward the earth at this moment?

23–19 A sound wave of frequency f_0 and wavelength λ_0 travels horizontally toward the right. It strikes and is reflected from a large, rigid, vertical plane surface, perpendicular to the direction of propagation of the wave and moving toward the left with a speed v.

a) How many positive wave crests strike the surface in a time interval t?

b) At the end of this time interval, how far to the left of the surface is the wave that was reflected at the beginning of the time interval?

c) What is the wavelength of the reflected waves, in terms of λ_0?

d) What is the frequency, in terms of f_0?

e) A listener is at rest at the left of the moving surface. How many beats per second does she hear as a result of the combined effect of the incident and reflected waves?

23–20 A man stands at rest in front of a large, smooth wall. Directly in front of him, between him and the wall, he holds a vibrating tuning fork of frequency f_0. He now runs toward the wall with a speed v. How many beats per second will he hear between the sound waves reaching him directly from the fork and those reaching him after being reflected from the wall?

23–21 The sound source of a ship's sonar system operates at a frequency of 50,000 Hz. The velocity of sound in water can be taken as 1450 m·s⁻¹.

a) What is the wavelength of the waves emitted by the source?

b) What is the difference in frequency between the directly radiated waves and the waves reflected from a whale traveling directly away from the ship at 6.95 m·s⁻¹?

CHALLENGE PROBLEMS

23–22 This problem asks you to perform an alternative derivation of the phenomenon of beats.

a) Consider two sound waves of the same amplitude but different frequencies f_1 and f_2. The displacements of the air as a function of time produced by the individual waves are $y_1 = A \sin (2\pi f_1 t)$ and $y_2 = A \sin (2\pi f_2 t)$. (For convenience take $x = 0$ in Eq. [23–1].) Apply the principle of superposition and the appropriate trigonometric identities to show that the resultant

displacement produced by the two simultaneous waves is

$$y = \left[2 A \cos 2\pi \left(\frac{f_1 - f_2}{2}\right)t\right] \sin 2\pi \left(\frac{f_1 + f_2}{2}\right)t.$$

b) From the result in (a), show that for f_1 and f_2 that are not too different, one will hear beats with a frequency of occurrence given by Eq. (23–12). What happens when $|f_1 - f_2|$ is large?

23–23 A source of sound waves, S, emitting waves of frequency f_0, is traveling toward the right in still air with a speed v_1. At the right of the source is a large, smooth, reflecting surface moving toward the left with a speed v_2. The velocity of the sound waves is c.

a) How far does an emitted wave travel in time t?

b) What is the wavelength of the emitted waves in front of (i.e., at the right of) the source?

c) How many waves strike the reflecting surface in time t?

d) What is the speed of the reflected waves?

e) What is the wavelength of the reflected waves?

f) What is the frequency of the reflected waves, as heard by a stationary listener?

g) Calculate a numerical value for the frequency in part (f), for $c = 345$ m·s^{-1}, $v_1 = 30$ m·s^{-1}, $v_2 = 60$ m·s^{-1}, and $f_0 = 1000$ Hz.

APPENDIX A
THE INTERNATIONAL SYSTEM OF UNITS

The Système International d'Unités, abbreviated SI, is the system developed by the General Conference on Weights and Measures and adopted by nearly all the industrial nations of the world. It is based on the mksa (meter-kilogram-second-ampere) system. The following material is adapted from NBS Special Publication 330 (1981 edition) of the National Bureau of Standards.

Quantity	Name of unit	Symbol	
SI Base Units			
length	meter	m	
mass	kilogram	kg	
time	second	s	
electric current	ampere	A	
thermodynamic temperature	kelvin	K	
luminous intensity	candela	cd	
amount of substance	mole	mol	
SI Derived Units			**Equivalent Units**
area	square meter	m^2	
volume	cubic meter	m^3	
frequency	hertz	Hz	s^{-1}
mass density (density)	kilogram per cubic meter	$kg \cdot m^{-3}$	
speed, velocity	meter per second	$m \cdot s^{-1}$	
angular velocity	radian per second	$rad \cdot s^{-1}$	
acceleration	meter per second squared	$m \cdot s^{-2}$	
angular acceleration	radian per second squared	$rad \cdot s^{-2}$	
force	newton	N	$kg \cdot m \cdot s^{-2}$
pressure (mechanical stress)	pascal	Pa	$N \cdot m^{-2}$
kinematic viscosity	square meter per second	$m^2 \cdot s^{-1}$	
dynamic viscosity	newton-second per square meter	$N \cdot s \cdot m^{-2}$	
work, energy, quantity of heat	joule	J	$N \cdot m$
power	watt	W	$J \cdot s^{-1}$
quantity of electricity	coulomb	C	$A \cdot s$
potential difference, electromotive force	volt	V	$W \cdot A^{-1}, J \cdot C^{-1}$
electric field strength	volt per meter	$V \cdot m^{-1}$	$N \cdot C^{-1}$
electric resistance	ohm	Ω	$V \cdot A^{-1}$
capacitance	farad	F	$A \cdot s \cdot V^{-1}$

Quantity	Name of unit	Symbol	Equivalent Units
magnetic flux	weber	Wb	$V \cdot s$
inductance	henry	H	$V \cdot s \cdot A^{-1}$
magnetic flux density	tesla	T	$Wb \cdot m^{-2}$
magnetic field strength	ampere per meter	$A \cdot m^{-1}$	
magnetomotive force	ampere	A	
luminous flux	lumen	lm	$cd \cdot sr$
luminance	candela per square meter	$cd \cdot m^{-2}$	
illuminance	lux	lx	$lm \cdot m^{-2}$
wave number	1 per meter	m^{-1}	
entropy	joule per kelvin	$J \cdot K^{-1}$	
specific heat capacity	joule per kilogram kelvin	$J \cdot kg^{-1} \cdot K^{-1}$	
thermal conductivity	watt per meter kelvin	$W \cdot m^{-1} \cdot K^{-1}$	
radiant intensity	watt per steradian	$W \cdot sr^{-1}$	
activity (of a radioactive source)	becquerel	Bq	s^{-1}
radiation dose	gray	Gy	$J \cdot kg^{-1}$
radiation dose equivalent	sievert	Sv	$J \cdot kg^{-1}$
SI Supplementary Units			
plane angle	radian	rad	
solid angle	steradian	sr	

DEFINITIONS OF SI UNITS

meter (m) The *meter* is the length equal to the distance traveled by light, in vacuum, in a time of 1/299,792,458 second.

kilogram (kg) The *kilogram* is the unit of mass; it is equal to the mass of the international prototype of the kilogram. (The international prototype of the kilogram is a particular cylinder of platinum-iridium alloy that is preserved in a vault at Sèvres, France, by the International Bureau of Weights and Measures.)

second (s) The *second* is the duration of 9,192,631,770 periods of the radiation corresponding to the transition between the two hyperfine levels of the ground state of the cesium-133 atom.

ampere (A) The *ampere* is that constant current that, if maintained in two straight parallel conductors of infinite length, of negligible circular cross section, and placed 1 meter apart in vacuum, would produce between these conductors a force equal to 2×10^{-7} newton per meter of length.

kelvin (K) The *kelvin*, unit of thermodynamic temperature, is the fraction 1/273.16 of the thermodynamic temperature of the triple point of water.

ohm (Ω) The *ohm* is the electric resistance between two points of a conductor when a constant difference of potential of 1 volt, applied between these two points, produces in this conductor a current of 1 ampere, this conductor not being the source of any electromotive force.

coulomb (C) The *coulomb* is the quantity of electricity transported in 1 second by a current of 1 ampere.

candela (cd) The *candela* is the luminous intensity, in a given direction, of a source that emits monochromatic radiation of frequency 540×10^{12} hertz and that has a radiant intensity in that direction of 1/683 watt per steradian.

mole (mol) The *mole* is the amount of substance of a system that contains as many elementary entities as there are carbon atoms in 0.012 kg of carbon 12. The elementary entities must be specified and may be atoms, molecules, ions, electrons, other particles, or specified groups of such particles.

newton (N) The *newton* is that force that gives to a mass of 1 kilogram an acceleration of 1 meter per second per second.

joule (J) The *joule* is the work done when the point of application of 1 newton is displaced a distance of 1 meter in the direction of the force.

watt (W) The *watt* is the power that gives rise to the production of energy at the rate of 1 joule per second.

volt (V) The *volt* is the difference of electric potential between two points of a conducting wire carrying a constant current of 1 ampere, when the power dissipated between these points is equal to 1 watt.

weber (Wb) The *weber* is the magnetic flux that, linking a circuit of one turn, produces in it an electromotive force of 1 volt as it is reduced to zero at a uniform rate in 1 second.

lumen (lm) The *lumen* is the luminous flux emitted in a solid angle of 1 steradian by a uniform point source having an intensity of 1 candela.

farad (F) The *farad* is the capacitance of a capacitor between the plates of which there appears a difference of potential of 1 volt when it is charged by a quantity of electricity equal to 1 coulomb.

henry (H) The *henry* is the inductance of a closed circuit in which an electromotive force of 1 volt is produced when the electric current in the circuit varies uniformly at a rate of 1 ampere per second.

radian (rad) The *radian* is the plane angle between two radii of a circle that cut off on the circumference an arc equal in length to the radius.

steradian (sr) The *steradian* is the solid angle that, having its vertex in the center of a sphere, cuts off an area of the surface of the sphere equal to that of a square with sides of length to the radius of the sphere.

SI Prefixes The names of multiples and submultiples of SI units may be formed by application of the prefixes listed in Table 1–1, page 6.

APPENDIX B
USEFUL MATHEMATICAL RELATIONS

ALGEBRA

$$a^{-x} = \frac{1}{a^x} \qquad a^{(x+y)} = a^x a^y \qquad a^{(x-y)} = \frac{a^x}{a^y}$$

Logarithms: If $\log a = x$, then $a = 10^x$. $\log a + \log b = \log (ab)$ $\log a - \log b = \log (a/b)$ $\log (a^n) = n \log a$

If $\ln a = x$, then $a = e^x$. $\ln a + \ln b = \ln (ab)$ $\ln a - \ln b = \ln (a/b)$ $\ln (a^n) = n \ln a$

Quadratic formula: If $ax^2 + bx + c = 0$, $x = \dfrac{-b \pm \sqrt{b^2 - 4ac}}{2a}$.

BINOMIAL THEOREM

$$(a + b)^n = a^n + na^{n-1}b + \frac{n(n-1)a^{n-2}b^2}{2!} + \frac{n(n-1)(n-2)a^{n-3}b^3}{3!} + \cdots$$

TRIGONOMETRY

In the right triangle ABC, $x^2 + y^2 = r^2$.

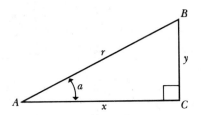

Definitions of the trigonometric functions: $\sin a = y/r$ $\cos a = x/r$ $\tan a = y/x$

Identities: $\sin^2 a + \cos^2 a = 1$ $\tan a = \dfrac{\sin a}{\cos a}$

$\sin 2a = 2 \sin a \cos a$ $\cos 2a = \cos^2 a - \sin^2 a = 2 \cos^2 a - 1$

$\sin \tfrac{1}{2}a = \sqrt{\dfrac{1 - \cos a}{2}}$ $\cos \tfrac{1}{2}a = \sqrt{\dfrac{1 + \cos a}{2}}$

$\sin (-a) = -\sin a$ $\sin (a \pm b) = \sin a \cos b \pm \cos a \sin b$

$\cos (-a) = \cos a$ $\cos (a \pm b) = \cos a \cos b \mp \sin a \sin b$

$\sin (a \pm \pi/2) = \pm\cos a$ $\sin a + \sin b = 2 \sin \tfrac{1}{2}(a + b) \cos \tfrac{1}{2}(a - b)$

$\cos (a \pm \pi/2) = \mp\sin a$ $\cos a + \cos b = 2 \cos \tfrac{1}{2}(a + b) \cos \tfrac{1}{2}(a - b)$

GEOMETRY

Circumference of circle of radius r: $C = 2\pi r$

Area of circle of radius r: $A = \pi r^2$

Volume of sphere of radius r: $V = 4\pi r^3/3$

Surface area of sphere of radius r: $A = 4\pi r^2$

Volume of cylinder of radius r and height h: $V = \pi r^2 h$

CALCULUS

Derivatives:

$$\frac{d}{dx} x^n = nx^{n-1}$$

$$\frac{d}{dx} \sin ax = a \cos ax$$

$$\frac{d}{dx} \cos ax = -a \sin ax$$

$$\frac{d}{dx} e^{ax} = ae^{ax}$$

$$\frac{d}{dx} \ln ax = \frac{1}{x}$$

Integrals:

$$\int x^n \, dx = \frac{x^{n+1}}{n+1}$$

$$\int \frac{dx}{x} = \ln x$$

$$\int \sin ax \, dx = -\frac{1}{a} \cos ax$$

$$\int \cos ax \, dx = \frac{1}{a} \sin ax$$

$$\int e^{ax} \, dx = \frac{1}{a} e^{ax}$$

$$\int \frac{dx}{\sqrt{a^2 - x^2}} = \arcsin \frac{x}{a}$$

$$\int \frac{dx}{\sqrt{x^2 + a^2}} = \ln (x + \sqrt{x^2 + a^2})$$

$$\int \frac{dx}{x^2 + a^2} = \frac{1}{a} \arctan \frac{x}{a}$$

$$\int \frac{dx}{(x^2 + a^2)^{3/2}} = \frac{1}{a^2} \frac{x}{\sqrt{x^2 + a^2}}$$

$$\int \frac{x \, dx}{(x^2 + a^2)^{3/2}} = -\frac{1}{\sqrt{x^2 + a^2}}$$

Power series (convergent for range of x shown):

$$\sin x = x - \frac{x^3}{3!} + \frac{x^5}{5!} - \frac{x^7}{7!} + \cdots \qquad \text{(all } x\text{)}$$

$$\cos x = 1 - \frac{x^2}{2!} + \frac{x^4}{4!} - \frac{x^6}{6!} + \cdots \qquad \text{(all } x\text{)}$$

$$\tan x = x + \frac{x^3}{3} + \frac{2x^5}{15} + \frac{17x^7}{315} + \cdots \qquad (|x| < \pi/2)$$

$$e^x = 1 + x + \frac{x^2}{2!} + \frac{x^3}{3!} + \cdots \qquad \text{(all } x\text{)}$$

$$\ln (1 + x) = x - \frac{x^2}{2} + \frac{x^3}{3} - \frac{x^4}{4} + \cdots \qquad (|x| < 1)$$

APPENDIX C

THE GREEK ALPHABET

Name	Capital	Lowercase	Name	Capital	Lowercase
Alpha	A	α	Nu	N	ν
Beta	B	β	Xi	Ξ	ξ
Gamma	Γ	γ	Omicron	O	o
Delta	Δ	δ	Pi	Π	π
Epsilon	E	ϵ	Rho	P	ρ
Zeta	Z	ζ	Sigma	Σ	σ
Eta	H	η	Tau	T	τ
Theta	Θ	θ	Upsilon	Υ	υ
Iota	I	ι	Phi	Φ	ϕ
Kappa	K	κ	Chi	X	χ
Lambda	Λ	λ	Psi	Ψ	ψ
Mu	M	μ	Omega	Ω	ω

APPENDIX D

PERIODIC TABLE OF THE ELEMENTS

Period	IA	IIA	IIIB	IVB	VB	VIB	VIIB	VIIIB	VIIIB	IB	IIB	IIIA	IVA	VA	VIA	VIIA	Noble gases
1	1 **H** 1.008																2 **He** 4.003
2	3 **Li** 6.941	4 **Be** 9.012										5 **B** 10.811	6 **C** 12.011	7 **N** 14.007	8 **O** 15.999	9 **F** 18.998	10 **Ne** 20.179
3	11 **Na** 22.990	12 **Mg** 24.305										13 **Al** 26.982	14 **Si** 28.086	15 **P** 30.974	16 **S** 32.064	17 **Cl** 35.453	18 **Ar** 39.948
4	19 **K** 39.098	20 **Ca** 40.08	21 **Sc** 44.956	22 **Ti** 47.90	23 **V** 50.942	24 **Cr** 51.996	25 **Mn** 54.938	26 **Fe** 55.847	27 **Co** 58.933	28 **Ni** 58.70	29 **Cu** 63.546	30 **Zn** 65.38	31 **Ga** 69.72	32 **Ge** 72.59	33 **As** 74.922	34 **Se** 78.96	35 **Br** 79.904	36 **Kr** 83.80
5	37 **Rb** 85.468	38 **Sr** 87.62	39 **Y** 88.906	40 **Zr** 91.22	41 **Nb** 92.906	42 **Mo** 95.94	43 **Tc** (99)	44 **Ru** 101.07	45 **Rh** 102.905	46 **Pd** 106.4	47 **Ag** 107.868	48 **Cd** 112.41	49 **In** 114.82	50 **Sn** 118.69	51 **Sb** 121.75	52 **Te** 127.60	53 **I** 126.905	54 **Xe** 131.30
6	55 **Cs** 132.905	56 **Ba** 137.33	57 **La** 138.905	72 **Hf** 178.49	73 **Ta** 180.948	74 **W** 183.85	75 **Re** 186.2	76 **Os** 190.2	77 **Ir** 192.22	78 **Pt** 195.09	79 **Au** 196.966	80 **Hg** 200.59	81 **Tl** 204.37	82 **Pb** 207.19	83 **Bi** 208.2	84 **Po** (210)	85 **At** (210)	86 **Rn** (222)
7	87 **Fr** (223)	88 **Ra** (226)	89 **Ac** (227)	104 **Rf(?)** (261)	105 **Ha(?)** (262)	106 (257)	107 (260)											

58 **Ce** 140.12	59 **Pr** 140.907	60 **Nd** 144.24	61 **Pm** (145)	62 **Sm** 150.35	63 **Eu** 151.96	64 **Gd** 157.25	65 **Tb** 158.925	66 **Dy** 162.50	67 **Ho** 164.930	68 **Er** 167.26	69 **Tm** 168.934	70 **Yb** 173.04	71 **Lu** 174.96
90 **Th** (232)	91 **Pa** (231)	92 **U** (238)	93 **Np** (239)	94 **Pu** (239)	95 **Am** (240)	96 **Cm** (242)	97 **Bk** (245)	98 **Cf** (246)	99 **Es** (247)	100 **Fm** (249)	101 **Md** (256)	102 **No** (254)	103 **Lr** (257)

For each element the average atomic mass of the mixture of isotopes occurring in nature is shown. For elements having no stable isotope, the approximate atomic mass of the most common isotope is shown in parentheses.

APPENDIX E

UNIT CONVERSION FACTORS

LENGTH

1 m = 100 cm = 1000 mm = 10^6 μm = 10^9 nm
1 km = 1000 m = 0.6214 mi
1 m = 3.281 ft = 39.37 in.
1 cm = 0.3937 in.
1 in. = 2.540 cm
1 ft = 30.48 cm
1 yd = 91.44 cm
1 mi = 5280 ft = 1.609 km
1 Å = 10^{-10} m = 10^{-8} cm = 10^{-1} nm
1 nautical mile = 6080 ft
1 light year = 9.461×10^{15} m

AREA

1 cm^2 = 0.155 in^2
1 m^2 = 10^4 cm^2 = 10.76 ft^2
1 in^2 = 6.452 cm^2
1 ft^2 = 144 in^2 = 0.0929 m^2

VOLUME

1 liter = 1000 cm^3 = $10^{-3} m^3$ = 0.03531 ft^3 = 61.02 in^3
1 ft^3 = 0.02832 m^3 = 28.32 liters = 7.477 gallons
1 gallon = 3.788 liters

TIME

1 min = 60 s
1 hr = 3600 s
1 da = 86,400 s
1 yr = 365.24 da = 3.156×10^7 s

ANGLE

1 rad = 57.30° = 180°/π
1° = 0.01745 rad = π/180 rad
1 revolution = 360° = 2π rad
1 rev·min^{-1} (rpm) = 0.1047 rad·s^{-1}

SPEED

1 m·s^{-1} = 3.281 ft·s^{-1}
1 ft·s^{-1} = 0.3048 m·s^{-1}
1 mi·min^{-1} = 60 mi·hr^{-1} = 88 ft·s^{-1}
1 km·hr^{-1} = 0.2778 m·s^{-1} = 0.6214 mi·hr^{-1}
1 mi·hr^{-1} = 1.466 ft·s^{-1} = 0.4470 m·s^{-1} = 1.609 km·hr^{-1}
1 furlong·$fortnight^{-1}$ = 1.662×10^{-4} m·s^{-1}

ACCELERATION

1 m·s^{-2} = 100 cm·s^{-2} = 3.281 ft·s^{-2}
1 cm·s^{-2} = 0.01 m·s^{-2} = 0.03281 ft·s^{-2}
1 ft·s^{-2} = 0.3048 m·s^{-2} = 30.48 cm·s^{-2}
1 mi·hr^{-1}·s^{-1} = 1.467 ft·s^{-2}

MASS

1 kg = 10^3 g = 0.0685 slug
1 g = 6.85×10^{-5} slug
1 slug = 14.59 kg
1 u = 1.661×10^{-27} kg
1 kg has a weight of 2.205 lb when g = 9.80 m·s^{-2}

FORCE

1 N = 10^5 dyn = 0.2248 lb
1 lb = 4.448 N = 4.448×10^5 dyn

PRESSURE

1 Pa = 1 N·m^{-2} = 1.451×10^{-4} lb·in^{-2} = 0.209 lb·ft^{-2}
1 bar = 10^5 Pa
1 lb·in^{-2} = 6891 Pa
1 lb·ft^{-2} = 47.85 Pa
1 atm = 1.013×10^5 Pa = 1.013 bar
      = 14.7 lb·in^{-2} = 2117 lb·ft^{-2}
1 mm Hg = 1 torr = 133.3 Pa

ENERGY

1 J = 10^7 ergs = 0.239 cal
1 cal = 4.186 J (based on 15° calorie)
1 ft·lb = 1.356 J
1 Btu = 1055 J = 252 cal = 778 ft·lb
1 eV = 1.602×10^{-19} J
1 kWh = 3.600×10^6 J

MASS–ENERGY EQUIVALENCE

1 kg $\leftrightarrow$ 8.988×10^{16} J
1 u $\leftrightarrow$ 931.5 MeV
1 eV $\leftrightarrow$ 1.073×10^{-9} u

POWER

1 W = 1 J·s^{-1}
1 hp = 746 W = 550 ft·lb·s^{-1}
1 Btu·hr^{-1} = 0.293 W

APPENDIX F
NUMERICAL CONSTANTS

FUNDAMENTAL PHYSICAL CONSTANTS

Name	Symbol	Value
Speed of light	c	2.9979×10^8 m·s^{-1}
Charge of electron	e	1.602×10^{-19} C
Gravitational constant	G	6.673×10^{-11} N·m^2·kg^{-2}
Planck's constant	h	6.626×10^{-34} J·s
Boltzmann's constant	k	1.381×10^{-23} J·K^{-1}
Avogadro's number	N_0	6.022×10^{23} molecules·mol^{-1}
Gas constant	R	8.314 J·mol^{-1}·K^{-1}
Mass of electron	m_e	9.110×10^{-31} kg
Mass of neutron	m_n	1.675×10^{-27} kg
Mass of proton	m_p	1.673×10^{-27} kg
Permittivity of free space	ϵ_0	8.854×10^{-12} C^2·N^{-1}·m^{-2}
	$1/4\pi\epsilon_0$	8.987×10^9 N·m^2·C^{-2}
Permeability of free space	μ_0	$4\pi \times 10^{-7}$ Wb·A^{-1}·m^{-1}

OTHER USEFUL CONSTANTS

Name	Symbol	Value
Mechanical equivalent of heat		4.186 J·cal^{-1} (15° calorie)
Standard atmospheric pressure	1 atm	1.013×10^5 Pa
Absolute zero	0 K	-273.15°C
Electronvolt	1 eV	1.602×10^{-19} J
Atomic mass unit	1 u	1.661×10^{-27} kg
Electron rest energy	mc^2	0.511 MeV
Energy equivalent of 1 u	Mc^2	931.5 MeV
Volume of ideal gas (0°C and 1 atm)	V	22.4 liter·mol^{-1}
Acceleration due to gravity (sea level, at equator)	g	9.78049 m·s^{-2}

ASTRONOMICAL DATA

Body	Mass, kg	Radius, m	Orbit radius, m	Orbit period
Sun	1.99×10^{30}	6.95×10^8	—	—
Moon	7.36×10^{22}	1.74×10^6	0.38×10^9	27.3 d
Mercury	3.28×10^{23}	2.57×10^6	5.8×10^{10}	88.0 d
Venus	4.82×10^{24}	6.31×10^6	1.08×10^{11}	224.7 d
Earth	5.98×10^{24}	6.38×10^6	1.49×10^{11}	365.3 d
Mars	6.34×10^{23}	3.43×10^6	2.28×10^{11}	687.0 d
Jupiter	1.88×10^{27}	7.18×10^7	7.78×10^{11}	11.86 y
Saturn	5.63×10^{26}	6.03×10^7	1.43×10^{12}	29.46 y
Uranus	8.61×10^{25}	2.67×10^7	2.87×10^{12}	84.02 y
Neptune	9.99×10^{25}	2.48×10^7	4.49×10^{12}	164.8 y
Pluto	5×10^{23}	4×10^5	5.90×10^{12}	247.7 y

ANSWERS TO ODD-NUMBERED PROBLEMS

CHAPTER 1

1–1   1.61 km

1–3   a) 0.403 mi·s^{-1} b) 648 m·s^{-1}

1–5 120 in^3

1–7 40.0 mi·gal^{-1}

1–9   a) 7.04×10^{-10} s   b) 5.11×10^{12} cycles·hr^{-1}
     c) 4.48×10^{26} cycles   d) 4.60×10^4 s

1–11   0.0038

1–13 $5.50 \times 10^3 \text{ kg·m}^{-3}$

1–15   Ten thousand

1–17   No; no

1–19 10^5

1–21 $\$1 \times 10^{15}$

1–23   7000

1–25   a) 5.0 cm, $-53.1°$ b) 13.0 cm, $-112.6°$
     c) 3.6 km, 123.7°

1–27   5.0 km, 36.9° E of S

1–29   a) 15.0 m, 53.3°   b) 25.9 m, 27.7° below $-x$-axis

1–31   a) $N_x = -1.0$ cm, $N_y = +1.0$ cm   b) 1.41 cm,
     135°

1–33   a) $A = 3.61$, $B = 2.24$ b) $3i + j$
     c) 3.16, 18.4° above $+x$-axis   d) $i + 5j$
     e) 5.10, 78.7° above $+x$-axis

1–35 $i \cdot i = j \cdot j = k \cdot k = 1$
     $i \cdot j = i \cdot k = 0$
     $j \cdot i = j \cdot k = 0$
     $k \cdot i = k \cdot j = 0$

1–37   130°

1–39 $i \times i = j \times j = k \times k = 0$
     $i \times j = k \quad j \times i = -k \quad k \times i = j$
     $i \times k = -j \quad j \times k = i \quad k \times j = -i$

1–41   a) $A \cdot B = 0$, $A \times B = 29\,k$ b) $A \cdot B = 58$,
     $A \times B = 0$

1–45   70.0 m, 26.3° E of N

1–47 $A = \sqrt{(x_2 - x_1)^2 + (y_2 - y_1)^2}$;
     $\theta = \arctan (y_2 - y_1 / x_2 - x_1)$

1–49   b) $\theta = 120°$ or 240°   c) $\sqrt{A^2 + B^2 - 2AB \cos \theta}$
     d) $\theta = 60°$ or 300°

1–51   14

1–53   90°

1–57   b) $C \cdot (A \times B)$

CHAPTER 2

2–1   a) 15.0 mi·hr^{-1} b) 22.0 ft·s^{-1}

2–3 5.0 cm·s^{-1}; -4.0 cm·s^{-1}

2–5 60 cm·s^{-1}

2–7   a) 0; 1.0 m·s^{-2}; 1.5 m·s^{-2}; 2.5 m·s^{-2}; 2.5 m·s^{-2};
     2.5 m·s^{-2}; 1.0 m·s^{-2}; 0. No. Yes, from 6 s to 12 s.
     b) 2.5 m·s^{-2}; 1.3 m·s^{-2}; 0

2–11   a) 3.95 ft·s^{-2} b) 620 ft

2–13   a) 5.0 m·s^{-1} b) 1.67 m·s^{-2}

2–15   a) 0, 6.25 m·s^{-2}, -11.2 m·s^{-2} b) 100 m, 230 m,
     320 m

2–17   a) 30.5 ft   b) 260 ft

2–19   a) 6000 m·s^{-1} b) 0.99   c) 1120 min

2–21   a) $x = 0.20t^3 - 0.010t^4$; $v = 0.60t^2 - 0.040t^3$
     b) 20 m·s^{-1}

2–23   a) 19.8 m·s^{-1} b) 4.04 s

2–25   a) 29.6 m·s^{-1} b) 39.6 m   c) 17.2 m·s^{-1}
     d) 50.0 m·s^{-2} e) 2.01 s   f) 29.7 m·s^{-1}

2–27   a) 48 ft·s^{-1} b) 36 ft   c) 0   d) 32 ft·s^{-2},
     downward   e) 80 ft·s^{-1}

2–29   a) 815 ft·s^{-2} b) 25.5 times longer   c) 1320 ft
     d) $a = 20.7\,g$; not consistent if you assume constant
     acceleration

2–31   a) 50 s   b) 150 s

2–33   a) $B + 3Ct^2$ b) $6Ct$

2–35   a) 9.0 m·s^{-1}; 11.0 m·s^{-1}; 13.0 m·s^{-1}
     b) 1.0 m·s^{-2} c) 8.0 m·s^{-1} d) 8.0 s
     e) 8.5 m   f) 1.06 s   g) 9.06 m·s^{-1}

2–37   a) 100 m   b) 20 m·s^{-1}

2–39 $x = (2 \text{ m·s}^{-2})t^2 + (0.75 \text{ m·s}^{-4})t^4$,
     $a = 4 \text{ m·s}^{-2} + (9 \text{ m·s}^{-4})t^2$

2–41   a) $A\omega \cos \omega t$ b) $-A\omega^2 \sin \omega t$
     c) $v = \pm \omega \sqrt{A^2 - x^2}$; max at $x = 0$, min at $x = \pm A$
     d) $-\omega^2 x$; max at $x = \pm A$, min at $x = 0$ e) A

f) $\pm\omega A$ g) $+\omega^2 A$

2-43 a) 11.3 m   b) 0.51 s   c) $v = 9.97\ m\cdot s^{-1}$, $a = -9.8\ m\cdot s^{-2}$

2-45 a) 4.0 m·s^{-2} b) 6.0 m·s^{-1} c) 4.5 m

2-47 2.54 s after first is dropped

2-49 a) 17.3 s   b) 421 m   c) 28.6 m·s^{-1}

2-53 a) 20 s, 120 m   b) 4.0 m·s^{-1} d) 40 s, 8.0 m·s^{-1} e) No   f) 5.66 m·s^{-1}; 28.3 s, 160 m

2-55 a) 25   b) 91

2-57 a) A b) A/B

CHAPTER 3

3-1 a) $v = -4i + 6tj$; $a = 6j$ b) $v = 18.4$ m·s^{-1}, 102.5°; $a = 6.0$ m·s^{-2}, 90°

3-3 a) $t = 0$: $x = 0$, $y = 0$
$t = 1$ s: $x = 20$ m, $y = -4.9$ m
$t = 2$ s: $x = 40$ m, $y = -19.6$ m
$t = 3$ s: $x = 60$ m, $y = -44.1$ m
$t = 4$ s: $x = 80$ m, $y = -78.4$ m
b) $v = 20$ m·s^{-1}i, $a = -9.8$ m·s^{-2}j
c) $v_x = 20.0$ m·s^{-1}, $v_y = -19.6$ m·s^{-1}; yes

3-5 a) 1.22 m   b) 2.00 m   c) $v_x = 4.00$ m·s^{-1}, $v_y = 4.90$ m·s^{-1}

3-7 a) 0.446 ft   b) 4.00 ft

3-9 a) 52.1 m   b) 3.26 s   c) 0.91 s and 5.61 s
d) $t = 0.91$ s: $v_x = 24.1$ m·s^{-1}, $v_y = 23.0$ m·s^{-1}
$t = 5.61$ s: $v_x = 24.1$ m·s^{-1}, $v_y = -23.0$ m·s^{-1}
e) 40.0 m·s^{-1}, 53° below horizontal

3-11 a) 66.5 m   b) 67.7 m·s^{-1} c) 410 m

3-13 0.0337 m·s^{-2}

3-15 a) 6.75 m·s^{-2}, upward   b) 8.38 s

3-17 a) 19.3° W of N   b) 274 km·hr^{-1}

3-19 a) 3.61 m·s^{-1}; 33.7° N of E   b) 333 s
c) 666 m

3-21 a) 11.7 m   b) 8.25 m·s^{-1}, 76.0° below x-axis
c) 4.0 m·s^{-2}, $-y$-direction   d) $t = 0$ e) $t = 0$ s, $x = 0$, $y = 19$ m; $t = 3.00$ s, $x = 6.0$ m, $y = 1.0$ m
f) 6.08 m at $t = 3.00$ s

3-23 a) $r(t) = [2\ \text{m·s}^{-1}t - 1\ \text{m·s}^{-3}t^3]i + [2.5\ \text{m·s}^{-2}t^2]j$; $a(t) = -6\ \text{m·s}^{-3}ti + 5\ \text{m·s}^{-2}j$ b) 5.0 m

3-25 a) 140 ft·s^{-1} b) 138 ft

3-27 a) 227 m·s^{-1} b) 908 m   c) $v_x = 182$ m·s^{-1}, $v_y = -185$ m·s^{-1}

3-33 a) 100 km·hr^{-1}, 36.9° W of S   b) 30.0° N of W

3-35 18.7 km·hr^{-1}, 31.0° E of N

3-37 17.8 m·s^{-1}

3-39 a) $[2v_0^2 \cos^2(\phi + \theta)/g \cos \theta][\tan(\phi + \theta) - \tan \theta]$
b) $\pi/4 - \theta/2$

3-41 $\Delta t = 0.5$ s: $a = 11.70$ m·s^{-2}, 54.2° from direction of v_1
$\Delta t = 0.1$ s: $a = 12.47$ m·s^{-2}, 82.8°
$\Delta t = 0.05$ s: $a = 12.49$ m·s^{-2}, 86.4°
Exact result is $a = 12.50$ m·s^{-2}, 90°

CHAPTER 4

4-1 $F_x = 34.6$ N (to right); $F_y = 20.0$ N (down)

4-3 109 lb, 18.9°

4-5 a) 0.102 kg   b) 78.4 N

4-7 $m = 5.31$ kg; $W = 19.6$ N

4-9 a) Earth's gravity (downward)   b) Gravitational force on earth, by bottle

4-11 a) 5.0 kg   b) 200 m

4-13 a) $x = 0.025$ m, $v = 0.010$ m·s^{-1} b) $x = 0.200$ m, $v = 0.020$ m·s^{-1}

4-15 6 Bmt

4-17 14.1 N, 135° counterclockwise from F_1

4-19 46.6 N, 60° clockwise from F_2

4-21 a) 2.2 m·s^{-2} b) 10.3 m·s^{-2}

4-23 a) $F_x = -60$ N·s^{-1}t, $F_y = 64$ N   b) 191 N, 160° counterclockwise from $+x$-axis

4-25 $r(t) = [(k_1/2m)t^2 + (k_2k_3/120m^2)t^5]i + [(k_3/6m)t^3]j$, $v(t) = [(k_1/m)t + (k_2k_3/24m^2)t^4]i + [(k_3/2m)t^2]j$

CHAPTER 5

5-3 a) 3.84 lb   b) 2.34 s

5-5 a) 46.5 m   b) 15.1 m·s^{-1} (33.8 mi·hr^{-1})

5-7 Low pressure: 0.0210; high pressure: 0.00449

5-9 a) 10 N   b) 20 N

5-11 a) 1540 N   b) 1.15°

5-13 a) $F_1 = 14.1$ N, $F_2 = 14.1$ N   b) 14.1 N

5-15 a) $w \sin \theta$ b) $2w \sin \theta$

5-17 a) $\mu_k(w_A + w_B)$ b) $\mu_k w_A$

5-19 a) Held back   b) 480 N

5-21 a) $\mu_k w/(\cos \theta - \mu_k \sin \theta)$ b) 1 tan θ

5-25 5.20 m·s^{-2}

5-27 a) 16,200 lb   b) 12,150 lb   c) 8100 lb
d) 4050 lb   e) Same magnitude as above, but opposite direction

5-29 a) 4.90 m·s^{-2} b) 3.20 m·s^{-2}

5-31 a) 3.27 m·s^{-2}, upward   b) 2.45 m·s^{-2}, downward
c) Yes (free-fall)

5-33 a) 0.98 m·s^{-2} b) 35.3 N

5-35 a) 3.92 m   b) 2.35 N before, 1.96 N after
c) 4.70 N before

5-37 $T = w$ each chain; $F = w/2\ w$

5-39 $w \tan \theta$

5-41 $\mu_s/(\mu_s + 1)$

5-43 a) 20 N   b) 30 N

5-45 a) 10.8 N   b) 6.47 N

5-47 0.251

5-49 a) 100 N, toward front of truck   b) 78.4 N, toward rear of truck

5-51 a) Yes   b) 1.58 m·s^{-2}, toward front of truck

5-53 a) 10.7 kg   b) 23.6 N, 83.2 N

5-55 $a_1 = 2m_2g/(4m_1 + m_2)$, $a_2 = m_2g/(4m_1 + m_2)$

5-57 a) 2.70 m·s^{-2} b) 112.5 N   c) 87.5 N

5-59 2.43 m above the floor

5-61 $a = g/\mu_s$

5-63 a) $\mu_k w/(\cos \phi + \mu_k \sin \phi)$ b) $\phi = 0°$, $P = 160$ N; $\phi = 10°$, $P = 152$ N; $\phi = 20°$, $P = 149$ N; $\phi = 30°$, $P = 150$ N; $\phi = 40°$, $P = 156$ N; $\phi = 50°$, $P = 169$ N; $\phi = 60°$, $P = 189$ N; $\phi = 70°$, $P = 223$ N; $\phi = 80°$, $P = 282$ N; $\phi = 90°$, $P = 400$ N
c) $\phi = \tan^{-1}(\mu_k)$; 21.8°

5-65 101 m

5-67 c) $v = 50$ m·s^{-1}, $m = 0.25$ kg, $T = 0.01$ s gives $F_0 = 940$ N

CHAPTER 6

6–1 22.4 m·s^{-1}

6–3 $7.97°$

6–5 $1.34 \text{ rev·min}^{-1}$

6–7 3.13 m·s^{-1}

6–9 a) 472 m   b) 6270 N

6–11 2.16

6–13 $6.18 \times 10^{24} \text{ kg}$

6–15 $3.71 \times 10^{-11} \text{ m·s}^{-2}$

6–17 $g_x = -10 \text{ m·s}^{-2}, g_y = +40 \text{ m·s}^{-2}$

6–19 $7.35 \times 10^3 \text{ m·s}^{-1}$

6–21 a) 97.1 min   b) 8.14 m·s^{-2}

6–23 a) 50 N   b) 46 N

6–25 a) 15.5 m·s^{-1} b) 4.95 m·s^{-1}

6–27 b) 0.331   c) No

6–29 a) $51.6°$ b) No   c) Bead rides at the bottom of the hoop ($\theta = 0°$)

6–31 a) $0.403w$ b) 4.49 s   c) $2.00w$ d) Hits ground 10 m from center of wheel

6–33 $2.58 \times 10^8 \text{ m}$

6–35 a) $g_x = g_y = 4.55 \times 10^{-10} \text{ m·s}^{-2}$
      b) $6.44 \times 10^{-12} \text{ N}$, 45° above x-axis

6–37 a) $m = 2.88 \times 10^{15} \text{ kg}, g = 0.00768 \text{ m·s}^{-2}$
      b) 6.20 m·s^{-1}; yes

6–39 1.41 hr

6–41 $(2GmM/a^2)[1 - x/\sqrt{x^2 + a^2}$, attractive and along line connecting point mass and center of the disk

6–43 $\tau_{\min} = \{[(\tan\theta - \mu_s)/(1 + \mu_s \tan\theta)]4\pi^2 h \tan\theta/g\}^{1/2}$;
      $\tau_{\max} = \{[(\tan\theta + \mu_s)/(1 - \mu_s \tan\theta)]4\pi^2 h \tan\theta/g\}^{1/2}$

CHAPTER 7

7–1 a) 4.0 J   b) −0.8 J

7–3 a) 270 ft·lb   b) −165 ft·lb   c) 0
     d) 105 ft·lb

7–5 a) 25 N; 50 N   b) 1.25 J; 5.00 J

7–7 a) −6 N   b) −48 N   c) −22.5 J

7–9 250 J

7–11 14.1 m·s^{-1}

7–13 a) $v_0^2/2\mu_k g$ b) 315 ft

7–15 a) $1.35 \times 10^4 \text{ m·s}^{-1}$ b) $5.45 \times 10^{-8} \text{ m}$

7–17 $2.98 \times 10^6 \text{ J}$

7–19 a) 261 N   b) 784 J

7–21 a) 96.7 ft·s^{-1} b) 96.7 ft·s^{-1}

7–23 −0.230 J

7–25 a) 2400 J   b) −470 J   c) 1415 J   d) 515 J

7–27 7.30 m·s^{-1}; just barely

7–29 a) 60 J; 15 J   b) 14.4 J

7–31 0.100 m

7–33 2.04 m

7–35 0.204

7–37 0.927 hp

7–39 a) $8.23 \times 10^3 \text{ J}$ b) 412 W

7–41 $1200 \text{ m}^3 \cdot \text{s}^{-1}$

7–43 3000 N

7–45 a) 814 W (1.09 hp)   b) 565 W (0.757 hp)
      c) 83.6 W (0.112 hp)

7–47 5.42 m·s^{-1}

7–49 $48.2°$

7–51 a) 4.43 m·s^{-1} b) 14.7 N

7–53 a) 0.272   b) −3.6 J

7–55 a) $7.84 \times 10^4 \text{ J}$ b) $1.60 \times 10^5 \text{ J}$
      c) $3.97 \times 10^3 \text{ W}$

7–57 a) $nRT \ln(V_2/V_1)$ b) $k(V_2^{1-\gamma} - V_1^{1-\gamma})/(1-\gamma)$

7–59 a) 24 J   b) 0   c) 16 J

7–61 $h/R_E = 0.01$; 63.8 km

7–63 $v = \sqrt{2gR_E[1 - R_E/(R_E + h)]}$

7–65 a) $40x^2 + 5x^3$ b) 5.87 m·s^{-1}

7–67 334 J

7–69 a) 3.87 m·s^{-1} b) 0.10 m

7–71 a) $U(x_0) = 0$ b) $v(x) = \pm\sqrt{(2\alpha/mx_0 x)(1 - x_0/x)}$
      c) $x = 2x_0$; $v_{\max} = \pm\sqrt{\alpha/2mx_0^2}$ d) 0
      e) $v(x) = \pm\sqrt{(2\alpha/9mx^2 x_0^2)(9xx_0 - 2x^2 - 9x_0^2)}$
      f) released at x_0: $x_{\min} = x_0$, $x_{\max} \to \infty$
      released at x_1: $x_{\min} = 3x_0/2$, $x_{\max} = 3x_0$

CHAPTER 8

8–1 a) $2.00 \times 10^5 \text{ kg·ms}^{-1}$ b) 40.0 m·s^{-1}
     c) 28.3 m·s^{-1}

8–3 a) 18.5 m·s^{-1}, to the right   b) 11.5 m·s^{-1}, to the left

8–5 2500 N; no

8–7 a) $At_2 + Bt_2^3/3$ b) $(t_2/m)(A + t_2^2/3)$

8–9 a) 0.20 m·s^{-1} b) 4780 J

8–11 a) $v_A = 22.0 \text{ m·s}^{-1}, v_B = 15.5 \text{ m·s}^{-1}$ b) 19.6%

8–13 a) 0.667 m·s^{-1} b) $1.33 \times 10^4 \text{ J}$ c) 1.0 m·s^{-1}

8–15 24.0 m·s^{-1}, 33.7° S of E

8–17 595 m·s^{-1}

8–21 $v_A = 26.0 \text{ m·s}^{-1}$; $v_B = 15.0 \text{ m·s}^{-1}$ at 60° from initial direction of A and 90° from final direction of A

8–23 a) 1.0 m·s^{-1} b) 0.75 J

8–25 0.898 m·s^{-1}

8–27 $4.62 \times 10^6 \text{ m}$ from center of earth

8–29 a) 250 N   b) Yes

8–31 a) $5.6 \times 10^{-53}\%$ b) 28.7%

8–33 b) $3.0 \times 10^{-3} \text{ s}$ c) 0.60 N·s   d) $2.0 \times 10^{-3} \text{ kg}$

8–35 $v = 0.417 \text{ m·s}^{-1}\mathbf{i} - 0.625 \text{ m·s}^{-1}\mathbf{j}$

8–37 Station wagon: 8.90 m·s^{-1}; truck: 18.4 m·s^{-1}

8–39 a) 0.163   b) 240 J   c) 0.320 J

8–41 0.300 m

8–43 20

8–45 a) 12.5 m·s^{-1}, 36.9° from initial direction of bullet and hence 126.9° from final direction of bullet
      b) No

8–47 a) 8.95%   b) 32.6%

8–49 $1.30 \times 10^6 \text{ m·s}^{-1}$ and $5.54 \times 10^5 \text{ m·s}^{-1}$;
      $v_r = 1.53 v_{Ba}$

8–51 b) $1/2 \, MV^2$

8–53 a) $1.39v_{Kr}$ b) $1.18v_r$ c) $2.57v_r$
      d) 3.12 km·s^{-1}

8–55 a) $F = -\lambda yg$, where y is the length of rope hanging over edge; $W = \lambda g l^2/18$ b) same W as in (a)

8–57 a) $v_{A1}m_A/(m_A + m_B)$ b) yes
      c) $u_{A1} = v_{A1}m_B/(m_A + m_B), u_{B1} = -v_{A1}m_A/(m_A + m_B)$, $P = 0$ d) $p_{B2} = -p_{B1}, p_{A2} = -p_{A1}; u_{A2} = -u_{A1}$, $u_{B2} = -u_{B1}$ e) $v_{A2} = -(2/3) \text{ m·s}^{-1}$, $v_{B2} = +(4/3) \text{ m·s}^{-1}$

8–59 b) 0.312 s^{-1} e) 8.55 g   f) 166 m·s^{-2}

CHAPTER 9

9–1   a) 1.5 rad   b) 1.57 rad; 90°   c) 1.2 m
9–3   a) $2.0 \text{ rad·s}^{-1} + (0.15 \text{ rad·s}^{-3})t^2$ b) 2.0 rad·s^{-1}
      c) 5.75 rad·s^{-1}; 3.25 rad·s^{-1}
9–5 $2b + 6ct$
9–7   a) -2.00 rev·s^{-2}; 58.3 rev   b) 3.33 s
9–9   7.5 s
9–11 a) 5.89 m·s^{-1} b) 229 rev·min^{-1}
9–13 a) 50.0 m·s^{-2} b) 5.0 m·s^{-1}; 50.0 m·s^{-2}
9–15 a) 0.15 m·s^{-2}; 0; 0.15 m·s^{-2} b) 0.15 m·s^{-2};
      0.628 m·s^{-2}; 0.646 m·s^{-2} c) 0.15 m·s^{-2};
      1.26 m·s^{-2}; 1.27 m·s^{-2}
9–17 a) 1.50 kg·m^2 b) 0.75 kg·m^2
9–19 0.138 kg·m^2
9–21 0.80 kg·m^2
9–23 a) 0.18 kg·m^2 b) 0.32 kg·m^2 c) 0.50 kg·m^2
9–25 $2Ma^2/3$
9–27 a) 80.0 ft·lb, counterclockwise   b) 69.3 ft·lb,
      counterclockwise   c) 40.0 ft·lb, counterclockwise
      d) 34.6 ft·lb clockwise   e) 0   f) 0
9–29 $(-0.10 \text{ N·m})\mathbf{k}$
9–31 0.589
9–33 a) 65.3 N   b) 16.2 m·s^{-1} c) 2.48 s
      d) 261 N
9–35 a) 19.6 N·m   b) 3.08×10^3 J   c) 3.08×10^3 J
9–37 a) 5.97×10^3 N·m   b) 2.39×10^4 N
      c) 62.8 m·s^{-1}
9–39 a) 5.88 N   b) 0.452 s   c) 27.7 rad·s^{-1}
9–41 $6.06 \times 10^{-8} \text{ kg·m}^2\text{·s}^{-1}$
9–43 $116 \text{ kg·m}^2\text{·s}^{-1}$
9–45 a) $43.0 \text{ kg·m}^2\text{·s}^{-1}$ b) 7.96 rad·s^{-1} c) Before:
      108 J; after: 171 J. Increase in K equals work done
      by man.
9–47 a) 12 rad·s^{-1} b) 0.027 J   c) 0.027 J
9–49 218 rev·s^{-1}
9–51 0.149 rad·s^{-1}
9–53 a) 99.1 min   b) 1.65×10^5 N·m
9–55 $0.371 \, MR^2$
9–57 a) 247 rev·min^{-1} b) 800 W
9–59 a) 5.42 rad·s^{-1} b) 5.42 m·s^{-1}
9–61 $v = \sqrt{2gd(m_B - \mu_k m_A)/(m_A + m_B + I/R^2)}$
9–63 a) 17.4 kg·m^2 b) -1.82 N·m   c) 91.6 rev
9–65 0.597 s
9–67 a) 10 N at block on table; 39 N at hanging block
      b) 0.145 kg·m^2
9–69 a) 1.97 m·s^{-2} b) 9.85 N
9–71 $a = F/(2M)$; $\mathcal{F} = F/2$
9–73 0.299 m
9–75 3000 J
9–77 2.94 rad·s^{-1} ($28.1 \text{ rev·min}^{-1}$)
9–79 $3MR^2/5$ (larger than $MR^2/2$)
9–81 a) $mv_1^2 r_1^2/r^3$ b) $1/2 \, mv_1^2[(r_1/r_2)^2 - 1]$
      c) $\Delta K = 1/2 \, mv_1^2[(r_1/r_2)^2 - 1]$; same
9–83 a) $a = \mu_k g$, $\alpha = -2\mu_k g/R$ b) $R^2 \omega_0^2/18 \mu_k g$
      c) $-\frac{1}{6} MR^2 \omega_0^2$

CHAPTER 10

10–1 Between the balls, 0.1 m from the 3 kg ball
10–3 $x = 0.0667$ m, $y = 0.0667$ m

10–5 1000 N; 0.8 m from the end where the 600 N force
      is applied
10–7 b) 2.50 m beyond pont B c) 2.88 m
10–9 a) $T_1 = 1.73w$, $T_2 = w$; $F = 2w$ at 30° above the
      horizontal
      b) $T_1 = 2.73w$, $T_2 = w$; $F = 3.35w$ at 45° above the
      horizontal
10–11 Top hinge: $F_{\text{horiz}} = -66.7$ N, $F_{\text{vert}} = 100$ N;
      Bottom hinge: $F_{\text{horiz}} = 66.7$ N, $F_{\text{vert}} = 100$ N
10–13 a) 2.71 N   b) 2.71 N   c) 2.71 N
10–15 a) 2.31 m   b) 213 N
10–17 $F_{\text{horiz}} = 60$ N, $F_{\text{vert}} = 80$ N
10–19 a) 60 N   b) 53.1°
10–21 a) 900 N   b) 727 N
10–23 a) 214 N   b) 186 N   c) 293 N
10–25 b) Yes; slides at 21.8°, tips at 26.6°   c) Tips first;
      tips at 26.6°, slides at 31.0°
10–27 a) A: 100 N; B: 700 N   b) 2 m
10–29 a) 390 N   b) 11,700 N   c) Slide:
      $0.3w/[\sin\theta - 0.3\cos\theta]$; tip:
      $0.9w/[1.8\sin\theta + 0.1\cos\theta]$; $\theta = 39.8°$
10–31 $X = 0$, $Y = 0.424a$

CHAPTER 11

11–1 $\omega = 31.4 \text{ rad·s}^{-1}$; $\tau = 0.2$ s
11–3 $\tau = 1.26$ s, $f = 0.796$ Hz, $\omega = 5.00 \text{ rad·s}^{-1}$
11–5 $\tau = 0.333$ s, $\omega = 18.8 \text{ rad·s}^{-1}$, $m = 0.563$ kg
11–7 a) 94.7 m·s^{-2}, 3.77 m·s^{-1} b) -56.8 m·s^{-2},
      $\pm 3.02 \text{ m·s}^{-1}$ c) 0.0369 s
11–9 $A = 1.22$ m, $\theta_0 = 1.41$ rad, $E = 74.0$ J,
      $x(t) = (1.22 \text{ m})\cos[(5 \text{ s}^{-1})t + 1.41]$
11–11 a) 98.0 N·m^{-1} b) 0.898 s   c) 1.27 s
11–13 0.0620 m
11–15 a) 39.2 J, 0, 0, 39.2 J   b) 9.8 J, 9.8 J, 19.6 J,
      39.2 J   c) 0, 0, 39.2 J, 39.2 J
11–17 0.190 m
11–19 a) $1.80 \times 10^{-7} \text{ kg·m}^2$ b) $2.84 \times 10^{-5} \text{ N·m·rad}^{-1}$
11–21 0.248 m
11–23 a) $7 \, mL^2/48$ b) $2\pi\sqrt{7L/12g}$
11–25 a) 4.24 Hz   b) 21.9 kg·s^{-1}
11–27 a) $7.11 \times 10^3 \text{ m·s}^{-2}$ b) 3.55×10^3 N
      c) 18.8 m·s^{-1}
11–29 a) 2.09 s   b) 0.0916 m   c) 0.0918
11–31 1.67 s
11–33 a) $\pm 0.314 \text{ m·s}^{-1}$ b) 0.493 m·s^{-2}, downward
      c) 0.333 s   d) 0.993 m
11–35 0.719 m
11–37 a) $\mathcal{N}(t) = m[g - (2\pi f)^2 A \cos(2\pi f t + \theta_0)]$
      b) $(2\pi f_b)^2 A$
11–39 a) $r = 0$, $\tau \to \infty$; $r = R/4$, $\tau = 1.04$ s; $r = R/2$,
      $\tau = 0.815$ s;
      $r = 3R/4$, $\tau = 0.827$ s; $r = R$, $\tau = 0.851$ s;
      $r = 3R/2$, $\tau = 0.941$ s
      c) $r/R = 0.707$; $\tau = 0.827$ s
11–41 a) 3.97 m   b) Stick of length 0.5 m, pivoted
      0.525 cm above its center
11–43 a) $l_1 = 0.35$ m, $l_2 = 0.25$ m   b) 0.993 s
11–45 a) $-2\pi\sqrt{L/g}(\Delta g/2g)$ b) $-\frac{1}{2}\Delta g/g$
      c) 9.7977 m·s^{-2}

11–47 a) $\sqrt{(1 + 3d^2/l^2)/(1 + 2d/l)}$, where $l = 1$ m is length of meterstick   b) $2l/3$

CHAPTER 12

12–1 4.80×10^{11} Pa

12–3 1.0×10^8 N·m^{-2}; 5.0×10^{-4}; 2.5×10^{-3} m; $-9.5 \times 10^{-3}\%$

12–5   a) Upper wire: 1.84×10^{-3}; lower wire: 1.23×10^{-3} b) Upper wire: 9.20×10^{-4} m; lower wire: 6.15×10^{-4} m

12–7   1.38 m

12–9 $B = 4.0 \times 10^9$ Pa; $k = 2.5 \times 10^{-10}$ Pa^{-1}

12–11 4.71×10^4 N

12–13   13.1 m·s^{-2}

12–15 1.66×10^{-3} m

12–17   a) 0.700 m to right of A
     b) 0.600 m to right of A

12–19 1.22×10^6 Pa

12–21   a) 1.82 m   b) Steel: 3.0×10^8 Pa; copper: 1.5×10^8 Pa   c) Steel: 1.5×10^{-3}; copper: 1.36×10^{-3}

12–23   a) $F \cos^2 \theta/A$ b) $F \sin 2\theta/2A$ c) 0°   d) 45°

12–25 $(A^2x - k_0V_0F)/FV_0$

12–27   a) 1.23×10^{-3} m   b) 20.0 N   c) 0.0245 J
     d) 0.0050 J   e) 0.0295 J   f) 0.0295 J

CHAPTER 13

13–1   a) 889 kg·m^{-3} b) Yes

13–3   a) 6.06×10^6 Pa   b) 1.07×10^5 N

13–5   a) 882 Pa   b) 3.33×10^3 Pa

13–7   a) 1.077×10^5 Pa   b) 1.037×10^5 Pa
     c) 1.037×10^5 Pa   d) 6.66×10^3 Pa
     e) 5.0 cm of Hg   f) 68 cm of water

13–9 2.04×10^{-4} m^3; 6.00×10^3 kg·m^{-3}

13–11   1.00 m^3

13–13   a) 118 Pa   b) 784 Pa   c) 0.680 kg

13–15   4.00 Pa

13–17   a) 6.37 m·s^{-1} b) 0.259 m

13–19   a) 88.9 m·s^{-1} b) 8.14×10^3 N

13–21 5.35×10^{-4} m^2

13–23 1.96×10^5 Pa

13–25   665 N, downward

13–27   a) 2.25 m·s^{-1} b) 0

13–29   a) 0.0533 m^3·s^{-1} b) 1.28×10^4 Pa
     c) 0.114 m^3·s^{-1}

13–31   167 m·s^{-1}

13–33   a) 22,400   b) Turbulent

13–35 1.96×10^4 N·m

13–37   a) 1.10×10^8 Pa   b) 1.085 kg·m^{-3}; 5.3% increase

13–39   a) 8.33×10^3 m^3 b) 9260 kg

13–41 8.29×10^6 kg; yes

13–43 7.77×10^{-5} m^3

13–45   0.10087 kg

13–47   25.9 N

13–49   7.02 N·m

13–51   0.10 m

13–53   0.0268 m^3·min^{-1}

13–55   107 m·s^{-1}

13–57 $3h_1$

13–59   a) $r = r_0\sqrt{v_0}/(v_0{}^2 + 2gy)^{1/4}$, where r_0 and v_0 are the radius and velocity at the opening and y is the distance the water has fallen   b) 0.765 m

13–61   a) 3.26 mm·s^{-1} b) 0.541 m·s^{-1}

13–63   a) 5.6×10^{-5} m^3·s^{-1} b) 0.56 m·s^{-1}; 1.12 m·s^{-1}; 2.80 m·s^{-1} c) 0.384 m; 0.336 m   d) 0.180 m
     e) 0.718 m   f) 1.12 m·s^{-1}, 2.24 m·s^{-1}, 5.60 m·s^{-1} g) Yes ($N_R = 355$)
     h) No ($N_R = 22{,}000$)

13–65   a) 88.2 N

13–67   a) la/g b) $\omega^2l^2/2g$ c) Yes; yes

13–69   a) $2\gamma \cos \theta/\rho xg$ b) 2.97 cm

13–71   a) $\sqrt{2gh}$ b) $p_1/\rho g - h$, where p_1 is atmospheric pressure

CHAPTER 14

14–1   a) Yes (104°F)   b) 37°   c) -40

14–3   90.18 K = 162.3°R

14–5 -286°C

14–7   79.97 m

14–9 1.5×10^{-5}(C°)$^{-1}$

14–11   2.509 cm

14–13   53.3°C

14–15 5.62×10^{-3} m^3

14–17   a) 1.0×10^{-5}(C°)$^{-1}$ b) 1×10^9 Pa

14–19 7.14×10^7 Pa

14–21   a) 1.44 cm^3 b) -3.89 kg·m^{-3}

14–23   a) 75.6°C   b) -63.2°C

14–25   20 cm; 10 cm

14–27 2.8×10^{-5}(C°)$^{-1}$

14–29 4.8×10^8 Pa

14–31 5.06×10^7 Pa

14–33   a) -0.012% b) 5.18 s   c) 1.93°C

14–35   b) 1.009×10^{-4}(C°)$^{-1}$

CHAPTER 15

15–1 1.88×10^4 J

15–3   a) 8380 J   b) 324°C   c) 42.3°C

15–5 1.05×10^6 J

15–7 3.01×10^3 J·kg^{-1}·(C°)$^{-1}$

15–9   3029 J, 724 cal, 2.87 Btu

15–11   20%

15–13   357 m·s^{-1}

15–15 3.53×10^3 W, 1.20×10^4 Btu·hr^{-1}

15–17   25.0°C

15–19   1.83 kg

15–21   108 grams

15–23   0.102 kg

15–25   170 m

15–27   a) 273 J   b) 3.41 J·mol^{-1}·K^{-1}
     c) 10.9 J·mol^{-1}·K^{-1}

15–29   a) 1.91×10^{11} J   b) 12.8 m on a side

15–31   a) 45.6 m^3 b) 8.75 m^3

15–33   2.99 grams

15–35   40.0°C

15–37 3.69×10^8 J·m^{-3}

CHAPTER 16

16–1 4.34×10^5 J
16–3 a) $-4.8°C$ b) 1.65 cal·s^{-1}·m^{-2}
16–5 1320 Btu = 1.39×10^6 J
16–7 a) -1000 C°·m^{-1} b) 38.5 J·s^{-1} c) 80 C°
16–9 a) 2.58 cm b) 2.46×10^3 Pa
16–11 1.89×10^6 J
16–13 196 W
16–15 0.408 cm^2
16–17 72.3 m^3
16–19 2.40×10^3 km^2 = 927 mi^2
16–21 a) 48 J·s^{-1} b) 25.8
16–23 7.61¢
16–25 Copper-steel junction: 9.48°C; steel-aluminum junction: 82.2°C
16–27 a) 4.73 J·s^{-1} b) Copper: 95.8%; steel: 4.2%
16–29 84.5°C
16–31 1.64 grams·hr^{-1}
16–33 a) $4\pi k(T_2 - T_1)ab/(b - a)$
b) $T(r) = T_2 - (T_2 - T_1)(b/r)[(r - a)/(b - a)]$
c) $2\pi kL(T_2 - T_1)/\ln(b/a)$
d) $T(r) = T_2 - (T_2 - T_1)[\ln(r/a)/\ln(b/a)]$
16–35 b) $T = 0°C$ throughout d) 31.4°C·cm^{-1}
e) 121 J·s^{-1} f) 0 g) 1.11×10^{-4} m^2·s^{-1}
h) $-10.9°C·s^{-1}$ i) 9.13 s j) Decrease
k) $-7.74°C·s^{-1}$

CHAPTER 17

17–1 3.0 atm
17–3 a) 50 moles b) 6.24×10^6 Pa = 61.6 atm
17–5 3.0×10^4 Pa
17–7 486°C
17–9 a) pM/RT b) 1.20 kg·m^{-3}
17–11 Solid + vapor in equilibrium (no liquid)
17–13 a) 0.125×10^5 Pa; solid → vapor
b) 33.9×10^5 Pa; solid → liquid, liquid → vapor
17–15 2.30 atm (33.8 lb·in^{-2})
17–17 12.2 cm, so piston has traveled 33.5 cm
17–19 a) 7.60×10^4 Pa b) 1.95 grams
17–21 a) 13.5 cylinders b) 5880 N c) 5440 N
17–23 a) 15.5 m·s^{-1}
b) $h_2 = 2.5$ m: 9.80 m·s^{-1}; $h_2 = 2.0$ m: 4.43 m·s^{-1}
c) 1.82 m
17–25 a) $n RT/(V-nb)^2 = 2an^2/V^3$; $2n RT/(V - nb)^3 = 6an^2/V^4$ b) $(V/n)_c = 3b$; $T_c = 8a/27bR$
c) $a/27b^2$ d) 8/3 e) H$_2$: 3.28; N$_2$: 3.44; O$_2$: 3.25; H$_2$O: 4.35
f) H$_2$: $T_c = 33.1$ K, $p_c = 12.9 \times 10^5$ Pa, $(V/n)_c = 79.8 \times 10^{-6}$ m^3·mol^{-1}
N$_2$: $T_c = 128$ K, $p_c = 34.1 \times 10^5$ Pa, $(V/n)_c = 117 \times 10^{-6}$ m^3·mol^{-1}
O$_2$: $T_c = 155$ K, $p_c = 50.5 \times 10^5$ Pa, $(V/n)_c = 95.4 \times 10^{-6}$ m^3·mol^{-1}
H$_2$O: $T_c = 646$ K, $p_c = 220 \times 10/^5$ Pa, $(V/n)_c = 91.5 \times 10^{-6}$ m^3·mol^{-1}

CHAPTER 18

18–1 0
18–3 2494 J

18–5 a) Yes b) No c) Negative
18–7 a) 4.0×10^4 J b) 8.0×10^4 J c) No
18–9 a) 60 J b) 70 J liberated c) ad: 50 J; db: 10 J
18–11 a) 365 J b) 0 c) Yes, 365 J liberated
18–13 $C_v = 25.2$ J·mol^{-1}·K^{-1}; $C_p = 33.5$ J·mol^{-1}·K^{-1}
18–15 736 K, 25.1 atm
18–17 a) 42.2 J b) 0
18–19 0.306 m^3
18–21 a) 0.174 m b) 204°C
18–23 a) 500 J; 0 b) 0; -500 J
18–25 b) 1.25×10^4 J c) 0 d) 1.25×10^4 J
e) 0.112 m^3
18–27 a) 1247 J; 4412 J; 3165 J b) 0; -3165 J; -3165 J c) 0

CHAPTER 19

19–1 a) 37.5% b) 5000 J c) 0.16 grams
d) 1.50×10^5 W = 201 hp
19–3 a) 25% b) 600×10^6 J·s^{-1}
19–5 a) $p_1 = 1.0$ atm, $V_1 = 2.46 \times 10^{-3}$ m^3;
$p_2 = 2.0$ atm, $V_2 = 2.46 \times 10^{-3}$ m^3
$p_3 = 1.0$ atm, $V_3 = 3.73 \times 10^{-3}$ m^3
b) 52 J
19–7 678 K
19–9 a) 8.76×10^5 J b) 1.75×10^5 J
c) 1.05×10^6 J
19–11 a) 319 K b) 20.2%
19–13 a) 1.84×10^7 J b) 1.65×10^6 J
19–15 47.4 J·K^{-1}
19–17 11.5 J·K^{-1}
19–19 a) 8.33 km^2 = 3.22 mi^2 b) 16.7 km^2 = 6.44 mi^2
19–21 15.8%
19–23 21 J·K^{-1}
19–25 $e = 1 - (6.96r^{-0.56} - 1)/1.4(4 - r^{0.4})$; 68.7%

CHAPTER 20

20–1 11.1 mol; 6.69×10^{24} molecules
20–3 a) 18 cm^3 b) 3.10×10^{-10} m c) Comparable
20–5 4095°C
20–7 1.006
20–9 a) 6.21×10^{-21} J b) 2.34×10^5 m^2·s^{-2}
c) 484 m·s^{-1} d) 2.57×10^{-23} kg·m·s^{-1}
e) 1.24×10^{-19} N f) 1.24×10^{-17} Pa
g) 8.15×10^{21} molecules
h) 2.44×10^{22} molecules i) $(v_x^2)_{av} = (v^2)_{av}/3$
20–11 24.9 J·mol^{-1}·K^{-1}; vibration is significant
20–15 2.45×10^6 molecules
20–17 a) $v_{rms} = 517$ m·s^{-1} b) $v_{x,rms} = 298$ m·s^{-1}
29–19 a) 1.24×10^{-20} kg b) 4.15×10^5 molecules
c) Sphere of radius 1.5×10^{-6} cm; no; no
20–21 b) 1.61×10^5 K; 1.00×10^4 K
20–23 a) $F(r) = (U_0/\sigma)[12(\sigma/r)^{13} - 6(\sigma/r)^7]$ b) $r_1 = \sigma$, $r_2 = 1.122\sigma$, $r_1/r_2 = 0.820$ c) $U_0/4$

CHAPTER 21

21–1 a) 10.8 m b) 282 Hz
21–3 a) 17.2 m; 0.0172 m b) 74.0 m; 0.0740 m
21–7 b) +x-direction

21-9 a) $49.5 \text{ m} \cdot \text{s}^{-1}$ b) 0.206 m

21-11 0.270 s

21-13 $35 \text{ m} \cdot \text{s}^{-1}$ greater at 57°C

21-15 b) $294 \text{ m} \cdot \text{s}^{-1}$ c) Large thermal conductivity, small λ

21-19 a) 0.50 Hz b) $3.14 \text{ rad} \cdot \text{s}^{-1}$ c) 3.14 m^{-1}
d) $y(x, t) = -(0.10) \sin (\pi(t - x))$, x and y in meters, t in seconds   e) $y(t) = -(0.10) \sin (\pi t)$, y in meters, t in seconds   f) $y(t) = -(0.10) \cos (\pi t)$, y in meters, t in seconds   g) $0.314 \text{ m} \cdot \text{s}^{-1}$
h) $+0.0707 \text{ m}$; $-0.222 \text{ m} \cdot \text{s}^{-1}$

21-21 0.452 s

21-23 $Y/100$

21-25 $c = 0.683 v_{\text{rms}}$

CHAPTER 22

22-5 a) $36.0 \text{ m} \cdot \text{s}^{-1}$ b) 64.8 N

22-7 a) 35.3 Hz b) 17.7 Hz

22-9 1290 Hz

22-11 a) $4960 \text{ m} \cdot \text{s}^{-1}$ b) $342 \text{ m} \cdot \text{s}^{-1}$

22-13 a) 170 N b) 12.2%

22-15 a) $375 \text{ m} \cdot \text{s}^{-1}$ b) 1.39

22-17 a) 0.657 m b) $56°C$

22-19 4.98 m, 1.15 m, 0.069 m

CHAPTER 23

23-1 a) 12.8 Pa (below pain threshold)   b) 514 Pa (well above pain threshold)

23-3 a) 58.7 b) 2.90×10^{-4}

23-5 $0.0325 \text{ W} \cdot \text{m}^{-2}$; 105 dB

23-9 443.6 Hz or 436.4 Hz

23-11 a) 0.63 m b) 0.75 m c) 548 Hz
d) 460 Hz e) 571 Hz f) 440 Hz
g) a: 0.65 m; b: 0.73 m; c: 546 Hz; d: 459 Hz; e: 569 Hz; f: 438 Hz

23-13 a) 0.286 Pa b) $2.53 \times 10^{-7} \text{ m}$ c) 50 m

23-15 a) 8.92 m b) 892 m

23-17 $9.89 \text{ m} \cdot \text{s}^{-1}$

23-19 a) $t(c + v)/\lambda_0$ b) $(c - v)t$
c) $\lambda_0[(c - v)/(c + v)]$ d) $f_0[(c + v)/(c - v)]$
e) $2v f_0/(c - v)$

23-21 a) 0.029 m b) 477 Hz

23-23 a) ct b) $(c - v_1)/f_0$ c) $f_0 t(c + v_2)/(c - v_1)$
d) c e) $f_0(c - v_2)(c - v_1)/(c + v_2)$
f) $[c(c + v_2)/(c - v_2)(c - v_1)]f_0$ g) 1560 Hz

INDEX

APPENDIX F
NUMERICAL CONSTANTS

FUNDAMENTAL PHYSICAL CONSTANTS

Name	Symbol	Value
Speed of light	c	2.9979×10^8 m·s^{-1}
Charge of electron	e	1.602×10^{-19} C
Gravitational constant	G	6.673×10^{-11} N·m^2·kg^{-2}
Planck's constant	h	6.626×10^{-34} J·s
Boltzmann's constant	k	1.381×10^{-23} J·K^{-1}
Avogadro's number	N_0	6.022×10^{23} molecules·mol^{-1}
Gas constant	R	8.314 J·mol^{-1}·K^{-1}
Mass of electron	m_e	9.110×10^{-31} kg
Mass of neutron	m_n	1.675×10^{-27} kg
Mass of proton	m_p	1.673×10^{-27} kg
Permittivity of free space	ϵ_0	8.854×10^{-12} C^2·N^{-1}·m^{-2}
	$1/4\pi\epsilon_0$	8.987×10^9 N·m^2·C^{-2}
Permeability of free space	μ_0	$4\pi \times 10^{-7}$ Wb·A^{-1}·m^{-1}

OTHER USEFUL CONSTANTS

Name	Symbol	Value
Mechanical equivalent of heat		4.186 J·cal^{-1} (15° calorie)
Standard atmospheric pressure	1 atm	1.013×10^5 Pa
Absolute zero	0 K	-273.15°C
Electronvolt	1 eV	1.602×10^{-19} J
Atomic mass unit	1 u	1.661×10^{-27} kg
Electron rest energy	mc^2	0.511 MeV
Energy equivalent of 1 u	Mc^2	931.5 MeV
Volume of ideal gas (0°C and 1 atm)	V	22.4 liter·mol^{-1}
Acceleration due to gravity (sea level, at equator)	g	9.78049 m·s^{-2}

ASTRONOMICAL DATA

Body	Mass, kg	Radius, m	Orbit radius, m	Orbit period
Sun	1.99×10^{30}	6.95×10^8	—	—
Moon	7.36×10^{22}	1.74×10^6	0.38×10^9	27.3 d
Mercury	3.28×10^{23}	2.57×10^6	5.8×10^{10}	88.0 d
Venus	4.82×10^{24}	6.31×10^6	1.08×10^{11}	224.7 d
Earth	5.98×10^{24}	6.38×10^6	1.49×10^{11}	365.3 d
Mars	6.34×10^{23}	3.43×10^6	2.28×10^{11}	687.0 d
Jupiter	1.88×10^{27}	7.18×10^7	7.78×10^{11}	11.86 y
Saturn	5.63×10^{26}	6.03×10^7	1.43×10^{12}	29.46 y
Uranus	8.61×10^{25}	2.67×10^7	2.87×10^{12}	84.02 y
Neptune	9.99×10^{25}	2.48×10^7	4.49×10^{12}	164.8 y
Pluto	5×10^{23}	4×10^5	5.90×10^{12}	247.7 y